Geotechnics for Hazard Mitigation

Geotechnics for Hazard Mitigation

Topic Editors

Mowen Xie
Yan Du
Yujing Jiang
Bo Li
Xuepeng Zhang

MDPI

Basel • Beijing • Wuhan • Barcelona • Belgrade • Novi Sad • Cluj • Manchester

Topic Editors

Mowen Xie
School of Civil and Resource Engineering
University of Science and Technology Beijing
Beijing
China

Yan Du
School of Civil and Resource Engineering
University of Science and Technology Beijing
Beijing
China

Yujing Jiang
School of Engineering
Nagasaki University
Nagasaki
Japan

Bo Li
College of Civil Engineering
Tongji University
Shanghai
China

Xuepeng Zhang
College of Energy and Mining Engineering
Shandong University of Science and Technology
Qingdao
China

Editorial Office
MDPI AG
Grosspeteranlage 5
4052 Basel, Switzerland

This is a reprint of the Topic, published open access by the journals *Applied Sciences* (ISSN 2076-3417), *GeoHazards* (ISSN 2624-795X), *Geosciences* (ISSN 2076-3263), *Remote Sensing* (ISSN 2072-4292), *ISPRS International Journal of Geo-Information* (ISSN 2220-9964) and *Materials* (ISSN 1996-1944), freely accessible at: https://www.mdpi.com/topics/5R6A0Y99W6.

For citation purposes, cite each article independently as indicated on the article page online and as indicated below:

Lastname, A.A.; Lastname, B.B. Article Title. *Journal Name* **Year**, *Volume Number*, Page Range.

ISBN 978-3-7258-6123-1 (Hbk)
ISBN 978-3-7258-6124-8 (PDF)
https://doi.org/10.3390/books978-3-7258-6124-8

Contents

About the Editors

Mowen Xie

Mowen Xie is a Professor and PhD supervisor at the School of Resources and Safety Engineering, University of Science and Technology Beijing. He earned his bachelor's degree in Engineering Geology and Hydrogeology from Hohai University (1985), his master's degree in Geotechnical Engineering from Wuhan University (1991), and his PhD in Environmental Geology and GIS from Kyushu University, Japan (2003). He previously served as Associate Professor at Wuhan University and as a JSPS Research Fellow at Kyushu University, before joining USTB, where he was selected for the Ministry of Education's "New Century Excellent Talents Program." His research focuses on geological disasters, geotechnical and environmental engineering, and GIS/remote sensing applications in hazard assessment. He has led multiple NSFC projects on rainfall-induced landslide prediction, slope instability in reservoir areas, and GIS-based geotechnical information systems. Prof. Xie has published over 60 papers and authored monographs including Information-Based Slope Engineering and Landslide Hazard Assessment Using GIS. He received the Metallurgical Science and Technology Award (Special Prize, 2006; First Prize, 2011) and serves as Vice Chair of the Ground Rock Engineering Committee of the Chinese Society for Rock Mechanics and Engineering. His work has significantly advanced early-warning technologies, slope stability evaluation, and disaster risk reduction.

Yan Du

Yan Du is an Associate Professor and master's supervisor at the School of Future City, University of Science and Technology Beijing. He obtained his PhD in Civil Engineering from USTB, completed postdoctoral research in Engineering Mechanics, and was a visiting scholar at Nagasaki University. He is also a certified national constructor in China. His research focuses on spaceborne and ground-based remote sensing, intelligent and green construction, and automated safety monitoring and early warning for lifeline projects. He has led or participated in more than ten NSFC, postdoctoral, and provincial projects, as well as over twenty enterprise collaborations. He has published over forty academic papers, applied for more than ten patents, and authored one monograph. His research achievements have received multiple provincial and ministerial-level awards, including First Prize in the Science and Technology Progress Award from the China Petroleum Engineering Construction Association (2019, 2021) and Second Prize from the Chinese Society for Optical Engineering (2020). He serves as a senior member of the Chinese Society for Optical Engineering, a youth committee member of the China Optics and Optoelectronics Manufacturers Association, and a young Editorial Board Member of several journals including the *International Journal of Mining Science and Technology*. Recognized as an Excellent Mentor for Innovation and Entrepreneurship in Beijing, he is committed to advancing research and education in civil and geotechnical engineering.

Yujing Jiang

Yujing Jiang is a Taishan Distinguished Professor at Shandong University of Science and Technology, leading the discipline of Safety Technology and Engineering and the Laboratory for Mine Disaster Prevention and Control. He received bachelor's and master's degrees in Mining Engineering from the Shandong Mining Institute and a PhD from Kyushu University, Japan (1993). He was Associate Professor at Kyushu University (1993–1999) and Professor at Nagasaki University (1999–2007), also serving as Deputy Director of its Urban Structure Research Center. He holds adjunct appointments at the Institute of Rock and Soil Mechanics, CAS, Tongji University, and Shandong University. His research spans environmental rock mechanics, resource and environmental systems engineering, and underground construction. He has pioneered work in soft rock nonlinear mechanics, top-coal caving interaction theory, and environmental mechanics of rock strata, developing generalized models of plastic zone formation and stability control under complex stress. He has designed advanced experimental systems for rock stability and joint shear–seepage coupling, recognized internationally. Honors include being a NSFC Distinguished Overseas Young Scholar (2000), Best Paper Awards (2002, 2007), the Invention Award of the Japan Society for Rock Mechanics (2007), being a Taishan Scholar Special Professor (2008), China's "Thousand Talents Plan" (2011), and a Foreign Member of the Engineering Academy of Japan (2018). He has led numerous national and international projects on landslide and debris flow monitoring and underground construction. He actively promotes international exchange, serving on editorial boards (*International Journal of Geomechanics, TUST*) and as a chair/committee member of major conferences. His contributions significantly advance geotechnical engineering and disaster prevention.

Bo Li

Bo Li is a Professor at the College of Civil Engineering, Tongji University. He received his PhD and gained international experience as Assistant Professor at Nagasaki University (2010–2015) and visiting scholar at the Royal Institute of Technology, Sweden (2012–2013). He was a Professor at Shaoxing University (2016–2021) before joining Tongji University in 2021. His research focuses on rock mechanics, geological disaster prevention, and multi-physical coupling in fractured rocks, emphasizing the shear behavior of rock joints, THMC processes, and hazard mechanisms in slopes and underground engineering. He has led numerous NSFC and international projects, including studies on ice-filled joint shear failure, coupled seepage and mass transport, precision grouting, and disaster early-warning for infrastructure. He has published extensively in leading journals (*Engineering Geology, Computers and Geotechnics, IJRMMS, Landslides, Underground Space*) with notable work on rockslide dynamics, discontinuous SPH modeling, and joint shear behavior. His achievements have earned recognition from ISRM (Rocha Medal nomination), the "Qian Qihu Award," and the Special Prize of Science and Technology Progress from CSRME. He serves as Secretary-General of the Technical Committee on Geological Engineering, CSRME. His career reflects strong contributions to rock fracture mechanics and innovative disaster prediction and mitigation methods for large-scale infrastructure.

Xuepeng Zhang

Xuepeng Zhang is an Associate Researcher and master's supervisor at the Shandong University of Science and Technology, where he also directs the Academic Office of the Key Laboratory for Mine Disaster Prevention and Control. He obtained his bachelor's in Engineering Mechanics (2012) and master's in Solid Mechanics (2015) at SDUST, and a PhD in Geotechnical Engineering from Nagasaki University, Japan (2019). He worked as a visiting researcher at Nagasaki University (2019–2020) and has been with SDUST since 2019, promoted to Associate Researcher in 2022. He was a visiting scholar at China University of Geosciences (Wuhan) in 2024–2025. His research covers shear properties of fractured rock masses, tunnel stability, the seismic design of underground structures, and the reinforcement of underground engineering systems. He has published over fifty papers in journals including *Automation in Construction, IJRMMS, RMRE, TUST*, and *Underground Space*. As first/corresponding author, he has 28 SCI/EI-indexed papers, 12 in JCR Q1, 7 in top journals, 1 ESI that is highly cited, and 1 selected for bilingual dissemination by CAST. He has authored three patents, two monographs, one textbook, and contributed to two CSRME standards. He has led six funded projects (NSFC, Shandong NSFC) and participated in many national/provincial projects. His received awards include the 2023 Frontier Award (JSRM), 2020 Outstanding Paper Award (JSRM), and 2020 Shandong Provincial Higher Education Award (Second Prize). He serves as a council member of the CSRME Underground Engineering Division, Science and Technology Commissioner of Qingdao, and as a young Editorial Board Member of the *Journal of Tunnel and Underground Engineering Hazard Prevention*. He is also a reviewer for leading journals and a member of ISRM, CSRME, JSRM, ARMA, and other societies.

Preface

Geohazard mitigation represents one of the most critical challenges confronting the global geoscience and engineering communities. In an era marked by climate change, accelerating urbanization, and the expansion of infrastructure into vulnerable regions, the threats posed by landslides, collapses, debris flows, and other slope-related disasters transcend national boundaries and demand coordinated international responses. This volume, *Geotechnics for Hazard Mitigation*, bridges fundamental research in rock and soil mechanics with applied methodologies for prediction, monitoring, and risk reduction. Its scope spans from quantitative slope stability assessment and early warning systems to engineered reinforcement solutions and innovative remote sensing applications.

The central aim of this specialized publication is to establish an international platform for sharing innovative methods and practical solutions that enhance our collective capacity to anticipate, manage, and mitigate geological hazards. The impetus for this work stems from the escalating risks associated with environmental change and development pressures in sensitive terrains worldwide. These challenges underscore the necessity for robust monitoring frameworks, data-driven risk evaluation, and resilient engineering strategies that are adaptive to diverse geographical and socio-economic contexts.

This volume is aimed at researchers, engineers, policymakers, and practitioners across the fields of geotechnical engineering, remote sensing, disaster risk management, and environmental resilience. By integrating diverse international perspectives and case studies, we hope to inspire further innovation, foster cross-disciplinary collaboration, and advance the development of effective, scalable hazard mitigation practices around the world.

We extend our sincere gratitude to all the authors, reviewers, and institutions whose contributions have made this publication possible. Special acknowledgment is given to the National Key Research and Development Program of China (grant nos. 2023YFC3081200, 2023YFC3081400) for its generous support, which has been instrumental in facilitating the research and collaboration reflected in this specialized publication.

Mowen Xie, Yan Du, Yujing Jiang, Bo Li, and Xuepeng Zhang
Topic Editors

International Journal of
Geo-Information

MDPI

Editorial

Recent Progress in and Future Perspectives on the Monitoring, Assessment, and Mitigation of Geological Disasters

Yan Du [1], Mengjia Lyu [2], Mowen Xie [2], Yujing Jiang [3], Bo Li [4] and Xuepeng Zhang [5,*]

1 Beijing Key Laboratory of Urban Underground Space Engineering, University of Science and Technology Beijing, Beijing 100083, China; mutulei@163.com
2 School of Resources and Safety Engineering, University of Science and Technology Beijing, Beijing 100083, China; lmj202410@163.com (M.L.); mowenxie@126.com (M.X.)
3 School of Engineering, Nagasaki University, Nagasaki 852-8521, Japan; jiang@nagasaki-u.ac.jp
4 College of Civil Engineering, Tongji University, Shanghai 200092, China; libo@usx.edu.cn
5 State Key Laboratory of Disaster Prevention and Ecology Protection in Open-pit Coal Mines, Shandong University of Science and Technology, Qingdao 266590, China
* Correspondence: zhangxuepeng0722@126.com; Tel.: +86-0532-86058039

1. Introduction

Among all the geological disasters (GDs), collapses, landslides, and debris flows are the most severe. Predicting geological disasters is one of the most effective means of reducing casualties and property losses. However, technologies for the early identification and warning of collapses have been difficult to implement effectively. Therefore, the scope of this research topic includes the quantitative identification of dangerous rocks, risk assessment of landslide instability precursors, quantitative modeling of movable solid sources in debris flow source areas, application studies on new monitoring and early warning technology systems, and the corresponding control and reinforcement measures.

This Special Issue brings together 25 cutting-edge research articles addressing critical challenges in geotechnical engineering, landslide dynamics, rock mechanics, and disaster resilience. As global climate change and urbanization intensify geohazard risks, from landslides and slope failures to seismic disasters, this collection offers innovative methodologies and practical insights to enhance prediction, monitoring, and mitigation strategies. The contributions span theoretical advancements, experimental validations, and real-world applications, reflecting a multidisciplinary effort to safeguard infrastructure, ecosystems, and communities. Taken together, the 25 articles in this Special Issue address these critical challenges, offering innovative methodologies and practical insights to advance prediction accuracy, monitoring capabilities, and mitigation effectiveness. Key themes and breakthroughs are discussed in the subsequent sections to provide a comprehensive understanding of the topic.

2. Slope Stability and Landslide Mechanisms

Several studies focus on novel approaches for evaluating slope stability. Jia et al. [1] propose using natural vibration frequency to rapidly assess perilous rock stability on soil slopes, linking buried depth to safety factors. Cao et al. [2] integrate UAV mapping, blasting tests, and numerical simulations to analyze multi-factor influences on open-pit mine slopes, revealing instability risks in excavation designs. Moresi et al. [3] compare GIS-based models (4SLIDE vs. SHALSTAB) for shallow landslide prediction, demonstrating 4SLIDE's superior accuracy in Mediterranean terrains. Komu et al. [4] model landslide runout distances in Türkiye's flysch formations, incorporating climate scenarios (RCP 4.5/8.5) to forecast future hazard zones.

 https://doi.org/10.3390/ijgi14080310

Collectively, these advances signify a transformative shift in slope stability research, moving from reactive post-failure analysis to proactive approaches leveraging real-time monitoring (e.g., vibration sensors and UAVs) and predictive modeling (e.g., climate-integrated runout simulations) [1]. This shift is characterized by the integration of multi-factor data (geotechnical, geophysical, and operational), which is essential for capturing complex instability mechanisms in engineered slopes [2], and crucially, by embedding climate uncertainty (e.g., RCP scenarios) into assessments, a necessity for infrastructure resilience under intensifying extreme weather [4].

Coupling AI (e.g., real-time sensor analytics), cross-scale modeling (micro-fracture to regional slope systems), and social vulnerability metrics will transform slope stability analysis from a technical exercise into a pillar of sustainable development.

3. Material Behavior Under Environmental Stress

Understanding soil and rock responses to environmental triggers is critical. Chen et al. [5] employ high-precision CT scanning to quantify pore evolution in tuff under frost salt weathering, revealing magnesium sulfate's severe erosive impact. Zhu et al. [6] develop a coupled damage model for deep expansive soil under freeze–thaw cycles, validated via triaxial tests. Guo et al. [7] review the geomechanical complexities of "block-in-matrix" rocks (Bimrocks), emphasizing the block proportion and welding degree as key failure controls. Valentino's [8] microstructural analysis of lateritic soils in Rwanda explains landslide triggers through mineralogical alterations unseen in conventional tests.

These studies illuminate three transformative shifts in geomechanics research:

1. From Macro to Micro/Nano Scales:

Advanced tools like high-resolution CT [5] and mineralogical mapping [8] expose hidden mechanisms, such as magnesium sulfate exhibiting a pore expansion rate approximately two times greater than that of sodium sulfate in tuff that dictates the macroscopic failure. On a different scale, in granular systems, arching structures that develop during excavation constitute a critical failure mode under stress redistribution. These have been effectively characterized through the optical and mechanical modeling of force chains [9]. This scale-bridging is redefining failure prediction;

2. From Single-Stress to Multi-Field Coupling:

Zhu et al.'s [6] freeze–thaw-loading damage model exemplifies a critical advance, quantifying how concurrent environmental and mechanical stresses (e.g., cyclic freezing and structural loads) accelerate material degradation, essential for infrastructure in climate-volatile zones [10];

3. From Homogeneous Assumptions to Heterogeneity Embracing:

Guo et al.'s [7] Bimrock analysis proves that block–matrix interfaces control stability. Similarly, Valentino's [8] discovery of 7.71% kaolinite porosity gaps in lateritic soils reveals why "identical" strata fail differentially, demanding stochastic modeling over simplistic averages.

The frontier demands:

1. AI-accelerated multi-scale modeling merging quantum-scale chemistry with continuum mechanics;
2. Field-deployable micro-sensors to capture real-time mineralogical changes;
3. Global degradation atlases mapping material vulnerabilities under projected climate scenarios [11].

Ultimately, decoding environmental stress responses transforms geomaterials from passive hazards into intelligently engineered assets—where every mineral alteration becomes a calculable risk variable.

4. Advanced Monitoring and Early Warning Systems

Innovative technologies enable real-time hazard detection. Chen et al. [12] introduce MEMS sensors with dual indicators (natural frequency and RMS velocity) to warn of rock collapses 3× faster than tilt-based systems. Chen et al. [13] utilize time-series InSAR (PS/SBAS) to monitor subsidence along China's Yangtze River, identifying urbanization and underground engineering as key drivers. Damiano et al. [14] characterize pyroclastic soils for landslide early warning, emphasizing liquefaction susceptibility's role in post-failure evolution. He et al. [15] validate an aerial remote sensing platform with real-time satellite transmission for disaster response, achieving <1 cm resolution imagery with <2 s latency.

Contemporary advanced geohazard monitoring is undergoing a paradigm shift, transitioning from post-event diagnosis to real-time early warning systems, as demonstrated by MEMS sensors enabling second-level rock collapse alerts [12] and satellite-linked drones transmitting centimeter-resolution imagery within 2 s [15]. This evolution integrates multi-scale observation networks, synergizing InSAR-based subsidence tracking [13], UAV remote sensing, and soil behavior characterization [14] to bridge microscale mechanisms with macroscale hazard evolution. Intelligent algorithms now supersede empirical thresholds, reducing false alarms by approximately 50% through dynamic indicators like coupled frequency–velocity indices, significantly enhancing warning precision [12]. These advancements deliver critical societal benefits: namely, ≤15 min warnings could reduce landslide mortality by 90% (UNDRR) and InSAR-based urban monitoring plays a critical role in mitigating considerable economic and infrastructural losses, while low-latency satellite transmission enables broader access for high-risk regions [16]. Future progress necessitates miniaturized sensors, AI-driven big data analytics, and standardized protocols to realize a planetary-scale hazard immune system—transforming disaster resilience into an ethical imperative through ubiquitous, intelligent safeguarding.

5. Earthquake and Flood Impact Assessment

Rapid post-disaster modeling saves lives. Miura et al. [17] create a near-real-time model for estimating seismic intensities and building losses in Bogotá, generating impact maps within 5 min. Annunziato et al. [18] reconstruct Libya's catastrophic Derna dam break using satellite/social media data, highlighting the critical 15–30 min evacuation window. Zhang et al. [19] apply Neural Radiance Fields (NeRFs) to rapidly generate 3D landslide models, reducing reconstruction time by 30% compared to photogrammetry for emergency response.

Advanced near-real-time modeling, demonstrated by seismic loss mapping within 5 min [17] and dam break flood reconstruction leveraging satellite/social media fusion [18], has significantly shortened critical assessment timelines, enabling proactive evacuation and resource allocation before secondary disasters escalate. This acceleration is underpinned by multi-source data integration (e.g., InSAR, crowdsourcing, and NeRF-generated 3D terrain), which deciphers cascading events like earthquake-triggered landslides [19], climate-induced mega-slides [20], or subsequent flooding through unified impact narratives. Crucially, physics-based scenario engineering now supports the preemptive stress testing of infrastructure resilience against compound hazards, such as post-quake dam breaches or sediment-driven floods, validated by mortality reduction rates where sub-30 min warnings reduce fatalities by 8% per minute in multi-hazard scenarios (World

Bank). To fully transform impact assessment into proactive risk interception, future work must prioritize cross-hazard algorithms that unify seismic, hydrologic, and geotechnical predictors to anticipate and mitigate domino effect disasters.

6. Risk Mapping and Infrastructure Resilience

Spatial and statistical tools enhance risk management. Lu et al. [21] combine the Analytic Hierarchy Process–Information Content Model (AHP-ICM) and entropy methods for collapse risk zoning in Huinan County, achieving 87.4% prediction accuracy (AUC). Nie et al. [22] couple information quantity with random forest to assess highway collapse risks, showing a reduction in high-risk areas when including emergency response capacity. Rouhana et al. [23] use network theory to evaluate Beirut's transportation fragility, revealing low redundancy and high disruption sensitivity.

Advanced geological disaster mapping has evolved from static zoning to dynamic intelligence, exemplified by machine learning-enhanced models achieving 87.4% prediction accuracy [21], field-tested DMT-based methods for sliding surface localization [24], and random forest-driven assessments that reduce high-risk areas by 4.66% through integrated emergency capacity metrics [22]. This paradigm shift quantifies systemic vulnerabilities, such as infrastructure redundancy deficits that paralyze 40% of urban emergency routes [23], highlighting critical interdependencies between human systems and infrastructure, particularly in regions where high landslide hazard aligns with limited resilience (UNDRR, 2023). The integration of social vulnerability indices ensures climate justice, directing resilience investments toward marginalized communities, while high-precision zoning helps to optimize global mitigation funding by significantly reducing false alarms (World Bank). Future frontiers demand digital twin cities for adaptive evacuation planning, AI-powered regulatory frameworks to enforce building compliance in high-risk zones, and Global South-centered protocols addressing informal settlement realities.

7. Synthesis and Future Directions

Collectively, the contributions in this Special Issue highlight a transformative shift toward integrated, data-driven approaches in geohazard mitigation. From the fusion of field data with AI models (e.g., Mercurio's [25] MARS model for earthquake-induced landslides) to the scaling of micro-scale structural insights to macro-scale slope stability assessments [8], and the integration of emerging technologies such as Neural Radiance Fields (NeRFs) with conventional monitoring frameworks [19], these studies present actionable and forward-looking frameworks. Yet, challenges remain in translating laboratory findings to field environments, capturing nonlinear site responses [17], and adapting predictive models to evolving climatic and anthropogenic pressures. Addressing these translational challenges remains critical. Recent experimental research has proposed high vibration amplitude and uncoordinated dynamic responses as key indicators for early warning of long-runout landslides [26]. Nonetheless, solutions such as MEMS-based early warning systems and real-time aerial monitoring platforms [27] are bridging the gap between theoretical advances and engineering practice.

The innovations presented in this Special Issue are particularly vital in mountainous regions experiencing accelerated infrastructure development. In the context of China's major national strategies—such as the Belt and Road Initiative—projects like the Sichuan–Tibet Railway, the China–Myanmar oil and gas pipeline, and the Wudongde Hydropower Station are advancing across geologically complex alpine canyon zones. The increasing prevalence of high and steep slope constructions has heightened both the frequency and intensity of rockfall hazards. However, continued reliance on passive mitigation—due to the lack of effective early warning technologies—often results in considerable economic losses and ecological impacts.

Against this backdrop, advancing robust, field-ready monitoring and early warning technologies is not only a scientific imperative but also a national strategic necessity. The convergence of cutting-edge geomechanical theories, AI-enhanced predictive modeling, real-time micro-sensing, and low-latency satellite–UAV platforms marks a paradigm shift in our capacity to detect, analyze, and mitigate geological risks. Looking ahead, realizing the full potential of these innovations—particularly in terrain-sensitive, infrastructure-intensive settings—will require sustained technological breakthroughs, cross-disciplinary collaboration, and equitable international knowledge exchange. Only through such collective efforts can we strengthen global resilience, safeguard critical infrastructure, and protect vulnerable communities in the face of escalating geohazard challenges.

Author Contributions: Writing—original draft preparation, Yan Du and Mengjia Lyu; writing—review and editing, Mowen Xie, Yujing Jiang, Bo Li and Xuepeng Zhang. All authors have read and agreed to the published version of the manuscript.

Funding: This research was funded by the National Key R&D Program of China, grant number 2023YFC3081400, and Open Fund Research Project Supported by the State Key Laboratory of Disaster Prevention and Ecology Protection in Open-pit Coal Mines (Open Fund Research Project), grant number DPEPM202502.

Data Availability Statement: Not applicable.

Acknowledgments: We thank Santos D. Chicas for their invaluable contributions to this paper.

Conflicts of Interest: The authors declare no conflicts of interest.

References

1. Jia, Y.; Song, G.; Wang, L.; Jiang, T.; Zhao, J.; Li, Z. Research on Stability Evaluation of Perilous Rock on Soil Slope Based on Natural Vibration Frequency. *Appl. Sci.* **2023**, *13*, 2406. [CrossRef]
2. Cao, H.; Ma, G.; Liu, P.; Qin, X.; Wu, C.; Lu, J. Multi-Factor Analysis on the Stability of High Slopes in Open-Pit Mines. *Appl. Sci.* **2023**, *13*, 5940. [CrossRef]
3. Moresi, F.V.; Maesano, M.; Di Cristofaro, M.; Scarascia Mugnozza, G.; Brunori, E. Evaluating Shallow Landslide Prediction Mapping by Using Two Different GIS-Based Models: 4SLIDE and SHALSTAB. *ISPRS Int. J. Geo-Inf.* **2025**, *14*, 144. [CrossRef]
4. Komu, M.P.; Nefeslioglu, H.A.; Gokceoglu, C. Modeling Shallow Landslide Runout Distance in Eocene Flysch Facies Using Empirical–Statistical Models (Western Black Sea Region of Türkiye). *ISPRS Int. J. Geo-Inf.* **2024**, *13*, 84. [CrossRef]
5. Chen, L.; Sun, Y.; Wang, C.; Sha, P.; Jin, H.; Liu, M.; Li, A. Study on the Characteristics of Pore Change in Tuff under the Frost and Salt Action Using High-Precision CT Scanning Equipment. *Appl. Sci.* **2023**, *13*, 10483. [CrossRef]
6. Zhu, Z.; Lin, B.; Chen, S. Study on Mechanical Properties of Deep Expansive Soil and Coupling Damage Model of Freeze–Thaw Action and Loading. *Appl. Sci.* **2023**, *13*, 11099. [CrossRef]
7. Guo, S.; Wei, Q.; Qi, S.; Xue, L.; Zheng, B.; Wang, H.; Li, J.; Song, S.; Liang, N.; Zou, Y.; et al. Research Progress on the Geomechanical Properties of Block-in-Matrix Rocks. *Materials* **2024**, *17*, 1167. [CrossRef]
8. Valentino, R.; Pizzati, M.; Mizero, J. Combined Effect of the Microstructure and Mechanical Behavior of Lateritic Soils in the Instability of a Road Cut Slope in Rwanda. *GeoHazards* **2024**, *5*, 559–581. [CrossRef]
9. Wang, M.; Zheng, J.; Xue, S. Mechanics and Stability of Force Chain Arch in Excavated Granular Material. *Appl. Sci.* **2024**, *14*, 2485. [CrossRef]
10. Dang, W.; Tao, K.; Fu, J.; Wu, B. Experimental Investigation on Unloading-Induced Sliding Behavior of Dry Sands Subjected to Constant Shear Force. *Appl. Sci.* **2025**, *15*, 401. [CrossRef]
11. Song, S.; Ruan, H.; Wei, J.; Niu, R.; Cheng, X.; Chen, B. The Influence of the Key Characteristics of Overburden Rock Structure on the Development Height of Water-Conducting Fracture in Yushenfu Coal Mine Area, China. *Appl. Sci.* **2024**, *14*, 10537. [CrossRef]
12. Chen, C.; Xie, M.; Du, Y.; Zhang, X. Dual-Indicator Micro-Electro-Mechanical System Monitoring Method for Rock Instability Early Warning. *Appl. Sci.* **2025**, *15*, 4210. [CrossRef]
13. Chen, Y.; Guo, L.; Xu, J.; Yang, Q.; Wang, H.; Zhu, C. Monitoring and Cause Analysis of Land Subsidence along the Yangtze River Utilizing Time-Series InSAR. *ISPRS Int. J. Geo-Inf.* **2024**, *13*, 230. [CrossRef]

14. Damiano, E.; de Cristofaro, M.; Brunzo, A.; Carrieri, G.; Iavazzo, L.; Netti, N.; Olivares, L. The Mechanical Characterization of Pyroclastic Deposits for Landslide Early Warning Systems. *Geosciences* **2023**, *13*, 291. [CrossRef]
15. He, X.; Xu, C.; Tang, S.; Huang, Y.; Qi, W.; Xiao, Z. Performance Analysis of an Aerial Remote Sensing Platform Based on Real-Time Satellite Communication and Its Application in Natural Disaster Emergency Response. *Remote Sens.* **2024**, *16*, 2866. [CrossRef]
16. Du, Y.; Lyu, M.; Liu, J.; Xie, M.; Jiang, Y.; Chen, H. Research and Prospects on Intelligent Monitoring and Early Warning of Brittle Failure-Type Geological Hazard. *J. Basic Sci. Eng.* **2025**, *33*, 755–769. [CrossRef]
17. Miura, H.; Matsuoka, M.; Reyes, J.C.; Pulido, N.; Hashimoto, M.; Riaño, A.C.; Hurtado, A.; Rincon, R.; García, H.; Lozano, C. Quick Estimation Model for Mapping Earthquake Impacts in Bogotá, Colombia. *ISPRS Int. J. Geo-Inf.* **2023**, *12*, 471. [CrossRef]
18. Annunziato, A.; Santini, M.; Proietti, C.; de Girolamo, L.; Lorini, V.; Gerhardinger, A.; Tucci, M. Modelling and Validation of the Derna Dam Break Event. *GeoHazards* **2024**, *5*, 504–529. [CrossRef]
19. Zhang, J.; Du, Y.; Jiang, Y.; Zhang, S.; Chen, H.; Shang, D. Exploring the Application of NeRF in Enhancing Post-Disaster Response: A Case Study of the Sasebo Landslide in Japan. *ISPRS Int. J. Geo-Inf.* **2025**, *14*, 218. [CrossRef]
20. Gargani, J. Influence of Relative Sea-Level Rise, Meteoric Water Infiltration and Rock Weathering on Giant Volcanic Landslides. *Geosciences* **2023**, *13*, 113. [CrossRef]
21. Lu, Z.; Yu, C.; Liu, H.; Zhang, J.; Zhang, Y.; Wang, J.; Chen, Y. Application of AHP-ICM and AHP-EWM in Collapse Disaster Risk Mapping in Huinan County. *ISPRS Int. J. Geo-Inf.* **2023**, *12*, 395. [CrossRef]
22. Nie, Z.; Lang, Q.; Zhang, Y.; Zhang, J.; Chen, Y.; Pan, Z. Risk Assessment of Landslide Collapse Disasters along National Highways Based on Information Quantity and Random Forest Coupling Methods: A Case Study of the G331 National Highway. *ISPRS Int. J. Geo-Inf.* **2023**, *12*, 493. [CrossRef]
23. Rouhana, F.; Jawad, D. A Spatial-Network Approach to Assessing Transportation Resilience in Disaster-Prone Urban Areas. *ISPRS Int. J. Geo-Inf.* **2025**, *14*, 261. [CrossRef]
24. Grabar, K.; Jug, J.; Bek, A.; Strelec, S. Comparison of the Piezocone Penetrometer (CPTU) and Flat Dilatometer (DMT) Methods for Landslide Characterisation. *Geosciences* **2024**, *14*, 64. [CrossRef]
25. Mercurio, C.; Calderón-Cucunuba, L.P.; Argueta-Platero, A.A.; Azzara, G.; Cappadonia, C.; Martinello, C.; Rotigliano, E.; Conoscenti, C. Predicting Earthquake-Induced Landslides by Using a Stochastic Modeling Approach: A Case Study of the 2001 El Salvador Coseismic Landslides. *ISPRS Int. J. Geo-Inf.* **2023**, *12*, 178. [CrossRef]
26. Du, Y.; Zhang, H.; Xie, M.; Jiang, Y.; Chicas, S.D.; Liu, J. A Possible Mechanism of High-Speed and Long-Distance Rockslides. *J. Earth Sci.* **2024**, *35*, 2158–2162. [CrossRef]
27. Liu, T.; Deng, J. Photogrammetry-Based 3D Textured Point Cloud Models Building and Rock Structure Estimation. *Appl. Sci.* **2023**, *13*, 4977. [CrossRef]

applied sciences

MDPI

Article

Research on Stability Evaluation of Perilous Rock on Soil Slope Based on Natural Vibration Frequency

Yanchang Jia [1,*], Guihao Song [1,*], Luqi Wang [2], Tong Jiang [1], Jindi Zhao [1] and Zhanhui Li [1]

1 College of Geosciences and Engineering, North China University of Water Resources and Electric Power, Zhengzhou 450046, China
2 School of Civil Engineering, Chongqing University, Chongqing 400030, China
* Correspondence: jiayanchang@ncwu.edu.cn (Y.J.); songgh0322@163.com (G.S.); Tel.: +86-18737132511 (Y.J.); +86-18436919556 (G.S.)

Abstract: Perilous rock instability on the soil slope brings a substantial threat to project operation and even people's lives. The buried depth of the perilous rock is a challenge to deal with and primarily determines its stability, and the indirect rapid identification of its buried depth is the key to its stability evaluation. The paper aims to find a new and quick method to measure the buried depth of perilous rock on the soil slope and to solve the hard-to-measure buried depth stability evaluation. When the damping ratio is less than one, and the deformation is linear elastic throughout the amplitude range, the potentially perilous rock vibration model may reduce to a multi-degree-of-freedom vibration one. By theoretical deduction, a quantitative relationship is established among the perilous rock mass, the basement response coefficient, the buried depth of the perilous rock, and the natural horizontal vibration frequency. In addition, the accuracy of this relationship is confirmed via numerous indoor experiments, showing that the horizontal vibration frequency of the perilous rock model in one dimension increases as the buried depth increases. Finally, based on the natural vibration frequency and guided by the limit balance model, a stability evaluation model of the perilous rock on the soil slope is constructed. Hence, the example shows that the method is feasible. The research findings are of vital significance for the stability evaluation of the perilous rock on the soil slope and give a novel approach and theoretical foundation for quick identification and monitoring.

Keywords: perilous rock on soil slope; vibration frequency; stability evaluation; buried depth of perilous rock; perilous rock model

1. Introduction

Perilous rocks on high and steep slopes are unstable, particularly when exposed to environmental or anthropogenic forces, such as precipitation, earthquake, or vibration [1–4]. The safety coefficient of the perilous rock stability on the soil slope refers to the ratio of the anti-overturning torque to the overturning one. When the ratio is greater than one, the slope is stable; when it is equal to one, the slope is in the limit equilibrium state; when it is less than one, the slope is damaged, and the perilous rock becomes unsteady. Then, under the action of gravity, these unstable rocks roll rapidly and seriously affect the project operation and even people's safety. In China, perilous rock is one of the common geological disasters in mountainous areas and along roads. Even though the rock mass involved in a perilous rock collapse is moderate, it features wide distribution, sudden eruption, fast speed, high energy, long movement distance, and difficulty in monitoring and early warning. In addition, it poses a serious threat to people's safety [5–9]. Especially with the implementation and advancement of the western development strategy and the "Belt and Road" initiative, many large-scale infrastructures are being constructed or have been built in southwestern China and along the Silk Road Economic Belt, where the risk of collapses and rockfalls may occur. For example, a building of the Sichuan–Tibet Railway will face engineering geological problems: the collapse of the shallow surface of the slope and the perilous rocks during construction and operation, due to the steep landform along the line, the complex geological environment conditions, and the development of unfavorable geological conditions, such as active fault zones and high ground stress [10,11]. It can be said that collapses and rockfalls are an unavoidable and important challenge in rock slope engineering, and scientific researchers and engineering technicians have paid much attention to it in recent years [12,13].

The buried depth of the perilous rock on the soil slope is an important parameter to evaluate its stability, which is hard to measure directly by instruments, so it brings great challenges for its identification, stability evaluation, and monitoring and early warning. To solve this problem, this paper, through theoretical deduction, model tests, and field tests, constructs a method based on dynamic characteristic parameters to identify the buried depth of perilous rock on the soil slope and then achieves its stability evaluation with the limit equilibrium theory. Research into the best way to assess the safety of a rock on the soil slope involves breaking the process down into qualitative stability assessments and quantitative numerical modeling approaches. It is difficult to achieve a quantitative evaluation of the stability of perilous rock by using qualitative research methods. The stability of the perilous rock may be quantitatively evaluated using the limit equilibrium and numerical simulation methods; however, the primary elements influencing the steadiness of perilous rock, such as its buried depth, size of the bond area, and bond strength of the perilous rock structures, are hard to grasp. Owing to current scientific and technological advancements, a technique has evolved for gauging the stability of the perilous rock based on its variation of dynamic characteristics. Amitrano et al. [14] analyzed the vibration before the collapse of the rock mass on the coast of Normandy in western France and concluded that vibration characteristics, such as vibration amplitude, can play a certain role in early monitoring and forewarning of a potentially fatal rockfall; Got et al. [15] concluded based on the displacement and vibration of a natural limestone cliff rock mass in the southeastern part of Vickers Mountains that spectrum analysis can be used as a supplement to the displacement velocity to provide early warning of perilous rock collapse.

Based on the vibration monitoring technology, Bottelin et al. [16] studied the dynamic response of four unstable rock masses in the Alps, and according to the measurements, there was a distinct energy peak at the resonance frequency of the perilous rock mass's vibration spectrum; Ma et al. [17], by comparing the safety factors of the test blocks, found that the vibration characteristic parameters, such as frequency, have a certain indication effect on the stability of the rock block. After analyzing the mechanical indicators, such as friction force, it was concluded that rock mass safety monitoring based on the vibration mode will play an important role in practical engineering [18–20]; then, Jia Yanchang et al. [21–27] obtained the change rule of the dynamic characteristic parameters of perilous rock with structural damage through tests and a theoretical analysis and realized the quantitative stability evaluation based on the vibration frequency. Numerous investigations have shown that for the stability assessment of the rock blocks it is reliable to use dynamic indicators. When it comes to collapse catastrophes, this is one of the most useful technological tools available. Taking perilous rock on the soil slope as the research object, the paper establishes a quantitative link among the rolling rock mass, the foundation response coefficient, and the buried depth of the perilous rock. Significant progress has been made in determining the natural vibration frequency in the horizontal direction theoretically and computing the buried depth of the perilous rock rapidly. Laboratory experiments confirm the accuracy of the theoretical model. Based on the natural vibration frequency, the stability evaluation model of the perilous rock on the soil slope is then established in conjunction with the limit equilibrium model, and the calculation example demonstrates the feasibility of this approach. The research findings are extremely significant for checking the precariousness of rocks on a slope and provide a novel approach and theoretical foundation for the quick identification and monitoring of perilous rocks.

2. Materials and Methods

2.1. Perilous Rock on Soil Slope and Research Method

A perilous rock appears when a mass of rock on a soil slope splits from its parent body but is still on the slope (see Figure 1), and it may cause a collapse disaster at any time due to gravity or other natural factors, such as rainfall, earthquake, wind action, and animal interference. After the perilous rock collapses, the rockfall often moves downward at a high speed along the slope, with an extremely powerful force of impact. The incidence of perilous rock collapse is connected to factors such as the external force, the nature of the slope on which the perilous rock is located, the rock's proclivity, and its buried depth. The perilous rock vibration model is simplified into a multi-freedom vibration model. The quantitative relationship among perilous rock mass, rock bottom response coefficient, perilous rock buried depth, and natural horizontal vibration frequency is established by theoretical inference. In the experiment, several perilous rock models with different masses and different buried depths were set up, and the vibration frequencies under the same buried depths and the buried depths under the same vibration frequencies were compared, researching the relationship between the buried depths of the perilous rock and the vibration frequencies.

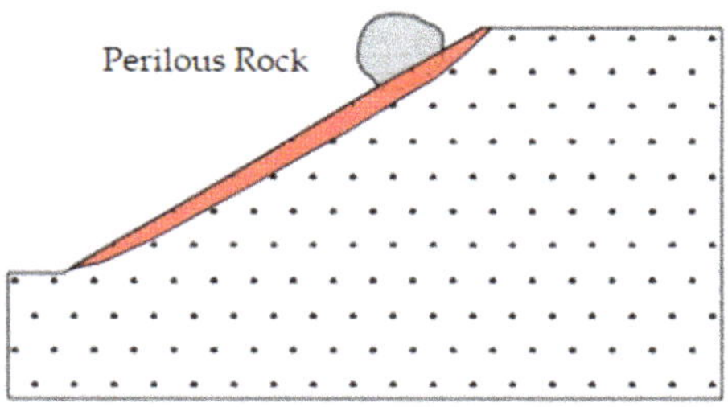

Figure 1. Schematic diagram of perilous rock on a soil slope.

2.2. Dynamic Characteristic Model of Perilous Rock on Soil Slope

Perilous rock formations will shake when prompted by external forces. In the case of the homogeneous isotropic foundation soil, when the damping ratio is smaller than one, the multi-degree-of-freedom rock model simplifies down to a single-degree-of-freedom model. Rock vibration models may be simplified by considering just a few degrees of freedom due to the linear elastic character of the deformation over the whole amplitude range. See Figure 2.

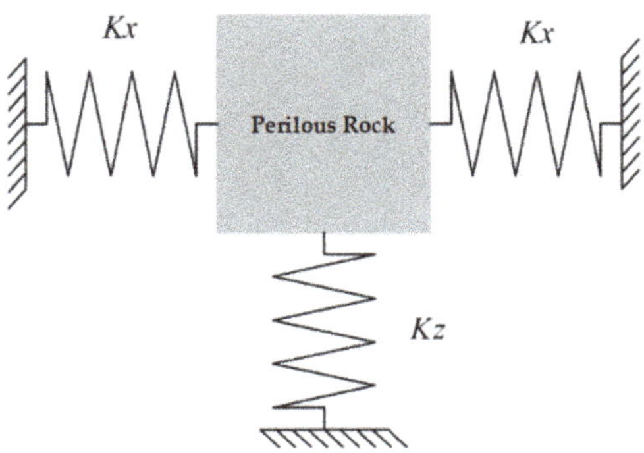

Figure 2. Dynamic characteristic model of perilous rock.

According to dynamic theory, when an outside force causes the rolling rock slope to vibrate, the vibrations travel parallel to the slope. The potentially fatal chasm between the rock and the starting point is initially fixed at x; the horizontal stiffness of the soil layer is K_x, and according to Hooke's law, the perilous rock is subjected to a force $K_x x$ moving in the opposite direction of the shift, calculated by using Newton's second law and the equations of horizontal mechanical equilibrium for the potentially perilous rock:

$$M\ddot{x} = -K_x x \tag{1}$$

$$p^2 = K_x/M \tag{2}$$

The frequency of the rock's natural vibrations, without any dampening, is denoted by P, and its mass, M, is the other key variable. Then, Equation (3) was applied for the free vibration of the perilous rock on the soil slope without any damping.

$$\ddot{x} + p^2 x = 0 \tag{3}$$

We let the equation at time t be solved as:

$$x = e^{xt} \tag{4}$$

We replaced the solution with Equation (3) to obtain the roots of the equations and identified equations:

$$s_{12} = \pm ip \tag{5}$$

$$s^2 + p^2 = 0 \tag{6}$$

Combined with the above equations, the equation's universal answer is:

$$x = C_1 e^{ipt} + C_2 e^{-ipt} \tag{7}$$

We applied Euler's formula to rewrite Formula (7):

$$x = C\cos pt + D\sin pt \tag{8}$$

The magnitudes of the integral constants C and D were defined by the beginning circumstances. By substituting $t = 0, x = x_0, \dot{x} = \dot{x}_0$ into Formula (8), we obtained the value of C and D as:

$$C = x_0, D = \dot{x}_0/p \tag{9}$$

$$x = x_p \cos pt + \sin pt\dot{x}/p \tag{10}$$

The trigonometric transformation of Formula (10) resulted in the following new form:

$$x = A\sin(pt + \alpha) \tag{11}$$

A is a constant, where:

$$A = \sqrt{x_0^2 + (\frac{\dot{x}_0}{p})^2}, \alpha = tg^{-1}\frac{px_0}{\dot{x}_0} \tag{12}$$

The system's vibration period is:

$$T = 2\pi/p = 2\pi\sqrt{M/K_x} \tag{13}$$

The system's vibration frequency is:

$$f = 1/T = \frac{1}{2\pi}\sqrt{\frac{K_x}{M}} \tag{14}$$

The relationship between the stiffness coefficient of the foundation soil K_x and the foundation reaction force coefficient k is:

$$K_x = khb \quad (15)$$

Equation (16) represents the horizontal vibration frequency at which a potentially perilous rock is most likely to oscillate after combining Equations (14) and (15):

$$f = \frac{1}{2\pi}\sqrt{\frac{khb}{M}} \quad (16)$$

Considering the ratio effect of the height of the perilous rock to its breadth parallel to the direction of vibration at its fundamental frequency, it becomes clear that the perilous rock on the soil slope operates against the damping force when the damping is weak. Over time, the amplitude of the vibrations will diminish as the system's mechanical energy is transformed into heat and eventually lost as heat. The resulting Formula (16) can be expressed as:

$$f_d = \frac{\sqrt{1-\xi^2}}{2\pi}\sqrt{\frac{(\frac{a}{H_s})^2}{(\frac{a}{H_s})^2+1}}\sqrt{\frac{khb}{M}} \quad (17)$$

where f_d is the damped system's fundamental vibration frequency (Hz); ξ is the damping ratio of the system (unitless); a is the perilous rock's dimension in the direction of the vibration, m; b is the perilous rock length in the vertical vibrational axis, m; M is the rock's bulk, kg; h is the perilous rock's buried depth, m; H is the height of the precarious rock, m; H_s is the vertical distance from base to top of perilous rock, m.

To further verify the correctness of Formula (17), the following part verifies whether the relationship among the buried depth, mass, and vibration frequency of the perilous rock is consistent with Formula (17) through the model test.

3. Model Test

3.1. Test Equipment

(1) Vibration meter
Specifically, a Microchip Cube wireless vibrometer manufactured by Beijing Beike Andi Technology Development Co., Ltd. (Beijing, China) was used in the experiment with a quick sampling rate, a high signal-to-noise ratio, and wireless data transmission. The frequency of the object's natural vibrations in all three dimensions was determined using the Fourier transform. The exact vibration meter parameters are depicted in Table 1.

Table 1. The performance indexes of Microchip Cube.

Parameter	Minimum	Maximum	Unit
range	−5.0	+5.0	g
sampling ratio	5.0	800	Hz
resolution	0.50	—	μg

(2) Vibration measurement
We attached the vibration detector to the model's summit and then used epoxy glue to connect the detector's fixed plate to the perilous rock block shown in the foreground. The installation of the layout sensor occurred after the epoxy resin had dried and the test model had been taken entirely. The velocity/acceleration in the three dimensions was activated after the hammer excited the wide frequency domain, measuring the object's time-domain representation in Figure 3. After obtaining a representation of the observed object's frequency domain by filtering and the Fourier transform (Figure 4), the measured object's natural vibration frequency could be determined.

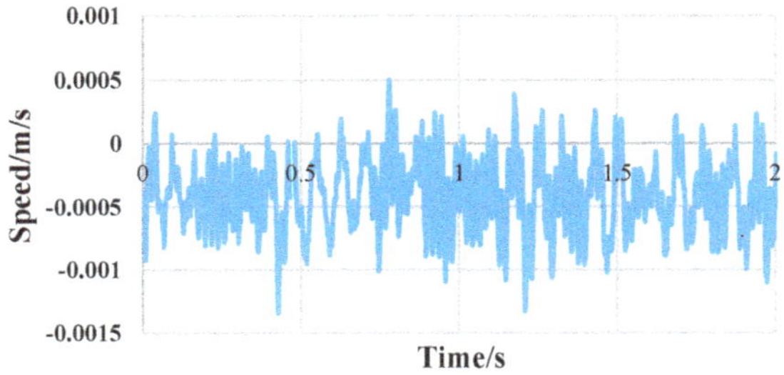

Figure 3. Vibration time-domain diagram.

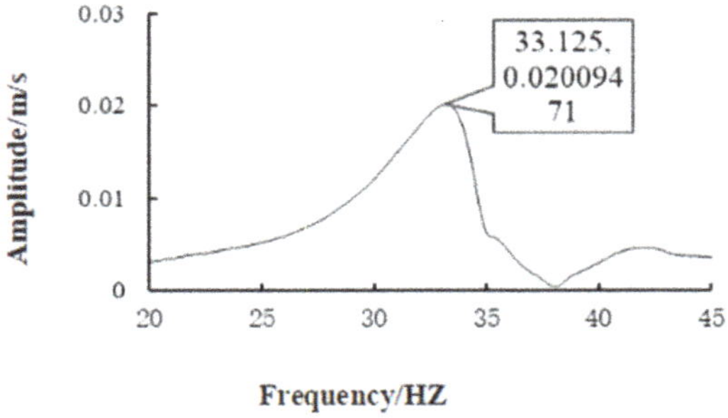

Figure 4. Vibration frequency-domain diagram.

3.2. Experimental Model

(1) Test model

In the experiment, cubes of different sizes and qualities were used to simulate perilous rocks. The dimensions of models A, B, and C were 0.3 m × 0.3 m × 0.4 m, 0.3 m × 0.3 m × 0.3 m, and 0.3 m × 0.3 m × 0.2 m, respectively, and the masses were 84.3 kg, 63.2 kg, and 42.5 kg, respectively. Figure 5 depicts the experiment's perilous rock model.

Figure 5. Perilous rock test models.

(2) Foundation soil

We took the soil on the open-top landslide body on the Dadu River National Road slope as the test foundation soil. The natural density, natural dry density, deformation modulus, and grain group content of the soil were obtained through conventional soil tests. Table 2 displays the standard properties of the foundation soil.

3.3. Test Plan

The models were buried on the soil slope to simulate perilous rocks, and 1/5, 2/5, 3/5, and 4/5 of the model's height governed the buried depth. The slope angle θ remained unchanged. We used a hammer with a wide frequency range to hit the model, and then a vibrometer. The acceleration time–history curve of the perilous rock model was obtained. We obtained the model's power spectrum by first applying a horizontal filter to the acceleration time–history curve and afterward by performing a Fourier transform on the filtered data. The test plan is displayed in Figure 6.

Table 2. Foundation soil parameters.

Test Factor	Average	Unit
natural density (γ)	1.92	g/cm^3
internal friction angle (φ)	20.0	°
coefficient of subgrade reaction (k)	6.5 × 104	kN/m^3
cohesion (C)	7.0	kN/m^2

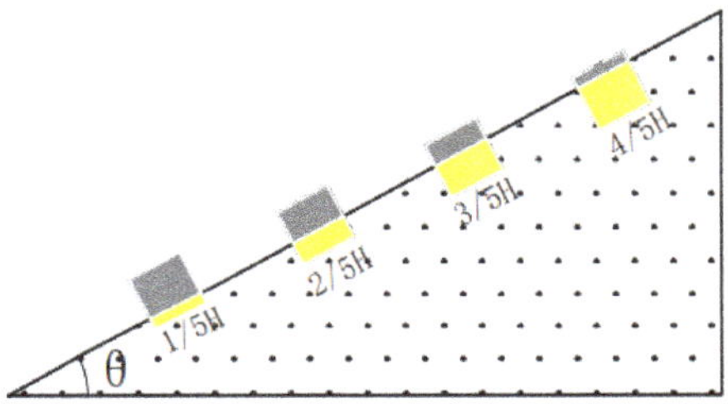

Figure 6. Test scheme.

3.4. Test Results

After the test, perilous rock models with various buried depths had their horizontal natural vibration frequencies measured. Table 3 displays the results of our statistical analysis.

Table 3. Experimental block horizontal vibration frequency.

Models	Buried Depth/cm	Measured Frequency/Hz	Buried Depth Calculation/cm	Error/%
A	8	13.6	7.49	6.37%
A	16	23.2	16.72	4.51%
A	24	31.8	24.61	2.52%
A	32	39.2	31.18	2.56%
B	6	16.5	6.34	5.69%
B	12	25.3	12.36	3.03%
B	18	32.6	17.51	2.73%
B	24	39.9	23.51	2.02%
C	4	17.7	3.84	3.92%
C	8	27.4	8.32	3.97%
C	12	35.4	12.82	6.83%
C	16	40.9	16.26	1.63%

The test findings reveal a maximum error of 6.83%, a minimum error of 1.63%, and an average error of 3.82% for the inferred buried depth of the perilous rocks.

3.5. Discussion

Figure 7 displays the experimentally determined connection among the buried depth of perilous rock, the vibration frequency, and the mass of perilous rock.

(1) Correlation between the buried depth of the perilous rock and the natural vibration frequency.

Figure 7 shows that the natural vibration frequency of the perilous rock grows nonlinearly as the buried depth increases.

By transforming Equation (17), the relationship between the buried depth of the perilous rock and the natural vibration frequency could be obtained as:

$$h = \frac{4\pi^2 M f_d^2 (a/H_s{}^2 + 1)}{kb(1-\xi^2)(a/H_s)^2} \quad (18)$$

Equation (18) demonstrates that as the buried depth of the perilous rock increases, the frequency of its natural vibrations grows steadily in a quadratic fashion. Additionally, the theoretical model has an average inaccuracy of 3.82% when comparing the experimentally observed buried depth of the perilous rock model to the predicted buried depth, showing that there is a strong theoretical relationship between the frequency of the natural vibrations and their buried depth, as proposed in this work. Unlike the manual method to measure the buried depth of the perilous rock, this paper, by measuring its natural vibration frequency, provides the theoretical basis and new technical support for its early identification, stability evaluation, and monitoring and early warning.

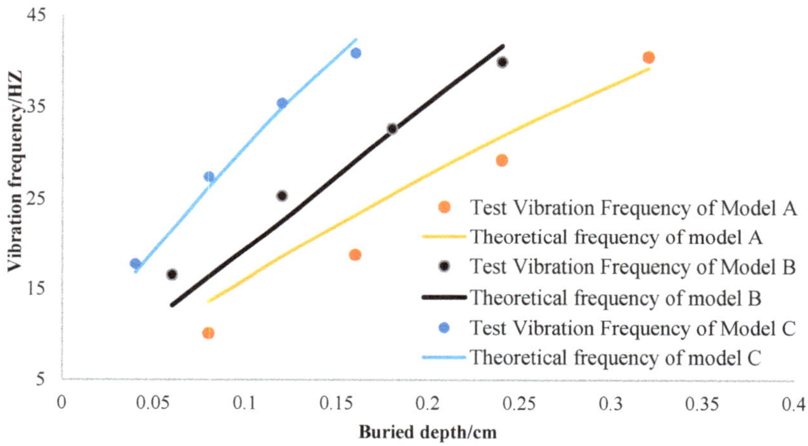

Figure 7. Relationship between natural vibration frequency and buried depth.

(2) Relationship between the natural vibration frequency and the perilous rock mass.

Figure 7 shows that, for a given natural vibration frequency, a lighter model has a shallower buried depth than a heavier one. When the models have the same natural vibration frequency, the buried depths of the models fall in the following order: A > B > C. If two rocks of equal mass are buried at the same depth, the vibration frequency of a perilous rock with a larger mass is lower than that with a smaller mass. The order of the natural vibration frequency of the models under the same buried depth is C > B > A.

Equation (17) leads to the following conclusion on the link between the perilous rock's bulk and its inherent vibration frequency:

$$M = \frac{kbh(1-\xi^2)(a/H_s)^2}{4\pi^2 f_d^2 (a/H_s{}^2 + 1)} \quad (19)$$

The natural vibration frequency reduces quadratically with the enhancing mass of the perilous rock, as shown by Formula (19). Data gathered from experiments involving the magnitude of perilous rocks and the frequency of natural vibrations confirm the accuracy of this result.

4. Perilous Rock Stability Evaluation

4.1. Force Analysis of Perilous Rock

Figure 8 depicts the results of a force study performed on the potentially perilous rock, where F is the friction force between the perilous rock and the soil, N; G is the gravity of the perilous rock, N; E_p is the passive earth pressure, N; E_a is the active earth pressure, N; F_e is the external force of the perilous rock, N; θ is the angle of the slope, °; Q is the water pressure at the bottom, N; z is the depth of the calculation point from the soil surface, m; δ is the friction angle between the perilous rock and the soil, where the value is φ, °.

$$E_p = 0.5\gamma h^2 \tan^2(45^\circ + \varphi/2)b \tag{20}$$

$$E_a = 0.5\gamma(h - z)^2 \tan^2(45^\circ - \varphi/2)b \tag{21}$$

$$z = 2c/(\gamma \tan(45^\circ - \varphi/2)) \tag{22}$$

$$F = G\sin\theta + F_e\cos\theta + E_a - E_p \tag{23}$$

For rolling stones in a rolling state:

$$M_{qf} = \left(\frac{H}{2} - h\right)G\sin\theta + \frac{a}{2}Q + \left(\frac{H}{2} - h\right)F_e\cos\theta + \frac{a}{2}F_e\sin\theta + \frac{1}{3}E_p h^2 \tag{24}$$

$$M_{kqf} = \frac{a}{2}G\cos\theta + \frac{1}{3}E_p h^2 \tag{25}$$

Among them, M_{qf} is the rolling motion state rolling rock overturning moment, and M_{kqf} is the rolling motion state rolling rock anti-overturning moment.

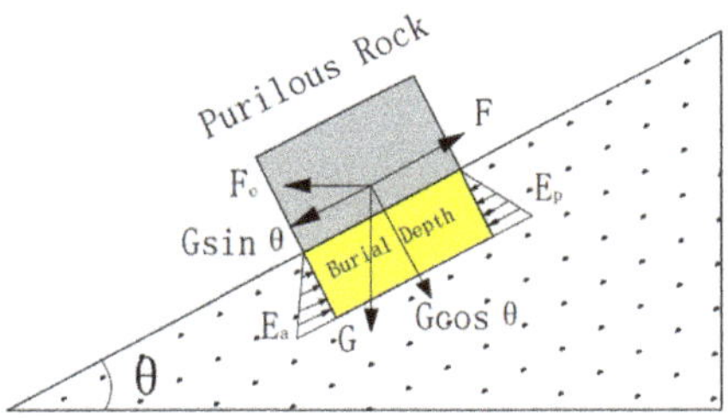

Figure 8. Force diagram of perilous rock.

4.2. Stability Evaluation Model

According to the limit equilibrium model, the safety factor K of the rolling rock under the condition of translational motion was obtained:

$$K = \frac{E_p + F + (ab + 2ah + 2bh)c}{E_a + G\sin\theta} \tag{26}$$

The safety factor K of the rolling stone under the condition of rolling motion state:

$$K = \frac{M_{kqf}}{M_{qf}} \tag{27}$$

Formulas (18), (20)–(23) were put into Formula (26), and Formulas (24) and (25) were put into Formula (27) to obtain the connection between the safety factor of the rolling stones and the natural vibration frequency under various motion states.

4.3. Example

During the impoundment of the Monkey Rock Reservoir in Ganzi Prefecture, Sichuan Province, there were obvious signs of deformation on the outer high retaining wall and the inner slope of the highway at the gully downstream of the open-top landslide. These deformations severely threaten people's safety and property within the S211 Provincial Highway and the downstream reservoir area. According to the site geological survey, the open-top landslide is about 14.5 km away from the Monkey Rock Dam and is located at the right bank of Dadu River in Kairao Village, Gezong Township, Danba County, and about 450 m upstream of Xihegou. The deformation body is about 460 m long along the river, with a top elevation of 2080 m, a bottom elevation of 1820 m or less, and a total volume of about four million cubic meters. The shallow surface of the open-top landslide is composed of colluvium–deluvial large boulders, block stones, and gravelly soil. The underlying surface is Silurian greenschist, containing phyllite. There is an interlayer compression fracture zone in the rock layer, and the surface cracks are mainly distributed in the middle and rear of the deformation body. The maximum boulder size is 40 m $\times$ 10 m $\times$ 15 m, and there are many cracks on the side surface, extending to the bottom of the boulder. The total volume of the boulders is about 150,000 cubic meters.

It is vital to evaluate the safety status of perilous rocks on the soil slope and issue an early warning. The main aspect of determining the stability of the perilous rock is to determine its buried depth. However, the shallow surface of the open-top landslide has complex terrain and a harsh geological environment, also with many unknown dangerous factors. Therefore, it is extremely dangerous to carry out manual operation, adding difficulties to directly measuring the buried depth of the perilous rock. The traditional surveying method is limited; hence, it is urgent to find a more practical measurement method to conform to the project's reality.

This paper establishes the evaluation model of the perilous rock on the soil slope. Based on the natural vibration frequency, the buried depth of the perilous rock can be determined quickly and efficiently, and its safety state can be evaluated. Firstly, the natural vibration frequency and the volume of perilous rocks in the horizontal direction were measured, and their vibration model was simplified into a multi-degree-of-freedom vibration model. Secondly, the buried depth was calculated according to the relationship between the natural vibration frequency of the perilous rock and its buried depth in the Formula (17). Finally, the safety factor was further determined according to the equilibrium conditions, and the safety state of the perilous rock was evaluated. Table 4 shows the parameter values of the two perilous rocks (see Figure 9) on the open-top landslide on the bank slope of the Dadu River Reservoir.

Table 4. Parameter indexes of the perilous rocks.

Length of Perilous Rock	/cm	Width/cm	Height/cm	Mass/kg	Slope Angle/°
A	3.33	5.21	7.23	327,386.50	25.00
B	2.85	4.89	4.56	165,866.65	25.00

The average monthly precipitation in the Dadu River Basin is 100–700 mm, and it reaches the maximum in July and August at nearly 700 mm. The basic earthquake intensity is VII degrees. On-site monitoring data show that the natural vibration frequency of perilous rock A is 3.95 Hz and that of perilous rock B is 4.28 Hz. The natural density of the foundation soil is 1.92 g/cm^3; the internal friction angle is 20.0°; the foundation reaction force coefficient is 6.5 $\times$ 104 kN/m^3, and the cohesion is 7.0k N/m^2. We substituted the parameters of perilous rock A and B into Formula (18) to calculate that the buried depth of perilous rock A is 2.18 m and that of perilous rock B is 1.05 m. According to Formula (26), the safety factor of perilous rock A is 1.11, and that of the rolling rock B is 1.11 under the condition of translational motion. According to Formula (27), the safety factor of rolling rock A is 2.33 and that of rolling rock B is 2.83 under rolling motion. Both perilous rocks are in a very stable condition.

(a) (b)

Figure 9. Perilous rocks. (**a**) A is a perilous rock on the open-top landslide; (**b**) B is a perilous rock on the open-top landslide near A.

4.4. Research Prospects

In the alpine canyon areas of southwestern China, a lot of perilous rocks exist near railways and highways. The government attaches great importance to investigating and cleaning them, with massive governance in progress. The research results are being piloted and popularized in the Monkey Rock and Baihetan Hydropower Station. The research is in the early stage. The next step is to take the remote non-contact laser vibration measuring instrument to replace the current contact wireless sensor. The perilous rock areas are difficult to reach by manpower in high mountain and canyon areas. Therefore, the remote non-contact laser vibration measuring instrument can be used to survey the natural vibration frequency of perilous rocks on the slope and ensure the safety of technicians. At the same time, the pilot application is strengthened more to further verify and improve the early identification model of the perilous rock proposed in this paper.

5. Conclusions

(1) This paper takes the perilous rock on the soil slope as the research object, assuming that the foundation soil is homogeneous and isotropic, the damping ratio is less than 1, and the deformation is linear elastic deformation in the amplitude range. The vibration model of the perilous rock on the slope was simplified into a vibration model with multiple degrees of freedom. Based on the dynamic theory, the quantitative relationship among the natural vibration frequency, mass, side length, deformation modulus of the foundation soil, and buried depth of the perilous rock on the soil slope was established; that is, the dynamic characteristic model of the perilous rock on the slope was set up.

(2) The indoor model tests were carried out with models of different mass, different side lengths, and different buried depths. The experimental findings demonstrate that as the buried depth of perilous rock increases the frequency of its natural vibrations grows steadily in a quadratic fashion. The natural amplitude vibrations decrease quadratically as the mass of potentially perilous rock increases. Table 3 shows that the dynamic characteristic model of the perilous rock developed in this research is accurate, as predicted by the tests.

(3) Combined with the dynamic characteristic model and limit equilibrium model of the perilous rock on the soil slope, the evaluation model of perilous rock stability based on the natural vibration frequency was established, and the model was verified to be correct by a field test. Based on the research results, the buried depth of the perilous rock can be calculated indirectly by measuring its natural vibration frequency. Its safety coefficient on the soil slope can be calculated. In this way, these calculations offer the theoretical basis and technical support for early identification, stability evaluation, and monitoring and early warning.

Author Contributions: Y.J. and L.W. contributed to the conception of the study and performed the experiment; Y.J. and G.S. were responsible for data collection; Y.J., G.S. and T.J. made significant edits to the manuscript; Y.J., L.W. and G.S. were in charge of making important revisions to the manuscript; J.Z. and Z.L. helped perform the analysis with constructive discussions. All authors have read and agreed to the published version of the manuscript.

Funding: This research was funded by the National Natural Science Foundation of China (No. U1704243) and the Natural Science Foundation granted by the Department of Education, Anhui Province (No. KJ2020A0235).

Institutional Review Board Statement: Not applicable.

Informed Consent Statement: Not applicable.

Data Availability Statement: Not applicable.

Conflicts of Interest: The authors declare no conflict of interest.

References

1. Hung, W.-Y.; Tran, M.-C.; Yeh, F.-H.; Lu, C.-W.; Ge, L. Centrifuge modeling of failure behaviors of sandy slope caused by gravity, rainfall, and base shaking. *Eng. Geol.* **2020**, *271*, 105609. [CrossRef]
2. Zhang, W.-G.; Meng, F.-S.; Chen, F.-Y.; Liu, H.-L. Effects of spatial variability of weak layer and seismic randomness on rock slope stability and reliability analysis. *Soil Dyn. Earthq. Eng.* **2021**, *146*, 106735. [CrossRef]
3. Pan, Y.; Wu, G.; Zhao, Z.; He, L. Analysis of rock slope stability under rainfall conditions considering the water-induced weakening of rock. *Comput. Geotech.* **2020**, *128*, 103806. [CrossRef]
4. Romeo, S.; Di Matteo, L.; Melelli, L.; Cencetti, C.; Dragoni, W.; Fredduzzi, A. Seismic-induced rockfalls and landslide dam following the October 30, 2016 earthquake in Central Italy. *Landslides* **2017**, *14*, 1457–1465. [CrossRef]
5. He, S.M.; Shen, J.; Wu, Y. Rock shed dynamic response to impact of rock-fall. *Rock Soil Mech.* **2011**, *32*, 781–788.
6. Cappadonia, C.; Cafiso, F.; Ferraro, R.; Martinello, C.; Rotigliano, E. Rockfall hazards of Mount Pellegrino area (Sicily, Southern Italy). *J. Maps* **2021**, *17*, 29–39. [CrossRef]
7. Jacklitch, C.; Shakoor, A.; Lund, W.R. Evaluation of Rockfall-Hazard Potential for Rockville, Utah, Following a 2013 Fatal Rockfall. *Environ. Eng. Geosci.* **2018**, *24*, 143–163. [CrossRef]
8. Zhang, J.; Liang, Z.; Feng, D.; Zhang, C.; Xia, C.; Tu, Y. Response of the buried steel pipeline caused by perilous rock impact: Parametric study. *J. Loss Prev. Process. Ind.* **2016**, *43*, 385–396. [CrossRef]
9. Jiang, N.; Li, H.-B.; Zhou, J.-W. Quantitative hazard analysis and mitigation measures of rockfall in a high-frequency rockfall region. *Bull. Eng. Geol. Environ.* **2021**, *80*, 3439–3456. [CrossRef]
10. Xue, Y.; Kong, F.; Yang, W.; Qiu, D.; Su, M.; Fu, K.; Ma, X. Main unfavorable geological conditions and engineering geological problems along Sichuan-Tibet railway. *Chin. J. Rock Mech. Eng.* **2020**, *39*, 445–468. [CrossRef]
11. Xu, H.; Zhou, T.Y.; Wang, X.-Y. Compression test and numerical simulation research on improved red beds subgrade fillers in Sichuan-Tibet Railway. *Rock Soil Mech.* **2021**, *42*, 2259–2268. [CrossRef]
12. Liu, G.; Li, J. Research on the Effect of Tree Barriers on Rockfall Using a Three-Dimensional Discontinuous Deformation Analysis Method. *Int. J. Comput. Methods* **2019**, *17*, 1950046. [CrossRef]
13. Wang, X.L.; Zhang, L.Q.; Zhang, Z.J.; Zhang, A.M.; Yun, G.R. Rockfall hazard analysis of slope at sutra caves of Shijing Mountain. *Rock Soil Mech.* **2012**, *33*, 191–196.
14. Amitrano, D.; Grasso, J.R.; Senfaute, G. Seismic precursory patterns before a cliff collapse and critical point phenomena. *Geophys. Res. Lett.* **2005**, *32*, 1–5. [CrossRef]
15. Got, J.-L.; Mourot, P.; Grangeon, J. Pre-failure behaviour of an unstable limestone cliff from displacement and seismic data. *Nat. Hazards Earth Syst. Sci.* **2010**, *10*, 819–829. [CrossRef]
16. Bottelin, P.; Jongmans, D.; Baillet, L.; Lebourg, T.; Hantz, D.; Levy, C.; Le Roux, O.; Cadet, H.; Lorier, L.; Rouiller, J.-D.; et al. Spectral Analysis of Prone-to-fall Rock Compartments using Ambient Vibrations. *J. Environ. Eng. Geophys.* **2013**, *18*, 205–217. [CrossRef]
17. Ma, G.C.; Sawada, K.; Yashima, A.; Saito, H. Experimental Study of the Applicability of the Remotely Positioned Laser Doppler Vibrometer to Rock-Block Stability Assessment. *Rock Mech. Rock Eng.* **2015**, *48*, 787–802. [CrossRef]
18. Du, Y.; Xie, M.; Jiang, Y.; Li, B.; Gao, Y. Methods for determining early warning indices Based on natural frequency monitoring. *Rock Soil Mech.* **2015**, *36*, 2284–2290. (In Chinese) [CrossRef]
19. Jia, B.; Wu, Z.; Du, Y. Real-time stability assessment of unstable rocks based on fundamental natural frequency. *Int. J. Rock Mech. Min. Sci.* **2019**, *124*, 104134. [CrossRef]
20. Du, Y.; Xie, M.; Jiang, Y.; Li, B.; Gao, Y.; Liu, Q. Safety monitoring experiment of unstable rock based on natural vibration frequency. *Rock Soil Mech.* **2016**, *37*, 3035–3040. (In Chinese) [CrossRef]
21. Jia, Y.; Xie, M.; Chang, S. A model for evaluation of stability of sliding- and falling-type dangerous rock blocks based on natural vibration frequency. *Rock Soil Mech.* **2017**, *38*, 2149–2156. (In Chinese) [CrossRef]

22. Jia, Y. *Study on Stability Model of Slope Dangerous Rock Mass based on Dynamic Characteristics*; University of Science and Technology Beijing: Beijing, China, 2018.
23. Jia, Y.; Jiang, T.; Liu, H. Stability evaluation model of toppling unstable rock based on natural vibration frequency. *J. Appl. Sci. Eng.* **2020**, *24*, 177–184. [CrossRef]
24. Jia, Y.; Song, G.; Jiang, T.; Yu, G.; Liu, H.; Pan, X. Buried depth calculation of the slope of the unstable rock based on natural vibration frequency. *Front. Earth Sci.* **2022**, *10*, 929825. [CrossRef]
25. Jia, Y.; Jiang, T.; Yin, Y.; Yu, G.; Wang, L. Experimental research on damage identification of topping dangerous rock structural surface based on dynamic characteristic parameters. *Front. Earth Sci.* **2022**, *10*, 942025. [CrossRef]
26. Zhang, W.; He, Y.; Wang, L.; Liu, S.; Meng, X. Landslide Susceptibility Mapping Using Random Forest and Extreme Gradient Boosting: A Case Study of Fengjie, Chongqing. *Geol. J.* **2023**. [CrossRef]
27. Liu, S.; Wang, L.; Zhang, W.; He, Y.; Pijush, S. A comprehensive review of machine learning-based methods in landslide susceptibility mapping. *Geol. J.* **2023**. [CrossRef]

geosciences

MDPI

Article

Influence of Relative Sea-Level Rise, Meteoric Water Infiltration and Rock Weathering on Giant Volcanic Landslides

Julien Gargani

Geops, CNRS, Université Paris-Saclay, 91405 Orsay, France; julien.gargani@universite-paris-saclay.fr

Abstract: Recent studies have shown that giant landslides correlate with climatic variations. However, the precise processes involved in this phenomenon need to be better defined. This study investigates the causes of giant landslides using a modeling approach. Here, I show that the effect of meteoric water infiltration could be distinguished from that of the sea level rise in triggering paleo-landslides. It is possible to identify the cause of coastal paleo-landslides based on the age of occurrence and comparison with climatic signals when glacial maxima are wetter than during interglacial periods, as in Polynesia and East Equatorial Africa, but not in other cases (Caribbean, Indonesia). The role of pore-pressure variations and sea water loading variations is discussed. The interaction between the relative sea level rise, pre-existing relief and deep weak structure due to the presence of highly weathered lavas may trigger the conditions for a large landslide. Highly weathered lavas have very low friction angles in volcanic islands. When volcanoes are still active, pressure fluctuations in the magma chamber caused by sea level lowering are expected to play a significant role in the destabilization of the relief. Competing processes in real cases make it difficult to distinguish between these processes.

Keywords: landslide; alteration; angle of friction; pore pressure; subsidence; sea level

1. Introduction

Landslides are one of the main processes that destroy relief and displace material. A dramatic example is the catastrophic Vajont landslide in Italy in 1963 in relation to water-level rise and rainfall increase [1]. Catastrophic landslides are often expected in areas of active seismicity [2] or in the context of significant volcanic activity [3,4]. Volcano slopes are a major concern in terms of civil engineering monitoring and risk prevention because important cities are built on potentially unstable slopes. In the case of active seismicity, the acceleration of the ground surface and subsurface may generate slope destabilization. From a theoretical perspective, the slope stability depends strongly on the mechanical properties of the rocks [5], on the geometry of weakness zones, on the loading and stress conditions [2,6–9] and on the presence of fluids [1,10–12]. In volcanic areas, the interaction between magma reservoirs, magma intrusions, and pre-existing faults influences the deformation of volcanic surfaces [13–15] and could generate landslides. Phreatomagmatic processes [16], pore pressure increases due to precipitation [17], and rift zone intrusions [13] are thought to influence slope instability on volcanic edifices. The study of coastal volcanism and volcanic islands is essential to a better understanding of volcanic edifices' growth, their dismantling mechanism and their activity [18]. Fundamental destabilization processes have been observed in context, such as in Mount Etna, where the slide occurred in response to the loading of the edifice on a weak basement [19], highlighting the significant role of gravitational spreading and sliding on volcanic edifices [20].

From a practical perspective, the present-day slope stability conditions of a volcanic edifice are influenced by factors such as its past eruptive history, possible landslides, faults and fractures, and alteration by fluid circulation, which could have created significant heterogeneities. Such heterogeneities may hide weaknesses and the potential occurrence of slope instabilities [21,22]. Thus the long geological history of old volcanic edifices is expected to produce complex geometries that are often difficult to understand in detail. A complex volcano can be defined by a succession of different eruptive phases that contribute to the building, destroying and re-building parts of the volcano again (Figure 1). On the flanks of volcanoes and in their internal structure, the rocks may consist of fresh or weathered massive lavas, tephra-fall deposits, ignimbrites or low cemented pyroclasts, among other rocks. This study focuses on volcano slope instabilities caused by non-eruptive processes. The volcano could be active or inactive, and the rocks composing the volcano are affected by phreatic activity (i.e., hot water and gas circulation) that would have weakened their mechanical properties. The eruptive history of a volcano, as well as the nature of the constituent rocks of its flanks, must be carefully considered when assessing the slope stability of volcanoes from a geotechnical perspective. These complex scenarios could lead to unfavorable conditions for the stability of the present-day slope and difficulties in taking the spatial heterogeneity of the mechanical parameters into account (Figure 1).

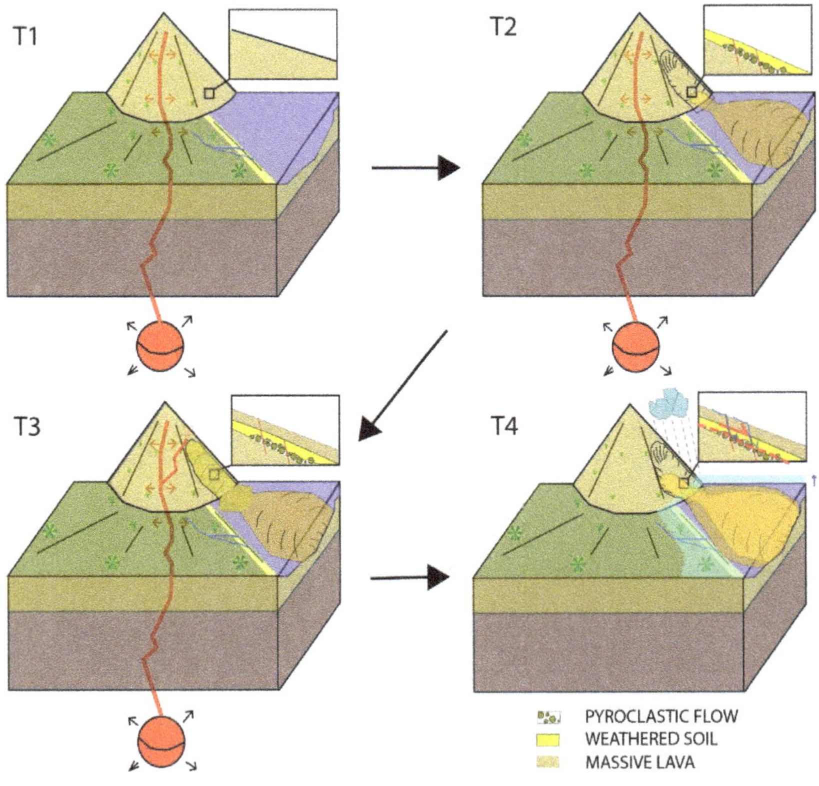

Figure 1. Complex evolution of volcanic edifices and locations of deep weakness zone, (**T1**) Young volcanic edifice composed of massive lavas, (**T2**) Volcanic edifice after a giant landslide caused by a volcanic process, where pyroclastic flows, as well as soil and lava weathering, occur, (**T3**) Filling of the landslide scar by thick volcanic lavas above the pyroclastic debris and the weathered soils, (**T4**) Deep rooting of the landslide into the weakness zone favored by meteoric water infiltration and/or sea level loading in the case of an old and complex volcano and/or the presence of deeply incised canyons.

It has been suggested that giant landslides could be correlated with climatic variations [16]. Some studies assume that giant landslides may have occurred during periods of a low stand [23], while others suggest that high stands or sea level rise are more favorable for causing slope instability [24]. This issue remains controversial and poorly studied, despite the expected sea level rise associated with climate warming.

The role of climatic conditions in triggering giant landslides on volcanoes is investigated in this study. The possible occurrence of a giant landslide on a volcano in relation to climatic variations is investigated from a theoretical perspective. The processes favoring instability that are discussed in this study are (i) Sea water loading with relative sea level rise, (ii) Pore pressure changes related to climatic variation, (iii) Rock and soil weathering, (iv) Weakness zone geometries, (v) Pressure fluctuations of the magma reservoir caused by sea level unloading.

2. Method

2.1. Slope Stability Model

In order to analyze the causes of stress that may trigger a giant landslide and to discuss the stability of a given relief, a 2D model was implemented. Two different geometries of the landslide were implemented to compare the role of the geometry of a weakness zone on the theoretical stability of a volcanic edifice. The first geometry (Figure 2A) is based on the Cullman wedge model [25]. Basically, a balance is achieved between the weight of the expected landslide and the resistance force generated by the rock properties [26]. The resistance shear stress along the sliding surface is given by $\tau_c = \sigma_n \tan\phi + C$ until the landslide occurs, where σ_n is the normal stress, ϕ is the friction angle of the slope-forming material, and C is the cohesion, and could be used to describe bedded rocks [27]. Considering the angle for which the effective cohesion is maximized for a value equal to $\beta/2 + \phi/2$, it is assumed that $\theta = \beta/2 + \phi/2$ [25,28]. This model allows us to calculate the maximum stable height for a simple geometry (Figure 2A):

$$H = 4C{\cdot}\sin\beta\cos\phi/(\rho_r\, g[1 - \cos(\beta - \phi)]) \quad (1)$$

where H is the maximum stable height of the slope, β is the hillslope gradient, ρ_r is the bulk rock density, and g is the gravitational acceleration. This model has been widely used in geomorphological studies to predict the maximum unfailed height of slopes [25]. In the theoretical calculation, vertical loading, pore pressure increase and volcanic rock properties with depth are included. Including the pore pressure U and considering the load of the sea water column of thickness H_m with a bulk density of the water ρ_m, the following can be obtained:

$$H^2 - 4H(C\sin\beta\cos\phi - U\sin\phi\cos\beta)/(\rho_r\, g\,[1 - \cos(\beta - \phi)]) + \rho_m H_m^2/\rho_r = 0 \quad (2)$$

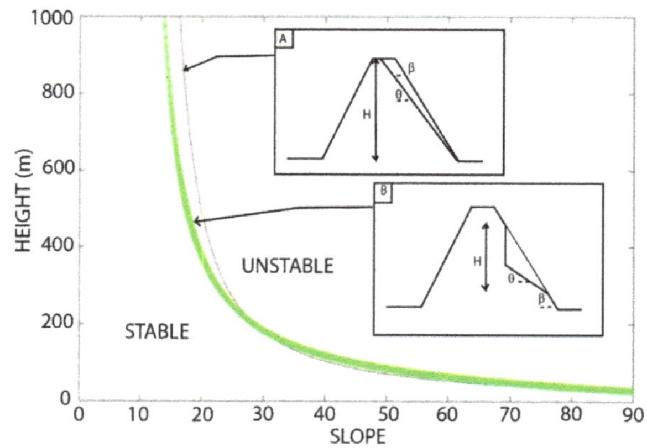

Figure 2. Effect of the initial slope β on the maximum stable relief H for two different landslide geometries. (**A**) Cullman wedge model, (**B**) concave geometry model. $\rho_c = 3000$ kgm^{-3}, C = 1 MPa, $\phi = 10^0$.

Equation (2) allows us to estimate the geometry of the maximum stable relief (height and slope) immediately before the collapse, as well as the geometry of the landslide. The thickness of the sea water column considered could include sea level variations related to Quaternary climate variations, but also the relative sea level rise due to subsidence.

Using the same method, it is also possible to calculate the maximum stable height for a second kind of landslide geometry (Figure 2B, concave geometry model) using the following equation:

$$H = H_1 + H_1 \cos\beta \sin(\beta/2 + \phi/2)/\sin(\beta/2 - \phi/2) \tag{3}$$

where H_1 can be calculated by resolving the following equation:

$$\begin{aligned} &H_1^2 - 4H_1 (C\cos\phi - U\sin\phi)/(\rho_r\, g[\sin\beta - \sin\phi]) \\ &+ \rho_m H_m^2 \sin(\beta/2 - \phi/2)/(\rho_r \sin\beta \cos\beta/2 + \phi/2) = 0 \end{aligned} \tag{4}$$

This study focuses on the role of pore pressure, water loading and mechanical properties of rocks on slope stability. Geotechnical approaches are also useful to evaluate the safety factor F_S = resisting forces/driving forces $\approx [\tau \times S]/[\rho_r\, V g \sin\theta]$ when the slope is destabilized only by its own weight [6,24,29].

2.2. Mechanical Properties of Volcanic Rocks

There are at least two ways to estimate the effective mechanical properties of rocks in a given geological context. The first method is to use experimental data that describe the mechanical behavior of a rock close to that of the studied area. The second is to propose hypotheses about the causes of failure (i.e., gravity, pore pressure, ground acceleration, eruption) and to estimate the effective values of the mechanical parameters that triggered the observed failure. Here, we verify that the mechanical properties of the rocks estimated using a modeling approach are not in contradiction with experimental data for volcanic rocks.

The parameterization of the model was conducted, assuming that the relief was close to critical stability. The effective mechanical properties of the rocks of the volcanic edifice considered in this numerical experiment were calibrated using some characteristics of a giant landslide that occurred in Tahiti 872 kyrs ago [30–33]. However, the investigation presented in this study is purely theoretical, and the geometries, as well as the rock heterogeneity, do not represent a real case.

The relief was assumed to have a total height of ~4500 m with an initial slope β of 8–12°. Two different geometries were tested to estimate the maximum stable relief H as a function of the initial slope (Figure 2). The effect of the precise geometry (Cullman wedge model vs. concave geometry model) of the landslide on the critical values of height and the initial slope of the stable relief is not negligible (Figure 2), but this study does not focus on the precise geometry of landslides.

The mechanical properties of volcanic rocks are strongly dependent on their alteration [34] and stress loading. The cohesion and angle of friction were estimated by Rodriguez-Losada et al. [5] for different volcanic rocks under different slope heights. Cohesion C can be obtained using the equation $C = kH^l$, where H is the slope height, and k and l are coefficients determined experimentally for different volcanic rocks [5]. For weathered, massive lavas, $k = 0.04$ and l varies from 0.41 to 0.60. For weakly cemented pyroclasts, $k = 0.05$ and $l = 0.5$. The angle of friction ϕ could also be estimated using the equation $\phi = a\, ln(H) + b$, where a and b are obtained experimentally. Rodriguez-Losada et al. [5] showed that very low friction angles of <13° are obtained at depths >1000 m for low cemented pyroclasts ($a = 6$; $b = 54$). A volcanic edifice composed of low cemented pyroclasts has cohesion values between 0.5 and 1 MPa at depths of 1500–3000 m (Figure 3A). At this depth, a theoretical volcanic edifice with weakly cemented pyroclasts has friction angles between 6° and 10° (Figure 3B).

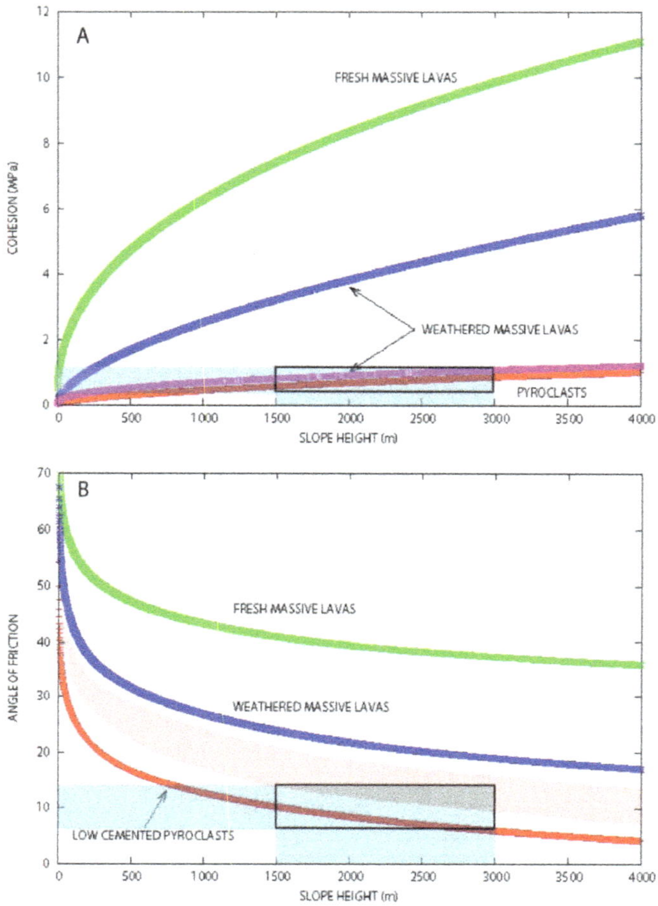

Figure 3. Effect of depth on mechanical properties. (**A**) Cohesion for low cemented pyroclasts (k = 0.05; l = 0.5) and fresh (k = 0.37; l = 0.60) and weathered (k = 0.04; l = 0.41 and 0.60) massive lavas, (**B**) Angle of friction for low cemented pyroclasts (a = 6; b = 54) and fresh (a = 5.6; b = 82) and weathered (a = 7; b = 75) massive lavas.

2.3. Modeling Hydroclimatic Variations

In this study, sea level loading and pore pressure were simulated as a function of climate evolution. The water column loading considered was estimated using a proxy. Sea level is known to change almost simultaneously with $\delta^{18}O$ variation [35,36]. A $\delta^{18}O$ curve [37] was used to simulate climate forcing. This curve was recalibrated to simulate Quaternary sea level variations reaching an amplitude of 120 m.

Hydroclimatic variations are also caused by Milankovitch astronomic cycles and can be simulated using insolation or $\delta^{18}O$ curves [38]. In this study, it is assumed that more humid conditions are capable of causing an increase in the effective pore pressure. The purpose of this assumption is to discuss the timing of potential giant landslides in relation to Quaternary climatic conditions (i.e., precipitation rates). The time required for the water to flow from the surface to a depth of 4.5 km was estimated to be ~150 days in Oregon [39]. In the model, the propagation is considered instantaneous, which is justified on a long time scale (>1 kyr). The amplitude of the variation of the pore pressure ΔP due to water infiltration into the crust is 0.01 MPa at Mt. Hood [39] and 2 MPa on the south flank of Kilauea volcano [17]. In this study, different values for the pore pressure variation were tested in the case where sea level effects are dominant ($0 < \Delta P < 0.125$ MPa) and in the case where pore pressure processes are dominant ($0 < \Delta P < 0.5$ MPa).

2.4. Pressure Variation in the Magma Reservoir

The loading fluctuation caused by sea level variation is $\Delta L_{SL}(t) = g\rho_m H_m(t)$.

Sea water unloading, which causes a decrease in lithostatic pressure at depth, enhances magma production [40]. Decompression favors partial melting of the mantle and magma release. The magma production rate DF/Dt at constant entropy S can be estimated as [40,41]:

$$DF/Dt = (\delta F/\delta P_M)_S\ (dP_M/dt - U \cdot \nabla P_M) \quad (5)$$

where P_M is the magma reservoir pressure, F is the melt fraction, t is the time, and U is the mean mantle upwelling rate.

The magma pressure variation $\delta\Delta P_M/\mathrm{dt}$ can be considered as a function of the sea-water unloading variation $\delta\Delta L_{SL}/dt$ under adiabatic conditions and for very high viscosity (i.e., η_c=10^{23} Pa.s) of the crustal rocks. In this case, $\delta\Delta P_M/\mathrm{dt} = -\delta\Delta L_{SL}/dt$.

When the viscous response is not negligible ($\eta_c < 10^{22}$ Pa.s; [40]), the equation is $\delta\Delta P_M/\mathrm{dt} + \Delta P_M\ (E_c/\eta_c) = -\delta\Delta L_{SL}/dt$, where E_c is the elastic modulus, and a delay of ~10 kyr after the forcing by sea water unloading is expected.

3. Results

The current question that we attempt to answer focuses on the thresholds and conditions that favor slope instability for volcanic edifices in relation to climate variations.

3.1. Influence of Water Column Loading on Slope Stability

Considering a simple geometry for the expected landslide (Figure 2A, Cullman wedge geometry), the role of sea level variation on slope stability was estimated using Equation (1). As a volcanic island subsides, there is an increase in relative sea level, consequently, an increase in water loading on the slope, which could be significant, enough to trigger a potential landslide.

Under the Cullman wedge geometry condition, an increase in loading by a water column of 120 m causes a slight decrease in the stability height (Figure 4C). The more the increase in water loading, the less the stability. Irrespective of the geometry of the sliding mass, the maximum height of the stable relief is affected by the increase in the loading by a water column of 400 m. A water column increase of 400 m is reached in less than 1 Myr for a moderate subsidence rate (~0.25 mm/yr) contemporaneously with a sea level high of +120 m during an interglacial period. In this case, slope instabilities are observed on higher slopes (>25°) but also on smaller slopes (Figure 4). The role of sea water loading could trigger landslides of smaller dimensions (150–300 m) for higher slopes (25–30°, Figure 4C, dark gray square).

3.2. Influence of the Pore Pressure on Slope Stability

Our model suggests that a significant effect of the pore pressure increase on the height of the stable relief could occur (Figure 4). The pore pressure increase significantly reduces the stability of the relief. As expected, the higher the pore pressure is, the lower the maximum height H before the giant landslide occurs. If the pore pressure increases by more than 1 MPa, the slope stability is significantly reduced (Figure 4A,B). A relief of more than 1500 m, with a slope of ~12°, could be affected by landslides under these conditions. Depending on the initial relief H, a pore pressure increase of 1 MPa could cause instabilities of more than 1000 m.

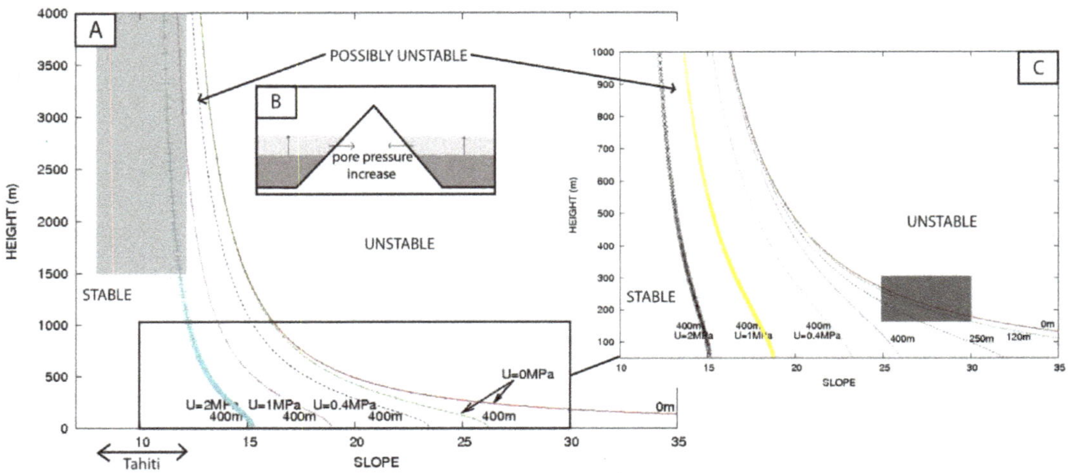

Figure 4. (**A**) Effect of increasing pore pressure and sea water loading on the relief stability. The light gray area represents the field where a theoretically stable relief changes to an unstable relief when submitted to an increase in the sea water column loading between 0 and 400 m and an increase in pore-water pressure between 0 MPa and 2 MPa. (**B**) Schematic representation of the effect of subsidence. (**C**) The details are as follows: a relief of ~200 m with a slope of $25° < \beta < 30°$ could be affected by water loading (dark gray square). ρ_c = 3000 kgm^{-3}, C = 1 MPa, ϕ = 10° and the geometry of the landslide is identical to those presented in Figure 2A.

3.3. Competing Influence of Sea Level Loading vs. Pore Pressure Variations over Time on Slope Stability

In this study, the influence of sea level loading and pore pressure variations over time was modeled in relation to meteoric water infiltration. The variation over time of the maximum height H of a volcanic edifice before the occurrence of a giant landslide is shown in Figure 5. The maximum stable height depends on the sea level and on the pore pressure value. Taking into account that the maximum sea level fluctuation during the last million years is ~120 m, and that the amplitude and the timing of the loading are known (Figure 5B), the variation in the maximum stable height H of a theoretical volcano before a giant landslide occurs was calculated. The absolute value of the maximum stable height H depends strongly on the mechanical properties of the rocks and the slope. This issue is not discussed in this study because the focus of this study is on the variation of stability caused by the climatic evolutions.

If sea water loading is the main process influencing landslides, then the higher the sea level is, the lower the stability will be. The maximum height H is expected to decrease as the sea level rises (Figure 5). With all other parameters being equal, the maximum stable height H of a volcanic edifice during interglacial periods is 50 m higher than during glacial periods considering the sea water loading is the main parameter influencing the system.

In the specific case of Polynesian climatic variation, where the climate is more humid during glacial maxima [42], as simulated here, the giant landslides generated by a pore pressure increase are expected to occur during glacial periods (Figure 5A). In the specific case of the Polynesian volcanic edifice, the period when the landslides are expected (i.e., the time when the maximum stable height H before a giant landslide is expected to occur) is different when the dominant process triggering the landslide is sea level loading or pore pressure loading (Figure 5A).

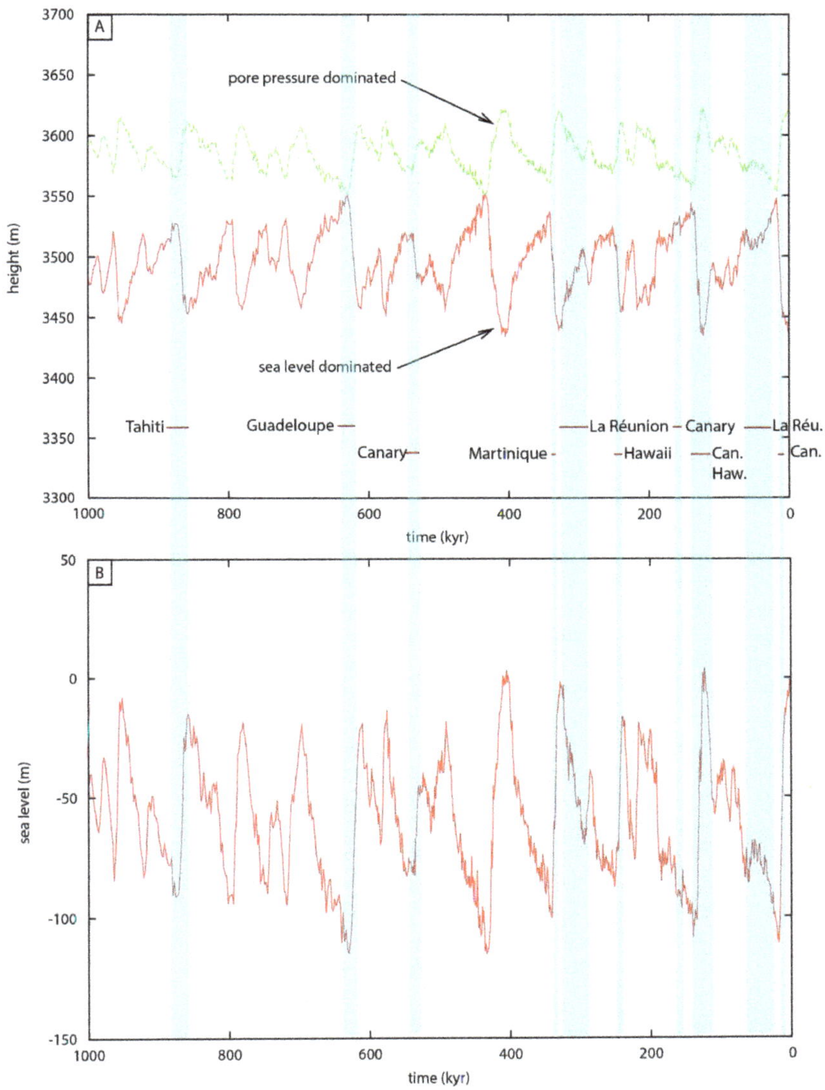

Figure 5. (**A**) Maximum height H before the landslide occurred. The geometry of the island is shown in Figure 4. The effects of sea level loading and pore pressure are simulated. The sea water loading depends on the variation in sea level, while the effect of the pore pressure depends on the fluctuation in precipitation. In the first case (sea level-dominated), the pore pressure is considered to range between 0 and 125 kPa depending on the climatic variation. In the second case (pore pressure-dominated), the pore pressure is assumed to range between 0 and 500 kPa depending on the climatic variation. The absolute values of H are indicative, and only the trends are interpreted in this study. $g = 9.81$ ms^{-2}, $C = 1$ MPa, $\phi = 10°$, $\rho = 2800$ kg/m^{-3}, $\beta = 7°$, and a subsidence rate of 0.25 mm/yr are considered. (**B**) Simulated sea level variations during the last million years using the δ^{18}O curve of Lisiecki and Raymo [37]. The light blue lines represent the giant landslides on volcanoes during the last 1 Ma of Table 1.

The influence of sea level unloading on magma reservoir pressure increases at depth is better correlated than the sea level loading with the occurrence of giant landslides (Figure 5).

This may indicate that, in many cases, the volcanoes were active (i.e., a magma reservoir was present at depth). If this process was the cause of the giant landslides, the viscosity of the crust was very high in many cases, except for La Réunion (290–320 kyr; [37]), the Canary Islands (134 ± 6; [3]) and Hawaii (127 ± 10; [43]), where a delay is observed with low sea level. The effect of sea level unloading influence on magma pressure rise is contemporaneous with the effect of pore pressure rise caused by meteoric water infiltration in Polynesia and East Africa (i.e., where precipitation increases during glacial periods).

4. Discussion

4.1. Rock Weathering and Slope Instability

Volcanic edifices are composed of massive lavas as well as non-massive lava, scoria, ash falls and pyroclastic flows, among others. These volcanic products can be highly weathered by the humid climate when they are exposed for thousands or millions of years. Laboratory experiments on weathered, massive lavas show that cohesion can be low (0.8–1 MPa) at depths ranging from 1500 m to 3000 m (Figure 3A).

The angle of friction of weathered, massive lavas is estimated from laboratory experiments [5] to range between 23° and 19° (Figure 3B). Lavas that have been extremely altered by rain and fluid circulation at depth for thousands of years are expected to have mechanical properties lower than those of fresh volcanic rocks. Furthermore, the effective mechanical properties (C and ϕ) of fractured and faulted reliefs are lower than those used for triaxial experiments on unfractured rocks. Effective mechanical properties consider the heterogeneities present in old and complex volcanic edifices, including low-cemented pyroclast and weathered soils. Soil layers can be observed trapped between volcanic lava flows or lava deposits [44]. These soils are expected to have reduced mechanical properties. A sliding bed with distributed soft and hard rock is a fundamental factor underlying translational landslides [21]. The model has been calibrated to be compatible with past giant landslide events (~1500–3000 m thick, initial slope β~8–12°, cohesion C of 0.3–1 MPa, Figure 2), such as the event that occurred in Tahiti, Society Archipelago (Figure 2) [32,33]. The values used here correspond to highly weathered rocks of an old and fractured volcanic edifice or low-cemented pyroclasts that could be interlayered at depth.

Rock mechanical properties C and ϕ decrease over time when exposed to weathering. The more weathering that occurs, the greater the decrease in C and ϕ. During humid periods, this decrease is greater than during dry periods (Figure 6A). In contrast, diagenetic processes are expected to improve the mechanical properties of rocks (Figure 6B). As pressure increases at depth, porosity and connectivity may decrease. In addition, mineralization at depth can also improve rock cohesion in certain cases.

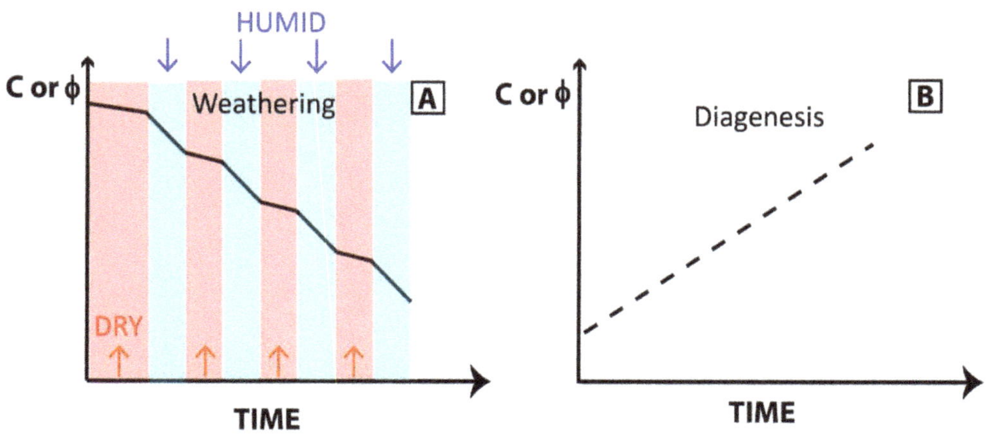

Figure 6. Influence on the mechanical rock properties C and ϕ of (**A**) Weathering (**B**) Diagenesis.

4.2. Weakness Zones and Geological Inheritance

The construction of volcanoes is often polygenic, and numerous heterogeneities exist on polygenic volcanic edifices [45], but also on monogenetic ones due to variations in eruptive dynamism. The complex and long volcano-tectonic history of long-lived volcanic edifices may explain why numerous fractures and significant weathering are observed. The highly fractured and heterogeneous material (basaltic flows and volcanic breccia) of volcanic edifices are often complex and difficult to model in detail at all scales. Layers of weathered soils interlayered between lavas flows can also be observed [44], as well as pyroclastic deposits. Low friction angle values were estimated or used in various studies for weak structures, effective rock properties, or faults [6,25,46–48].

Studies using theoretical models simplify the complexity of the real world [49]. The effective mechanical properties implemented in modeling differ from experimental properties in the laboratory. Pre-existing zones of weakness at depth are present in many volcanic edifices but are different in each case. The real geometry of a specific volcanic edifice or a specific landslide is not the aim of this study. A general flat isosceles trapezoid model is considered. This model does not permit us to forecast exactly when a large landslide will occur, but only to discuss the possibility of distinguishing the influence of meteoric water infiltration from sea water loading/unloading using climatic correlation from a theoretical perspective.

4.3. Sea Level Variation Effect on Landslides and Volcanic Activity

The role of sea level rise in triggering landslides has been suggested in the case of the Vajont landslide (1963, Italy) [1,7]. As the sea level rises, (i) Water loading increases (ii) Water infiltrates the rocks. The loading caused by sea level rises could cause slope instability if the slope is close to equilibrium immediately prior to the rise in sea level. Furthermore, a sea level rise may cause erosion at the base of the cliff, which could promote cliff retreat, rock fall and landslides [50]. However, the effect of loading forces on the base of a slope can also favor slope stability. A vertical loading force at the base of the volcano can cause an increase in stability by counteracting the potential movements of the relief. In other words, when the sea level drops at the base of a slope, it could favor instability (Figure 7A). This result is suggested by the decrease of the safety factor F_S as the loading decreases. When a significant loading is above the center of mass of the potential landslide, it causes the destabilization of the slope by triggering rock failure (Figure 7B).

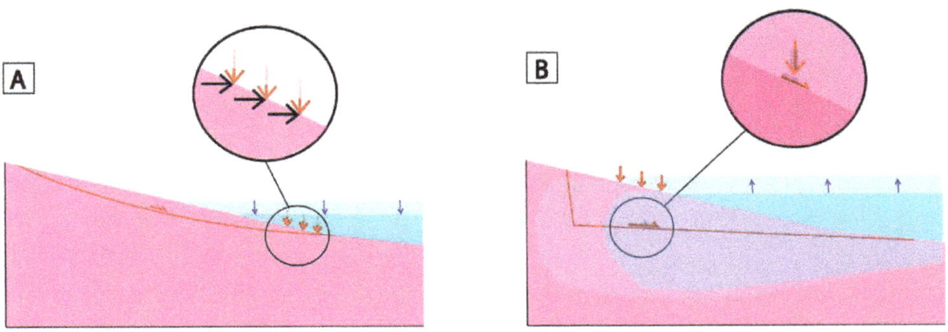

Figure 7. Influence of the sea level variation with respect to the position of the center of mass of the potential sliding area: (**A**) Slope instability caused by sea level lowering when the load is located at the base of the slope, (**B**) slope instability caused by a sea level rise above the center of mass with seawater infiltration.

Giant landslides in volcanic areas appear to be correlated with climatic variations [23]. Fluctuation in sea level loading is one of the potential causes of giant landslides [12]. However, indirect interactions are also possible in volcanic areas. The unloading associated with sea level lowering could change the pressure on the magma reservoir [40]. A decrease in pressure has implications as magma flows into and out of the reservoir. If the connection to the deeper magma source remains open, one consequence is an increased rate of replenishment with primitive magma [51]. In addition, reduced load allows the eruption of denser magma that would otherwise be trapped at shallow depths [52].

The potential interaction between sea level changes and landslides, as well as landslides and volcanic eruptions [53], could cause difficulties in interpretation. If the magma reservoir is shallow, a landslide could cause dike intrusion, but if the magma reservoir is deeper, dike intrusion after a landslide is more difficult.

4.4. Pore Pressure Variation

Another process that can cause landslides is pore pressure increase. Fluids are known to be a triggering and driving factor for landslides [11]. Various mechanisms could explain a pore pressure increase of ~1 MPa. First, a significant amount of water must enter the system. For example, unsaturated rocks of porosity n in an area of volume V could become saturated if a significant volume nV infiltrates. The volume of water could be higher if a dense fracture network exists. Meteoric water infiltration at a depth of 1000 m under lithostatic pressure could cause an increase in pore pressure. An increase of more than 1 MPa in the pore pressure at depth due to meteoric water is not unrealistic and has been proposed in previous studies (e.g., [17]).

An alternative mechanism that could also lead to a significant increase in pore pressure is the progressive collapse of the base of the volcanic edifice. In this scenario, the slow creep of the edifice on a pre-existing weak structure can locally trigger overpressure conditions that reach high pressures [8]. In a volcanic context, hydrothermal processes are also expected to play a role in the pore pressure increase and rock weathering at depth.

The pore pressure can change in relation to meteoric water infiltration in highly fractured rocks. In this case, meteoric water infiltration increases as the precipitation rate increases. Consequently, climatic variations are expected to influence the pore pressure variations in highly fractured rocks. The correlation between climatic variations and giant landslides may be caused by pore pressure increases at depth.

4.5. Climate Variation and Correlation with Giant Landslides

Precipitation increased during glacial maxima in Polynesia [42], Mexico [54], and eastern equatorial Africa [55] but decreased in NW Europe [56], the Caribbean [57] and Indonesia [58,59].

When humid conditions occurred during warmer phases, the potential effect of water infiltration on pore pressure acted to increase instability during warmer interglacial periods, similar to sea level loading. In this case, interglacial phases are expected to cause more slope instability. However, it is difficult to distinguish the influence of sea level loading from meteoric water infiltration at depth on giant paleo-landslides in this case.

In contrast, when precipitation increased during glacial maxima, the effects of sea level loading and depth water infiltration on giant paleo-landslides were not simultaneous. In this case, it is theoretically possible to discriminate the process that plays the main role in triggering giant landslides. For instance, in Polynesia, increased meteoric water infiltration occurred during glacial maxima and may have increased slope instability during glacial maxima, whereas sea level rise caused slope instability during interglacials (Figure 5A). The age of giant landslides (Table 1) may be a good argument for distinguishing between these two processes, but only in Polynesia, Mexico or eastern equatorial Africa, and not in NW Europe, the Caribbean or Indonesia.

In this study, I suggest that a climatic origin of landslides is possible, even if a volcano is extinct, if there are pre-existing weakness zones at depth that cause low effective cohesion and internal friction. Seismic or volcanic processes are not necessarily directly responsible for giant landslides, although these processes may have played a role in creating weak zones in these areas in the past. In 70% of cases, giant landslides occurred when the sea level was rising, regardless of the precipitation rate. Because of the potential delay caused by viscous processes (i.e., 10 kyr of delay for a viscosity less than 10^{22} Pa.s), it could also be correlated with the magma chamber unloading due to sea level lowering. This result could be influenced by the data set considered, and further studies on giant landslides on volcanic islands would improve this interpretation.

Table 1. Giant landslide ages in volcanic areas during the last 1 Ma.

Volcano	Age (ka)	References
Canary Islands (El Hierro, El Golfo)	10–17	[60]
La Réunion	20–68	[61]
Hawaii (Alika phase 1)	112	[43]
Hawaii (Alika phase 2)	127 ± 10	[43]
Canary Islands (El Hierro, El Golfo)	134 ± 6	[3]
Hawaii (Southern Lanai)	135	[62]
Canary Islands (Tenerife)	150–170	[63] [64]
Hawaii (Southern Lanai)	240	[62]
La Réunion	290–320	[61]
Martinique	337 ± 5	[65]
Canary Islands (La Palma)	537 ± 8	[66] [67]
Guadeloupe	629 ± 13	[68]
Hawaii (Haleakala, Hana)	860	[69]
Tahiti-Nui (north)	872 ± 10	[30]

4.6. Discussion of Giant Landslides Real Cases

This theoretical study allows us to describe the causes of giant landslides more precisely and could be used to discuss real cases. For the first time, the reason why all giant landslides caused by climate variation do not occur during interglacial maximum could be explained. For example, the Tahiti volcano (Society Archipelago, French Polynesia, Pacific Ocean) is a shield volcano, extinct since 230 kyr, with a mean slope of 6–10°, subsiding at a rate of 0.39–0.15 mm/yr (e.g., [33]). At least two major listric discontinuities caused by two giant landslides that occurred 872 ± 10 kyrs ago and 850–650 kyrs B.P. have been described at a depth > 1000 m [30,32,33]. In the South Pacific, glacial periods of time are wetter than interglacial ones. The Tahiti giant landslide, dated at 872 ± 10 kyrs B.P., could have been triggered by a pore pressure increase because it occurred during a glacial period (Figure 5). Nevertheless, if it occurred during the warming phase, the landslide could have been triggered by the sea water unloading that caused a magma reservoir pressure increase with a delay of several thousand years. A more accurate age is necessary to propose a unique interpretation of the Tahiti giant landslide.

La Réunion volcano (i.e., Piton de la Fournaise) is an active shield volcano, composed of basaltic lavas, formed from solidified lava flows accumulated over hundreds of thousands of years, and located in the Indian Ocean above a hot spot. Several landslides occurred on La Réunion volcano slopes at 290–320 kyr and 20–68 kyr B.P. [61]. The eastern flank of the volcano is unstable and is in the initial stages of failure. In La Réunion, heavy precipitations are known in the present day and also during the interglacial time. The past giant landslides that occurred at 290–320 kyr and 20–68 kyr B.P. during cooling periods of times (Figure 5) could have been triggered by sea water unloading, causing magma reservoir pressure increase with a delay of several thousand years.

The Guadeloupe volcano (i.e., La Soufrière), Lesser Antilles, in the Caribbean, is an active stratovolcano [68] composed of lavas flows, debris flows, pyroclastics gathering ash, pumice, and scoria deposits [34]. It is 1467 m high, and. several eruptions, as well as landslides, have been described. The main landslide is 629 ± 13 kyrs old [68]. Concerning the Guadeloupe landslides, they occurred during a glacial phase or during the warming phase (Figure 5). The glacial phases are drier in the Caribbean and exclude pore pressure increases. If it occurred during the glacial phase, the potential cause of the Guadeloupe giant landslide is the magma reservoir pressure increase associated with sea water unloading. Nevertheless, if the giant landslide occurred during the warming phase, the causes could be the sea level loading increase or/and the water infiltration increase. The stratovolcano of La Martinique (i.e., Montagne Pelée), Lesser Antilles, in the Caribbean, which occurred 337 ± 5 kyrs ago [65] during the glacial phase or during the warming phase (Figure 5) has the same potential causes.

4.7. *Small Landslides vs. Giant Landslides*

The size of the expected landslide depends on the initial slope of the relief. Locally, as in a deeply incised canyon of volcanic islands, significant slopes can be observed. In the case of a significant slope (>25°), the critical value for the height of the stable relief is less than 300 m (Figure 4C). Consequently, smaller landslides can be triggered locally. When rotational landslides occur, significant slopes develop in the upstream part. The resulting relief may be close to the instability conditions that favor the development of new landslides and other retrogressive erosion processes. It can be difficult to distinguish the relief associated with numerous "small" landslides from that caused by giant landslides [32,70].

Erosion and repetition of small landslides could gradually reduce the loading. As previously shown, loading affects the maximum height expected before a landslide occurs. Reducing the load increases the maximum height before a landslide occurs and thus reduces the potential occurrence of a landslide. Consequently, the potential occurrence of a giant landslide could be reduced by the occurrence of small landslides and relief erosion. Previous studies have shown that the probability of small landslides occurring is higher than that of giant landslides (e.g., [9]). Nevertheless, giant landslides have also been described, suggesting that there are specific conditions that favor giant landslides.

5. Conclusions

The probability of a large landslide on an old volcanic edifice in an area where no recent volcanic and/or seismic activities have been observed in the last thousand years is not null. Large landslides caused by non-volcanic processes could occur on old volcanic islands (>200 kyrs). In fact, our modeling suggests that a large landslide (>1500 m deep) could occur due to an increase in pore pressure (~0.5 MPa) or sea level variation. These processes could be discriminated against in areas where glacial maxima are wetter than interglacial periods, but not in the other cases, based only on temporal correlation. This finding is also a consequence of the mechanical properties of highly weathered or fractured volcanic rocks in weakness zones (low cohesion < 1 MPa, and angle of friction < 13°) at depth in old weathered volcanic edifices. Sea level unloading can be responsible for giant landslides, which cause pressure increases in the magma reservoirs at depth and magma release. When the viscous response of the magma reservoir is not negligible ($\eta_c < 10^{22}$ Pa.s), a delay of 10 kyrs could occur. During the last million years, several giant landslides in tropical areas have been correlated with sea level unloading during glacial periods of uplift, suggesting that this effect is a driving mechanism in many cases.

Funding: This research received no external funding.

Data Availability Statement: No new data were created or analyzed in this study.

Acknowledgments: The three anonymous reviewers are acknowledged for their suggestions and the time dedicated to the review process. The academic editor is also acknowledged.

Conflicts of Interest: The authors declare no conflict of interest.

References

1. Muller-Salzburg, L. The Vajont catastrophe—A personnal review. *Eng. Geol.* **1987**, *24*, 423–444. [CrossRef]
2. Keefer, D.K. The importance of earthquake-induced landslides to long-term slope erosion and slope-failure hazards in seismically active regions. *Geomorphology* **1994**, *10*, 265–284. [CrossRef]
3. Carracedo, J.C.; Day, S.J.; Guillou, H.; Pérez Torrado, F.J. Giant Quaternary landslides in the evolution of La Palma and El Hierro, Canary Islands. *J. Volcanol. Geotherm. Res.* **1999**, *94*, 169–190. [CrossRef]
4. Blahůt, J.; Balek, J.; Klimeš, J.; Rowberry, M.; Kusák, M.; Kalina, J. A comprehensive global database of giant landslides on volcanic islands. *Landslides* **2019**, *16*, 2045–2052. [CrossRef]
5. Rodriguez-Losada, J.A.; Hernandez-Gutérrez, L.E.; Olalla, C.; Perucho, A.; Serrano, A.; Eff-Darwich, A. Geomechanical parameters of intact rocks and rock masses from the Canary Islands: Implications on their flank stability. *J. Volcanol. Geotherm. Res.* **2009**, *182*, 67–75. [CrossRef]
6. Cala, M.; Flisiak, J. Slope stability with FLAC and limit equilibrium methods. In *FLAC and Numerical Modeling in Geomechanics*; Billaux, D., Rachez, X., Detournay, C., Hart, R., Eds.; Balkema: Rotterdam, The Netherlands, 2001; pp. 111–114.
7. Kilburn, C.R.J.; Petley, D.N. Forecasting giant, catastrophic slope collapse: Lessons from Vajont, Northern Italy. *Geomorphology* **2003**, *54*, 21–32. [CrossRef]
8. Veveakis, E.; Vardoulakis, I.; Di Toro, G. Thermoporomechanics of creeping landslides: The 1963 Vaiont slide, northern Italy. *J. Geophys. Res.* **2007**, *112*, F03026. [CrossRef]
9. Urgeles, R.; Camerlinghi, A. Submarine landslides of the Mediterrranean Sea: Trigger mechanisms, dynamics, and frequency-magnitude distribution. *J. Geophys. Res.* **2013**, *118*, 2600–2618. [CrossRef]
10. Crozier, M.J. Deciphering the effect of climate change on landsliding activity: A review. *Geomorphology* **2010**, *in press*. [CrossRef]
11. Cappa, F.; Guglielmi, Y.; Viseur, S.; Garambois, S. Deep fluids can facilitate rupture of slow-moving giant landslides as a result of stress transfer and frictional weakening. *Geophys. Res. Lett.* **2014**, *41*, 61–66. [CrossRef]
12. Aslan, G.; De Michele, M.; Raucoules, D.; Bernardie, S.; Cakir, Z. Transient motion of the largest landslide on earth, modulated by hydrological forces. *Sci. Rep.* **2021**, *11*, 10407. [CrossRef]
13. Le Corvec, N.; Walter, T.R. Volcano spreading and fault interaction influenced by rift zone intrusions: Insights from analogue experiments analyzed with digital image correlation technique. *J. Volcanol. Geotherm. Res.* **2009**, *183*, 170–182. [CrossRef]
14. Gargani, J.; Geoffroy, L.; Gac, S.; Cravoisier, S. Fault slip and Coulomb stress variations around a pressured magma reservoir: Consequences on seismicity and magma intrusion. *Terra Nova* **2006**, *18*, 403–411. [CrossRef]
15. Hampel, A.; Hetzel, R. Slip reversals on active normal faults related to the inflation and deflation of magma chambers: Numerical modeling with application to the Yellowstone-Teton region. *Geophys. Res. Lett.* **2008**, *35*, L07301. [CrossRef]
16. McMurtry, G.M.; Watts, P.; Fryer, G.J.; Smith, J.R.; Imamura, F. Giant landslides, mega-tsunamis, and paleo-sea level in the Hawaiian Islands. *Mar. Geol.* **2004**, *203*, 219–233. [CrossRef]
17. Cervelli, P.; Segall, P.; Johnson, K.; Lisowski, M.; Miklius, A. Sudden aseismic fault slip on the south flank of Kilauea volcano. *Nature* **2002**, *415*, 1014–1018. [CrossRef]
18. De Ritis, R.; Passaro, S.; Pensa, A. Editorial for Special Issue "Present and Past Submarine Volcanic Activity (1)". *Geosciences* **2022**, *12*, 458. [CrossRef]
19. Nicolosi, I.; Caracciolo, F.; Branca, S.; Ventura, G.; Chiappini, M. Volcanic conduit migration over a basement landslide at Mount Etna (Italy). *Sci. Rep.* **2014**, *4*, 5293. [CrossRef]
20. Murray, J.B.; van Wyke de Vries, B. Basement sliding and the formation of fault systems on Mt. Etna volcano. *J. Volcanol. Geotherm. Res.* **2022**, *428*, 107573. [CrossRef]
21. Du, Y.; Xie, M.; Jia, J. Stepped settlement: A possible mechanism for translational, landslides. *Catena* **2019**, *187*, 104365. [CrossRef]
22. Du, Y.; Chen, C. Data Mining for Landslide Genetic Mechanism Analysis in the Yunnan Province of China. *Geotech. Geol. Eng.* **2022**, *40*, 5631–5642. [CrossRef]
23. Quidelleur, X.; Hildenbrand, A.; Samper, A. Causal link between Quaternary paleoclimatic changes and volcanic islands evolution. *Geophys. Res. Lett.* **2008**, *35*, L02303. [CrossRef]
24. Gargani, J.; Bache, F.; Jouannic, G.; Gorini, C. Slope destabilization during the Messinian Salinity Crisis. *Geomorphology* **2014**, *213*, 128–138. [CrossRef]
25. Bigot-Cornier, F.; Montgomery, D.R. Valles Marineris landslides: Evidence for a strength limit to Martian relief? *Earth Planet. Sci. Lett.* **2007**, *260*, 179–186. [CrossRef]
26. Du, Y.; Xie, M.-W.; Jiang, Y.-J.; Li, B.; Chicas, S. Experimental Rock Stability Assessment Using the Frozen–Thawing Test. *Rock Mech. Rock Eng.* **2017**, *50*, 1049–1053. [CrossRef]
27. Zhang, L.; Niu, F.; Liu, M.; Wang, Z.; Wang, J.; Dong, T. Fracture characteristics and anisotropic strength criterion of bedded sandstone. *Front. Earth Sci.* **2022**, *10*, 879332. [CrossRef]
28. Champel, B.; van der Beck, P.; Mugnier, J.-L.; Leturny, P. Growth and lateral propagation of fault-related folds in the Siwaliks of western Nepal: Rates, mechanisms, and geomorphic signature. *J. Geophys. Res.* **2002**, *107*, 2111. [CrossRef]
29. Hürlimann, M.; Garcia-Piera, J.O.; Ledesma, A. Causes and mobility of large volcanic landslides: Application to Tenerife, Canary Islands. *J. Volcanol. Geotherm. Res.* **2000**, *103*, 121–134. [CrossRef]
30. Hildenbrand, A.; Gillot, P.-Y.; Le Roy, I. Volcano-tectonic and geochemical evolution of an oceanic intra-plate volcano: Tahiti-Nui (French Polynesia). *Earth Planet. Sci. Lett.* **2004**, *217*, 349–365. [CrossRef]

31. Hildenbrand, A.; Gillot, P.-Y.; Marlin, C. Geomorphological study of long-term erosion on a tropical volcanic ocean island: Tahiti-Nui (French Polynesia). *Geomorphology* **2008**, *93*, 460–481. [CrossRef]
32. Gargani, J. Modelling the mobility and dynamics of a large Tahitian landslide using runout distance. *Geomorphology* **2020**, *370*, 107354. [CrossRef]
33. Gargani, J. Isostatic Adjustment, Vertical Motion Rate Variation and Potential Detection of Past Abrupt Mass Unloading. *Geosciences* **2022**, *12*, 302. [CrossRef]
34. Navelot, V.; Géraud, Y.; Favier, A.; Diraison, M.; Corsini, M.; Lardeaux, J.-M.; Verati, C.; Mercier de Lépinay, J.; Legendre, L.; Beauchamps, G. Petrophysical properties of volcanic rocks and impacts of hydrothermal alteration in the Guadeloupe Archipelago (West Indies). *J. Volcanol. Geotherm. Res.* **2018**, *360*, 1–21. [CrossRef]
35. Gargani, J.; Rigollet, C. Mediterranean Sea level variations during the Messinian Salinity Crisis. *Geophys. Res. Lett.* **2007**, *34*, L10405. [CrossRef]
36. Gargani, J.; Moretti, I.; Letouzey, J. Evaporite accumulation during the Messinian Salinity Crisis: The Suez Rift Case. *Geophysical Research Letters* **2008**, *35*, L02401. [CrossRef]
37. Lisiecki, L.E.; Raymo, M.E. A Pliocene-Pleistocene stack of 57 globally distributed benthic ^{18}O records. *Paleoceanography* **2005**, *20*, 1003.
38. Gargani, J.; Stab, O.; Cojan, I.; Brulhet, J. Modelling the long-term fluvial erosion of the river Somme during the last million years. *Terra Nova* **2006**, *18*, 118–129. [CrossRef]
39. Saar, M.O.; Manga, M. Seismicity induced by seasonal groundwater recharge at Mt. Hood, Oregon. *Earth Planet. Sci. Lett.* **2003**, *214*, 605–618. [CrossRef]
40. Sternai, P.; Caricchi, L.; Garcia-Castellanos, D.; Jolivet, L.; Sheldrake, T.E.; Castelltort, S. Magmatic pulse driven by sea-level changes associated with the Messinian salinity crisis. *Nat. Geosci.* **2017**, *10*, 783–787. [CrossRef]
41. Jull, M.; McKenzie, D. The effect of deglaciation on mantle melting beneath Iceland. *J. Geophys. Res.* **1996**, *101*, 21815–21828. [CrossRef]
42. Saez, A.; Valero-Garces, B.L.; Giralt, S.; Moreno, A.; Bao, R.; Pueyo, J.J.; Hernandez, A.; Casas, D. Glacial Holocene slimate changes in SE Pacific. The Raraku Lake sedimentary record (Easter Island, 27°S). *Quat. Sci. Rev.* **2009**, *28*, 2743–2759. [CrossRef]
43. McMurtry, G.M.; Herrero-Bervera, E.; Cremer, M.; Resig, J.; Sherman, C.; Smith, J.R.; Torresan, M.E. Stratigraphic constraints on the timing and emplacement of the Alika 2 giant Hawaiian submarine landslide. *J. Volcanol. Geotherm. Res.* **1999**, *94*, 35–58. [CrossRef]
44. Hevia-Cruz, F.; Hildenbrand, A.; Sheldon, N. *Enhanced Weathering of Volcanic Islands during Glacial-Interglacial Transitions: Insights from Well-Dated Paleosols Spanning the Last Million Year in Sao Miguel Island, Eastern Azores*; Conférence "Climat et impacts": Orsay, France, 2022.
45. Hildenbrand, A.; Gillot, P.-Y.; Bonneville, A. Offshore evidence for a huge landslide of the northern flank of Tahiti-Nui (French Polynesia). *Geochem. Geophys. Geosystems* **2006**, *7*, Q03006. [CrossRef]
46. Got, J.-L.; Monteiller, V.; Monteux, J.; Hassani, R.; Okubo, P. Deformation and rupture of the oceanic crust may control growth of Hawaiian volcanoes. *Nature* **2008**, *451*, 453–456. [CrossRef] [PubMed]
47. Egholm, D.L.; Clausen, O.R.; Sandiford, M.; Kristensen, M.B.; Korstgard, J.A. The mechanics of clay smearing along fault. *Geology* **2008**, *36*, 787–790. [CrossRef]
48. Abers, G.A. Slip on shallow-dipping normal faults. *Geology* **2009**, *37*, 767–768. [CrossRef]
49. Gargani, J. *Poincaré, le Hasard et L'étude des Systèmes Complexes*; L'Harmattan: Paris, France, 2012; 124p.
50. Ye, F.Y.; Barriot, J.P.; Carretier, S. Initiation and recession of the fluvial knickpoints of the Island of Tahiti (French Polynesia). *Geomorphology* **2013**, *186*, 162–173. [CrossRef]
51. Pinel, V.; Jaupart, C. The effect of edifice load on magma ascent beneath a volcano. *Philos. R. Soc. London Ser. A* **2000**, *358*, 1515–1532. [CrossRef]
52. Pinel, V.; Jaupart, C. Magma chamber behavior beneath a volcanic edifice. *J. Geophys. Res.* **2003**, *108*, 2072. [CrossRef]
53. Longpré, M.A.; Troll, V.R.; Walter, T.R.; Hansteen, T.H. Volcanic and geochemical evolution of the Teno massif, Tenerife, Canary Islands: Some repercussions of giant landslides on ocean island magmatism. *Geochem. Geophys. Geosyst.* **2009**, *10*, Q12017. [CrossRef]
54. Ganeshram, R.S.; Pedersen, T.F. Glacial-interglacial variability in upwelling and bioproductivity off NW Mexiso: Implications for Quaternary paleoclimate. *Paleoceanography* **1998**, *13*, 634–645. [CrossRef]
55. Chiang, J.C.H. The Tropics of Paleoclimate. *Ann. Rev. Earth Planet. Sci.* **2009**, *37*, 263–297. [CrossRef]
56. Guiot, J.; Pons, A.; de Beaulieu, J.L.; Reille, M. A 140000 year continental climate reconstruction from two European pollen records. *Nature* **1989**, *338*, 309–313. [CrossRef]
57. Curtis, J.H.; Brenner, M.; Hodell, D.A. Climate change in the circum-Caribbean (Late Pleistocene to Present) and implications for Regional Biogeography. In *Biogeography of the West Indies*; Charles, A.W., Florence, E.S., Eds.; CRC Press: Boca Raton, FL, USA, 2001; 608p. [CrossRef]
58. Costa, K.M.; Russel, J.M.; Vogel, H.; Bijaksana, S. Hydrological connectivity and mixing of Lake Towati, Indonesia in response to paleoclimatic changes over the last 60,000 years. *Palaeogeogr. Palaeoclimatol. Palaeoecol.* **2015**, *417*, 467–475. [CrossRef]
59. Russell, J.M.; Vogel, H.; Konecky, B.L.; King, J.W. Glacial forcing of central Indonesian hydroclimate since 80000 yr B.P. *Proc. Natl. Acad. Sci. USA* **2014**, *111*, 5100–5105. [CrossRef]

60. Gee, M.J.; Watts, A.B.; Masson, D.G.; Mitchell, N.C. Landslides and the evolution of El Hierro in the Canary Islands. *Mar. Geol.* **2001**, *177*, 271–293. [CrossRef]
61. Lenat, J.F.; Labazuy, P. Growth and collapse of the Reunion Island volcanoes. *Bull. Volcanol.* **2008**, *70*, 717–742.
62. Rubin, K.H.; Fletcher, C.H., III; Sherman, C. Fossiliferous Lanai deposits formed by multiple events rather than a single giant tsunami. *Nature* **2000**, *408*, 675–681. [CrossRef]
63. Hunt, J.E.; Wynn, R.B.; Masson, D.G.; Talling, P.J.; Teagle, D.A.H. Sedimentological and geochemical evidence for multistage failure of volcanic island landslides: A case study from Icod landslide on north Tenerife, Canary Islands. *Geochem. Geophys. Geosystems* **2011**, *12*, Q12007. [CrossRef]
64. Masson, D.G.; Watts, A.B.; Gee, M.J.R.; Urgeles, R.; Mitchell, N.C.; Le Bas, T.P.; Canals, M. Slope failures on the flanks of the western Canary Islands. *Earth-Sci. Rev.* **2002**, *57*, 1–35. [CrossRef]
65. Quidelleur, X.; Samper, A.; Boudon, G.; Le Friant, A.; Komorowski, J.C. Radiometric dating of large volume flank collapses in the Lesser Antilles Arc. *Eos Trans. AGU* **2004**, *85*.
66. Guillou, H.; Carracedo, J.C.; Duncan, R.A. K-Ar, 40Ar/39Arages and magnetostratigraphy of Brunhes and Matuyama lava sequencesfrom La Palma Island. *J. Volcanol. Geotherm. Res.* **2001**, *106*, 175–194. [CrossRef]
67. Groom, S.; Barfod, D.N.; Millar, I.; Downes, H. The Cumbre Nueva collapse (La Palma, Canary Islands): New age determinations and evidence of an isotopic excursion. *J. Volcanol. Geotherm. Res.* **2022**, *433*, 107708. [CrossRef]
68. Samper, A.; Quidelleur, X.; Lahitte, P.; Mollex, D. Timing ofeffusive volcanism and collapse events within an oceanic arc island:Basse Terre, Guadeloupe archipelago (Lesser Antilles Arc). *Earth Planet. Sci. Lett.* **2007**, *258*, 175–191. [CrossRef]
69. Moore, J.G.; Clague, D.A. Volcano growth and evolution of the island of Hawaii. *Geol. Soc. Am. Bull.* **1992**, *104*, 1471–1484. [CrossRef]
70. Gargani, J. Relative sea level and abrupt mass unloading in Barbados during the Holocene. *Geomorphology* **2022**, *413*, 108353. [CrossRef]

applied sciences

MDPI

Article

Photogrammetry-Based 3D Textured Point Cloud Models Building and Rock Structure Estimation

Tiexin Liu [1] and Jianhui Deng [2,*]

[1] Department of Civil Engineering, Dalian Maritime University, Dalian 116026, China
[2] State Key Laboratory of Hydraulics and Mountain River Engineering, College of Water Resource & Hydropower, Sichuan University, Chengdu 610065, China
* Correspondence: jhdeng@scu.edu.cn; Tel.: +86-139-8095-9251

Abstract: Trace lines on the outcrop of a rock mass are usually the primary data source for the estimation of rock structure. It is important to obtain the data of trace lines precisely. Photogrammetry is well suited to finish this task. However, this is mainly conducted by commercial software, and not every researcher has easy access to the method of digital photogrammetry. This study aims to provide researchers with a low-cost method of building a photogrammetry-based textured 3D point cloud model (FMBPM) and display the applicability of the method to estimating the rock structure of rock masses. In the FMBPM, a digital single-lens reflex camera with a prime lens and a total station are the necessary hardware employed to capture images and measure the coordinates of feature points. A coordinate transformation means of converting model coordinates to physical coordinates was introduced. A program for calculating a joint orientation based on the coordinates of inflection points on the trace line of the joint was developed. A section of a rock slope was selected as a case to show the procedures and the practicability of the FMBPM. The textured 3D point cloud model of the rock slope was successfully built, and the rock structure of the rock slope was analyzed using the joint disk model generated based on the trace lines extracted from the point cloud model. The results show that: (1) the precision of the point coordinates of the textured 3D point cloud model could achieve 3.96 mm, taking the data of the total station as the reference; (2) the rock structure of the slope is good, according to the value of the rock quality designation; (3) the new method is applicable in engineering practices.

Keywords: photogrammetry; rock structure; coordinates transformation; joint disk model

1. Introduction

The mechanical and hydraulic properties of rock masses are very complex, considering the existence of joints [1]. Rock structure directly reflects the development of joints, and it is an essential index in rock mass classification systems, such as RMR, Q, and RMi [2]. The geometrical parameters of joints, such as size, orientation, density, and spacing, significantly influence the rock structure of a rock mass [3,4]. Due to the lightproof characteristic of rock masses, most of the parameters are inferred according to the information of joints on outcrops rather than directly measured or observed. Trace lines are the intersection lines of joints and an exposed surface of a rock mass, and they are usually the primary data source for the estimation of rock structure [5]. Consequently, it is essential to obtain the data of trace lines precisely.

There are two main methods to obtain the data of trace lines: contact field investigations and non-contact remote sensing investigations [6]. A geological compass and a tape are the main measuring equipment in a contact field investigation, and the investigation results of joints can be acquired immediately. The method is applicable when the number of joints is small and the site of joints can be easily reached by surveyors. In recent years, rock engineering has become so large that many areas of rock exposures are manually inaccessible. Yet, the detailed rock structure information of the inaccessible areas is also urgently needed to ensure the stability of the engineering. Consequently, remote sensing techniques are being used increasingly, which permits a safe, fast, comprehensive, and accurate acquisition of information on inaccessible rock exposures and increases engineer safety. Digital photogrammetry and laser scanning are two main methods of remote sensing techniques related to rock engineering, both of which can generate high-precision 3D point clouds of research objects [7–10]. Laser scanning methods generate a point cloud model using time-of-flight technology [11]. In comparison, digital photogrammetry methods calculate the spatial coordinates of a point based on its planar geometric position in different photographs. Many studies related to rock structure extracting and geological hazard analysis have been conducted based on the two methods. Chen et al. [6] presented a method for evaluating the development of the joints on rock tunnel faces based on a 3D point cloud model obtained by photogrammetry. Kaminski et al. [12] estimated the dynamics and sizes of landslides by comparing digital terrain models, which are generated photogrammetrically from stereo pairs of images. Cheng et al. [13] demonstrated the effectiveness of Unmanned Aerial Vehicle photogrammetry in the remote sensing and assessment of landslide behavior through a case study of a landslide that occurred in Guizhou, China. Nappo et al. [14] provided a tool for the semi-automatic damage assessment of roads in landslide-affected areas to support the risk analysis and planning of mitigation measures based on photogrammetry. Buyer et al. [15] determined the block size and shape distributions according to a textured digital surface model of an investigated outcrop. Menegoni et al. [16] detected joints on the surface of a rock slope based on digital photogrammetry and a remotely piloted aircraft system. Liu et al. [5] searched the dangerous parts of a rock slope using block theory based on photogrammetry. Lato et al. [17] established an online shared repository of photogrammetry data and light detection and ranging (LiDAR) for researchers. Sturzenegger et al. [18] investigated the applicability of photogrammetry and LiDAR to the derivation of joint intensity, mean trace length, and block size.

Compared with laser scanning, photogrammetry has the advantages of lower equipment cost and better portability [19,20]. However, the process of generating point cloud models is conducted by commercial software, in most studies, after rock mass photographs are collected; thus, not every researcher has easy access to the method of digital photogrammetry. This study aims to provide researchers with a low-cost method of building a photogrammetry-based textured 3D point cloud model (FMBPM) of the exposures of rock masses and to illustrate the applicability of the method to estimating the rock structure of rock masses. The paper is organized as follows: (1) the primary geological conditions of a rock slope at the study site are introduced, a section of which is selected to display the processes and to verify the applicability of the FMBPM; (2) the processes of collecting photographs and generating textured 3D point cloud models are proposed; (3) a coordinate transformation means of converting model coordinates to physical coordinates is supplied; (4) the accuracy of the textured 3D point cloud model is verified, and trace lines of the model are extracted; (5) a joint disk model is built for estimating the rock structure of the section of the rock slope.

2. The Point Cloud Model Generating for the Study Site

The study site is located in Dalian City, Liaoning Province, northeast of China. The city has plenty of rain, sunshine, and lush vegetation. The risk of rockfalls is the primary potential geological disaster influencing the stability of the study site. A rock slope at the site is adjacent to the Fengcai road and faces the Longwangtang reservoir (Figure 1), a section of which was selected as the case study to illustrate the processes of the FMBPM and to verify the applicability of the FMBPM to rock structure estimation. The studied area of the slope covers a rectangular region with dimensions of approximately 29 m in length and 7 m in height. The joints in the slope are well-developed, and the rock structure of the slope is generally blocky.

Figure 1. Location of the study site.

A NIKON D7100 digital single-lens reflex camera with a 35 mm lens and a South NTS-362R total station were employed to capture images and to measure the coordinates of feature points. The maximum pixel value of the NIKON D7100 is 24,000,000 (4000 in height and 6000 in width). The precision of the distance measurement of the total station is $\pm(3 + 2 \times 10 - 6D)$ mm. D is the distance between the total station and a measuring point, and it is not greater than 30 m in this case. Therefore, the precision of the distance is less than 3.6 mm in theory. The accuracy of angle measurement is 2 s, which means the error of the coordinates of a measuring point is theoretically less than 0.291 mm when D is less than 30 m.

Digital photogrammetry calculates the spatial coordinates of a point based on its planar geometric positions in different photographs. Hence, the same feature point of the slope should be presented in at least two images. A total of 194 images of the slope were taken from different camera positions. The textured 3D point cloud model of the slope was generated by Meshroom, which is a free, open-source 3D reconstruction software (https://alicevision.org/#meshroom, accessed on 20 December 2021). The pipeline of photogrammetry of Meshroom was used. The processes of generating textured 3D point cloud models of study subjects from images in the pipeline are highly automatic. The steps related to image processing, such as feature extracting, image matching, feature matching, meshing, and texturing, were all done automatically. Users can modify the degree of fineness of a 3D point cloud model; the finer the point cloud model, the longer it takes to generate. The textured 3D point cloud model of the study site includes 3,566,492 vertices and 7,127,652 faces, and it took about 15 h to create the model. The CPU model of the computer is an Intel(R) Core(TM) i7-9700 with 3.00 GHz, and the graphics card model is Nvidia Quadro P1000 (4 GB). It is worth noting that the surface of the rock slope where it is not directly covered by vegetation can be modeled by adding camera positions (Figure 2).

The covered part of the surface must be exposed in images taken from the added camera positions. Figure 3 is the 2D and 3D comparison image of rock structure identification.

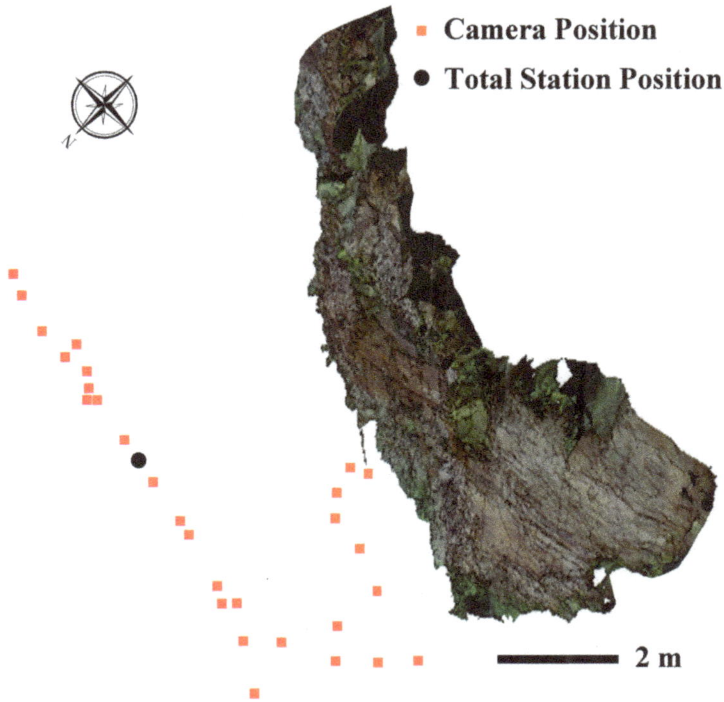

Figure 2. Positions of the camera and total station. (Note that the coordinate origin and coordinate system of the study area can be determined after setting up the total station).

Figure 3. Rock structure identification 2D and 3D comparison image.

3. Coordinate Transformation

It should be noted that in Meshroom, only the relative positions of points are accurate, rather than the physical coordinates; therefore, point cloud models built by Meshroom must be scaled and rotated before conducting measurements. Meshlab is open-source software for processing and editing 3D point cloud models (https://www.meshlab.net/#download, accessed on 12 January 2022), and it provides a series of tools for editing, cleaning, healing, inspecting, rendering, texturing, and converting meshes. The postprocessing of coordinate transformation and trace line extraction were finished in Meshlab after the textured 3D point cloud model had been built in Meshroom.

The coordinates of at least four feature points on the slope should be measured simultaneously in Meshlab and in the field (denoted as ***CF*** and ***CT***, respectively) to obtain the transformation matrix. ***CF*** can be directly extracted in Meshlab, but ***CT*** needs to be measured with a total station in the field. The processes of the transformation can be summarized by Equation (1).

$$\boldsymbol{Tranmat} = \begin{bmatrix} \boldsymbol{R} \cdot S & \boldsymbol{T} \\ \boldsymbol{O} & 1 \end{bmatrix} \tag{1}$$

In Equation (1), ***Tranmat*** is a 4-by-4 matrix, which represents the total transformation of the 2 sets of paired points; ***R*** is a 3-by-3 matrix representing the rotation transformation; *S* is a scalar, which represents the scale factor; ***T*** is a 3-by-1 matrix, which signifies the translation; ***O*** is a 1-by-3 zero matrix. The matrixes of ***R*** and ***T***, and the scale factor of *S* can be solved according to ***CF*** and ***CT***, which are the coordinates of two sets of paired points (Equations (2) and (3)). Each set of the points can be represented by a 4-by-3 matrix. The *N*th row in the matrixes consists of the coordinates of the *N*th point (*N* is not greater than 4).

$$\boldsymbol{CF} = \begin{bmatrix} x_{F1} & y_{F1} & z_{F1} \\ x_{F2} & y_{F2} & z_{F2} \\ x_{F3} & y_{F3} & z_{F3} \\ x_{F4} & y_{F4} & z_{F4} \end{bmatrix} \tag{2}$$

$$\mathbf{CT} = \begin{bmatrix} x_{T1} & y_{T1} & z_{T1} \\ x_{T2} & y_{T2} & z_{T2} \\ x_{T3} & y_{T3} & z_{T3} \\ x_{T4} & y_{T4} & z_{T4} \end{bmatrix} \tag{3}$$

The centroids of ***CF*** and ***CT*** can be calculated by Equations (4) and (5), respectively. The mean values of each set of coordinates should be translated to the origin of its own coordinate system; this can be done by subtracting from the point coordinates of the centroid. Equations (6) and (7) correspond to this process.

$$\boldsymbol{CFCEN}_{1j} = \frac{1}{4}\sum_{i=1}^{4} \boldsymbol{CF}_{ij} \tag{4}$$

$$\boldsymbol{CTCEN}_{1j} = \frac{1}{4}\sum_{i=1}^{4} \boldsymbol{CT}_{ij} \tag{5}$$

$$\boldsymbol{CFC}_{ij} = \boldsymbol{CF}_{ij} - \begin{bmatrix} \boldsymbol{CFCEN} \\ \boldsymbol{CFCEN} \\ \boldsymbol{CFCEN} \\ \boldsymbol{CFCEN} \end{bmatrix}_{ij} \tag{6}$$

$$\boldsymbol{CTC}_{ij} = \boldsymbol{CT}_{ij} - \begin{bmatrix} \boldsymbol{CTCEN} \\ \boldsymbol{CTCEN} \\ \boldsymbol{CTCEN} \\ \boldsymbol{CTCEN} \end{bmatrix}_{ij} \tag{7}$$

In Equations (4)–(7), j ranges from 1 to 3; ***CFCEN*** and ***CTCEN*** are the mean values of ***CF*** and ***CT***, respectively; ***CFC*** and ***CTC*** are the coordinate matrixes of ***CF*** and ***CT***, respectively, after translation to the origin of their own respective coordinate system. The scale factor can be calculated by Equations (8)–(10).

$$S_F = \left(\frac{1}{4} \sum_{i=1}^{4} \sum_{j=1}^{3} CFC_{ij}^2 \right)^{0.5} \tag{8}$$

$$S_T = \left(\frac{1}{4} \sum_{i=1}^{4} \sum_{j=1}^{3} CTC_{ij}^2 \right)^{0.5} \tag{9}$$

$$S = \frac{S_T}{S_F} \tag{10}$$

In Equations (8)–(10), S_F and S_T are the size factors of ***CF*** and ***CT***, respectively.

$$\boldsymbol{CFCS}_{ij} = \frac{\boldsymbol{CFC}_{ij}}{S_F} \tag{11}$$

$$\boldsymbol{CTCS}_{ij} = \frac{\boldsymbol{CTC}_{ij}}{S_T} \tag{12}$$

In Equations (11) and (12), ***CFCS*** and ***CTCS*** are the matrixes normalized by size factors of ***CF*** and ***CT***, respectively. The rotation matrix of ***R*** can be computed by Equations (13) and (14) [21].

$$\boldsymbol{H} = \boldsymbol{CFCS}^{\mathrm{T}}\boldsymbol{CTCS} \tag{13}$$

$$\boldsymbol{R} = \left(\boldsymbol{H}^{\mathrm{T}}\boldsymbol{H}\right)^{0.5}\boldsymbol{H}^{-1} \tag{14}$$

In Equations (13) and (14), ***H*** is a temporary matrix for the calculation of ***R***. The translation matrix is calculated by Equations (15) and (16).

$$\boldsymbol{W}_{ij} = \boldsymbol{CT}_{ij} - \left(S \cdot \sum_{k=1}^{3} \left(\boldsymbol{R}_{jk}\boldsymbol{CF}^{\mathrm{T}}{}_{ki}\right) \right)^{\mathrm{T}} \tag{15}$$

$$\boldsymbol{T}_{j1} = \frac{1}{4}\sum_{i=1}^{4} \boldsymbol{W}_{ij} \tag{16}$$

In Equations (15) and (16), ***W*** is a temporary translation matrix for the calculation of ***T***. The coordinates of point cloud models generated by Meshroom can be converted to a physical size according to the transformation matrix of ***Tranmat*** in Meshlab. The corresponding MATLAB code for the calculation of ***Tranmat*** is provided in Appendix A. The four control points on the slope for the calculation of ***Tranmat*** are shown in Figure 4. Because ***Tranmat*** is a 4-by-4 matrix, the coordinates of a point in Meshroom should be modified to the form of $(x, y, z, 1)$ before conducting the transformation. The textured 3D point cloud model of the rock slope at the physical scale is shown in Figure 4.

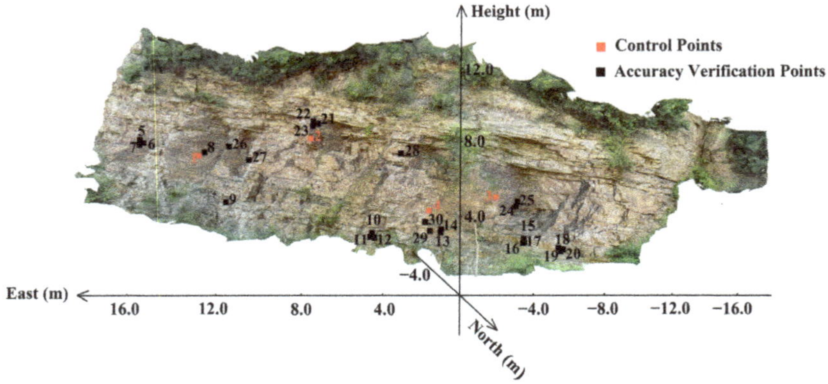

Figure 4. Textured 3D point cloud model of the rock slope at the physical scale.

4. The Accuracy Verification of the 3D Point Cloud Model

The accuracy of the textured 3D point cloud model includes two aspects: length accuracy and angle accuracy. The exposed surface of the slope fluctuates obviously. It is inconvenient to conduct measurements with a compass and a tape; therefore, the accuracy verification is conducted with a total station. The length accuracy is verified directly by comparing the distance between two points obtained by Meshlab and the total station. The angle accuracy is verified by comparing the orientations of joints obtained by the two methods. The orientations are calculated according to the normal vectors of the joint planes, which can be solved by at least three non-collinear points. Taking Figure 5 as an example, P_1, P_2, and P_3 are the inflection points of a joint, and they are non-collinear. A fitting plane (shown as a blue dash circle in Figure 5) can be obtained by the least square method with the coordinates of the three inflection points, and then the normal vector ($\boldsymbol{n}$ (x_n, y_n, z_n)) of the fitting plane can be calculated. The dip and dip direction (corresponding to β and α, respectively, in Figure 5) of the fitting plane can be inferred based on the normal vector of $\boldsymbol{n}$. The dip direction is defined as the angle between north and the horizontal projection of the normal vector, which ranges from 0 to 360. The dip is the dihedral angle between the horizontal plane and the joint, which runs from 0 to 90.

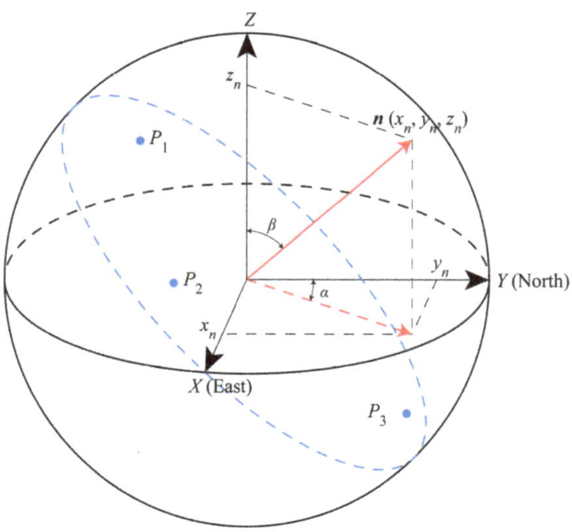

Figure 5. Relationship between the normal vector of a joint and its corresponding orientation.

As can be seen from the above introduction, the measurement accuracy of the total station is vital for the accuracy verification of the textured 3D point cloud model. An outdoor experiment was carried out to obtain the actual measurement accuracy of the total station and to verify the applicability of the calculation method of joint orientations. The main equipment used in the experiment included a total station, tripod, compass, tape, vernier caliper, and measuring pad. Seven reflective stickers were attached to the measuring pad, denoted as A, B, C, D, E, F, and G. Figure 6 displays the main equipment used in the experiment. The outdoor experiment was conducted as the following steps:

(1) Identify the specific model and brand of the total station to be tested and make sure that it is properly calibrated and in good working condition.
(2) Choose a suitable outdoor site for the experiment, taking into account factors such as the terrain, weather conditions, and availability of reference points.
(3) Set up the total station at a fixed position and level it carefully. Use a tripod and make sure that the instrument is stable and properly oriented.
(4) Place the measuring pad at a known distance from the total station and mark it clearly with a reflective target or prism.
(5) Take multiple measurements of the measuring pad using the total station, varying the horizontal and vertical angles and the distance. Record the measurements carefully and accurately, taking into account any sources of error or uncertainty.
(6) Repeat the measurements at different times of day or under different weather conditions to assess the effect of environmental factors on the accuracy of the total station.
(7) Compare the measured distances and angles with the actual values of the measuring pad obtained from a reference source with a vernier caliper.
(8) Calculate the errors and uncertainties in the measurements and analyze the results to determine the actual measurement accuracy of the total station under the given conditions.

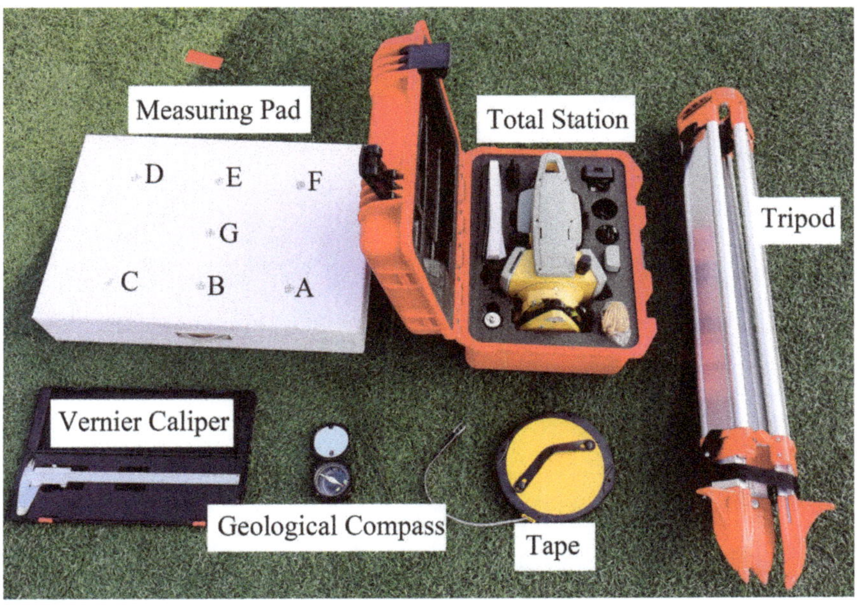

Figure 6. Main equipment used in the outdoor experiment.

The measuring pad was placed 10 m away from the total station in the directions of east, west, south, and north. In the direction of west, 4 different distances of 10 m, 30 m, 50 m, and 100 m were selected. A total of seven different combinations of direction and distance were employed to verify the accuracy of the total station. Table 1 shows the results of the outdoor experiment. From the table, the following points can be drawn. First, the absolute values of the dip direction differences for the 2 measuring methods range from 0.20 to 1.90 degrees, and the mean value at the distance of 10 m is 0.69 degrees. The value generally rises following the increase of the measuring distance. Second, the absolute values of the dip differences range from 0.08 to 1.25 degrees, and the mean value of 10 m is 0.315 degrees. The value is smaller compared with the data of the dip direction difference at the same distance. The same trend of the value with distance can also be observed here. Third, the absolute values of the distance differences range from 0.32 to 1.33 mm, and the mean value at 10 m is 0.51 mm. The value basically increases following the rise of the measuring distance. Fourth, because the distance between the slope surface and the total station is not greater than 30 m, the absolute values of the differences of the dip direction, dip, and distance are not higher than 1.19, 0.83 degrees, and 0.82 mm, respectively (taking the largest error values at the distances of 10 and 30 m). The results mean that the accuracy of the total station is high enough as the reference for verifying the accuracy of the 3D point cloud model, and the method of calculating the orientations of joints according to the coordinates of points is applicable.

Table 1. Results of the outdoor experiment for the accuracy verification of the total station.

No.		1	2	3	4	5	6	7
Distance between the total station and the measuring pod (m)		10	10	10	10	30	50	100
Orientation obtained by compass (°)	Dip direction	277	9	93	192	285	274	313
	Dip	53	53	54	51	50	50	54
Orientation calculated from coordinates (°)	Dip direction	277.20	10.19	93.78	192.59	286.09	275.90	314.88
	Dip	52.63	53.67	54.14	51.08	50.83	50.55	55.25
Distances obtained by vernier caliper (mm)	AB				174.62			
	BC				178.54			
	CD				246.22			
	DE				175.02			
	EF				173.64			
	FA				240.10			
	AG				211.66			
	GD				217.22			
	CG				215.44			
	GF				209.22			
Distances calculated from coordinates (mm)	AB	174.39	174.58	174.03	174.59	173.92	174.11	172.47
	BC	178.32	177.82	179.29	178.43	178.30	178.02	176.86
	CD	245.71	246.10	245.32	246.45	246.28	244.90	249.10
	DE	175.01	174.88	174.83	174.90	174.89	175.24	175.44
	EF	173.23	173.27	173.63	173.19	173.08	172.57	173.98
	FA	239.96	240.31	239.07	241.05	239.90	240.12	239.61
	AG	212.09	212.62	210.08	212.70	211.36	211.44	213.91
	GD	217.07	217.09	217.77	217.10	217.42	216.48	216.26
	CG	214.79	213.35	216.82	215.17	215.27	214.66	215.01
	GF	208.78	209.78	208.04	208.93	208.40	209.00	207.49
Absolute values of orientation differences (°)	Dip direction	0.20	1.19	0.78	0.59	1.09	1.90	1.88
	Dip	0.37	0.67	0.14	0.08	0.83	0.55	1.25
Absolute values of distance differences (mm)	AB	0.23	0.04	0.59	0.03	0.70	0.51	2.15
	BC	0.22	0.72	0.75	0.11	0.24	0.52	1.68
	CD	0.51	0.12	0.90	0.23	0.06	1.32	2.88
	DE	0.01	0.14	0.19	0.12	0.13	0.22	0.42
	EF	0.41	0.37	0.01	0.45	0.56	1.07	0.34
	FA	0.14	0.21	1.03	0.95	0.20	0.02	0.49
	AG	0.43	0.96	1.58	1.04	0.30	0.22	2.25
	GD	0.15	0.13	0.55	0.12	0.20	0.74	0.96
	CG	0.65	2.09	1.38	0.27	0.17	0.78	0.43
	GF	0.44	0.56	1.18	0.29	0.82	0.22	1.73
Mean value (mm)		0.32	0.53	0.82	0.36	0.34	0.56	1.33

The coordinates of a total of 30 feature points (corresponding to the numbers 1 to 30) on the slope (see Figure 4) were measured in Meshlab (after conducting the coordinate transformation) and in the field simultaneously. Feature points 1 to 4 are control points that were used for calculating the transformation matrix, while feature points 5 to 30 were employed for the accuracy verification of the 3D point cloud model. The mean value of the distances of the feature points 5 to 30 between the 2 sets of coordinates is 3.96 mm. Tables 2 and 3 list the length and the angle accuracy verification results, respectively, for the 3D point cloud model. As can be seen from the 2 tables, the absolute values of the length, dip, and dip direction differences calculated based on the 2 sets of coordinates are 1.57 mm, 0.38 degrees, and 0.42 degrees, respectively; therefore, the accuracy of the textured 3D point cloud model is high enough for the engineering application. It should be noted that the accuracy listed above is taking the coordinates obtained by the total station as a reference. The real accuracy of a textured 3D point cloud model is controlled by the following factors: the accuracy of the total station, the number and the quality of the images of the rock mass, the grid density of the point cloud, and the fineness of the texture of the rock mass.

Table 2. Results of the length accuracy verification for the textured 3D point cloud model.

No.	Start Point Series Number	Endpoint Series Number	Length Calculated Based on Coordinates Extracted from Meshlab (m)	Length Calculated Based on Coordinates Extracted from Total Station (m)	Absolute Value of Difference of Two Lengths (mm)
1	5	18	20.9986	21.0002	1.6
2	6	19	20.9473	20.9466	0.7
3	7	20	21.0564	21.0564	0.0
4	8	21	7.5396	7.5405	0.9
5	9	22	6.3363	6.3389	2.6
6	10	23	5.5685	5.5653	3.2
7	11	24	7.2291	7.2286	0.5
8	12	25	7.2948	7.2932	1.6
9	13	26	11.5674	11.5693	1.9
10	14	27	9.8953	9.8936	1.7
11	15	28	6.6601	6.6637	3.7
12	16	29	4.3543	4.3546	0.3
13	17	30	4.6618	4.6601	1.7
Mean value					1.57

Table 3. Results of the angle accuracy verification for the textured 3D point cloud model.

No.	Point Series Numbers	Orientation Calculated Based on Coordinates Extracted from Meshlab (°)		Orientation Calculated Based on Coordinates Obtained from Total Station (°)		Absolute Value of Difference of Two Orientations (°)	
		Dip	Dip Direction	Dip	Dip Direction	Dip	Dip Direction
1	5, 6, 7	68.2	157.2	67.7	157.5	0.5	0.3
2	10, 11, 12	68.7	203.1	68.4	203.7	0.3	0.6
3	15, 16, 17	86.5	157.2	86.7	157.2	0.2	0.0
4	18, 19, 20	78.9	342.3	79.6	341.8	0.7	0.5
5	21, 22, 23	71.2	153.5	71.4	152.8	0.2	0.7
Mean value						0.38	0.42

5. The Rock Structure Analysis for the Rock Slope

Rock structure is a composite indicator for reflecting the development of joints in a rock mass, which can be evaluated by joint spacing, density, orientation, and size. In most cases, this information has to be inferred according to the trace lines exposed on the surface of a rock mass. Trace lines are the intersection lines of joints and the surface of a rock mass. Trace lines can be clearly identified in point cloud models. The study involved a manual process of extracting trace lines by identifying multiple inflection points along a single trace line, connecting them, and obtaining the final trace line. Joint models, which simulate the development of joints in rock masses, can be built based on this information. No obvious difference between the strike length and the dip length of joints was observed by Robertson [22] through analyzing a mass of joint data collected in the field. This finding suggests that joints are approximately equidimensional. Baecher et al. [23] developed a joint disk model, in which joints are simplified into equidimensional planar disks. Due to its good applicability, the joint disk model is widely used in the aspects of simulating the mechanical and hydraulic behavior of rock masses [24–27].

The geometry of joint disks in a model is mainly governed by three parameters: orientation, diameter, and density. Because the number of joints in a rock mass is often very large, the parameters of the joints used in a model are generated by the group. Four main steps are needed to build a joint disk model for a rock mass. First, joints collected on the outcrop of the rock mass are divided into different groups. Second, joint density is calculated according to the relative position of joints and sampling windows on the outcrop. Third, the diameters of joint disks are inferred based on the distribution of the trace lengths. Fourth, the orientations of joint disks are generated on the basis of the joint density and the orientation distribution of joints on the outcrop. It should be noted that all the steps are executed based on the data of joints collected on the outcrop, and steps 2 to 4 are carried out by the joint group.

A total of 2993 traces on the rock slope were artificially extracted from the textured 3D point cloud model in Meshlab. The study involved a manual process of extracting trace lines by identifying multiple inflection points along a single trace line, connecting them, and obtaining the final trace line. The software currently cannot automatically discriminate between lines and areas in 3D point cloud models, and this work was done manually instead. It is preferred that joints be automatically extracted and identified in the future. The lengths of the traces range from 0.01 to 12.16 m. Figure 7 shows the distribution of the traces on the slope. Each of the traces consists of dozens of inflection points, the coordinates of which are the basis of the joint parameter calculation. The orientation (including the dip and dip direction) of a joint can be inferred based on at least three non-collinear points on the joint.

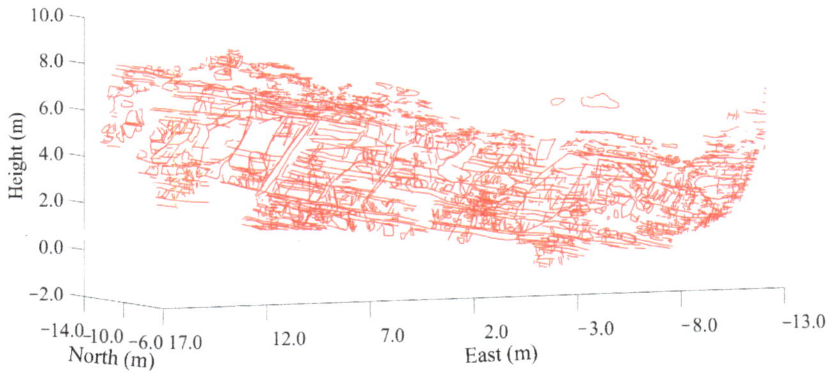

Figure 7. Trace lines extracted from the textured 3D point cloud model.

Joints were grouped according to their orientations. The orientation is composed of two parameters: dip and dip direction. Fuzzy C-means was used to execute the grouping when the number of joint sets ranged from 2 to 8 [28]. The number of joint sets was determined according to a simplified Xie–Beni index, shown as Equation (17) [29]. The index has the advantages of simple programming, high precision, and good practicability [26]. The number of joint sets that minimize the index is the optimum number of joint sets.

$$SXB_K = \frac{\sum_{i=1}^{N}\sum_{vj=1}^{K} I_{ivj}\left(\arccos\left|x_i x_{vj} + y_i y_{vj} + z_i z_{vj}\right|\right)^2}{\min_{vj\neq vl}\left(\arccos\left|x_{vj} x_{vl} + y_{vj} y_{vl} + z_{vj} z_{vl}\right|\right)^2} \tag{17}$$

where SXB_K is the simplified Xie–Beni set validity index when the N joints are divided into K sets. (x_i, y_i, z_i) are coordinates of the unit normal vector of joint i. (x_{vj}, y_{vj}, z_{vj}) and (x_{vl}, y_{vl}, z_{vl}) are coordinates of the unit normal central vectors of sets vj and vl, respectively. I_{ivj} is a Boolean value, which is equal to 1 if the joint i belongs to the set vj; otherwise, it is equal to 0. Joints were grouped into three sets. Figure 8 is the poles diagram of the grouping results for the 2993 traces. One pole represents the orientation of one joint, and the angle between the pole and North represents the dip direction, while the distance between the pole and the circle center represents the magnitude of the dip. Because the surface of a rock mass is opaque, the joint density and diameters of joint disks were inferred using 3D topographic sampling windows, which allow consideration of joint traces within buffer zones [18]. Due to the fact that extracted traces in the 3D point cloud model are three-dimensional curves, and the actual slope surface is also a three-dimensional surface, the buffer zone method is adopted in this paper for trace information statistics. This method converts the linear boundary of the circular measuring window into a surface boundary, counts the trace information within and intersecting with the buff zone boundary, and then calculates the distribution parameters of the joint diameter and density. Finally, a joint disk model can be generated based on these parameters. Figure 9 shows the traces of joints by the set and the employed 3D topographic sampling windows. Because the dimension of the studied slope is greater in the horizontal direction, several topographic sampling windows were placed at different locations of the slope (shown as Figure 9), and the average results were used. The area density (ρ_A) and the mean trace length of joints (l_t) in a sampling window were inferred with Equations (18) and (19) [30].

$$\rho_A = \frac{m}{0.5\pi {d_c}^2} \tag{18}$$

$$l_t = \frac{\pi d_c n}{4m} \tag{19}$$

where m is the number of joint endpoints inside a circular sampling window; n is the number of joint intersections with the circular scanline; d_c is the diameter of the circular sampling window. The mean diameter of joint disks was estimated with Equation (20) [31].

$$E({d_i}^2) = 1.5E(l_t^2) \tag{20}$$

where d_i is the diameter of a joint disk. The volume density of joints (ρ_A) was inferred with Equation (21) [32].

$$\rho_v = \frac{\rho_A}{E(d_i)\cdot\sin(\gamma)} \tag{21}$$

where γ is the angle between a topographic sampling window and the mean joint plane. The number of joint disks in each of the joint sets can be determined according to the volume density. The orientation and diameter distributions of the joint disks are shown in Figure 10. The parameters of the joint disk model built for the rock slope are listed in Table 4. The space distribution of the joint disks in each joint set and the final joint disk model are shown in Figure 11.

The rock quality designation (RQD), which is defined as the percentage of the lengths of intact rock core pieces that are not shorter than 0.1 m, of the rock slope was calculated according to the joint disk model. A total of 300 equally spaced boreholes were drilled in the joint disk model, with 100 boreholes distributed on each of the planes of *xy*, *yz*, and *zx*. Figure 12 displays the layout of the boreholes in the model. The lengths of core pieces between two adjacent joints on the borehole lines were measured to calculate the value of RQD. Figure 13 shows the distribution of the lengths of intact rock core pieces. The average value of RQD is equal to 97.05%. According to Figures 12 and 13 and Table 4, the following points can be concluded. (1) The joints in set 1 are generally subhorizontal, with the mean dip value of 14.7°, while the joints in sets 2 and 3 are nearly vertical, with the mean dip values of 76.6° and 81.5°, respectively. (2) The sizes of the joints in set 1 are the largest in all 3 joint sets, with a mean length of 0.67 m. (3) The rock structure of slope is good, according to the value of RQD. (4) The method of photogrammetry-based rock structure analysis proposed in this study is applicable in real rock engineering.

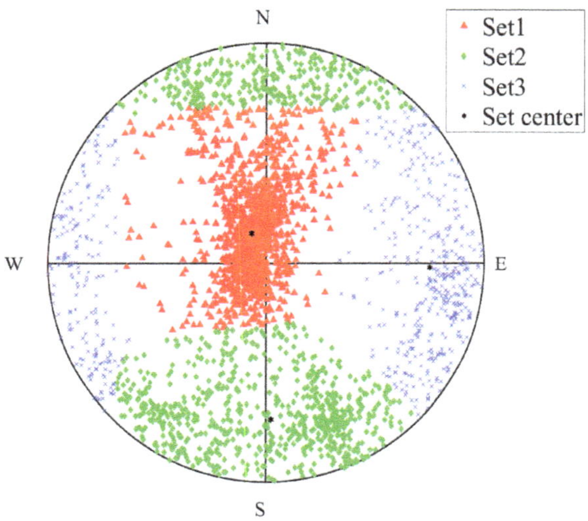

Figure 8. Poles diagram of the grouping results of the 2993 traces.

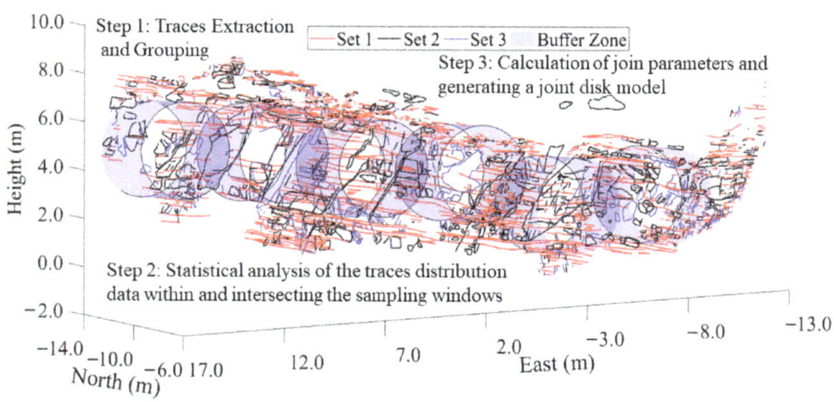

Figure 9. Traces of joints shown by set and the employed 3D topographic sampling windows.

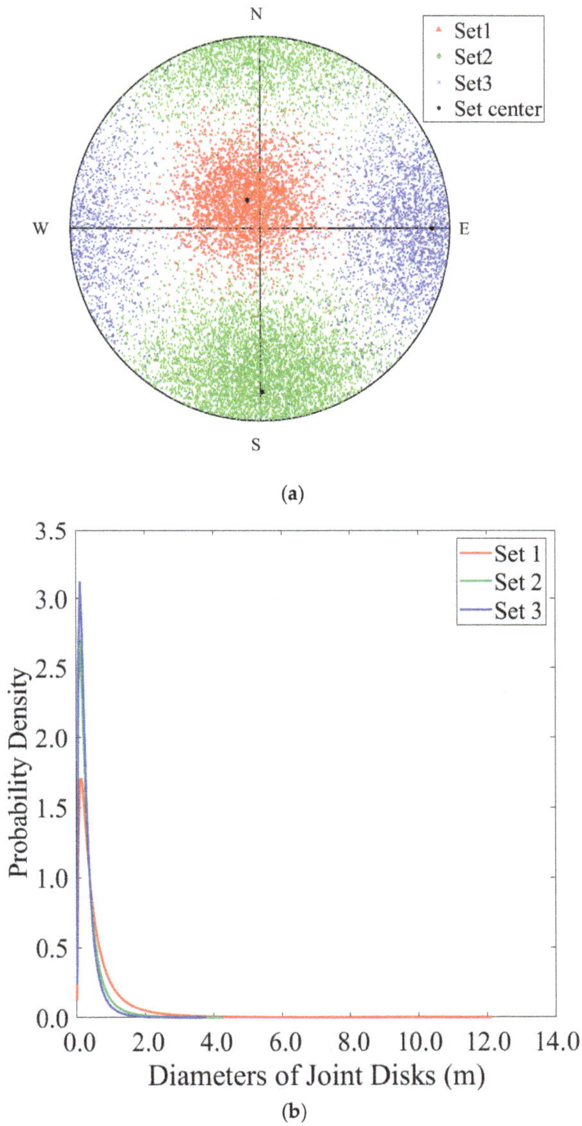

Figure 10. Parameters of the three joint sets used in the joint disk model. (**a**) Poles diagram of the joint disk orientations used in the joint disk model. (**b**) Diameter distributions of the three joint sets used in the joint disk model.

Table 4. Parameters used in the joint disk model of the rock slope.

Joint Set Series	Volume Density (m^{-3})	Orientation (Fisher Distribution)			Diameter (Lognormal Distribution)	
		Dip Direction (°)	Dip (°)	K_F	Mean Value (m)	Variance (m^2)
1	3.02	335.6	14.7	12.0	0.67	1.28
2	7.86	179.4	76.6	7.1	0.38	0.22
3	3.99	90.1	81.5	8.1	0.31	0.12

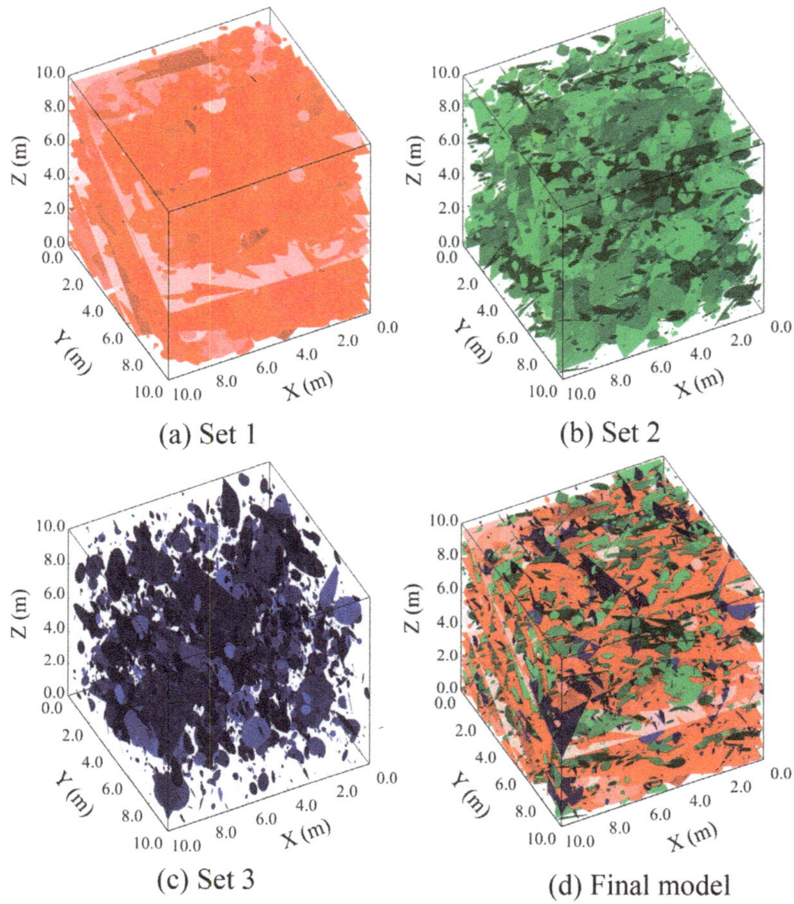

Figure 11. Joint disk models of the rock slope.

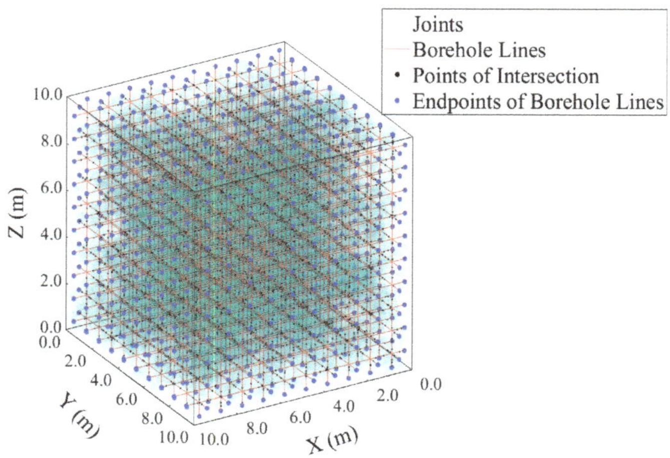

Figure 12. Layout of the boreholes in the joint disk model.

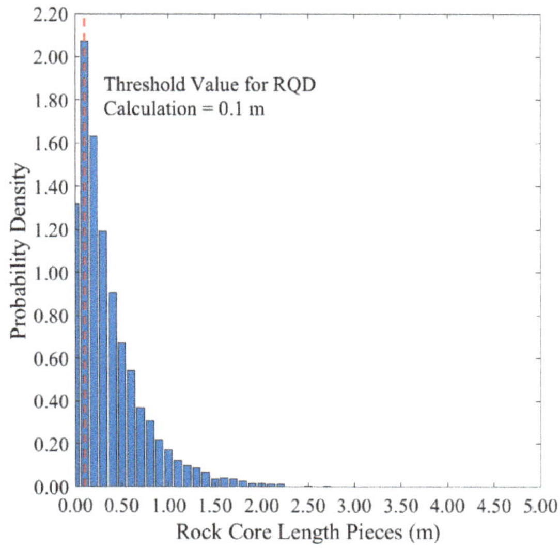

Figure 13. Distribution of the lengths of intact rock core pieces of the joint disk model.

Despite the promising results of this methodology in the study of rock mass structure, there are several limitations that need to be addressed. A main limitation is the computational resources required to generate 3D point cloud models, which can be a bottleneck when dealing with large outcrops. Additionally, the accuracy of the method heavily depends on the number photos used to generate 3D texture models, which can be a challenging task in some field scenarios. Furthermore, the generated joint models can only provide information about the geometry of the joints, and not their mechanical properties, which are crucial for slope stability analysis. To overcome these limitations, there are several approaches that can be taken. Firstly, we suggest improving the hardware configuration of the computer. Additionally, if feasible, a large rock outcrop can be divided into several smaller models. Secondly, we recommend using unmanned aerial vehicles to capture more photos to solve the issue of insufficient photo data. Moreover, unmanned aerial vehicles can overcome some limitations of certain field conditions, such as inaccessibility or safety concerns. Finally, we recommend conducting indoor mechanical experiments to obtain the mechanical parameters of joints. If conducting mechanical experiments is difficult, the mechanical parameters of joints can be obtained from geological reports.

6. Summary and Conclusions

A low-cost method of obtaining photogrammetry-based, textured, 3D point cloud models was provided with two open-source software programs, Meshroom and Meshlab. A coordinate transformation means of converting model coordinates to physical coordinates was introduced, and the corresponding computer code was supplied. The length accuracy and angle accuracy of a textured 3D point cloud model obtained with the method were verified, taking the data of a total station as the reference. The accuracy of the total station was illustrated by an outdoor experiment. A section of a rock slope located in Dalian city was selected as the case study to represent the processes of textured 3D point cloud model building and the applicability of the method to rock structure estimation. The findings are summarized as follows.

(1) The accuracy of the total station is enough as a reference for verifying the accuracy of the 3D point cloud model. The accuracy of the point coordinates of the 3D point cloud model could achieve 3.96 mm. The accuracy of the dip, dip direction, and length are 0.315 degrees, 0.69 degrees, and 0.51 mm, respectively, at a distance of 10 m.

(2) The accuracy of the 3D point cloud model is high enough for engineering applications, taking the coordinates obtained by the total station as a reference. The accuracies of the length, dip, and dip direction are 1.57 mm, 0.38 degrees, and 0.42 degrees, respectively.
(3) The method of building a photogrammetry-based 3D point cloud model is applicable. A total of 2993 joint traces of the rock slope were successfully extracted, and the rock structure was estimated with a joint disk model built according to the traces.
(4) The rock structure of the slope is good, according to the value of the rock quality designation. The joints of the slope can be grouped into three sets. The joints in set 1 are generally subhorizontal, with a mean dip value of 14.7°, while the joints in sets 2 and 3 are nearly vertical, with mean dip values of 76.6° and 81.5°, respectively.

Author Contributions: Methodology, T.L.; Writing—review & editing, J.D. All authors have read and agreed to the published version of the manuscript.

Funding: This research was funded by National Natural Science Foundation of China grant number No. U19A2098. The APC was funded by No. U19A2098.

Institutional Review Board Statement: Not applicable.

Informed Consent Statement: Not applicable.

Data Availability Statement: Not applicable.

Acknowledgments: The authors thank Zheng Hongchun and Han Ju for their contributions to the data collection and revision processes.

Conflicts of Interest: The authors declare that they have no known competing financial interests or personal relationships that could have appeared to influence the work reported in this paper.

Appendix A

```
function [Tranformation_matrix ] = Transformation_Matrix(C_From, C_To)
% C_From is a 4-by-3 row vector. Coordinates of reference points in Meshroom.
% C_To is a 4-by-3 row vector. Coordinates of reference points measured by a total station.
% Tranformation_matrix is a 4*4 matrix. It is used for transforming
% coordinates from Meshroom to Meshlab.
A_From = mean(C_From);
A_To = mean(C_To);
C_From_MN = C_From-A_From;
C_To_MN = C_To-A_To;
sFrom = sqrt(sum(C_From_MN(:).^2)/4);
sTo = sqrt(sum(C_To_MN(:).^2)/4);
s = sTo/sFrom; % Scale factor
C_From_MN = C_From_MN/sFrom;
C_To_MN = C_To_MN/sTo;
H = C_From_MN'*C_To_MN;
R = (H'*H)^0.5*inv(H);
A = s*R;
theory_To = (A*C_From')';
T = mean(C_To - theory_To);
Tranformation_matrix(4, 4) = 1;
for i = 1:3
  for j = 1:3
     Tranformation_matrix(i, j) = A(i, j);
  end
end
Tranformation_matrix(1:3, 4) = T';
Tranformation_matrix(4, 1:3) = [0, 0, 0];
End
```

References

1. Wu, X.; Jiang, Y.; Gong, B.; Guan, Z.; Deng, T. Shear performance of rock joint reinforced by fully encapsulated rock bolt under cyclic loading condition. *Rock Mech. Rock Eng.* **2019**, *52*, 2681–2690. [CrossRef]
2. Tzamos, S.; Sofianos, A. A correlation of four rock mass classification systems through their fabric indices. *Int. J. Rock Mech. Min. Sci.* **2007**, *44*, 477–495. [CrossRef]
3. Bahaaddini, M.; Sharrock, G.; Hebblewhite, B.K. Numerical investigation of the effect of joint geometrical parameters on the mechanical properties of a nonpersistent jointed rock mass under uniaxial compression. *Comput. Geotech.* **2013**, *49*, 206–225. [CrossRef]
4. Ban, L.; Zhu, C.; Qi, C.; Tao, Z. New roughness parameters for 3D roughness of rock joints. *Bull. Eng. Geol. Environ.* **2019**, *78*, 4505–4517. [CrossRef]
5. Liu, T.; Deng, J.; Zheng, J.; Zheng, L.; Zhang, Z.; Zheng, H. A new semi-deterministic block theory method with digital photogrammetry for stability analysis of a high rock slope in China. *Eng. Geol.* **2017**, *216*, 76–89. [CrossRef]
6. Chen, J.Y.; Huang, H.W.; Zhou, M.L.; Chaiyasarn, K. Towards semi-automatic discontinuity characterization in rock tunnel faces using 3D point clouds. *Eng. Geol.* **2021**, *291*, 106232. [CrossRef]
7. Deb, D.; Hariharan, S.; Rao, U.M.; Ryu, C.-H. Automatic detection and analysis of discontinuity geometry of rock mass from digital images. *Comput. Geosci.* **2008**, *34*, 115–126. [CrossRef]
8. Fujii, Y.; Fodde, E.; Watanabe, K.; Murakami, K. Digital photogrammetry for the documentation of structural damage in earthen archaeological sites: The case of Ajina Tepa, Tajikistan. *Eng. Geol.* **2009**, *105*, 124–133. [CrossRef]
9. Fujii, Y.; Shogaki, T.; Miyakawa, M. Photogrammetric documentation and non-invasive investigation of a stone dry dock, the Yokosuka arsenal dry dock no. 1, Japan. *Eng. Geol.* **2018**, *234*, 122–131. [CrossRef]
10. Guo, J.; Liu, S.; Zhang, P.; Wu, L.; Zhou, W.; Yu, Y. Towards semi-automatic rock mass discontinuity orientation and set analysis from 3D point clouds. *Comput. Geosci.* **2017**, *103*, 164–172. [CrossRef]
11. Sturzenegger, M.; Stead, D. Close-range terrestrial digital photogrammetry and terrestrial laser scanning for discontinuity characterization on rock cuts. *Eng. Geol.* **2009**, *106*, 163–182. [CrossRef]
12. Kaminski, M.; Zientara, P.; Krawczyk, M. Electrical resistivity tomography and digital aerial photogrammetry in the research of the "Bachledzki Hill" active landslide–in Podhale (Poland). *Eng. Geol.* **2021**, *285*, 106004. [CrossRef]
13. Cheng, Z.; Gong, W.P.; Tang, H.M.; Juang, C.H.; Deng, Q.; Chen, J.; Ye, X. UAV photogrammetry-based remote sensing and preliminary assessment of the behavior of a landslide in Guizhou, China. *Eng. Geol.* **2021**, *289*, 106172. [CrossRef]
14. Nappo, N.; Mavrouli, O.; Nex, F.; Van, C.; Gambillara, R.; Michetti, A.M. Use of UAV-based photogrammetry products for semi-automatic detection and classification of asphalt road damage in landslide-affected areas. *Eng. Geol.* **2021**, *294*, 106363. [CrossRef]
15. Buyer, A.; Aichinger, S.; Schubert, W. Applying photogrammetry and semiautomated joint mapping for rock mass characterization. *Eng. Geol.* **2020**, *264*, 105332. [CrossRef]
16. Menegoni, N.; Giordan, D.; Perotti, C.; Tannant, D. Detection and geometric characterization of rock mass discontinuities using a 3D high-resolution digital outcrop model generated from RPAS imagery-Ormea rock slope, Italy. *Eng. Geol.* **2019**, *252*, 145–163. [CrossRef]
17. Lato, M.; Kemeny, J.; Harrap, R.M.; Bevan, G. Rock bench: Establishing a common repository and standards for assessing rockmass characteristics using LiDAR and photogrammetry. *Comp. Geosci.* **2013**, *50*, 106–114. [CrossRef]
18. Sturzenegger, M.; Stead, D.; Elmo, D. Terrestrial remote sensing-based estimation of mean trace length, trace intensity and block size/shape. *Eng. Geol.* **2011**, *119*, 96–111. [CrossRef]
19. Laribi, A.; Walstra, J.; Ougrine, M.; Seridi, A.; Dechemi, N. Use of digital photogrammetry for the study of unstable slopes in urban areas: Case study of the El Biar landslide, Algiers. *Eng. Geol.* **2015**, *187*, 73–83. [CrossRef]
20. Mora, P.; Baldi, P.; Casula, G.; Fabris, M.; Ghirotii, M.; Mazzioni, E.; Pesci, A. Global positioning systems and digital photogrammetry for monitoring of mass movements: Application to the Ca' di Malta landslide (northern Apennines, Italy). *Eng. Geol.* **2003**, *68*, 103–121. [CrossRef]
21. Kabsch, W. A Solution for the Best Rotation to Relate Two Sets of Vectors. *Acta. Cryst. A* **1976**, *32*, 922–923. [CrossRef]
22. Robertson, A. The interpretation of geological factors for use in slope theory. In *Planning Open Pit Mines: Proceedings of the Symposium on the Theoretical*; SAIMM: Johannesburg, South Africa, 1970; pp. 55–71.
23. Baecher, G.B.; Lanney, N.S.; Einstein, H.H. Statistical descriptions of rock properties and sampling. In Proceedings of the 18th US Symposium on Rock Mechanics (USRMS), Golden, CO, USA, 22–24 June 1977; OnePetro: Richardson, TX, USA, 1977.
24. Lei, Q.; Latham, J.P.; Tsang, C.F. The use of discrete fracture networks for modelling coupled geomechanical and hydrological behaviour of fractured rocks. *Comput. Geotech.* **2017**, *85*, 151–176. [CrossRef]
25. Liu, T.; Zheng, J.; Deng, J. A new iteration clustering method for rock discontinuity sets considering discontinuity trace lengths and orientations. *Bull. Eng. Geol. Environ.* **2021**, *80*, 413–428. [CrossRef]
26. Liu, T.; Jiang, A.; Deng, J.; Zheng, J.; Zhang, Z. A new multiple-factor clustering method considering both box fractal dimension and orientation of joints. *J. Rock Mech. Geotech. Eng.* **2021**, *14*, 366–376. [CrossRef]
27. Zhang, W.; Lan, Z.; Ma, Z.; Tan, C.; Que, J.; Wang, F. Determination of statistical discontinuity persistence for a rock mass characterized by non-persistent fractures. *Int. J. Rock Mech. Min. Sci.* **2020**, *126*, 104177. [CrossRef]

28. Hammah, R.E.; Curran, J.H. Fuzzy cluster algorithm for the automatic identification of joint sets. *Int. J. Rock Mech. Min. Sci.* **1998**, *35*, 889–905. [CrossRef]
29. Ma, G.; Xu, Z.; Zhang, W.; Li, S. An enriched K-means clustering method for grouping joints with meliorated initial centers. *Arab. J. Geosci.* **2015**, *8*, 1881–1893. [CrossRef]
30. Mauldon, M.; Dunne, W.M.; Rohrbaugh, M.B. Circular scanlines and circular windows: New tools for characterizing the geometry of fracture traces. *J. Struct. Geol.* **2001**, *23*, 247–258. [CrossRef]
31. Nie, Z.; Chen, J.; Zhang, W.; Chun, T. A new method for three-dimensional fracture network modelling for trace data collected in a large sampling window. *Rock Mech. Rock Eng.* **2019**, *53*, 1145–1161. [CrossRef]
32. Warburton, P.M. A stereological interpretation of joint trace data. *Int. J. Rock Mech. Min. Sci. Geomech. Abstr.* **1980**, *17*, 181–190. [CrossRef]

International Journal of
Geo-Information

MDPI

Article

Predicting Earthquake-Induced Landslides by Using a Stochastic Modeling Approach: A Case Study of the 2001 El Salvador Coseismic Landslides

Claudio Mercurio [1,*], Laura Paola Calderón-Cucunuba [1], Abel Alexei Argueta-Platero [1,2], Grazia Azzara [1], Chiara Cappadonia [1], Chiara Martinello [1], Edoardo Rotigliano [1] and Christian Conoscenti [1]

1 Dipartimento di Scienze della Terra e del Mare (DiSTeM), University of Palermo, Via Archirafi 22, 90123 Palermo, Italy; laurapaola.calderoncucunuba@unipa.it (L.P.C.-C.); abel.argueta@ues.edu.sv (A.A.A.-P.); grazia.azzara@unipa.it (G.A.); chiara.cappadonia@unipa.it (C.C.); chiara.martinello@unipa.it (C.M.); edoardo.rotigliano@unipa.it (E.R.); christian.conoscenti@unipa.it (C.C.)

2 Escuela de Posgrado y Educación Continua, Facultad de Ciencias Agronómicas, University of El Salvador, Final de Av. Mártires y Héroes del 30 Julio, San Salvador 1101, El Salvador

* Correspondence: claudio.mercurio@unipa.it

Abstract: In January and February 2001, El Salvador was hit by two strong earthquakes that triggered thousands of landslides, causing 1259 fatalities and extensive damage. The analysis of aerial and SPOT-4 satellite images allowed us to map 6491 coseismic landslides, mainly debris slides and flows that occurred in volcanic epiclastites and pyroclastites. Four different multivariate adaptive regression splines (MARS) models were produced using different predictors and landslide inventories which contain slope failures triggered by an extreme rainfall event in 2009 and those induced by the earthquakes of 2001. In a predictive analysis, three validation scenarios were employed: the first and the second included 25% and 95% of the landslides, respectively, while the third was based on a *k*-fold spatial cross-validation. The results of our analysis revealed that: (i) the MARS algorithm provides reliable predictions of coseismic landslides; (ii) a better ability to predict coseismic slope failures was observed when including susceptibility to rainfall-triggered landslides as an independent variable; (iii) the best accuracy is achieved by models trained with both preparatory and trigger variables; (iv) an incomplete inventory of coseismic slope failures built just after the earthquake event can be used to identify potential locations of yet unreported landslides.

Keywords: landslide susceptibility; multivariate adaptive regression splines (MARS); GIS; earthquake; earthquake-induced landslides; rainfall-induced landslides; El Salvador; Central America

1. Introduction

In El Salvador, CA, landslides are among the most destructive natural processes and can cause mass fatalities and devastation in the built environment [1]. They may occur either due to heavy rains (rainfall-induced landslides) or as a secondary event of an earthquake (earthquake-induced landslides). The extreme rainfall events that frequently affect the Central American region are indeed responsible for activating gravitational phenomena consisting of shallow and fast-moving flow landslides that may cause severe economic damage and fatalities [2]. Additionally, seismic events can produce hundreds to thousands of gravitational phenomena which may cause more damage and victims than the earthquake itself, a situation that occurs in countries often affected by high-intensity phenomena. Indeed, among the various secondary effects of a seismic event (i.e., tsunamis, fires and liquefaction phenomena), 71.1% of deaths worldwide result from landslides [3]. Landslides usually occur during or shortly after a seismic event and have different spatial distributions according to the slope gradient, lithology, earthquake magnitude and hypocentral and epicentral distances [4–9]. A number of studies observed a correlation between earthquake intensity and landslide occurrence [10,11]. Keefer identified a critical Richter magnitude of 4.0 and observed that the area affected by gravitational phenomena progressively grows up to 500,000 km^2 for an earthquake with a magnitude of 9.2 [10]. However, this correlation may present exceptions. A landslide could indeed be initiated even by a weak shaking if a slope failure is imminent prior to the seismic event or if the material involved includes weakly cemented and/or incoherent rocks. Volcanic rocks, which are widely distributed across the country, are among these lithologies, as they are usually characterized by a higher mobility than non-volcanic rocks due to differences in their material properties such as granularity, collapsibility, and water content [12–16]. These geological conditions can significantly increase the frequency of gravitational phenomena triggered and the areas affected by landslides when compared to areas that experienced comparable magnitudes [17,18].

Damage due to coseismic landslides is closely related to economic development and the ability to prevent and reduce disasters. Indeed, developed countries have the economic and scientific resources to plan and manage structures designed to contain blocks of rock and stop or divert both slow and rapid flows [19,20]. They may also activate direct or indirect protection and alarm systems to favor a prompt and rapid general evacuation. Projecting effective defensive engineering structures and alarm systems requires knowing where landslides are more likely to occur. This information, which is also fundamental for land use planning aimed at mitigating landslide risk, is provided by landslide susceptibility maps.

Landslide susceptibility reflects the relative probability of slope failure occurrence in a given area [21,22]. The mapping of landslide susceptibility can be achieved by using various methods, among which the stochastic approach is the most popular choice [23–28]. This methodology is based on the hypothesis that "the past is the key to the future" [29–31]. Therefore, future landslides are more likely to be triggered under the same environmental conditions that have produced past and present slope failures. The first step of landslide susceptibility modeling using a stochastic approach is to select a mapping unit, which usually corresponds to square grid cells derived from digital elevation models (DEMs) or to slope units limited by streams and water divides. Then, a value of the dependent and independent variables is assigned to each mapping unit. The dependent variable is usually dichotomous (0: absence of landslide; 1: presence of landslide). The values of the selected independent variables (also known as covariates or predictors), which serve as proxies for the preparatory factors that mainly control landslide occurrence in the study area, are usually derived from DEMs and from already available thematic maps, such as geological or land use maps. Finally, multivariate statistical techniques or machine learning algorithms are usually employed to model the relationships between the landslide spatial distribution and the variability of the predictors to calculate a probability value of landslide occurrence for each mapping unit. The most common statistical and machine learning methods employed for this aim are logistic regression [22,28,32–34], discriminant analysis [30,35,36], support vector machines [37,38] or random forests [39]. Comprehensive reviews of the most popular methods for landslide susceptibility zonation can be found in [30,40–42]. Landslide susceptibility zonation usually relies on independent variables that reflect only preparatory factors, assuming that the geographical distribution of the landslide inventories is not affected by the pattern of the trigger (either rainfall or earthquake), which is thus considered spatially homogeneous. However, if a homogenous trigger pattern can be assumed for single and relatively small areas, this hypothesis does not apply when modeling large areas or multiple study areas that may experience different intensities of the trigger. In other words, when landslide susceptibility is assessed at a regional scale or in different areas spread over a large territory, it is necessary to consider not only variables that express preparatory factors but also variables that represent the pattern and intensity of the trigger in relation to the return period.

This paper focuses on predicting the landslides triggered by two earthquakes of magnitudes 7.7 and 6.6 which occurred in El Salvador in January and February 2001, respectively, and triggered thousands of slope failures and caused more than 1150 fatalities.

The main aim of this study is to evaluate the ability of rainfall- and earthquake-induced landslide susceptibility models to predict the geographical distribution of the coseismic slope failures that occurred in 2001 under three calibration and validation scenarios. The first is a scenario in which most landslides are known (i.e., 75%), and the second is a scenario in which only a minor portion of the gravitational phenomena that occurred are identified (i.e., 5%). A comparable and more than acceptable predictive ability of these scenarios might suggest that an inventory of slope failures built just after the triggering event, even if incomplete, may help rapidly generate reliable landslide susceptibility maps. To further evaluate our approach, we evaluated a third scenario based on a *k*-fold spatial cross-validation [37,43,44].

The experiment was performed in fourteen sectors of El Salvador, accounting for a total area of 397 km^2. Eight sectors (92 km^2) were strongly hit by the January earthquake, whereas most of the landslides triggered by the February earthquake occurred in the other six sectors (305 km^2). The modeling strategy was based on both rainfall- and earthquake-triggered landslide susceptibility models. These were based on statistical relationships between landslide inventories and multiple predisposing and triggering factors, established using multivariate analysis regression splines (MARS), which demonstrated excellent landslide predictive ability in previous papers [22,45–50]. The first model was calibrated using an inventory of thousands of landslides triggered on 7–8 November 2009 by the combined action of Hurricane Ida and the low-pressure system 96E in a sector extending 287 km^2 along the slopes of the San Vicente Volcano, El Salvador [49]. The rainfall-triggered model, which was trained by eleven preparatory factors, was applied to all fourteen study areas to verify its ability to predict the position of the coseismic landslides that occurred in January and February 2001. On the other hand, the earthquake-induced landslide predictive models were calibrated using both preparatory and trigger variables, with the latter reflecting the intensity of ground shaking.

The article is organized as follows. Section 2 describes the available data and the methods employed to map the coseismic landslides and to calibrate and validate the predictive models. Section 3 describes the inventory of the coseismic landslides and shows the results of models' training and testing. The results are discussed in Section 4, and the main conclusions of the study are reported in Section 5.

2. Materials and Methods

2.1. Study Areas

The complex geodynamic context in which the small country of El Salvador is located is characterized by the convergence of the Cocos plate and the Caribbean plate through the Middle America Trench at a rate of about 7 cm/year [51] (Figure 1a). This active tectonic environment is responsible for several active volcanoes in the country which are aligned in the WNW–ESE direction and have frequently erupted and deposited widespread and poorly consolidated tephra in many parts of the country [13]. These lithologies, in combination with acid pyroclastites and acid and basic effusive rocks, ensure that volcanic rocks represent 95% of the outcropping lithologies in the country (Figure 1b).

The combination of the geodynamic contexts responsible for medium and high earthquakes, steep slopes due to rugged volcanic ranges, incoherent rocks and a subtropical climate characterized by heavy storms, cause a high probability of the occurrence of earthquake- and rainfall-induced landslides in most of the country. The consequences of these gravitational phenomena are devastating for the country, which is the smallest country in Central America but has the highest population density (310 inhabitants/km^2). Indeed, the fertility of the soils around the volcanoes, which are oriented WNW–ESE due to the tectonic assessment [17], has generated a situation in which a large part of the population lives in the areas of the country most susceptible to the gravitational phenomena (Figure 1b). As a confirmation of this, El Salvador has the highest ratio of destroyed houses to affected risks in Central America [52]. This work was carried out in fourteen sectors along the volcanic chain (Figure 2), which were among the areas most affected by the two intense earthquakes that occurred in the first two months of 2001.

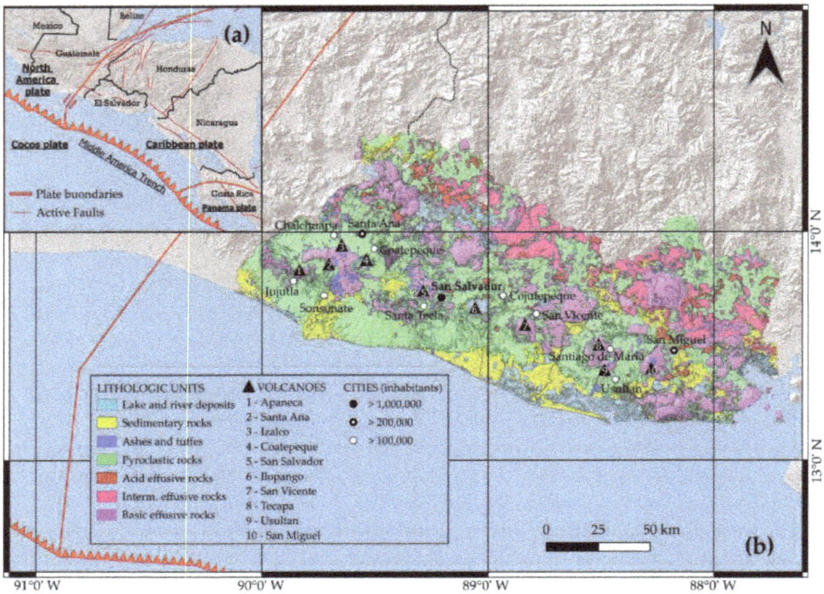

Figure 1. Location and geological setting of the study area. El Salvador is located in Central America, where the Cocos Plate is subducted beneath the Caribbean Plate into the Middle American Trench (**a**); major cities and geological setting of El Salvador, mainly characterized by the volcano chain and by the large outcropping of volcanic rocks (**b**).

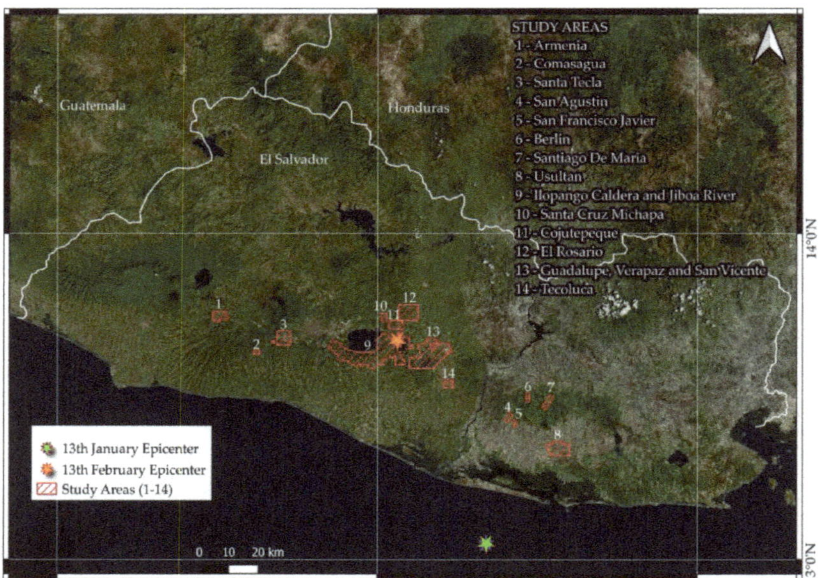

Figure 2. Location of the fourteen study areas: sectors 1–3 and 4–8 were heavily affected by the earthquake of 13 January 2001; sectors 9–14 were severely affected by the earthquake of 13 February 2001.

The first eight study areas, numbered from 1 to 8, were selected because of the high number of landslides triggered by the January 2001 earthquake. Sectors 1 to 3 are in the southwestern part of the country, close to the San Salvador volcano, whereas sectors 4 to 8 are located in the southeastern portion of El Salvador. Sectors 9 to 14 are placed in the southern central part of the country and were selected because they were close to the epicenter of the February 2001 earthquake and were strongly affected by coseismic landslides. Table 1 reports the extent, average and standard deviation of the altitude and slope angle and the main outcropping lithology of all the study areas.

Table 1. Extent, average and standard deviation of altitude and slope angle and the main outcropping lithology classes. GEO1: incoherent soils/rock (alluvial, marine and lacustrine deposits); GEO2: soft rocks (intermediate basic effusive rocks and subordinate pyroclastites); GEO3: hard rocks (intermediate basic effusive rocks); GEO4: medium rocks (acid effusive and acid intermediate rocks); GEO5: very soft soils (Tierra Blanca: acidic pyroclastites, subordinate volcanic epiclastites and acid effusive rocks); GEO6: soft soils (quaternary sedimentary deposits); GEO7: hard soils (acid pyroclastic rocks; volcanic epiclastites); GEO8: medium soils (volcanic epiclastites and pyroclastites; locally effusive basic–intermediate rocks).

Sector	Area [km²]	Altitude [m a.s.l.]		Slope Angle [°]		Two Main Lithology Classes	
		Mean	Std. Dev.	Mean	Std. Dev.		
1	15.41	577.73	64.19	15.10	11.22	GEO8 (83.92%)	GEO3 (15.39%)
2	2.31	1039.00	50.69	25.17	10.02	GEO3 (84.42%)	GEO8 (15.58%)
3	23.89	975.86	70.16	14.71	12.96	GEO8 (42.38%)	GEO5 (26.70%)
4	3.98	286.16	41.14	20.51	10.71	GEO8 (100.00%)	-
5	2.66	287.80	31.01	16.38	9.41	GEO8 (98.44%)	GEO2 (1.56%)
6	3.89	1008.35	85.91	12.11	9.23	GEO2 (72.80%)	GEO8 (27.20%)
7	8.29	862.50	70.54	11.05	8.79	GEO2 (50.97%)	GEO8 (49.03%)
8	31.42	78.34	22.40	3.29	4.28	GEO8 (95.32%)	GEO6 (1.81%)
9	155.32	618.41	122.58	24.23	12.43	GEO8 (30.17%)	GEO7 (27.98%)
10	5.32	685.48	28.97	16.46	12.00	GEO7 (56.19%)	GEO5 (41.14%)
11	13.79	781.87	74.31	16.09	10.14	GEO3 (51.69%)	GEO7 (31.39%)
12	34.80	699.61	51.11	16.30	10.01	GEO3 (10.39%)	GEO7 (73.24%)
13	87.65	833.88	365.10	17.93	13.55	GEO5 (36.26%)	GEO2 (31.78%)
14	8.84	266.62	39.17	5.12	5.55	GEO7 (90.25%)	GEO3 (8.57%)

2.2. The 2001 Earthquakes

El Salvador has been severely affected by medium- and high-intensity earthquakes that have triggered numerous and devastating landslides. Bommer et al. [17] report that the country is experiencing earthquakes from two major seismic sources. The first source of seismicity originates from deep earthquakes in the Benioff–Wadati zones of the subducted Cocos plate beneath the Caribbean plate, which are characterized by relatively high magnitude values (often with surface wave magnitude, Ms, and moment magnitude, Mw, values of >6.0), capable of involving large areas of El Salvador. Numerous seismic phenomena with this genesis were witnessed in the last century, many of which were particularly destructive and had dramatic impacts on the country. Among these phenomena were the September 1915 event, Ms 7.8, which caused destruction in the country's western area, and the Ms 7.3 earthquake in June 1982, which caused hundreds of earthquake-induced landslides and had 49 victims. The second source of seismicity is a zone of upper-crustal earthquakes that coincides with the array of volcanoes and is orientated WNW–ESE [17]. These intraplaque earthquakes occur at relatively lower hypocentral depths. Since the main Salvadoran cities are located just along the volcanoes' alignment, it is not hard to determine the catastrophic consequences of the seismic events that can occur, often with more significant damage than the seismic phenomena originating in Benioff–Wadati zones. Among the most significant events of this origin in the last century are the earthquake of 8 June 1917 (Ms 6.7), which had 101 victims, the 20 December 1936 (Ms 6.1) event, which resulted in more than 200 victims, and the 10 October 1986 event (Ms 7.5), which produced more than 1500 fatalities.

On 13 January 2001, the country was hit by a violent earthquake, originating from the first source of seismicity, at a depth of 60 km with a magnitude of 7.7 Mw (Figure 3). This event caused thousands of gravitational phenomena, injuring 5565, causing 844 fatalities and destroying over 100,000 buildings. The differences in damage and fatalities between this event and the one in 1982, which was of a similar magnitude, are related to the increase in the elements at risk. Indeed, a rapid expansion of urban areas occurred during this time span, which continued to grow mainly toward the flanks of the volcanoes, as revealed by the difference between the topographical maps of the country dating back to the year 1981 and aerial photos from 2001.

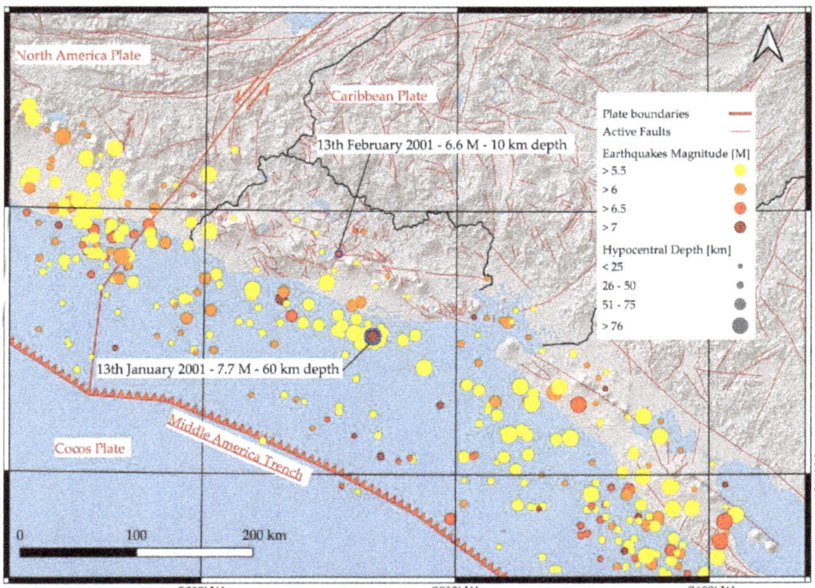

Figure 3. Main earthquakes (Mw > 5.5) that affected El Salvador. The earthquakes that hit the country on 13 January and 13 February 2001 are highlighted.

Exactly one month later, on 13 February 2001, El Salvador experienced another intense earthquake, which increased the destruction experienced by the country. This event, originating from the second source of seismicity, had a Mw of 6.6, with a hypocenter placed at a depth of 10 km. This earthquake resulted in 315 fatalities and injured 3399 people due to ground-shaking and landslides (Figure 3).

Nearly 45,000 homes were destroyed and 16,000 were seriously damaged. This earthquake affected areas comparable to those affected by the 1936 event, which was of the same genesis. Once again, the increased risk factors have led to more damage and more victims.

Most of the landslides triggered by the 2001 earthquakes were shallow debris flows; however, there were also numerous slides and debris and rock falls, according to Cruden and Varnes' classification of landslides [53]. Moreover, both earthquakes triggered a few large and deep landslides that were highly damaging.

There are several doubts about whether these two earthquakes were related. However, some studies [17,54] reported that normal faulting subduction earthquakes in Central America tend to be followed by either large thrust events or shallow intraplate events within four years. This trend can also be observed in El Salvador, where earthquakes belonging to the second type of seismic genesis have occurred even after a month, as in the case of the earthquakes of 2001. To mention other examples of this trend, Bommer et al. [17] cited subduction earthquakes that occurred in 1915, 1932 and 1982 and were followed by intra-crustal earthquakes dated 1917, 1936 and 1986, respectively.

El Salvador is in a tropical humid climate system characterized by a dry season from November to April and a wet season from May to October in which about 75% of the annual precipitation falls. The earthquakes of 2001 occurred in the first two months of the year and thus during the dry season, as confirmed by the precipitation data recorded close to the study areas. Furthermore, the cumulative precipitation in the 12 months preceding the earthquakes of 2001 was lower than the annual average recorded at the same gauge stations in the 29 years before the earthquakes (Table 2).

Table 2. Precipitation in the 12 months preceding the seismic events of 2001 and the average annual precipitation in 1971–1999 in the fourteen areas under consideration.

Sector	1	2	3	4	5	6	7	8	9	10	11	12	13	14
Rainfall [mm] March 2000–February 2001	1166	1614	1614	1571	1571	1571	1571	1571	1632	1632	1632	1632	1632	1632
Average rainfall [mm] 1971–1999	1657	1847	1847	1650	1650	1650	1650	1650	1836	1836	1836	1836	1836	1836

In addition, November 2000 was the last month that experienced rainfall, which was lower than the average, and unlike the averages of previous years, the following months until the earthquakes of 2001 were without rain. These data suggest that the water content did not play a decisive role in the genesis of gravitational phenomena due to the absence of or low water pore pressure in static conditions and the presence of a relatively high apparent cohesion in pyroclastic soils due to suction.

2.3. Mapping Strategy

Landslide inventories are valuable tools for investigating landscape susceptibility to these phenomena. When investigating the slope's response to a specific triggering event, the archive must exclude the landslides that occurred before and after that event. In stochastic and empirical assessment of landslide susceptibility, the presence or absence of a gravitational phenomenon within a selected mapping unit is employed as dependent variable. Thus, the accuracy and degree of completeness of the inventory largely affect the predictive performance of the models [47,55–58].

The 2001 landslide inventories available to date are incomplete, as they only include the main gravitational phenomena; therefore, the first step of this study was to prepare a complete inventory of the slope failures triggered by the two earthquake events in the study areas. To this aim, aerial photos provided by the CNR (Centro Nacional de Registros de El Salvador—Instituto Geográfico y del Catastro Nacional) were employed. Regarding the January earthquake, 136 aerial photos covering an area of 91.85 km^2 were analyzed, while 313 aerial photos acquired over an area of 305.72 km^2 were employed to identify the landslides triggered by the February earthquake (Figure 4).

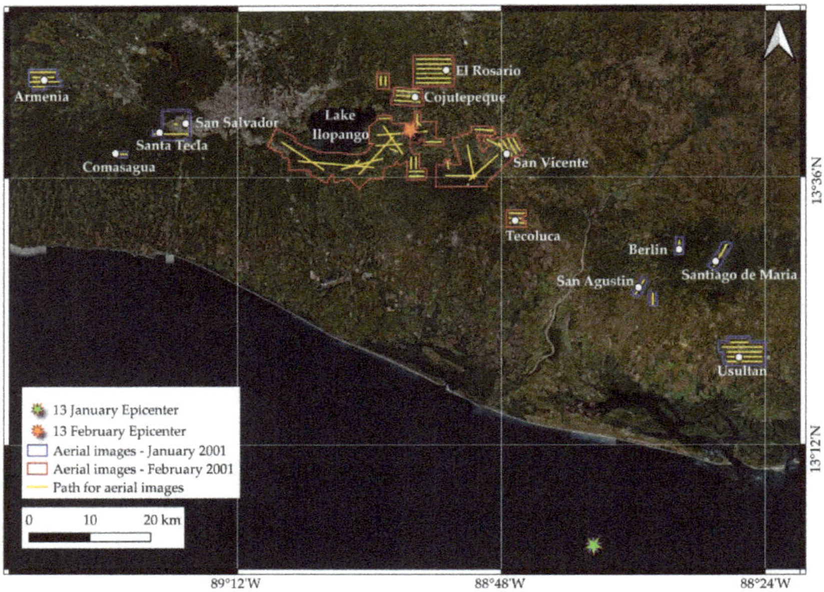

Figure 4. Positions of the epicenters, paths and areas covered by the aerial images employed to map the landslides that occurred in January and February 2001.

Recognizing and mapping landslides in tropical areas where vegetation recovers rapidly requires aerial/satellite images taken not long after the event. The employed aerial photos were taken in the days just after the earthquakes, when the displaced materials had not yet been removed or covered by vegetation. The aerial images cover the areas of the country that were the most heavily affected by the 2001 earthquakes. These correspond to sectors of the eastern and western parts of the country in the case of the January earthquake, and to the central sector in the case of the February earthquake. To exclude slope failures triggered by a previous seismic or rainy event from the analysis, we analyzed a SPOT-4 satellite image dated 27 November 2000 (Scene ID: 4 611-323 00-11-27 16:42 2M) with 10 m spatial resolution (European Space Agency). In addition, in the areas covered by the February 2001 aerial photos, SPOT-4 satellite images from 21 January (Scene ID: 1 611-323 01-01-21 16:44 1P) were also analyzed to identify those landslides triggered by the January earthquake. It is worth noting that the aerial and satellite images allowed us to detect only landslides due to slope collapses; thus, landslide-producing fracturing and tension cracks from short coseismic slides were not included in the inventories.

In this study, the landslides were recognized, and their bodies were mapped as vector polygons. Then, the highest point of each landslide crown was identified, assuming that the environmental properties of these points, known as LIPs (landslide identification points) [23,59–61], reflect the pre-failure conditions that favored the initiation of the landslides.

2.4. Dependent and Independent Variables

Landslide susceptibility assessment involves dividing the study area into mapping units, which usually correspond to grid cells or slope units [29,62,63]. As grid cell partition is largely and effectively employed in landslide susceptibility mapping [8,64,65], this strategy was employed in our experiment. The study areas were resampled into 10 m pixels resulting from rescaling a 5 m cell DEM from the European Space Agency (ESA). To each of these pixels, a value for the dependent and independent variables was assigned. The cell size was selected to find a compromise between the spatial resolution and computation time needed to train and test the models [64,66], taking into account that a 10 m cell partition was successfully employed in previous studies [33,67].

In this study, the dependent variable is dichotomous, taking the value 1 when a pixel hosts one or more LIPs and the value 0 when it does not intersect a LIP or a landslide body. Independent variables were selected according to data availability and their supposed direct or indirect relationships with the spatial distribution of the slope failures that occurred in the study areas [68]. The following eleven predictors, which reflect preparatory factors, were selected: Northness (N), Eastness (E), Catchment Area (C_AR), Convergence Index (C_IND), Downslope Curvature (D_CUR), Elevation (ELE), Geology (GEO), Normalized Difference Vegetation Index (NDVI), Slope (SLO), Topographic Positioning Index (TPI), Topographic Wetness Index (TWI) and Upslope Curvature (U_CUR). N and E were obtained by calculating the cosine and sine of slope aspect, respectively. The NDVI was calculated from images dated 1 January 2001 and 2 February 2001 to analyze the areas affected by the January and February 2001 events, respectively. The terrain attributes were extracted from the DEM, which is dated 2017 and is thus more recent than the 2001 earthquake events. Therefore, to avoid training and testing the models with topographic conditions altered by the slope failures, the landslide points coincided with the highest points of each crown (i.e., LIP) and the non-landslide points were randomly selected outside the landslide areas. Additionally, two trigger (i.e., depending on the earthquake characteristics) variables were also employed as covariates, namely: Peak Ground Accelerations (PGAs) and the distance from the hypocenter (DIST). The first was obtained by rasterizing the PGA maps from both earthquakes from the USGS ShakeMap website [69], while the second by producing a distance raster from the epicenter coordinates known through the USGS website.

The absence of multicollinearity among the predictors was evaluated by calculating the Variance Inflation Factor (VIF) (e.g., [32,70]) and by applying the rule of ten ([26,71] and references therein). As all the predictors exhibited a VIF value < 10, they were used to build the landslide predictive models.

Table 3 summarizes the variables employed to model the landslide spatial distribution.

Table 3. Independent variables used to model rainfall- and earthquake-induced landslides.

Independent Variable	Application for Rainfall-Induction/Earthquake-Induction Landslide Susceptibility (Y: Yes, N: No)	Type	Reference
N Northness	Y/Y	Preparatory	Geospatial Data Abstraction Library QGis
E Eastness	Y/Y	Preparatory	Geospatial Data Abstraction Library QGis
C_AR Catchment Area	Y/Y	Preparatory	SAGA GIS
C_IND Convergence Index	Y/Y	Preparatory	SAGA GIS
D_CUR Downslope Curvature	Y/Y	Preparatory	SAGA GIS
ELE Elevation	Y/Y	Preparatory	10-m DEM resampled from a 5-m DEM from ESA
DIST Epicentral Distance	N/Y	Trigger	USGS website
GEO Geology	Y/Y	Preparatory	Geological map from Schmidt-Thomé (modified)
NDVI Normalized Difference Vegetation Index	Y/Y	Preparatory	Terra/MODIS Satellite NASA
PGA Peak Ground Acceleration	N/Y	Trigger	USGS Shakemap
SLO Slope	Y/Y	Preparatory	Geospatial Data Abstraction Library QGis
TPI Topographic Positioning Index	Y/Y	Preparatory	Geospatial Data Abstraction Library QGis
TWI Topographic Wetness Index	Y/Y	Preparatory	SAGA GIS
U_CUR Upslope Curvature	Y/Y	Preparatory	SAGA GIS

2.5. Modeling Technique

The likelihood of landslide occurrence at each 10 m grid cell of the study area was calculated using multivariate adaptive regression splines (MARS; [72]) as a modeling technique. MARS is a non-parametric regression method that is useful for handling complex data and is able to model non-linear relationships between dependent and independent variables, splitting their range into an optimized number of linear branches. The extreme values of these intervals, called knots, allow an optimal linear function known as a basis function (BF). The result is a weighted sum of terms which includes a basis function or the product of two or more of them. The MARS regression function is expressed as (1):

$$y = f(x) = \alpha + \sum_{i=1}^{N} \beta_i h_i(x) \tag{1}$$

where y is the dependent variable predicted by the function $f(x)$, α is the constant, each $h_i(x)$ is a BF and β_i its coefficients, and N is the number of functions.

The calibration and validation of the MARS models were carried out by using the following R software packages: "raster" [73], "usdm" [74], "splitstackshape" [75], "pROC" [76], "ROCR" [77], "caret" [78], "sperrorest" [43] and "earth" [79].

2.6. Calibration and Validation Strategy

In this study, the landslide predictive models were calibrated and validated using five datasets (Figure 5; Table 4), which correspond to the fourteen study sectors (JAN, FEB and JF), to the area of the San Vicente (SV) volcano landslide inventory prepared by Mercurio et al. [49] and to the intersection between the latter and sectors 9 and 13 (FSE).

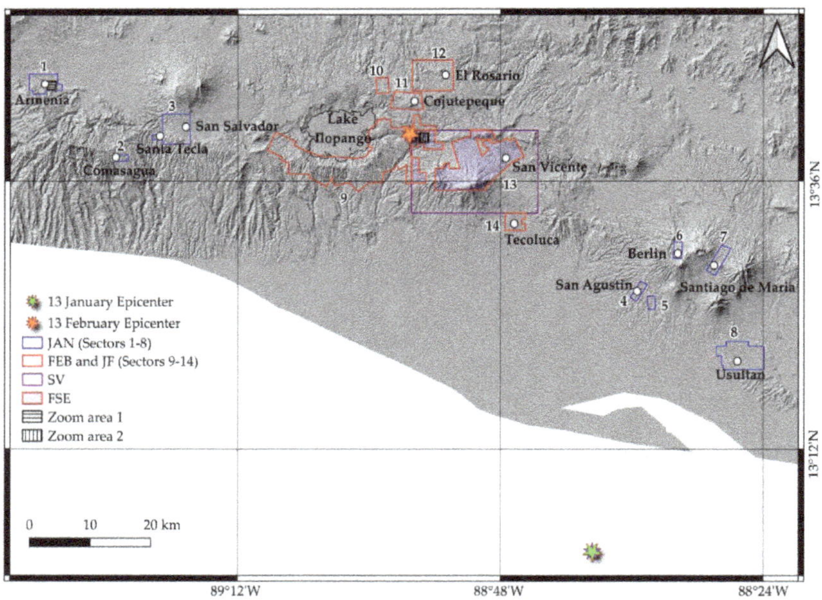

Figure 5. Locations of the datasets used for the landslide susceptibility modeling.

These datasets are matrices in which rows correspond to the pixels of a specific area, and columns report the values of the dependent and independent variables calculated for each pixel. The dataset of rainfall-triggered landslides (i.e., SV) includes only preparatory covariates, whereas those of the coseismic landslides (i.e., JAN, FEB, JF and FSE) report the values of both preparatory and trigger variables.

Table 4. Datasets used to calibrate and validate the landslide predictive models.

Dataset Code	Area	Trigger	Landslides Date
SV	San Vicente area (Mercurio et al., 2021 [49])	Rainfall	November 2009
JAN	Sectors 1 to 8	Earthquake	January 2001
FEB	Sectors 9 to 14	Earthquake	February 2001
JF	Sectors 9 to 14	Earthquake	January 2001
FSE	Intersections between FEB and SV areas	Earthquake	February 2001

The MARS predictive models were prepared using different calibration datasets or sets of predictors (Table 5). The models named M1 were trained with the dataset SV, which includes rainfall-induced landslides and eleven independent variables reflecting preparatory factors. The M1 model calibrated with the dataset SV was then transferred to all fourteen study sectors of El Salvador, calculating a probability (PSV) of landslide occurrence for each pixel of these areas. The PSV values were then employed as new predictors when training the M3 model. The M2, M3 and M4 models were calibrated with the earthquake-induced landslides of January and February 2001. The variables PGA and DIST were used in the M2 model as the only covariates and in the M3 model in addition to PSV. The M4 model contains both preparatory and triggering independent variables.

Table 5. Inventories and independent variables used in M1, M2, M3 and M4 models.

Inventory for Calibration	Model			
	M1	M2	M3	M4
SV (rainfall induction)	X			
JAN, JF, FEB and FSE (earthquake induction)		X	X	X
Independent variables				
PSV			X	
Ten preparatory variables	X			X
Two trigger variables		X	X	X
Area/Inventory for validation				
JAN, JF, FEB and FSE (earthquake induction)	X	X	X	X

The predictive ability of the M1 models was evaluated by comparing the PSV values and the presence or absence of coseismic landslides in the datasets JAN, FEB, JF and FSE.

The MARS models were calibrated and validated using three strategies. The first and second strategies consisted of the following steps. First, we randomly selected ten subsets from each dataset, which included all the event cells and the same number of non-event cells. The subsets of cells extracted from the SV dataset were used to train ten M1 models. Second, we extracted two calibration samples from the ten subsets of all the datasets, excluding SV, by randomly selecting 75% and 5% of the event cells, respectively. The remaining event cells (25% and 95%, respectively) were included in the validation samples. Both calibration and validation samples were completed by randomly selecting the same number of non-event cells. Thus, this second step produced twenty calibration and twenty validation samples from the datasets JAN, FEB, JF and FSE. The two ratios of calibration/validation samples were chosen in order to evaluate the predictive performance of the MARS models under two different scenarios: in the first scenario, the position of a large part (i.e., 75%) of the earthquake-induced landslides is known; the second scenario simulates the situation just after the earthquake occurrence, when only a very small part (i.e., 5%) of the landslides had been identified. The probability of landslide occurrence was calculated for each cell of the validation samples by averaging the score obtained from the ten MARS model runs (one for every calibration sample). This strategy was applied to increase the stability of the landslide predictions and mitigate the rare-events issue [22,32,80,81].

To further evaluate the reliability of the models, a k-fold spatial cross-validation (CV) strategy was also adopted [44,82]. According to this approach, the models are trained for k-1 combined samples at time and validated using the remaining sample. This procedure is repeated k times. Unlike the classic k-fold cross-validation approach, a sample is extracted from each of k different areas. In this study, five (k = 5) different sampling areas were identified by applying the k-means clustering algorithm. Similarly, to the first and second calibration and validation strategy, the spatial CV was repeated ten times in order to assess the robustness of the models.

For all the validation samples, a receiver operating characteristics (ROC) curve [42,48,71,83] was prepared using the average MARS score to predict the value of the dependent variable at each pixel (0: LIP absence; 1: LIP presence). The ROC curve plots sensitivity (true positive rate) against 1—specificity (false positive rate). Then, the area under the ROC (AUC) curves were calculated, thus obtaining ten AUC values for the prediction of the datasets JAN, FEB, JF and FSE for each of the models M1 to M4 and validation scenarios. This procedure led to a total of 480 AUC values. An AUC value close to 0.5 reveals a random prediction, whereas a value of 1.0 reflects a perfect ability to discriminate between event and non-event pixels. AUC values larger than 0.7, 0.8 and 0.9 were considered as acceptable, excellent and outstanding, respectively [84]. The Wilcoxon signed-rank test was employed to identify significant differences between the predictive skill performance of the models by setting the level of significance at 0.01.

2.7. Landslide Susceptibility Maps

The models M1 to M4 were used to generate a landslide susceptibility map for each of the four datasets. To this aim, we employed the MARS models calibrated with the samples including 75% of the LIPs. For each pixel of the datasets JAN, FEB, JF and FSE, a probability (P) of landslide occurrence was calculated for the models M1, M2, M3 and M4 by averaging the score of their ten repetitions. Since the models were trained on samples including the same number of LIP and non-LIP locations, P should be interpreted as the spatial relative probability of landslide occurrence.

3. Results

3.1. The 2001 Earthquake-Induced Landslide Inventories

The landslide inventory of the dataset JAN was obtained by analyzing eight sets of aerial photos taken a few days after the earthquake of 13 January 2001. These images cover eight inhabited sectors, extending for 91.85 km^2 (Figure 6), which experienced the most significant damage due to the seismic shaking and the consequent landslides. These were mainly rotational slides, debris/earth flows and earth/rock falls.

The inventory consists of 997 LIPs, each hosted by a 10 m pixel. The percentage of event cells (i.e., cells hosting at least one LIP) compared to the total number of pixels (1,260,811) of the study area is equal to 0.08%, while there are 9556 cells intersecting the landslide polygons, corresponding to 0.76% of the total pixels. The JF inventory includes the LIPs and landslides caused by the January earthquake in the areas covered by the aerial images taken after the February event (sectors 9 to 14, counted from west to east, are Ilopango south and Rio Jiboa, S. Cruz Michapa, Cojutepeque, El Rosario, San Vicente and Tecoluca, respectively; red polygons of Figure 6). These landslides were identified by analyzing images from the SPOT-4 satellite dated 21 January 2001. In these sectors, with relatively low January PGA values, 123 LIPs were mapped, each placed in one 10 m cell and accounting for 0.002% of the total number of cells of the area under examination. When landslide bodies are considered, unstable pixels cover 0.2% of the area.

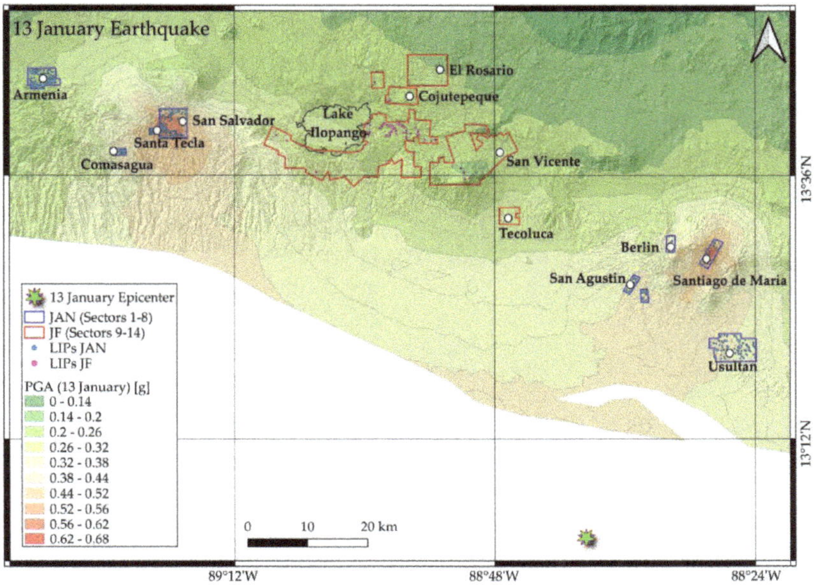

Figure 6. PGA values and LIPs related to the seismic event of Mw 7.7 on 13 January 2001 in the fourteen study areas.

When the JAN and JF inventories are merged, the percentages of LIPs and landslide pixels in relation to the total area (sectors 1 to 14) are significantly reduced (0.017 and 0.30, respectively) due to the low PGA values within sectors 9 to 14. However, the landslides triggered in these sectors, despite being about one tenth of the number of the JAN landslides, cover a significantly higher number of pixels (Table 6). This can be explained considering that the steepest slopes are characterized by thick deposits of poorly consolidated late Pleistocene and Holocene Tierra Blanca rhyolitic tephras that erupted from Ilopango caldera, an area particularly susceptible to the initiation of gravitational phenomena [47,61]. In fact, in that region, the January earthquake resulted in a low number of debris flows and rotational slides; however, they had a large extent. This is also demonstrated by the number of gravitational phenomena that were larger than 10,000 m^2: two and twenty-two in the JAN and the JF datasets, respectively. However, two gravitational phenomena, which both occurred in sector 3 (Las Colinas neighborhood of Santa Tecla, west of the capital San Salvador), caused the worst damage and the largest number of victims. One of these was a complex landslide (rotational slide and earth flows) that covered an area of about 200,000 m^3 in which approximately 585 people lost their life. This landslide, which sadly become famous for the devastation it generated, was about 800 m long and 150 m wide, with a 50 m high escarpment activated on the north side of El Balsamo Ridge, which is composed of the andesitic cinders and some interbedded tephra of the El Balsamo Formation [17]. The second event, located approximately 2.5 km from the first landslide, was a debris flow about 1500 m long which also began from the north side of El Balsamo Ridge and reached and destroyed a sector of the Carrera Panamericana, the most crucial road in the country which connects the capital with the western part of El Salvador. The highway interruption hindered the arrival of aid, creating enormous damage to the population. However, the absence of small landslides in sectors 9 to 14 compared to that of sectors 1 to 8 (41.42%) could also be related to the different resolution of the images used for mapping, as phenomena with an extension of less than 100 m^2 are hardly detectable in a satellite analysis with a 10 m resolution.

Table 6. Amounts and characteristics of the landslides triggered by the January earthquake in sectors 1 to 14.

Sector	Area of the Sector (km²)	N° of LIPs	% of Event Cells	N° of Landslides Extending <100 m²; (%)	N° of Landslides Extending 100–1000 m²; (%)	N° of Landslides Extending 1000–10,000 m²; (%)	N° of Landslides Extending >10,000 m²; (%)	Landslides' Average Extent in m²	% of Landslide Cells
1	15.41	245	0.10	84 (34.29)	143 (58.37)	18 (7.35)	0 (0)	297.69	1.24
2	2.31	128	0.52	67 (52.34)	57 (44.53)	4 (3.13)	0 (0)	190.91	3.16
3	23.89	296	0.10	107 (36.15)	159 (53.72)	28 (9.46)	2 (0.68)	870.79	1.45
4	3.98	67	0.07	38 (56.72)	29 (43.28)	0 (0)	0 (0)	145.82	0.39
5	2.66	43	0.15	27 (62.79)	14 (32.56)	2 (4.65)	0 (0)	151.94	0.74
6	3.89	8	0.02	1 (12.50)	6 (75.00)	1 (12.50)	0 (0)	328.59	0.19
7	8.29	17	0.01	1 (5.88)	16 (94.12)	0 (0)	0 (0)	267.09	0.06
8	31.42	193	0.05	88 (45.60)	103 (56.37)	2 (1.04)	0 (0)	116.19	0.31
JAN (1–8)	91.85	997	0.08	413 (41.42)	527 (52.86)	55 (5.52)	2 (0.20)	407.13	0.76
9	155.32	93	0.0025	0 (0)	19 (20.43)	55 (59.14)	19 (20.43)	8744.50	0.27
10	5.32	1	0.0018	0 (0)	0 (0)	0 (0)	1 (100)	19,571.39	0.44
11	13.79	23	0.0136	0 (0)	8 (34.78)	15 (65.22)	0 (0)	2079.28	0.35
12	34.80	0	0	0 (0)	0 (0)	0 (0)	0 (0)	0	0
13	87.65	6	0.004	0 (0)	1 (16.67)	3 (50.00)	2 (33.33)	10,263.87	0.05
14	8.84	0	0	0 (0)	0 (0)	0 (0)	0 (0)	0	0
JF (9–14)	305.72	123	0.002	0 (0)	28 (22.77)	73 (59.34)	22 (17.89)	7660.25	0.20
JAN + JF (1–14)	397.57	1120	0.017	413 (36.88)	555 (49.56)	128 (11.42)	24 (2.14)	3585.67	0.30

The seismic phenomenon of 13 February triggered thousands of landslides in the country's central sector. Many landslides occurred where slopes were covered by thick and slightly consolidated deposits of the rhyolitic tephras of the Tierra Blanca Formation, which originated from eruptions from the Ilopango Caldera.

Landslide mapping was carried out using aerial photos taken a few days after the event in areas that experienced a very high PGA and suffered the most significant damage (sectors 9 to 14, Figure 7). The landslides that were also included in the JF inventory were discarded except for those that have been reactivated.

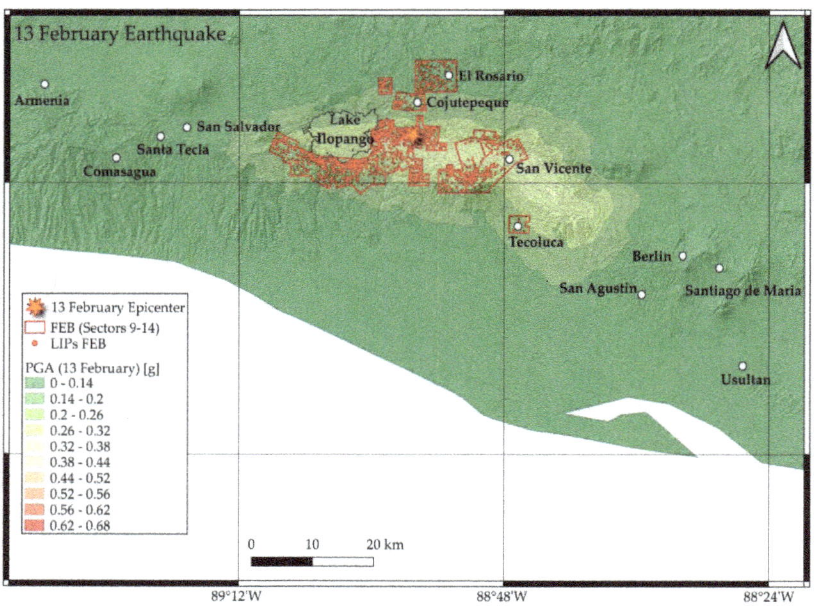

Figure 7. PGA values of the seismic event of 13 February 2001 and LIPs of the landslides triggered in the FEB study areas.

The FEB inventory includes 5371 LIPs intersecting 5344 pixels. The percentage of event cells of the FEB dataset is 0.09%. The extent of the landslide bodies is 123,488 pixels, which accounts for 2.12% of the study area. This seismic event triggered landslides similar to those of the JAN dataset, but their frequency and size were higher (Table 7).

Table 7. Amounts and characteristics of the gravitational phenomena of landslides belonging to sectors 9 to 14, related to the February earthquake.

Sector	Area of the Sector (km^2)	N° of LIPs	% of Event Cells	N° of Landslides Extending <100 m^2; (%)	N° of Landslides Extending 100–1000 m^2; (%)	N° of Landslides Extending 1000–10,000 m^2; (%)	N° of Landslides Extending >10,000 m^2; (%)	Landslides' Average Extent in m^2	% of Landslide Cells
9	155.32	3650	0.10	427 (11.70)	2145 (58.77)	955 (26.16)	123 (3.37)	1694.63	2.47
10	5.32	186	0.33	32 (17.20)	119 (63.98)	34 (18.28)	1 (0.54)	789.50	4.68
11	13.79	279	0.16	91 (32.62)	151 (54.12)	37 (13.26)	0 (0)	554.75	1.58
12	34.80	209	0.06	6 (2.87)	144 (68.90)	57 (27.27)	2 (0.96)	1093.83	1.16
13	87.65	1029	0.07	172 (16.72)	563 (54.71)	267 (25.95)	27 (2.62)	1500.60	1.56
14	8.84	18	0.02	8 (44.44)	10 (55.56)	0 (0)	0 (0)	126.75	0.08
FEB (9–14)	305.72	5371	0.092	736 (13.70)	3132 (58.31)	1350 (25.13)	153 (2.85)	1538.26	2.12

Among the many large-scale phenomena, large rotational flows of debris were mapped in the area close to the epicenter, corresponding to areas near the Jiboa River, which fortunately have low population densities. As a matter of fact, thousands of gravitational phenomena have been detected in the Tierra Blanca Formation (GEO5), some of which caused the damming of the river for hundreds of meters, causing the formation of an artificial lake that could have caused a high hydrogeological risk downstream.

The FSE inventory was generated by overlapping the areas of the datasets FEB and SV. The intersection extended for 107 km^2 and included the entirety of sector 13 and 12.5% of sector 9 in the FEB study areas (Figure 8). The FSE dataset contains 1587 LIPs pertaining to the landslides triggered by the February earthquake, which were mainly earth/debris flows.

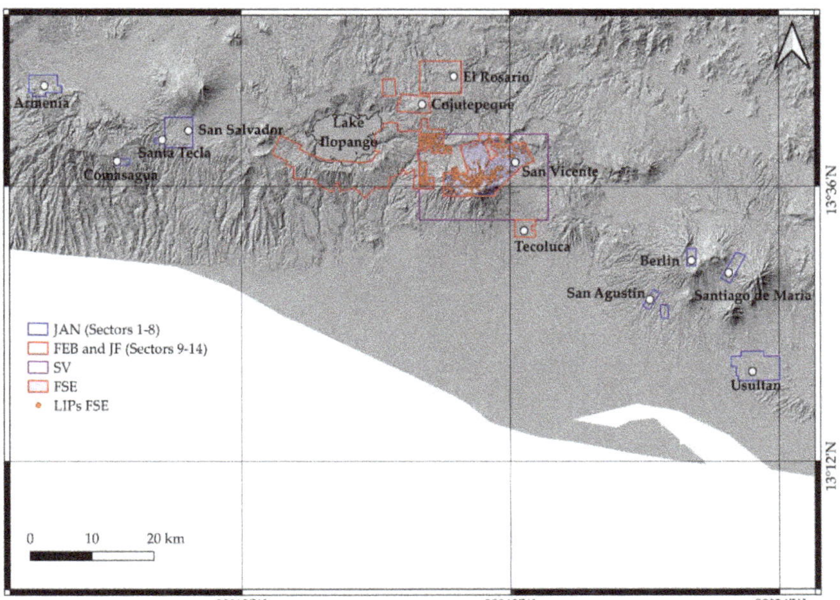

Figure 8. LIPs of the landslides triggered by the seismic event on 13 February 2001 in the intersection between FEB and SV study areas (FSE).

The conditional density plots of Figure 9 reveal empirical relationships between landslide inventories and independent variables. These plots show the relative frequency of non-LIP (0) and LIP (1) pixels within the predictor classes. A higher frequency of LIPs occurs in steeper slopes and where the upslope and downslope curvatures are convex and concave, respectively. The N and E plots show that landslide density was higher on the north- and east-facing slopes of the datasets JF and FSE, respectively. Landslides are very rare at high-elevation JF positions, while they are more numerous where the elevation is high in the JAN dataset. As expected, the density of February LIPs decreases with the distance from the earthquake epicenter; surprisingly, the opposite occurs for the LIPs triggered by the January seismic event. A relatively high landslide frequency is observed in hard rocks (JAN), very soft soils (FEB and JF) and hard soils (FSE), as well as in upper (positive TPI) and lower (negative TPI) slopes. The density of LIPs is slightly higher than the average in intermediate PGA classes of the datasets FEB, JF and FSE.

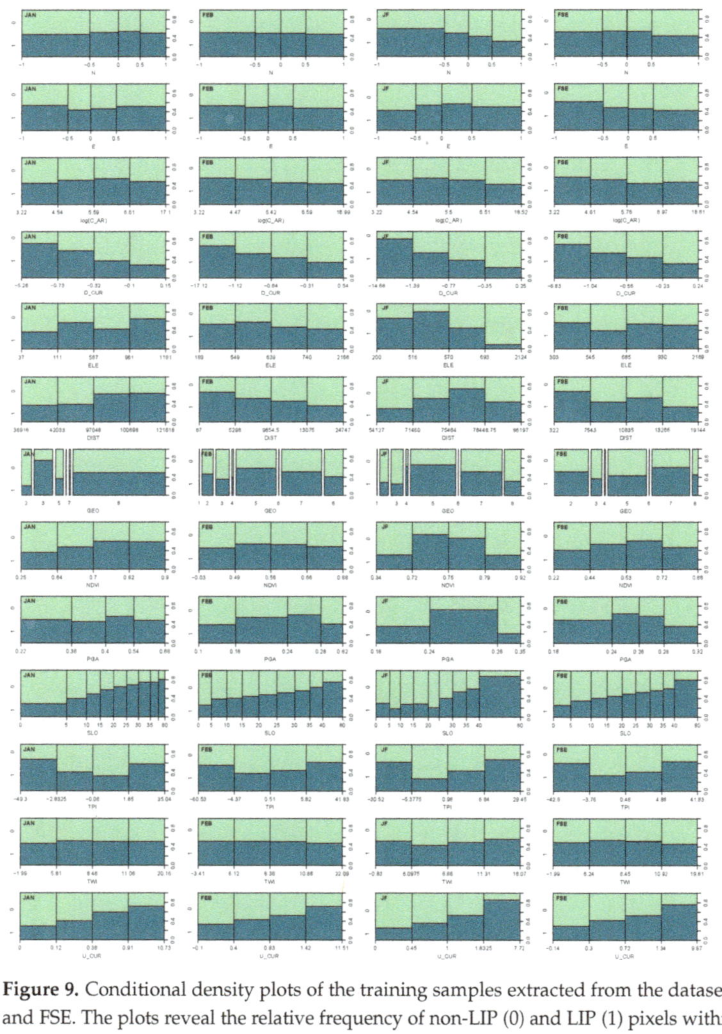

Figure 9. Conditional density plots of the training samples extracted from the datasets JAN, FEB, JF and FSE. The plots reveal the relative frequency of non-LIP (0) and LIP (1) pixels within the predictor classes. The width of the bars shows the relative frequency of each class. Lithology units: (1) GEO1 incoherent soils/rocks, (2) GEO2 soft rocks, (3) GEO3 hard rocks, (4) GEO4 medium rocks, (5) GEO5 very soft soils, (6) GEO6 soft soils, (7) hard soils and (8) GEO8 medium soils.

3.2. The Ida/96E 2009 Inventory

The SV inventory [49] covers an area of 287 km^2 around the flanks of the San Vicente Volcano and includes the LIPs of 5609 landslides triggered between 7 and 8 November 2009 [85] by an extreme rainfall event. This event was caused by the combined action of Hurricane Ida and the low-pressure system 96E, which produced 350 mm in 24 h in an area of about 400 km^2 located between Ilopango Lake and the San Vicente Volcano (Figure 8). The rainfall triggered numerous and large flows that caused fatalities and massive damage, with an economic loss of approximately a quarter of USD one billion [86].

3.3. Predictive Performance of the MARS Models

The inventories and variables used for calibrating and validating the M1, M2, M3 and M4 models are shown on Table 5. The predictive skill of the models, measured on the validation samples of the datasets JAN, FEB, JF and FSE, are revealed by the boxplots in Figure 10. These reflect, for each model and dataset, the variability and skewness of the ten AUC values calculated by applying the calibration and validation strategies described in Section 2.6. The skewness displays, for each group of ten AUCs, their first quartile (Q1), median and third quartiles (Q3), the minimum and maximum values within 1.5 times the interquartile range (IQR) above Q3 and below Q1 and eventually outliers outside 1.5 of the IQR. Moreover, Table 8 shows the average and standard deviation values of the AUC groups. Furthermore, the average ROC curves calculated for each group of ten model runs are shown in Figure 11.

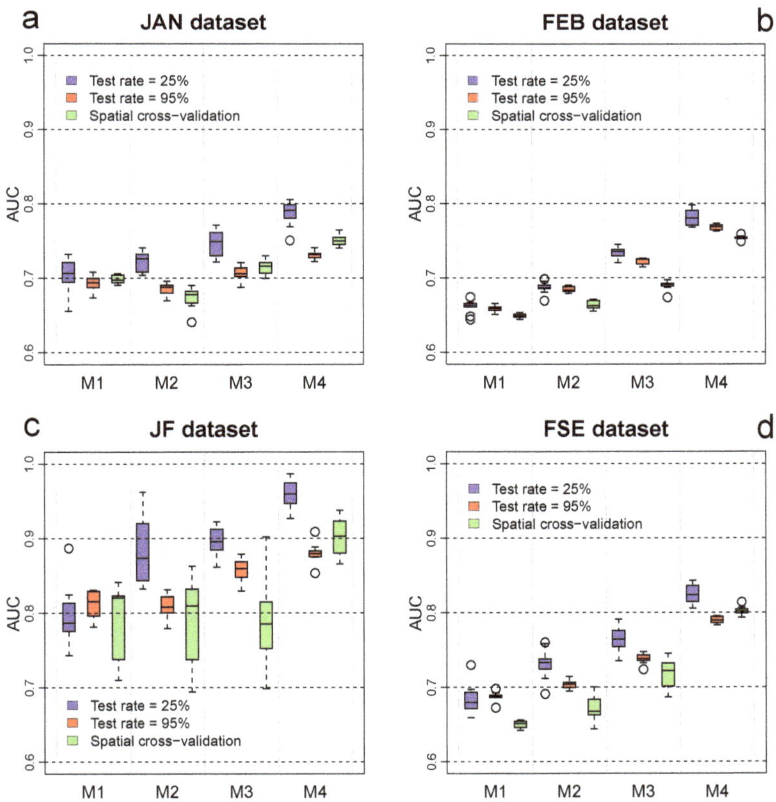

Figure 10. Box plots of the AUC values achieved by the M1, M2, M3 and M4 models, measured on the validation samples of the datasets JAN (**a**), FEB (**b**), JF (**c**) and FSE (**d**) for the three validation scenarios (Scenario 1: test rate 25%; Scenario 2: test rate: 95%; Scenario 3: 5-fold spatial CV).

Table 8. AUC mean and standard deviation (SD) calculated for the four models and datasets, with the three validation scenarios (Scenario 1: test rate 25%; Scenario 2: test rate: 95%; Scenario 3: 5-fold spatial CV).

Scenario 1	**AUC_JAN mean**	**AUC_JAN SD**	**AUC_FEB mean**	**AUC_FEB SD**	**AUC_JF mean**	**AUC_JF SD**	**AUC_FSE mean**	**AUC_FSE SD**
M1_75	0.7	0.026	0.661	0.009	0.795	0.041	0.685	0.02
M2_75	0.723	0.013	0.687	0.008	0.88	0.042	0.731	0.02
M3_75	0.747	0.017	0.734	0.007	0.896	0.018	0.765	0.016
M4_75	0.787	0.017	0.781	0.01	0.958	0.021	0.824	0.012
Scenario 2	AUC_JAN mean	AUC_JAN SD	AUC_FEB mean	AUC_FEB SD	AUC_JF mean	AUC_JF_ SD	AUC_FSE mean	AUC_FSE SD
M1_05	0.693	0.01	0.658	0.004	0.812	0.018	0.688	0.006
M2_05	0.685	0.008	0.684	0.004	0.809	0.015	0.704	0.006
M3_05	0.706	0.009	0.722	0.004	0.857	0.016	0.739	0.007
M4_05	0.731	0.006	0.768	0.003	0.88	0.014	0.79	0.005
Scenario 3	AUC_JAN mean	AUC_JANS D	AUC_FEB mean	AUC_FEB SD	AUC_JF mean	AUC_JF_ SD	AUC_FSE mean	AUC_FSE SD
JAN_M1	0.699	0.006	0.649	0.003	0.794	0.052	0.65	0.005
JAN_M2	0.673	0.014	0.664	0.006	0.787	0.06	0.671	0.017
JAN_M3	0.714	0.01	0.69	0.006	0.792	0.068	0.718	0.019
JAN_M4	0.75	0.008	0.754	0.003	0.901	0.025	0.802	0.006

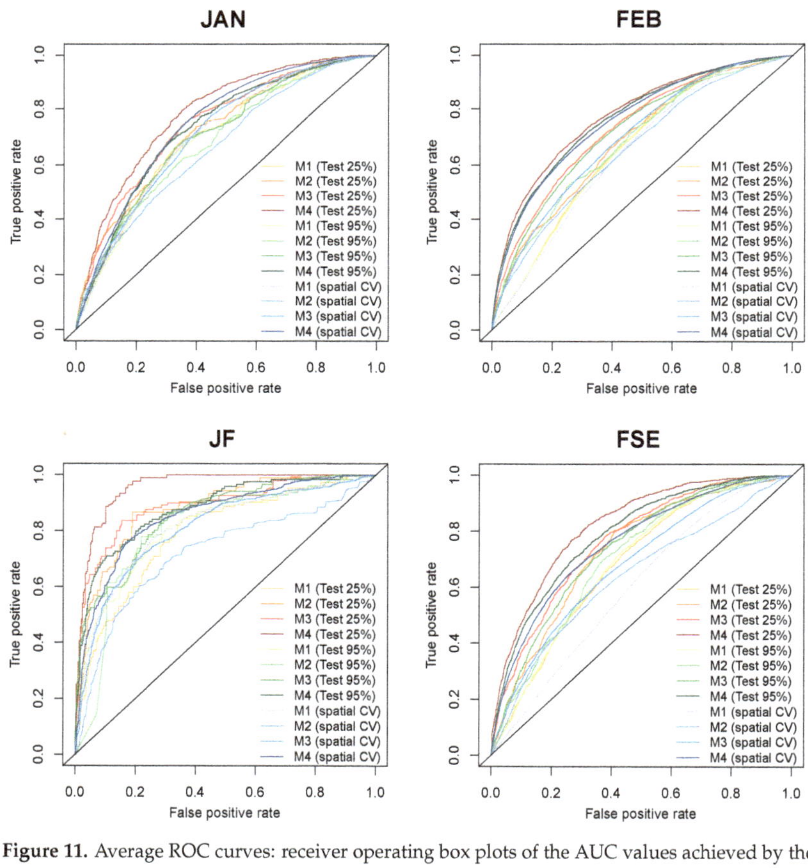

Figure 11. Average ROC curves: receiver operating box plots of the AUC values achieved by the M1, M2, M3 and M4 models, measured on the validation samples of the datasets JAN, FEB, JF and FSE for the validation Scenarios 1 (test rate 25%) and 2 (test rate: 95%).

The M1 models, calibrated with all rainfall-induced landslides that occurred near the San Vicente volcano and then transferred to the study areas, exhibit average AUC values ranging from 0.65 to 0.81. Overall, the accuracy of these models is poorer than that of the other models, except for what was observed in the datasets JAN and JF, where the M1 models perform similarly to the M2 models or even better when validation is carried out with 95% of the landslides or by applying the spatial CV. The average performance of the M1 models is not acceptable for the datasets FEB and FSE, whereas it is almost acceptable in the dataset JAN [84]. The M1 models show the best performance in the dataset JF, achieving a mean AUC in the range of 0.79–0.81.

The M2 models, which include as predictors the variables PGA and DIST, achieved mean AUC values in the range of 0.66–0.88. Their performance is better than the performance exhibited by the M1 models except for the dataset JAN when the validation scenarios 2 (test rate = 95%) and 3 (spatial CV) are applied. In these cases, as well as in the datasets FEB and FSE (with spatial CV), the predictive skills of the M2 models are not acceptable. On the other hand, their ability to discriminate between stable and unstable cells ranges from acceptable to excellent for the datasets FSE (with Scenarios 1 and 2) and JF.

The validation of the M3 models revealed mean AUC values in the range of 0.69–0.90, significantly higher than those achieved by the M1 and M2 models in all datasets except when applying spatial CV to the JF dataset. The average AUC achieved by the M3 models, which were fitted using the covariates PGA, DIST and PSV, ranges from acceptable to excellent, with lower values revealed by the spatial CV in the datasets FEB, JF and FSE.

The M4 models were trained with all independent variables available, excluding PSV. The validation results reveal a further improvement in the predictive ability of the M4 models compared to the predictive ability exhibited by the other models. The average AUC values, which are in the range of 0.73–0.96, exceed the acceptability threshold in the datasets JAN, FEB and FSE, achieving excellent to outstanding levels when the models are applied to the JF datasets.

The best predictive ability of the four tested models is observed in the dataset JF (mean AUC: 0.79–0.96), which contains the landslides triggered in sectors 9 to 14 by the January earthquake. Conversely, the models exhibited some inaccuracy in discriminating between stable and unstable cells of the JAN and FEB datasets, showing mean AUC values in the ranges of 0.67–0.79 and 0.65–0.78, respectively.

Overall, when the learning samples include 75% of the observed landslides, models M2 to M4 exhibit a significantly better accuracy (mean AUC: 0.69–0.96) than the accuracy achieved when they are calibrated using only 5% of the landslides (mean AUC: 0.68–0.88) or when applying the 5-fold spatial CV (mean AUC: 0.66–0.90). However, except for the dataset JF, the differences in mean AUC values are small and insignificant when the M2 models are fitted with 5% of the FEB landslides. Moreover, regarding the M1 models, there is no significant difference in accuracy when validation is carried out with 25% and 95% of the landslides. As expected, larger differences in AUCs are observed when comparing validation Scenarios 1 (test rate = 25%) and 3 (spatial CV), since the latter involves the transfer of the models between different areas.

Validation samples containing 95% of the landslides reveal a more stable predictive skill (SD: 0.003–0.018) than the predictive skill obtained when validating the models with the 25% of the event cells (SD: 0.007–0.042). The performance of the models is also stable when the spatial CV is carried out in JAN, FEB and FSE (SD: 0.003–0.019), but the AUC values are rather dispersed in JF (SD: 0.025–0.068), where Scenarios 1 (SD: 0.018–0.042) and 2 (SD: 0.014–0.018) also exhibit variability in their performances.

The ROC curves of Figure 11 confirm that predictive skill increases from the models M1 to M4. Since the FEB dataset contains the largest number of LIPs (5371), the ROC curves prepared for this dataset are smoother. Conversely, those calculated for JF are scattered, as they are derived from samples including a few landslides (123).

3.4. Comparision of the Landslide Susceptibility Maps

Figure 12 shows an example of the landslide susceptibility maps generated by models M1, M2, M3 and M4 for two zoom areas (Figure 5).

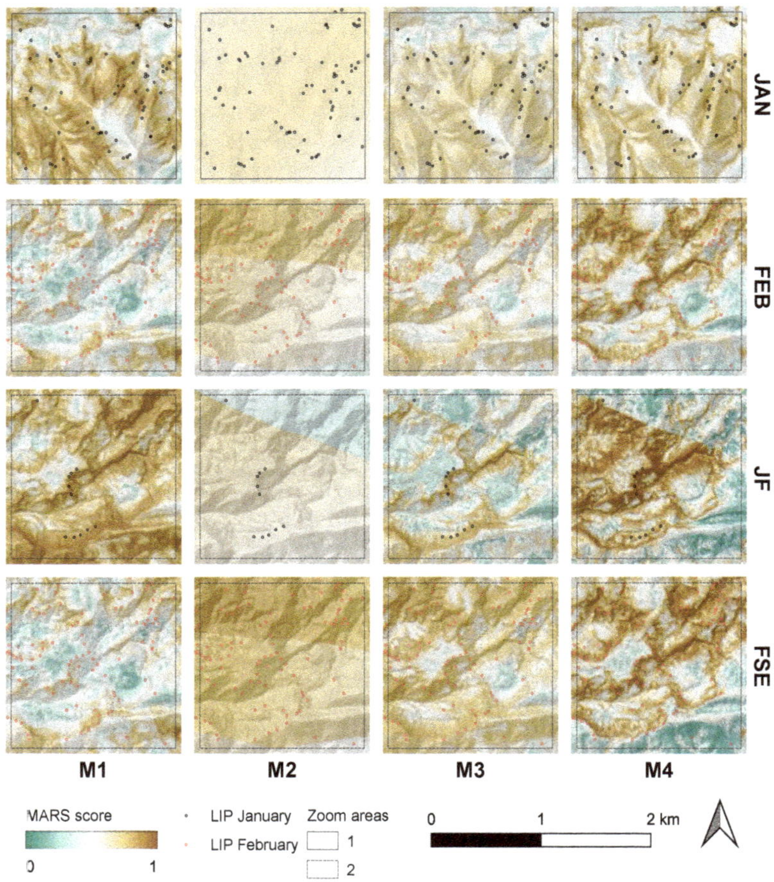

Figure 12. Examples of the landslide susceptibility maps extracted from the zoom areas 1 (dataset JAN) and 2 (datasets FEB, JF and FSE).

As they are landslide-prone areas, MARS score values tend to be high. Overall, maps generated by the M2 models show abrupt changes related to the pattern of PGA values, while the other maps are smoother except for the M3 and M4 maps of the dataset JF. Figures 13 and 14 reveal the values of the true positive rate (TPR), true negative rate (TNR), false positive rate (FPR) and false negative rate (FNR) calculated for the zoom areas of Figure 12 and for the entire landslide susceptibility maps, respectively. These metrics were calculated by setting a probability cut-off value of 0.5. In the zoom areas, the maps exhibit a very high rate of true positive but also a very high FPR (Figure 13). This is particularly evident for the M2 and M3 models. On the other hand, the entire susceptibility maps show a better discrimination ability, with markedly lower values of FPR, even if a decrease in true positives is also observed (Figure 14). These histograms confirm that the best ability to discriminate landslide pixels is achieved for the dataset JF by the models M2 (TPR: 0.886; TNR: 0.750), M3 (TPR: 0.894; TNR: 0.797) and M4 (TPR: 0.976; TNR: 0.856).

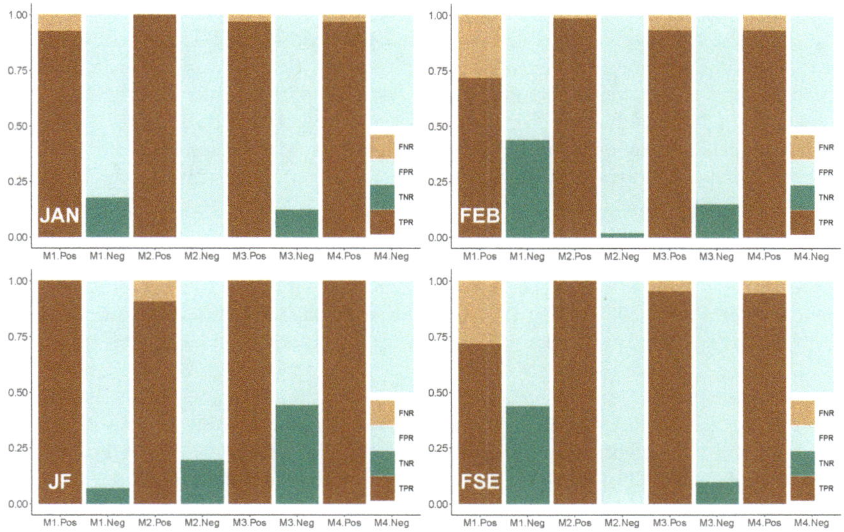

Figure 13. TPR, TNR, FPR and FNR values calculated for the landslide susceptibility maps in the zoom areas 1 (dataset JAN) and 2 (datasets FEB, JF and FSE).

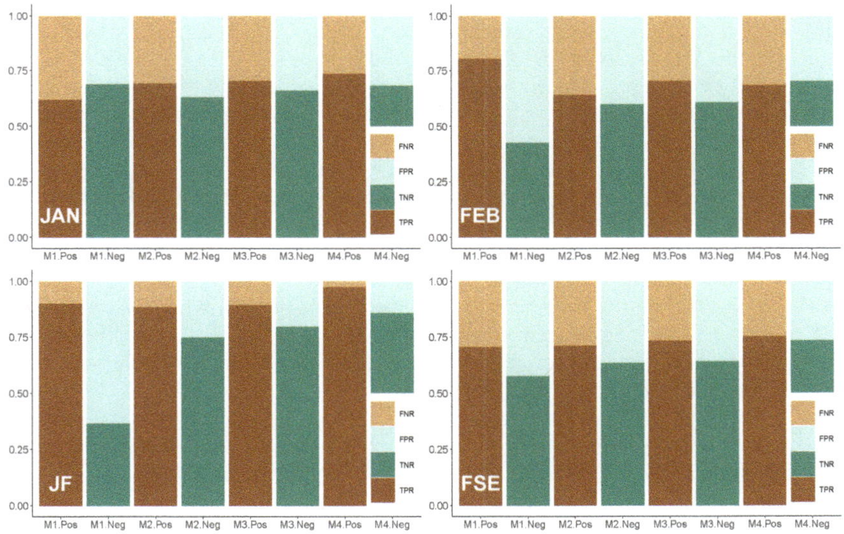

Figure 14. TPR, TNR, FPR and FNR values calculated for the entire landslide susceptibility maps of the datasets JAN, FEB, JF and FSE.

4. Discussion

In January and February of 2001, El Salvador was hit by two intense seismic phenomena of different natures which produced thousands of gravitational phenomena. Both 2001 earthquakes triggered similar landslides, but their distribution varied because of different earthquake source parameters, such as magnitude, epicentral distance and hypocentral depth. Indeed, earthquakes with large hypocentral depths tend to produce a relatively low number of gravitational phenomena spread over large areas. In contrast, phenomena that occur at a relatively low hypocentral depth usually produce many landslides concentrated in areas close to the epicenter [10]. The seismic event of January, with a hypocentral depth of 60 km, triggered 1200 gravitational phenomena across 398 km^2, whereas 5371 landslides, spread over 305 km^2, were induced by the February event, which had a hypocentral depth of 10 km. The difference in hypocentral depth caused different spatial distributions of the PGA and MMI (modified Mercalli intensity). In fact, their values decreased from the hypocenter to a greater distance in the January event compared to the February event. The slow decay of PGA values, which is observed for earthquakes with a deep and offshore genesis, triggers gravitational phenomena over larger areas than earthquakes with the most superficial genesis. Conversely, the latter are characterized by a higher concentration of slope failures around the hypocenter.

In this work, four different MARS models (M1 to M4) were produced using different predictor sets and landslide inventories, the latter containing the slope failures triggered by the Ida/96E event of November 2009 (M1) and those induced by the earthquakes of January and February 2001 (M2 to M4). Two different calibration and validation scenarios were employed. In the first scenario, three quarters of the observed landslides were used for calibrating the models M2 to M4, whereas the remaining quarter was included in the validation samples. This scenario is frequently used to produce and validate landslide susceptibility models [65,80,87,88]. In the second scenario, the learning samples contained only 5% of the earthquake-induced landslides, using the remaining 95% of the LIPs for validation. Additionally, to further investigate the predictive skill of the models, a third validation approach based on a 5-fold spatial cross-validation (CV) was applied.

The AUC values calculated for each model, dataset, and validation scenario, which are summarized by Table 8 and the boxplots of Figure 10, reveal that the MARS algorithm provides reliable predictions of the spatial distribution of earthquake-induced landslides. Indeed, 73% of the 320 AUC values AUC that were computed by applying the two test rates (25 and 95%) exceed the threshold of acceptability (AUC > 0.7), and a quarter of them even reveal the excellent predictive skill of the models (AUC > 0.8). Regarding the spatial CV, these percentages decrease to 50% and 21%, respectively. This is related to the data-driven nature of the adopted approach, which may generate a very good ability to predict unknown events if calibration and validation samples are extracted from the same area, even when time partition is performed [61]; however, models tend to lose predictive skill when transferred between different areas. As confirmation of this, the models trained with only 5% of the event pixels but which were spatially randomly distributed over all the sectors of the datasets achieve, in general, a better performance, with 68% of the model repetitions achieving an AUC value > 0.7.

When comparing the performance of the models M2 to M4, which were assessed with the two different percentages of validation LIPs (i.e., 25 and 95%), it is worth noting that the mean AUC gap is never greater than 0.08 and is0.04 on average. These results suggest that reliable predictive models of earthquake-triggered landslides can be obtained even when only a small portion of the slope failures have been identified. Therefore, in this situation, which is typical of the first moments immediately after a triggering event, predictive models such as those used in this work can help to quickly identify slopes where unmapped mass movements may have occurred. Indeed, even after over twenty years, only one fifth of the landslides identified in this work were already known. The latter were the main landslides that had hit the inhabited centers and which had been mapped immediately after the events to aid civil protection operations. The difference in the AUC values between the first validation scenario (test rate = 25%) and the 5-fold spatial CV are slightly higher (max: 0.10; mean: 0.05), which is acceptable if we consider that spatial transfer is very challenging for statistical landslide susceptibility zonation.

A general growing trend of predictive skill from M1 to M4 can be observed except in the datasets JAN and JF, where the predictive performance of the M1 and M2 models is similar. Conversely, when landslides triggered by the February earthquake are considered (i.e., datasets FEB and FSE), the M2 models' predictive ability is clearly better than the ability achieved by the M1 models, which do not achieve acceptable accuracy. This could be explained by a weaker control of the earthquake variables (i.e., PGA and DIST) on the spatial distribution of the January landslides due to a higher hypocentral depth and the largest distance between the epicenter and observed slope failures with respect to the February events. This hypothesis is confirmed by Figure 9, which reveals that the frequency of January LIPs increases with the distance from the earthquake epicenter and shows no relationship with the PGA. On the other hand, the higher accuracy of the M3 models compared to the M2 models, which is observed in all four datasets, reveal that a rainfall-induced landslide susceptibility model, even if calibrated outside the validation area, can significantly improve the predictive ability of a model containing only trigger variables. However, the fact that the M4 models achieve the best accuracy highlights the need to include preparatory variables measured in the same area in which validation is performed.

The AUC values also show that the models' predictive skill observed in the dataset FSE is better than that achieved in the dataset FEB. Based on these results, we could infer that rainfall-triggered landslide susceptibility models help more in predicting earthquake-induced slope failures when focusing on the same area in which the models were calibrated.

The landslide susceptibility maps of Figure 12 appear smoother as the number of predictors and/or training LIPs increases. Indeed, models with only two predictors (i.e., M2) and/or those that are fitted with few landslide pixels (i.e., M2, M3 and M4 calibrated for JF) produce more heterogenous maps. On the other hand, the M1 models, which include ten predictors and were calibrated with 5609 LIPs, and the M3 (three predictors) and M4 (twelve predictors) models fitted on JAN (997 LIPs), FEB (5371 LIPs), JF (123 LIPs) and FSE (1587 LIPs) generate smoother landslide susceptibility maps. As suggested by Steger et al. [56], landslide susceptibility maps with abrupt changes, such as those generated for JF by the M2, M3 and M4 models, are characterized by poor geomorphic plausibility and applicability and may be influenced by biased relationships between independent and dependent variables. Conversely, the other landslide susceptibility maps, which exhibit smoother prediction patterns, are more easily readable and thus might be more useful for land planning purposes, even though they derive from models with a slightly lower predictive skill.

The high number of false positives observed in the zoom areas of the landslide susceptibility maps (Figure 13) can be reasonably explained by considering that the models were fitted with a 1:1 sampling ratio of positives to negatives and that in these areas, even though they are among the most prone to landsliding, event pixels are very rare compared to the number of non-event pixels. Further research could be carried out to explore if increasing the ratio of negative to positive cases or changing the cut-off value could decrease the FPR while keeping the value of the TPR high. However, when considering entire maps (Figure 14), the number of false positives decreases significantly, although the TPR remains relatively high in most cases. The FPR and FNR values calculated in our study are comparable with those found in other studies that applied empirical models to model the spatial distribution of earthquake-induced landslides. Nowicki Jessee et al. [28], who developed a global empirical model for the near-real-time assessment of seismically induced landslides, found FPR and FNR values in the ranges of 0.06–0.45 and 0.02–0.47, respectively. These values changed in relation to the predicted probability, with higher error rates calculated for the average probability classes. In our study, FPR values are in the ranges 0.31 (M1)–0.37 (M2), 0.30 (M4)–0.58 (M1), 0.14 (M4)–0.63 (M1) and 0.27 (M4)–0.42 (M1), for the datasets JAN, FEB, JF and FSE, respectively. Regarding the FNR, the observed values are markedly lower, being in the ranges 0.26 (M4)–0.38 (M1), 0.20 (M1)–0.36 (M2), 0.02 (M4)–0.11 (M2) and 0.25 (M4)–0.29 (M1).

Some limitations must be considered when interpreting the results of our study. First, due to the lack of area photos covering larger areas, it was not possible to evaluate the presence/absence of landslides in a wider range of epicentral distances and PGA values. This would probably have led to an increase in the weight of these variables in the spatial distribution of landslides, thus determining a better predictive capacity of the models. Second, due to the resolution and spectral limits of the aerial photos that were visually analyzed in this study, we could not identify and consider permanent deformations of the slopes and fractures with little or no displacement, which are also important effects of an earthquake and may lead to severe infrastructural damage [89,90]. Unfortunately, the lack of aerial photos could not be remedied by using satellite images as their spatial resolution is lower. Furthermore, due to the coarser resolution of the SPOT-4 satellite images with respect to the resolution of the aerial photos employed to map the slope failures of the other datasets, the JF dataset includes only landslides extending more than 100 m^2. This limitation probably did not allow for the identification of smaller landslides and therefore resulted in a significantly lower number of LIPs in the JF dataset than in the other datasets.

5. Concluding Remarks

Based on the results of this experiment, the following main conclusions can be drawn for the areas of El Salvador (CA) that were most affected by the earthquakes of 2001.

- Overall, the MARS algorithm provides reliable predictions of the spatial distribution of earthquake-induced landslides.
- MARS models calibrated with rainfall-induced landslides (i.e., M1 models), even if exported from other areas, are able to predict landslides distributed over vast areas and caused by deep earthquakes, such as the one in January 2001, with an acceptable accuracy.
- When MARS models are based only on seismic covariates (i.e., M2 models), including as a predictor the probability of occurrence of rainfall-induced landslides (i.e., M3 models), even when transferred from other areas, a significant increase inaccuracy can be observed.
- The models containing both preparatory and trigger variables (i.e., M4 models) achieve the best accuracy in predicting the spatial distribution of earthquake-induced landslides.
- Even when only a small part of the landslides produced by an earthquake is known, as is usually the case shortly after the event, it is possible to use them to calibrate models, such as those used in this work, that help to identify slopes where yet unreported landslides may have occurred.

The results of this experiment and the main conclusions listed above suggest that the approach presented in this work can be useful for predicting the distribution of landslides caused by earthquakes such as those that occurred on Salvadoran territory in January and February of 2001. Seismic events with similar characteristics occur quite frequently in El Salvador due to the continuous subduction of the Cocos Plate beneath the Caribbean Plate, which generates severe offshore earthquakes, resulting in intraplate phenomena in a relatively short time. Therefore, it is useful to provide the country with landslide susceptibility maps that are also calibrated with slope failures induced by extreme rain events, which occur with greater frequency than earthquake-induced landslides. Indeed, as observed in our experiment, these maps can be used as a basis to identify slopes in which a future seismic event is most likely to cause landslides. On the other hand, in the case that a high magnitude earthquake occurs and the positions of some landslides caused by the same event are known, the models based on predisposing variables can be integrated with trigger variables such as PGAs and epicentral distances in order to obtain predictive maps that may help local civil protection to identify sites potentially affected by others and unknown earthquake-induced landslides.

Author Contributions: Conceptualization, Claudio Mercurio, Edoardo Rotigliano and Christian Conoscenti; methodology, Claudio Mercurio, Laura Paola Calderón-Cucunuba, Abel Alexei Argueta-Platero, Grazia Azzara, Chiara Cappadonia, Chiara Martinello, Edoardo Rotigliano and Christian Conoscenti; software, Claudio Mercurio, Laura Paola Calderón-Cucunuba, Abel Alexei Argueta-Platero, Chiara Martinello and Christian Conoscenti; validation; Claudio Mercurio, Chiara Cappadonia, Edoardo Rotigliano and Christian Conoscenti; formal analysis, Claudio Mercurio, Laura Paola Calderón-Cucunuba, Abel Alexei Argueta-Platero, Grazia Azzara, Chiara Martinello and Christian Conoscenti; investigation; Claudio Mercurio, Grazia Azzara, Chiara Martinello, Chiara Cappadonia and Christian Conoscenti; resources, Claudio Mercurio, Abel Alexei Argueta-Platero,

Grazia Azzara, Chiara Martinello and Christian Conoscenti; data curation, Claudio Mercurio, Laura Paola Calderón-Cucunuba, Grazia Azzara, Chiara Martinello, Chiara Cappadonia and Christian Conoscenti; writing—original draft preparation, Claudio Mercurio, Laura Paola Calderón-Cucunuba and Christian Conoscenti; writing—review and editing, Claudio Mercurio, Laura Paola Calderón-Cucunuba, Chiara Martinello and Christian Conoscenti; visualization, Claudio Mercurio, Abel Alexei Argueta-Platero, Grazia Azzara, Chiara Cappadonia and Christian Conoscenti; supervision, Chiara Cappadonia, Edoardo Rotigliano and Christian Conoscenti; project administration, Chiara Cappadonia, Edoardo Rotigliano and Christian Conoscenti; funding acquisition, Christian Conoscenti. All authors have read and agreed to the published version of the manuscript.

Funding: This research was developed in the framework of the CASTES project (AID 11726), funded by the Italian Agency for Development Cooperation (AICS) and coordinated by Prof. C.C. (Christian Conoscenti) "https://castes.agronomia.ues.edu.sv (accessed on 2 March 2023)".

Data Availability Statement: The landslide inventories, digital elevation model and geologic layers adopted for this study can be acquired from any of the authors. The NDVI values obtained can be downloaded from the website MODIS Satellite from modis.gsfc.nasa.gov/data (accessed on 2 March 2023). Peak ground accelerations and epicentral distance values can be generated by the rasterization of layers available from the USGS Shakemap website "earthquake.usgs.gov/data/shakemap (accessed on 2 March 2023)".

Conflicts of Interest: The authors declare no conflict of interest.

References

1. Rose, W.I.; Bommer, J.J.; López, D.L.; Carr, M.J.; Major, J.J. *Natural Hazards in El Salvador*; Geological Society of America: Boulder, CO, USA, 2004.
2. Quesada-Román, A.; Campos-Durán, D. Natural Disaster Risk Inequalities in Central America. *Pap. Appl. Geogr.* **2022**, *9*, 36–48. [CrossRef]
3. Marano, K.D.; Wald, D.J.; Allen, T.I. Global Earthquake Casualties Due to Secondary Effects: A Quantitative Analysis for Improving Rapid Loss Analyses. *Nat. Hazards* **2010**, *52*, 319–328. [CrossRef]
4. Agnesi, V.; Carrara, A.; Macaluso, T.; Monteleone, S.; Pipitone, G.; Sorriso-Valvo, M. Preliminary Observations of Slope Instability Phenomena Induced by the Earthquake of November 1980 on the Upper Valley of Sele River. *Geol. Appl. Idrogeol.* **1982**, *17*, 79–93.
5. Agnesi, V.; Carrara, A.; Macaluso, T.; Monteleone, S.; Pipitone, G.; Sorriso-Valvo, M. Typological and Morphological Features of Slope Instability Phenomena Induced by the 1980 Earthquake, Upper Sele Valley, Southern Italy. *Geol. Appl. Idrogeol.* **1983**, *18*, 309–341.
6. Carrara, A.; Agnesi, V.; Macaluso, T.; Monteleone, S.; Pipitone, G. Slope Movements Induced by the Southern Italy Earthquake of November 1980. *IAEG* **1986**, *2*, 237–250.
7. Agnesi, V.; Camarda, M.; Conoscenti, C.; Di Maggio, C.; Serena Diliberto, I.; Madonia, P.; Rotigliano, E. A Multidisciplinary Approach to the Evaluation of the Mechanism That Triggered the Cerda Landslide (Sicily, Italy). *Geomorphology* **2005**, *1–2*, 101–116. [CrossRef]
8. Rotigliano, E.; Cappadonia, C.; Conoscenti, C.; Costanzo, D.; Agnesi, V. Slope Units-Based Flow Susceptibility Model: Using Validation Tests to Select Controlling Factors. *Nat. Hazards* **2012**, *61*, 143–153. [CrossRef]
9. Du, Y.; Xie, M.; Jia, J. Stepped Settlement: A Possible Mechanism for Translational Landslides. *CATENA* **2020**, *187*, 104365. [CrossRef]
10. Keefer, D.K. Landslides Caused by Earthquakes. *Geol. Soc. Am. Bull.* **1984**, *95*, 406–421. [CrossRef]
11. Martino, S.; Prestininzi, A.; Romeo, R.W. Earthquake-Induced Ground Failures in Italy from a Reviewed Database. *Nat. Hazards Earth Syst. Sci.* **2014**, *14*, 799–814. [CrossRef]
12. Legros, F. The Mobility of Long-Runout Landslides. *Eng. Geol.* **2002**, *63*, 301–331. [CrossRef]
13. Hayashi, J.N.; Self, S. A Comparison of Pyroclastic Flow and Debris Avalanche Mobility. *J. Geophys. Res. Solid Earth* **1992**, *97*, 9063–9071. [CrossRef]
14. Xie, M.; Du, Y.; Jiang, Y.; Li, B.; Zhang, X.; Gargani, J. Influence of Relative Sea-Level Rise, Meteoric Water Infiltration and Rock Weathering on Giant Volcanic Landslides. *Geosciences* **2023**, *13*, 113. [CrossRef]
15. Du, Y.; Xie, M.w.; Jiang, Y.j.; Li, B.; Chicas, S. Experimental Rock Stability Assessment Using the Frozen–Thawing Test. *Rock Mech. Rock Eng.* **2017**, *50*, 1049–1053. [CrossRef]
16. Du, Y.; Lu, Y.; Xie, M.; Jia, J. A New Attempt for Early Warning of Unstable Rocks Based on Vibration Parameters. *Bull. Eng. Geol. Environ.* **2020**, *79*, 4363–4368. [CrossRef]
17. Bommer, J.; Benito, B.; Ciudad-Real, M.; Lemoine, A.; López-Menjívar, M.; Madariaga, R.; Mankelow, J.; Méndez de Hasbun, P.; Murphy, W.; Nieto-Lovo, M.; et al. The El Salvador Earthquakes of January and February 2001: Context, Characteristics and Implications for Seismic Risk. *Soil Dyn. Earthq. Eng.* **2002**, *22*, 389–418. [CrossRef]

18. Jibson, R.W.; Crone, A.J.; Harp, E.L.; Baum, R.L.; Major, J.J.; Pullinger, C.R.; Escobar, C.D.; Martínez, M.; Smith, M.E. Landslides Triggered by the 13 January and 13 February 2001 Earthquakes in El Salvador. *Spec. Pap. Geol. Soc. Am.* **2004**, *375*, 69–88. [CrossRef]
19. Chen, H. Controlling Factors of Hazardous Debris Flow in Taiwan. *Quat. Int.* **2006**, *147*, 3–15. [CrossRef]
20. Larsen, M.C.; Wieczorek, G.F. Geomorphic Effects of Large Debris Flows and Flash Floods, Northern Venezuela, 1999. *Z. Fur Geomorphol. Suppl.* **2006**, *145*, 147–175.
21. Brabb, E.E. Innovative Approaches to Landslide Hazard and Risk Mapping. In Proceedings of the IVth International Conference and Field Workshop in Landslides, Tokyo, Japan, 23–31 August 1985; Volume 1, pp. 17–22.
22. Vargas-Cuervo, G.; Rotigliano, E.; Conoscenti, C. Prediction of Debris-Avalanches and -Flows Triggered by a Tropical Storm by Using a Stochastic Approach: An Application to the Events Occurred in Mocoa (Colombia) on 1 April 2017. *Geomorphology* **2019**, *339*, 31–43. [CrossRef]
23. Lombardo, L.; Cama, M.; Maerker, M.; Rotigliano, E. A Test of Transferability for Landslides Susceptibility Models under Extreme Climatic Events: Application to the Messina 2009 Disaster. *Nat. Hazards* **2014**, *74*, 1951–1989. [CrossRef]
24. Khalaj, S.; BahooToroody, F.; Mahdi Abaei, M.; BahooToroody, A.; De Carlo, F.; Abbassi, R. A Methodology for Uncertainty Analysis of Landslides Triggered by an Earthquake. *Comput. Geotech.* **2020**, *117*, 103262. [CrossRef]
25. Bahootoroody, F.; Khalaj, S.; Leoni, L.; De Carlo, F.; Di Bona, G.; Forcina, A. Reliability Estimation of Reinforced Slopes to Prioritize Maintenance Actions. *Int. J. Environ. Res. Public Health* **2021**, *18*, 373. [CrossRef] [PubMed]
26. Conoscenti, C.; Martinello, C.; Alfonso-Torreño, A.; Gómez-Gutiérrez, Á. Predicting Sediment Deposition Rate in Check-Dams Using Machine Learning Techniques and High-Resolution DEMs. *Environ. Earth Sci.* **2021**, *80*, 380. [CrossRef]
27. Cappadonia, C.; Cafiso, F.; Ferraro, R.; Martinello, C.; Rotigliano, E. Rockfall Hazards of Mount Pellegrino Area (Sicily, Southern Italy). *J. Maps* **2021**, *17*, 29–39. [CrossRef]
28. Nowicki Jessee, M.A.; Hamburger, M.W.; Allstadt, K.; Wald, D.J.; Robeson, S.M.; Tanyas, H.; Hearne, M.; Thompson, E.M. A Global Empirical Model for Near-Real-Time Assessment of Seismically Induced Landslides. *J. Geophys. Res. Earth Surf.* **2018**, *123*, 1835–1859. [CrossRef]
29. Carrara, A.; Cardinali, M.; Guzzetti, F.; Reichenbach, P. *Gis Technology in Mapping Landslide Hazard*; Springer: Dordrecht, The Netherlands, 1995; pp. 135–175.
30. Guzzetti, F.; Reichenbach, P.; Cardinali, M.; Galli, M.; Ardizzone, F. Probabilistic Landslide Hazard Assessment at the Basin Scale. *Geomorphology* **2005**, *72*, 272–299. [CrossRef]
31. Furlani, S.; Ninfo, A. Is the Present the Key to the Future? *Earth Sci. Rev.* **2015**, *142*, 38–46. [CrossRef]
32. Heckmann, T.; Gegg, K.; Gegg, A.; Becht, M. Sample Size Matters: Investigating the Effect of Sample Size on a Logistic Regression Susceptibility Model for Debris Flows. *Nat. Hazards Earth Syst. Sci.* **2014**, *14*, 259–278. [CrossRef]
33. Conoscenti, C.; Ciaccio, M.; Caraballo-Arias, N.A.; Gómez-Gutiérrez, Á.; Rotigliano, E.; Agnesi, V. Assessment of Susceptibility to Earth-Flow Landslide Using Logistic Regression and Multivariate Adaptive Regression Splines: A Case of the Belice River Basin (Western Sicily, Italy). *Geomorphology* **2015**, *242*, 49–64. [CrossRef]
34. Allison, P.D. Logistic Regression Using the SAS System: Theory and Application; SAS Institute Inc.: Cary, CA, USA. Available online: http://www.sciepub.com/reference/248083 (accessed on 12 December 2022).
35. Guzzetti, F.; Reichenbach, P.; Ardizzone, F.; Cardinali, M.; Galli, M. Estimating the Quality of Landslide Susceptibility Models. *Geomorphology* **2006**, *81*, 166–184. [CrossRef]
36. Carrara, A.; Crosta, G.; Frattini, P. Comparing Models of Debris-Flow Susceptibility in the Alpine Environment. *Geomorphology* **2008**, *94*, 353–378. [CrossRef]
37. Brenning, A. Spatial Prediction Models for Landslide Hazards: Review, Comparison and Evaluation. *Nat. Hazards Earth Syst. Sci.* **2005**, *5*, 853–862. [CrossRef]
38. Tien Bui, D.; Tuan, T.A.; Klempe, H.; Pradhan, B.; Revhaug, I. Spatial Prediction Models for Shallow Landslide Hazards: A Comparative Assessment of the Efficacy of Support Vector Machines, Artificial Neural Networks, Kernel Logistic Regression, and Logistic Model Tree. *Landslides* **2016**, *13*, 361–378. [CrossRef]
39. Vorpahl, P.; Elsenbeer, H.; Märker, M.; Schröder, B. How Can Statistical Models Help to Determine Driving Factors of Landslides? *Ecol. Modell.* **2012**, *239*, 27–39. [CrossRef]
40. Yilmaz, I. Landslide Susceptibility Mapping Using Frequency Ratio, Logistic Regression, Artificial Neural Networks and Their Comparison: A Case Study from Kat Landslides (Tokat—Turkey). *Comput. Geosci.* **2009**, *35*, 1125–1138. [CrossRef]
41. Reichenbach, P.; Rossi, M.; Malamud, B.D.; Mihir, M.; Guzzetti, F. A Review of Statistically-Based Landslide Susceptibility Models. *Earth-Sci. Rev.* **2018**, *180*, 60–91. [CrossRef]
42. Frattini, P.; Crosta, G.; Carrara, A. Techniques for Evaluating the Performance of Landslide Susceptibility Models. *Eng. Geol.* **2010**, *111*, 62–72. [CrossRef]
43. Brenning, A. Spatial Cross-Validation and Bootstrap for the Assessment of Prediction Rules in Remote Sensing: The R Package Sperrorest. In Proceedings of the 2012 IEEE International Geoscience and Remote Sensing Symposium, Munich, Germany, 22–27 July 2012; pp. 5372–5375. [CrossRef]
44. Ruß, G.; Brenning, A. Spatial Variable Importance Assessment for Yield Prediction in Precision Agriculture. *Lect. Notes Comput. Sci.* **2010**, *6065*, 184–195. [CrossRef]

45. Gutiérrez, Á.G.; Schnabel, S.; Lavado Contador, J.F. Using and Comparing Two Nonparametric Methods (CART and MARS) to Model the Potential Distribution of Gullies. *Ecol. Modell.* **2009**, *220*, 3630–3637. [CrossRef]
46. Gutiérrez, Á.G.; Conoscenti, C.; Angileri, S.E.; Rotigliano, E.; Schnabel, S. Using Topographical Attributes to Evaluate Gully Erosion Proneness (Susceptibility) in Two Mediterranean Basins: Advantages and Limitations. *Nat. Hazards* **2015**, *79*, 291–314. [CrossRef]
47. Rotigliano, E.; Martinello, C.; Agnesi, V.; Conoscenti, C. Evaluation of Debris Flow Susceptibility in El Salvador (CA): A Comparisobetween Multivariate Adaptive Regression Splines (MARS) and Binary Logistic Regression (BLR). *Hung. Geogr. Bull.* **2018**, *67*, 361–373. [CrossRef]
48. Conoscenti, C.; Agnesi, V.; Cama, M.; Caraballo-Arias, N.A.; Rotigliano, E. Assessment of Gully Erosion Susceptibility Using Multivariate Adaptive Regression Splines and Accounting for Terrain Connectivity. *Land Degrad. Dev.* **2018**, *29*, 724–736. [CrossRef]
49. Mercurio, C.; Martinello, C.; Rotigliano, E.; Argueta-Platero, A.A.; Reyes-Martínez, M.E.; Rivera-Ayala, J.Y.; Conoscenti, C. Mapping Susceptibility to Debris Flows Triggered by Tropical Storms: A Case Study of the San Vicente Volcano Area (El Salvador, CA). *Earth* **2021**, *2*, 66–85. [CrossRef]
50. Martinello, C.; Cappadonia, C.; Conoscenti, C.; Rotigliano, E. Landform Classification: A High-Performing Mapping Unit Partitioning Tool for Landslide Susceptibility Assessment—A Test in the Imera River Basin (Northern Sicily, Italy). *Landslides* **2022**, *19*, 539–553. [CrossRef]
51. Dewey, J.W.; Suárez, G. Seismotectonics of Middle America. *Neotectonics N. Am.* **1991**, *1*, 309–321. [CrossRef]
52. Campos-Durán, D.; Quesada-Román, A. Intensive and Extensive Risk in Central America Between 1990 and 2015. *Anuário Inst. Geociências UFRJ* **2017**, *40*, 234–249. [CrossRef]
53. Cruden, D.M.; Varnes, D.J. Landslide Types and Processes. In *Landslides Investigation and Mitigation, Transportation Research Board, Special Report No. 247*; Turner, A.K., Shuster, R.L., Eds.; Scientific Research: 1996; National Research Council: Washington, DC, USA, 1996; pp. 36–75. Available online: https://www.scirp.org/(S(czeh2tfqyw2orz553k1w0r45))/reference/ReferencesPapers.aspx?ReferenceID=1855371 (accessed on 10 March 2023).
54. Lomnitz, C.; Elizarrarás, S.R. El Salvador 2001: Earthquake Disaster and Disaster Preparedness in a Tropical Volcanic Environment. *Seismol. Res. Lett.* **2001**, *72*, 346–351. [CrossRef]
55. Galli, M.; Ardizzone, F.; Cardinali, M.; Guzzetti, F.; Reichenbach, P. Comparing Landslide Inventory Maps. *Geomorphology* **2008**, *94*, 268–289. [CrossRef]
56. Steger, S.; Brenning, A.; Bell, R.; Glade, T. The Propagation of Inventory-Based Positional Errors into Statistical Landslide Susceptibility Models. *Nat. Hazards Earth Syst. Sci.* **2016**, *16*, 2729–2745. [CrossRef]
57. Steger, S.; Brenning, A.; Bell, R.; Glade, T. The Influence of Systematically Incomplete Shallow Landslide Inventories on Statistical Susceptibility Models and Suggestions for Improvements. *Landslides* **2017**, *14*, 1767–1781. [CrossRef]
58. Martinello, C.; Mercurio, C.; Cappadonia, C.; Hernández Martínez, M.Á.; Reyes Martínez, M.E.; Rivera Ayala, J.Y.; Conoscenti, C.; Rotigliano, E. Investigating Limits in Exploiting Assembled Landslide Inventories for Calibrating Regional Susceptibility Models: A Test in Volcanic Areas of El Salvador. *Appl. Sci.* **2022**, *12*, 6151. [CrossRef]
59. Rotigliano, E.; Agnesi, V.; Cappadonia, C.; Conoscenti, C. The Role of the Diagnostic Areas in the Assessment of Landslide Susceptibility Models: A Test in the Sicilian Chain. *Nat. Hazards* **2011**, *58*, 981–999. [CrossRef]
60. Cama, M.; Lombardo, L.; Conoscenti, C.; Agnesi, V.; Rotigliano, E. Predicting Storm-Triggered Debris Flow Events: Application to the 2009 Ionian Peloritan Disaster (Sicily, Italy). *Nat. Hazards Earth Syst. Sci.* **2015**, *15*, 1785–1806. [CrossRef]
61. Rotigliano, E.; Martinello, C.; Hernandéz, M.A.; Agnesi, V.; Conoscenti, C. Predicting the Landslides Triggered by the 2009 96E/Ida Tropical Storms in the Ilopango Caldera Area (El Salvador, CA) Optimizing MARS-Based Model Building and Validation Strategies. *Environ. Earth Sci.* **2019**, *78*, 210. [CrossRef]
62. Carrara, A.; Guzzetti, F. *Geographical Information Systems in Assessing Natural Hazards*; Springer: Berlin/Heidelberg, Germany, 1995.
63. Hansen, A. Landslide Hazard Analysis. In *Slope Instability*; Brunsden, D., Prior, D.B., Eds.; John Wiley and Sons: New York, NY, USA, 1984; pp. 523–602. Available online: http://www.sciepub.com/reference/209033 (accessed on 1 October 2022).
64. Cama, M.; Lombardo, L.; Conoscenti, C.; Rotigliano, E. Improving Transferability Strategies for Debris Flow Susceptibility AssessmentApplication to the Saponara and Itala Catchments (Messina, Italy). *Geomorphology* **2017**, *288*, 52–65. [CrossRef]
65. Martinello, C.; Cappadonia, C.; Conoscenti, C.; Agnesi, V.; Rotigliano, E. Optimal Slope Units Partitioning in Landslide Susceptibility Mapping. *J. Maps* **2020**, *17*, 152–162. [CrossRef]
66. Martinello, C.; Cappadonia, C.; Rotigliano, E. Investigating the Effects of Cell Size in Statistical Landslide Susceptibility Modelling for Different Landslide Typologies: A Test in Central–Northern Sicily. *Appl. Sci.* **2023**, *13*, 1145. [CrossRef]
67. Garosi, Y.; Sheklabadi, M.; Pourghasemi, H.R.; Besalatpour, A.A.; Conoscenti, C.; Van Oost, K. Comparison of Differences in Resolution and Sources of Controlling Factors for Gully Erosion Susceptibility Mapping. *Geoderma* **2018**, *330*, 65–78. [CrossRef]
68. Shao, X.; Xu, C. Earthquake-Induced Landslides Susceptibility Assessment: A Review of the State-of-the-Art. *Nat. Hazards Res.* **2022**, *2*, 172–182. [CrossRef]
69. Wald, D.J.; Worden, C.B.; Thompson, E.M.; Hearne, M. ShakeMap Operations, Policies, and Procedures. *Earthq. Spectra* **2022**, *38*, 756–777. [CrossRef]
70. Jebur, M.N.; Pradhan, B.; Tehrany, M.S. Optimization of Landslide Conditioning Factors Using Very High-Resolution Airborne Laser Scanning (LiDAR) Data at Catchment Scale. *Remote Sens. Environ.* **2014**, *152*, 150–165. [CrossRef]

71. Conoscenti, C.; Rotigliano, E. Predicting Gully Occurrence at Watershed Scale: Comparing Topographic Indices and Multivariate Statistical Models. *Geomorphology* **2020**, *359*, 107123. [CrossRef]
72. Friedman, J.H. Multivariate Adaptive Regression Splines. *Ann. Stat.* **1991**, *19*, 1–67. [CrossRef]
73. Hijmans, R.J. *Geographic Data Analysis and Modeling, [R Package Raster Version 3.6-3]*; University of Davis: Davis, CA, USA, 2022; Available online: https://cran.r-project.org/web/packages/raster/index.html (accessed on 18 April 2023).
74. Naimi, B. *Package "USDM" Type Package Title Uncertainty Analysis for Species Distribution Models*; University of Helsinki: Helsinki, Finland, 2022; Available online: https://cran.r-project.org/web/packages/usdm/usdm.pdf (accessed on 18 April 2023).
75. Mahto, A. *Package "SPLITSTACKSHAPE" Type Package Title Stack and Reshape Datasets After Splitting Concatenated Values*; University of Santa Barbara: Santa Barabara, CA, USA, 2018; Available online: https://cran.r-project.org/web/packages/splitstackshape/splitstackshape.pdf (accessed on 18 April 2023).
76. Robin, X.; Turck, N.; Hainard, A.; Tiberti, N.; Lisacek, F.; Sanchez, J.C.; Müller, M. PROC: An Open-Source Package for R and S+ to Analyze and Compare ROC Curves. *BMC Bioinform.* **2011**, *12*, 77. [CrossRef]
77. Sing, T.; Sander, O.; Beerenwinkel, N.; Lengauer, T. ROCR: Visualizing Classifier Performance in R. *Bioinformatics* **2005**, *21*, 3940–3941. [CrossRef] [PubMed]
78. Kuhn, M.; Wing, J.; Weston, S.; Williams, A.; Keefer, C.; Engelhardt, A.; Cooper, T.; Mayer, Z.; Kenkel, B.; R Core Team. Package "CARET" Title Classification and Regression Training Description Misc Functions for Training and Plotting Classification and Regression Models. 2018, p. 2. Available online: https://cran.r-project.org/web/packages/caret/caret.pdf (accessed on 16 November 2022).
79. Milborrow, S. Notes on the Earth Package. University of Cape Town: South Africa, 2021. Available online: https://cran.r-project.org/web/packages/earth/earth.pdf (accessed on 18 April 2023).
80. Van Den Eeckhaut, M.; Hervás, J.; Jaedicke, C.; Malet, J.P.; Montanarella, L.; Nadim, F. Statistical Modelling of Europe-Wide Landslide Susceptibility Using Limited Landslide Inventory Data. *Landslides* **2012**, *9*, 357–369. [CrossRef]
81. Svoray, T.; Michailov, E.; Cohen, A.; Rokah, L.; Sturm, A. Predicting Gully Initiation: Comparing Data Mining Techniques, Analytical Hierarchy Processes and the Topographic Threshold. *Earth Surf. Process. Landf.* **2012**, *37*, 607–619. [CrossRef]
82. Goetz, J.N.; Brenning, A.; Petschko, H.; Leopold, P. Evaluating Machine Learning and Statistical Prediction Techniques for Landslide Susceptibility Modeling. *Comput. Geosci.* **2015**, *81*, 1–11. [CrossRef]
83. Javidan, N.; Kavian, A.; Pourghasemi, H.R.; Conoscenti, C.; Jafarian, Z.; Rodrigo-Comino, J. Evaluation of Multi-Hazard Map Produced Using MaxEnt Machine Learning Technique. *Sci. Rep.* **2021**, *11*, 6496. [CrossRef]
84. Hosmer, D.W.; Lemeshow, S. *Applied Logistic Regression*; John Wiley & Sons, Inc.: Hoboken, NJ, USA, 2000; ISBN 9780471722144.
85. Avila, L.A.; Cangialosi, J. *Tropical Cyclone Report Hurricane Ida (AL112009)*; National Hurricane Center: Miami, FL, USA, 2010; 19p. Available online: https://www.nhc.noaa.gov/data/tcr/AL112009_Ida.pdf (accessed on 16 November 2022).
86. Avila, R. *Síntesis de Los Informes de Evaluación Técnica de Las Lluvias Del 7 y 8 de Noviembre 2009 En El Salvador: Análisis Del Impacto Físico Natural y Vulnerabilidad Socio Ambiental*; Report Comisión Técnico Científica; Coordina el Ministerio de Medio Ambiente y Recursos Naturales, MARN: San Salvador, El Salvador, 2010; Available online: https://www.academia.edu/42170439/_Síntesis_de_los_informes_de_evaluación_técnica_de_las_lluvias_del_7_y_8_de_noviembre_2009_en_El_Salvador_Análisis_del_impacto_físico_natural_y_vulnerabilidad_socio_ambiental_ (accessed on 16 November 2022).
87. Clerici, A.; Perego, S.; Tellini, C.; Vescovi, P. A GIS-Based Automated Procedure for Landslide Susceptibility Mapping by the Conditional Analysis Method: The Baganza Valley Case Study (Italian Northern Apennines). *Environ. Geol.* **2006**, *50*, 941–961. [CrossRef]
88. Costanzo, D.; Chacón, J.; Conoscenti, C.; Irigaray, C.; Rotigliano, E. Forward Logistic Regression for Earth-Flow Landslide Susceptibility Assessment in the Platani River Basin (Southern Sicily, Italy). *Landslides* **2014**, *11*, 639–653. [CrossRef]
89. Liu, Z.; Shao, J.; Xu, W.; Chen, H.; Shi, C. Comparison on Landslide Nonlinear Displacement Analysis and Prediction with Computational Intelligence Approaches. *Landslides* **2014**, *11*, 889–896. [CrossRef]
90. Tao, H.; Zhang, M.; Gong, L.; Shi, X.; Wang, Y.; Yang, G.; Lei, S. The Mechanism of Slope Instability Due to Rainfall-Induced Structural Decay of Earthquake-Damaged Loess. *Earthq. Res. Adv.* **2022**, *2*, 100137. [CrossRef]

applied sciences

MDPI

Article

Multi-Factor Analysis on the Stability of High Slopes in Open-Pit Mines

Hui Cao [1,*], Gaotong Ma [1], Peng Liu [2], Xiushan Qin [3], Chunping Wu [1] and Jin Lu [1]

1 School of Civil and Resources Engineering, University of Science and Technology Beijing, Beijing 100083, China
2 State Key Laboratory for Geomechanics and Deep Underground Engineering, China University of Mining and Technology, Beijing 100083, China
3 BGRIMM Technology Group, Beijing 100160, China
* Correspondence: cao_h@ustb.edu.cn; Tel.: +86-132-6941-5686

Abstract: During the production of open-pit mines, the stability of slopes can be affected by various factors such as structural surfaces, production blasting vibrations, and mining areas. In this study, the researchers focused on the slope of the open-pit mine at Yinshan and employed UAV mapping technology to conduct an on-site geological engineering investigation. Information on the yield, trace length, spacing, and density of the structural surface of the south slope was obtained. The researchers also carried out vibration blasting tests in combination with the production blasting activities in the mine to determine the blasting vibration attenuation law and whether the blasting vibration speed met safety specifications. Additionally, numerical simulation methods were used to examine the influence of the mining area on the stability of the current slope and the designed excavation slope. The slope stability was evaluated using the limit equilibrium method, and the researchers separately discussed the influence of self-weight load and self-weight load plus blasting vibration force on the stability of the high slope of the open pit. The results showed the following: (1) The rock mass structural plane in the south slope of the mining area was mainly dominated by a medium-large dip structural plane, and three faults and joint fissures in the investigation area combined to form cutting and sliding surfaces in the rock mass that were prone to collapse and sliding. (2) The maximum blasting vibration speed met safety requirements. (3) There was no large range of plastic zone damage in the entire slope, and the overall stability of the slope was good. (4) The present slope was relatively stable when considering only self-weight stress and the blasting vibration force. However, there was a certain risk of instability in the design of the excavation slope.

Keywords: open-pit mine; slope stability analysis; drone geological survey; attenuation law of blasting vibration; goaf area

1. Introduction

With the increasing mining depth of medium and large open-pit mines, a number of high and steep slopes are inevitably formed. The stability of these slopes poses a huge hidden danger to the daily functioning of mines and to the safety of the lives and property of construction personnel [1,2]. There are many unfavorable factors affecting the safety and stability of open-pit mine slopes [3–5]. First of all, the traditional method of geological surveying requires manual standing at the foot of the open-pit slope and using geological survey tools such as a compass and measuring tape to measure and record on site, which requires the participation of many people and requires a large survey area, resulting in low survey efficiency. The general height of the slope of the open-pit mine is large, and the height of the slope after merging sections is often as high as 30 m, while manual measurements can only be carried out within a very limited range of 2–3 m at the top of the slope. The structural surface measured is not very representative. On-site surveyors stand under the slope for a long time, but the upper part of the slope is affected by high wind, self-weight, and production activities, and there is a risk of falling pumice, which poses a great threat to the safety of on-site surveyors, often impeding the ability to conduct geological investigations. Secondly, the risks posed to slope safety by blasting vibrations generated in the process of mine production cannot be ignored [6–8]. As open-pit mining becomes increasingly challenging, its mining costs are rising, prompting many mines to transition from open-pit mining to combined open-pit and underground mining. Underground mining destroys the original geological structure, and the influence of the leftover goaf on slope stability needs to be studied [9–12]. Therefore, it is necessary to study the influence of the goaf on slope stability. In conjunction with the current mining situation, the research and evaluation of mine slope stability is a major concern for ensuring the safety of mine production. By carrying out investigation and test analysis on the slope of the excavated area, the current slope stability situation and the potential landslide risk in the lower mining process are further analyzed and evaluated. This study provides beneficial recommendations and guidance for the subsequent safe working of the mine, and it makes a significant contribution to the safety, damage prevention, and control of the mine slope, as well as to efficient deep mining.

The study of slope stability is based on the study of the rock mass structure. The development degree of the structural plane seriously affects the stability of the slope rock mass. Although traditional geological survey methods, such as the survey line method and the window statistics method, are more accurate, the efficiency is low and the workload is large [13]. In order to solve these problems, the method of using three-dimensional laser scanning [14,15] and UAV mapping technology for geological surveying and structural plane information acquisition has gradually emerged; this method provides significant benefits to slope engineering geological surveys. Among them, aerial survey technology with small UAVs has been rapidly applied, and has a broad prospect of widely implemented applications in major engineering construction, mining resource development, land monitoring, disaster emergency and treatment, geological survey and mapping, digital city construction, et cetera [16–18]. In 2014, Yathunanthan Vasuki et al. [19] studied the measurement data of rock based on UAV aerial images and developed a photogrammetric dataset that can be used to generate the rock surface by UAV aerial images. C. Michele et al. [20] used UAV aerial images to generate a digital grid of the rock mass and obtained the geometric parameters of the grid, and further compared the differences between the geometric parameters collected by the UAV and the actual project. Jin et al. [21] obtained the digital elevation model of a slope by using UAV aerial photography and evaluated the stability of the slope model using FLAC3D 6.0 (Itasca Consulting Group, Minneapolis, MN, USA) software.

Most open-pit mines frequently produce blasting operations, and blasting seismic waves have a great impact on slope stability. A large number of experts and scholars have conducted relevant research on the impact of blasting vibrations on slopes. Zhou et al. [22] studied the influence of blasting amplitude and frequency on slope stability, based on the established slope instability mutation theory. Wu et al. [23] studied the safety control standard value of blasting vibrations through analyzing slope stability during blasting. Zhang et al. [24] discussed the weakening law of blasting damage based on the disturbance factor D value in the Hoek–Brown strength criterion, and they discussed the influence of the D value on the slope safety factor. Cui et al. [25] used PFC3D 6.0 (Itasca Consulting Group, Minneapolis, MN, USA) software to simulate the variation in internal cracks and the displacement of the slope with blasting height. With increases in mining depth, the operational costs of open-pit mines are also increasing. Most mines have changed from open-pit mining to underground mining or open-pit combined mining. The goaf left by underground mining seriously threatens the stability of the slope. Therefore, Yang et al. [26] conducted research based on the strength reduction theory and elastic plastic mechanics; they used FLAC3D software to study the influence of the underground goaf on slope stability. Li et al. [27] studied the influence of the goaf on the instability of red clay slopes using a numerical simulation and the limit equilibrium method. Sun et al. [28] explored the characteristics and evolution of slope slip under different mining sequence combinations of open-pit and underground combined mining. Shi et al. [29] carried out an analysis and evaluation of the potentially dangerous area division, slope stability characteristics, and isolation thickness of an open-pit slope according to the theoretical deduction of the goaf and the results of borehole detection.

A large number of experts and scholars have made fruitful studies on various factors affecting slope stability respectively, but due to the complex environment of open-pit mines, there are more factors that may cause slope instability, whether they are the preferred structure plane, production blasting vibration, or underground mining area. In order to solve this problem, all kinds of factors that may cause slope instability damage need to be analyzed.

2. Background of the Project and Technical Route

2.1. Overview of Regional Project

This study is based on the copper-gold open-pit mine in the ninth district of Yinshan Mining Co., Ltd., Jiangxi Copper Industry Group. The site is located in the northern suburbs of Dexing City, Jiangxi Province, at the northwest foot of the Damaoshan branch of the Huaiyu Mountains (Figure 1). It is a typical hilly mountain landform. The mining area is 2.7 km long from north to south and 2.15 km wide from east to west, with an area of about 4.36 km^2 and a production capacity of 5000 t/d. The Neoproterozoic Zhangcun Group is widely exposed in the mining area, followed by the Mesozoic Lower Cretaceous Daguding Formation and a small amount of the Shixi Formation. The Zhangcun Group is the basement strata of the mining area, which is composed of metamorphic rocks. The Daguding Formation is a continental volcanic rock series covering the Zhangcun Group, which is distributed in the southwest of the mining area and near the Xishan crater. Its distribution is characterized by a northeast zonal distribution. The Zhangcun Group and the Daguding Formation are the main ore-bearing surrounding rocks in the mining area. The Shixi Formation is unconformably covered by the Zhangcun Group and the Daguding Formation, and a small amount is distributed in the southwest of the mining area. The main lithology of the mining area includes phyllite, quartz porphyry, and Quaternary residual deposits. Phyllite is the most important and widely distributed rock mass in the mining area, mainly comprising sericite phyllite and mixed with quartz. The quartz porphyry is a small dyke, which is nearly east–west to northeast-east, with an inclination angle of 80–90°, a length of 300–800 m, a width of 2–20 m, and a depth of more than 500 m. The Quaternary is the Quaternary residual slope and alluvial loose accumulation layer, widely distributed in the mining area. The lithofacies are mainly clay, mixed with breccia such as slate and phyllite, mostly gravel with a diameter of less than 2 cm. The surrounding rock of the ore body is mainly phyllite, quartz porphyry, and a small amount of dacite porphyry. The surrounding rock is relatively stable. The factors affecting the stability of the surrounding rocks are the development of phyllite schist and faults in local sections.

2.2. Technology Lines

The technical route of this study is shown in Figure 2.

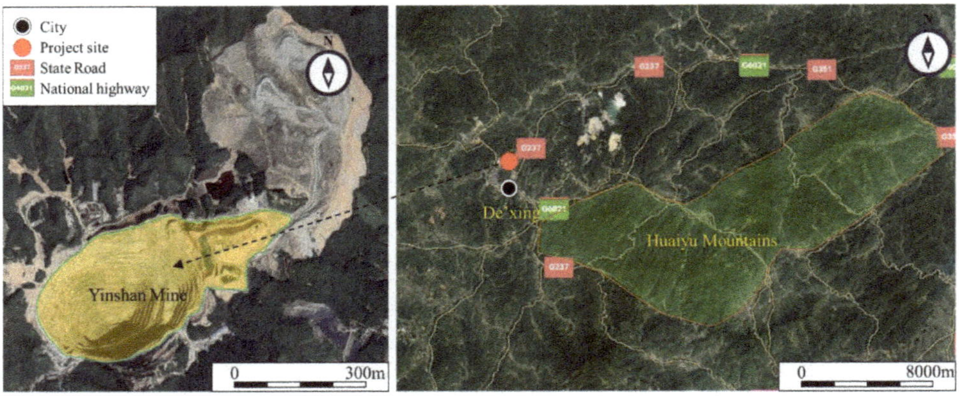

Figure 1. Project location diagram.

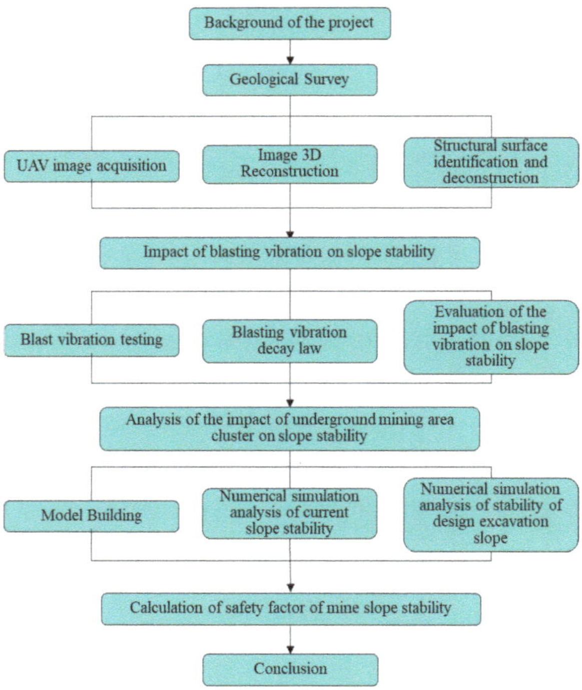

Figure 2. Research technology lines.

3. Geological Survey

3.1. UAV Technology

Due to the influence of the geological structure and external conditions, a large number of fracture structural surfaces with different directions and scales are left in the rock mass. The fractured structural surfaces are the parts of the rock mass with the lowest strength and the weakest resistance to deformation, and their existence leads to significant weakening and strong anisotropy of the overall mechanical properties of the rock mass. Rock mechanics emphasizes that the deformation and damage of the rock mass is in general the deformation and damage of the structural faces and their networks, mainly through the tension and shear deformation of the structural faces. In order to fully understand the appearance quality as well as distribution state of rock joints in each area of the mine, and to derive the dominant orientation of structural face production in the investigated area, this paper uses UAV mapping technology to conduct on-site geological investigation of slopes. UAV mapping geological surveying is a new non-contact slope engineering geological survey method, the principle of which is to use a digital camera carried on the UAV to collect aerial images by remote control, and finally use professional interpretation software to interpret the collected images. Compared with conventional geological survey methods, the UAV mapping geological survey method has the advantages of high degree, low cost, and high safety. The flight platform used in this survey is a DJI brand (Shenzhen, China) Phantom 4 Pro V2.0 quadrotor UAV, the 20-megapixel FC6310 camera on the UAV is used for image acquisition, and the 3D reconstruction system based on Sfm algorithm is used to reconstruct the aerial images [30]. Finally, the 3D reconstructed point cloud model is used for intelligent recognition and information resolution of the structural surface [31] to obtain information on the yield, location, trace length, spacing, and density of the structural surface.

3.2. Scope and Content of the Survey

Due to the influence of the geological structure and external conditions on the south slope of the stope, historically, small landslides have occurred on the south slope in December 2014 and June 2016. Therefore, geological surveying work has been carried out in this area to gather information about its structural plane and to provide the basis for slope stability evaluation and future disaster prevention and control. The survey is divided into two areas, namely, survey area I and survey area II, as shown in Figure 3. The geological survey includes (1) the yield of the structural surface, including its tendency and dip angle, (2) the location and scale of the traces of the structural surface, and (3) the spacing and density of the structural surface.

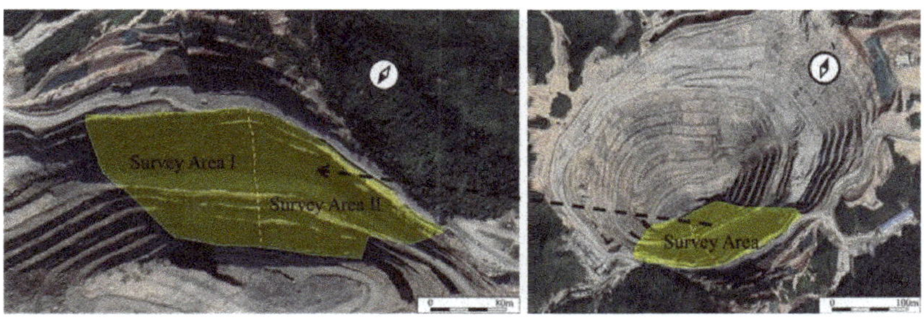

Figure 3. UAV Geological survey area.

3.3. Analysis Result

According to the on-site survey, the survey area was determined by the UAV mapping method, of which six ground control target sites were set up in the survey areas I and II. There were a total of eight steps in the effective area of the point cloud generated in area I, with a length of about 230 m and a height of about 180 m, and a total of seven steps in the effective area of the point cloud generated in survey area II.

The point cloud data of the investigation area I is shown in Figure 4. The numbers 1–9 in Figure 4 represent the number of faults or joints. Two large-scale interlayer faults have developed in the slope of survey area I, accompanied by several joint fissures. Fault 1 has an exposure length of about 120 m, and an occurrence of 290–320°∠45–55°. The lower part forms a wedge with joint 8 (335–355°∠60–70°, outcrop length of about 40 m), causing a rock landslide there. Fault 2 has an outcrop length of about 105 m, and the occurrence is 280–300°∠60–70°. We surveyed the lower part of the mud: a section about 2–3 m thick. The joints 3, 5, 6, and 9 are in the same group of joints: the occurrence is 295–320°∠50–60°, and the exposure length is 13 m, 18 m, 16 m, and 29 m, respectively. The joints 4 and 7 are in the same group of joints: the occurrence is 355–025°∠50–60°, and the exposure length is 10 m and 14 m, respectively.

Figure 4. Investigation area I point cloud data.

The point cloud data of the investigation area II is shown in Figure 5. The numbers 1–3 in Figure 5 represent the numbers of faults or joints. A large-scale interlayer fault has developed in slope II of the survey area, accompanied by several joint fissures of the same occurrence. The exposure length of fault 1 is about 148 m, and the attitude is 300°∠50°. The exposure length of joint 2 is 57 m, and that of joint 3 is 60 m.

Figure 5. Investigation area II point cloud data.

There are large faults with several joint fissures in investigation areas I and II, and the average joint spacing is about 5–30 m. The average joint density is 0.033–0.2/m. DIPS software was used to map the nodal strike rosettes of the two out survey areas, as shown in Figure 6.

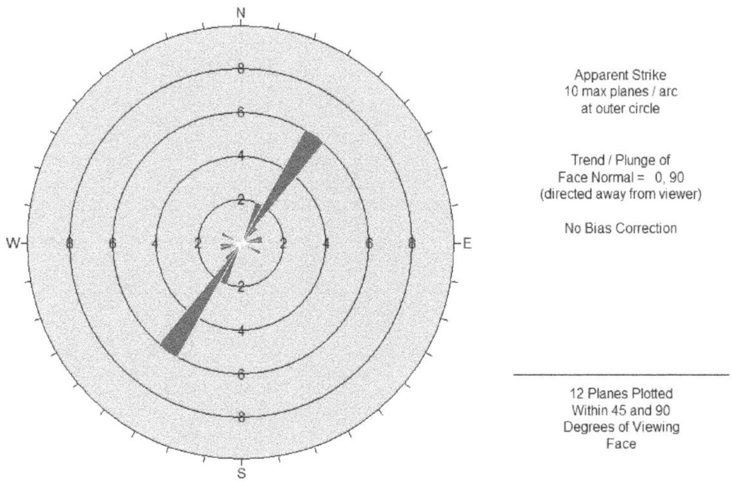

Figure 6. Rose diagram of rock joint surface strike.

4. Analysis of the Influence of Blasting Vibrations on Slope Stability

Blasting operations are frequent in most open-pit mining processes. The blasting of seismic waves is an important factor affecting slope stability. Due to the existence of structural planes in the rock and soil of the slope, when the vibration wave generated by blasting passes through the weak structural plane, the reflection of the vibration wave at the structural plane is enhanced, which may cause the slope to be destroyed at the structural plane. On the other hand, the blasting vibration wave exerts a dynamic load on the weak structural plane. When the load exceeds the bearing capacity of the structural plane, the slope is destroyed. At the same time, due to the heterogeneity and discontinuity of the slope rock mass, the vibration wave causes a certain relative displacement of the internal structure of the slope rock mass, which then causes the slope to be destroyed.

4.1. Blasting Vibration Test

4.1.1. Testing Method

Usually, the bumping and shaking of all nearby objects caused by blasting is called the blasting seismic effect. When the distance is more than 150 times the radius of the explosive package, the intensity of the vibration wave caused by the blasting is weakened; this means that the particles can only undergo elastic vibration and cannot cause rock mass damage. This is called an elastic wave. A seismic wave is an elastic wave, which is composed of a body wave and a surface wave. The body wave is divided into a longitudinal wave and a transverse wave. The longitudinal wave has the characteristics of low amplitude, short duration, and high speed, while the transverse wave has a long duration, and its amplitude is larger than that of the longitudinal wave. The surface wave is formed by the multiple reflection and superposition of the body wave on the free surface. Its propagation speed is slower than that of the body wave, but the carrying energy is larger. The seismic damage caused by blasting is mainly the effect of a surface wave. When the distance is short, the longitudinal wave, shear wave, and surface wave arrive almost at the same time, so it is difficult to identify the type of wave. When the distance is large, the three waves begin to separate, and can therefore be identified, and then the parameters of the blasting vibration are determined. This vibration monitoring was conducted using the Mini series blasting vibration meter (Figure 7) produced by Chengdu Tai Ce Technology Co., Ltd., (Chengdu, China) which facilitates the online monitoring of blasting vibration velocity and frequency.

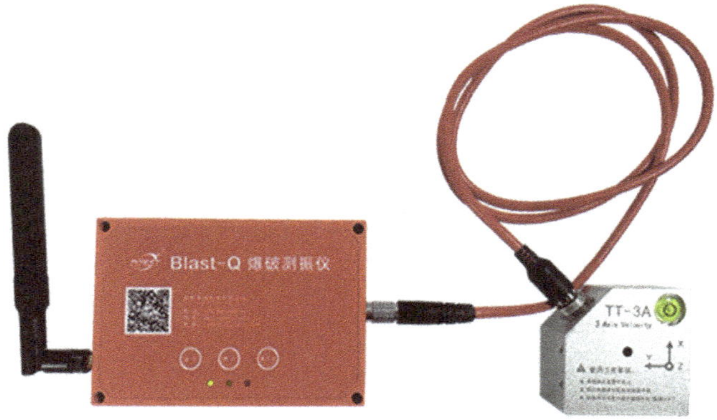

Figure 7. Composition of blasting vibration meter.

4.1.2. Vibration Test Scheme Design

According to the time sequence of on-site production blasting, the blasting vibration monitoring is divided into four blasting areas: A, B, C, and D. Five monitoring points are set up in each blasting area. The blasting area and vibration monitoring points are shown in Figure 8.

Figure 8. Layout of measuring points in the blasting area and vibration measuring area.

The center coordinates of each blasting area are shown in Table 1. The coordinate system used for all coordinates in Table 1 is China Geodetic Coordinate System 2000 (CGCS2000), and the unit of coordinates is m. The blasting area A is located on the −36 m platform on the north side of the stope, and its maximum one-stage charge is 410 kg. The vibration parameters of blasting area A are monitored by the A1–A5 vibration points. The blasting area B is located on the −48 m platform on the southwest side of the stope. The maximum one-stage charge is 250 kg. The vibration parameters of blasting area B are monitored by the B1-B5 vibration points. The blasting area C is located on the −48 m platform on the east side of the stope. The maximum one-stage charge is 400 kg. The blasting area D is located on the −60 m platform on the east side of the stope. The vibration parameters of blasting area C and blasting area D are monitored by the C1–C5 vibration measuring points. The coordinates of each vibration point are shown in Table 2.

Table 1. Center coordinates of the blasting area.

Blast Area	Blasting Zone Center Coordinates		
	X	Y	Z
A	566,644.7281	3,201,628.888	−31.3655
B	566,624.6358	3,201,261.0618	−42
C	566,965.4442	3,201,443.7396	−42
D	566,890.1092	3,201,446.9961	−56

Table 2. Vibration point coordinates.

Blast Area	Maximum One-Stage Charge (kg)	Measuring Point	Distance between the Center of Blasting Area and the Measuring Point (m)		
			Level	Vertical	Straight Line
A	410	A1	242.46	1.13	242.47
		A2	223.98	1.28	223.98
		A3	210.07	1.13	210.08
		A4	24.64	11.92	27.37
		A5	46.11	12.05	47.66
B	250	B1	140.22	114.62	181.11
		B2	226.98	113.15	253.62
		B3	256.25	113.81	289.56
		B4	393.24	113.74	409.36
		B5	416.99	114.42	432.40
C	400	C1	155.16	113.79	192.41
		C2	149.96	114.04	188.39
		C3	149.66	113.90	188.08
		C4	148.56	113.52	186.97
		C5	158.02	113.59	194.61
D	380	C1	225.14	127.79	258.88
		C2	224.09	128.04	258.09
		C3	224.90	127.90	258.73
		C4	223.96	127.52	257.71
		C5	232.47	127.59	265.18

4.1.3. Blasting Vibration Monitoring Results

According to the *Technical Code for the Safety Monitoring of Slopes of Metallic and Non-metallic Open-Pit Mines* (AQ2063-2018), the monitoring points of blasting vibration velocity should be set at the foot of the slope in the main sliding direction of the slope, and there should be more than three monitoring points. The velocity of the particle in the three directions of X, Y, and Z should be monitored at the same time, and the maximum value in the three directions should be taken as the maximum value of the particle. Monitoring accuracy should be less than 0.001 cm/s. After monitoring, 13 groups of effective data for the peak vibration velocity, acceleration, and main vibration frequency of the particles were obtained. The monitoring results of each blasting vibration are shown in Table 3.

Table 3. Blasting vibration monitoring data.

Measuring Point	Peak Vibration Velocity in the X Direction (cm/s)	Peak Vibration Velocity in the Y Direction (cm/s)	Peak Vibration Velocity in the Z Direction (cm/s)	Superposition Vibration Velocity (cm/s)
A1	0.8315	0.0255	0.4773	0.2453
A2	2.9327	6.9657	7.8745	8.8547
A3	0.4719	0.6274	0.2629	0.6832
A4	0.0731	0.4694	0.0057	0.4706
B2	0.0315	0.3421	0.2483	0.4014
B3	0.4348	0.4670	0.2861	0.5179
B4	0.1762	0.1338	0.0974	0.2008
B5	0.0052	0.1298	0.0014	0.1298
C1	0.1260	0.0075	0.6600	0.6605
C2	1.0147	0.9307	1.1884	1.4475
C3	0.9227	1.5859	1.4036	2.0280
C4	0.0916	0.9070	0.9894	1.0875
C5	0.3434	0.9278	1.0621	1.2176

For specific seismic wave propagation conditions, the particle vibration velocity is related to the charge weight and the distance from the particle to the center of the explosion source. The relationship of peak particle vibration velocity with charge weight, distance, site factor K, and the attenuation index α conforms to Sadaovsk's formula:

$$V = K(\frac{Q^{\frac{1}{3}}}{R})^{\alpha} = K\rho^{\alpha} \quad (1)$$

In Formula (1), V is the peak vibration velocity of particle, and the unit is cm/s. Q is the amount of explosive, and the unit is kg. R is the distance between the measuring point and the center of the explosion source, and the unit is m. K is a coefficient related to rock properties, blasting methods, and other factors, namely the site factor. α is the seismic wave attenuation index, related to geological conditions. ρ is the proportional dose.

Because V and ρ in the above attenuation formula are not linear, it is necessary to convert the formula into a linear relationship for regression, so as to obtain the corresponding K and α values. By taking logarithms on both sides of the equal sign of the formula, the following linear form is obtained:

$$lgV = \alpha lg\rho + lgK \quad (2)$$

When the effective data are sufficient, the above equation is regressed by the least square method of mathematical statistics.

4.2. Evaluation of the Influence of Blasting Vibrations on Slope Stability

After eliminating the obvious data noise, the least square method is used to analyze Equation (2), and the coefficient K and attenuation coefficient α, which are related to the terrain and geological conditions between the blasting point and the measuring point, are fitted. The fitting results are shown in Figure 9. The site coefficient K = 328.02 and the attenuation index α = 1.809 are calculated. The formula for the vibration velocity attenuation of blasting vibration particles in the monitoring area of the Yinshan Mine is:

$$V = 328.02 \times (\frac{\sqrt[3]{Q}}{R})^{1.809} \quad (3)$$

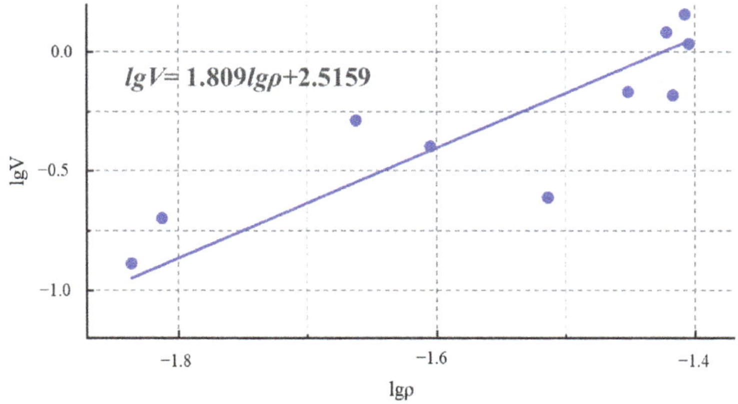

Figure 9. Fitting curve of the particle velocity of the blasting vibration.

Three blasting vibration tests were carried out on the blasting vibration measuring points. The following conclusions are drawn from the analysis. According to China's "Technical Code for Non-coal Open-pit Slope Engineering" (GB51016-2014) and *Technical Code for Safety Monitoring of Slope of Metallic and Non-metallic Open-Pit Mines* (AQ2063-2018), the particle vibration velocity of the slope should be less than 24 cm/s. According to the monitoring results, the maximum one-stage charge of the three blastings is 410 kg, and the maximum blasting vibration is measured to be 7.8745 cm/s in the vertical direction (vector combined velocity 8.8547 cm/s), which meets the requirements of the safety specifications.

5. Analysis of the Influence of the Underground Goaf Area Group on Slope Stability

5.1. Model Building

Due to the increasing depth of mining and the increasing difficulty of mining in large and medium-sized open-pit mines at home and abroad, the cost of open-pit mining has also increased. Therefore, more and more open-pit mines have begun to shift from open-pit mining to a combination of open-pit mining and underground mining. Underground mining is carried out above a certain mining depth, and open-pit mining is still used below the depth. The influence of the goaf left by underground mining on the stability of the slope cannot be ignored. The mined-out area near the slope of the Yinshan open-pit mine is mainly distributed in the south of the stope, which contains a total of thirteen partitions. The maximum width of the mined-out area is about 3 m, the maximum length is about 75 m, and the maximum height is about 35 m. In order to study the influence of the mined-out area on the slope stability of the Yinshan Mine, this chapter uses 3Dmine 2021 (Beijing, China) mining engineering software and Midas GTS NX 2021 (Seoul, Republic of Korea) software to complete the establishment of a three-dimensional slope model. The three-dimensional model is shown in Figure 10. FLAC3D three-dimensional numerical simulation software is used to analyze the influence of the mined-out area on the stability of the Yinshan open-pit slope under the conditions of the mined-out area group. FLAC3D is a fast 3D Lagrangian analysis program that uses explicit Lagrangian algorithms and hybrid discrete partitioning techniques to simulate the 3D mechanical properties of geotechnical or other geotechnical materials [32]. The analysis process is shown in Figure 11.

Considering the size effect, the size of the calculation model is 800 m × 1410 m × 774 m, and a total of 181,272 nodes and 1,032,056 meshes are generated. The final calculation model is shown in Figure 12. The calculation is divided into two phases: on the one hand, the influence of the goaf on slope stability under the current slope conditions is analyzed, while on the other hand, the influence of the goaf on slope stability in the later mining process is analyzed. Since the mined-out area group is located in the southern slope of the stope, the main control lithology is quartz porphyry and phyllite, so the numerical calculation model strata are dominated by these two kinds of lithology. The mechanical parameters of the rock mass are shown in Table 4.

Table 4. Mechanical parameters of the rock mass.

Lithologic Characters	Compressive Strength (MPa)	Poisson Ratio	Force of Cohesion (MPa)	Angle of Internal Friction (°)	Elastic Modulus (GPa)	Tensile Strength (MPa)
Quartz porphyry	114.06	0.341	1.806	35.75	36	1.06
Phyllite	64.43	0.334	1.001	23.25	24.9	0.93

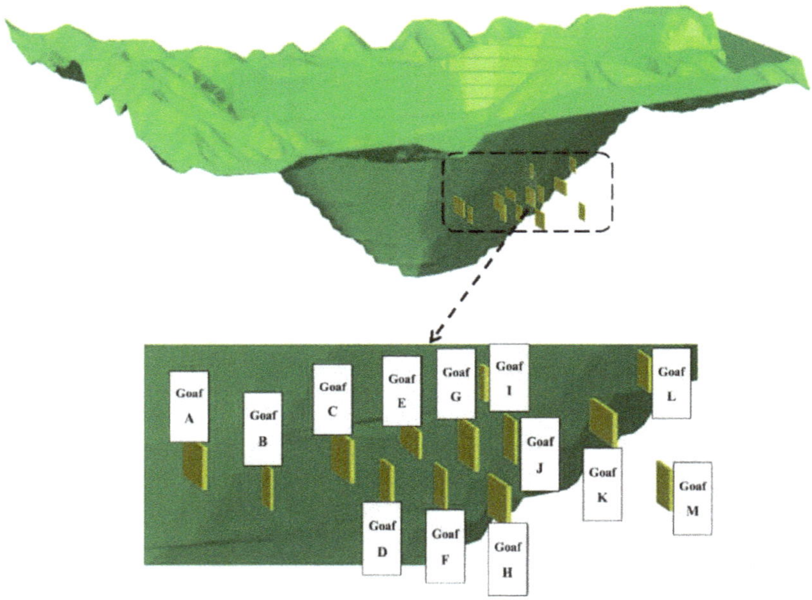

Figure 10. Three-dimensional model of the relative position between mine and goaf.

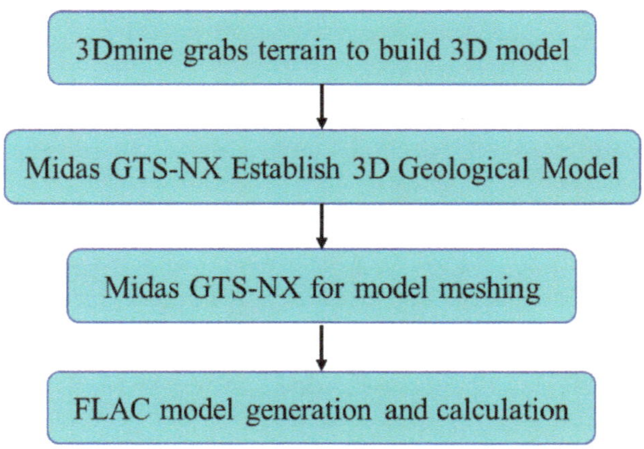

Figure 11. Modeling process.

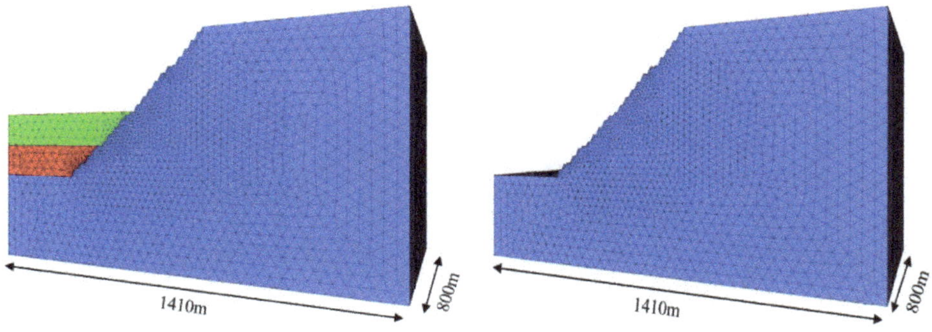

(a) **(b)**

Figure 12. Three-dimensional slope model. (**a**) Present slope mesh model. (**b**) Design excavation slope mesh model.

The model calculation considers the self-weight stress field and the initial stress in the three directions of X, Y, and Z according to the ratio of horizontal stress to vertical stress, and it solves the equilibrium under this condition. According to the selected mechanical parameters of rock mass, the elastic constitutive model is used to quickly balance the model. After the initial stress field is formed by solving the balance, the displacement is initialized, and then the Mohr–Coulomb constitutive model is used to complete the balance again. After that, the Null constitutive model is used for the subsequent excavation calculation. The boundary conditions are set as follows: the top of the model is a free surface, with normal constraints applied to the surrounding boundaries and full constraints applied to the bottom.

5.2. Numerical Simulation Analysis of Slope Stability

5.2.1. Numerical Simulation Analysis of the Current Slope Stability

Because there are many existing mined-out areas, the relative position of each mining area to the slope of the quarry is not the same, and the interaction between the mined-out areas and the influence degree of each mined-out area on the slope stability is not the same. This paper mainly studies the mined-out areas H and I, which have the most obvious influence on the current slope and the designed excavation slope.

The numerical calculation results of the profile of the current slope H mining hollow area and I mining hollow area are shown in Figure 13. The existence of I mining hollow area above H mining hollow area makes the plastic zone around the upper left corner of H mining hollow area extend upward to the bottom of the current slope, and the overall stability of the slope is less affected by the plastic zone penetration. However, in order to avoid further inducing the plastic zone to be penetrated in a large scale in the later mining, the mining area measures should be advanced to deal with the mining hollow area I and H, focusing on the stability of the upper left area of the H mining hollow area. From the vertical displacement cloud map, it can be seen that the vertical displacement gradually increases from the inside of the slope to the slope surface, and the displacement from the top of the slope to the foot of the slope also gradually increases. The maximum vertical displacement of the slope surface appears at the broken foot of the slope, which is 7 cm, and the impact of H and I mining areas on the internal displacement of the slope is limited. There is no large vertical displacement around the two mining areas.

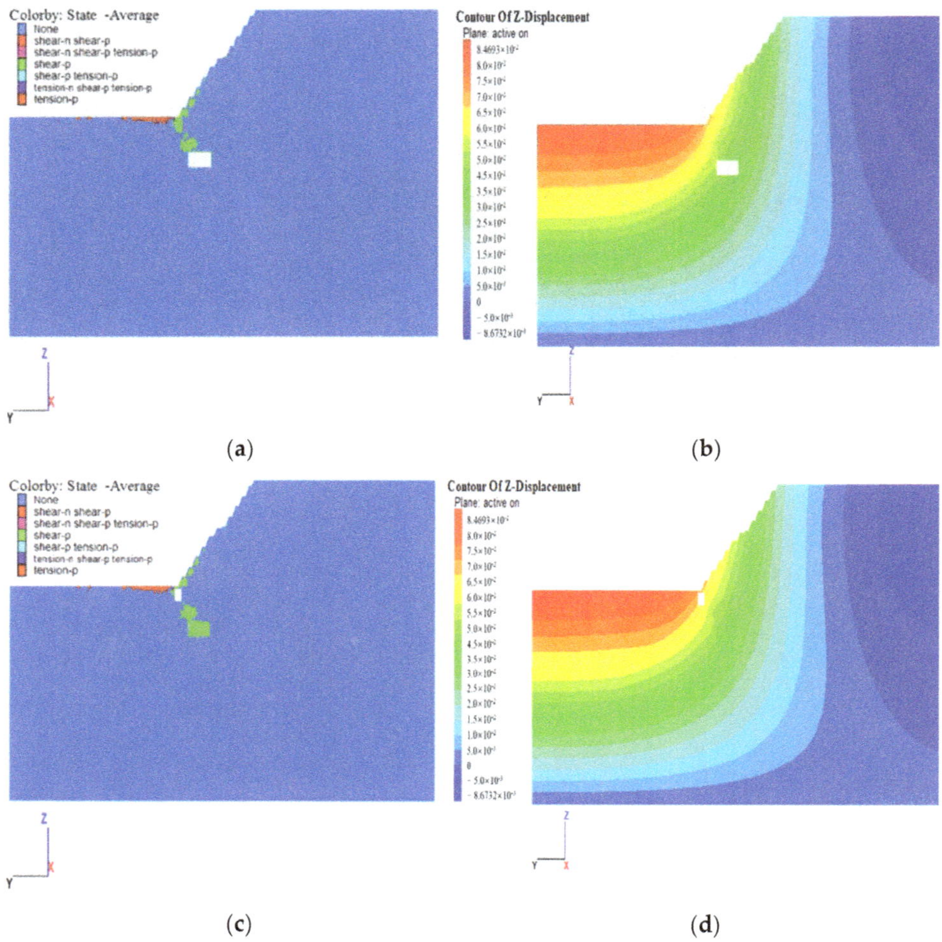

Figure 13. Numerical calculation result cloud of the current slope mining area H and mining area I profile. (**a**) H mining area plastic zone. (**b**) Vertical displacement cloud map of H mining area. (**c**) I mining area plastic zone. (**d**) Vertical displacement cloud map of I mining area.

5.2.2. Design Excavation Slope Stability Numerical Simulation Analysis

The results of the numerical calculation of the profile of H and I mining area of the design excavation slope are shown in Figure 14. The H and I mining areas interact with each other, making a plastic zone appearing around the two mining areas, and the plastic zone extends upward and penetrates the step slope. The plastic zone gathers at the foot of the design excavation slope, while the plastic zone does not appear or appears in a small amount at other locations. The vertical displacement of the designed excavation slope has increased compared with the current slope, and the maximum displacement appears at the slope surface near the foot of the slope, reaching 10 cm.

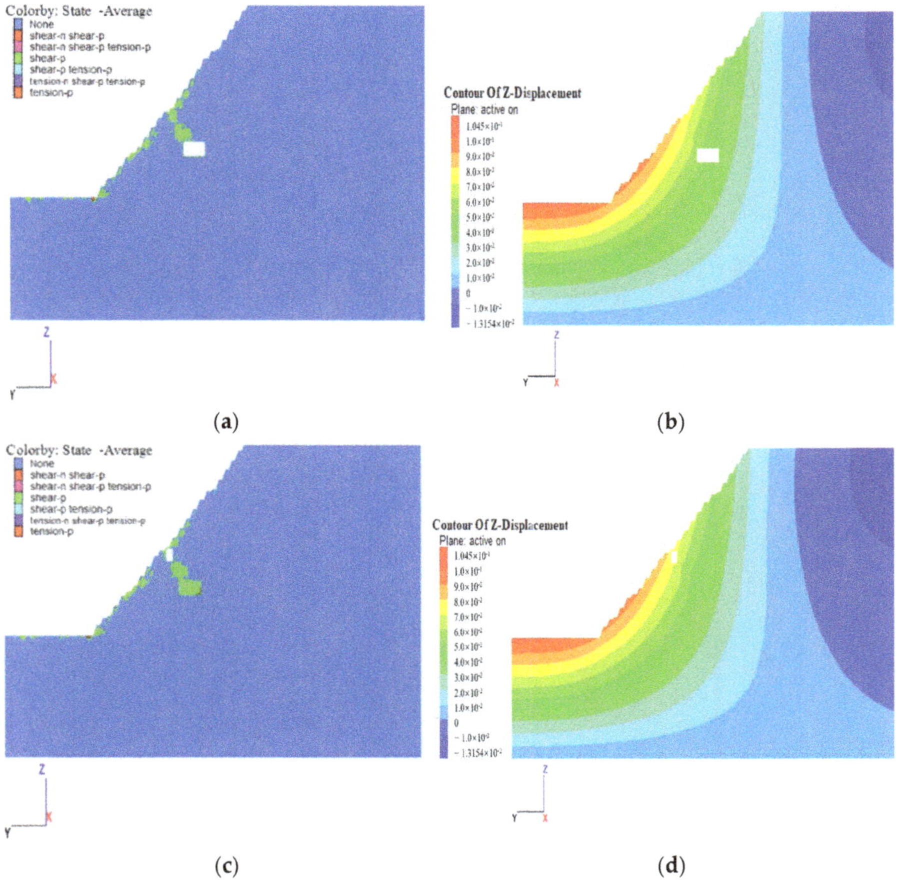

Figure 14. Numerical calculation result cloud of the design excavation slope mining area H and mining area I profile. (**a**) H mining area plastic zone. (**b**) Vertical displacement cloud map of H mining area. (**c**) I mining area plastic zone. (**d**) Vertical displacement cloud map of I mining area.

6. Safety Factor Calculation for the South Slope of the Open Pit

6.1. Stability Analysis Method

The limit equilibrium analysis method is one of the most commonly used methods for slope stability research. The limit equilibrium method includes the Fellenius method, the simplified Bishop method, the simplified Janbu method, and the Spencer method. The technical code for building slope engineering (GB50330-2013) recommends the simplified Bishop method for calculating slope stability. At the same time, a large number of practices have proved that the Bishop method has high accuracy, and it has become the most popular safety factor calculation method in engineering. Therefore, this paper uses the simplified Bishop method to analyze the slope stability of each partition of the Yinshan open-pit mine, using Rocscience Slide 6.0 software (Toronto, ON, Canada).

6.2. Calculation Load Combination and Calculation Method

According to the specific conditions of the Yinshan open-pit mine, two load combinations were analyzed and designed for this slope analysis. Load combination I is the self-weight of the slope, and load combination II is the self-weight of the slope + the blasting vibration force. In this analysis, the horizontal blasting vibration force of each slice can be calculated according to the following formula:

$$F_i' = \frac{\alpha_i \beta_i W_i}{g} \tag{4}$$

$$\alpha_i = 2\pi f V_i \tag{5}$$

In Formulas (4) and (5), F_i' is the horizontal equivalent static of the blasting vibration force in kN. α_i is the maximum horizontal acceleration of the blasting vibration particle in m/s^2, its value is shown in Table 5. β_i is the blasting vibration force coefficient, which may take values from 0.1–0.3. W_i is the weight of the slice in kN. g is the gravitational acceleration in m/s^2. f is the blasting vibration frequency in Hz. V_i is the horizontal vibration velocity of the particle at the center of gravity. The calculation method is shown in Formula (1), and the unit is m/s.

Table 5. Maximum horizontal acceleration data for blast vibration.

Measuring Point	Maximum Horizontal Acceleration of a Mass (cm/s^2)
A1	0.4063
A2	2001.3306
A3	186.2939
A4	56.9758
B2	85.2876
B3	134.2878
B4	19.2972
B5	31.4129
C1	61.8013
C2	243.3945
C3	571.9120
C4	306.6573
C5	324.5797

When considering the influence of blasting vibration force in slope stability software calculations, it is necessary to determine the influence coefficient of the blasting vibration force level ε :

$$\varepsilon = \frac{\alpha_i \beta_i}{g} \tag{6}$$

According to the calculation of the blasting vibration test results, the average blasting vibration force level influence coefficient of each zone is between 0.01 and 0.03, and the maximum value of 0.03 is used to calculate the slope stability under the action of the blasting vibration force.

6.3. Safety Factor Calculation of Mine Slope Stability

The section selected for the calculation of the safety factor of the mine slope is shown in Figure 15. The calculated safety factor of the current slope is shown in Figure 16. The safety factor of the designed excavation slope is shown in Figure 17. The calculated safety factor and the planned excavation slope parameters are shown in Table 6. According to the relevant provisions of China's *Non-ferrous Metal Mining Design Specification* (GB50771-2012) and *Technical Specification for Slope Engineering of Non-Coal Open-Pit Mines* (GB51016-2014), safety factor 1.20 is selected as the design stability safety factor. Combining the above limit equilibrium calculation results, it can be seen that:

(1) The safety factors of the current slope under combined load I and load combination II are 1.473 and 1.409, respectively, which meet the requirements of safety factor 1.20 specified in the code, meaning that the overall slope is stable.

(2) The safety factor of the designed excavation slope is 1.130 when only the self-weight load is considered, and the safety factor after considering the self-weight load and blasting vibration is 1.081, which does not meet the requirements of the safety factor of 1.20. Because of the final slope angle or slope height in this area, and because faults and goafs occur under some slopes in this area, there is a certain risk of instability.
(3) The bottom of the final slope of the Yinshan open-pit mine is affected by the fault and the original underground mined-out area. The slope rock mass structure is broken, and prone to wedge failure and local collapse.

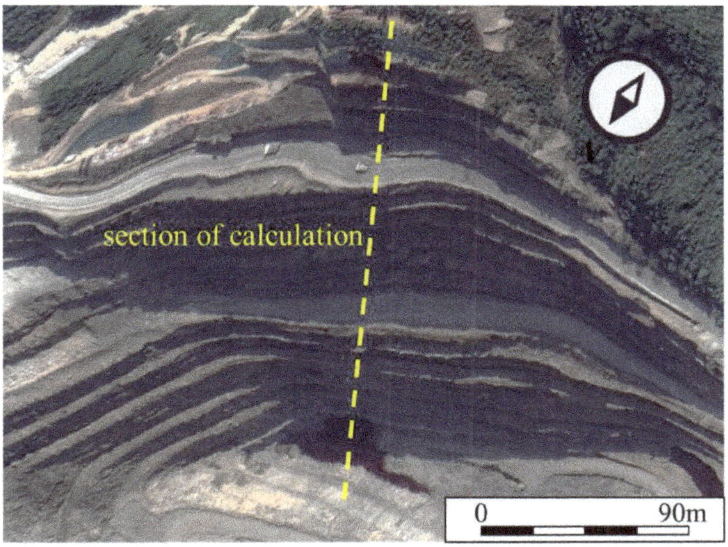

Figure 15. Safety factor calculation section.

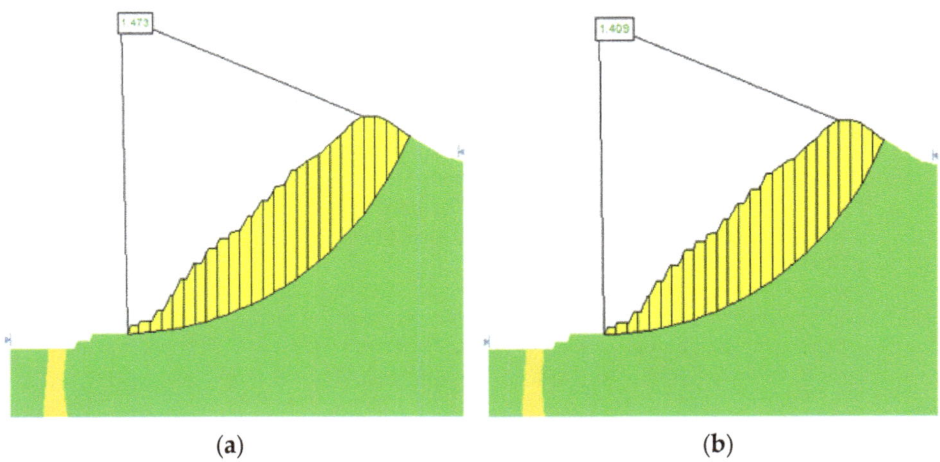

Figure 16. Current slope safety factor. (**a**) Load combination I. (**b**) Load combination II.

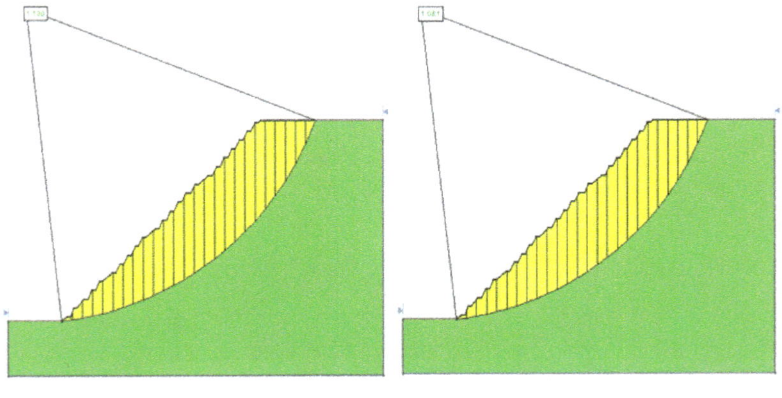

(**a**) (**b**)

Figure 17. Planned excavation slope safety factor. (**a**) Load combination I. (**b**) Load combination II.

Table 6. Planned excavation slope safety factor.

Present Safety Slope Coefficient		Planned Excavation Slope Safety Factor		Planned Excavation Slope Parameters	
Load Combination I	Load Combination II	Load Combination I	Load Combination II	Final Slope Angle (°)	Maximum Height (m)
1.473	1.409	1.130	1.081	44.8	540

7. Discussion

This paper presents the results of a geological survey conducted on the southern slope of the open-pit slope of the Yinshan Mine, utilizing UAV mapping technology and the digital structural surface recognition method. Additionally, a production blasting vibration test was performed to analyze the influence of blasting vibration on the rock mass of the slope. Using FLAC3D software, the impact of the underground goaf on the stability of the slope was analyzed, and the limit balance analysis method was applied to evaluate the status quo of the open-pit slope and the stability of the design excavation slope. The following results were obtained:

(1) The southern slope of the mining area primarily exhibits a medium-large inclination angle structural surface. The development of thousands of rock fragments in each area was traced from the statistical results of joint fractures in the entire measurement site, where the average joint spacing was found to be 5–30 m, and the average joint density was 0.033–0.2/m.

(2) A total of three blasting vibration tests were conducted using the blasting vibration monitoring system, in accordance with the specification. Based on the monitoring results, the maximum one-stage charge of the three blasts was 410 kg, and the measured value did not exceed 24 cm/s. The analysis of the measured data allowed for the blasting vibration attenuation law for the Yinshan Mine to be obtained, and the vibration velocity attenuation law formula was fitted.

(3) According to the three-dimensional finite element analysis, the entire slope exhibits no large range of plastic zone damage, with only a small range of plastic zone generated on the step slope above the location of the empty area, which is mainly shear failure.

(4) A typical section was selected to carry out the limit equilibrium analysis. Load combination I (considering only the self-weight stress) and load combination II (which includes the blasting vibration force) of the current slope meet the safety factor of 1.20, and the overall slope is stable. However, the designed excavation slope is near the

critical value of the safety factor specified in the code under the conditions of load combination I and load combination II, and, thus, poses a certain risk of instability.

8. Conclusions and Recommendations

In this paper, we analyzed the engineering geology of the mining area of Yinshan open pit in detail, and carried out on-site engineering geological investigation work by using small low-speed UAV and image reconstruction and interpretation. We carried out a production blasting vibration test in the Yinshan Mine, obtained the blasting vibration attenuation law, and analyzed the impact of blasting vibration on the slope rock. We analyzed the stability of the slope of the mining site by using FLAC3d software and the limit equilibrium method. Based on the analysis of the data, the following conclusions and recommendations were made on the stability of the south slope of the open pit of the Yinshan Mine:

(1) The main lithology of the mining area includes phyllite, quartz porphyry, and Quaternary residual deposits. The rock integrity of the slope ranges from "broken" to "more complete". There are three faults in the geological investigation area, which are combined with joints and fissures, constituting cutting and sliding surfaces in the rock mass, and easily forming wedges and other conditions favorable to collapse and sliding. It is suggested to pay close attention to the distribution of the newly revealed fault fragmentation zone and dense structure surface, further analyze and summarize the destabilization and damage characteristics of the slope, and, if necessary, take reinforcement measures for the potentially dangerous slope body or partially slow down the slope.

(2) The maximum one-stage charge of the three blasts was 410 kg, and the maximum blast vibration was measured as 7.8745 cm/s in vertical direction (vector combined velocity 8.8547 cm/s), which did not exceed 24 cm/s and met the safety requirements. It is recommended that pre-cracking blasting and light surface blasting technology should be used when blasting against the gang slope adjacent to the final boundary of the quarry or the permanent slope, optimizing blasting parameters and further strengthening the research on vibration reduction control technology.

(3) According to the results of numerical simulation analysis, the top plastic zones of mining area H and mining area I are penetrated with the upper step slope, suggesting that damage may occur to the slope at local locations due to the impact of the mining area. The influence of the dense area of the empty zone is more obvious. According to the result of the ultimate equilibrium analysis, the current slope meets the requirements of the safety code when considering only the self-weight and taking into account the blasting vibration force load, and the slope stability is good. However, the design excavation slope is close to the critical value of the safety coefficient specified in the code under both combinations of load, and there is the risk of instability. It is suggested that the final slope angle of some slopes should be optimally adjusted during the design mining process, and relevant technical measures should be taken to over-detect and timely deal with the potential mining void area to effectively guarantee production safety.

Author Contributions: Software, P.L. and X.Q.; Investigation, C.W.; Writing—original draft, H.C. and G.M.; Writing—review & editing, J.L. All authors have read and agreed to the published version of the manuscript.

Funding: This research was funded by the Fundamental Research Funds for the Central Universities, grant number 06500229.

Institutional Review Board Statement: Not applicable.

Informed Consent Statement: Not applicable.

Data Availability Statement: The original contributions presented in the study are included in the article. Further inquiries can be directed to the corresponding authors.

Conflicts of Interest: The authors declare no conflict of interest.

References

1. Yang, T.H.; Zhang, F.C.; Yu, Q.L.; Cai, M.F.; Li, H.Z. Research situation of open-pit mining high and steep slope stability and its developing trend. *Rock Soil Mech.* **2011**, *32*, 1437–1451+1472. [CrossRef]
2. Du, S.G. Method of equal accuracy assessment for the stability analysis of large open-pit mine slopes. *Chin. J. Rock Mech. Eng.* **2018**, *37*, 1301–1331. [CrossRef]
3. Yang, T.H.; Wang, H.; Dong, X.; Liu, F.Y.; Zhang, P.H.; Deng, W.X. Current situation, problems and countermeasures of intelligent evaluation of slope stability in open pit. *J. China Coal Soc.* **2020**, *45*, 2277–2295. [CrossRef]
4. Zhong, C.B.; Zhou, Y.M.; Gao, P.T. The Influencing Factors and Control of Slope Stability in Open-pit Mine of Qinghai-Tibet Plateau. *Saf. Coal Mines* **2012**, *43*, 203–206. [CrossRef]
5. Wang, J.Q.; Li, J.; Li, Q.; Chen, L. Analysis of influence factors of high slope stability of loess: Taking the Baojixia Water Division Project for example. *Rock Soil Mech.* **2009**, *30*, 2114–2118. [CrossRef]
6. Sun, P.C.; Lu, W.B.; Lei, Z.; Chen, M.; Li, R.Z.; Li, F.Q. Blasting vibration response and control of high rock slopes of thin mountain. *Chin. J. Geotech. Eng.* **2021**, *43*, 877–885. [CrossRef]
7. Hu, Y.G.; Wu, X.X.; Zhao, G.; Liu, M.S.; Li, P.; Lu, W.B. Determination of safety control standard for high rock slopes under blasting vibration. *Chin. J. Rock Mech. Eng.* **2016**, *35*, 2208–2216. [CrossRef]
8. Zhou, H.Y.; Chi, E.A.; Ouyang, T.Y.; Yu, H.K.; Gao, Z.H. Analysis of Stability of Open Slope under Blasting Load. *Blasting* **2021**, *38*, 80–87. [CrossRef]
9. Zhao, J.J.; Xiao, J.G.; Xiang, X.Q.; Huang, R.Q.; Wang, Y.C.; Shi, W.B. Failure mechanism numerical simulation of mining landslide with gentle bedding coal strata. *J. China Coal Soc.* **2014**, *39*, 424–429. [CrossRef]
10. Wang, C.Y.; Li, J.P.; Liu, W.; Zhang, Q. Study on the Stability and Timeliness of Slope Mining by the Combination of Open Pit and Underground. *Min. Res. Dev.* **2017**, *37*, 107–110. [CrossRef]
11. Xie, L.K.; Xiong, D.Y.; Yang, T.H.; Wan, C.C.; Hu, J.J. Analysis on disaster impact of concealed goaves underneath open-pit slope and its treatment. *Chin. J. Nonferrous Met.* **2020**, *30*, 2505–2512.
12. Yin, G.Z.; Li, X.S.; Li, Y.J. Simulation on the deformation and failure response features and stability of a slope from open pit mining to underground mining under the effecting of excavation goaf by the floor friction model. *Chin. J. Eng.* **2012**, *34*, 231–238. [CrossRef]
13. Jia, S.G.; Jin, A.B.; Zhao, Y.Q. Application of UAV oblique photogrammetry in the field of geology survey at the high and steep slope. *Rock Soil Mech.* **2018**, *39*, 1130–1136. [CrossRef]
14. Dong, X.J.; Huan, R.Q. Application of 3D laser scanning technology to geologic survey of high and steep slope. *Chin. J. Rock Mech. Eng.* **2006**, *2006*, 3629–3635. [CrossRef]
15. Huan, J.; Shi, Y.C.; Ji, F. Discussion on the Application of 3-D Laser Scanning Technology to the Investigation of High Slop Perilous Rockmass. *J. Yangtze River Sci. Res. Inst.* **2013**, *30*, 45–49. [CrossRef]
16. Yang, L.L. Application of Aerial Photogrammetry Based on Light and Small Unmanned Aerial Vehicle in Geometric Information Survey of High and Steep Slope. Master's Thesis, Southwest Jiaotong University, Chengdu, China, 3 May 2017.
17. Fang, K. Application of aerial photography technology in topographic map repairing and surveying engineering. *Fujian Build. Mater.* **2018**, *5*, 47–48.
18. Wang, G.; Jiang, R.B.; Xiao, H.H.; Zhang, D. Research on slope reconstruction technique based on UAV oblique photogrammetry. *China Min. Mag.* **2017**, *26*, 158–161.
19. Vasuki, Y.; Holden, E.J.; Kovesi, P.; Peter, K.; Steven, M. Semi-automatic mapping of geological Structures using UAV-based photogrammetric data: An image analysis approach. *Comput. Geosci.* **2014**, *69*, 22–32. [CrossRef]
20. Michele, C.; Rita, A. Advanced 3D Photogrammetric Surface Reconstruction of Extensive Objects by UAV Camera Image Acquisition. *Sensors* **2018**, *18*, 2815. [CrossRef]
21. Jin, A.B.; Chen, S.J.; Zhao, A.Y.; Sun, H.; Zhang, Y.S. Numerical simulation of open-pit mine slope based on unmanned aerial vehicle photogrammetry. *Rock Soil Mech.* **2021**, *42*, 255–264. [CrossRef]
22. Zhou, Z.H.; Chen, Z.H.; Wang, J.M.; Zhang, L.F.; Nian, G.Q. Catastrophe analysis of open-pit slope stability under blasting load. *Rock Soil Mech.* **2020**, *41*, 849–857. [CrossRef]
23. Wu, X.X.; Rao, Y.; Hu, Y.G.; Cai, C.Z. Determination method of control standard under blasting vibration based on stop stability. *Eng. Blasting* **2021**, *27*, 11–18. [CrossRef]
24. Zhang, H.; Wu, Z.X.; Xu, S.; Wu, C.F.; Lv, Q. Influence of Blasting Damage Zone on Slope Stability based on Hoek-brown Criterion. *Blasting* **2022**, *39*, 134–139.
25. Cui, T.J.; Ma, Y.D.; Wang, L.G. Simulated study on the influence of the slope blasting height on the slope stability. *J. Saf. Environ.* **2017**, *17*, 896–900. [CrossRef]
26. Yang, X.J.; Geng, Q.; Liu, C.K.; Peng, W.Y.; Song, Z.G.; Zhao, D.D. Numerical Analysis for the Influence of Lower Goaf on the Slope Stability. *Min. Res. Dev.* **2018**, *38*, 71–74. [CrossRef]
27. Li, W.; Ren, P.; Li, Q.Y.; Guo, X.F. Study on instability mechanism of red clay slope under the influence of goat. *Saf. Coal Mines* **2021**, *52*, 237–240. [CrossRef]
28. Sun, S.G.; Zhang, Y.J.; Zhang, Y.H.; Guo, W.C. Research on Influencing Mechanism of the Synchronous Mining of Open-pit and Underground on Slope Deformation. *Met. Mine* **2016**, *2016*, 58–62.

29. Shi, F.W.; Xiao, Y.T.; Yao, G.H.; Cao, H. Safety Evaluation and Countermeasures of the Open-pit Mning under the Goaf Occurrence. *Mod. Min.* **2019**, *35*, 130–134.
30. Westoby, M.J.; Brasington, J.; Glasser, N.F.; Hambrey, M.J.; Reynolds, J.M. 'Structure-from-Motion' photogrammetry: A low-cost, effective tool for geoscience applications. *Geomorphology* **2012**, *179*, 300–314. [CrossRef]
31. Kong, D.H.; Wu, F.Q.; Saroglou, C. Automatic identification and characterization of discontinuities in rock masses from 3D point clouds. *Eng. Geol.* **2020**, *265*, 105442. [CrossRef]
32. Xie, J.H.; Yu, B.X.Z.H.; Zhang, Y.H. Brief introduction of numerical simulation software flac and its application on geosciences. *Geol. Explor.* **2005**, *2*, 77–80.

applied sciences

MDPI

Article

Study on the Characteristics of Pore Change in Tuff under the Frost and Salt Action Using High-Precision CT Scanning Equipment

Lilei Chen [1], Yue Sun [1], Chao Wang [1], Peng Sha [1], Huijun Jin [2,3], Minghao Liu [2] and Anyuan Li [1,3,*]

1 Key Laboratory of Rock Mechanics and Geohazards of Zhejiang Province, Shaoxing University, Shaoxing 312000, China; chenlilei2020@126.com (L.C.); sunyue_usx@163.com (Y.S.); wangchao_usx@126.com (C.W.); shapeng@usx.edu.cn (P.S.)

2 State Key Laboratory of Frozen Soils Engineering, Northwest Institute of Eco-Environment and Resources, Chinese Academy of Sciences, Lanzhou 730000, China; hjjin@nefu.edu.cn (H.J.); liuminghao@lzb.ac.cn (M.L.)

3 School of Civil Engineering, Institute of Cold-Regions Science and Engineering, Northeast Forestry University, Harbin 150040, China

* Correspondence: anyuanli1013@163.com

Abstract: Using high-precision CT scanning equipment, two series of tests on frost and salt weathering were conducted to investigate the characteristics of pore change in tuff. Experiments on frost and salt aging were performed with pieces of tuff from the same area of southeast China. One set of tuff samples was soaked in saturated sodium sulfate or magnesium sulfate solutions for 60 days. Another set of tuff samples were subjected to 60 freeze–thaw cycles after being submerged in saturated sodium sulfate or magnesium sulfate solutions for 48 h. Our study demonstrates that processes such as salt erosion and freeze–thaw affect the pore evolution of tuffs significantly. Tuff lost 1.56% of its mass after being submerged in magnesium sulfate solutions for 60 days, while tuff submerged in sodium sulfate solutions gained a negative 0.33% of its mass. After 60 freeze–thaw cycles, the mass loss of tuff samples immersed in sodium sulfate, magnesium sulfate, and distilled water solutions was 3.52%, 3.58%, and 3.82%, respectively. The average porosity of the magnesium sulfate and sodium sulfate test groups increased by 6.59% and 4.14%, respectively, when the number of days of salt erosion was extended from 10 to 60 days. The average porosity of tuff samples immersed in magnesium persulfate and sodium sulfate solutions increased by 2.25% and 2.18%, respectively, as the number of freeze–thaw cycles went from 10 to 60.

Keywords: tuff; frost weathering; salt weathering; pore; high-precision CT

1. Introduction

The frequency and intensity of extreme weather events (typhoons, heavy rainfall, etc.) are increasing in southeast China against the backdrop of global warming. This has accelerated the weathering process of stone and caused a large amount of deterioration on the surface of many ancient stone bridges (Figure 1). The deterioration of historical stone bridges in coastal regions can be attributed to two primary factors: salt weathering and frost weathering [1,2]. On the one hand, the salt that is present in the ocean gradually seeps into the rock throughout the typhoon and rainy season through the micro-cracks. The process of salt weathering results in the generation of crystallization pressure by the salt, leading to the deterioration of the stone [3,4]. On the other hand, when liquid water freezes into ice in the cracks inside rocks in winter, a 9% volume expansion induced by the phase transition of water to ice can put enough pressure on the rock to worsen the fracture and eventually destroy the boulder [5–8].

Figure 1. Surface weathering and deterioration of tuff in the Bazi Bridge in Shaoxing City.

During the past 30 years, year-round field observations or monitoring of rock deformation have revealed that air temperature, moisture content, and salt weathering are significant environmental factors that cause stone degradation in the natural environment [8–12]. The researchers carried out a number of indoor simulation tests in order to undertake a quantitative analysis of the effects of freeze–thaw cycles and salinity on the processes of rock weathering [13]. Typically, frost or salt weathering of a particular stone or rock is evaluated quantitatively by measuring the changes in sample weight, compressive strength, porosity, and other physical parameters under different experimental conditions [14–18]. For example, Ref. [19] performed static and impact mechanical tests on both weathered and un-weathered red sandstones. Furthermore, they developed a mechanical property decay model that incorporated the strain rate effect to better predict the dynamic mechanical degradation of sedimentary rocks following long-term frost weathering. Based on the idea that water freezing and moving through micro-pores induces micro-pore rupture, Ref. [20] developed a micromechanical model of micro-pores, provided an analytical approach to forecast the degradation degree of micro-pores, and established a finite element model for fracture analysis. Twenty dry–wet cycles at temperatures above 0 °C in a laboratory showed that samples of deionized water and sodium chloride remained intact, but those of sodium sulfate crumbled [21]. Even though these studies have advanced our understanding of rock weathering, there have been few quantitative investigations on the evolution of interior pores during coupled frost and slat weathering actions [22].

Porosity is one of the key indicators used to evaluate the extent of rock weathering [23]. Mercury injection test results suggested that micro-pores smaller than 0.05~0.1 μm in radius play a significant role in salt weathering of eight Japanese building stone varieties [24]. However, rocks can have pores of a wide variety of sizes and shapes, which leads to a wide range of possible variations, which in turn results in pores that are exceedingly complicated and irregular in structure. As a result, it is challenging to perform quantitative research on rocks in order to investigate their internal pore structure [25]. Due to their non-destructive nature and ability to reconstruct the 3D spatial structure of objects, X-ray computed microtomography or micro-CT have been widely used for observing and quantifying the change in the internal structure of materials under different conditions in recent years [26]. For instance, based on the results of CT detection, Ref. [27] found that the strength and density of rock samples decreased after freezing compared to before freezing. Using interior CT scanning technology to scan and model rock samples before and after freeze–thaw, Ref. [15] investigated the mechanism of damage influence of freezing speed and temperature on the internal structure of rock and the evolution of internal structure damage. Although these studies have laid the framework for the analysis of the internal pore evolution of rocks, there have been few 3D spatial structure evolution analyses of internal pores under coupled frost and salt action.

In this research, a series of freeze–thaw cycles and salt erosion tests were performed on tuff samples. Then, we used industrial CT with high precision to scan and reconstruct the three-dimensional spatial structure of tuff samples under various test conditions. Furthermore, a quantitative analysis was conducted on the evolution of the internal pore structure of tuff under various experimental conditions. The aim of this analysis was to establish a theoretical foundation for the restoration of deteriorating ailments observed in ancient stone tuff bridges.

2. Methods and Materials

The test rock came from Jinhua City, Zhejiang Province. It was mostly made of tuff. All rocks are carved into 100 mm in height and 50 mm in diameter cylindrical specimens. Two distinct series of tests were conducted to examine the pore changes and evolution that emerges from the salt and frost weathering processes in tuff samples. The first batch of tuff samples were immersed in saturated sodium or magnesium sulfate solutions for 60 days to simulate salt weathering. The experimental tuff samples were treated as follows: tuff samples were soaked in one of two saturated salt solutions or distilled water at 25 °C and oven-dried at 120 °C for 24 h. The second group of tuff samples were immersed for two days in saturated solutions of sodium sulfate, magnesium sulfate, and distilled water. Then, each sample was placed in a cabinet set to −20 °C. After 12 h, the samples were taken out of the cabinet and left to defrost for an additional 12 h at room temperature (25 °C). The freeze–thaw cycle was repeated until sixty cycles had been completed. All of the samples in the first and second group were weighted and X-rayed every 10 days or freeze-thaw cycles. (Figure 2).

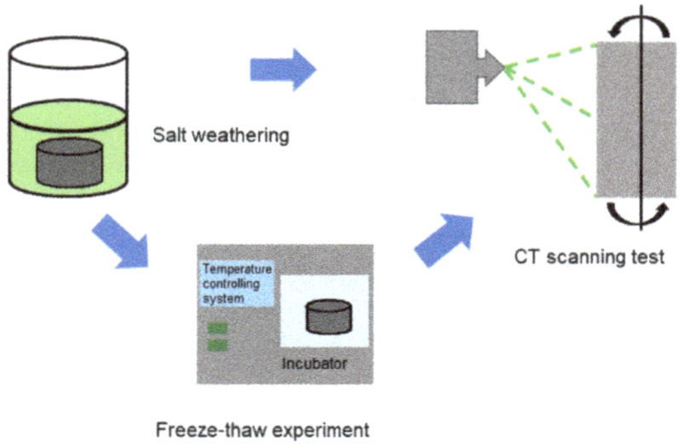

Figure 2. Schematic diagram of test system of frost and salt weathering of tuff.

A high-precision (5 µm) micro-CT scanning system (GE manufactures) including an X-ray detector, carrier rotating stage, two ray emitters, and a data processing system was used to scan tuff samples (Figure 3). Each sample group was divided into 1200 cross-sections for a 45-minute scan to assure 3D reconstruction accuracy. VG Studio Max, one of the most advanced CT image post-processing software platforms, was used to precisely measure, analyze, and calculate 3D volume data sets from the 1200 images captured by the scanner [25]. For the analysis of the evolution of the 3D spatial structure, we only selected the middle portion of the specimen (20 mm in diameter, 60 mm in height) to reduce the influence of the specimen boundary on the results (Figure 3). Since the CT scanner has a spatial resolution of 0.03 mm × 0.03 mm and a scanning layer thickness of 0.08 mm, each pixel cell in the image represents the porosity of 0.03 × 0.03 × 0.08 mm^3 cube cells. All of the scanned images were analyzed with the use of the software VG Studio Max, which performed operations such as binaryzation, segmentation, and normalization [25]. Pores were represented in the scanned images by gray values that were more than eight, while particle skeletons were represented by the other values. To quantitatively investigate the change and evolution in the pores, it was necessary to define the porosity of the samples. The calculation of porosity involved the division of the volume of identified pores by the overall volume of the selected region within the three-dimensional model of the specimen.

The determination of the porosity of a sample's cross-section involves calculating the ratio of the total area occupied by recognized pores to the entire area of the specimen section. In our paper, we only focused on the microscopic scale of the internal pores of pumice [28]. In total, four experimental runs were conducted with diverse salt and freeze–thaw cycles to examine the pore change in tuff during frost and salt weathering processes (Table 1).

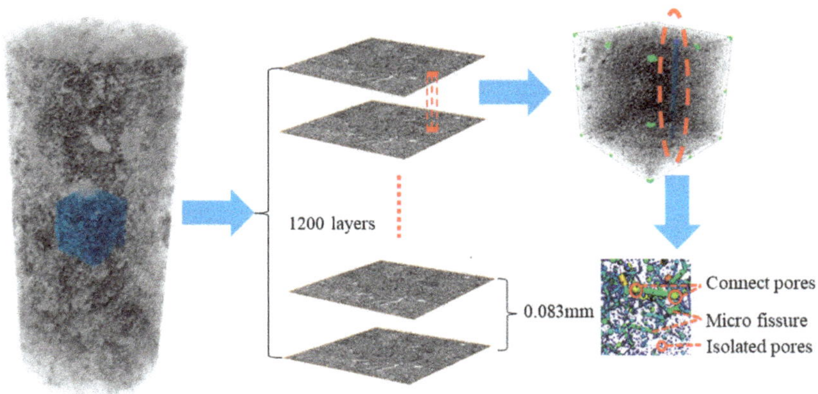

Figure 3. Schematic diagram of CT scanning and 3D reconstruction of tuff sample.

Table 1. Initial experiment conditions.

Rock Type	Sample Size (mm)	Experiment Content	Experiment Condition	Test Number	Initial Mass (g)
Tuff	φ50 × 100	Salt weathering (25 °C)	Distilled water	N1	482.12
			Sodium sulfate	N2	485.05
			Magnesium sulfate	N3	484.13
		Coupled frost and salt weathering (−20~20 °C)	Distilled water and freeze–thaw cycles	N4	484.50
			Sodium sulfate and freeze–thaw cycles	N5	487.60
			Magnesium sulfate and freeze–thaw cycles	N6	483.00

3. Results and Discussion

Among the two forms of salt weathering, $MgSO_4$ caused the greatest amount of damage and mass loss to tuff samples. Figure 4 demonstrates that as the time of erosion increased, the mass of tuff samples soaked in distilled water and sodium sulfate changed only slightly, whereas the mass loss of tuff samples soaked in magnesium sulfate reached ~10 g. The rate of mass loss was 1.56%. The results suggest that the presence of magnesium sulfate has a notable impact on both the expansion and dissolution of the tuff sample. The primary factor contributing to this phenomenon can be attributed to the process of salt crystallization [29]. The mass of the samples in each group of tests decreased as the freeze–thaw cycles increased in the second series of tuff samples. It revealed that the freeze–thaw cycle caused significant damage to the tuff. After 60 freeze–thaw cycles, the mass loss of tuff in the distilled water, sodium sulfate, and magnesium sulfate groups were 18.5 g, 17.2 g, and 17.3 g, respectively. The mass loss rates of tuff in the solutions of distilled water, sodium sulfate, and magnesium sulfate were 3.82%, 3.52%, and 3.58%, respectively. The observed mass loss rates of specimens immersed in sodium sulfate and magnesium sulfate solutions were marginally lower compared to those immersed in distilled water. This discrepancy may be attributed to the presence of salt crystals adhering to the specimen's surface, potentially obstructing certain micro-pores, and consequently reducing moisture infiltration into the rock specimen [30,31].

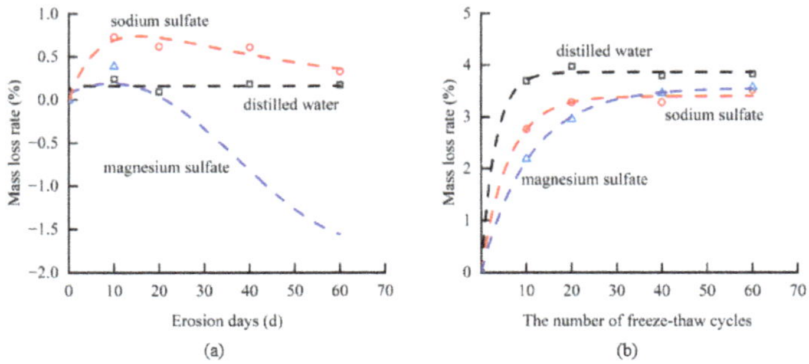

Figure 4. Mass change in tuff samples under frost and salt weathering (**a**) salt weathering (**b**) coupled salt and frost weathering.

The evolution of the cross-sectional morphology of tuff as a result of salt weathering is depicted in Figure 5. After 10 days, the outer edge of the cross-section does not change significantly. When the number of days of erosion reached 40, the outer margin of the cross-section appeared fragmented due to the strong chemical reaction between the magnesium sulfate solution and the mineral element in tuff, which caused partial dissolution of the outer edge. After 60 days, the cross-section of tuff that had been soaked in sodium sulfate solution revealed severe damage along its upper edge, and a fracture appeared in the lower right region (Figure 5f). Almost the whole edge of the cross-section of tuff that had been soaked in magnesium sulfate solution was damaged after 60 days of erosion (Figure 5c).

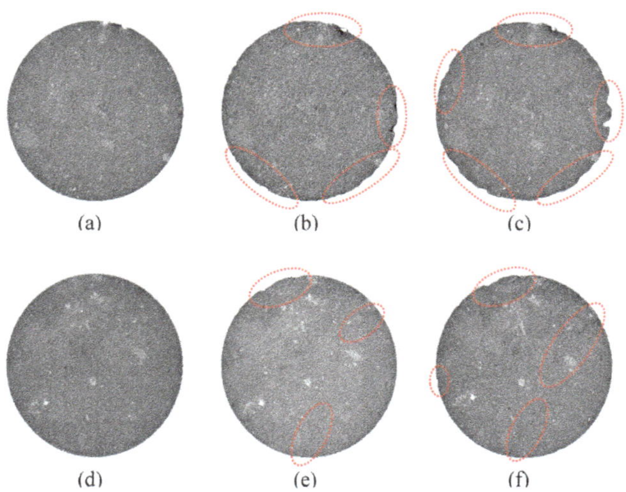

Figure 5. The evolution of the cross-section over time under salt weathering (**a**–**c**) samples soaked into magnesium sulfate solution for 10, 40, 60 days; (**d**–**f**) samples soaked into sodium sulfate solution for 10, 40, 60 days.

To analyze the change and evolution in pore characteristics quantitatively, all layers of scanning images were used to reconstruct the 3D model of the central pore distribution of tuff specimen. In Figure 6, the spectrum of porosity from 0 to 0.3 is represented by the color bar, with red representing locations with high porosity and blue representing regions with low porosity. The pore distribution of the sample exhibits notable heterogeneity in Figure 6. The primary factor contributing to this phenomenon is the diverse composition of the tuff. With an increase in the number of days and freeze–thaw cycles, the areas of high porosity in the specimen increase in each group, indicating that frost and salt weathering have a significant impact on the development of internal pores in the tuff.

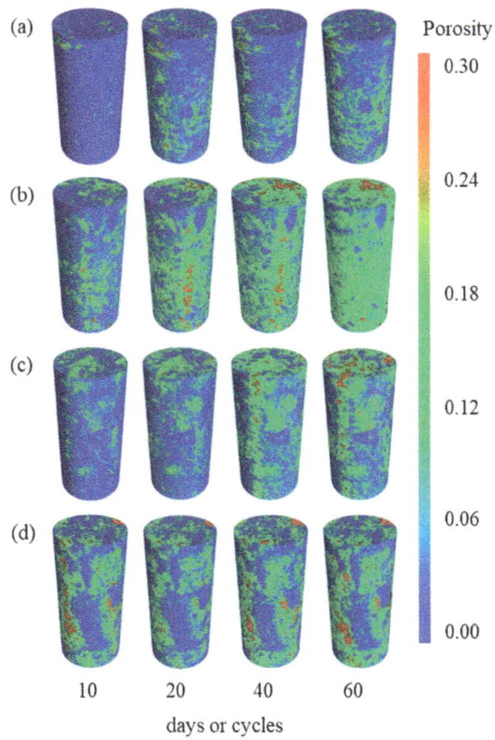

Figure 6. 3D Pore distribution of tuff (**a**) N2 group, (**b**) N3 group, (**c**) N5 group, and (**d**) N6 group.

To quantify the cross-section pore characteristics, we calculated the porosity of each cross-section of the tuff during the salt and frost weathering process (Figure 7). With an increase in erosion days and freeze–thaw cycles, the porosity of the entire tuff cross-section progressively increased. When the number of erosion days increased from 10 to 60, the maximal porosity change in all cross-sections in the sodium sulfate and magnesium sulfate groups was 3.42% and 5.54%, respectively. As the number of freeze–thaw cycles increased from 10 to 60, the greatest value of porosity change in the sodium sulfate and magnesium sulfate groups was 3.77% and 3.35%, respectively. The average porosity of all the cross-sections in the magnesium sulfate and sodium sulfate test groups increased by 6.59% and 4.14%, respectively, when the number of days of salt erosion was extended from 10 days to 60 days. The average porosity of all the cross-sections in the magnesium sulfate and sodium sulfate test groups increased by 2.25% and 2.18%, respectively, as the number of freeze–thaw cycles increased from 10 to 60.

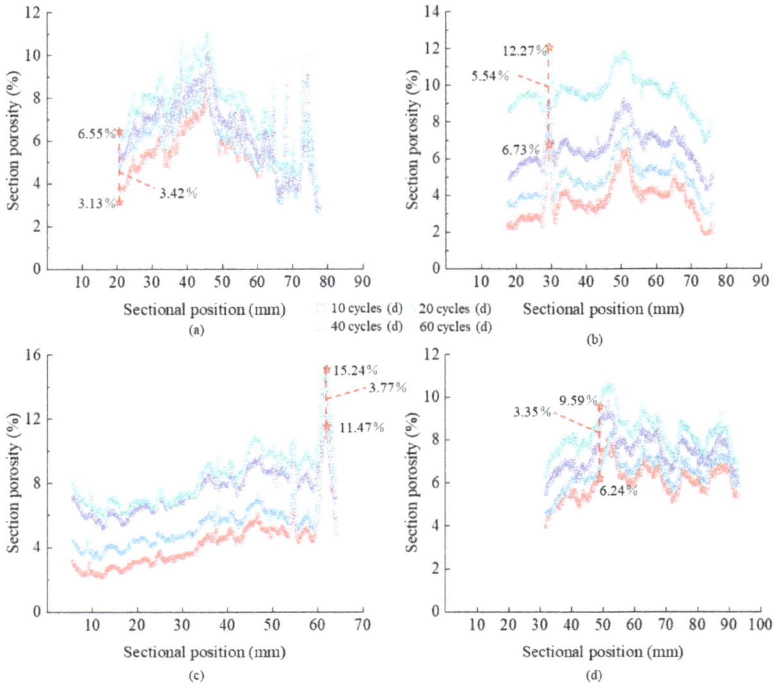

Figure 7. Cross-section porosity (**a**) N2 group, (**b**) N3 group, (**c**) N5 group, and (**d**) N6 group.

To examine and describe the variations in pore properties of the tuff, a statistical analysis was performed on the distribution of porosity obtained from the CT scanning images of three distinct cross-sections inside each sample. These cross-sections were selected for analysis because of their noticeable differences in porosity, as illustrated in Figures 8–11. Porosity is categorized into three distinct ranges: high porosity area, medium porosity area, and low porosity area, which correspond to values of 0.18~0.3, 0.06~0.18, and less than 0.06, respectively. As shown in Figure 8, after 10 days of erosion, the cross-sections at all three locations are dominated by areas with low porosity. The central local area of the cross-section at Z = 24 mm changed from a low porosity zone to an elongated high porosity zone band as the erosion days increased from 10 to 20. At Z = 46 mm and Z = 74 mm, there are numerous high porosity regions in the area to the lower right of the cross-section and in the area directly above the cross-section, respectively. As the number of erosion days increased from 10 to 60, the distribution of low porosity areas decreased significantly on the 24 mm, 46 mm, and 74 mm cross-sections, while the distribution of high porosity area increased significantly. The porosity of the section increased from 5.22%, 8.49%, and 8.96% to 8.14%, 11.05%, and 10.11%, respectively (Figure 8a). It showed that the saturated sodium sulfate solution corroded the tuff, destroyed its internal pore structure, and increased its porosity.

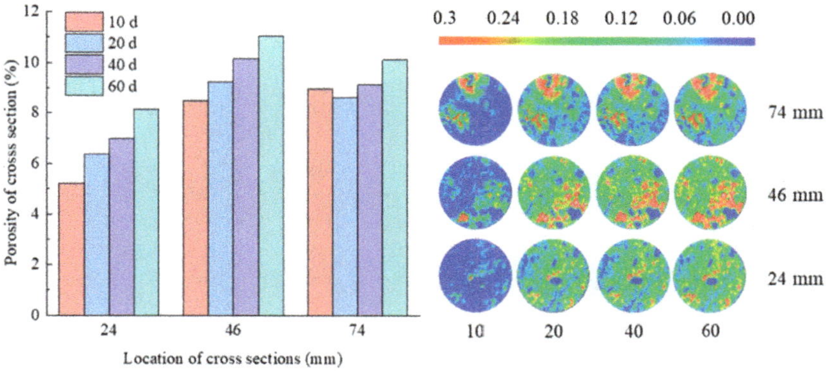

(a) (b)

Figure 8. (**a**) Porosity of cross-sections and (**b**) porosity distribution on the cross-sections at Z = 24, 46, 74 mm with 10, 20, 40, 60 days in N2 group.

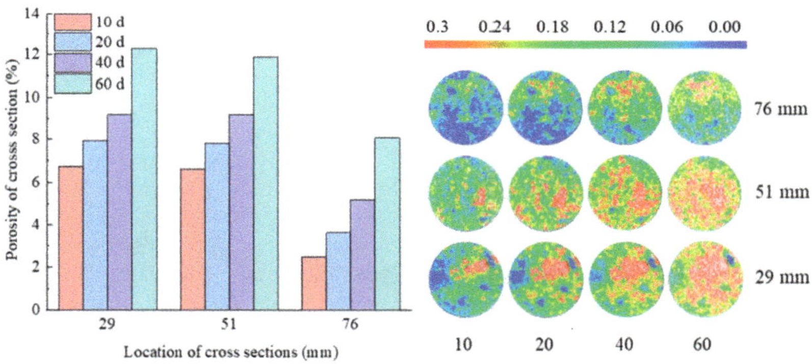

(a) (b)

Figure 9. (**a**) Porosity of cross-sections and (**b**) porosity distribution on the cross-sections at Z = 29, 51, 76 mm with 10, 20, 40, 60 days in N3 group.

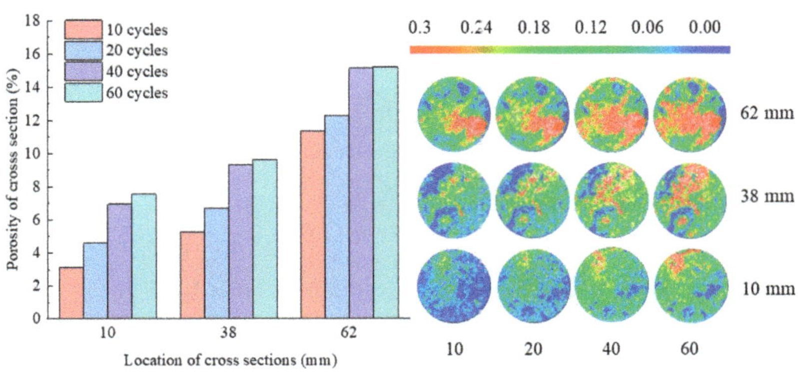

(a) (b)

Figure 10. (**a**) Porosity of cross-sections and (**b**) porosity distribution on the cross-sections at Z = 10, 38, 62 mm with 10, 20, 40, 60 days in N5 group.

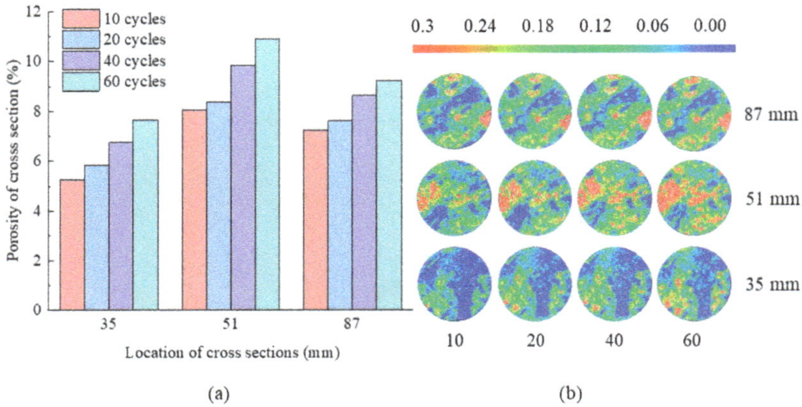

Figure 11. (**a**) Porosity of cross-sections and (**b**) porosity distribution on the cross-sections at Z = 35, 51, 87 mm with 10, 20, 40, 60 days in N6 group.

Figure 9 shows that after 10 days of erosion, the cross-sections at Z = 29 mm and Z = 51 mm were predominantly medium porosity zone, while the cross-section at Z = 76 mm was mostly low porosity zone. The cross-sections of the three sites exhibited a tendency toward steady increases in high porosity area and decreases in low porosity area as the erosion days rose from 10 to 20. After 60 days of erosion, the majority of the cross-section's medium porosity zones at Z = 29 mm and Z = 51 mm progressively merged and evolved into the high porosity zone, covering practically the whole section with only sporadic low porosity patches. From bottom to top, the cross-sectional porosity of the three places increased from 6.73%, 6.63%, and 2.49%, to 12.27%, 11.9%, and 8.09%, respectively. It showed that the saturated magnesium sulfate solution had a severe corrosive impact on tuff, gradually expanding the microscopic pores within the rock and increasing porosity.

As shown in Figure 10, after 10 freeze–thaw cycles, the majority of the Z = 10 mm and Z = 38 mm cross-sections consist of low and medium porosity regions, with only a few regions containing high porosity parts. In the Z = 62 mm cross-section, the lower right area had almost entirely high porosity areas, while the other regions are dominated by medium porosity areas. When the number of freeze–thaw cycles increases from 10 to 60, the upper left region of the cross-section at Z = 10 mm transforms from a low porosity region to an elongated band of high porosity. At Z = 38 mm, the middle and upper regions of the cross-section become labyrinthine high porosity regions. At Z = 62 mm, the central and lower regions of the cross-section are occupied with zones of high porosity. The porosity of the three cross-sections increased from 3.15%, 5.26%, and 11.4%, respectively, to 7.57%, 9.63%, and 15.24%, respectively. It demonstrates that the porosity of the tuff progressively increased as the number of freeze–thaw cycles increased. This is due to the volume enlargement caused by the phase transition of ice water during the freeze–thaw cycle, which expanded the pores within the rock. When the number of freeze–thaw cycles increased from 40 to 60, there was little change in the porosity distribution of the specimen's cross-section, indicating that freeze–thaw erosion tended to be stable. Concurrently, a fraction of the saline solution infiltrated the sample by pore seepage [30,31], resulting in the deposition of salt crystals that have the potential to obstruct certain pores and fortify the pore space within the tuff during successive freeze–thaw cycles.

Figure 11 shows that after 10 freeze–thaw cycles, punctate high porosity plaques emerge on the lower left border of the cross-section at Z = 35 mm; the majority of the remaining regions are low and medium porosity regions. At Z = 51 mm, a large high porosity zone appeared in the left part of the cross-section, and a definite band-like high porosity region appeared in the middle region. The lower right of the cross-section at Z = 87 mm shows lumpy areas of high porosity, whereas the middle and upper sections have punctate areas of high porosity. The lower left part of the cross-section at Z = 35 mm evolved from a low porosity zone to a punctate high porosity region as the freeze–thaw cycle increases from 10 to 60. The high porosity zone on the left side of the cross-section at Z = 51 mm gradually expanded, and a punctate high porosity region appears below; the high pore zone on the lower right of the cross-section at Z = 87 mm marginally increased. It demonstrates that freeze–thaw cycles have an influence on the interior pores of tuff. The porosity of the three cross-sections increased from 5.26%, 8.05%, and 7.26% to 7.66%, 10.89%, and 9.24%, respectively.

Frost and salt weathering are two of the primary causes of pore expansion, weathering, and rock deterioration [32,33]. In nature, free Mg^{2+}, SO_4^{2-}, and other minerals in soil and running water slowly infiltrate the surface layer of the tuff through seepage and diffusion, where they react with the mineral components on the surface, creating gypsum and other compounds and causing volume expansion [30,31]. This process creates microscopic fissures on the surface of the tuff, which are visible macroscopically as the erosion interface forms irregular curves. Ions continue to penetrate the specimen's microscopic pores over time, driving the expansion and distortion of the pores within the tuff. The volume expansion generated by the ice-water phase transition accelerates pore expansion within the tuff. This study examined the process of pore formation and evolution in tuff subjected to frost and salt weathering by a series of indoor tests. In our preliminary study, we employed CT scanning equipment to investigate the alterations in pore characteristics of tuff quantitatively. Our findings indicate that the porosity of the samples exhibits an increase in response to both freeze–thaw cycles and the duration of salt erosion, as shown by [24]. However, the degradation and failure mechanisms of tuff, resulting from the combined effects of frost and salt weathering, are highly intricate [34]. The present analysis did not account for the variations in temperature, water, and stress fields inside the tuff caused by freeze–thaw cycles and salt erosion due to the limitations of the experimental constraints. In further investigations, it is important to examine a quantifiable relationship between the physical characteristics, such as density, initial porosity, and microstructural changes in tuff under coupled frost and salt weathering [30,31].

4. Conclusions

In this research, high-precision CT scanning equipment was used to study the erosion process of saturated magnesium sulfate, sodium sulfate solution, and freeze–thaw cycles on tuff. We quantitatively examined the pore changes in tuff under frost and salt weathering.

(1) Saturated magnesium sulfate is more corrosive to tuff than sodium sulfate when erosion days reach 60; the quality loss rate of tuff immersed in a saturated magnesium sulfate solution is 1.56%. Sodium sulfate solution increases the mass of tuff by 0.33%.
(2) Freeze–thaw cycles have a significant influence on tuff's weathering. After 60 freeze thaw cycles, the mass loss rates of tuff in solutions of distilled water, sodium sulfate, and magnesium sulfate were 3.82%, 3.52%, and 3.58%, respectively.
(3) From 10 to 60 erosion days, the maximal and average porosity changes in all cross-sections in the sodium sulfate and magnesium sulfate groups were 3.42% and 5.54%, 4.14%, and 6.59%, respectively. The largest and average values of porosity change in the sodium sulfate and magnesium sulfate groups were 3.77% and 3.35%, and 2.18% and 2.25%, respectively, as the number of freeze–thaw cycles rose from 10 to 60.

Author Contributions: Conceptualization, H.J. and A.L.; Software, Y.S.; Formal analysis, Y.S. and A.L.; Investigation, L.C. and C.W.; Data curation, L.C. and C.W.; Writing—original draft, L.C. and A.L.; Writing—review & editing, P.S. and M.L.; Funding acquisition, P.S. and M.L. All authors have read and agreed to the published version of the manuscript.

Funding: This study has been partially funded by the Second Tibetan Plateau Scientific Expedition and Research Program (Grant No. 2019QZKK0904), the Program of the State Key Laboratory of Frozen Soil Engineering (Grant No. SKLFSE-ZT-202110), and the Science and Technology Plan of Gansu Province (22JR5RA059).

Institutional Review Board Statement: Not applicable.

Informed Consent Statement: Not applicable.

Data Availability Statement: Not applicable.

Acknowledgments: The authors thank the anonymous reviewers and editors for their constructive comments on this manuscript.

Conflicts of Interest: The authors declare no conflict of interest.

References

1. Walder, J.S.; Hallet, B. The physical basis of frost weathering: Toward a more fundamental and unified perspective. *Arct. Alp. Res.* **1986**, *18*, 27. [CrossRef]
2. Matsuoka, N.; Murton, J. Frost weathering: Recent advances and future directions. *Permafr. Periglac. Process.* **2008**, *19*, 195–210. [CrossRef]
3. Oguchi, C.T.; Yu, S. A review of theoretical salt weathering studies for stone heritage. *Prog. Earth Planet. Sci.* **2021**, *8*, 1–23. [CrossRef]
4. Yang, H.; Chen, C.; Ni, J.; Karekal, S. A hyperspectral evaluation approach for quantifying salt-induced weathering of sandstone. *Sci. Total. Environ.* **2023**, *885*, 163886. [CrossRef]
5. Trenhaile, A.S.; Mercan, D.W. Frost weathering and the saturation of coastal rocks. *Earth Surf. Process. Landf.* **1984**, *9*, 321–331. [CrossRef]
6. Williams, R.B.G.; Robinson, D.A. Frost weathering of rocks in the presence of salts—A review. *Permafr. Periglac. Process.* **1991**, *2*, 347–353. [CrossRef]
7. Peter, M.; Andersen, J.L.; Nixon, F.C.; Etzelmüller, B.; Westermann, S.; Fredin, O. Near-surface temperatures and potential for frost weathering in blockfields in Norway and Svalbard. *Earth Surf. Process. Landf.* **2023**, *48*, 940–955. [CrossRef]
8. Matsuoka, N. Frost weathering and rockwall erosion in the southeastern Swiss Alps: Long-term (1994–2006) observations. *Geomorphology* **2008**, *99*, 353–368. [CrossRef]
9. Goudie, A.S.; Viles, H.A.; Parker, A.G. Monitoring of rapid salt weathering in the central Namib Desert using limestone blocks. *J. Arid. Environ.* **1997**, *37*, 581–598. [CrossRef]
10. Matsuoka, N. Direct observation of frost wedging in alpine bedrock. *Earth Surf. Process. Landf.* **2001**, *26*, 601–614. [CrossRef]
11. Schnepfleitner, H.; Sass, O.; Fruhmann, S.; Viles, H.; Goudie, A. A multi-method investigation of temperature, moisture and salt dynamics in tafoni (Tafraoute, Morocco). *Earth Surf. Process. Landf.* **2016**, *41*, 473–485. [CrossRef]
12. Kellerer-Pirklbauer, A. Potential weathering by freeze-thaw action in alpine rocks in the European Alps during a nine year monitoring period. *Geomorphology* **2017**, *296*, 113–131. [CrossRef]
13. Matsuoka, N. Mechanisms of rock breakdown by frost action: An experimental approach. *Cold Reg. Sci. Technol.* **1990**, *17*, 253–270. [CrossRef]
14. Park, J.; Hyun, C.U.; Park, H.D. Changes in microstructure and physical properties of rocks caused by artificial freeze–thaw action. *Bull. Eng. Geol. Environ.* **2015**, *74*, 555–565. [CrossRef]
15. Yang, G.S.; Zhang, Q.S.; Ren, J.X.; Pu, Y. Study on the effect of freezing rate on the damage CT values of TongChuan sandstone. *Chin. J. Rock Mech. Eng.* **2004**, *23*, 4099–4104.
16. Jia, H.; Liu, Q.; Xiang, W.; Zhang, W.; Lang, L. Damage evolution model of saturated sandstone under freeze-thaw cycles. *Chin. J. Rock Mech. Eng.* **2013**, *32*, 3049–3055.
17. Shen, Y.; Yang, G.; Rong, T.L.; Jia, H.; Wang, M.; Liu, H. Localized damage effects of quasi-sandstone with single fracture and fracture behaviors of joint end under cyclic freezing and thawing. *Chin. J. Rock Mech. Eng.* **2017**, *36*, 562–570.
18. Zhang, H.; Wang, Y. Multi-scale analysis of damage evolution of freezing-thawing red sandstones. *Rock Soil Mech.* **2022**, *43*, 2103–2114.
19. Wang, P.; Xu, J.; Liu, S.; Liu, S.; Wang, H. A prediction model for the dynamic mechanical degradation of sedimentary rock after a long-term freeze-thaw weathering: Considering the strain-rate effect. *Cold Reg. Sci. Technol.* **2016**, *131*, 16–23. [CrossRef]
20. Hori, M.; Morihiro, H. Micromechanical analysis on deterioration due to freezing and thawing in porous brittle materials. *Int. J. Eng. Sci.* **1998**, *36*, 511–522. [CrossRef]
21. Williams, R.B.G.; Robinson, D.A. Weathering of sandstone by the combined action of frost and salt. *Earth Surf. Process. Landf.* **1981**, *6*, 1–9. [CrossRef]
22. Sun, Q.; Dong, Z.; Jia, H. Decay of sandstone subjected to a combined action of repeated freezing–thawing and salt crystallization. *Bull. Eng. Geol. Environ.* **2019**, *78*, 5951–5964. [CrossRef]
23. Angeli, M.; Bigas, J.-P.; Benavente, D.; Menéndez, B.; Hébert, R.; David, C. Salt crystallization in pores: Quantification and estimation of damage. *Environ. Geol.* **2007**, *52*, 205–213. [CrossRef]
24. Yu, S.; Oguchi, C.T. Role of pore size distribution in salt uptake, damage, and predicting salt susceptibility of eight types of Jap-anese building stones. *Eng. Geol.* **2010**, *115*, 226–236. [CrossRef]
25. Li, B.; Zhang, G.; Ma, W.; Liu, M.; Li, A. Damage mechanism of sandstones subject to cyclic freeze–thaw (FT) actions based on high-resolution computed tomography (CT). *Bull. Eng. Geol. Environ.* **2022**, *81*, 374. [CrossRef]

26. Deprez, M.; Kock, T.D.; Schutter, G.D.; Cnudde, V. A review on freeze-thaw action and weathering of rocks. *Earth-Sci. Rev.* **2020**, *203*, 103143. [CrossRef]
27. Lai, Y.; Wu, Z.; Zhu, Y.; Liao, Q. CT Analysis of Frost Damage of the Surrounding Rocks of a Tunnel in the Daban Mountain. *J. Glaciol. Geocryol.* **2000**, *22*, 206–210.
28. Wang, J.; Gao, J.; Liu, L. Porosity characteristics of sandstone by X-ray CT scanning system. *Acta Pet. Sin.* **2009**, *30*, 887–897.
29. Desarnaud, J.; Bonn, D.; Shahidzadeh, N. The Pressure induced by salt crystallization in confinement. *Sci. Rep.* **2016**, *6*, 30856. [CrossRef]
30. Santhanam, M.; Cohen, M.D.; Olek, J. Mechanism of sulfate attack: A fresh look: Part 1: Summary of experimental results. *Cem. Concr. Res.* **2002**, *32*, 915–921. [CrossRef]
31. Santhanam, M.; Cohen, M.D.; Olek, J. Mechanism of sulfate attack: A fresh look: Part 2. Proposed mechanisms. *Cem. Concr. Res.* **2003**, *33*, 341–346. [CrossRef]
32. Benavente, D.; Martinez-Martinez, J.; Cueto, N.; Ordoñez, S.; Garcia-Del-Cura, M.A. Impact of salt and frost weathering on the physical and durability properties of travertines and carbonate tufas used as building material. *Environ. Earth Sci.* **2018**, *77*, 147. [CrossRef]
33. Sato, M.; Hattanji, T. A laboratory experiment on salt weathering by humidity change: Salt damage induced by deliquescence and hydration. *Prog. Earth Planet. Sci.* **2018**, *5*, 84. [CrossRef]
34. Qazi, M.J.; Salim, H.; Doorman, C.A.W.; Jambon-Puillet, E.; Shahidzadeh, N. Salt creeping as a self-amplifying crystallization process. *Sci. Adv.* **2019**, *5*, eaax1853. [CrossRef]

geosciences

Article

The Mechanical Characterization of Pyroclastic Deposits for Landslide Early Warning Systems

Emilia Damiano [1,*], Martina de Cristofaro [1], Antonia Brunzo [1], Goffredo Carrieri [2], Luisa Iavazzo [1], Nadia Netti [3] and Lucio Olivares [1]

[1] Department of Engineering, University of Campania "Luigi Vanvitelli", 81031 Aversa, Italy; martina.decristofaro@unicampania.it (M.d.C.); antonia.brunzo@unicampania.it (A.B.); ing.iavazzo@gmail.com (L.I.); lucio.olivares@unicampania.it (L.O.)

[2] Italian Air Force, Viale dell'Università, 00118 Rome, Italy; goffredo.carrieri@aeronautica.difesa.it

[3] Department of Economics, Management and Institutions, University of Napoli "Federico II", 80138 Naples, Italy; nadia.netti@unina.it

* Correspondence: emilia.damiano@unicampania.it; Tel.: +39-081-5010262

Abstract: Broad mountainous areas in the western Campania (southern Italy), where young pyroclastic deposits extensively outcrop, frequently experience rainfall-induced slope movements of different degrees of mobility, causing heavy damage and fatalities. Such landslides cannot be easily mitigated, and the implementation of physically based early warning systems is still not able to predict the post-failure evolution of slope movements and the exposed areas at risk. This paper is devoted to overcoming this limit. To this end, the mechanical characterization of pyroclastic soil, carried out through an extensive laboratory testing program, is presented and compared with those of two other ashy soils of different depositional mechanisms. The results show that the depositional mode influences soil properties; to begin with, it affects the unsaturated shear strength, whose intercept of cohesion is up to 5 kPa higher in ashes of flow deposition than in airfall ash deposits. The saturated undrained soil response allowed for the identification of different levels of susceptibility to the liquefaction of pyroclastic deposits, which is one of the main factors governing the post-failure evolution of landslides. Gathering all the acquired information, including saturated and unsaturated soil shear strength, permeability function, and water retention curves, into a soil database, it was possible to present all data under a unitary framework. Finally, the implementation of the proposed flowchart for a simplified assessment of post-failure evolution to be employed in regional early warning systems can enhance our knowledge of the areas at risk.

Keywords: rainfall-induced landslide; pyroclastic soils; unsaturated soil mechanics; liquefaction; early warning systems

1. Introduction

A significant part of the Campanian slopes (southern Italy) is covered by young deposits arising from the recent explosive activity of the Somma–Vesuvius and Phlaegrean Fields volcanic centers present in the area. They have generated pyroclastic deposits of different thicknesses and characteristics depending on the distance from the eruptive centers and on deposition mechanisms (airfall, flow, or surge deposits), which mantled steep slopes present in a wide area around the Campanian Plain. Over the years, these covers were subjected to numerous shallow landslides due to infiltration processes following prolonged and intense rainfalls. The stability of these shallow slopes is mainly due to the unsaturated conditions of the covers: the presence of matric suction causes an apparent cohesive intercept, which increases the soil shear resistance [1]. During rainwater infiltration, the increase in soil weight and the decrease in matric suction may lead to slope instabilities [2–4]. Some of them evolved into catastrophic flowslides able to run for kilometers even over flat areas, destroying entire population centers and causing numerous fatalities [2,5].

Such landslides cannot be easily mitigated owing to their complexity and their large areal extension; therefore, one of the strategies used to face them is the implementation of physically based early warning systems capable of simulating the triggering phenomena [6–8]. However, these systems are available for analysis at the slope scale, and usually, they are not able to predict the evolution of subsequent movements. To overcome this limitation, a first attempt was made in [9], where the authors suggested developing a soil database (SDB) for implementation in regional landslide early warning systems, based on advanced constitutive models. In addition to the main morphological characteristics of the slopes, the information in the database includes data on the mechanical properties of the soils in both saturated and partially saturated conditions as well as the assessment of their susceptibility to liquefaction, which is considered to play a role in the evolution of landslides into flowslides [10,11]. The database only includes airfall deposits lying on limestone bedrock involved in previous catastrophic flowslides. In many cases, however, the rapid landslides that affect some shallow covers around Naples do not evolve into flowslides. In this case, the sliding mass shows less mobility and, usually, stops along the slope or at its toe. This kind of phenomenon can involve pyroclastic deposits formed through different deposition mechanisms, like pyroclastic flows, lying on a bedrock of different nature. The aim of this work is to extend the knowledge framework by analyzing also the pyroclastic deposits involved in slope movements that do not eevolve into flowslides.

In particular, this paper shows the results of an extensive laboratory experimental program performed in both saturated and unsaturated conditions on a flow deposit present along a hill near the city of Naples and recurrently involved in rainfall-induced landslides with no flow-like characteristics. The obtained results contribute to the enrichment of the database first proposed by [9]. Moreover, by comparing the characteristics of the deposit herein analyzed with those of deposits involved in flowslides (Sarno and Cervinara), it is possible to highlight the similarities and dissimilarities between the two types of soil deposits and enhance the comprehension of the mechanisms that govern the post-failure evolution of shallow rainfall-induced slope movements. Finally, this study aims to gather all the acquired information in a flowchart for a simplified assessment of the post-failure evolution of landslides.

The SDB and the proposed flowchart provide the features necessary to improve the existing EWSs whose reliability depends on adopting realistic hydro-geotechnical models (e.g., for unsaturated soils) that enable researchers to detect triggering thresholds, thus minimizing the risk of false or late alarms. Although the extension of such a database to the regional scale requires a major effort, which may be considered costly for the community, undoubtedly, many economic benefits are derived from the use of well-calibrated physically based models in the implementation of EWSs.

2. Background

From a geomorphological point of view, the Campania region is very complex and can be divided into macro-areas depending on the nature of the bedrocks (volcanic, carbonate, terrigenous, and alluvial) [12] (Figure 1a).

The deposits covering these areas consist of pyroclastic materials, accumulated after the eruptions of the volcanic centers of Roccamonfina, Phlaegrean Fields, and Somma-Vesuvius. They can be categorized into air-fall, flow, or surge deposits, and although originating from the same volcanic eruptions, they take on different physical and mechanical characteristics depending on the distance from the eruptive centers and transport, accumulation, and deposition processes. Airfall deposits are formed by pyroclastic particles directly ejected from the crater in the form of a sustained eruptive column that, after wind transport,, fall down mantling slopes and plains. Due to transport and deposition modes, the particles have sharp edges, as they are not abraded by dragging mechanisms. By contrast, surge and flow deposits are generated as a result of the sedimentation of eruptive mixtures composed of solid particles and gases that flow along the sides of the volcano.

These particles have rounded edges due to being subjected to abrasion they undergo during sliding. Surge deposits have less heterogeneous and smaller granule sizes than those constituting flow deposits. Surge deposits often contain small, rounded structures, called pisolites, comprising ash particles. The substantial difference between flow and surge deposits is the different percentage of the aeriform component, which is greater in surge deposits [13].

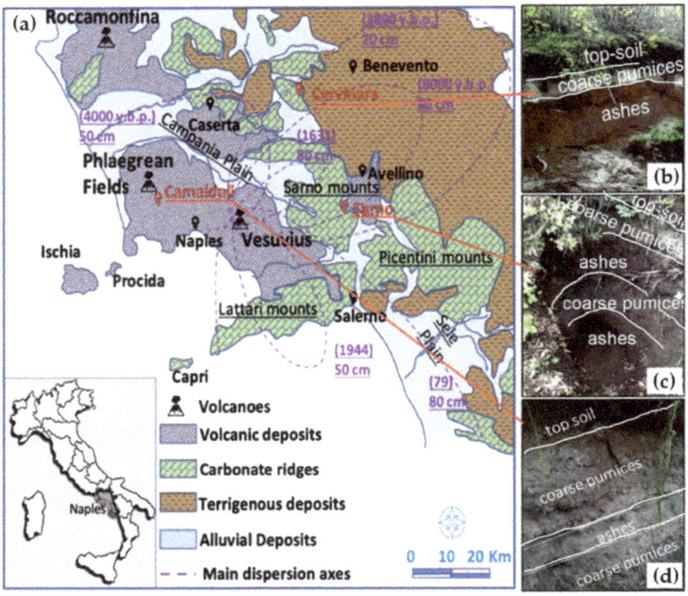

Figure 1. (**a**) Geomorphological contexts of the pyroclastic deposits in Campania, Italy; schematic stratigraphy at the location of the study sites: (**b**) Cervinara; (**c**) Sarno; (**d**) Camaldoli.

In recent years, the scientific community has shown considerable interest in the landslide events of Campania along the slope covered by airfall pyroclastic deposits that evolved into flowslides. Notably, significant attention has been given to the Sarno and Cervinara sites (Figure 2), where catastrophic flowslides occurred in 1998 and 1999, respectively. In this geomorphological context, defined henceforth as "carbonatic context", slope failures involve airfall pyroclastic deposits in a primary position lying on fractured limestone bedrock [14]. Covers consist of alternating layers of loose ash and pumice: they are regularly layered, with thicknesses ranging between 0.5 and 3 m in the detachment zone and inclination between 30 and 40°.

In this paper, instead, our focus will also be on the slopes surrounding the city of Naples in the Phlegrean area, where landslides are characterized by the accumulation of the landslide body at the toe of the slopes (Figure 2). Along these slopes, numerous landslides occur about every year, which, although characterized by a degree of mobility lower than flowslides, cause engineering problems to surrounding structures and infrastructure, as they are very close to highly urbanized areas. The covers involved are generally in primary position; however, differently from those involved in flowslides, they consist of flow or surge pyroclastic deposits. In this "volcanic context", the nature of bedrock is different; it is mainly constituted by lava, tuff, and weakly cemented pyroclastic materials, as well as slope inclinations in the detachment areas, which reach a value as high as 60° (Camaldoli and Agnano hills) (Figure 2). Moreover, the covers show a less regular stratigraphic sequence: alternations of paleosoils and loose soils (ash, pumice, and scoriae) (Figure 1d) reach tens of meters on the top of the slope. In this case, shallow landslides develop in the most superficial portion of the soil cover, with a thickness ranging from tens of cm to almost

1–2 m [15]. According to [14,16], landslides triggered in the volcanic context usually evolve into small slides or debris avalanches and propagate, with a travel distance usually no more than 300 m, while in the carbonatic context, large flowslides or debris avalanches develop, which generally propagate over distances as long as several kilometers.

Figure 2. Landslides case history in pyroclastic covers (Campania region).

Hence, the assessment of the post-failure evolution remains a key point for hazard mapping in these areas. In this regard, and considering the simplified hypothesis of homogeneous infinite slope in cohesionless soil, Ref. [3] implemented a flowchart for a simplified assessment of landslide evolution, pointing out that the post-failure evolution of a landslide is strictly related to the degree of saturation at the onset of instability and the susceptibility of the involved soil to liquefaction. Along steep slopes, characterized by an inclination almost equal to the friction angle of the soil ($\alpha \cong \phi'$), after intense precipitation, failure occurs in conditions close to saturation ($Sr \cong 1$). Under these conditions, in soil deposits susceptible to liquefaction, the evolution of the landslide into a flowslide is associated with the undrained instability (static liquefaction) that occurs in saturated loose granular soils under shear stress. In soil deposits not susceptible to liquefaction, the mechanisms underlying the post-failure evolution associated with the development of undrained response are essentially debris avalanches or slides. On very steep slopes ($\alpha >> \phi'$), failure occurs when the soil is far from a saturated condition ($Sr << 1$). In this case, the slope movement evolves into a debris avalanche or slide, independently from the eventual susceptibility to the liquefaction of the involved soils because the partially saturated condition prevents the development of an undrained response. Nevertheless, a comprehensive investigation of the mechanical properties of the soil cover is needed to effectively evaluate the triggering and post-failure mechanisms of rainfall-induced landslides in pyroclastic deposits.

To this aim, an extensive experimental program on pyroclastic soil, taken from a flow deposit located in the surrounding of Naples, was carried out to frame the saturated soil response within the steady-state theory and define soil susceptibility to the liquefaction of the deposit. Partial saturation conditions were also analyzed to assess the influence of the degree of saturation on soil shear strength and evaluate the water retention curve and permeability function. The experimental results shed light on the possibility of enriching the database for early warning system implementation [17,18] by adding the data of soil belonging to a pyroclastic deposit of flow or surge origin lying on volcanic bedrock to the database.

3. Materials and Methods

The Camaldoli slope was chosen as representative of a flow deposit generated near the eruptive center of Phlegrean Fields. The hill, located to the north of the city, is the highest relief of Naples. The south-facing side is characterized by steep slopes of tuff rocks, while the north-facing side presents gentler slopes covered by a wooded area. As reported by [15], the hill can be ideally subdivided into four different parts: the top high plain, characterized by local acclivity higher than 60°, bounded at its lower limit by the vertical cliff of tuff rock; the main slope with an inclination between 30° and 50°; the foot slope and the basal plain where the steepness becomes lower. Detachment areas are usually present at the highest altitudes of the slope above the vertical tuff cliff, where the thickness of the cover is very small.

The study area is in the upper part of the south-facing slope near the "Camaldoli hermitage" in an area characterized by slope angles ranging between 40 and 60°. Here, the slope is covered by a soil deposit tens of meters thick lying on volcanic bedrock consisting of yellow Neapolitan tuff.

The undisturbed samples used in the experimental program were taken in a 1.5 m deep recess cut along a subvertical escarpment excavated for the construction of a road that crosses the slope at a depth of about 4–6 m from the top. In this way, it was possible to identify and collect soils in primary deposition. To minimize disturbances, the soil specimens were directly taken with thin-wall metal samplers (Figure 3) having the same size as laboratory equipment used for studying the specimens (38 mm and 68 mm).

Figure 3. Example of manual undisturbed sampling.

The mean values of the physical properties of the sampled soils are reported in Table 1 in terms of the maximum diameter of soil d_{max}, the specific unit weight of the soil particles γ_s, the unit weight of soil volume γ, porosity n, and the degree of saturation S_r. The soil is characterized by a low specific unit weight of about 25 kN/m^3, probably due to the presence of internal voids of the soil particles, by porosity ranging between 53% and 55% and a saturation degree ranging between 24% and 46% related to the weather conditions.

Table 1. Main physical properties of Camaldoli ash.

d_{max} (mm)	γ_s (kN/m^3)	γ (kN/m^3)	n (%)	S_r (%)
6	24.9	12.0	53–55	24–46

The grain size distribution of the soil, illustrated in Figure 4, shows that it can be classified as silty sand. The uniformity coefficient is about 10, showing a quite uniform particle size distribution. Visual observation allows for the recognition of particles with rounded edges because of the abrasion that occurs during the formation of flow deposits.

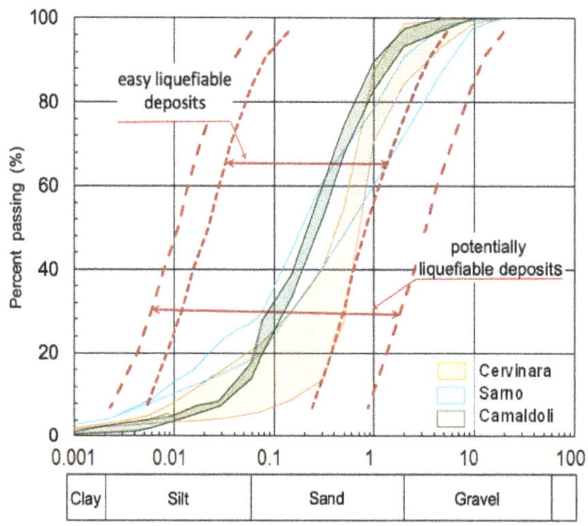

Figure 4. Grain size distribution of Camaldoli ash compared with other liquefiable ash deposits of Campania and with the bounds of liquefiable natural soil deposits.

In the same Figure 4, the grain size distribution of the Camaldoli ash is compared with the bounds of natural deposits susceptible to liquefaction under monotonic and cyclic loading [19] as well as with the bounds of Cervinara and Sarno deposits. It should be noted that the Camaldoli ash is included in the range of easily liquefiable deposits. Therefore, from this preliminary analysis, the soil can be defined as potentially liquefiable.

To define the mechanical properties of the soil in both saturated and unsaturated conditions, a comprehensive laboratory testing program was carried out, which consisted of the following tests:

- Consolidated isotropically drained and undrained (CID and CIU) triaxial tests on saturated specimens to evaluate soil shear strength and susceptibility to liquefaction;
- Constant head permeability test to determine the saturated hydraulic conductivity;
- Suction-controlled triaxial tests (SCTX) on undisturbed specimens to define hydraulic conductivity function and evaluate the effects of partial saturation on shear strength;
- Infiltration/evaporation method for the evaluation of the soil water retention curve (SWRC).

4. Results

4.1. Mechanical Properties in Saturated Conditions

4.1.1. Saturated Shear Strength and Undrained Response

The mechanical response of the Camaldoli ash was investigated through CID and CIU tests performed on undisturbed natural soil samples saturated by adopting the experimental technique proposed by [20]. After soil saturation, specimens were isotropically consolidated under mean effective stresses between 25 kPa and 300 kPa. The void ratios at the end of the consolidation phases ranged between 1.1 and 1.2 (Table 2) [21].

Table 2. Triaxial tests in saturated conditions: state parameters at the end of the consolidation stage and the applied mean effective stress [21].

Test ID	γ (kN/m^3)	γ_d (kN/m^3)	w (%)	n (%)	e	p'$_0$ (kPa)
C_Cid_1	16.9	11.5	46.5	53.6	1.16	73.4
C_Cid_2	17.0	11.7	45.4	53.1	1.13	99.38
C_Cid_3	17.0	11.8	44.4	52.7	1.11	147.66
C_Cid_4	16.7	11.6	42.9	53.3	1.14	149.86
C_Cid_5	16.7	11.4	47.2	54.4	1.19	25.85
C_Cid_6	16.9	11.7	45	53.1	1.13	229.5
C_Cid_7	16.6	11.2	48	54.9	1.22	49.56
C_Cid_8	16.9	11.5	46.8	53.7	1.16	125.67
C_Ciu_1	17.1	11.9	43.1	52.1	1.09	297.46
C_Ciu_2	16.8	11.6	44.2	53.2	1.14	34.69
C_Ciu_3	17.0	11.7	44.7	52.8	1.12	69.97
C_Ciu_4	17.1	11.9	44	52.3	1.10	220.11
C_Ciu_5	16.9	11.8	43.5	52.7	1.12	126.69
C_Ciu_6	17.0	11.8	44.2	52.7	1.12	99.83
C_Ciu_7	17.0	11.9	42.8	52.3	1.10	148.01
C_Ciu_8	17.0	11.9	42.6	52	1.09	199.99

The results of both types of tests are reported in Figure 5 in terms of stress paths (the p′–q plane in Figure 5a), the evolution of deviatoric stress (the ε_a–q plane in Figure 5b,c), the evolution of volumetric strain during CID test (the ε_a–ε_v plane in Figure 5d), and the development of excess pore pressure during undrained CIU test (the ε_a–Δu plane in Figure 5e).

In drained conditions, the soil shows a fragile behavior (Figure 5a,b): the mobilized strength first reaches a peak value and then stabilizes at a lower critical state value. The corresponding strength parameters are as follows:

- A friction angle equal to 38° and cohesion of 15 kPa at peak;
- A friction angle of 38° and cohesion of 0 kPa at the critical state.

The soil has a high friction angle and a relatively low effective cohesion in the investigated stress field.

In undrained conditions (green lines in Figure 5a,c), the stress paths clearly show that, with an increase in deviatoric stress, partial instability occurs, with an initial decrease in the mean effective stress and a subsequent progressive increase (transformation phase point) until a larger final value of deviatoric stress is reached under a high level of strain. The soil, therefore, shows a potentially non-liquefiable behavior.

The results are reported in the compression plane (log p′–e) in Figure 6. In detail, the points represent the final state of the specimens in CID tests and both the initial and final conditions of the specimens in CIU tests. The final conditions fall under a narrow range of values (highlighted in red in Figure 6), which allows for the identification of the steady-state conditions and the position of the steady-state line (SSL) in the compression plane. The SSL is almost straight and is characterized by a slope slightly higher than the isotropic compression line, shown in the same figure (green line).

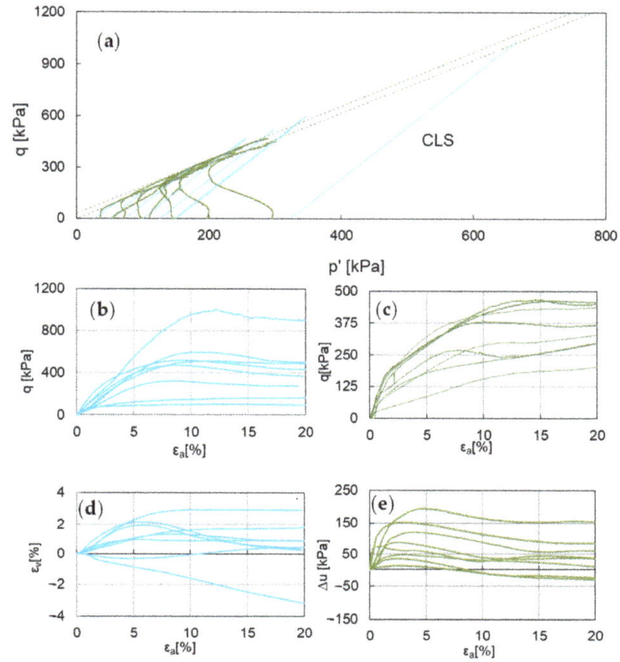

Figure 5. Triaxial test results: (**a**) stress plane; (**b**) results of CID tests in εa–q plane; (**c**) results of CIU tests in εa–q plane; (**d**) results of CID tests in εa–εv plane; (**e**) results of CIU tests in εa–Δu plane.

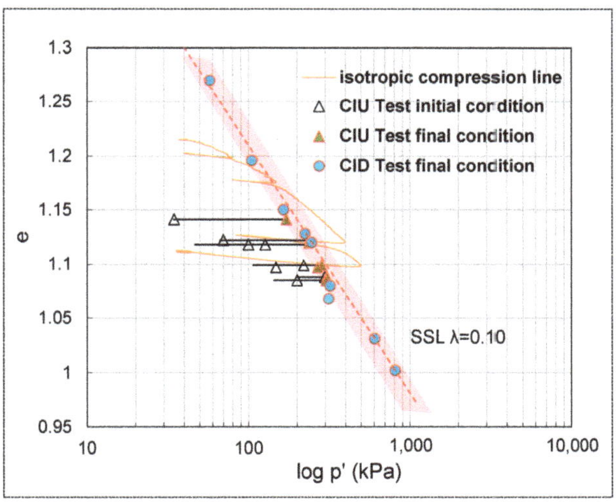

Figure 6. Triaxial test results: compression plane.

In framing the undrained soil behavior through the lens of the steady-state theory [22], we can use the SSL as a discriminating factor to verify soil susceptibility to liquefaction of natural soil deposits. For Camaldoli ash, the points representing the initial conditions of the natural deposit, characterized by void ratios around 1.1 and mean stress values of tens of kPa, are located well below the SSL (Figure 6). This suggests that the undrained behavior of the Camaldoli natural deposit is stable [23]. Thus, the experimental results show that, under undrained deformation, the Camaldoli ash, which has a particle size distribution typical of potentially liquefiable natural deposits, is not subjected to an increase in pore pressure to be able to diminish the soil shear resistance, and therefore, it is not liquefiable.

4.1.2. Saturated Hydraulic Conductivity

The saturated conductivity of the soil, k_{sat}, was investigated through constant head permeability tests on natural samples. The results are shown in Figure 7. The values of k_{sat} are reported as a function of void ratio (Figure 7a) and the mean applied effective stress (Figure 7b). For a low state of stress, close to the in situ stress, the saturated permeability is about 10^{-5} m/s, which progressively decreases by an order of magnitude for applied effective stress as high as 700 KPa (Figure 7b).

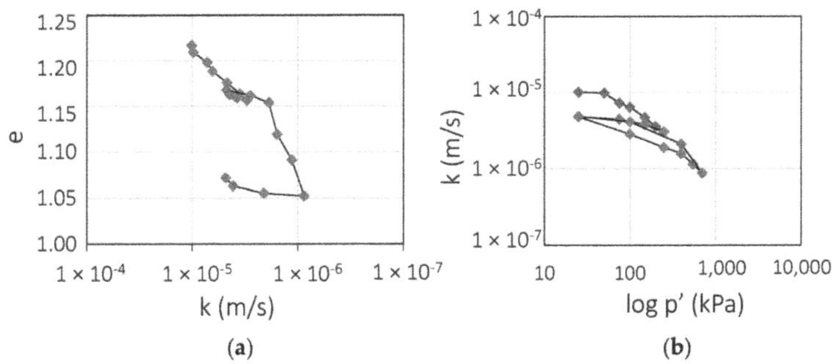

Figure 7. Saturated conductivity as a function of (**a**) void ratio and (**b**) mean effective stress.

4.2. *Mechanical Properties in Unsaturated Conditions*

Unsaturated soil properties play a key role in affecting the rate of rainwater infiltration and variation in the slope safety factor during a rainfall event. Therefore, to set up a physically based EWS, the proper characterization of the hydraulic and mechanical properties of unsaturated soil is crucial. The functional relationships used to interpret the experimental results, the corresponding parameters, and the procedures used in the following sections are extensively discussed in [1].

4.2.1. Shear Strength under Unsaturated Conditions (Sr < 1)

The experimental program was carried out in a suction-controlled triaxial cell (SCTX) on undisturbed specimens 70 mm in diameter by applying a mean net stress (T) of 50 and 100 kPa, and suction values ranging between 5 and 50 kPa (Table 3). The triaxial cell was equipped for internal measurements of axial deformation, carried out using LVDT transducers, radial mean deformation, and water volume exchanges through level transducers and a double burette system. Inside these instruments, both air flow and pore water flow were initiated through the upper and lower surfaces of the specimen to keep the imposed suction level constant, which was applied using the axis translation technique. After an equalization phase, needed to obtain a uniform suction distribution inside the specimen, an isotropic consolidation stage at constant suction was performed. Finally, the failure stage was induced by keeping the mean net stress and suction constant and progressively increasing the axial stress with a rate of 1.5 kPa/h to ensure a drained loading condition.

The test results are shown in the stress plane ($p–u_a$, q) in Figure 8a. In the same Figure, the critical state line corresponding to the saturated condition of the material (dashed line) is also reported. Thus, from the comparison, the mechanical effect of suction on the soil strength can be revealed. All the points representing failure in unsaturated conditions are located well above the failure envelope of the saturated soil, showing that suction in these soils plays a significant role.

Table 3. Triaxial tests in unsaturated conditions: initial conditions and applied state of stress.

SCTX	γ (kN/m^3)	γ_d (kN/m^3)	n (%)	e	$S_{r,f}$	$p-u_a$ (kPa)	u_a-u_w (kPa)
C_USP_1	15.2	11.8	52.50	1.104	0.64	50	5
C_USP_2	14.4	11.8	52.45	1.104	0.48	50	15
C_USP_3	13.6	11.8	52.47	1.105	0.34	50	35
C_USP_4	13.5	11.6	53.61	1.154	0.36	50	50
C_USP_5	14.2	11.8	52.46	1.105	0.44	100	5
C_USP_6	12.8	11.1	55.48	1.247	0.30	100	15
C_USP_7	15.4	11.4	54.96	1.192	0.74	100	35
C_USP_8	13.4	11.7	52.87	1.123	0.33	100	50

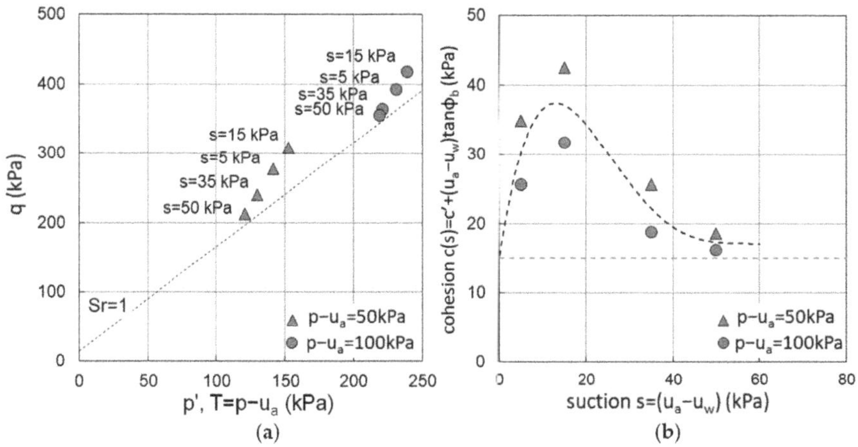

Figure 8. Unsaturated shear strength: (**a**) stress plane; (**b**) intercept of cohesion as a function of suction.

The results of the SCTX tests were also interpreted by using the extension of the Mohr–Coulomb criterion to the unsaturated soils proposed by [1], assuming a constant effective friction angle of 37°. The results, in terms of apparent cohesion as a function of matric suction, are illustrated in Figure 8b. The apparent cohesion, and hence the shear strength of the soil, increases at low matric suction until it reaches a peak for suction of about 15–18 kPa, and then it rapidly decreases. For values of suction higher than 50 kPa, the effect of suction on shear strength approaches a residual value ranging between 15 and 18 kPa, which should correspond to its effective value. Although the trends of the relationship between suction and apparent cohesion are similar for the two groups of specimens consolidated at different levels of mean net stresses, they are characterized by different magnitudes: the lower the mean net stress, the higher the effect of suction on shear strength. This can be ascribed to the different particle arrangements induced by the higher mean net stress that modifies the SWRC, especially at low suction values [24].

Nevertheless, the lowest values of the intercept of cohesion are sufficient to justify the stability of the slope with an inclination also higher than 40°.

4.2.2. Soil Water Retention Curve and Conductivity Function

To retrieve the SWRC of the Camaldoli ash in a range of suction from 0 up to 80 kPa, of primary interest for our research, an experimental set-up based on the evaporation method proposed by [25] was used. A sketch of the experimental apparatus is illustrated in Figure 9. The test was performed on a cubic sample of soil (10 × 10 × 10 cm^3), reconstituted with an initial porosity of 0.55, by using two mini-tensiometers (jet-fill, Soil Moisture type) with ceramic tips located at different heights, and a TDR probe inserted horizontally at the mid-height of the specimen. Mini-tensiometers measure suction up to 100 kPa with an accuracy of 1 kPa. The TDR probe was used to measure the volumetric water content of the soil. It is based on the correlation between electric conductivity and water content. Taking advantage of the dielectric properties of the soil, the travel time of the electromagnetic pulse along the metal probes buried into the soil is measured and correlated to the volumetric water content [26]. An estimation of the volumetric water content was also carried out using gravimetric measurements by assuming a constant porosity in the soil sample (no significant volumetric changes and shrinkage occurred during the test).

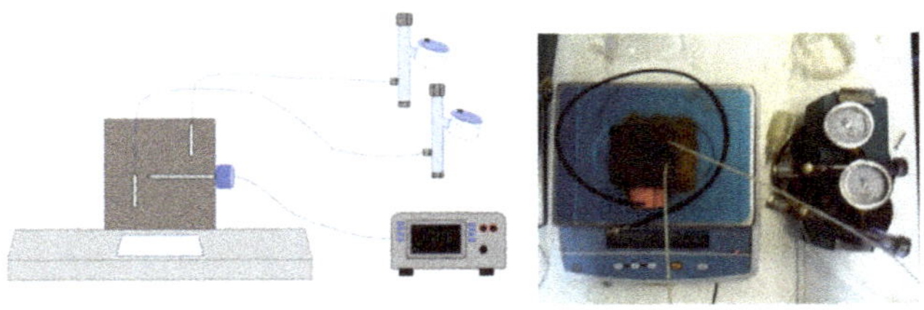

Figure 9. Sketch of the apparatus used for determination of the SWRC.

The test consisted of monitoring a series of evaporation and infiltration stages starting from an initial saturated condition. As there was no significant difference found by coupling the water content with matric suction measured at the most superficial or the deepest tensiometer, the two suction measurements were averaged to obtain the representative suction incurred by the specimen. The wetting and drying paths of the soil during the various phases are reported in Figure 10 by coupling suction measurements with the volumetric water content obtained using mass measurements. The experimental points obtained by coupling suction measurements with volumetric water content values retrieved with the TDR probe are also reported.

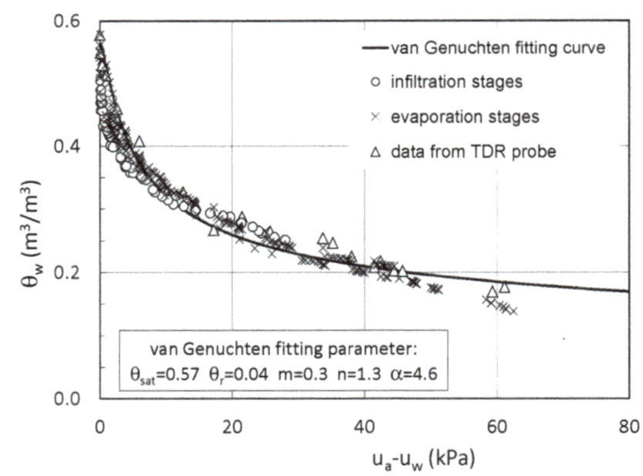

Figure 10. Soil water retention curve of the Camaldoli ash.

The Camaldoli ash has a negligible hydraulic hysteresis in the investigated range of suction, as the wetting and drying paths partly overlap. The obtained SWRC is typical of coarse soil, characterized by a very small air-entry value (in the order of a few kPa), a steep front in the transition zone, and a flatter trend approaching the residual zone of unsaturation. The experimental points were fitted by using the van Genuchten model [27]. The fitting parameters (α, n, m) are reported in Figure 10.

The hydraulic conductivity function (HCF) was analyzed using the SCTX tests. Each value was determined by analyzing the transient flow induced during the equalization phase of an imposed suction level by using the Kunze expression [28]. Experimental values (green dots) are shown in Figure 11, along with the saturated conductivity values (white triangles). The same figure also shows another estimation of the unsaturated conductivities performed with the expression proposed by [29] based on the suction and saturation degrees measured at the end of each equalization phase. Finally, the trend of the permeability function determined with Gardner's function [30], which better approximates the experimental data, is depicted as a dashed line in Figure 11. In the investigated field of suction, hydraulic conductivity greatly reduces by about three orders of magnitude, reaching a value of about 1×10^{-8} m/s. For suction lower than 30 kPa, the data show a greater dispersion.

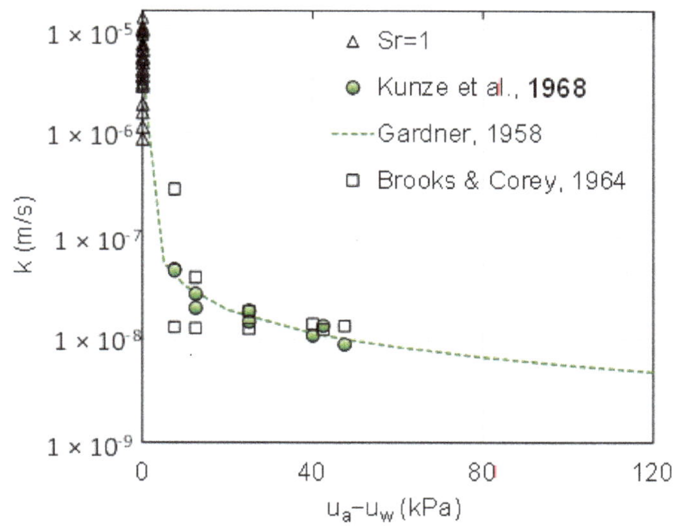

Figure 11. Hydraulic conductivity function of the Camaldoli ash (relationships for data interpretation from [28–30].

5. Discussion

The experimental results related to the Camaldoli soil can be incorporated into the soil database (SDB) on pyroclastic materials initially proposed by [9] for airfall deposition covers by adding information regarding volcanic ash belonging to a pyroclastic deposit of flow origin. The enriched SDB is reported in Figure 12, where data from the three study areas of Camaldoli, Cervinara, and Sarno are reported. In addition, to facilitate comparison and comments on the properties of these deposits, in Figure 13, the SWRCs (Figure 13a), HCFs (Figure 13b), apparent cohesion trends (Figure 13c), and undrained paths in the compression plane (Figure 13d) are shown.

The SDB is organized in the form of sheets where both geographical and geotechnical characteristics of a soil deposit are reported, together with an evaluation of the reliability of the most utilized literature expressions that were used to fit the experimental data.

The first section of the sheet indicates the site's geographical coordinates and some geomorphological data such as the average slope angle, the mean thickness of the cover, and the position within the stratigraphic sequence of the investigated layer from which undisturbed soil samples were taken and analyzed.

In the second section, the physical and state properties of the soil are reported. For the Camaldoli flow pyroclastic ash, the grain size distribution is located within the range of potentially liquefiable soil, and this information is highlighted in red in the SDB (Figure 12a) as a first indication of susceptibility to the liquefaction of the soil deposit.

In the third section of the sheet, the mechanical characteristics of the soil in saturated conditions are shown. In detail, the critical state parameters and the undrained behavior of the soil, evaluated through CIU triaxial tests on natural specimens consolidated in the range of stress acting in situ, are indicated. Once again, this latter information is highlighted in red in the case of an unstable soil response (liquefiable soil), yellow in the case of a partially unstable soil response, or green in the case of a stable undrained response.

Camaldoli pyroclastic soil

Geographical information and layer data

Latitude	Longitude	Average slope angle	Thickness cover	Thickness soil layer	Depth of the soil layer min	Depth of the soil layer max
N	E	[°]	[m]	[m]	[m]	[m]
40° 51' 30"	14° 11' 32"	50°	>10	0.5–2.00	1	1.5

Physical and state properties

Gs	n	γ_d	γ_{sat}
	%	[kN/m^3]	[kN/m^3]
2.49	54	10.7	16.6

Grain size distribution of liquefiable soil [yes or no]	yes

Saturated shear strength and critical state parameters

c'	φ'	M	λ
[kPa]	[°]		
10–15	37	1.53	0.1

Undrained unstable response in CTX tests [yes or no]	no

Unsaturated shear strength

mod. after Fredlund and Rahardjo 1993

c'	φ'	c(u$_a$-u$_w$)
[kPa]	[°]	[kPa]
10–15	37	*See Fig.8*

Evaluation of reliability	high

Lu et al. 2010

$\sigma_s=-(u_a-u_w)/[1+(\alpha(u_a-u_w))^n]^m$

α; n; m van Genuchten parameters

Evaluation of reliability	low

Permeability functions

Kunze et al. 1968

$k= 1/\alpha^2 \cdot L^2/V \cdot [Q_o/\Delta u_o] \cdot \rho_w \cdot g \cdot b_\alpha$

b_α from analysis of suction equalization (SCTX)

ks [S$_r$=1]	k$_P$	L$_P$	$\alpha=L_P/K_P \cdot k/L$
[m/s]	[m/s]	[mm]	
4.50 × 10^{-6}	1 × 10^{-7}	5	

Evaluation of reliability	high

Brooks and Corey 1964

$k=ks$ if $(u_a-u_w)\leq(u_a-u_w)_b$

$k=ks \cdot S_e^{\eta}$ if $(u_a-u_w)\leq(u_a-u_w)_b$

$\eta=2+3\lambda$; λ=pore size distribution index

ks [S$_r$=1]	(ua–uw)$_b$	λ	η
[m/s]	[kPa]	[-]	[-]
4.50 × 10^{-6}	1	0.45	3.35

Evaluation of reliability	low

Gardner 1958

$k=ks/[1+a\,[(u_a-u_w)/\rho g]^n]$

ks [S$_r$=1]	a	n
[m/s]	[-]	[-]
4.50 × 10^{-6}	1	0.87

Evaluation of reliability	medium

Water retention curves

van Genuchten 1980

$(\theta-\theta_r)/(\theta_s-\theta_r)=[1/(1+(\alpha h)^n)]^m$

h=pressure head assumed positive [m]

θ_s	θ_r	α	n	m
		[1/m]		
0.57	0.04	4.6	1.3	0.3

Evaluation of reliability	high

(a)

Figure 12. *Cont.*

Cervinara pyroclastic soil							
Geographical information and layer data							
Latitude	Longitude	Average slope angle	Thickness cover	Thickness soil layer	Depth of the soil layer min	Depth of the soil layer max	
N	E	[°]	[m]	[m]	[m]	[m]	
41° 00′ 28″	14° 38′ 38″	38°	2.2	0.6–1.4	0.8	1.8	

Physical and state properties			
Gs	n	γ_d	γ_{sat}
	%	[kN/m^3]	[kN/m^3]
2.58	69	8.0	14.8
Grain size distribution of liquefiable soil [yes or no]		yes	
Saturated shear strength and critical state parameters			
c′	ϕ'	M	λ
[kPa]	[°]		
0	38	1.55	0.19
Undrained unstable response in CTX tests [yes or no]		yes	
Unsaturated shear strength			
mod. after Fredlund and Rahardjo 1993			
c′	ϕ'	$c(u_a-u_w)$	
[kPa]	[°]	[kPa]	
		if $(u_a-u_w)\leq 1$	0
0	38	if $(u_a-u_w)>1$	$2.4\ln(u_a-u_w)$
Evaluation of reliability		high	
Lu et al. 2010			
$\sigma_s=-(u_a-u_w)/[1+(\alpha(u_a-u_w))^n]^m$ α; n; m van Genuchten parameters			
Evaluation of reliability		medium	

Permeability functions				
Kunze et al. 1968				
$k= 1/\alpha^2\cdot L^2/V\cdot[Q_o/\Delta u_o]\cdot\rho_w\cdot g\cdot b_\alpha$				
b_α from analysis of suction equalization (SCTX)				
ks [S_r=1]	k_P	L_P	$\alpha=L_P/K_P\cdot k/L$	
[m/s]	[m/s]	[mm]		
3.70×10^{-7}	1×10^{-7}	5		
Evaluation of reliability				high
Brooks and Corey 1964				
$k=ks$ if $(u_a-u_w)\leq(u_a-u_w)_b$				
$k=ks[(u_a-u_w)_b/(u_a-u_w)]^\eta$ if $(u_a-u_w)\leq(u_a-u_w)_b$				
$\eta=2+3\lambda$;		λ=pore size distribution index		
ks [S_r=1]		$(ua-uw)_b$	λ	η
[m/s]		[kPa]	[-]	[-]
3.70×10^{-7}		1	0.39	3.17
Evaluation of reliability				low
Gardner 1958				
$k=ks/[1+a\,[(u_a-u_w)/\rho g]^n]$				
ks [S_r=1]			a	n
[m/s]			[-]	[-]
3.70×10^{-7}			1	0.8
Evaluation of reliability				medium
Water retention curves				
van Genuchten 1980				
$(\theta-\theta_r)/(\theta_s-\theta_r)=[1/(1+(\alpha h)^n)]^m$ h=pressure head assumed positive [m]				
θ_s	θ_r	α	n	m
		[1/m]		
0.7	0.17	1	0.47	1.5
Evaluation of reliability				high

(b)

Figure 12. *Cont.*

Sarno pyroclastic soil							
Geographical information and layer data							
Latitude	Longitude	Average slope angle	Thickness cover	Thickness soil layer	Depth of the soil layer min	Depth of the soil layer max	
N	E	[°]	[m]	[m]	[m]	[m]	
40° 50′ 45″	14° 36′ 42″	40°	3.5	0.8–2.0	1	2.6	

Physical and state properties			
Gs	n	γ_d	γ_{sat}
	%	[kN/m^3]	[kN/m^3]
2.61	68	8.3	15.5
Grain size distribution of liquefiable soil [yes or no]		yes	

Saturated shear strength and critical state parameters			
c′	ϕ'	M	λ
[kPa]	[°]		
0	37	1.51	0.12
Undrained unstable response in CTX tests [yes or no]		yes	

Unsaturated shear strength			
mod. after Fredlund and Rahardjo 1993			
c′	ϕ'	$c(u_a-u_w)$	
[kPa]	[°]	[kPa]	
		if $(u_a-u_w)\le 1$	0
0	37	if $(u_a-u_w)>1$	$1.8\ln(u_a-u_w)$
Evaluation of reliability		high	
Lu et al. 2010			
$\sigma_s=-(u_a-u_w)/[1+(\alpha(u_a-u_w))^n]^m$			
α; n; m van Genuchten parameters			
Evaluation of reliability		medium	

Permeability functions				
Kunze et al. 1968				
$k= 1/\alpha^2 \cdot L^2/V \cdot [Q_o/\Delta u_o] \cdot \rho_w \cdot g \cdot b_\alpha$				
b_α from analysis of suction equalization (SCTX)				
ks [S_r=1]	k_P	L_P	$\alpha=L_P/K_P \cdot k/L$	
[m/s]	[m/s]	[mm]		
6 × 10^{-7}	1 × 10^{-7}	5		
Evaluation of reliability				high
Brooks and Corey 1964				
$k=ks$ if $(u_a-u_w)\le(u_a-u_w)_b$				
$k=ks[(u_a-u_w)_b/(u_a-u_w)]^\eta$ if $(u_a-u_w)\le(u_a-u_w)_b$				
$\eta=2+3\lambda$;		λ=pore size distribution index		
ks [S_r=1]	$(ua-uw)_b$	λ	η	
[m/s]	[kPa]	[-]	[-]	
6 × 10^{-7}	1.1	0.45	3.35	
Evaluation of reliability				low
Gardner 1958				
$k=ks/[1+a\,[(u_a-u_w)/\rho g]^n]$				
ks [S_r=1]		a	n	
[m/s]		[-]	[-]	
6 × 10^{-7}		1	0.87	
Evaluation of reliability				medium

Water retention curves				
van Genuchten 1980				
$(\theta-\theta_r)/(\theta_s-\theta_r)=[1/(1+(\alpha h)^n)]^m$				
h=pressure head assumed positive [m]				
θ_s	θ_r	α	n	m
		[1/m]		
0.7	0.18	0.9	0.45	1.4
Evaluation of reliability				high

(c)

Figure 12. Database of soil characteristics: (**a**) Camaldoli flow deposit; (**b**) Cervinara airfall deposit; (**c**) Sarno airfall deposit (literature expressions from [1,27–31]).

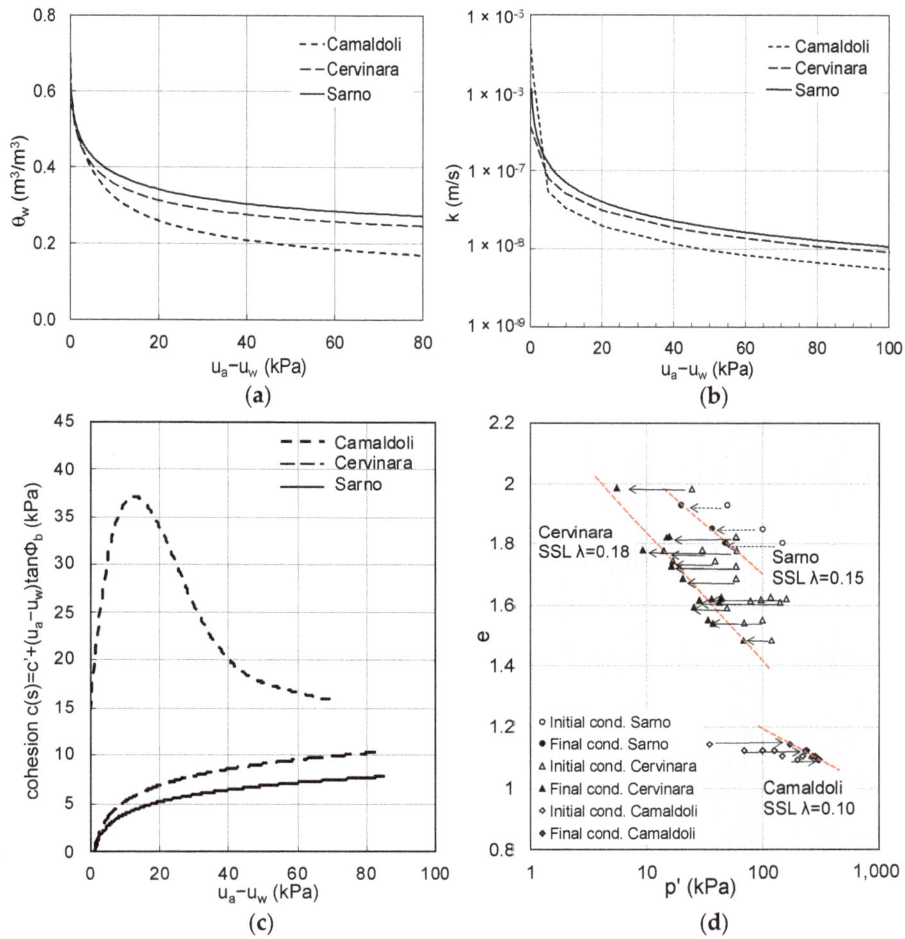

Figure 13. Comparison of the mechanical properties of three investigated pyroclastic deposits: (**a**) SWRCs; (**b**) HCFs; (**c**) apparent cohesion trends; and (**d**) undrained paths in the compression plane.

Afterward, a section that synthetizes the unsaturated shear strength in terms of the functional relationship between suction and apparent cohesion is provided. When a simple unique trend for this relationship is not found, as is the case for the Camaldoli ash, the information in the box reports the experimental points illustrated in the figure. In this section, the reliability of the expression proposed by [32], which is widely used in the literature for the evaluation of the unsaturated shear strength of granular soils, is also determined.

Finally, the last section of the SDB reports the hydraulic properties of the soil. The parameters of various expressions reported in the literature for soil water retention curves and the hydraulic conductivity function that best fits the experimental data are indicated therein.

Based on the data collected in the SDB and illustrated in Figure 13, the properties of the Camaldoli flow deposit are different from those of the Cervinara and Sarno airfall deposits. Although the three deposits are characterized by similar trends of hydraulic properties (Figure 13a,b), greater differences arise when we look at the shear strength in both unsaturated and saturated states. Indeed, under unsaturated conditions, the soil from the flow deposit shows a trend of apparent cohesion intercept very different from that of the airfall ash deposits, and they are also characterized by higher values (Figure 13c).

Despite the similar grain size distribution of the three deposits, which has a great influence on the relationship between suction and unsaturated shear strength [31,32], the significant difference observed in the experimentally obtained functions must be attributed to the dissimilar formation of the two kinds of deposits.

Analyzing the mechanical properties of the soils under saturated conditions, it can be seen that the effective friction angles of the three deposits are very similar ranging between 37 and 38°, but the flow ash deposit presents an effective cohesion of few kPa, as reported in the SBD (Figure 12). Considering the soil response under undrained loading, framed within the steady-state theory and illustrated in the compression plane of Figure 13d, various behaviors can be recognized. Firstly, the locations of the SSLs are very different, which is related to the dissimilar void ratios of the natural deposits (around 1.9 for fall deposits and 1.1 for flow deposits). The SSLs of the airfall ash deposits localize in the upper part of the compression plane, presenting a high inclination ($\lambda = 0.19$ for Cervinara and $\lambda = 0.12$ for Sarno), whereas the SSL of the flow ash deposit is located in the lower part and characterized by a lower inclination. More importantly, the undrained responses of the three deposits are significantly different. In the range of stresses that the soils may potentially experience at failure in situ after soil saturation, the undrained paths of the airfall deposits show an unstable behavior, differently from the flow deposit, which shows a stable behavior (de Cristofaro et al., 2022). Flow or surge deposits, characterized by lower porosity than airfall ash deposits, do not show a tendency toward liquefaction under the actual in situ stress level. In fact, by evaluating the state parameter ψ defined by [33] using the state conditions at the end of the consolidation stage of the CIU test in the range of p' representing in situ conditions ($20 < p' < 50$ kPa; $e = 1.14$ for Camaldoli; $e = 1.98$ for Cervinara; $e = 1.93$ for Sarno), it can be seen that it assumes a negative value for the Camaldoli ash soil (about -0.20), which represents a non-liquefiable deposit, whereas it assumes positive values ($0.13 \div 0.31$) for both the Cervinara and Sarno soils, indicative of a liquefiable deposit.

Again, these characteristics can be attributed to the different deposition modes of the materials: in the case of a flow deposit, a medium porosity soil structure results in a stable behavior in an undrained loading process. In contrast, the eolian transport over long distances, followed by slow fall deposition in the carbonatic context, allows for the formation of an open-spaced metastable soil structure characterized by porosity values as high as 70%, which leads to an unstable undrained response and a loss of shear resistance.

Assessment of the Post-Failure Evolution of Landslides in Pyroclastic Covers

Data collected in the SDB constitute the basic and essential information required not only to implement a physically based model for the analysis of the infiltration process and analyze the slope stability in a boundary value problem but also for forecasting the post-failure evolution of landslides. Indeed, data presented in the SDB of Figure 12, along with the flowchart illustrated in Figure 14, can be used in an EWS as a supporting tool for the identification of the post-failure evolution of rainfall-induced landslides in pyroclastic shallow covers of Campania. The flowchart is organized into two sections. The first concerns checking the degree of saturation of soil at the onset of instability as it is the discriminator between the potential development of a drained or undrained soil response, whereas the second allows for the assessment of the post-failure evolution of the sliding mass based on the information reported in the SDB.

The flowchart, originally proposed by [3] for cohesionless soils, was modified to extend the analysis to granular covers characterized by small values of cohesion, as in the case of the Camaldoli hill. In the simplified assumption of infinite slope and cohesionless soil, the degree of saturation at failure can be easily determined by comparing the acclivity of the slope with the effective friction angle: for slope inclination less or equal to the friction angle of the soil (gentle or steep slope) failure occurs in saturated conditions, whereas for slope inclination higher than the friction angle (very steep slope), failure involves partially saturated soil. However, in the case of soils with effective cohesion, it is not possible to assess the degree of saturation at failure simply by comparing the friction angle with the slope angle. Moreover, differently from the case of cohesionless soil, the safety factor is no longer independent from the depth, decreasing with an increase in depth and introducing another variable, the depth of the sliding surface, to the equilibrium analysis. Therefore, the critical slope angle (α_{crit}) is introduced instead of the slope angle (α), defined as the slope inclination at which failure occurs under saturated conditions. This critical angle must be determined case by case, eventually assuming a simplified hypothesis. Obviously, for homogeneous cohesionless soils, the critical slope angle is equal to the friction angle of the material.

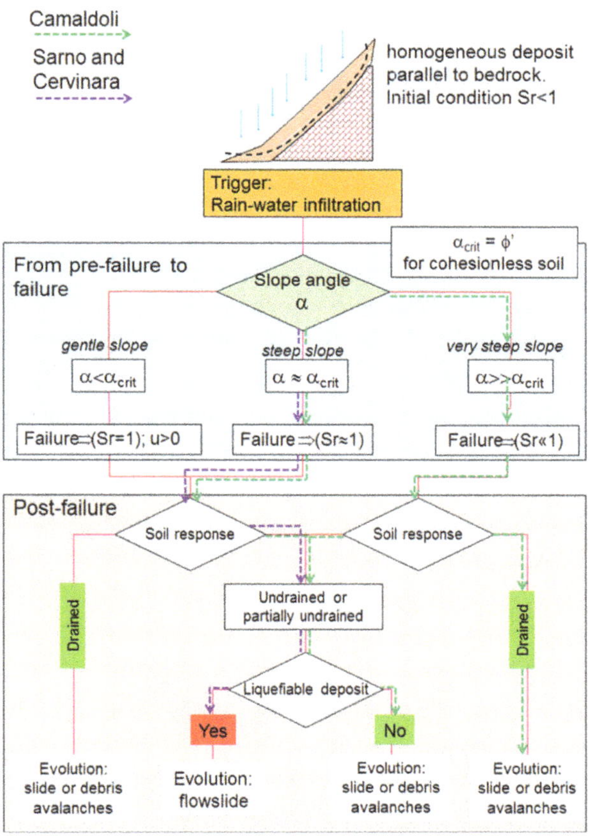

Figure 14. Flowchart for simplified assessment of the post-failure evolution of landslides in granular soils (red box highlights the potential post-failure evolution in flowslide; green boxes denote evolution in slide or debris avalanches).

Here, an example of the application of this method for the three investigated sites is proposed. The analysis starts with the evaluation of the degree of saturation of soil at failure. In the case of Cervinara and Sarno cohesionless soil deposits, which have effective friction angles close to the inclination of the slopes ($\phi' \approx \alpha_{crit}$), failure occurs in an almost saturated condition. In the flowchart, the path to be followed (purple line) is the one corresponding to a steep slope ($\alpha \approx \alpha_{crit}$).

In the case of Camaldoli hill, as the soil presents an intercept of effective cohesion, the first step consists of the evaluation of α_{crit}. Based on the observation of detachment areas, which shows that the sliding surfaces are located at about 1.5 m of depth, α_{crit} is determined by performing stability analysis at different inclinations of the slope, in the simplified hypothesis of a constant suction profile and by using the peak state parameters ($\phi' = 38°$ and $c' = 15$ kPa). The analysis results show that it corresponds to a slope inclination of 65°: In the presence of α lower than α_{crit}, slope failure occurs in saturated conditions; for α higher than α_{crit}, the mobilization of the soil deposit occurs in soil which is still in unsaturated conditions. Along the Camaldoli hill, slope instabilities mainly involve the upper part of the slope above the vertical tuff cliff where soil deposits present local acclivity higher than 60°. Thus, instability may occur regardless of whether the soil is saturated or not: the potential paths to follow in the flowchart (green lines) correspond to both the last two branches related to the steep slope ($\alpha \approx \alpha_{crit}$) and very steep slope ($\alpha \approx \alpha c_{rit}$).

Afterward, the assessment of the post-failure evolution of the landslide can be carried out by looking at the second part of the flowchart. If the post-failure movement occurs in a drained condition, landslide phenomena evolve into slide or debris avalanche; if it occurs under undrained or partially undrained conditions, the evolution of the landslide is related to soil susceptibility to liquefaction.

In the case of the Sarno and Cervinara deposits, as evaluated through infiltration tests in small-scale slopes reconstituted in a multi-instrumented flume [34,35], in the very early post-failure stage, soils exhibit an undrained response, as evidenced by the high excess pore pressures observed. Thus, we can infer that the path to follow in the flowchart is the one corresponding to an undrained or partially undrained response, which, in the presence of such liquefiable deposits, leads to the prediction of evolution into a flowslide.

Different is the case of the Camaldoli slope. Indeed, although we cannot predict if a drained or undrained condition will occur at the onset of failure because the deposit is non-liquefiable, the slope movement will evolve into a slide or a debris avalanche.

In fact, as reported by [14], a significant feature of most landslides occurring in pyroclastic deposits in the volcanic context is the absence of any signs of liquefaction.

6. Conclusions

A wide mountainous area in Campania (southern Italy) is threatened by rainfall-induced landslides, which mainly affect the granular pyroclastic covers and evolve into rapid slides, debris avalanches, or catastrophic flowslides deeply affecting urban developments at the foot slope. For these densely urbanized areas, knowledge of both the landslide susceptibility of the territory and the type of evolution of the landslide mass, which can impact the exposed areas in a very different way, is mandatory.

To this end, in this paper, an extensive testing program was developed to investigate both unsaturated and saturated mechanical properties and evaluate soil susceptibility to liquefaction of volcanic ash belonging to a flow deposit, frequently involved in debris avalanches. Moreover, the properties of other types of ash belonging to airfall deposits, involved in flowslides, were compared to highlight the similarities and dissimilarities between the two types of deposits and explain the different patterns of the post-failure evolution of the landslide mass. Based on the experimental results, a soil database for hydro-geotechnical models was presented, which contains all the data needed for an advanced constitutive soil modeling to implement a physically based model for forecasting the initiation of landslides, thus improving the effectiveness of early warning systems. In addition, a framework was proposed for the assessment of landslide evolution under a simplified hypothesis that can be used by stakeholders for land management and the protection of areas at risk.

Author Contributions: Conceptualization, E.D. and L.O.; methodology, E.D. and L.O.; formal analysis, all authors; investigation, M.d.C., L.I., G.C. and A.B.; data curation, M.d.C. and L.I.; writing—original draft preparation, E.D. and L.I.; writing—review and editing, all authors; supervision, E.D. and L.O.; funding acquisition, L.O. All authors have read and agreed to the published version of the manuscript.

Funding: This research was funded by the Program VALERE: "VAnviteLli pEr la RicErca", DDG n. 516–24/05/2018.

Data Availability Statement: The data used in this research work are available upon request from the corresponding authors.

Conflicts of Interest: The authors declare no conflict of interest.

References

1. Fredlund, D.G.; Rahardjo, H. Shear Strength Theory. In *Soil Mechanics for Unsaturated Soils*; Wiley: New York, NY, USA, 1993; pp. 217–259. [CrossRef]
2. Cascini, L.; Cuomo, S.; Guida, D. Typical source areas of May 1998 flow-like mass movements in the Campania region, Southern Italy. *Eng. Geol.* **2008**, *96*, 107–125. [CrossRef]
3. Damiano, E.; Olivares, L. The role of infiltration processes in steep slope stability of pyroclastic granular soils: Laboratory and numerical investigation. *Nat. Hazards* **2010**, *52*, 329–350. [CrossRef]
4. Sorbino, G.; Nicotera, M.V. Unsaturated soil mechanics in rainfall-induced flow landslides. *Eng. Geol.* **2013**, *165*, 105–132. [CrossRef]
5. De Vita, P.; Di Clemente, E.; Rolandi, M.C.P. Engineering geological models of the initial landslides occurred on April 30, 2006, at Monte Vezzi (Ischia Island, Italy). *Ital. J. Eng. Geol. Environ.* **2007**, *2*, 119–141.
6. Olivares, L.; Damiano, E.; Mercogliano, P.; Picarelli, L.; Netti, N.; Schiano, P.; Savastano, V.; Cotroneo, F.; Manzi, M.P. A simulation chain for early prediction of rainfall-induced landslides. *Landslides* **2014**, *11*, 765–777. [CrossRef]
7. Cascini, L.; Cuomo, S.; Pastor, M.; Sorbino, G. Modeling of Rainfall-Induced Shallow Landslides of the Flow-Type. *J. Geotech. Geoenviron. Eng.* **2010**, *136*, 85–98. [CrossRef]
8. Marino, P.; Santonastaso, G.F.; Fan, X.; Greco, R. Prediction of shallow landslides in pyroclastic-covered slopes by coupled modeling of unsaturated and saturated groundwater flow. *Landslides* **2021**, *18*, 31–41. [CrossRef]
9. Olivares, L.; Damiano, E.; Netti, N.; de Cristofaro, M. Geotechnical properties of two pyroclastic deposits involved in catastrophic flowslides for implementation in early warning systems. *Geosciences* **2019**, *9*, 24. [CrossRef]
10. Olivares, L.; Picarelli, L. Shallow flowslides triggered by intense rainfalls on natural slopes covered by loose unsaturated pyroclastic soils. *Geotechnique* **2003**, *53*, 283–287. [CrossRef]
11. Picarelli, L.; Olivares, L.; Lampitiello, S.; Darban, R.; Damiano, E. The undrained behaviour of an air-fall volcanic ash. *Geosciences* **2020**, *10*, 60. [CrossRef]
12. de Falco, M.; Di Crescenzo, G.; Santo, A. Volume estimate of flow-type landslides along carbonatic and volcanic slopes in Campania (Southern Italy). *Nat. Hazards* **2012**, *61*, 51–63. [CrossRef]
13. Picarelli, L.; Evangelista, A.; Rolandi, G.; Paone, A.; Nicotera, M.V.; Olivares, L.; Scotto di Santolo, A.; Lampitiello, S.; Rolandi, M. Mechanical Properties of Pyroclastic Soils in Campania Region. In Proceedings of the 2nd International Workshop on Characterization and Engineering Properties of Natural Soils, Singapore, 29 November–1 December 2006; Volume 4, pp. 2331–2384.
14. Picarelli, L.; Olivares, L.; Damiano, E.; Darban, R.; Santo, A. The effects of extreme precipitations on landslide hazard in the pyroclastic deposits of Campania Region: A review. *Landslides* **2020**, *17*, 2343–2358. [CrossRef]
15. Calcaterra, D.; Coppin, D.; de Vita, S.; Di Vito, M.; Orsi, G.; Palma, B.; Parise, M. Slope processes in weathered volcaniclastic deposits within the city of Naples: The Camaldoli Hill case. *Geomorphology* **2007**, *87*, 132–157. [CrossRef]
16. de Falco, M.; di Crescenzo, G.; Santo, A. Empirical relationships to predict the runout of flow-type landslides in pyroclastic deposits in Campania region (Southern Italy). *Rend. Online Soc. Geol. Ital.* **2013**, *24*, 86–88.
17. Zona, R.; de Cristofaro, M.; Esposito, L.; Ferla, P.; Palladino, S.; Totaro, E.; Olivares, L.; Minutolo, V. Early warning system for landslide risk and SHM by means of reinforced optic fiber in lifetime strain analysis. In Proceedings of the 5th International Conference on Internet of Things, Big Data and Security, Prague, Czech Republic, 7–9 May 2020. [CrossRef]
18. Minutolo, V.; Cerri, E.; Coscetta, A.; Damiano, E.; De Cristofaro, M.; Di Gennaro, L.; Esposito, L.; Ferla, P.; Mirabile, M.; Olivares, L.; et al. NSHT: New smart hybrid transducer for structural and geotechnical applications. *Appl. Sci.* **2020**, *10*, 4498. [CrossRef]
19. Hunter, G.; Fell, R. Prediction of impending failure of embankments on soft ground. *Can. Geotech. J.* **2003**, *40*, 209–220. [CrossRef]
20. Olivares, L.; Picarelli, L. Occurrence of flowslide in soils of pyroclastic origin and considerations for landslide hazard mapping. In Proceedings of the Fourteenth Southeast Asian Geotechnical Conference, Hong Kong, China, 14 December 2001.
21. Carrieri, G. Caratterizzazione di Depositi Piroclastici Saturi e Parzialmente Saturi. Ph.D. Thesis, Seconda Università degli Studi di Napoli, Caserta, Italy, 2013. [CrossRef]
22. Poulos, S.J.; Castro, G. Liquefaction Evaluation Procedure. *J. Geotech. Eng.* **1985**, *111*, 772–792. [CrossRef]
23. de Cristofaro, M. Essential Elements to Define Early Warning Systems in Volcanic Deposits. Ph.D. Thesis, University of Campania 'Luigi Vanvitelli', Caserta, Italy, 2022.
24. Lee, I.-M.; Sung, S.-G.; Cho, G.-C. Effect of stress state on the unsaturated shear strength of a weathered granite. *Can. Geotech. J.* **2005**, *42*, 624–631. [CrossRef]
25. Schindler, U.; Müller, L. Simplifying the evaporation method for quantifying soil hydraulic properties. *J. Plant Nutr. Soil Sci.* **2006**, *169*, 623–629. [CrossRef]
26. Topp, G.C.; Davis, J.L.; Annan, A.P. Electromagnetic determination of soil water content: Measurements in coaxial transmission lines. *Water Resour. Res.* **1980**, *16*, 574–588. [CrossRef]

27. van Genuchten, M.T. A Closed-form Equation for Predicting the Hydraulic Conductivity of Unsaturated Soils. *Soil Sci. Soc. Am. J.* **1980**, *44*, 892–898. [CrossRef]
28. Kunze, R.J.; Uehara, G.; Graham, K. Factors Important in the Calculation of Hydraulic Conductivity. *Soil Sci. Soc. Am. J.* **1968**, *32*, 760–765. [CrossRef]
29. Brooks, R.; Corey, A. *Hydraulic Properties of Porous Media*; Hydrology Papers; Colorado State University: Fort Collins, CO, USA, 1964.
30. Gardner, W.R. Some steady-state solutions of the unsaturated moisture flow equation with application to evaporation from a water table. *Soil Sci.* **1958**, *85*, 228–232. [CrossRef]
31. Lu, N.; Godt, J.W.; Wu, D.T. A closed-form equation for effective stress in unsaturated soil. *Water Resour. Res.* **2010**, *46*. [CrossRef]
32. Lu, N.; Likos, W.J. Suction Stress Characteristic Curve for Unsaturated Soil. *J. Geotech. Geoenviron. Eng.* **2006**, *132*, 131–142. [CrossRef]
33. Been, K.; Jefferies, M.G. A state parameter for sands. *Géotechnique* **1985**, *35*, 99–112. [CrossRef]
34. Darban, R.; Damiano, E.; Minardo, A.; Olivares, L.; Picarelli, L.; Zeni, L. An Experimental Investigation on the Progressive Failure of Unsaturated Granular Slopes. *Geosciences* **2019**, *9*, 63. [CrossRef]
35. Damiano, E.; Darban, R.; Olivares, L.; Picarelli, L. An investigation on progressive failure in granular slopes leading to flow-like landslides. In Proceedings of the 13th International Symposium on Landslides, Cartagena, Colombia, 15–19 June 2020.

International Journal of
Geo-Information

Article

Application of AHP-ICM and AHP-EWM in Collapse Disaster Risk Mapping in Huinan County

Zengkang Lu [1], Chenglong Yu [1,*], Huanan Liu [2], Jiquan Zhang [3], Yichen Zhang [1], Jie Wang [1] and Yanan Chen [1]

1 College of Jilin Emergency Management, Changchun Institute of Technology, Changchun 130012, China; 20221311204@stu.ccit.edu.cn (Z.L.); zhangyc@ccit.edu.cn (Y.Z.); 20221311210@stu.ccit.edu.cn (J.W.); chenyn061@nenu.edu.cn (Y.C.)
2 School of Prospecting and Surveying, Changchun Institute of Technology, Changchun 130021, China; liuhuanan@ccit.edu.cn
3 School of Environment, Northeast Normal University, Changchun 130024, China; zhangjq022@nenu.edu.cn
* Correspondence: yuchenglong@ccit.edu.cn; Tel.: +86-1350-088-8781

Abstract: Collapses are one of the most common geological disasters in mountainous areas, which easily damage buildings and infrastructures and bring huge property losses to people's production and life. This paper uses Huinan County as the study area, and with the help of a geographic information system (GIS) based on the formation principle of natural disaster risk, the information content method (ICM), the analytical hierarchy process (AHP), and the analytical hierarchy process–information content method (AHP-ICM) model are applied to hazard mapping, and the analytical hierarchy process-entropy weight method (AHP-EWM) model is applied to exposure, vulnerability and emergency responses, and recovery capability mapping. A risk mapping model for collapse disasters was also constructed using these four elements. Firstly, an inventory map of 52 landslides was compiled using remote sensing interpretation, field verification, and comprehensive previous survey data. Then, the study area mapping units were delineated using the curvature watershed method in the slope unit, and 21 indicators were used to draw the collapse disaster risk zoning map by considering the four elements of geological disaster risk. The prediction accuracy of the three hazard mapping models was verified using the receiver operating characteristic (ROC) curve, and the area under the curve (AUC) results of the AHP, ICM, and AHP-ICM models were 80%, 85.7%, and 87.4%, respectively. After a comprehensive comparison, the AHP-ICM model is the best of the three models in terms of collapse hazard mapping, and it was applied to collapse risk mapping with the AHP-EWM model to produce a reasonable and reliable collapse risk zoning map, which provides a basis for collapse management and decision making.

Keywords: risk mapping; GIS; analytical hierarchy process; information content method; spatial analysis

1. Introduction

Geological hazards such as collapses and landslides often occur in mountainous areas and can pose a major threat to the safety of life, property, and transport infrastructure in areas of intense human activity [1,2]. A collapse is a phenomenon in which a mountain, earth or rock deposit, or building loses its stable balance and collapses partially or completely due to geological action or external forces, mostly among steep-angled mountains [3], and may be caused by factors such as heavy rainfall [4], seismic activity [5], and human engineering activities [6] (mining, slope excavation, etc.). To better identify areas where collapse disasters may occur and to take effective prevention and control measures, risk mapping of collapse disasters is essential [7–9].

In recent years, driven by the development of mountainous cities, many road construction projects require collapse hazard and risk evaluation to protect people's lives and properties [10–12]. According to statistics, in 2017 alone, more than 2800 geological disasters occurred in China, resulting in hundreds of deaths and economic losses of more than CNY 1 billion. Among them, repeated collapse disasters occurred in some mountainous areas, destroying a large amount of infrastructure and farmland and causing a large number of rural people to lose their production and livelihood. Although the size of rock masses is relatively small, these rock falls can instantly cause huge losses and hazards due to their suddenness and the huge impact they contain, especially along roads [13].

At present, great progress has been made in the work of collapse disaster risk evaluation, which mainly consists of three major categories: qualitative analysis, quantitative analysis, and a combination of both [14–19]. The analytical hierarchy process (AHP) is a more widely used method for qualitative analysis of collapse disaster risk evaluation [18,20,21], which is a multi-objective decision-making method with a high degree of subjectivity for determining the degree of contribution of each collapse indicator to collapse risk. Quantitative analysis is a method of analysing problems using mathematical models, which takes objective data as input and applies tools such as mathematical and statistical analysis to calculate the results and analyse them, such as the information content method (ICM) [22], frequency ratio method (FRM) [17,23], entropy weight method (EWM) [24], and logistic regression (LR) [25,26]. Using only one method is too subjective or relies too much on mathematical models and yields somewhat one-sided weight values. However, combining the two types of weight-defining methods can make up for the single deficiency and achieve the integration of subjective and objective and qualitative and quantitative. In this study, AHP is combined with ICM and EWM, respectively, based on the advantages of different methods and ArcGIS 10.8.1 to investigate the riskiness of collapse disasters in Huinan County.

In collapse risk mapping, the selection of a suitable mapping unit is a key step. At present, more and more scholars choose slope units based on the characteristics of collapses [22,27], while the hydrological analysis method is more cumbersome in GIS extraction, and the method cannot identify the horizontal plane correctly. To give a more complete picture of the slope where the collapse site is located and thus be able to carry out a more practical and reasonable division and improve the model prediction accuracy, the curvature watershed method [25] was chosen for the unit division of the mapping unit in this study.

The purpose of this study is to provide a basis for the establishment of the riskiness model of collapse in Huinan County, Tonghua City, Jilin Province. Based on the 1:50,000 geological hazard survey in Huinan County, a total of 52 collapse disaster sites in the study area were counted. The curvature watershed method was used to divide the entire study area into units and then apply the formation principle of natural disaster risk [28]. Evaluation indexes were selected from four standard layers: disaster hazard, exposure, vulnerability, and emergency response and recovery capability of the disaster bearers, and correlation analysis was performed on the selected indicators. Then, the AHP, ICM, and integrated AHP and ICM methods were used to establish the hazard mapping models, and the AHP-EWM method was used to establish the exposure, vulnerability, and emergency response and recovery capability mapping models. Finally, the predictive ability of the three hazard mapping models was verified by using the receiver operating characteristic (ROC) curve, and the optimal model was selected and plotted with the AHP-EWM model. The risk map of Huinan County was created, and risk mapping of collapse disasters in Huinan County was performed. The results of the study will provide guidance for risk management and disaster prevention and mitigation in Huinan County and provide a reference for policymakers.

2. Data and Methodology

2.1. Overview of the Study Area

The research area is in Huinan County, Tonghua City, Jilin Province (Figure 1), with a total area of about 2277 km^2 and geographical coordinates at 125°58′–126°45′ E and 42°16′–42°54′ N. The terrain is characterised by high southeast and low northwest points, with the highest point at 1222 m and the lowest point at 187 m above sea level. The climate of the study area is a north temperate continental monsoon, with an average annual precipitation of 696.3 mm according to meteorological statistics from 1998 to 2014, with precipitation mainly concentrated in the period of June to August. Field investigations show that the landslides are mainly located along the Huinan–Jingyu highway and railway, as well as in the mining areas of Chaoyang Town, Huinan Town, Fumin Town, etc. The types of slopes forming the landslides are mainly artificial rocky slopes, followed by natural rocky slopes. The geotechnical types are more complex, mainly basalt (β), granite (γ), hard clastic rock (Q + Z), softer clastic rock (K + J), limestone (∈ + O + Z), metamorphic rock (As) and clay and gravel double-layered soil (Q). Possible destabilising factors are rainfall, frost heave, and human damage, with collapse mostly occurring during the rainfall process.

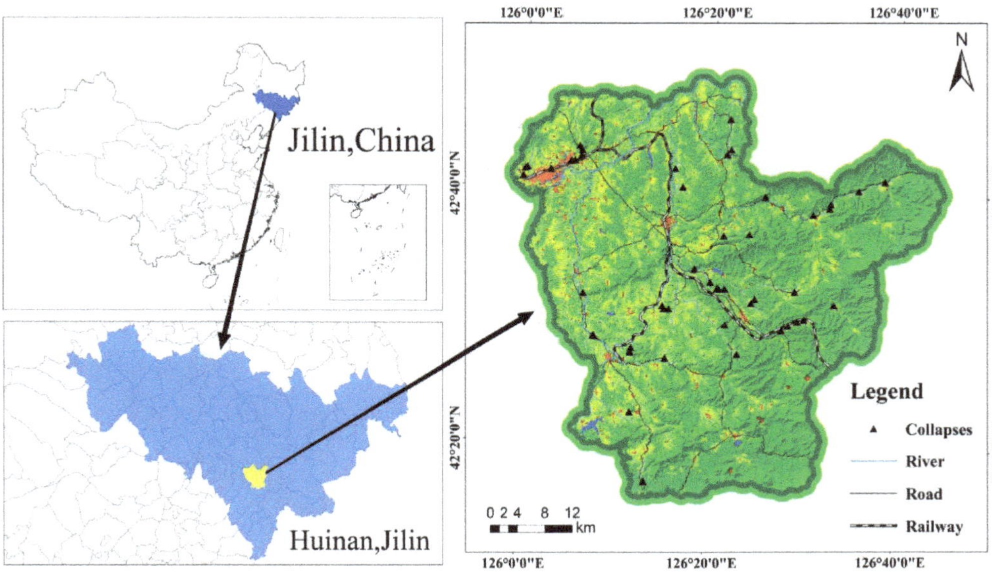

Figure 1. Location map of the study area.

2.2. Collapse Evaluation Indicators

2.2.1. Establishment of the Indicator System

This study is based on the evaluation indicators selected from the formation principle of natural disaster risk. For disaster hazards, in addition to reviewing the relevant literature, we also combined the geological characteristics of the study area and the mechanism of the occurrence of collapse hazards and filtered out the high-correlation evaluation indicators with the occurrence of collapses after comparison. For the exposure, vulnerability, emergency response, and recovery capability of the disaster acceptor, indicators related to population, economy, and education, as well as infrastructure, were considered from the perspective of human society and according to the population, economic, and social situation of the study area.

In summary, 21 evaluation indicators were selected for the study to apply to the collapse risk mapping, as follows:

Selection of Hazard Indicators

The hazard contains 11 indicators, namely slope angle, slope aspect, multi-year average precipitation, distance from the fault, distance from the river, distance from the road, landform type, vegetation type, NDVI, lithology, and mining point density. Slope angle is a crucial factor when examining the potential for collapses in an area, and where the slope angle is large, the risk of collapses is higher due to the increased gravitational effect of the mountain and reduced resistance [29]. The slope aspect is one of the topographic factors, and differences in the slope aspect can result in differences in the direction of light exposure, light intensity, and hours of light on the slope [30]. Rainfall is a primary external factor of geological disasters in the region. Both surface water and groundwater's perennial scouring and erosion effects on geotechnical bodies will reduce the mechanical properties of geotechnical bodies [31], providing conditions for the formation of disasters such as collapse and landslides. The severity, distribution, and type of geological disasters are closely related to lithology, and lithology is an intrinsic factor in the development or occurrence of geological disasters [32]. The distance from the fault reflects the influence of geological formations on the formation and development of collapse disasters in many ways. The development of faults reduces the overall joint strength of the rock mass, resulting in a reduction in shear strength, which makes the rock mass vulnerable to destabilisation and damage under the action of external factors [18]. Distance from the river reflects the gradient relationship between rivers and the possible occurrence of collapses, with hills close to rivers being more prone to collapse disasters [25,33]. Distance from the road and mining point density [33] represent the intensity of human engineering activities, which can cause the movement of geotechnical bodies and groundwater transport under the influence of human activities and often cause geological disasters when beyond a stable state. Topography is the main controlling factor in determining the extent of geological disasters, and the presence of certain topographical conditions must be accompanied by a corresponding type of geological disaster. To a certain extent, a certain topography necessarily corresponds to a certain type of geological disaster. Lush vegetation can effectively reduce the degree of influence of the external environment on geological disasters [34,35], and geological disasters are less likely to be caused in low human activity areas with lots of vegetation. The spatial distribution of the 11 hazard indicators is shown in Figure 2.

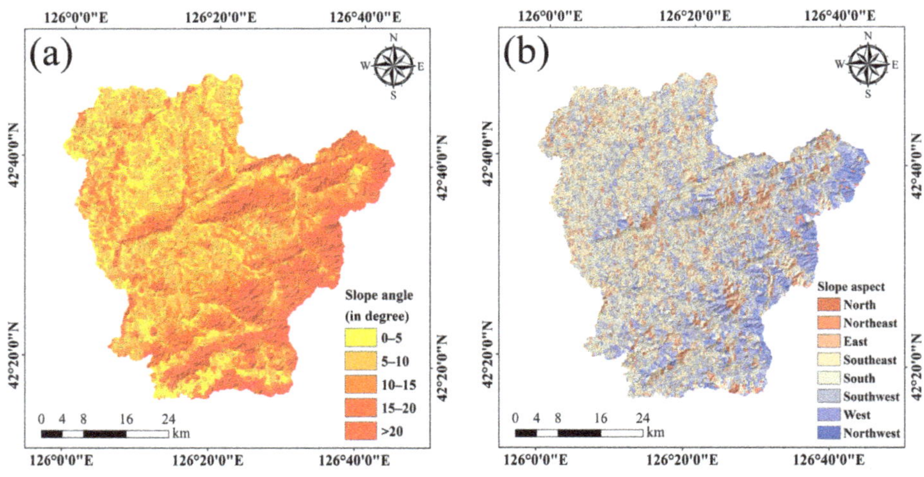

Figure 2. *Cont.*

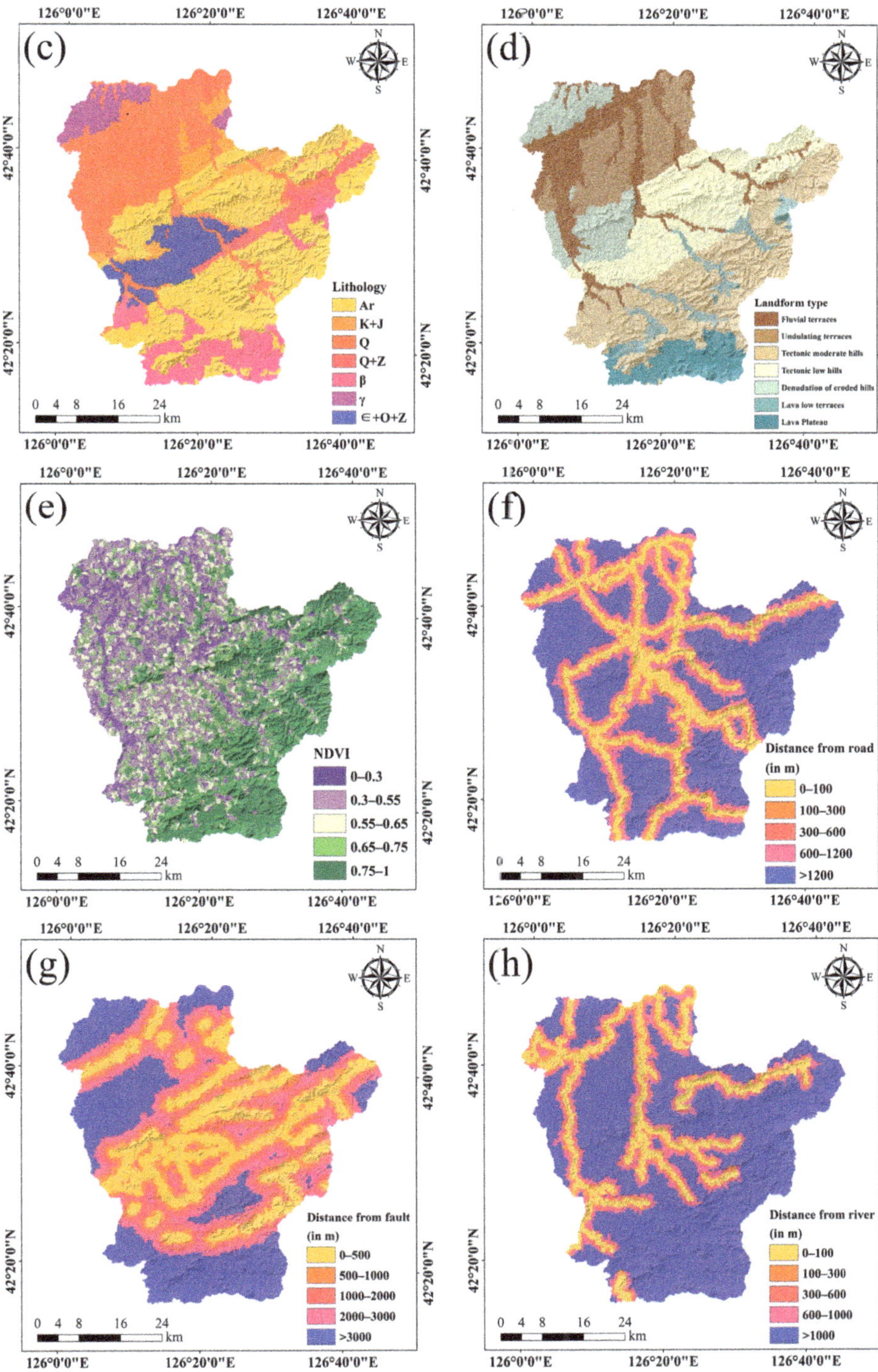

Figure 2. *Cont.*

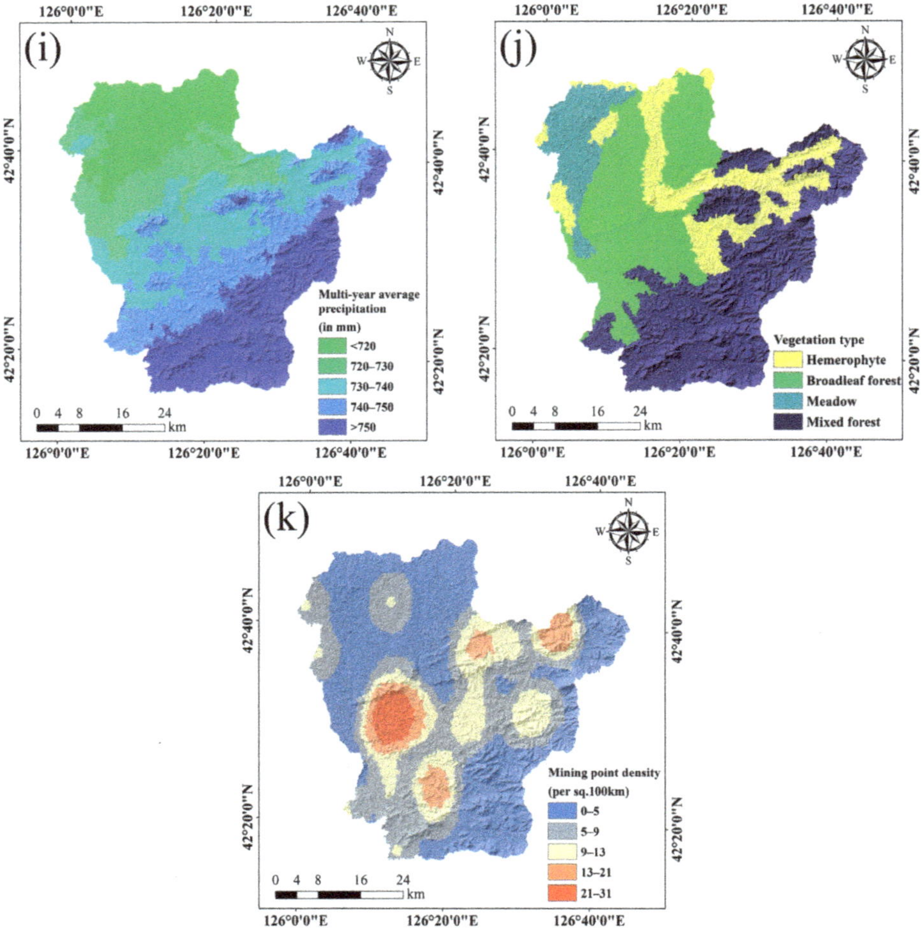

Figure 2. Hazard indicator maps: (**a**) slope angle; (**b**) slope aspect; (**c**) lithology; (**d**) landform type; (**e**) NDVI; (**f**) distance from road; (**g**) distance from fault; (**h**) distance from river; (**i**) multi-year average precipitation; (**j**) vegetation type; and (**k**) mining point density.

Selection of Exposure Indicators

Exposure refers to the economic, social, and natural environmental systems that are vulnerable to geological disasters, including agriculture, human living conditions, and the ecological environment. In this study, exposure mapping selected indicators such as population density, road density, housing density, and GDP are used. Population density is calculated by dividing the population of a township by its area. Areas with high population density have greater pressure on the living environment, leading to higher corresponding exposure [36], which is more severely affected by collapses. GDP is a measure of economic development [37], which is taken from a kilometre grid dataset of the spatial distribution of GDP in China, and the higher the GDP of the study area, the higher its exposure. Housing density and road density, which are key indicators of the urban development scale [38], are important factors regarding exposure, whereby dense housing and roads reflect greater exposure. The spatial distribution of the four exposure indicators is shown in Figure 3.

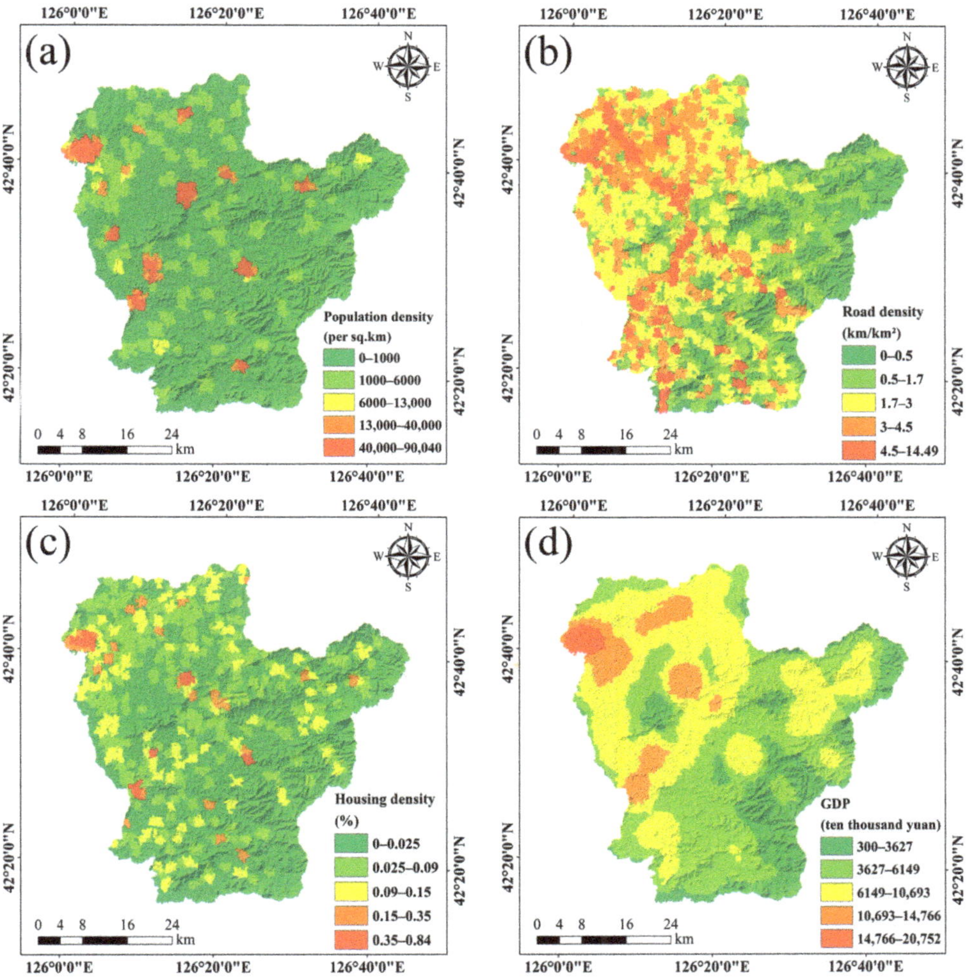

Figure 3. Exposure indicator maps: (**a**) population density; (**b**) road density; (**c**) housing density; and (**d**) GDP.

Selection of Vulnerability Indicators

Vulnerability refers to the degree of damage and loss to the disaster bearers in Huinan County due to potential disaster factors, including population, property, and ecosystems. The level of human vulnerability depends on factors such as age and awareness of disaster preparedness [28]. The selected vulnerability indicators in this study include the proportion of vulnerable populations, education status, and residential buildings. The vulnerable population includes people aged 0–17 and over 60 years old, as well as those with poor health who are more susceptible to disasters [28,38]. Education increases awareness of disaster preparedness, leading to reduced vulnerability. Residential buildings are human property and are more prone to collapse disasters when located in close proximity to mountainous areas [37]. The spatial distribution of these three vulnerability indicators is displayed in Figure 4.

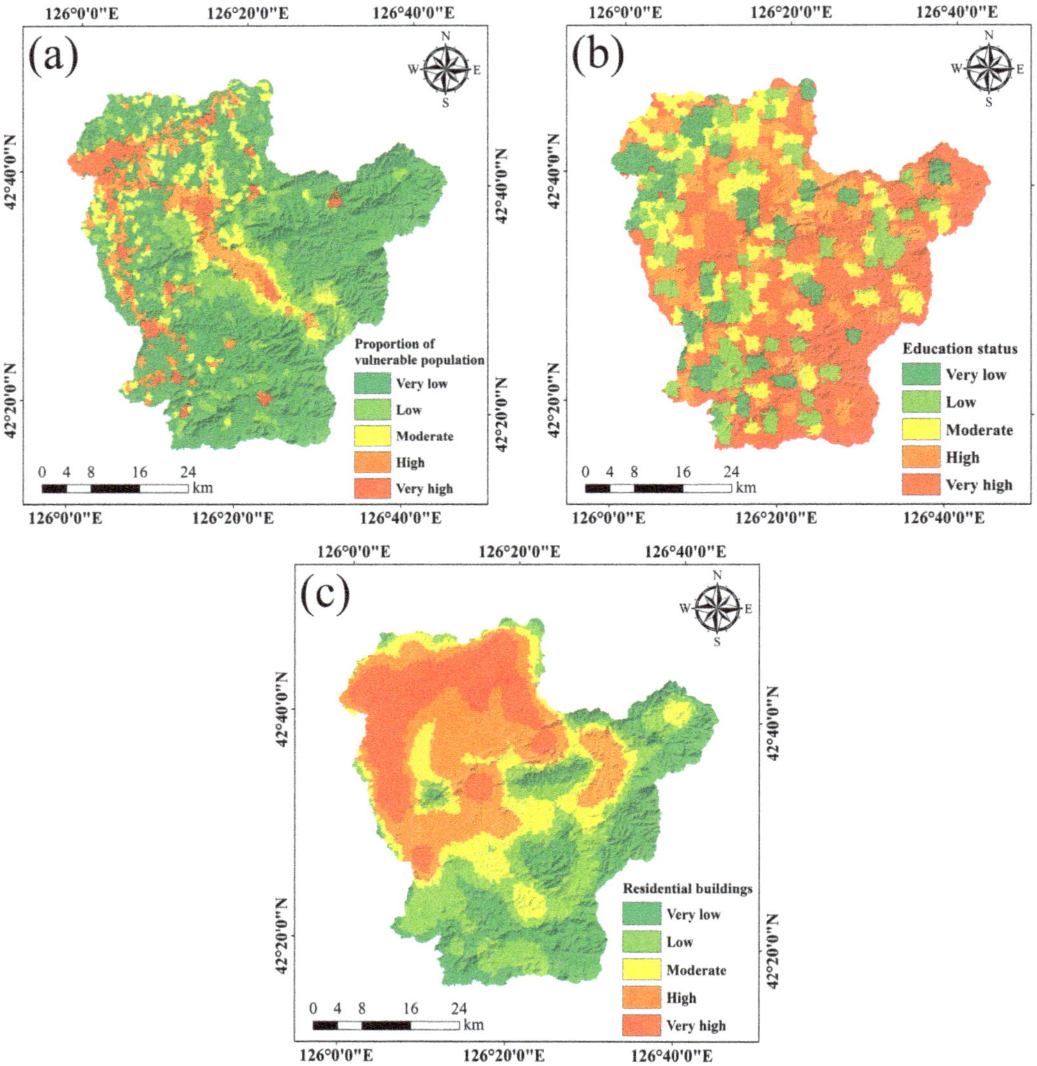

Figure 4. Vulnerability indicator maps: (**a**) proportion of vulnerable population; (**b**) education status; and (**c**) residential buildings.

Selection of Emergency Response and Recovery Capability Indicators

Emergency response and recovery capability refer to a series of disaster prevention measures taken to reduce the damage caused when a disaster bearer is affected by a collapse disaster. The level of local economic development and the number of emergency response agencies reflect the level of emergency response and recovery capability. In this study, education investment, local financial revenue, and the capacity of relief agencies are selected as indicators of emergency response and recovery capability. The spatial distribution of these three indicators is presented in Figure 5.

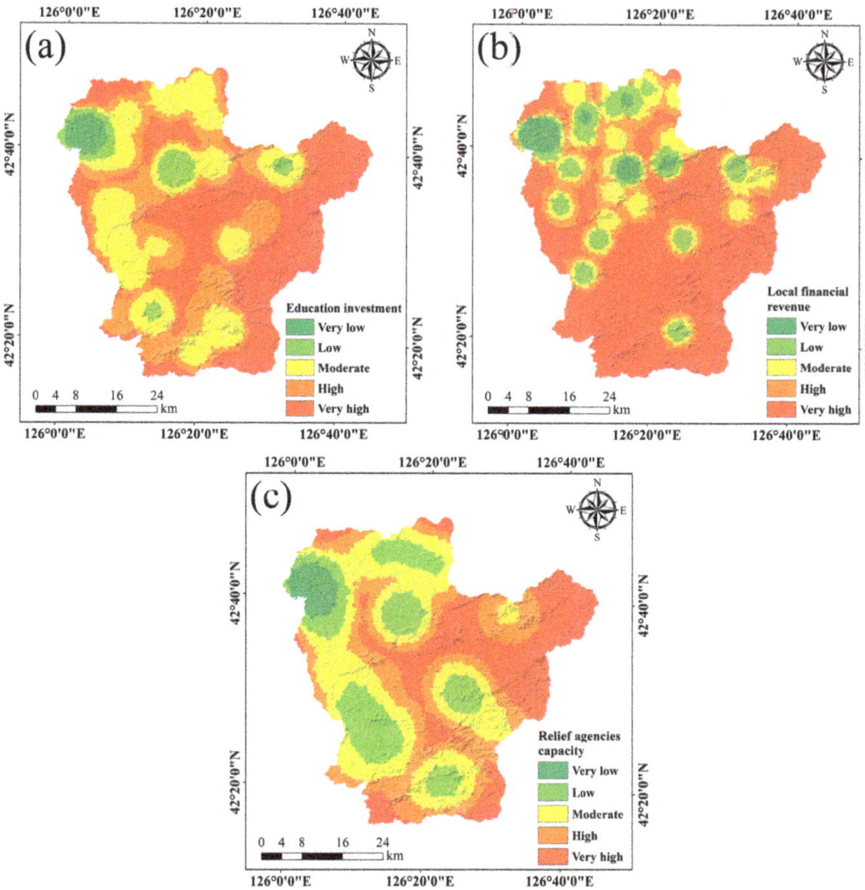

Figure 5. Emergency response and recovery capability indicator maps: (**a**) education investment; (**b**) local financial revenue; and (**c**) relief agencies' capacity.

2.2.2. Multicollinearity Analysis of Evaluation Indicators and Results

There may be statistical covariance between the initially selected collapse evaluation indicators, which can lead to the vulnerability model not being able to accurately analyse the true relationship between the evaluation indicators and collapses. Before building the model, this paper uses tolerance (TOL) and variance inflation factor (VIF) to test for factor covariance to ensure that the indicators are independent of each other [39]. The VIF is calculated as follows:

$$VIF = \frac{1}{1 - R_i^2} \quad (1)$$

where R_i is the negative correlation coefficient of an evaluation indicator to other evaluation indicators. It is generally accepted that when the VIF value is greater than 10, there is strong co-collinearity between the factors, and the variable should be removed from the model.

Multiple covariance analysis was performed on the evaluation indicators by VIF and TOL to ensure that each evaluation indicator was independent of the other, and if the VIF of an evaluation indicator was greater than 10, it was excluded. As can be seen from the results in Table 1, the VIF values of all indicators were less than 10, indicating that each indicator was independent of the other and did not need to be excluded.

Table 1. Variance inflation factor (VIF) values and tolerance (TOL) for evaluation indicators.

Indicator	TOL	VIF
Local financial revenue	0.16	6.248
Distance from road	0.179	5.602
Multi-year average precipitation	0.266	3.755
Vegetation type	0.34	2.942
Education investment	0.351	2.849
Slope aspect	0.369	2.707
Lithology	0.369	2.712
Education status	0.432	2.312
Population density	0.477	2.095
Landform type	0.56	1.787
Road density	0.578	1.731
Distance from river	0.606	1.65
Proportion of vulnerable population	0.627	1.595
Housing density	0.75	1.333
GDP	0.796	1.256
Relief agencies' capacity	0.8	1.251
NDVI	0.846	1.182
Slope angle	0.898	1.114
Distance from fault	0.906	1.104
Mining point density	0.913	1.095
Residential buildings	0.947	1.055

2.3. Data Collection

This study constructs a collapse risk mapping indicator system for Huinan County from four elements: hazard, exposure, vulnerability, and emergency response and recovery capability. The data mainly include vector and raster data from remote sensing, meteorology, and basic geographic information, as well as attribute data related to population and economy. The data for lithology, distance from fault, landform type, mining point density, housing density, road density, and residential buildings were obtained from the Report on Geological Disaster Investigation and Mapping in Huinan County, Jilin Province, where the distance from fault was processed using the Euclidean Distance, the mining point density and residential buildings were processed using the Kernel Density, and the housing density and road density were processed through the Calculate Geometry of the Attribute Table. The slope angle and slope aspect are derived from DEM digital elevation data (ASTER GDEM 30 M) from the Geospatial Data Cloud, which can be extracted directly by ArcGIS 10.8.1. The data on distance from the river and distance from the road were obtained from the National Catalogue Service for Geographic Information and then processed using Euclidean Distance. Multi-year average precipitation was obtained directly from the Resource and Environment Science and Data Center of the Chinese Academy of Science. Vegetation type was obtained from the databox, and NDVI data were obtained from Landsat 8 OLI_TIRS in the Geospatial Data Cloud, then processed and converted to raster data using ENVI 5.3 software. Population density and proportion of vulnerable populations are available directly from the Worldpop Hub, with a resolution of 100 m selected. GDP data are available directly from the National Earth System Science Data Center, with a resolution of 1 km. Data on education status, education investment, local financial revenue, and relief agencies' capacity were obtained from POI from Planning Cloud (http://guihuayun.com/ (accessed on 15 February 2023)) for point extraction, and then Kernel Density was processed using ArcGIS 10.8.1, while the data used for analysing these evaluative indicators were obtained from the Tonghua Statistical Yearbook. Table 2 shows the selection of each indicator and its source, and Figure 6 shows the technical route and methodology.

Table 2. Indicators selected for the collapse risk model and their sources.

Sl. No.	Indicator	Data Types	Resolution/Year	Data Source
Hazard indicators				
1	Lithology	Raster data	1:50,000	Report on Geological Disaster Investigation and Mapping in Huinan County, Jilin Province
2	Distance from fault	Vector data	1:50,000	Report on Geological Disaster Investigation and Mapping in Huinan County, Jilin Province
3	Slope angle	Raster data	30 m	https://www.gscloud.cn (accessed on 30 February 2023)
4	Slope aspect	Raster data	30 m	https://www.gscloud.cn (accessed on 30 February 2023)
5	Landform type	Vector data	1:50,000	Report on Geological Disaster Investigation and Mapping in Huinan County, Jilin Province
6	Distance from river	Vector data	1:1,000,000	https://www.webmap.cn/ (accessed on 9 March 2023)
7	Multi-year average annual precipitation	Raster data	1 km	https://www.resdc.cn/ (accessed on 15 March 2023)
8	Vegetation type	Vector data	1:1,000,000	https://www.databox.store (accessed on 11 February 2023)
9	NDVI	Raster data	30 m	Landsat 8 OIL_TIRS
10	Distance from road	Vector data	1:1,000,000	https://www.webmap.cn/ (accessed on 9 March 2023)
11	Mining point density	Vector data	1:50,000	Report on Geological Disaster Investigation and Mapping in Huinan County, Jilin Province
Exposure indicators				
1	Population density	Raster data	100 m	https://www.worldpop.org/ (accessed on 14 February 2023)
2	Housing density	Vector data	1:50,000	Report on Geological Disaster Investigation and Mapping in Huinan County, Jilin Province
3	Road density	Vector data	1:50,000	Report on Geological Disaster Investigation and Mapping in Huinan County, Jilin Province
4	GDP	Raster data	1 km	http://www.geodata.cn/ (accessed on 16 March 2023)
Vulnerability indicators				
1	Proportion of vulnerable population	Raster data	100 m	https://www.worldpop.org/ (accessed on 14 February 2023)
2	Education status	Attribute data	2015–2019	Tonghua Statistical Yearbook
3	Residential buildings	Vector data	1:50,000	Report on Geological Disaster Investigation and Mapping in Huinan County, Jilin Province
Emergency response and recovery capability indicators				
1	Education investment	Attribute data	2015–2019	Tonghua Statistical Yearbook
2	Local financial revenue	Attribute data	2015–2019	Tonghua Statistical Yearbook
3	Relief agencies' capacity	Attribute data	2015–2019	Tonghua Statistical Yearbook

All data for this study were processed in ArcGIS 10.8.1, and the geographical coordinates were GCS_WGS_1984. The data analysis was carried out in the projection coordinates of WGS 1984 UTM Zone 52 N, and the cell size of all indicators was adjusted to 100 m when doing the analysis, so the total number of cells for each indicator map was 227,121.

2.4. Mapping Unit

The division of mapping units is the first and most important basis for the evaluation of the susceptibility, hazard, and risk of disasters in large-scale areas [40]. Its purpose is to extract the actual area into the unit, which in turn makes the results of the collapse risk mapping more accurate. Before doing the analysis, the data of all evaluation indicators are extracted into the mapping unit, and then GIS technology is used to conduct spatial overlay analysis of all evaluation indicators and visualise the data so that the results obtained will be more representative of the corresponding regions in reality. There are five main types of mapping units: grid unit, regional unit, watershed unit, uniform condition unit, and slope unit [41,42].

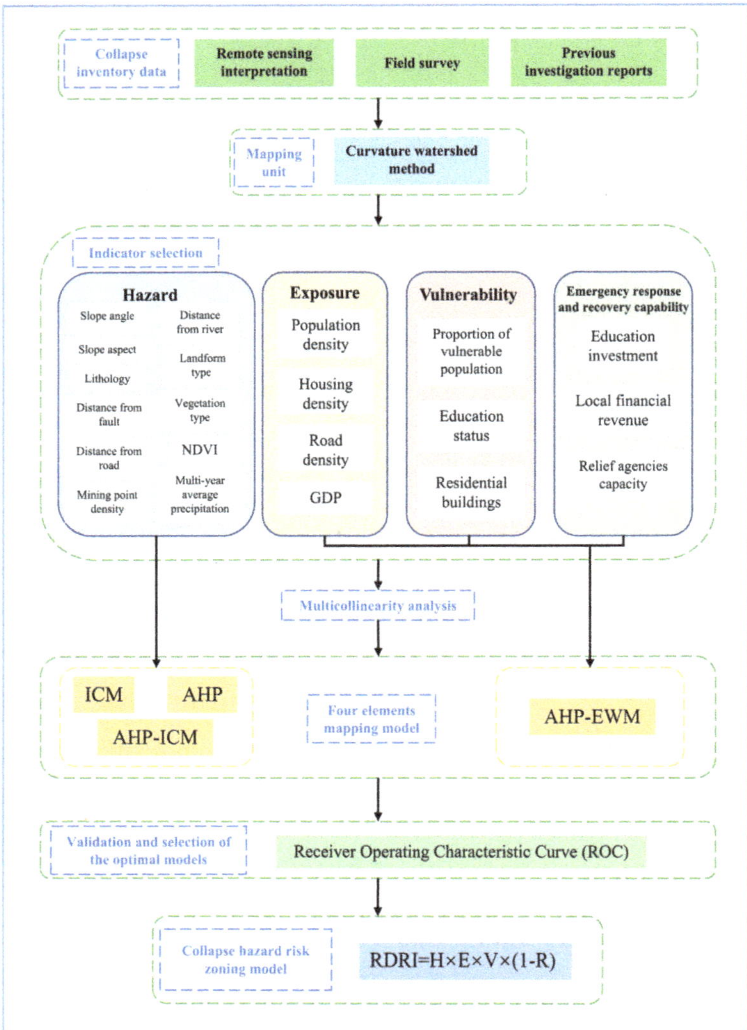

Figure 6. Flow chart of collapse risk mapping.

The accuracy of the mapping unit delineation reflects the fact that the study area is delineated as accurately as possible on ArcGIS so that a more accurate analysis of the study area can be carried out. Therefore, the correct choice of mapping unit delineation is essential for collapse risk evaluation. Previously, grid units were chosen as mapping units for collapse disaster evaluation [43]; however, the disadvantage of this approach is that it does not reflect the geological characteristics of the study area, which is divided into a certain number of regular grids, and after extracting information about the study area onto grid units, the unit is almost completely irrelevant to information about the spatial topography of the study area. Therefore, the slope unit was selected as the mapping unit in this paper, and it was divided into slope units using the hydrological analysis module in

ArcGIS 10.8.1 for the 90 m resolution DEM data of the study area. Currently, slope units include the hydrological analysis method and the curvature watershed method [18,44]. The hydrological analysis method uses DEM data with inverted DEM data to extract ridge lines and valley lines in the study area. The disadvantages of this method are the complexity of the operation, the need to control the appropriate flow thresholds and the resolution of the DEM, and the difficulty of distinguishing between horizontal and sloping terrain. The curvature watershed method is through the plane curvature and profile curvature of great and small values, reflecting the changes in slope and slope direction, and thus can identify the ridge line and valley line well, including the horizontal terrain and inclined terrain between the boundary; the method is not only simple to operate, but in the later stages of manual modification of the unreasonable units, greatly reduces the amount of work. The delineation accuracy is only related to the resolution of the DEM [25]. In this paper, the curvature watershed method is chosen to divide the study area into slope units. Figure 7 shows the detailed slope unit division process, and Figure 8 shows the slope unit delineation results of the study area. After comparing the results based on the slope unit division with the real terrain, the study area was divided into 10,096 units, and the geometric calculations yielded an area of 1.37 km^2 for the largest unit divided and 0.005 km^2 for the smallest unit.

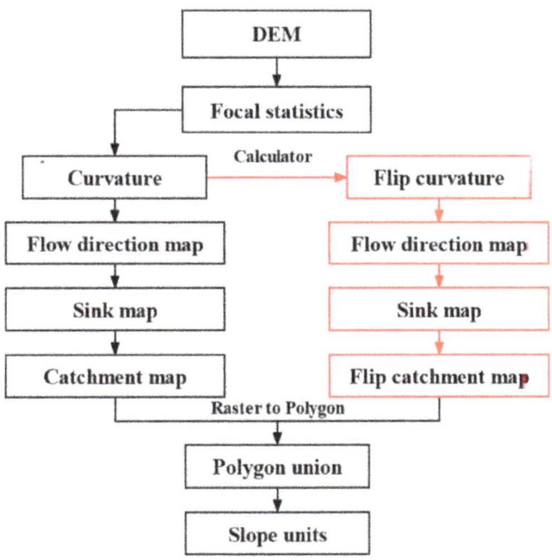

Figure 7. Slope unit classification process based on the curvature watershed method.

2.5. Collapse Inventory

The precise location and extent of the collapse are crucial in the process of collapse risk mapping [45]. For any collapse risk mapping, one of the important processes is collapse inventory, which relates to the location, number, extent of impact, and intensity of activity of collapses in an area. In this paper, remote sensing interpretation, field investigation, and synthesis of previous survey information are used to create a collapse inventory map. Information on collapses in the study area was obtained from the geological disaster survey and mapping report of Huinan County, Jilin Province. A total of 52 collapses were identified in this study using available information (Figure 8). According to statistics, nearly half of the collapse disaster potential sites (20) in the study area have formed multiple mountain-cut slopes due to the construction of roads and the intense folk mining activities in and around the chert and limestone quarries.

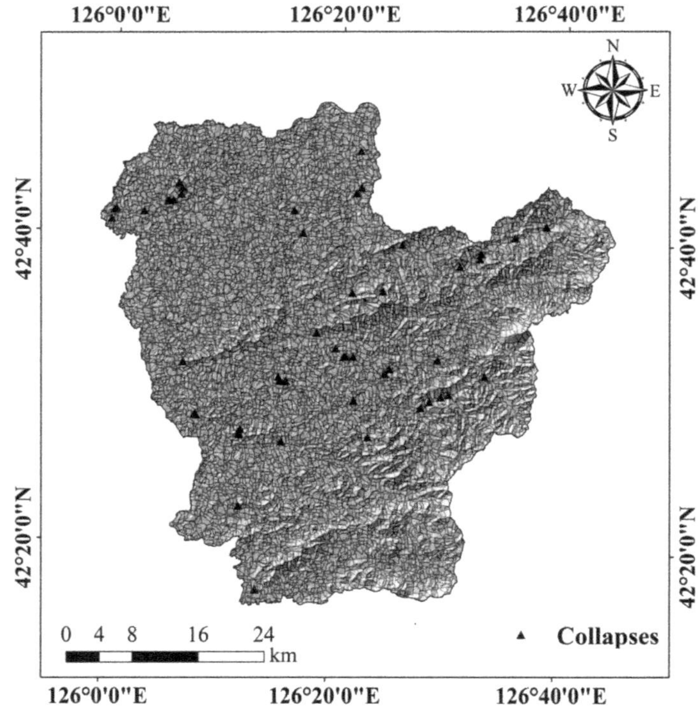

Figure 8. Slope unit division and collapse inventory in the study area.

2.6. Collapse Mapping Model

2.6.1. Hazard Mapping Model

Information Content Model (ICM)

The *ICM* is one of the basic concepts of information theory, which was proposed by Claude E. Shannon. According to the information theory, the amount of information can be measured by probability. *ICM* is a statistical analysis and prediction method that has been widely used in geohazard risk assessment [17,18,46]. The method calculates the *ICM* value of each indicator based on known information about the collapse and its evaluation indicators; the smaller the *ICM* value, the less likely it is to cause collapse disasters. The formula is calculated as follows:

$$I(X_i, H) = \ln \frac{N_i/N}{S_i/S} \tag{2}$$

where X_i is the evaluation indicator, $I(X_i, H)$ is the information value of each evaluation indicator X_i affecting the occurrence of collapse, N is the total number of collapse disasters in the study area, N_i is the number of collapses contained in each evaluation indicator X_i, S is the total number of rasters in the study area, and S_i is the total number of rasters in each evaluation indicator X_i. Then, the *ICM* values were assigned to each evaluation indicator map to reclassify them. Finally, the collapse hazard index (*CHI*) was obtained by superimposing each evaluation indicator map. The calculation formula is as follows:

$$CHI_{ICM} = \sum ICM_i \tag{3}$$

where *ICM* refers to the evaluation indicator map reclassified by the informative values, and *i* denotes hazard indicators from the 1st to the 11th.

Analytical Hierarchy Process (AHP)

AHP is a multi-criteria decision-making method proposed by Thomas L. Saaty [47]. The weights of the indicators are determined by building a hierarchical model, decomposing them into a series of levels and criteria, and then determining the relative importance between the elements of each level by comparing them two by two. The relative importance of these indicators is assessed on a scale of 1–9 from low to high. The AHP method consists of three steps: firstly, constructing an affiliation model for the highest level of decision making (target level), the middle level (criterion level), and the bottom level (indicator level); secondly, establishing the corresponding judgement matrix; and thirdly, calculating the weights of each factor and testing the random consistency of each factor [20,48,49].

Establishing a system hierarchy: According to the specific performance and causes of collapse disasters in Huinan County, reasonable evaluation indicators were selected after combining the relevant situations in the study area with the analysis of the research results of Huinan County, analysing the main influencing elements of collapse disasters occurring in the study area, such as the natural environment and geological elements, etc. The hierarchy model and specific factors were built in the process of the study (Figure 9).

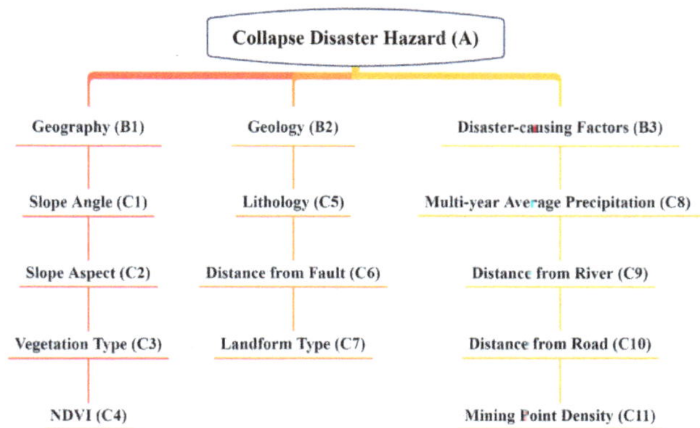

Figure 9. Schematic diagram of the structure of the AHP of collapse disaster.

Establishing a judgement matrix: The structural model of the collapse disaster analytical hierarchy process constructed in Figure 9 allows for the analysis of factors and the accurate representation of the affiliation problem of factors. In accordance with the 1~9 scale method and other relevant contents, there will be differences in the hierarchical aspects between the factors. After specifying their weights, the judgement matrices are built for the A–B and B–C layers, respectively. The judgement matrix can be expressed as follows:

$$X = (x_{ij})_{n\times n} = \begin{bmatrix} x_{11} & x_{12} & \cdots & x_{1n} \\ x_{21} & x_{22} & \cdots & x_{2n} \\ \cdots & \cdots & \cdots & \cdots \\ x_{n1} & x_{n2} & \cdots & x_{nn} \end{bmatrix} \tag{4}$$

where X is the judgement matrix; x_{ij} is the result of the comparison of the importance of factor i and factor j, and has the following property:

$$a_{ij} = \frac{1}{a_{ji}} \tag{5}$$

Relative weights and consistency test: In order to check the reasonableness of the weight determination method, a consistency test is required. The maximum eigenvalue λmax of the matrix is utilised while the corresponding eigenvectors are obtained, and the consistency of the judgment matrix is verified according to Equations (6) and (7).

$$CI = \frac{\lambda_{\max} - n}{n - 1} \quad (6)$$

$$CR = \frac{CI}{RI} \quad (7)$$

where *CI* is the consistency indicator, and the smaller the value, the lower the degree of inconsistency; λmax is the maximum eigenvalue of the judgment matrix; *n* is the order; *CR* is the consistency ratio; and *RI* is the random consistency indicator.

Based on the principles of *AHP*, the relative weights of each hazard indicator are calculated, and the collapse hazard index (*CHI*) is calculated as follows:

$$CHI_{AHP} = \sum_{i=1}^{n} (W_i \times W_{ij}) \quad (8)$$

where W_i is the weight of evaluation index *i*, W_{ij} is the *i*-th evaluation indicator and the *j*-th sub-class weight, and *n* is the number of hazard evaluation indicators, with $n = 11$ in this study.

AHP-ICM Model

AHP-ICM is a combination of subjective and objective methods, which avoids the limitations of a single method and improves the comprehensiveness and accuracy of collapse hazard evaluation. This method has been widely used in the risk evaluation of geologic disasters [18,19,24,50]. After determining the sub-classes to which the hazard evaluation indicators belong, the information value of each sub-class is calculated using the *ICM*, and the weights of each evaluation indicator are obtained using the *AHP*, and the two are coupled for calculation. In this study, the *AHP-ICM* model is used for collapse hazard calculation, and the specific formula is

$$CI_{ij} = W_i \times I_{ij} \quad (9)$$

where CI_{ij} is the combined information value of the sub-class *j* in evaluation indicator *i*, W_i is the weight of each evaluation indicator obtained by *AHP*, and I_{ij} is the information value of the *i*th evaluation indicator and the *j*th sub-class.

2.6.2. Exposure, Vulnerability, and Emergency Responses and Recovery Capability Mapping Model

Entropy Weighting Method (EWM)

The entropy weighting method is an objective weight-defining method that is used to characterise the relative intensity of each indicator in the evaluation system of a target system. The entropy method does not have any subjective analysis for the calculation of individual indicator weights, relying only on objective indicator data to calculate the magnitude of the indicator weights [24,51]. The specific steps are as follows:

With *m* objects, each with *n* evaluation indicators, *a* judgment matrix is constructed.

$$A = (a_{ij})_{m*n} (i = 1, 2, \ldots, m; j = 1, 2, \ldots, n) \quad (10)$$

where a_{ij} is the value of the *j*th indicator for the *i*-th object.

The judgment matrix is normalised to obtain the normalised judgment matrix $D = (d_{ij})_{mn}$.

$$d_{ij} = \frac{r_{ij} - r_{\min}}{r_{\max} - r_{\min}} \quad (11)$$

$$d_{ij} = \frac{r_{\max} - r_{ij}}{r_{\max} - r_{\min}} \quad (12)$$

where $r_{\max}$ is the highest indicator value under different objects for the same indicator and $r_{\min}$ is the lowest indicator value under different objects for the same indicator.

For a situation with m objects and n indicators, the entropy of the evaluation indicator can be determined as

$$S_j = -\frac{\sum_{i=1}^{m} f_{ij} \ln f_{ij}}{\ln m} \quad (i = 1,2,\ldots,m; j = 1,2,\ldots,n) \quad (13)$$

where $f_{ij} = \frac{1+d_{ij}}{\sum_{i=1}^{m} (1+d_{ij})}$.

The entropy weight vector W of the evaluation indicators is

$$W = (\omega_j)_{1\times n} \quad (14)$$

where ω_j is the entropy weight of the jth evaluation indicator, and $\omega_j = \frac{1-S_j}{n-\sum_{j=1}^{n} S_j}$.

AHP-EWM Model

Firstly, the sub-classes of exposure, vulnerability, and emergency response and recovery capability evaluation factors are determined; their weights are calculated using the AHP, then the weights of the evaluation factors are calculated using the EWM, and finally, the two are coupled to derive the combined weights of each sub-class in each evaluation factor. In this study, the method is used for the calculation of exposure, vulnerability, and emergency response and recovery capability. The specific formulae are as follows:

$$S_{ij} = W_{ij} \times W_e \quad (15)$$

where S_{ij} is the combined weight value of sub-class j in evaluation indicator i. o is the i-th evaluation indicator and the weight of the j-th sub-class calculated by hierarchical analysis. r is the weight of each evaluation indicator calculated by the entropy weight method.

2.6.3. Collapse Risk Mapping Model

In this study, using the formation principle of natural disaster risk, the four elements of geohazard disaster, the selected index system, and the integrated weighted analysis method were used to establish the collapse disaster risk index model, whose formula can be expressed as follows [28]:

$$RDRI = H \times E \times V \times (1 - R) \quad (16)$$

where *RDRI* represents the collapse disaster risk index; a higher value means a higher level of collapse disaster risk. *H*, *E*, *V*, and *R* represent the collapse disaster hazard, exposure, vulnerability, and emergency response and recovery capability indicator indices, respectively.

3. Results and Analysis of the Hazard Mapping Model

3.1. Results of the Model

3.1.1. Results of the Information Content Model (ICM)

There is a close relationship between the collapse disaster and the information content (IC) value; when IC > 0, the probability of collapse is greater than the average value, i.e., the probability of collapse is high; conversely, the probability of collapse is low in the reverse case; if IC = 0, the probability of occurrence is of an average value. The results are shown in Table 3.

Table 3. Spatial relationship between each collapse evaluation indicator and collapse.

Indicator	Class	Collapse Count	Total Count	ICM
Slope angle	0–5	5	32,898	−0.4099
	5–10	12	68,537	−0.2682
	10–15	20	51,643	0.5256
	15–20	12	38,386	0.3115
	>20	3	35,653	−1.0010
Slope aspect	North	0	1296	0.0000
	Northeast	1	12,326	−1.0375
	East	2	25,050	−1.0535
	Southeast	14	48,,583	0.2300
	South	12	51519	0.0172
	Southwest	19	46,038	0.5892
	West	3	30,587	−0.8477
	Northwest	1	11,718	−0.9869
Multi-year average precipitation	<720	5	39,165	−0.5841
	720–730	11	35,490	0.3029
	730–740	21	44,830	0.7159
	740–750	10	48,315	−0.1009
	>750	5	59,321	−0.9993
Lithology	Q	6	62,666	−0.8718
	γ	11	10,226	1.5472
	K + J	3	12,000	0.0879
	Ar	10	90,418	−0.7276
	Q + Z	2	10,027	−0.1379
	β	5	23,725	−0.0829
	∈ + O + Z	15	18,059	1.2886
Distance from fault	0–500	18	56,250	0.3348
	500–1000	10	27,686	0.4559
	1000–2000	7	44,551	−0.3765
	2000–3000	11	28,746	0.5136
	>3000	6	69,888	−0.9809
Landform type	Fluvial terrace	6	34,499	−0.2749
	Undulating terrace	5	30,078	−0.3201
	Denudation of eroded hill	9	23,872	0.4988
	Tectonic low hill	22	49,696	0.6594
	Tectonic moderate hill	6	61,991	−0.8610
	Lava low terrace	2	10,751	−0.2076
	Lava plateau	2	16,234	−0.6197
Distance from river	0–100	32	33,744	1.4212
	100–300	3	10,748	0.1981
	300–600	7	17,049	0.5841
	600–1000	6	19,992	0.2707
	>1000	4	145,588	−2.1202
Distance from road	0–100	33	44,349	1.1787
	100–300	3	15,005	−0.1355
	300–600	4	19,397	−0.1046
	600–1200	7	33,720	−0.0979
	>1200	5	114,650	−1.6582
Vegetation type	Hemerophyte	18	42,454	0.6162
	Broadleaf forest	15	69,491	−0.0589
	Meadow	7	21,029	0.3742
	Mixed forest	12	94,147	−0.5857
NDVI	0–0.3	20	32,809	0.9793
	0.3–0.55	4	24,824	−0.3513
	0.55–0.65	21	50,226	0.6022
	0.65–0.75	4	38,058	−0.7786
	0.75–1	3	81,204	−1.8241

Table 3. *Cont.*

Indicator	Class	Collapse Count	Total Count	ICM
Mining point density	0–5	14	104,598	−0.5369
	5–9	15	66,194	−0.0103
	9–13	14	35,987	0.5301
	13–21	7	14,501	0.7459
	21–31	2	5830	0.4044
Total area		52	227,121	

As can be seen from Table 3, the ICM values for slope angles of 10–15° and 15–20° are 0.5256 and 0.3115, respectively, and according to previous studies, the higher the slope angle, the greater the probability of collapse occurring [29]. According to field surveys, within these two gradations, collapse is more likely to occur because of the steep cliffs or steep slopes within a small area formed by road cutting and engineering construction. The ICM values are 0.23 and 0.5892 for southeast- and southwest-facing slopes, respectively. For the multi-year average precipitation, the ICM values are 0.3029 and 0.7159 for 720–730 mm and 730–740 mm, respectively, with the higher precipitation concentrated in areas of the study area with less human activity, where the geotechnical structure is intact, the vegetation cover is larger and subject to precipitation erosion, and the erosion effect is less. For lithology, the ICM values for the granite rock group (γ) and the hard limestone rock group ($\in + O + Z$) are 1.5472 and 1.2886, respectively. As both γ weathering fissures and primary fissures are more developed, the upper part is strongly weathered and the rock integrity is poor, while $\in + O + Z$ consists of medium-thick laminated pure limestone, sandstone and siltstone, with weakly developed karst, generally dominated by solution gaps and small caves, and in micro-landscapes of mostly steep cliffs and steep slopes, so these two rock groups are most prone to collapse. The degree of development of collapse disasters in Huinan County is obviously influenced by faults. Fractures are developed in the area, with more structural surfaces and locally weak surfaces or weak zones, with low rock strength and easy weathering and stripping. For landform type, the ICM values for denudation of eroded hill and tectonic low hill landform areas are 0.4988 and 0.6594, respectively. These two areas are more prone to collapse due to more residential settlements, developed transportation and strong human activities such as road cutting, deforestation, and mining. In mountainous areas close to rivers, the occurrence of collapses is mainly located within 100 m of the river due to the erosive action of the river. In this study area, as the construction of the road required extensive slope cutting, the collapse occurred mainly within 100 m of the road. The ICM values for hemerophyte and meadow were 0.6162 and 0.3742, respectively, with higher ICM values for NDVI within 0.65, indicating that areas with low vegetation cover are prone to collapse. Where the mining point density is high, collapse is more likely to occur due to the long-term disturbance of the geotechnical body. The results are shown in Figure 10a.

3.1.2. Results of the Analytical Hierarchy Process (AHP)

The degree of merit of the model was judged by CR based on the comparison between variables and indicators, and the consistency of the judgment matrix was accepted if CR < 0.1 [20,52–54]. Then, the weights Wi normalised by the criterion layer to the indicator layer (B–C) were obtained, and the weights of the target layer to the indicator layer (A–C) were obtained after normalising them again under the target layer (A), and since all CRs were less than 0.1, all judgment matrices met the consistency requirements and were considered to be reasonably weighted. The specific data content is shown in Tables 4 and 5. Based on the results, the multi-year average precipitation has the highest weighting of 0.4119, followed by distance from road and lithology at 0.1924 and 0.1118, respectively. This indicates that the multi-year average precipitation and distance from road and lithology are the main factors contributing to collapse in the study area. Based on Equation (8), the

AHP-generated collapse hazard map was derived by combining Tables 4 and 5 to calculate CHI, as shown in Figure 10b.

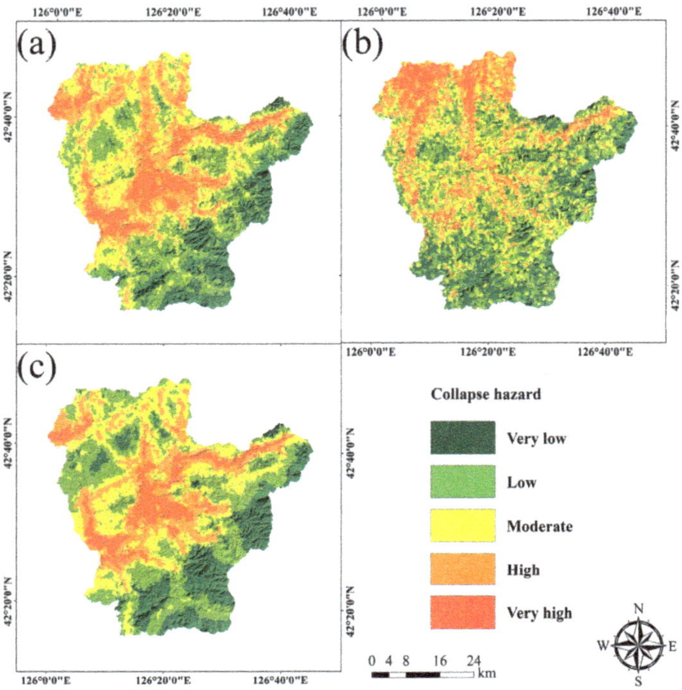

Figure 10. Results of hazard mapping: (**a**) information content method (ICM) method; (**b**) analytical hierarchy process (AHP) method; and (**c**) analytical hierarchy process–information content method (AHP-ICM) method.

Table 4. Judgement matrix of evaluation indicators and their weights.

Target Layer (A)	Criterion Layer (B)	A–B Judgement Matrix			A–B Weight	Indicator Layer (C)	B–C Judgement Matrix				B–C Weight	A–C Weight (Wi)
Collapse hazard (1.0000)	Geography	1	1/3	1/7	0.081	Slope angle	1	3	3	1/2	0.2947	0.0239
						Slope aspect	1/3	1	1/3	1/5	0.0781	0.0063
						Vegetation type	1/3	3	1	1/3	0.1537	0.0124
						NDVI	2	5	3	1	0.1537	0.0383
	Geology	3	1	1/5	0.1884	Lithology	1	3	3		0.5936	0.1118
						Distance from fault	1/3	1	1/2		0.1571	0.0296
						Landform type	1/3	2	1		0.2493	0.047
	Disaster-causing factors	7	5	1	0.7306	Multi-year average precipitation	1	5	3	7	0.5638	0.4119
						Distance from river	1/5	1	1/3	3	0.1178	0.0861
						Distance from road	1/3	3	1	5	0.2634	0.1924
						Mining point density	1/7	1/3	0.2	1	0.0550	0.0402

Table 5. Judgement matrix for evaluation indicators' sub-classes and their weights.

Indicator	Class	Judgement Matrix					Weight (W_{ij})
Slope angle	0–5	1	1/3	1/3	1/4	1/3	0.06386
	5–10	3	1	1/3	1/3	1/3	0.10497
	10–15	3	3	1	1/3	3	0.25278
	15–20	4	3	3	1	3	0.41551
	>20	3	3	1/3	1/3	1	0.16289

Table 5. *Cont.*

Indicator	Class	Judgement Matrix								Weight (W_{ij})
Slope aspect	North	1	1	1/5	1/7	1/8	1/9	1/3	1/2	0.02632
	Northeast	1	1	1/3	1/5	1/7	1/6	1/2	1	0.03591
	East	5	3	[illegible]	1/3	1/4	1/4	2	2	0.09014
	Southeast	7	5	3	1	1/2	2	4	4	0.22180
	South	8	7	4	2	1	3	5	7	0.33640
	Southwest	9	6	4	1/2	1/3	1	2	3	0.17164
	West	3	2	1/2	1/4	1/5	1/2	1	4	0.07541
	Northwest	2	1	1/2	1/4	1/7	1/3	1/4	1	0.04237
Vegetation type	Hemerophyte	1	4	1/2	5					0.37499
	Broadleaf forest	1/4	1	1/2	1					0.12538
	Meadow	2	2	1	3					0.39248
	Mixed forest	1/5	1	1/3	1					0.10715
NDVI	0–0.3	1	1	1/3	1/3	1/4				0.08391
	0.55–0.65	1	1	1/3	1/2	2				0.13792
	0.3–0.55	3	3	1	1	4				0.35182
	0.65–0.75	3	2	1	1	3				0.30628
	0.75–1	4	1/2	1/4	1/3	1				0.12007
Lithology	Q	1	1/2	4	1	2	3	1/5		0.12901
	γ	2	1	7	1	3	4	1/2		0.21439
	K + J	1/4	1/7	1	1/3	1/2	3	1/5		0.05090
	Ar	1	1	3	1	2	4	1/4		0.14705
	Q + Z	1/2	1/3	2	1/2	1	2	1/3		0.08317
	β	1/3	1/4	1/3	1/4	1/2	1	1/3		0.04333
	∈ + O + Z	5	2	5	4	3	3	1		0.33215
Distance from fault	0–500	1	3	1/2	1/2	2				0.19789
	500–1000	1/3	1	1/3	1/2	1/2				0.08911
	1000–2000	2	3	1	2	2				0.34454
	2000–3000	2	2	1/2	1	1				0.20961
	>3000	1/2	2	1/2	1	1				0.15885
Landform type	Fluvial terrace	1	1/3	1/5	1/4	1/2	3	4		0.08124
	Undulating terrace	3	1	1/2	1	2	4	3		0.18835
	Denudation of eroded hill	5	2	1	3	5	4	4		0.34318
	Tectonic low hill	4	1	1/3	1	3	2	5		0.19120
	Tectonic moderate hill	2	1/2	1/5	1/3	1	3	2		0.09903
	Lava low terrace	1/3	1/4	1/4	1/2	1/3	1	1		0.05027
	Lava plateau	1/4	1/3	1/4	1/5	1/2	1	1		0.04673
Multi-year average precipitation	<720	1	3	4	3	4				0.44926
	720–730	1/3	1	2	1	1/2				0.13347
	730–740	1/4	1/2	1	1/2	1/3				0.07666
	740–750	1/3	1	2	1	1/2				0.13347
	>750	1/4	2	3	2	1				0.20713
Distance from river	0–100	1	3	2	2	4				0.37497
	100–300	1/3	1	1/2	1/2	2				0.12081
	300–600	1/2	2	1	1	3				0.21536
	600–1000	1/2	2	1	1	3				0.21536
	>1000	1/4	1/2	1/3	1/3	1				0.07350
Distance from road	0–100	1	3	2	1/2	4				0.28286
	100–300	1/3	1	1/2	1/4	2				0.10469
	300–600	1/2	2	1	1	3				0.21437
	600–1200	2	4	1	1	3				0.32492
	>1200	1/4	1/2	1/3	1/3	1				0.07316
Mining point density	0–5	1	3	2	3	0.5				0.27758
	5–9	1/3	1	1/2	1/2	1/2				0.09473
	9–13	1/2	2	1	1/3	1/2				0.12500
	13–21	1/3	2	3	1	1/3				0.16494
	21–31	2	2	2	3	1				0.33774

3.1.3. Results of the AHP-ICM Model

Based on the results of the actual survey in the study area, a combination of subjective and objective methods was used to determine the hazard of collapse in Huinan County, taking into account the geographic and geological environment and external factors that contribute to the disaster. The comprehensive information content (Table 6) amount of each sub-class in each hazard evaluation indicator was calculated using Equation (9), and then the sub-class of each evaluation indicator was assigned a value using a raster calculator, finally superimposed, and the natural breaks classification method was used to divide it into five hazard classes: very low, low, medium, high, and very high, to obtain a hazard zoning map, and the results are shown in Figure 10c. As can be seen from Table 6, the CIs for multi-year average precipitation (730–743 mm), distance from road (0–100 m), and lithology (γ) are the largest at 0.29487, 0.22677, and 0.17298, respectively, which causes the very high hazard zones to be mainly located in the central part, in the northwestern part, along the northeastern road, and in the west-southwestern part of the study area, which makes the collapse disaster prone to occur in this part of the area. The main reason is that these areas are located in the middle and low mountainous belt and are affected by the long-term frequent economic activities of human beings. According to the statistics of the number of disaster points, it is concluded by comparison that the collapse disasters in Huinan County are mainly distributed in the tectonic low hill geomorphic area and hilly area, and the lithology is mostly limestone, granite, and metamorphic rock groups, which have low vegetation coverage, and the human activities are the strongest in these areas. In tectonic mesas and high terraces, human activity is less, elevation is greater, vegetation cover is greater, and collapse disasters are least developed.

Table 6. Combined information content for each collapse evaluation indicator's sub-classes.

Indicator	Class	CI	Indicator	Class	CI
Slope angle	0–5	−0.00980	Distance from river	0–100	0.12236
	5–10	−0.00641		100–300	0.01706
	10–15	0.01256		300–600	0.05029
	15–20	0.00744		600–1000	0.02330
	>20	−0.02392		>1000	−0.18255
Slope aspect	North	0.00000	Distance from road	0–100	−0.31904
	Northeast	−0.00654		100–300	−0.01884
	East	−0.00664		300–600	−0.02012
	Southeast	0.00145		600–1200	−0.02608
	South	0.00011		>1200	0.22677
	Southwest	0.00371	Vegetation type	Hemerophyte	0.00764
	West	−0.00534		Broadleaf forest	−0.00073
	Northwest	−0.00628		Meadow	0.00464
Multi-year average precipitation	<720	−0.24059		Mixed forest	−0.00726
	720–730	0.12476	NDVI	0–0.3	0.03751
	730–740	0.29487		0.3–0.55	0.02307
	740–750	−0.04157		0.55–0.65	−0.01345
	>750	−0.41161		0.65–0.75	−0.02982
Lithology	Q	−0.09747		0.75–1	−0.06986
	γ	0.17298	Mining point density	0–5	−0.02159
	K + J	0.00983		5–9	−0.00041
	Ar	−0.08135		9–13	0.02131
	Q + Z	−0.01542		13–21	0.02999
	β	−0.00926		21–31	0.01626
	∈ + O + Z	0.14407	Distance from fault	0–500	0.00991
Landform type	Fluvial terrace	−0.01292	Distance from fault	500–1000	0.01350
	Undulating terrace	−0.01505		1000–2000	−0.01114
	Denudation of eroded hill	0.02344		2000–3000	0.01520
	Tectonic low hill	0.03099		>3000	−0.02904
	Tectonic moderate hill	−0.04047		0–500	0.00991
	Lava low terrace	−0.00976			
	Lava plateau	−0.02913			

3.2. Validation of the Hazard Mapping Model

The results obtained from the ICM, AHP, and AHP-ICM models were validated to guarantee the accuracy of the collapse hazard mapping model, and then the optimal model was selected for the study area collapse risk mapping. In this study, the three models were validated by applying the receiver operating characteristic (ROC) curve in Origin 2022 software. The curve is a two-dimensional plot and is widely used in the validation of two-dimensional classification models [18,19,25], and the evaluation process is simple and easy to operate. The ROC uses the area under the curve (AUC) to directly determine the accuracy of the model. The results are intuitive, with AUC values between 0.5 and 1; the closer the result is to 1, the more accurate the model and the better the performance. In this study, ROC curves were plotted based on the principle of ROC curves with an equal number of randomly selected collapse and non-collapse points. The ROC results in Figure 11 show that the AHP-ICM model has the highest accuracy (AUC = 87.4%), followed by the ICM model (AUC = 85.6%) and the AHP (AUC = 80%).

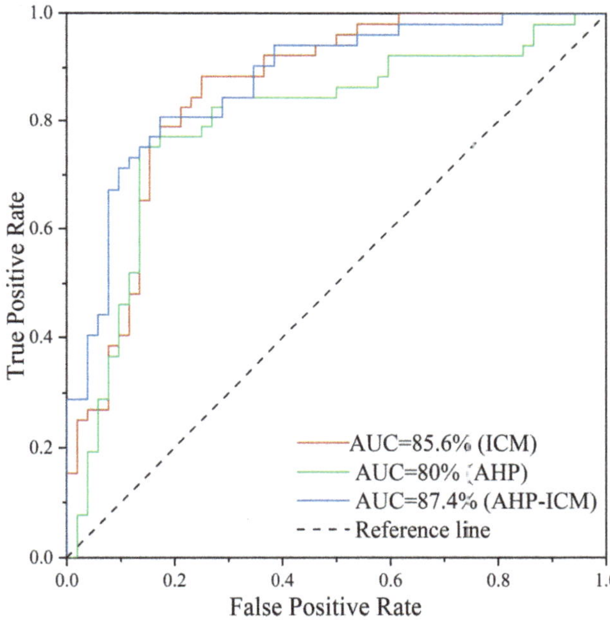

Figure 11. Receiver operating characteristic (ROC) curve.

3.3. Comparison of Hazard Mapping Models

Three models were used to compare the hazard mapping in this study: the ICM model, the AHP model, and the AHP-ICM model. The accuracy of the three models was verified by ROC curves, and the results in Figure 11 show that the AHP-ICM model has the highest accuracy. In addition, the results of the mapping of the three models are compared in more detail according to the five hazard classes, and the statistical results are shown in Figure 12.

The collapse hazard map produced by the natural breaks classification method should meet the following requirements: (1) The areas classified as high and very high hazard should cover as small an area of the study area as possible. (2) The areas with high and very high hazards should contain as many known collapse points as possible. The results show that the AHP model has the smallest total proportion of very high and high hazard classes (28.03%), but it contains the fewest known collapse points (37). It is followed by the AHP-ICM model with a total proportion of 28.88% of very high and high classes and 44 known collapse points, and then the ICM model with a total proportion of 32.51% of very high and high classes and 48 known collapse points. The last model is the ICM model, with a total of 32.51% of very high and high grades and 48 known collapse points. Although the ICM model contains the highest number of collapse points in areas with very high and high hazard classes, their total percentage is the highest of the three models, while the AHP-ICM model contains seven more known collapse points than the AHP model with a total percentage of very high and high hazard classes of only 0.85% more than the AHP model. In terms of disaster point density, the highest density in areas with very high and high hazard classes is found in the AHP-ICM model (6.69 per 100 km^2), followed by the ICM model (6.48 per 100 km^2) and the AHP model (5.8 per 100 km^2).

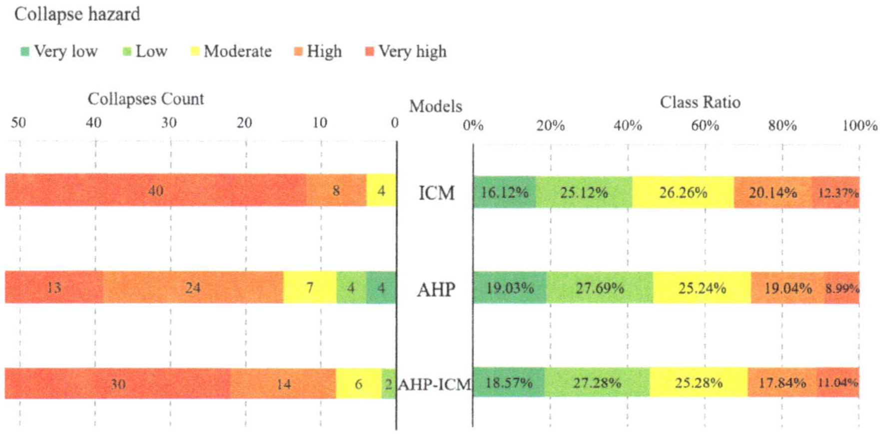

Figure 12. Statistics of the results of the collapse hazard mapping.

By comprehensively analysing the ROC curves, the percentage of high and very high hazard classes of the models, and the disaster point density, this study determines the degree of superiority of the three hazard mapping models, of which the optimal model is the AHP-ICM model, which is used in the study of collapse risk mapping (see Section 5).

4. Results of Exposure, Vulnerability, and Emergency Response and Recovery Capability Mapping

This study also used a combination of subjective and objective methods to analyse the exposure, vulnerability, and emergency response and recovery capability mappings. Weighting of the 10 evaluation indicators in the exposure, vulnerability, and emergency response and recovery capability indicator evaluation system was performed using EWM using Excel 2021 software; then, using AHP, judgment matrices were constructed to calculate the subjective weights of each of the five sub-classes in the evaluation indicators, and a total of 10 judgment matrices were constructed to derive the weight values of each sub-class; finally, the weights (We) of the evaluation indicators are coupled with the sub-class weights (W_{ij}) in the evaluation indicators using Formula (15) to obtain the sub-class comprehensive weights S. The results of the overlay are shown in Figure 13. The results show that the high and very high exposure areas are mostly concentrated in densely populated and built-up areas, the reason being that these areas have increased the scale of engineering and construction activities due to economic development and farming and reclamation, increasing the potential for collapse disasters to occur and threaten nearby areas. High and very high vulnerability areas are mainly located in rural and township areas close to the mountains, the reason being that the vulnerable population and dwellings are more numerous in this area and are most vulnerable to collapse disasters, and they are less likely to recognise the approaching danger or move slowly when a collapse disaster occurs. The closer to urban and rural areas, the more relief agencies are available, and the more money is invested in disaster management, the better their ability to prevent and mitigate disasters.

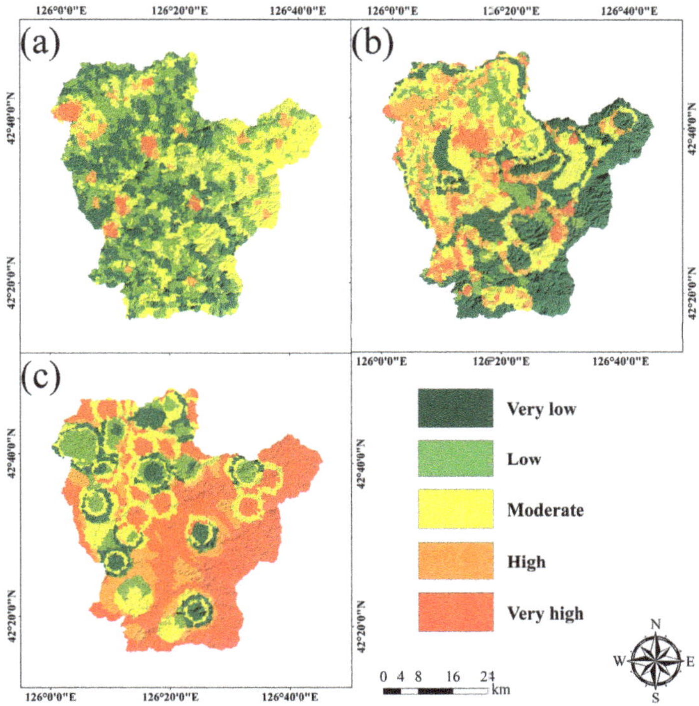

Figure 13. Results of the study area mapping based on the AHP-EWM model: (**a**) exposure overlay; (**b**) vulnerability overlay; and (**c**) emergency response and recovery capability overlay.

5. Results of Risk Mapping

Based on the above analysis, the optimal model for collapse hazard mapping is the AHP-ICM model. The final four elements of overlay maps are as follows: hazard mapping based on the AHP-ICM model, exposure overlay, vulnerability overlay, and emergency response and recovery capability mapping based on the AHP-EWM model.

Combining the four elements obtained by overlay, the collapse disaster risk index was calculated according to Equation (16) and classified using the natural breaks classification method into four classes: very high, high, moderate, and low, to obtain a collapse disaster risk zoning map, as shown in Figure 14. Of these, the very high-risk area accounts for 6.06% and contains 20 collapse points, mainly in the central, northwestern, southwestern, and northeastern parts of the study area along the roads. The very high-risk areas are mainly caused by the construction of the road, which causes many cut slopes, and by the mining and cultivation of the land, which destroys a lot of vegetation. The high-risk zone covers 30.07% of the area and contains 28 landslides, which are caused by the low topography and low vegetation cover in these areas and by the long-term rainfall and the incompatibility of human engineering construction with the geological environment. The low- and medium-risk areas, which together account for 63.87% of the study area, contain four collapse points and are mainly located in areas of high elevation and low elevation areas where human engineering activities are less frequent.

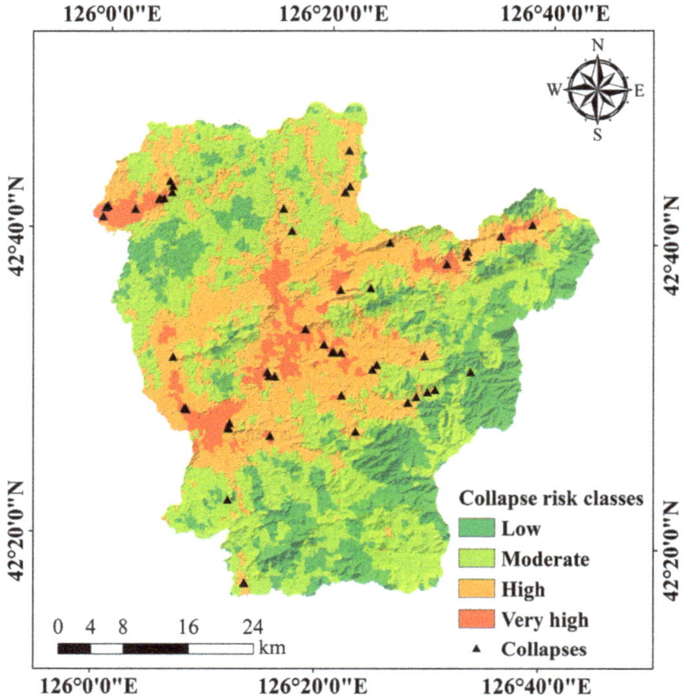

Figure 14. Collapse disaster risk zoning map.

6. Discussion

6.1. Importance and Significance of This Study

Huinan County is located in a semi-mountainous area, with large changes in topography and geomorphology, various types of rock and soil bodies, and development of local geological formations, superimposed on the role of strong human engineering activities such as road building, urban construction, and mineral resource extraction, geological disasters are very well developed, especially under the influence of extreme abnormal climate, which triggers or aggravates collapse disasters, so disaster prevention and control are still facing great pressure. Therefore, risk assessment and mapping of collapse disasters are of great practical significance. In this study, based on the formation principle of natural disaster risk, the hazard mapping model of collapse disaster was established using the ICM, AHP, and AHP-ICM, and the mapping model of exposure, vulnerability, and emergency response and recovery capability was established using AHP-EWM. The risk mapping model combines multiple evaluation indicators and comprehensively considers multiple dimensions of collapse risk, which is scientific and operable. This study provides a comprehensive assessment and accurate identification of the risk of collapses in Huinan County, which provides an important reference for collapse disaster management and decision making and has far-reaching practical significance for the prevention and mitigation of geological disasters in mountainous areas. In addition, the methods and models used in this study have some value and potential for application in risk assessment and mapping studies in other regions. It is worth mentioning that previous studies have mainly focused on the evaluation of disaster hazard, exposure, and vulnerability, while few studies have been conducted on the evaluation of emergency response and recovery capability, and there are few sources of data and ways to visualise their data. In this study, three socio-economic indicators, namely education investment, local financial revenue, and relief agencies' capacity, are used as indicators of emergency response and recovery capability, which play a key role in collapse risk mapping.

6.2. Comprehensive Evaluation of Hazard Mapping Model

The ICM method is a relatively objective evaluation model that establishes a suitable evaluation indicator system and its sub-class through the actual survey results of the study area and then calculates the information value of each indicator's sub-class, which reduces the influence of subjectivity [24,44] but ignores the relationship between the sub-class of evaluation indicators and the occurrence of collapse. The AHP method is a subjective weighting method [20,54], and the assignment of its weights requires sufficient experience and level. Because the established hierarchy and calculation of weights are more complicated, it is easy to calculate improper weights in a certain sub-class, which is difficult to avoid, so this leads to a lower accuracy of collapse hazard mapping. The AHP-ICM method, as a combined subjective and objective method [18,19,50], compared with a single method, can ensure both the risk mapping objective and scientific soundness but also allows for subjective analysis based on the actual situation in the study area. There have been researchers using the AHP-ICM method in the landslide sensitivity mapping study. Wang et al. [18] used the AHP-ICM method to carry out sensitivity analyses of 41 landslides, and more accurate results were obtained, which verified the correctness and reasonableness of the method. Du et al. [19] used the AHP-ICM method to analyse the landslide sensitivity mapping of 799 landslides in the Eastern Himalayan zone of Tibet, and the results showed that the AHP-ICM method has a high prediction accuracy. Ma et al. [24] compared the AHP-ICM method with other combined methods in landslide sensitivity mapping and found that the prediction accuracy of the AHP-ICM method was similar to the other methods. In the same field, although the AHP-ICM method still has a certain gap compared with existing machine learning methods (Support Vector Machine, Random Forest, Artificial Neural Networks, etc.), the AHP-ICM method is much easier to compute and operate and only needs to extract the relevant landslide data of the evaluation indicators in the GIS, and then the corresponding weights of the indicators can be derived by calculating them directly in the Excel 2021. Machine learning methods, on the other hand, are more demanding in terms of data quality and feature selection, requiring adequate data preparation and model training processes, while the models are less explanatory, making it difficult to explain the key features identified by the models.

6.3. Limitations and Perspectives of This Study

This study considers as comprehensively as possible the evaluation indicators related to collapse disasters from a four-element perspective; however, the data used are based on existing datasets and may be incomplete and inaccurate for economic and education-related data. In future work, more high-quality data can be further collected to improve the accuracy of the findings. In future collapse risk mapping, more indicators related to collapse risk should be considered, and earthquakes have a certain impact on it. There has been no seismic activity in this study area in 40 years, and seismic data are difficult to obtain, so the impact of earthquakes is not considered in this paper. In areas with high seismic impacts, it is important to incorporate seismic factors into collapse risk mapping in order to better predict and mitigate the occurrence of collapse disasters. For the impact of indicators on collapse, the distance category should be further improved based on the survey results to narrow the buffer zone and improve the accuracy of collapse risk zoning. For areas at high risk of collapse, the prevention, monitoring, and emergency rescue of collapse disasters should be strengthened, and the potential risk of collapse disasters should be reduced by means of land planning, engineering measures, and warning and education.

7. Conclusions

In this paper, the curvature watershed method is used to make mapping units, which are used to extract all evaluation indicators. Then, the ICM, AHP, and AHP-ICM models are applied to hazard mapping, the accuracy of the three models is verified by the ROC curve, the optimal model is selected, the zoning of exposure, vulnerability, and emergency response and recovery capability made by applying the AHP-EWM model is combined, the collapse disaster risk index model used in the risk index is calculated, and finally, the collapse risk zoning map was drawn using GIS. This led to the following conclusions.

Based on the evaluation indicator system determined by the four elements of natural disaster risk theory, TOL and VIF were used to analyse the covariance among the indicators, and it was verified that there was independence among all 21 indicators, indicating that the established evaluation indicator system was reasonable.

The ICM model, AHP model, and AHP-ICM model were compared in the collapse hazard mapping, and they were validated by the ROC curve, yielding AUC values for the three models: ICM model (85.6%), AHP model (80%), and AHP-ICM model (87.4%). After a comprehensive comparison, the percentage of the ICM, AHP, and AHP-ICM models with very high and high hazard ratings are 28.03%, 28.88%, and 32.51%, respectively, and the number of known collapses they contain are 48, 37, and 44, respectively, with disaster point densities of 4.48, 5.8, and 6.69 per km^2, respectively; the results show that the AHP-ICM model is the optimal model for collapse hazard mapping in Huinan County.

The AHP-EWM model was used to map the exposure, vulnerability, and emergency response and recovery capability of the collapse disaster bearers. The results show that very high exposure and high exposure areas are mostly concentrated in densely populated and built-up areas, very high vulnerability and high vulnerability areas are mainly located in rural and township areas close to the mountain, and areas with high emergency response and recovery capability are mostly concentrated in urban and township areas.

According to the results of the collapse risk mapping, the very high-risk zone accounts for 6.06% of the study area, containing 20 collapse points and a disaster point density of 14.49 per 100 km^2; the high-risk zone covers 30.07% of the area, containing 28 collapse points and a disaster point density of 4.09 per 100 km^2. Very high-risk areas and high-risk areas are mainly located in (1) Northwestern Huinan County (Chaoyang Township), (2) the central part (Huinan Town–Sansonggang Town–Fumin Town), (3) the southwest (look-alike whistle town), and (4) the northeast (Shidaohe Town). When preventing and controlling collapse disaster sites or sections, decision makers should consider these four areas first in order to reduce disaster losses and casualties.

Author Contributions: Conceptualisation, Zengkang Lu; data curation, Zengkang Lu and Jie Wang; formal analysis, Zengkang Lu and Chenglong Yu; methodology, Zengkang Lu; writing—original draft, Zengkang Lu; funding acquisition, Yichen Zhang and Huanan Liu; writing—review and editing, Chenglong Yu, Yanan Chen and Jiquan Zhang. All authors have read and agreed to the published version of the manuscript.

Funding: This research was funded by the Key Scientific and Technology Research and Development Program of Jilin Province (grant No. 20220203002SF), Jilin Science and Technology Program (grant No. 20230203130SF), Jilin Education Program (grant No. JJKH20230722KJ).

Data Availability Statement: The codes and data for this article are freely available at https://www.gscloud.cn/ (accessed on 30 February 2023), https://www.webmap.cn/ (accessed on 9 March 2023), https://www.resdc.cn/ (accessed on 15 March 2023), https://www.databox.store (accessed on 11 February 2023), https://www.worldpop.org/ (accessed on 14 February 2023), and http://www.geodata.cn/ (accessed on 16 March 2023).

Conflicts of Interest: The authors declare no conflict of interest.

References

1. Xu, Z.H.; Wang, W.Y.; Lin, P.; Wang, X.T.; Yu, T.F. Buffering Effect of Overlying Sand Layer Technology for Dealing with Rockfall Disaster in Tunnels and a Case Study. *Int. J. Geomech.* **2020**, *20*, 04020127. [CrossRef]
2. Xie, R.; Fan, W.; Liu, B.; Shen, D. Dynamic behavior and vulnerability analysis of bridge columns with different cross-sectional shapes under rockfall impacts. *Structures* **2020**, *26*, 471–486. [CrossRef]
3. Ritchie, A.M. Evaluation of Rockfall and Its Control. *Highw. Res. Rec.* **1963**, *17*, 13–28.

4. Rana, H.; Babu, G.L.S. Regional back analysis of landslide events using TRIGRS model and rainfall threshold: An approach to estimate landslide hazard for Kodagu, India. *Bull. Eng. Geol. Environ.* **2022**, *81*, 160. [CrossRef]
5. Guo, S.; Liu, Y.; Zhang, P.; Zhu, R.; Qian, Y. Insight into the critical morphological characteristics of earthquake-induced sequential rock avalanches in weathered-fractured rock cliffs. *Front. Earth Sci.* **2023**, *10*, 1101246. [CrossRef]
6. Loupasakis, C. Contradictive mining–induced geocatastrophic events at open pit coal mines: The case of Amintaio coal mine, West Macedonia, Greece. *Arab. J. Geosci.* **2020**, *13*, 582. [CrossRef]
7. Saroglou, C. GIS-Based Rockfall Susceptibility Zoning in Greece. *Geosciences* **2019**, *9*, 163. [CrossRef]
8. Zhou, X.; Wu, W.; Lin, Z.; Zhang, G.; Chen, R.; Song, Y.; Wang, Z.; Lang, T.; Qin, Y.; Ou, P.; et al. Landslide risk zoning in Ruijin, Jiangxi, China. *Nat. Hazards Earth Syst. Sci. Discuss.* **2020**, 1–21. [CrossRef]
9. Wen, H.; Hu, J.; Zhang, J.; Xiang, X.; Liao, M. Rockfall susceptibility mapping using XGBoost model by hybrid optimized factor screening and hyperparameter. *Geocarto Int.* **2022**, *37*, 16872–16899. [CrossRef]
10. Katz, O.; Reichenbach, P.; Guzzetti, F. Rock fall hazard along the railway corridor to Jerusalem, Israel, in the Soreq and Refaim valleys. *Nat. Hazard* **2010**, *56*, 649–665. [CrossRef]
11. Lee, J.; Barbato, M.; Lee, D.K. Rockfall Hazard Analysis Based on the Concept of Functional Safety with Application to the Highway Network in South Korea. *Rock Mech. Rock Eng.* **2021**, *54*, 6633–6647. [CrossRef]
12. Tanoli, J.I.; Chen, N.; Ullah, I.; Qasim, M.; Ali, S.; Rehman, Q.U.; Umber, U.; Jadoon, I.A.K. Modified "Rockfall Hazard Rating System for Pakistan (RHRSP)": An Application for Hazard and Risk Assessment along the Karakoram Highway, Northwest Pakistan. *Appl. Sci.* **2022**, *12*, 3778. [CrossRef]
13. Antoniou, A.A.; Lekkas, E. Rockfall susceptibility map for Athinios port, Santorini Island, Greece. *Geomorphology* **2010**, *118*, 152–166. [CrossRef]
14. Copons, R.; Vilaplana, J.M. Rockfall susceptibility zoning at a large scale: From geomorphological inventory to preliminary land use planning. *Eng. Geol.* **2008**, *102*, 142–151. [CrossRef]
15. Prina Howald, E.; Abbruzzese, J.M.; Grisanti, C. An approach for evaluating the role of protection measures in rockfall hazard zoning based on the Swiss experience. *Nat. Hazards Earth Syst. Sci.* **2017**, *17*, 1127–1144. [CrossRef]
16. Xie, W.; Li, X.; Jian, W.; Yang, Y.; Liu, H.; Robledo, L.F.; Nie, W. A Novel Hybrid Method for Landslide Susceptibility Mapping-Based GeoDetector and Machine Learning Cluster: A Case of Xiaojin County, China. *ISPRS Int. J. Geo-Inf.* **2021**, *10*, 93. [CrossRef]
17. Shu, H.; Guo, Z.; Qi, S.; Song, D.; Pourghasemi, H.; Ma, J. Integrating Landslide Typology with Weighted Frequency Ratio Model for Landslide Susceptibility Mapping: A Case Study from Lanzhou City of Northwestern China. *Remote Sens.* **2021**, *13*, 3623. [CrossRef]
18. Wang, F.; Xu, P.; Wang, C.; Wang, N.; Jiang, N. Application of a GIS-Based Slope Unit Method for Landslide Susceptibility Mapping along the Longzi River, Southeastern Tibetan Plateau, China. *ISPRS Int. J. Geo-Inf.* **2017**, *6*, 172. [CrossRef]
19. Du, G.; Zhang, Y.; Yang, Z.; Guo, C.; Yao, X.; Sun, D. Landslide susceptibility mapping in the region of eastern Himalayan syntaxis, Tibetan Plateau, China: A comparison between analytical hierarchy process information value and logistic regression-information value methods. *Bull. Eng. Geol. Environ.* **2018**, *78*, 4201–4215. [CrossRef]
20. Bhagya, S.B.; Sumi, A.S.; Balaji, S.; Danumah, J.H.; Costache, R.; Rajaneesh, A.; Gokul, A.; Chandrasenan, C.P.; Quevedo, R.P.; Johny, A.; et al. Landslide Susceptibility Assessment of a Part of the Western Ghats (India) Employing the AHP and F-AHP Models and Comparison with Existing Susceptibility Maps. *Land* **2023**, *12*, 468. [CrossRef]
21. Zhou, S.; Chen, G.; Fang, L.; Nie, Y. GIS-Based Integration of Subjective and Objective Weighting Methods for Regional Landslides Susceptibility Mapping. *Sustainability* **2016**, *8*, 334. [CrossRef]
22. Deng, H.; Wu, X.; Zhang, W.; Liu, Y.; Li, W.; Li, X.; Zhou, P.; Zhuo, W. Slope-Unit Scale Landslide Susceptibility Mapping Based on the Random Forest Model in Deep Valley Areas. *Remote Sens.* **2022**, *14*, 4245. [CrossRef]
23. Panchal, S.; Shrivastava, A.K. A Comparative Study of Frequency Ratio, Shannon's Entropy and Analytic Hierarchy Process (AHP) Models for Landslide Susceptibility Assessment. *ISPRS Int. J. Geo-Inf.* **2021**, *10*, 603. [CrossRef]
24. Ma, Z.; Qin, S.; Cao, C.; Lv, J.; Li, G.; Qiao, S.; Hu, X. The Influence of Different Knowledge-Driven Methods on Landslide Susceptibility Mapping: A Case Study in the Changbai Mountain Area, Northeast China. *Entropy* **2019**, *21*, 372. [CrossRef]
25. Sun, X.; Chen, J.; Li, Y.; Rene, N.N. Landslide Susceptibility Mapping along a Rapidly Uplifting River Valley of the Upper Jinsha River, Southeastern Tibetan Plateau, China. *Remote Sens.* **2022**, *14*, 1730. [CrossRef]
26. Lin, L.; Lin, Q.; Wang, Y. Landslide susceptibility mapping on a global scale using the method of logistic regression. *Nat. Hazards Earth Syst. Sci.* **2017**, *17*, 1411–1424. [CrossRef]
27. Sun, X.; Yu, C.; Li, Y.; Rene, N.N. Susceptibility Mapping of Typical Geological Hazards in Helong City Affected by Volcanic Activity of Changbai Mountain, Northeastern China. *ISPRS Int. J. Geo-Inf.* **2022**, *11*, 344. [CrossRef]
28. Zhang, J.-Q.; Liang, J.-D.; Liu, X.-P.; Tong, Z.-J. GIS-Based Risk Assessment of Ecological Disasters in Jilin Province, Northeast China. *Hum. Ecol. Risk Assess.* **2009**, *15*, 727–745. [CrossRef]
29. Hamza, T.; Raghuvanshi, T.K. GIS based landslide hazard evaluation and zonation—A case from Jeldu District, Central Ethiopia. *J. King Saud Univ. Sci.* **2017**, *29*, 151–165. [CrossRef]
30. Pellicani, R.; Argentiero, I.; Spilotro, G. GIS-based predictive models for regional-scale landslide susceptibility assessment and risk mapping along road corridors. *Geomat. Nat. Hazards Risk* **2017**, *8*, 1012–1033. [CrossRef]

31. Pradhan, A.M.S.; Kim, Y.-T. Rainfall-Induced Shallow Landslide Susceptibility Mapping at Two Adjacent Catchments Using Advanced Machine Learning Algorithms. *ISPRS Int. J. Geo-Inf.* **2020**, *9*, 569. [CrossRef]
32. Tan, Q.; Huang, Y.; Hu, J.; Zhou, P.; Hu, J. Application of artificial neural network model based on GIS in geological hazard zoning. *Neural Comput. Appl.* **2020**, *33*, 591–602. [CrossRef]
33. Wu, C.; Zhang, Y.; Zhang, J.; Chen, Y.; Duan, C.; Qi, J.; Cheng, Z.; Pan, Z. Comprehensive Evaluation of the Eco-Geological Environment in the Concentrated Mining Area of Mineral Resources. *Sustainability* **2022**, *14*, 6808. [CrossRef]
34. Yuan, X.; Liu, C.; Nie, R.; Yang, Z.; Li, W.; Dai, X.; Cheng, J.; Zhang, J.; Ma, L.; Fu, X.; et al. A Comparative Analysis of Certainty Factor-Based Machine Learning Methods for Collapse and Landslide Susceptibility Mapping in Wenchuan County, China. *Remote Sens.* **2022**, *14*, 3259. [CrossRef]
35. Kadavi, P.; Lee, C.-W.; Lee, S. Application of Ensemble-Based Machine Learning Models to Landslide Susceptibility Mapping. *Remote Sens.* **2018**, *10*, 1252. [CrossRef]
36. Do, C.; Kuleshov, Y. Multi-Hazard Tropical Cyclone Risk Assessment for Australia. *Remote Sens.* **2023**, *15*, 795. [CrossRef]
37. Ke, K.; Zhang, Y.; Zhang, J.; Chen, Y.; Wu, C.; Nie, Z.; Wu, J. Risk Assessment of Earthquake–Landslide Hazard Chain Based on CF-SVM and Newmark Model—Using Changbai Mountain as an Example. *Land* **2023**, *12*, 696. [CrossRef]
38. Duan, C.; Zhang, J.; Chen, Y.; Lang, Q.; Zhang, Y.; Wu, C.; Zhang, Z. Comprehensive Risk Assessment of Urban Waterlogging Disaster Based on MCDA-GIS Integration: The Case Study of Changchun, China. *Remote Sens.* **2022**, *14*, 3101. [CrossRef]
39. Mwakapesa, D.S.; Lan, X.; Nanehkaran, Y.A.; Mao, Y. Landslide susceptibility mapping using O-CURE and PAM clustering algorithms. *Front. Environ. Sci.* **2023**, *11*, 271. [CrossRef]
40. Zhang, T.; Fu, Q.; Quevedo, R.P.; Chen, T.; Luo, D.; Liu, F.; Kong, H. Landslide Susceptibility Mapping Using Novel Hybrid Model Based on Different Mapping Units. *KSCE J. Civ. Eng.* **2022**, *26*, 2888–2900. [CrossRef]
41. Ba, Q.; Chen, Y.; Deng, S.; Yang, J.; Li, H. A comparison of slope units and grid cells as mapping units for landslide susceptibility assessment. *Earth Sci. Inf.* **2018**, *11*, 373–388. [CrossRef]
42. Guzzetti, F.; Carrara, A.; Cardinali, M.; Reichenbach, P. Landslide hazard evaluation: A review of current techniques and their application in a multi-scale study, Central Italy. *Geomorphology* **1999**, *31*, 181–216. [CrossRef]
43. Deng, N.; Li, Y.; Ma, J.; Shahabi, H.; Hashim, M.; de Oliveira, G.; Chaeikar, S.S. A comparative study for landslide susceptibility assessment using machine learning algorithms based on grid unit and slope unit. *Front. Environ. Sci.* **2022**, *10*, 1009433. [CrossRef]
44. Yu, C.; Chen, J. Application of a GIS-Based Slope Unit Method for Landslide Susceptibility Mapping in Helong City: Comparative Assessment of ICM, AHP, and RF Model. *Symmetry* **2020**, *12*, 1848. [CrossRef]
45. Hong, H.; Liu, J.; Bui, D.T.; Pradhan, B.; Acharya, T.D.; Pham, B.T.; Zhu, A.X.; Chen, W.; Ahmad, B.B. Landslide susceptibility mapping using J48 Decision Tree with AdaBoost, Bagging and Rotation Forest ensembles in the Guangchang area (China). *Catena* **2018**, *163*, 399–413. [CrossRef]
46. He, H.; Hu, D.; Sun, Q.; Zhu, L.; Liu, Y. A Landslide Susceptibility Assessment Method Based on GIS Technology and an AHP-Weighted Information Content Method: A Case Study of Southern Anhui, China. *ISPRS Int. J. Geo-Inf.* **2019**, *8*, 266. [CrossRef]
47. Saaty, T.L. Modeling unstructured decision problems—The theory of analytical hierarchies. *Math. Comput. Simul.* **1978**, *20*, 147–158. [CrossRef]
48. Gao, R.; Wang, C.; Liang, Z.; Han, S.; Li, B. A Research on Susceptibility Mapping of Multiple Geological Hazards in Yanzi River Basin, China. *ISPRS Int. J. Geo-Inf.* **2021**, *10*, 218. [CrossRef]
49. Roccati, A.; Paliaga, G.; Luino, F.; Faccini, F.; Turconi, L. GIS-Based Landslide Susceptibility Mapping for Land Use Planning and Risk Assessment. *Land* **2021**, *10*, 162. [CrossRef]
50. Cao, C.; Wang, Q.; Chen, J.; Ruan, Y.; Zheng, L.; Song, S.; Niu, C. Landslide Susceptibility Mapping in Vertical Distribution Law of Precipitation Area: Case of the Xulong Hydropower Station Reservoir, Southwestern China. *Water* **2016**, *8*, 270. [CrossRef]
51. Li, X. TOPSIS model with entropy weight for eco geological environmental carrying capacity assessment. *Microprocess. Microsyst.* **2021**, *82*, 103805. [CrossRef]
52. Kincal, C.; Kayhan, H. A Combined Method for Preparation of Landslide Susceptibility Map in Izmir (Türkiye). *Appl. Sci.* **2022**, *12*, 9029. [CrossRef]
53. Saha, A.; Villuri, V.G.K.; Bhardwaj, A. Development and Assessment of GIS-Based Landslide Susceptibility Mapping Models Using ANN, Fuzzy-AHP, and MCDA in Darjeeling Himalayas, West Bengal, India. *Land* **2022**, *11*, 1711. [CrossRef]
54. Saha, A.; Villuri, V.G.K.; Bhardwaj, A.; Kumar, S. A Multi-Criteria Decision Analysis (MCDA) Approach for Landslide Susceptibility Mapping of a Part of Darjeeling District in North-East Himalaya, India. *Appl. Sci.* **2023**, *13*, 5062. [CrossRef]

applied sciences

MDPI

Article

Study on Mechanical Properties of Deep Expansive Soil and Coupling Damage Model of Freeze–Thaw Action and Loading

Zhuliang Zhu, Bin Lin * and Shiwei Chen

School of Civil Engineering and Architecture, Anhui University of Science and Technology, Huainan 232001, China; zhuzl1999@163.com (Z.Z.)
* Correspondence: linbin8910@163.com

Abstract: This study is primarily intended to present a damage constitutive equation under the combined action of confining pressures and freeze–thaw cycles by subjecting deep expansive clay to the consolidated undrained triaxial tests. We study the influence of the numbers of freeze–thaw cycles on various mechanical indexes of soil by using the TSZ-2 instrument (fully automatic triaxial instrument). As the number of freeze–thaw cycles increases, the ultimate peak stress of the soil decreases, and then, the effect of the freeze–thaw effect on the shear strength gradually weakened. By combining the expression method of the damage variable under the action of loading alone with the expression method under the action of freeze–thaw cycles alone, we brought in the damage evolution equation to obtain the damage constitutive equation under the combined action of confining pressures and freeze–thaw cycles. The stress values under three confining pressures (100 kPa, 200 kPa, and 300 kPa) can be determined by using the final damage constitutive model. The measured data with a water content of 17% and six freeze–thaw cycles were compared with the theoretical data. The actual strength values were 118.4 kPa, 152.3 kPa, and 184.1 kPa, and the theoretical strength values were 120 kPa, 150 kPa, and 186 kPa. The fitting degree of the strength value was as high as 99%, which verifies the feasibility of this model. This study can serve as an available reference for well wall construction and disaster prediction in deep coal mining.

Keywords: frozen soil; deep clay; triaxial tests; freeze–thaw cycles; damage model

1. Introduction

Coal resources make up a large proportion of the world's energy supply. Currently, coal accounts for 27% of the primary energy consumption worldwide [1]. Some countries around the world have an increasing demand for energy resources [2,3], while coal resources cannot be substituted by alternative sources of energy. The extraction of shallow coal reserves has been insufficient to satisfy the rising demand, and consequently, there has been a gradual increase in the excavation depths of mines worldwide [4,5]. With the increase in the depth of mining, the geological conditions are becoming more complex, landslide disasters caused by various reasons are prone to occur [6–9], and the mining environment is becoming increasingly unfavorable [10]. So far, artificial ground freezing has become the way to prevent water inrush and well gushing in the process of shaft sinking in deep alluvium and fractured rock strata [11–13]. The artificially chilled brine passes through the freezing pipe, and consequently, the frozen wall is formed around the shaft. The frozen wall is a useful protective screen for resisting earth pressure and preventing groundwater infiltration into the well. The construction complexity of the frozen wall inevitably increases with soil thickness and depth of freezing, resulting in more frequent disasters caused by the freezing and melting process of soil [14,15]. Therefore, it is imperative to study the mechanical properties and damage progression of deeply frozen soil.

From the perspective of damage mechanics, the failure of soil resulting from the application of external loading is a transition from primary to secondary soil structure.

The evolution of damage can be seen as a blend of undisturbed soil and soil that is wholly damaged [16–18]. Kachanov et al. [19] first proposed the concept of damage mechanics in the study of creep fracture and cited the concept of the 'continuity factor'. Shen [20] pointed out that, when the external loading is applied to the soil, the soil can be regarded as the superposition of the undisturbed soil and the damaged soil, and he introduced the concept of a damage ratio and established the damage constitutive model of structural clay. Tong et al. [21] used cement to improve the soil, studied the damage evolution of the improved soil, and established the damage constitutive model of cement-improved soil under uniaxial compression, based on the study of the evolution law. Zhu et al. [22] used Weibull distribution to simulate the whole process of damage evolution, coupled stress field, temperature field, and water field and fitted the stress–strain curves of frozen sand at different temperatures to obtain the model. Zhang et al. [23] conducted the rapid loading test, consolidation test, and triaxial compressive strength test of the natural sedimentary structural clay in Lianyungang and studied the relationship between the damage variable and disturbance degree of soil. Shi et al. [24] proposed the damage variable considering the cutting effect of the ice lenses and established the corresponding elastoplastic damage constitutive model of the frozen soil. Sun et al. [25] proposed an elastoplastic damage constitutive model based on super/sub-loading yield surface and verified the model by observing the mechanical behavior of frozen soil and frozen saline soil through the triaxial compression test. He et al. [26] established a new composite structure fractional constitutive model by establishing the creep relationship between the bonding unit and the friction unit based on the microscopic damage mechanism of frozen soil. The above research shows that scholars have limited research on the damage evolution of soil to only from a single direction, and there is almost no research on deep soil.

Therefore, in this paper, deep expansive clay was studied, and the damage evolution of soil under the combined action of loading and freeze–thaw cycles was studied in more depth. The conventional triaxial test and triaxial shear tests under the freeze–thaw cycles of deep expansive clay were carried out, and the influence of the number of freeze–thaw cycles on various mechanical indexes of soil was studied. The expression method of the damage variable under the action of loading alone and the expression method under the action of freeze–thaw cycles alone were coupled and brought into the damage evolution equation to obtain the damage constitutive equation under the combined action of confining pressures and freeze–thaw cycles and compared with the actual stress–strain curve. This study can provide some reference value for well wall construction and disaster prediction in deep coal mining.

2. Experiments Design

2.1. Sample Preparation

The deep expensive clay samples were obtained from detection hole of an air shaft in a mining area of Anhui, east China. The basic physical properties of the soil are listed in Table 1, and the soil particle size distribution was calculated according to ASTMD422-63 [27] and listed in Table 2.

Table 1. Basic physical property parameters of soil.

Natural Moisture Content (%)	Natural Density (g·cm^{-3})	Dry Density (g·cm^{-3})	Specific Gravity	Liquid Limit (%)	Plastic Limit (%)	Free Expansion Rate (%)
19.31	1.965	1.647	2.467	39.33	18.77	58.54%

Table 2. Particle size grading table.

Particle Size (mm)	Particle Size Ratio (%)
<0.075	12.23
0.075~0.25	37.97
0.25~0.5	30.19
0.5~1.0	13.64
1.0~2.0	5.7

The soil sample preparation method was as follows: the bulk disturbed soil was crushed by the soil crusher and sieved through a 2 mm standard sieve. The soil samples were dried in an oven at 105 °C for 8 h and cooled to room temperature in a sealed container. According to the natural water content of the soil, the experimental water content was set reasonably. A certain amount of water was added evenly to the soil samples by a sprayer to achieve the moisture content targets of 14%, 17%, 21%, and 24%. In order to ensure even distribution of water in the soil, soil samples were then sealed in a closed container for 24 h. After that, some soil samples were taken for moisture content testing to ensure that the target moisture content was achieved. Five equal amounts of soil were added to the standard mold, the sample being compacted after each addition. In order to reduce the friction between the soil samples and the mold, Vaseline was used to coat around the mold. Finally, a number of soil samples with an inner diameter of 39.1 mm and a height of 80 mm were produced. In addition, the soil samples were immediately sealed with plastic film [28,29].

2.2. Experimental System

As shown in Figure 1, the test instrument was the TSZ-2 fully automatic triaxial instrument produced by the Nanjing Soil Instrument Factory. The axial load of the upper and lower limit device adopts the force sensor; the measurement range is 0–30 kN, and the accuracy is ±1%. The freeze–thaw cycle system is a high–low temperature and humidity constant test chamber produced by Taiwan, China. It has the function of PID automatic calculus and high precision.

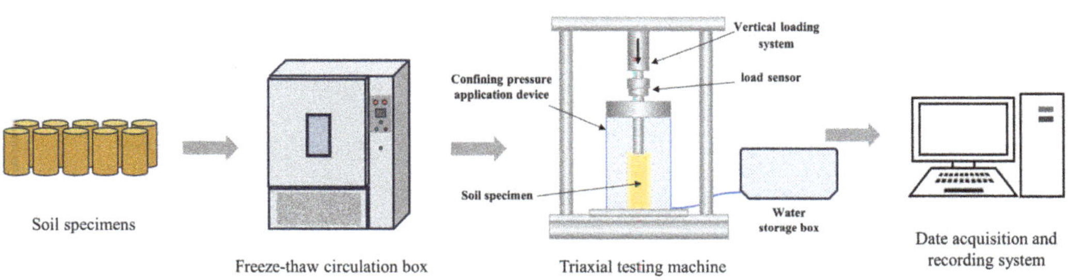

Figure 1. Schematic diagram of triaxial shear test under freeze–thaw cycles.

2.3. Experimental Program

The triaxial consolidation undrained method was used in the test. The axial loading rate was set to 0.8 mm/min during the test, and the confining pressure was controlled at 100 kPa, 200 kPa, and 300 kPa, respectively. If the stress–strain curve had no peak value, the stress corresponding to 15% strain was the ultimate stress. After completing the conventional triaxial test, we placed the prepared soil samples in the high–low temperature and humidity constant test chamber produced for freeze–thawing and the time of freezing and thawing was controlled at 4 h to ensure that 3 freeze–thaw cycles was 1 test cycle. Five test cycles of freeze–thaw cycles, i.e., 0, 3, 6, 9, and 12, were considered. After the freeze–thaw cycle, the triaxial shear test under freeze–thaw cycles was carried out. The test method and confining pressure setting were the same as the conventional triaxial test.

3. Experiments Results and Analysis

The stress–strain curve obtained from the consolidation undrained test is shown in Figure 2. From the stress–strain curve of deep clay under different water contents, it can be seen that the lower the water content, the greater the compressive strength of the soil. The compression curve of deep clay under four water contents undergoes a linear elastic stage and a nonlinear stage. In the case of low water content, the samples under triaxial confining pressure were all strain softening. The higher the water content, the later the peak point appeared. The soil sample with low water content had a peak point before the strain was 20%, indicating that the curve of deep clay showed a strain-hardening type when the water content gradually increased. At the same time, it can be seen from the figure that the stress–strain curve of deep expansive clay under high confining pressure tended to strain hardening, while low confining pressure showed strain softening.

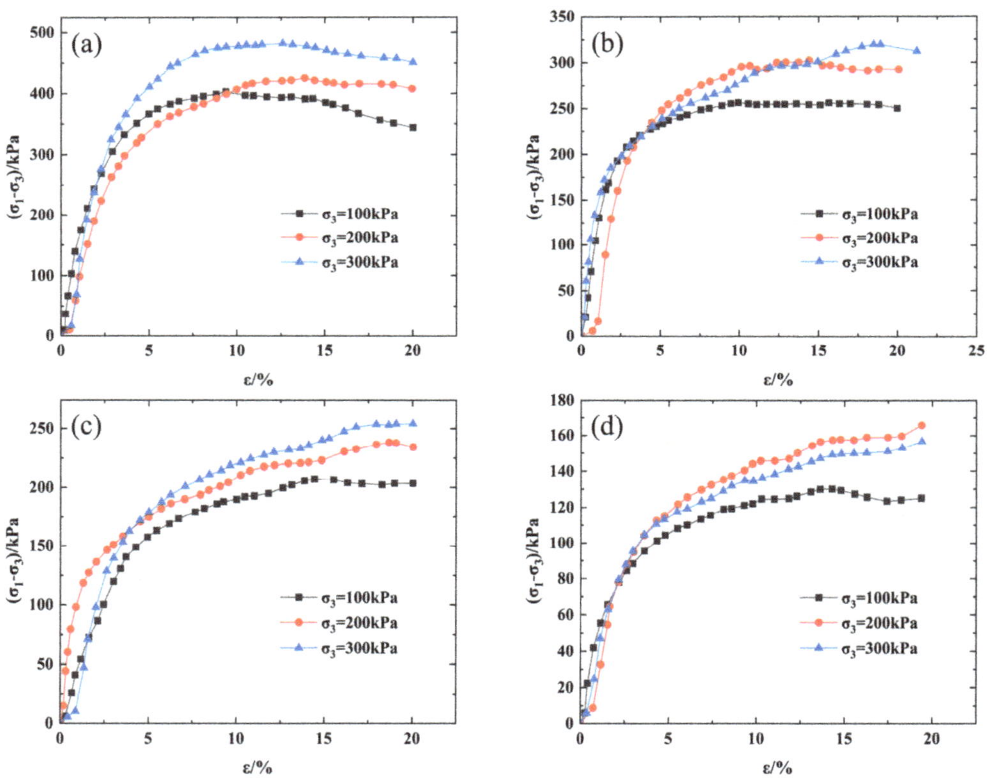

Figure 2. The stress–strain curve under different moisture contents: (**a**) ω =14%, (**b**) ω = 17%, (**c**) ω = 21%, and (**d**) ω = 24%.

Figure 3 illustrates that under the same confining pressure, the growth rate of stress with strain gradually decreased as the freeze–thaw times increased. Meanwhile, the ultimate peak stress also decreased as a result of the freeze–thaw cycles. The initial stress was linearly proportional to the strain during triaxial compression and, then, gradually became a nonlinear relationship because of internal damage. Under different freeze–thaw cycles, the stress–strain curves of triaxial tests all showed strain hardening. Under the three confining pressures, the relationship between compressive strength and number of freeze–thaw cycles is shown in the Figure 4. It can be seen that the change rules were basically similar under the three confining pressures. As the number of freeze–thaw cycles increased, the compressive strength decreased significantly. In addition, the compressive strength value decreased significantly when the number of freeze–thaw cycles was small. When the freeze–thaw cycles reached a certain number, the compressive strength remained basically stable, and the change was small. It can be concluded that as the number of freeze–thaw cycles increased, the effect of the freeze–thaw effect on the compressive strength gradually weakened.

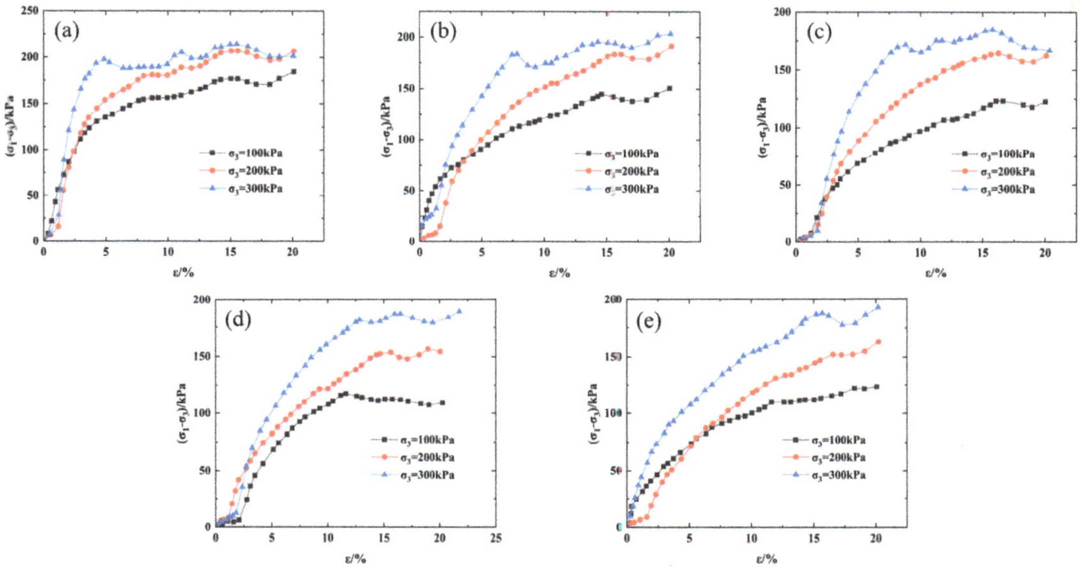

Figure 3. Stress–strain curves under different numbers of freeze–thaw cycles: (**a**) 0 time, (**b**) 3 times, (**c**) 6 times, (**d**) 9 times, and (**e**) 12 times.

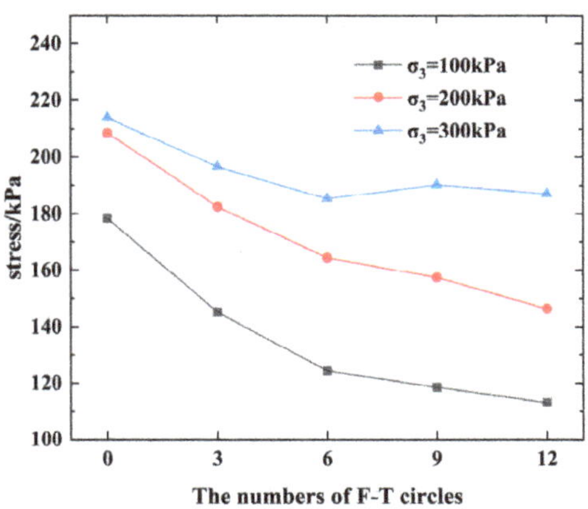

Figure 4. The relationship between compressive strength and number of freeze–thaw cycles.

The value of cohesion c and internal friction angle φ are the basic physical index of the triaxial test. The Mohr stress circle envelope was drawn by the limit stress and strain of each freeze–thaw cycle, and the value of the internal friction angle and cohesion was calculated. The Mohr stress circle envelope in several states is shown in Figure 5.

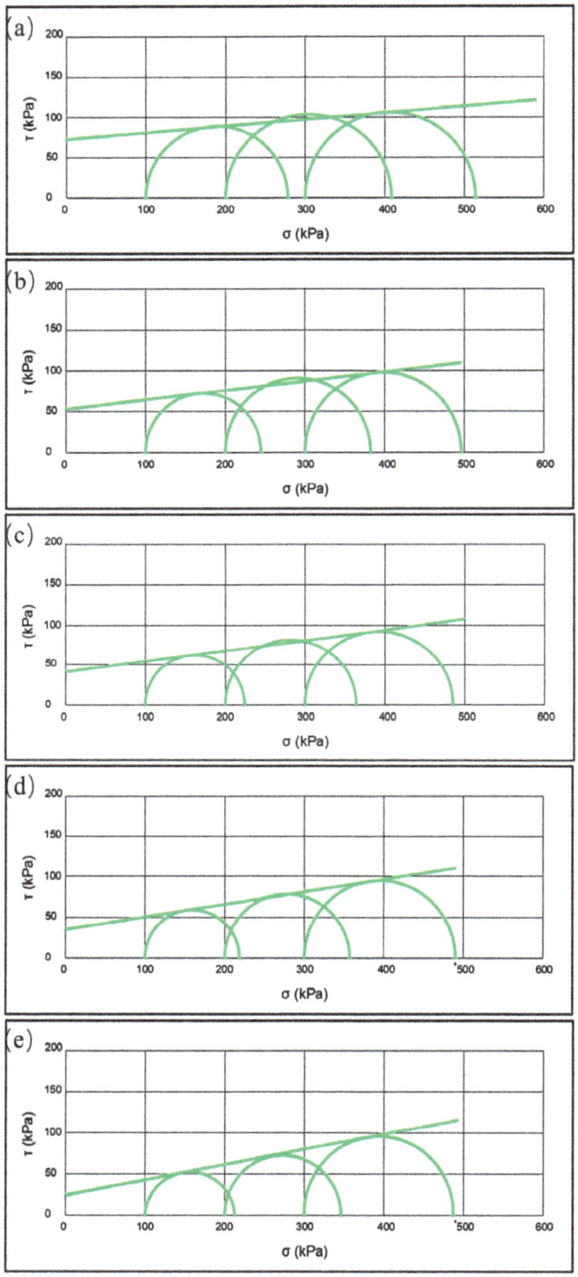

Figure 5. The Mohr stress circle under different numbers of freeze–thaw cycles: (**a**) 0 time, (**b**) 3 times, (**c**) 6 times, (**d**) 9 times, and (**e**) 12 times.

It can be seen from Figure 6 that as the number of freeze–thaw cycles increased, the cohesion tended to decrease, while the internal friction angle tended to increase. For small numbers of freeze–thaw cycles (within six cycles), the freeze–thaw cycles greatly affected both cohesion and the internal friction angle. When the number of freeze–thaw cycles reached roughly nine times, cohesion and the internal friction angle generally stabilized, indicating little impact from freeze–thaw cycles on these shear-strength indices.

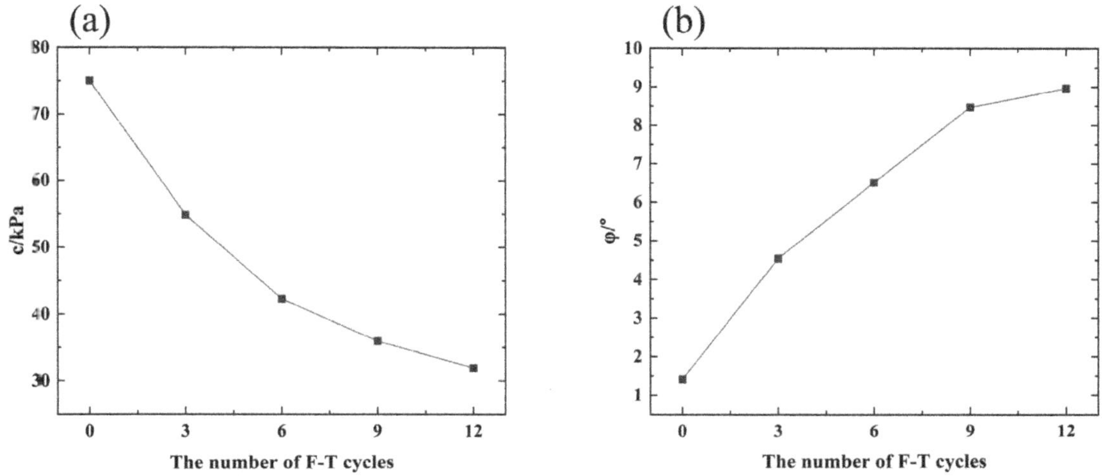

Figure 6. The change diagram of mechanical index with the number of freeze–thaw cycles. (**a**) The cohesion, (**b**) The internal friction angle.

4. Damage Constitutive Model of Deep Expansive Clay under Freeze–Thaw Cycles

The soil subjected to freeze–thaw cycles is damaged under the combined action of its own structural defects and internal microcracks and external loads during the compression process, which will lead to the propagation and interconnection of damage so that the macroscopic mechanical properties of freeze–thaw soil are significantly weakened. In addition, due to the repeated freeze–thaw action, small cracks in the soil can develop, which has a huge impact on the structure of the soil. How to properly reflect the damage law of soil under the combined action of loading and freeze–thaw cycles, the key is to select the appropriate damage variable. Many scholars had selected micro-benchmarks for damage analysis under loading [30,31]; however, there were some defects. The reason is that the initial soil damage has a great degree of randomness and random evolution. If the constitutive relationship of traditional methods is used, the randomness of soil in the damage process cannot be well reflected. Some scholars had cited the idea of probability and statistics [32], which provided a new way to solve such problems.

4.1. Establishment of Damage Variables

The effective stress is the main factor affecting the strain of the soil body. When the material damage occurs, its constitutive relationship only needs to change the stress to the effective stress, which is the famous Lenaitre's strain equivalence hypothesis, so as to establish the basic equation of the damage constitutive model of clay expansive soil:

$$\sigma_{ij} = \sigma^*(1 - D) = E_{ijkl}\varepsilon_{kl}(1 - D) \quad (1)$$

where E_{ijkl} is the tensor component of material elastic stiffness, σ_{ij} is the stress tensor component, σ^* is the effective stress tensor component, ε_{ij} is the strain tensor component, and D is the damage variable.

In soil mechanics, soil is often considered a continuous medium from a macroscopic perspective. Due to the dispersion of soil particles, soil is also considered to be a porous medium. However, there are many kinds of soil particles and cement in it, and the existence of interlocking between soil particles makes the soil have good integrity and structure.

Therefore, it is considered that the soil is continuous during a loading process of little structural damage, so the following assumptions are put forward:

1. When the sample is subjected to the loading, macroscopically, it can be regarded as an isotropic microelement, while at the microscopic level it contains some of the basic information that causes damage, and at this point, it can be regarded as a nonhomogeneous microscopic material.
2. The clay microelement can be regarded as linearly elastic before damage is produced under loading. At this time, its strength follows the Hooke law, and the nonlinearity of stress–strain results from the generation of material damage. The initial tangential modulus can also be replaced by the elastic modulus of the undamaged material.
3. The damage caused by microelements is considered to be strength damage in clay materials.

At the microscopic perspective, the ratio of the area in a damaged unit to the total cross-sectional area is called the damage variable. Macroscopically, it can be defined as the ratio of the number of self-damaged units, N_f, to the total number of cells, N. The damage equation is expressed as follows:

$$D = \frac{N_f}{N} \tag{2}$$

The former scholars demonstrated that the Weibull distribution is more appropriate to describe the pattern of strength distribution in clay under freeze–thaw cycles [33,34]. By introducing elemental intensity as a random variable and combining it with the Weibull distribution function, the following can be obtained:

$$\begin{cases} P(F^*) = \frac{\alpha}{F_0}\left(\frac{F^*}{F_0}\right)^{m-1} \exp\left[-\left(\frac{F^*}{F_0}\right)^m\right], & F^* \geq 0 \\ 0 & F^* \leq 0 \end{cases} \tag{3}$$

The damage variable D can be expressed as follows:

$$D = \frac{N_f}{N} = \int_{-\infty}^{F} P(x)dx = 1 - \exp\left[-\left(\frac{F^*}{F_0}\right)^m\right] \tag{4}$$

where m and F_0 are parameters of the Weibull distribution. They are determined by the mechanical properties of the clay.

4.2. Determination of the Clay Microelement Strength

The Druck–Prager damage criterion can be used to calculate the microelement strength of clay, and the basic form of clay microelement damage is selected as follows:

$$F^* = \alpha I_1^* + \sqrt{J_2^*} \tag{5}$$

In this formula,

$$I_1^* = \sigma_1^* + \sigma_2^* + \sigma_3^* \tag{6}$$

$$J_2^* = \frac{1}{6}\left[(\sigma_1^* - \sigma_2^*)^2 + (\sigma_2^* - \sigma_3^*)^2 + (\sigma_1^* - \sigma_3^*)^2\right] \tag{7}$$

The damage evolution equation can be obtained by substituting Equation (5) into Equation (4), resulting in the following:

$$D = 1 - \exp\left[-\left(\frac{\alpha I_1^* + \sqrt{J_2^*}}{F_0}\right)^m\right] \tag{8}$$

Equation (9) is obtained based on the previous assumptions about the damage variables and then combined with the Lemaitre strain equivalence principle as:

$$\frac{1}{1-D} = \exp\left[\left(\frac{\alpha I_1 + \sqrt{J_2}}{F_0(1-D)}\right)^m\right] \quad (9)$$

where

$$I_1 = \sigma_1 + \sigma_2 + \sigma_3 \quad (10)$$

$$J_2 = \frac{1}{6}\left[(\sigma_1 - \sigma_2)^2 + (\sigma_2 - \sigma_3)^2 + (\sigma_1 - \sigma_3)^2\right] \quad (11)$$

By $F = \alpha I_1 + \sqrt{J_2}$, we get:

$$\frac{1}{D} = \exp\left[\left(\frac{F}{F_0(1-D)}\right)^m\right] \quad (12)$$

When $D = 0$, the soil is not damaged; $F = 0$ is a prerequisite that the equation mentioned above is valid.

During triaxial compression, the clay was subjected to stresses in three dimensions. There is $\alpha > 0$ in the Druck–Prager damage criterion; a necessary condition is followed as:

$$\sigma_1 = \sigma_2 = \sigma_3 = 0 \quad (13)$$

By analyzing the stress–strain curve, it can be seen that the stress–strain can be regarded as a linear relationship at the beginning of axial stress loading [35]. As the stress increases, the slope starts to get smaller and smaller at a certain point, which is nonlinear. It is possible to identify the specific point in the stress–strain curve for clay at which damage occurs. The stress value that corresponds to this point can be considered as the damage stress. So, Equation (13) derived above is unreasonable. If the stress is below the stress threshold corresponding to a certain point, the stress–strain curve shows a linear relationship, and it can be inferred that no damage happens in the clay at this point. Once the stress point surpasses the damage threshold value, damage emerges within the clay body, and it intensifies throughout the loading process.

Combining Equation (1) with Equation (9), the stress–strain relationship of the soil under triaxial stress can be expressed as:

$$\sigma = E\varepsilon_1 \exp\left[-\left(\frac{F^*}{F_0}\right)^m\right] + 2\mu\sigma_3 \quad (14)$$

$$\sigma = E\varepsilon_1 \exp\left[-\left(\frac{\alpha I_1 + \sqrt{J_2}}{F_0}\right)^m\right] + 2\mu\sigma_3 \quad (15)$$

where E is the elastic modulus of soil, ε_1 is the axial strain of soil, μ is the Poisson ratio of soil, and σ_3 is the surrounding pressure of soil.

When considering the damage caused by freeze–thaw action, the freeze–thaw damage variable can be depicted by using a macroscopic scale. The damage variable incurred after a certain number of freeze–thaw experiences can be expressed as:

$$D_n = 1 - \frac{E_n}{E_0} \quad (16)$$

where E_0 is the elastic modulus of the soil without freeze–thaw cycles, and E_n is the elastic modulus of the soil with freeze–thaw cycles of n times.

Zhang et al. [36] proposed that the concept of the benchmark damage state is that the soil is initially damaged under the external loading. As the damage intensifies, the constitutive relationship of the soil damage under the combined effects of freeze–thaw action and loading can be defined as:

$$\sigma = E\varepsilon(1 - D_m) \quad (17)$$

$$D_m = D + D_n + DD_n \quad (18)$$

where D_m is the damage variable under the combined loading and freeze–thaw action; D is the damage variable under loading; D_n is the damage variable under freeze–thaw actions alone.

Substituting Equations (16) and (18) into Equation (8), we get:

$$D_m = 1 - \frac{E_n}{E_0} \exp\left[-\left(\frac{\alpha I_1 + \sqrt{J_2}}{F_0} \right)^m \right] \quad (19)$$

Uniting Equations (15), (17), and (19), we get the damage constitutive equation under freeze–thaw actions and loading as:

$$\sigma = E_n \varepsilon \exp\left[-\left(\frac{\alpha I_1 + \sqrt{J_2}}{F_0} \right)^m \right] + 2\mu\sigma_3 \quad (20)$$

4.3. Determination of the Parameters of the Damage Constitutive Equation

According to the identity $\sqrt{J_2} = sI_1 + t$, we can get the parameters of this identity in Table 3.

Table 3. The linear relationship coefficients of $\sqrt{J_2} \sim I_1$.

The Content of Moisture ω (%)	s	t	R^2
14	0.2254	132.145	0.9845
17	0.1647	112.014	0.9756
21	0.1225	97.014	0.9874
24	0.1014	80.257	0.9565

According to Table 3 above, the relationship of s and t and the content of moisture ω are obtained as follows:

$$s = 0.6745 \exp(-0.0474\omega) \quad (21)$$

$$t = 589.47\omega^{-0.48} \quad (22)$$

The damage conditions of deep expansive clay can be expressed as follows:

$$G = \sqrt{J_2} - sI_1 - t \quad (23)$$

Combining Equations (21), (22), and (23), we get the damage conditions of deep expansive clay; that is:

$$G = \sqrt{J_2} - 0.6745 \exp(-0.0474\omega) I_1 - 589.47\omega^{-0.48} \quad (24)$$

According to the damage threshold value relationship, the damage evolution equation of the damage variable can be obtained as:

$$D = \begin{cases} 0 & G < 0 \\ 1 - \exp\left[-\left(\frac{\alpha I_1^* + \sqrt{J_2^*} - F_G^*}{F_0} \right)^m \right] & G \geq 0 \end{cases} \quad (25)$$

where F_G^* is the strength value of deep expansive clay of the damage valve.

By combining Equations (1), (5), (20), (24), and (25), the damage constitutive equation of deep expansive clay, considering the damage threshold value under freeze–thaw cycle conditions, can be expressed as:

$$\varepsilon = \begin{cases} \frac{\sigma - 2\mu\sigma_3}{E_n} & G < 0 \\ \frac{\sigma - 2\mu\sigma_3}{E_n} \exp\left[\left(\frac{(\alpha I_1 + \sqrt{J_2} - (\alpha + 0.6745e^{-0.0474\omega}) I_{1G} - 589.47\omega^{-0.48} P_a) E\varepsilon_1}{F_0(\sigma - 2\mu\sigma_3)}\right)^m\right] & G \geq 0 \end{cases} \tag{26}$$

where E_n and μ are the elastic modulus and the Poisson ratio, respectively, under freeze–thaw cycles of n times.

The above expression can be simplified as follows:

$$\text{In}a + m\text{In}b = \text{In}\left[-\text{In}\left(\frac{\sigma - 2\mu\sigma_3}{E_n\varepsilon}\right)\right] \tag{27}$$

where

$$a = \left(\frac{1}{F_0}\right)^m \tag{28}$$

$$b = \frac{\left[\alpha I_1 + \sqrt{J_2} - \left(\alpha + 0.6745e^{-0.0474\omega}\right) I_{1G} - 589.47\omega^{-0.48} P_a\right] E_n\varepsilon_1}{(\sigma - 2\mu\sigma_3)} \tag{29}$$

The above expression can be simplified further as follows:

$$Y = mX + c \tag{30}$$

where $Y = \text{In}\left[-\text{In}\left(\frac{\sigma - 2\mu\sigma_3}{E_n\varepsilon}\right)\right]$; $X = \text{In}b$; $c = \text{In}a$.

The values of m and c can be obtained by linearly fitting the data measured in the experiment to the above formula. According to $F_0 = \exp(-\frac{c}{m})$, F_0 can be obtained as follows in the table. The Poisson ratio of the deep expansive clay is taken to be 0.45, and then, based on the Druck–Prager expression of failure criterion, $\alpha I_1 + \sqrt{J_2} = k$, the value of α and k can be obtained by linearly fitting.

After the mechanical parameters of the damage model under the combined action of freeze–thaw cycles and loading were determined, they were listed in Table 4; the stress values under three confining pressures (100 kPa, 200 kPa, and 300 kPa) can be determined by using the final damage constitutive model. The measured data with a water content of 17% and six freeze–thaw cycles were compared with the theoretical data. As shown in Figure 7, when the confining pressure was 100 kPa, 200 kPa, and 300 kPa, the actual strength values were 118.4 kPa, 152.3 kPa, and 184.1 kPa, respectively, and the theoretical strength values are 120 kPa, 150 kPa, and 186 kPa. The fitting degree of the strength value was as high as 99%. The theoretical curve was basically consistent with the measured curve. The model can reasonably predict the damage evolution of soil under the combined action of loading and freeze–thaw cycles.

Table 4. The mechanical parameters of the damage model under the combined action of freeze–thaw cycles and loading.

The Number of Freeze–Thaw Cycles	σ_3 = 100 kPa			σ_3 = 200 kPa			σ_3 = 300 kPa		
	E_n (MPa)	m	F_0	E_n (MPa)	m	F_0	E_n (MPa)	m	F_0
0	647	0.384	0.457	674	0.341	0.445	660	0.316	0.471
3	584	0.324	0.387	485	0.387	0.381	543	0.384	0.249
6	502	0.318	0.754	534	0.345	0.241	526	0.365	0.426
9	424	0.336	0.714	496	0.319	0.674	507	0.314	0.874
12	388	0.329	0.646	403	0.327	0.429	469	0.352	0.773

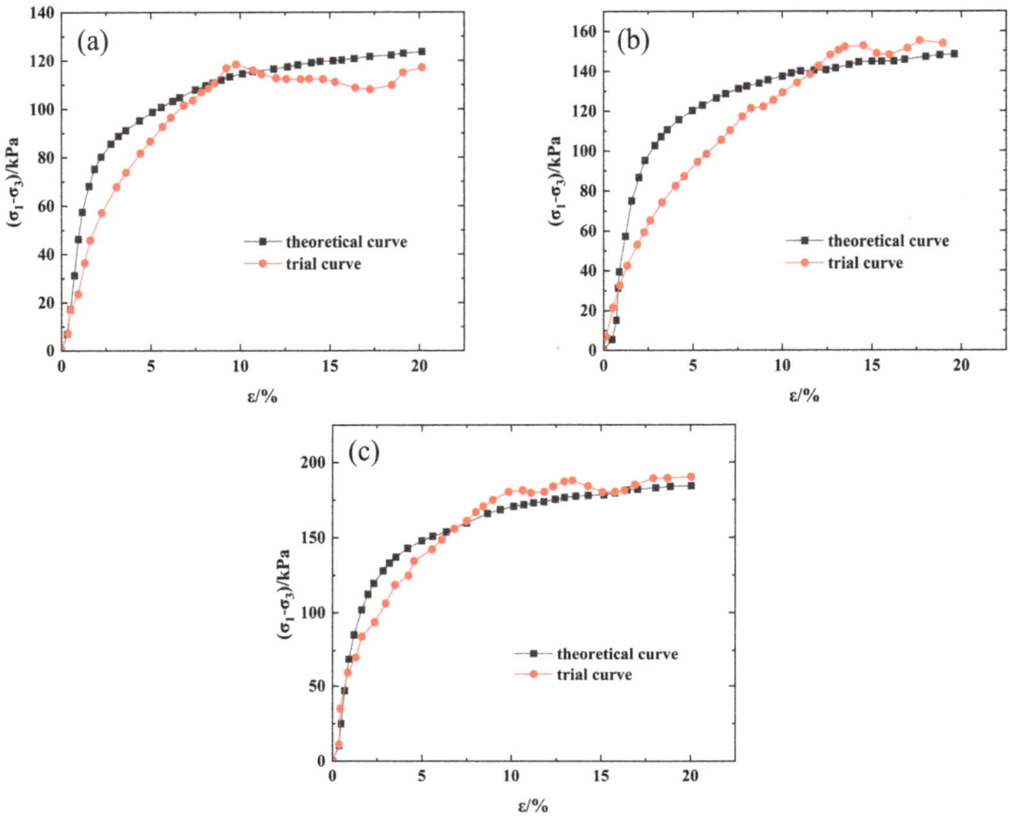

Figure 7. The comparison of the damage theoretical curve and test curve of deep expansive clay under different confining pressures. (**a**) 100 kPa, (**b**) 200 kPa, and (**c**) 300 kPa.

5. Discussion

In the early stage of this study, the conventional triaxial test and the triaxial test under freeze–thaw cycles were carried out on the deep expansive clay samples, and the mechanical properties of the soil were analyzed.

With the increase in the number of freeze–thaw cycles, the internal cracks of the soil will continue to expand with the expansion of water ice, which has a huge impact on the structure of the soil. How to properly reflect the damage law of soil under the combined action of loading and freeze–thaw cycles, the key is to select the appropriate damage variable. Many scholars had selected micro-benchmarks for damage analysis under loading [30,31]; however, there were some defects. The reason is that the initial soil damage has a great degree of randomness and random evolution. If the constitutive relationship of traditional methods is used, the randomness of soil in the damage process cannot be well reflected.

The reason is that the changes of microstructure and mechanical properties belong to micro and macro aspects, respectively. In order to study this problem fundamentally, it is necessary to combine macro research methods with micro research methods. In order to combine macro and micro research methods, macro and micro experiments must be carried out simultaneously. On this basis, this paper reasonably processes the test data by using probability and statistics methods, introduces appropriate damage variables, and establishes a damage constitutive theory combining the macro and micro.

Based on the research experience of previous scholars, this paper further studies the various mechanical properties of deep expansive clay and establishes the corresponding damage constitutive equation. However, due to time and other factors, the article still has shortcomings and needs further improvement:

1. This paper mainly studies some properties of deep expansive clay, but it does not take into account the great differences in particle size, mineral composition, and structure of expansive clay at different depths. We can further explore the influence of these factors on the mechanical properties of soil.
2. Whether the artificial soil can represent the deep environment and whether it is different from the deep in situ soil samples needs further discussion.
3. Due to the limited amount of undisturbed soil and the test period and test equipment and other factors, the test sample used in the article is slightly insufficient; it can be supplemented by some relevant tests to make the conclusions of the article more reliable.

6. Conclusions

In this study, the conventional triaxial test and the triaxial test under freeze–thaw cycles were carried out on the deep expansive clay samples, and the mechanical properties of the soil were analyzed. Based on the test, the damage classification and research methods were discussed, and the damage constitutive equation of the combined action of freeze–thaw cycles and loading was determined by establishing appropriate damage variables.

The major conclusions of this study are as follows:

1. In conventional triaxial test, the compressive strength of deep expansive clay gradually decreased with an increase in the water content. At this time, the stress–strain curve of clay under high confining pressure tended toward strain hardening, while low confining pressure shows strain softening.
2. In the triaxial shear test under freeze–thaw cycles, the growth rate of stress with strain gradually decreased as the number of freeze–thaw cycles increased. Moreover, the ultimate peak stress also decreased as a result of the freeze–thaw cycles. Under different freeze–thaw cycles, the stress–strain curves of the triaxial tests all showed strain hardening; as the number of freeze–thaw cycles increased, the cohesion tended to decrease, while the internal friction angle tended to increase.
3. Based on Lenaitre's strain equivalence hypothesis and the Druck–Prager damage criterion, the parameters of the damage constitutive equation are calculated and determined, and the calculated data are substituted into the final damage constitutive equation for verification. The fitting degree between the calculated strength value and the theoretical strength value is as high as 99%.
4. The damage constitutive equation can reasonably predict the damage evolution of soil under the combined action of loading and freeze–thaw cycles. This study can serve as an available reference for well wall construction and disaster prediction in deep coal mining.

Author Contributions: Conceptualization, Z.Z. and S.C.; methodology, Z.Z. and S.C.; validation, Z.Z.; formal analysis, Z.Z.; investigation, S.C.; resources, B.L.; data curation, Z.Z. and S.C.; writing—original draft preparation, Z.Z. and S.C.; writing—review and editing, Z.Z.; visualization, Z.Z.; supervision, Z.Z.; project administration, B.L.; funding acquisition, B.L. All authors have read and agreed to the published version of the manuscript.

Funding: This research was funded by the University Doctoral Program Special Project, grant number 200803610004. The study was supported by the Natural Science Foundation of Anhui Province, grant number 2008085ME143.

Data Availability Statement: The data used to support the findings of this study are available from the corresponding author upon request.

Conflicts of Interest: The authors declare no conflict of interest.

References

1. Thomas, L.P. 3—Global Standardization of Coal Resources and Reserves. In *The Coal Handbook*, 2nd ed.; Osborne, D., Ed.; Woodhead Publishing: Sawston, UK, 2023; Volume 1, pp. 53–83. ISBN 978-0-12-824328-2.
2. Jianxiong, M.; Tong, H. 13—Coal Resources, Production, and Use in China. In *The Coal Handbook*, 2nd ed.; Osborne, D., Ed.; Woodhead Publishing: Sawston, UK, 2023; Volume 2, pp. 431–454. ISBN 978-0-12-824327-5.
3. Singh, A.K.; Bhattacharya, S. 11—Coal Resources of India and Trends in Utilization. In *The Coal Handbook*, 2nd ed.; Osborne, D., Ed.; Woodhead Publishing: Sawston, UK, 2023; Volume 2, pp. 321–359. ISBN 978-0-12-824327-5.
4. Zou, J.; Wu, K.; Zhang, X.; Zhu, J.; Zhou, Z.; Zheng, F.; Xie, H.; Jiao, Y.-Y. Effective Evaluation of Deep-Hole Blasting for Controlling Strong Tremors Induced by Deep Coal Mining-A Case Study. *Int. J. Rock Mech. Min. Sci.* **2022**, *159*, 105211. [CrossRef]
5. Zhan, H.; Liu, S.; Wu, Q.; Li, Y.; Qi, K.; Zhang, X. Quantitative Prediction of the Impact of Deep Extremely Thick Coal Seam Mining on Groundwater. *Process Saf. Environ. Prot.* **2023**, *178*, 511–527. [CrossRef]
6. Damiano, E.; Cristofaro, M.; Brunzo, A.; Carrieri, G.; Iavazzo, L.; Netti, N.; Olivares, L. The Mechanical Characterization of Pyroclastic Deposits for Landslide Early Warning Systems. *Geosciences* **2023**, *13*, 291. [CrossRef]
7. Du, Y.; Xie, M.; Jiang, Y.; Li, B.; Chicas, S. Experimental Rock Stability Assessment Using the Frozen–Thawing Test. *Rock Mech. Rock Eng.* **2017**, *50*, 1049–1053. [CrossRef]
8. Du, Y.; Xie, M.; Jia, J. Stepped Settlement: A Possible Mechanism for Translational Landslides. *CATENA* **2020**, *187*, 104365. [CrossRef]
9. Mercurio, C.; Calderón-Cucunuba, L.P.; Argueta-Platero, A.A.; Azzara, G.; Cappadonia, C.; Martinello, C.; Rotigliano, E.; Conoscenti, C. Predicting Earthquake-Induced Landslides by Using a Stochastic Modeling Approach: A Case Study of the 2001 El Salvador Coseismic Landslides. *ISPRS Int. J. Geo-Inf.* **2023**, *12*, 178. [CrossRef]
10. Zhao, Y.; Yang, T.; Liu, H.; Wang, S.; Zhang, P.; Jia, P.; Wang, X. A Path for Evaluating the Mechanical Response of Rock Masses Based on Deep Mining-Induced Microseismic Data: A Case Study. *Tunn. Undergr. Space Technol.* **2021**, *115*, 104025. [CrossRef]
11. Levin, L.; Golovatyi, I.; Zaitsev, A.; Pugin, A.; Semin, M. Thermal Monitoring of Frozen Wall Thawing after Artificial Ground Freezing: Case Study of Petrikov Potash Mine. *Tunn. Undergr. Space Technol.* **2021**, *107*, 103685. [CrossRef]
12. Hegemann, J. A New Concept for Sinking Freeze Shafts into Great Depths. *Eng. Geol.* **1981**, *18*, 385–393. [CrossRef]
13. Hentrich, N.; Franz, J. About the Application of Conventional and Advanced Freeze Circle Design Methods for the Ust-Jaiwa Freeze Shaft Project, Vertical and Decline Shaft Sinking. In Proceedings of the International Mining Forum 2015, Cracow, Poland, 25–26 February 2015.
14. Farquharson, L.M.; Romanovsky, V.E.; Cable, W.L.; Walker, D.A.; Kokelj, S.V.; Nicolsky, D. Climate Change Drives Widespread and Rapid Thermokarst Development in Very Cold Permafrost in the Canadian High Arctic. *Geophys. Res. Lett.* **2019**, *46*, 6681–6689. [CrossRef]
15. Chen, H.; Gao, X.; Wang, Q. Research Progress and Prospect of Frozen Soil Engineering Disasters. *Cold Reg. Sci. Technol.* **2023**, *212*, 103901. [CrossRef]
16. Kruyt, N.P.; Rothenburg, L. Micromechanical Definition of the Strain Tensor for Granular Materials. *J. Appl. Mech* **1996**, *63*, 706–711. [CrossRef]
17. Christoffersen, J.; Mehrabadi, M.M.; Nemat-Nasser, S. A Micromechanical Description of Granular Material Behavior. *J. Appl. Mech.* **1981**, *48*, 339. [CrossRef]
18. Chang, C.S.; Misra, A.; Sundaram, S.S. Micromechanical Modelling of Cemented Sands under Low Amplitude Oscillations. *Géotechnique* **1990**, *40*, 251–263. [CrossRef]
19. Kachanov, L.M. Time of the Rupture Process under Creep Conditions. *Izv. Akad. Nauk. SSSR Otd. Teckhnicheskikh Nauk.* **1958**, *8*, 26–31.
20. Zhujiang, S. An Elasto-plastic Damage Model for Cemented Clays. *Chin. J. Geotech. Eng.* **1993**, *15*, 21–28.
21. Tong, X.; Gong, X.; Jiang, Y. The elastic-plastic damage model of cement reinforced soil. *Eng. Mech.* **2002**, *19*, 3338.
22. Zhu, Z.; Wuning, J.; Ma, W. Study on constitutive model of frozen soil based on damage and numerical simulation of water, heat and force coupling. *Chin. Sci.* **2010**, 758–772.
23. Zhang, T.; Xu, H.; Deng, Y. The correlation between structural soft clay damage variable and perturbation. *Rock Soil Mech.* **2015**, *36*, 7.
24. Shi, S.; Zhu, F.; Feng, D.; Zhang, F. Mechanical Behaviors and Elastoplastic Damage Constitutive Model for Heterogeneous Frozen Soil. *Geomech. Energy Environ.* **2023**, *35*, 100490. [CrossRef]
25. Sun, K.; Tang, L.; Zhou, A.; Ling, X. An Elastoplastic Damage Constitutive Model for Frozen Soil Based on the Super/Subloading Yield Surfaces. *Comput. Geotech.* **2020**, *128*, 103842. [CrossRef]
26. He, J.; Niu, F.; Jiang, H.; Jiao, C. Fractional Viscoelastic-Plastic Constitutive Model for Frozen Soil Based on Microcosmic Damage Mechanism. *Mech. Mater.* **2023**, *177*, 104545. [CrossRef]
27. *ASTM D422-63*; Standard Test Method for Particle-Size Analysis of Soils 1. ASTM International: West Conshohocken, PA, USA, 2007.
28. Li, B.; Zhu, Z.; Ning, J.; Li, T.; Zhou, Z. Viscoelastic–Plastic Constitutive Model with Damage of Frozen Soil under Impact Loading and Freeze–Thaw Loading. *Int. J. Mech. Sci.* **2022**, *214*, 106890. [CrossRef]
29. Wang, D.; Ma, W.; Niu, Y.; Chang, X.; Wen, Z. Effects of Cyclic Freezing and Thawing on Mechanical Properties of Qinghai–Tibet Clay. *Cold Reg. Sci. Technol.* **2007**, *48*, 34–43. [CrossRef]
30. Ning, J.; Zhu, Z. The constitutive model of frozen soil and coupling problems. *J. Mech.* **2007**, *39*, 7.

31. Lai, Y.; Li, S.; Gao, Z.; Chang, X. Stochastic Damage Constitutive Model for Warm Frozen Soil under Uniaxial Compression and Its Strength Distribution. *Glacier* **2007**, *6*, 969–976.
32. Xu, H.; Liu, Y.; Zhao, G.; Ye, N. Time-Varying Reliability Analysis Based on Improved Toughness Exhaustion Model and Probability Density Evolution Method to Predict Fatigue Damage Life. *Eng. Fract. Mech.* **2022**, *276*, 108930. [CrossRef]
33. Lai, Y.; Li, S.; Qi, J.; Gao, Z.; Chang, X. Strength Distributions of Warm Frozen Clay and Its Stochastic Damage Constitutive Model. *Cold Reg. Sci. Technol.* **2008**, *53*, 200–215. [CrossRef]
34. Li, S.; Lai, Y.; Zhang, S.; Liu, D. An Improved Statistical Damage Constitutive Model for Warm Frozen Clay Based on Mohr–Coulomb Criterion. *Cold Reg. Sci. Technol.* **2009**, *57*, 154–159. [CrossRef]
35. Song, F. Finite Element Analysis of Seinfeld Gravite. Ph.D. Thesis, Chang'an University, Xi'An, China, 2006.
36. Zhang, H. Discussion on the expansion characteristics of lotus rock. *J Xi'an Univ. Archit. Technol. Nat. Sci. Ed.* **2010**, *42*, 4.

International Journal of *Geo-Information*

MDPI

Article

Quick Estimation Model for Mapping Earthquake Impacts in Bogotá, Colombia

Hiroyuki Miura [1,*], Masashi Matsuoka [2], Juan C. Reyes [3], Nelson Pulido [4], Mitsufumi Hashimoto [5], Andrea C. Riaño [3], Alvaro Hurtado [3], Raul Rincon [6], Helber García [7] and Carlos Lozano [8]

1 Graduate School of Advanced Science and Engineering, Hiroshima University, Hiroshima 739-8527, Japan
2 Department of Architecture and Building Engineering, Tokyo Institute of Technology, Yokohama 226-8501, Japan; matsuoka.m.ab@m.titech.ac.jp
3 Department of Civil and Environmental Engineering, Universidad de los Andes, Bogotá DC 11711, Colombia; jureyes@uniandes.edu.co (J.C.R.); ac.riano384@uniandes.edu.co (A.C.R.); ai.hurtado77@uniandes.edu.co (A.H.)
4 National Research Institute for Earth Science and Disaster Resilience, Ibaraki 305-0006, Japan; nelson@bosai.go.jp
5 Kozo Keikaku Engineering Inc., Tokyo 164-0001, Japan; mitsu@kke.co.jp
6 Department of Civil and Environmental Engineering, Rice University, Houston, TX 77005, USA; raul.rincon@rice.edu
7 Servicio Geológico Colombiano (SGC), Bogotá DC 111321, Colombia; hgarcia@sgc.gov.co
8 Instituto Distrital de Gestión de Riesgos y Cambio Climático (IDIGER), Diagonal 47 Number 77A-09 Int. 11, Bogotá DC 111071, Colombia; clozano@idiger.gov.co
* Correspondence: hmiura@hiroshima-u.ac.jp

Abstract: Early disaster responses in damaged areas after a large earthquake are indispensable for stakeholders to assess and grasp the impacts such as building and infrastructure damage and disrupted community functionality as soon as possible. This study introduces a quick estimation model for mapping seismic intensities and building losses in Bogotá, the capital city of Colombia. The model uses ground motion records in the seismic network, soil maps of average shear-wave velocity in the upper 30 m (Vs30) with site amplifications, building inventory, and vulnerability functions for all building types. The spatial distribution of ground motion intensities, including spectral accelerations, was estimated by interpolating the observed seismic intensities with the Vs30-based site amplifications. The losses (repair cost) for all the buildings were evaluated by integrating the estimated spectral accelerations, the building inventory, and the vulnerability functions. The spatial distributions of seismic intensities and building losses can be computed within a few minutes immediately after triggering earthquake motions in the seismic network. The proposed model demonstrates evaluations of the impacts for the Mw6.0 earthquake that occurred on December 2019 and an earthquake scenario with Mw7.0 from an active fault near the Bogotá region.

Keywords: seismic intensity; building loss; vulnerability; Bogotá

1. Introduction

Identification of the spatial distribution of seismic intensities and building damage immediately after large earthquakes is crucial for the administrators of the national and local governments to understand the impacts and to define emergency response strategies. For this purpose, the ShakeMap system has been developed to rapidly and automatically generate seismic intensity maps after earthquakes by spatially interpolating observed ground motion intensities [1–3]. Similarly, early ground motion estimation systems have been implemented in some countries and cities, such as Japan and European countries [4–9].

Estimation of building damage distribution is important because of the human losses and the economic impacts that structural damage and failure can produce. Building damage and/or loss maps have been implemented in some seismic monitoring systems [4,7,8,10].

Up-to-date building inventory data and fragility/vulnerability functions required for damage mapping have been developed in each region for predicting building damage due to real and scenario earthquakes [11–17]. Such seismic shaking and building damage mapping systems, however, require a real-time seismic monitoring network, detailed soil amplification data, building inventory data, and fragility/vulnerability functions suitable for each region. The number of countries and cities where such quick estimation systems have been implemented is still limited because of the huge costs and investments needed for preparing the input data.

Bogotá, the capital city of Colombia, is one of the largest metropolitan cities in Latin America with a population of approximately eight million and a total number of buildings of over 800,000. Bogotá has been affected by magnitude 6–7 class crustal earthquakes triggered by active faults in the Eastern Andes range at intervals of approximately 100 years [18]. The largest event in this region assessed from the historical earthquake records was the Ms 7.1 in August 1917 [18]. Such a large earthquake has not been recorded near Bogotá since 1917 when Bogotá experienced the most damaging earthquake to date. Therefore, we should consider countermeasures against destructive magnitude 7 class earthquakes that will occur in the near future.

In order to estimate damage information for emergency responses in earthquake crises, the Laboratory of Automatic Seismic Instrumentation (LISA, Laboratorio de Instrumentación Sísmica Automática) has been developed in Bogotá [19,20]. The LISA produces shakemaps and building damage maps with block-level resolution including economic losses of the damaged buildings. The typical size of the blocks in Bogotá was several thousand to tens of thousands of square meters. The obtained shakemaps and damage maps by LISA can be sent to designated city officials by email and short message service (SMS) within five minutes after the earthquake. Although the system is a convenient and valuable tool to immediately assess the spatial distributions of ground shakings and building damage, the LISA still has limitations and uncertainties, especially in estimating shakemaps as follows:

- No seismic event information (such as the one provided by the National Seismic Network of Colombia [21]) was incorporated into the automatic shakemap calculations of LISA [20], hindering its timely and accurate release of information.
- The seismic motion intensities on the ground surface were estimated based on a single ground motion record observed at a bedrock site within Bogotá, and, therefore, the system may not properly reflect the large spatial variability expected in near-source ground motions, due to finite fault ruptures of potential M6-7 class earthquakes nearby.
- The geotechnical model used for detailed seismic response analysis of soils at a block level (~hundreds of meters), was interpolated by using a probabilistic approach from the geotechnical information available at 23 boreholes (including measurements of shear-wave velocity, Vs) within Bogotá (every ~3 km) [20].

Significant discrepancies might be found between the probabilistically determined models and more realistic models investigated by dense geophysical explorations. Since the Vs models strongly control the amplitudes of the seismic motions on the ground surface, a more realistic Vs model in Bogotá has been expected for accurate seismic hazard evaluations. Furthermore, the shakemaps in the LISA did not reflect the seismic intensities observed at other ground surface sites in Bogotá. More accurate shakemaps can be produced by adding a large number of ground motion records observed at multiple sites available in the seismic monitoring network currently in operation.

To address these issues, we have conducted very dense geophysical explorations (every ~1 km) based on single and array microtremor measurements as well as existing gravity measurements (~1 km), for developing more accurate three-dimensional (3D) Vs structures in the Bogotá basin [22,23]. We included empirical models of site amplification factors for peak ground accelerations (PGA), peak ground velocities (PGV), and spectral accelerations (Sa) that are dependent on the average shear-wave velocities in the upper 30 m (Vs30) obtained in geophysical explorations [24,25]. More site-specific and reliable shakemaps can be created by using observed ground motion data at multiple sites and the developed site amplification factors. Based on the previous studies, this study introduces a quick estimation model for mapping earthquake impacts, such as shakemaps and building damage maps, including building economic losses, based on seismic observation data, the Vs30 map, and building vulnerabilities.

2. Development of Quick Estimation Model

2.1. Accelerograph Network in Bogotá

Figure 1 shows an overview of the developed quick ground motion and building loss estimation model for Bogotá. The model is based on ground motion records observed by the seismic network operated by The Instituto Distrital de Gestión de Riesgos y Cambio Climático (IDIGER, District Institute of Risk Management and Climate Change) in Bogotá named Red de Acelerógrafos de Bogotá (RAB) [26]. Seismic observations at the ground surface in Bogotá have been operated by IDIGER at 29 sites, as shown in Figure 2. The strong motion seismographs named BASALT, OBSIDIAN, K2, ETNA-2, and ETNA in Kinemetrics Inc. have been used by the IDIGER in the seismic network (see Table 1). At the 25 stations where the BASALT, OBSIDIAN, and ETNA-2 equipment are installed, the observed seismic data are transmitted to the IDIGER headquarters by broadband telemetry system immediately after triggering a tremor. Peak ground accelerations (PGA), peak ground velocities (PGV), and acceleration response spectra (Sa) are calculated from the transmitted records, and they are used in the following quick estimations.

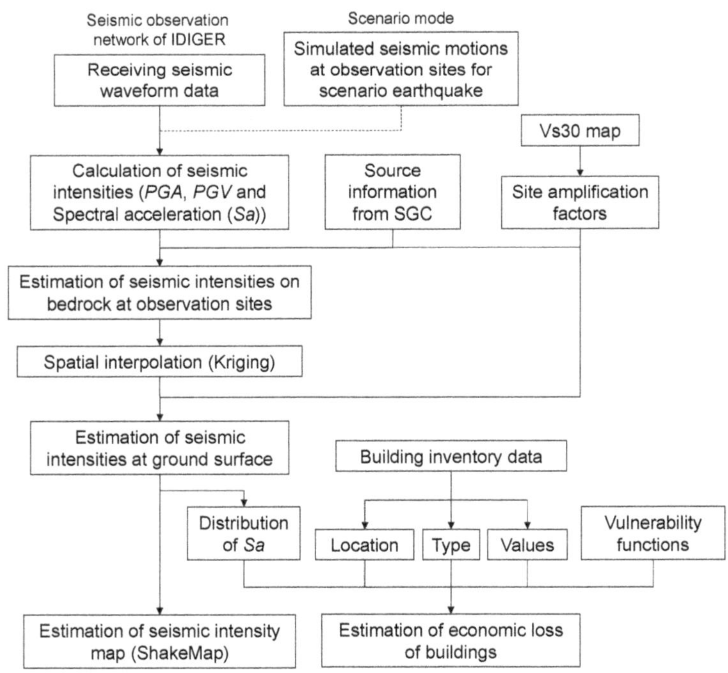

Figure 1. Diagram for quick estimation model for shakemap and economic loss of buildings developed for Bogotá, Colombia.

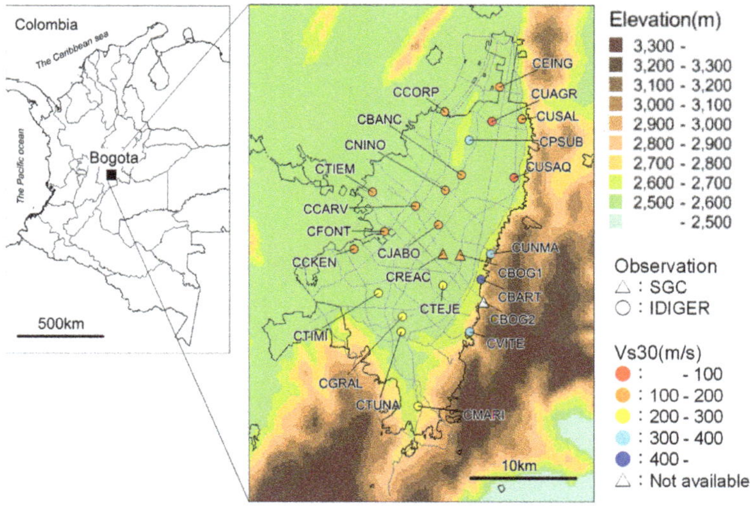

Figure 2. Seismic ground observation network by IDIGER in Bogotá with their Vs30 values. The map shows the distribution of the observation stations with topographical map.

Table 1. Seismic motion observation sites by IDIGER in Bogotá.

ID	Site	Longitude (deg.)	Latitude (deg.)	Elevation (m)	Vs30 (m/s)	Type of Seismograph
1	CARAN	−74.1128	4.6456	2555	161	ETNA-2
2	CARTI	−74.1234	4.5469	2569	333	BASALT
3	CBANC	−74.0790	4.7085	2552	123	ETNA-2
4	CBART	−74.0618	4.6200	2671	425	BASALT
5	CBOSA	−74.1922	4.6065	2552	220	BASALT
6	CCARV	−74.1188	4.6823	2556	120	BASALT
7	CCKEN	−74.1724	4.6459	2548	173	BASALT
8	CCORP	−74.0940	4.7619	2554	98	ETNA-2
9	CDIOS	−74.0884	4.5899	2583	-	ETNA-2
10	CEING	−74.0460	4.7836	2562	97	ETNA-2
11	CFLOD	−74.1464	4.7297	2557	112	ETNA
12	CFONT	−74.1456	4.6609	2546	157	BASALT
13	CGRAL	−74.1301	4.5879	2566	255	OBSIDIAN
14	CJABO	−74.0993	4.6665	2554	100	ETNA-2
15	CLAGO	−74.1003	4.7180	2552	-	ETNA-2
16	CMARI	−74.1171	4.5120	2689	257	BASALT
17	CNIÑO	−74.0931	4.6962	2555	109	ETNA-2
18	CPSUB	−74.0726	4.7379	2588	139	BASALT
19	CSMOR	−74.1701	4.5746	2783	398	BASALT
20	CTEJE	−74.0951	4.6146	2566	214	ETNA-2
21	CTIEM	−74.1560	4.6944	2552	123	ETNA
22	CTIMI	−74.1510	4.6083	2559	186	BASALT
23	CTUNA	−74.1311	4.5752	2563	257	BASALT

Table 1. *Cont.*

ID	Site	Longitude (deg.)	Latitude (deg.)	Elevation (m)	Vs30 (m/s)	Type of Seismograph
24	CTVCA	−74.0847	4.7179	2652	142	BASALT
25	CUAGR	−74.0527	4.7541	2561	93	ETNA
26	CUNMA	−74.0539	4.6416	2679	333	ETNA
27	CUSAL	−74.0267	4.7558	2567	114	ETNA-2
28	CUSAQ	−74.0339	4.7062	2565	100	ETNA-2
29	CVITE	−74.0717	4.5752	2777	555	ETNA-2

The Servicio Geológico Colombiano (SGC, Colombian Geological Survey) has developed a nationwide seismic network (RSNC) for monitoring activities of earthquakes and volcanos since 1993 [21,27,28]. When a tremor is detected by the seismic network, the source information, such as location, magnitude, and depth estimated from the observed waveforms, is immediately distributed to registered people and organizations by email service. The obtained seismic data in IDIGER are linked with the source information provided by email service from the RSNC in the model based on the registered GPS clock.

2.2. Vs30 Map and Site Amplification Factors

Miura et al. [24] found that long-period (approximately 2 s) surface waves were significantly predominant in ground motions observed within the Bogotá basin in cases of shallow earthquakes due to the basin-edge effect of the deep sedimentary basin. On the contrary, body waves directly propagated from the earthquake source were dominant in deep earthquakes. These authors also found strong correlations between Vs30s and site amplification factors and proposed a Vs30-dependent site amplification factor model of Sa, PGA, and PGV for surface wave type and body wave type [24,25]. Figure 3 shows the Vs30-dependent site amplification factors of Sa for surface wave and body wave types for several Vs30s proposed in reference [24]. It must be mentioned that the seismic records used had PGA values smaller than 80 cm/s^2, thus strong shaking amplification effects, such as nonlinear ground responses, still require further attention.

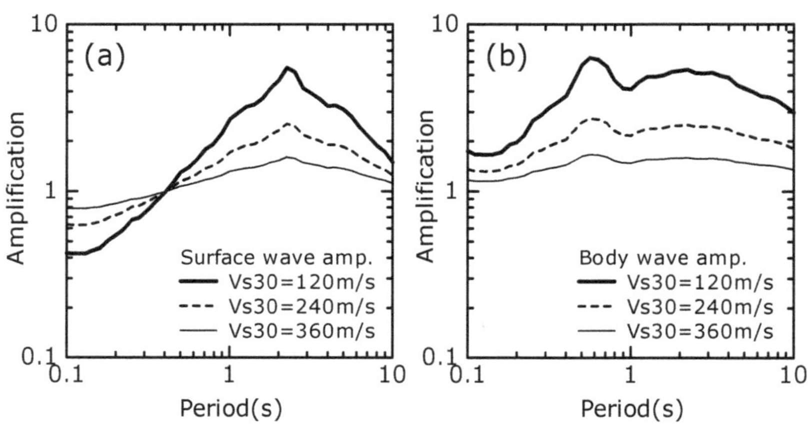

Figure 3. Vs30-dependent site amplification factors for (**a**) surface waves and (**b**) body waves in Bogotá proposed in Miura et al. [24].

Although Vs30s represent characteristics of very shallow surface soil, Miura et al. [24] observed that the ground motions in Bogotá can be characterized by the Vs30s, and the developed Vs30-based site amplification factors accurately reproduced the ground motion characteristics observed in Bogotá's basin. Given this observation, we used the Vs30 model in [24] as a proxy of site amplification factors on the quick loss estimation model. The Vs30 map included in the system uses data from three-dimensional (3D) velocity structures developed for the region [22]. For the development of the 3D model, array measurements and single-site observations of microtremors were performed at approximately 300 and 800 sites, respectively, in and around the city of Bogotá and 800 sites in the surrounding areas [23,24]. The shear-wave velocity profiles at the observation sites were estimated by inversion analysis of the microtremor-derived phase velocities. In addition, the 3D material model was constructed to characterize properties up to the seismic bedrock with a shear-wave velocity of approximately 3000 m/s. Figure 4 shows the Vs30 map with a spatial resolution of 250 m in Bogotá. Since Bogotá is located on a large basin with a long sedimentation process of lacustrine fluvio-glacial deposits, the surface soil condition is very soft. As shown in Figure 4, the Vs30 values in most of the city are smaller than 300 m/s, and small values of Vs30s with 200 m/s or less are widely distributed.

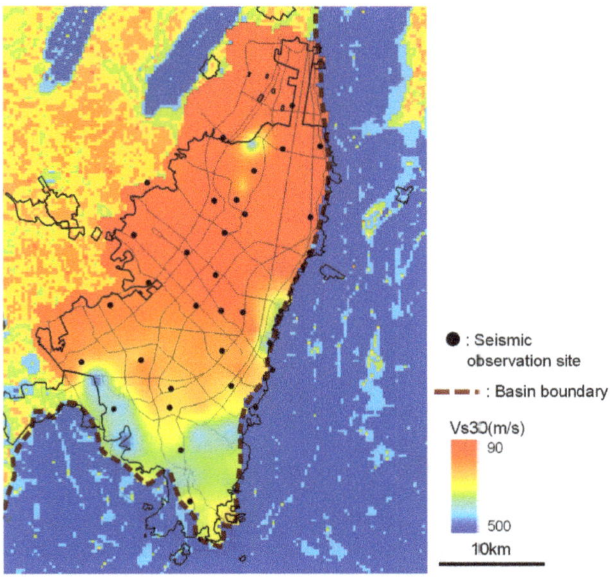

Figure 4. Vs30 map in Bogotá (after Pulido et al. [23]). Points and dotted line indicate the locations of seismic observation sites and the basin boundary, respectively.

Figure 5 shows the schematic diagram for estimating shakemaps from observation records and site amplification factors obtained from the Vs30 map. First, the seismic intensities at the bedrock of the observation stations are estimated by de-amplifying the seismic intensities observed at the surface with the site amplification factors. Then, the surface wave amplification factors are applied for shallow earthquakes (less than 150 km depth) whereas the body wave amplification factors are used for deeper earthquakes (larger than 150 km depth) based on the source information estimated by the RSNC. Next, the distributions of the seismic intensities on the bedrock are obtained by spatial interpolation based on the simple Kriging [29], since the technique was also adopted in a similar quick estimation model of strong ground motion maps for Japan [6]. The distributions of the seismic intensities at the surface are estimated from the spatially interpolated seismic intensities on the bedrock and the Vs30 map-based site amplification factors. With this, the system outputs the maps of PGA, PGV, and Sa for the period of 0.1 to 10.0 s as shakemaps.

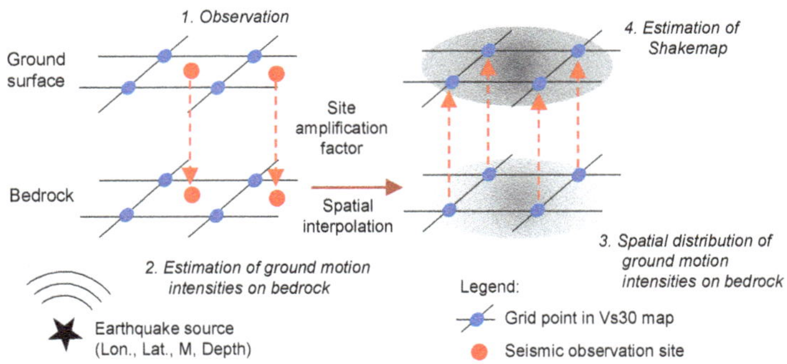

Figure 5. Schematic diagram for estimation of shakemap based on observation seismic data and site amplification factors. Step 1 to 4 in the figure indicates procedures for observations and estimations of ground motion intensities at ground surface and bedrock.

Note that explicit nonlinear soil behavior expected during strong shaking, in the rare case of a large earthquake near Bogotá, was not considered in the shakemaps because the developed amplification models were based on seismic records whose PGA is smaller than 80 cm/s^2. However, the observed strong motion records that possibly experienced nonlinearity will be also included for the shakemap calculations, and, therefore, the effect of nonlinearity in ground motions would be incorporated to a first order. Anyway, this limitation may be addressed in the future by extending the amplification models with analytical strong motion approaches.

2.3. Building Inventory and Vulnerability Functions

Building inventory databases and building fragility/vulnerability functions are required for estimating citywide building damage [30]. Detailed databases are available in Bogotá as part of the "Spatial Data Infrastructure of the Capital District" [31]; these databases consist of cadastral geographic information layers at 'lot-level' and 'construction-level'. The lot-level inventory data includes lot area and corresponding social-economic class. The number of lots in the existing database is approximately 890,000. Additionally, the construction-level data characterizes the building's footprints and are modeled by 1.6 million polygons, which include data about the number of stories and the information required to define the buildings' use. These existing databases do not include building typology that is the most important factor for evaluating seismic capacity of buildings; hence, a structural typology assignation model was used for this purpose.

A matrix-like algorithm was used to assign the building typologies in the exposure inventory. The algorithm assigns the most likely structural typology based on the building use, number of floors, socioeconomic level, and slope of the terrain. All possible combinations were considered in the assignation model. These characteristics were found to create spatially distributed clusters in dense cities, which relate well with the building typology clusters. For example, high-rise buildings with commercial usage (e.g., offices) located on high socio-economic levels are commonly found in the financial sector, which usually are reinforced concrete dual-systems; likewise, unreinforced, and confined masonry are observed to be dominant typologies in residential low-rise buildings located in low-income areas (probably with moderate-to-high slope terrain). The algorithm was developed in previous studies at the Universidad de los Andes (Uniandes), Colombia, using cadastral databases, web-based inspection of thousands of buildings, and in-person validation for hundreds of buildings and adapted to the current study's particularities [32]. Table 2 shows the building typologies in the developed inventory data. The building typologies from the tree-like model resulted in 23 classes. Structural types were classified as Adobe, MSC (unreinforced masonry bearing walls), MR (reinforced masonry), PCRD (concrete moment frame), PCRM (concrete frame with unreinforced masonry infill walls), and others. The design level (low, moderate, and high) was assigned to the PCRD and PCRM. The distribution of the classified structural types is illustrated in Figure 6a. The structural period in Table 2 indicate the fundamental vibration period for each building typology estimated from the typical height and the structural type [16,17,33]. The number of buildings in each class is also shown in Table 2. Approximately 90% of the buildings are low-rise dwellings with 1–3 stories, and the MSC type is predominant in Bogotá city. Figure 6b shows the distribution of the building heights; high-rise buildings are observed in the northeast part of the city.

Table 2. Building typologies and their characteristics in building inventory data of Bogotá, Colombia.

No.	Structural Code	Description	Height	Stories	Typical Height (m)	Structural Period (s)	Design Level	Number	Percentage (%)
1	ADOBE	Adobe	Low	1–2	6.1	0.50	-	19,193	1.20
2	MSC1_3	Unreinforced masonry bearing walls	Low	1–3	4.6	0.35	-	1,186,283	73.97
3	MSC4_5		Mid	4–5	10.7	0.50	-	26,913	1.68
4	MR1_3	Reinforced masonry bearing walls	Low	1–3	6.1	0.35	-	206,205	12.86
5	MR4_5		Mid	4–5	15.2	0.56	-	15,645	0.95
6	PCRDMI1_3	Concrete moment frames	Low	1–3	6.1	0.40	Low	253	0.02
7	PCRDMO1_3		Low	1–3	6.1	0.40	Moderate	14,943	0.93
8	PCRDMI4_5		Mid	4–5	15.2	0.75	Low	64	0.00
9	PCRDMO4_5		Mid	4–5	15.2	1.45	Moderate	5190	0.32
10	PCRDMO6_12		High	6–12	36.6	0.56	Moderate	3308	0.21
11	PCRM_DMI4_5	Concrete frame with unreinforced masonry infill walls	Mid	4–5	15.2	0.56	Low	10,299	0.64
12	PCRM_DMO4_5		Mid	4–5	15.2	0.56	Moderate	9904	0.62
13	PCRM_DES4_5		Mid	4–5	15.2	0.56	High	1390	0.09
14	PCRM_DMI6_12		High	6–12	36.6	1.09	Low	588	0.04
15	PCRM_DMO6_12		High	6–12	36.6	1.09	Moderate	4476	0.28
16	PCRM_DES6_12		High	6–12	36.6	1.09	High	1365	0.09
17	PCRM_DMO12_20		High	12–20	56.0	1.76	Moderate	490	0.03
18	PrFC4_5	Precast concrete tilt-up walls	Mid	4–5	15.2	0.56	-	12,553	0.78
19	SC6_12	Reinforced concrete frames and concrete shear walls	High	6–12	36.6	1.09	-	574	0.04
20	SC_20		High	12+	65.0	2.01	-	47	0.00
21	PRT_CER_MUR	Reinforced concrete frames and steel truss girder (Warehouses)	-	All	4.6	0.32	-	11,058	0.69
22	BOD_PEQ	Steel light frame	-	All	4.6	0.40	-	42,023	2.62
23	BOD_GRN		-	All	4.6	0.40	-	30,948	1.93
Total								1,603,712	100.00

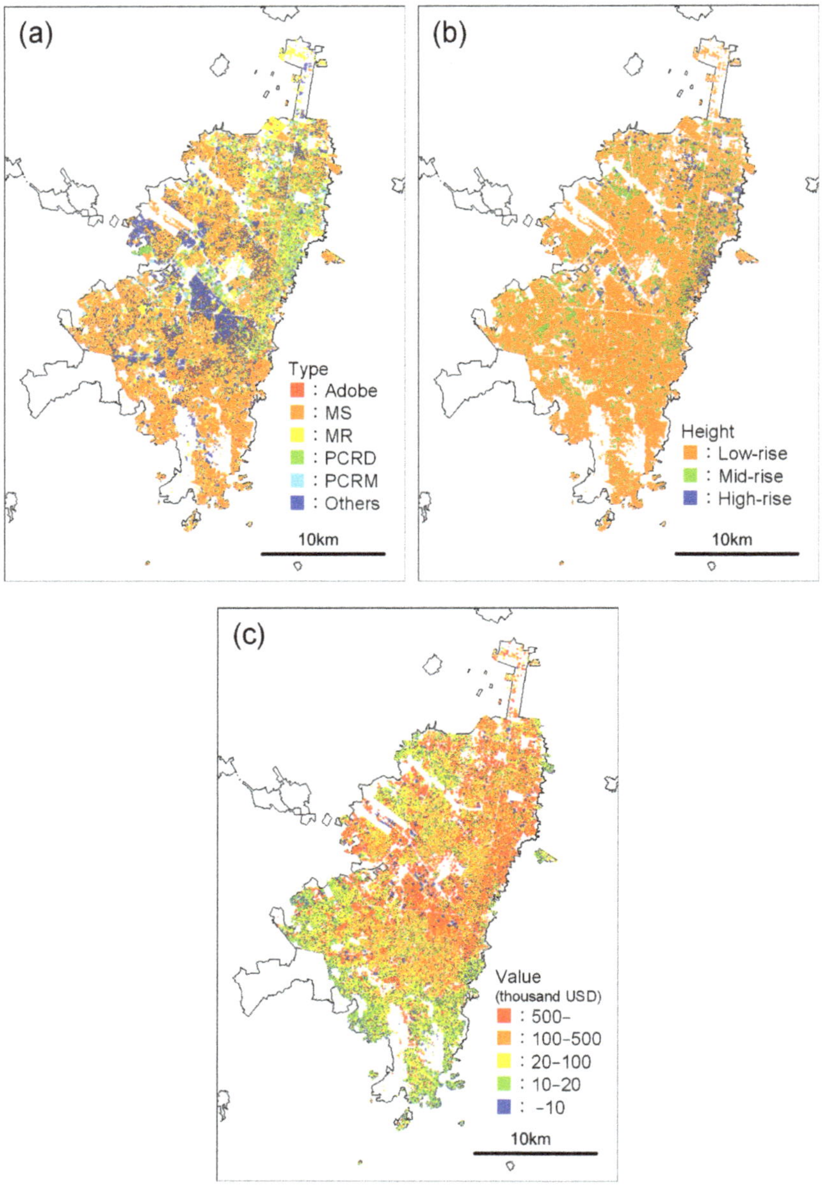

Figure 6. Building inventory data. (**a**) Building typologies (see Table 2), (**b**) Building heights, and (**c**) Building values in USD.

Another important characteristic that existing building databases do not contain is the building replacement cost. Considering that the replacement costs intent is to relate the building's damage to the expected economic losses, a cost estimation model is built into this study based on factors that capture the variation in the building's costs produced by phenomena such as predominant construction material, construction quality, construction age, variation in finishing costs, and quality, among other variables. Information that supports the abovementioned properties is not easy to collect, thus a construction price index (per square meter) is defined for different combinations of (1) building uses,

(2) socio-economic levels, (3) locality, and (4) building height, assuming that these factors jointly influence the buildings' replacement cost at Bogotá. The last factor was used only to differentiate low-to-midrise constructions from high-rise buildings. The price index model was constructed using available information from the public contracts databases [34], international construction surveys [35], commercial building costs databases [36], surveying the commercial costs of different building configurations in each of the 20 localities in Bogotá [36,37], and the use, structure-height, and socio-economic classification according to the cadastral database [31]. The building replacement cost was obtained by multiplying the corresponding price index per use by each constructed area minus the relative land value associated for each building into a lot unit. Here, it was assumed that the estimated building value corresponds to the replacement cost including materials, supplies, labors, tools, equipment, machines, and transportations and excludes the following costs: land acquisition, relocating utilities, construction technology equipment, and commercial aspects. The residential and commercial uses account for more than 90% of the regional exposed value, and the total building inventory cost was appraised at 720 billion COP (in 2019), a cost validated against the Real Estate Census (in 2019) reported in [38]. Figure 6c shows the obtained distribution of the building replacement costs in thousand USD under the currency exchange rate as of 2019 (1 USD = 3333.3 COP).

A unique vulnerability function was assigned to each building typology defined in the exposure database. The vulnerability functions represent the relationship between seismic intensity and loss value, which is expressed as total repair cost normalized by building replacement value (i.e., the mean damage ratio, MDR). The vulnerability functions were adapted from the previous studies [16,33], performing adjustments to represent the local characteristics of the building portfolio, as shown in Figure 7. The horizontal axis indicates spectral acceleration at the typical fundamental period of vibration of the building type (shown also in Table 2); the vertical axis corresponds to the MDR in percentage. More details on the exposure database and vulnerability curves for the city of Bogotá can be found in [33]. In this model, it was assumed that the MDRs in the vulnerability functions were zero when the MDRs were smaller than 0.1. This decision was based on the assumption that owners would not repair slight damage, such as hair cracks on non-structural walls.

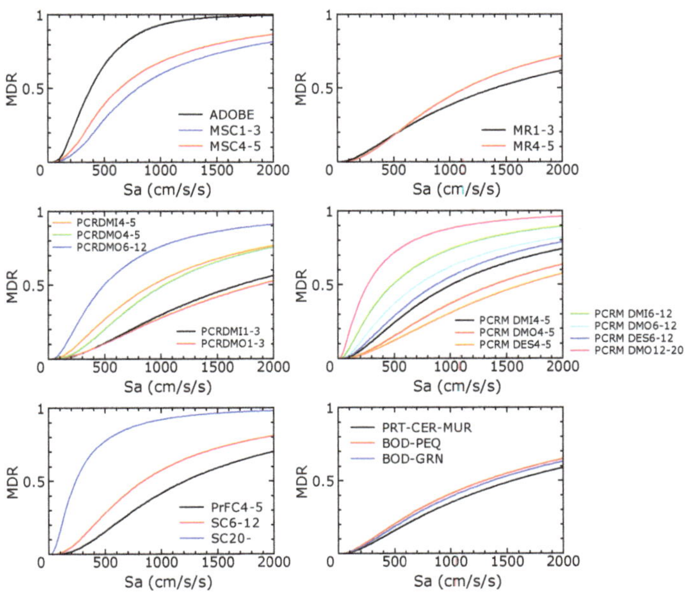

Figure 7. Vulnerability functions for building types in Bogotá.

2.4. Shakemaps and Building Loss Maps

The building loss is calculated first using the estimated spectral acceleration at ground level and the vulnerability curve to obtain the predicted MDR; then, the expected losses are computed as the building replacement cost multiplied by the MDR. The estimated building losses are aggregated by the 250 m mesh map, and the total loss in each mesh is illustrated into a building loss map.

The shakemaps and building loss maps can be obtained within several minutes after receiving seismic data transmitted from the observation stations. The city administrators and operators can confirm the estimation results displayed on internet browsers. The system automatically produces shakemaps and building loss maps in raw image file (RAW format) and compressed image file (PNG format) with the georeferencing information in HDR format. The color scales of the maps are automatically given according to the maximum and minimum values of the estimations.

3. Shakemaps for the Mesetas Earthquake on 24 December 2019

This chapter introduces an example for the estimation of shakemaps for a real earthquake. An earthquake with magnitude Mw6.0 occurred in the Mesetas district approximately 100 km south of Bogotá at 14:03 (local time) on 24 December 2019 [39]. The epicenter was located at a longitude of −74.18 degrees and a latitude of 3.46 degrees. The focal depth was about 13 km according to the source information by the SGC. This earthquake was the largest crustal earthquake in the last 20 years in Colombia [40]. The seismographs at 12 stations of the IDIGER seismic network were triggered in the earthquake.

Figure 8 shows the pasteups of the observed acceleration and velocity waveforms in Bogotá. Since the CMARI seismic station is located near the epicenter, large amplitudes were observed in accelerations and velocities. Large amplitudes were observed also at CUSAQ whereas the site is located far from the epicenter. As shown in Figure 4, the northeast area of the Bogotá basin around CUSAQ is covered with very low Vs30s, indicating that very soft soils with thick deposits exist in the area. Due to such very soft soils, large amplifications produced stronger ground shakings at the site. On the contrary, very small amplitudes were observed at CBART. Since CBART is located at the rock site of the eastern edge of the basin, very small amplifications were expected at the site. Figure 9 shows the acceleration response spectra at typical sites with their Vs30 values. As described above, smaller spectral accelerations were observed at CBART with the Vs30 of over 500 m/s. The responses increased as the Vs30s decreased. Especially, large amplitudes at approximately 1–2 s were observed at CUSAQ. Such strong shakings at longer periods confirm large amplifications due to soft soils, as shown in Figure 3. These results indicate that the site conditions strongly control the amplitudes of the ground shakings within the basin, and accurate soil models would be important for reliable shakemaps.

After receiving the observed waveform data, the developed system produced shakemaps and building loss maps within one minute. Although the building inventory data included more than 1.6 million, building damage and their repair costs were calculated with a very short computation time. The surface wave amplification models were automatically adopted in calculating the seismic intensities since the focal depth was only 13 km. Figure 10 shows the shakemaps of the estimated PGAs, PGVs, and spectral accelerations of typical periods for the earthquake. The southern areas show larger seismic intensities in the PGA map and the short-period spectral accelerations for the period of 0.35 s and 0.5 s because the source of the earthquake was closer. However, the northeast areas also show larger intensities especially in the PGV and longer period spectral accelerations, such as 1.0 s and 2.0 s, because of the strong amplification due to the soft soil conditions in the area. These shakemaps show the site conditions typically affected the seismic intensities, especially for longer periods.

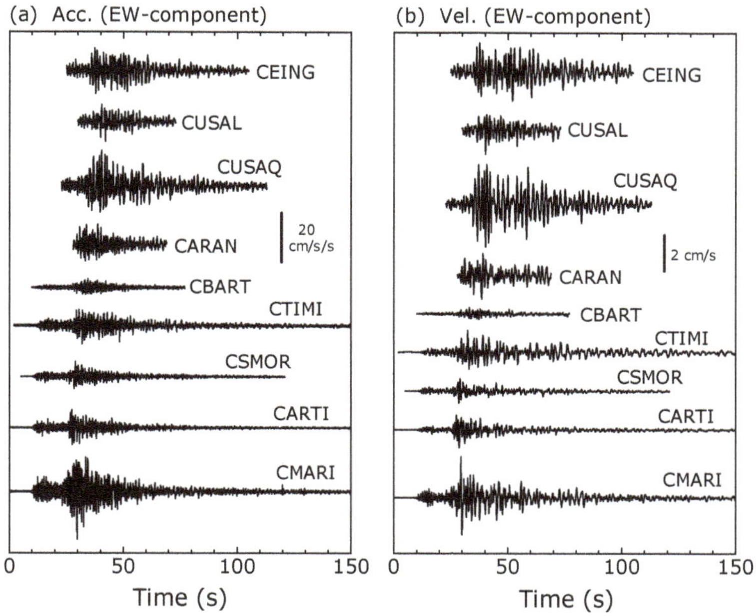

Figure 8. Observed time histories of EW component for (**a**) accelerations and (**b**) velocities in the 2019 Mesetas earthquake.

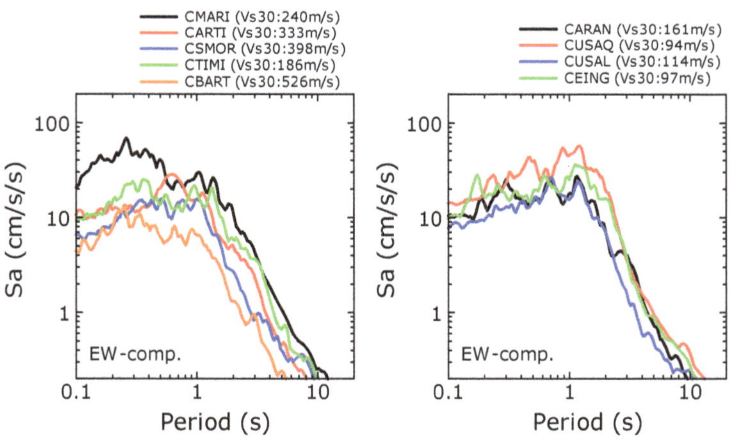

Figure 9. Spectral accelerations at typical sites observed in the 2019 Mesetas earthquake.

Since the observed and estimated seismic intensities were not so large and the estimated MDRs were smaller than 0.1 in all the buildings, no building loss was estimated by the proposed system in the earthquake. Whereas some damage to weak buildings was reported in the epicentral region [40], structural damage to buildings was not reported in Bogotá. The estimated building losses correspond to the actual situation.

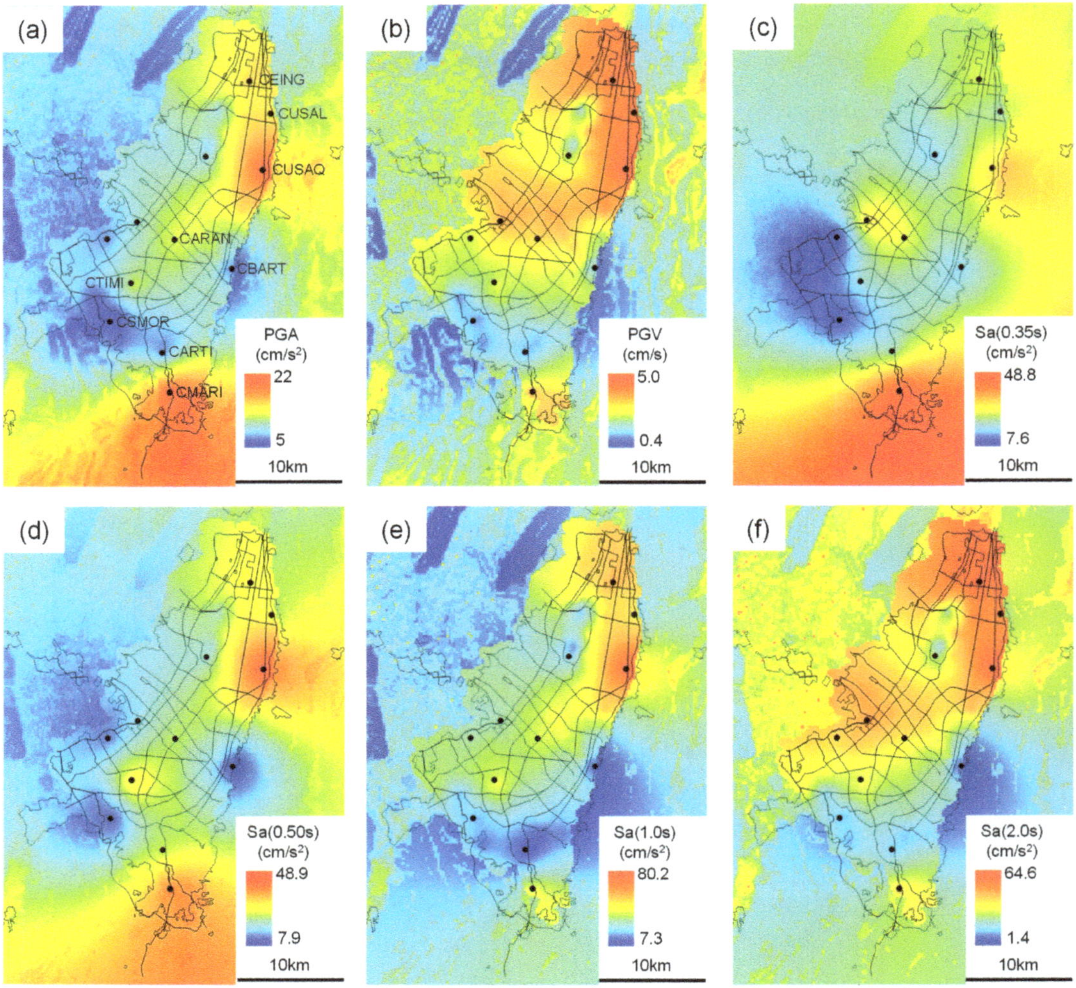

Figure 10. Spatial distribution of (**a**) PGA (**b**) PGV and spectral accelerations at (**c**) 0.35 s, (**d**) 0.50 s, (**e**) 1.0 s and (**f**) 2.0 s activated by the 2019 Mesetas earthquake.

4. Shakemaps and Building Losses for an Mw 7.0 Scenario Earthquake

As of 2023, a damaging earthquake has not been recorded in and around Bogotá since the development of the quick damage estimation system. To demonstrate economic building losses for a large earthquake, numerical simulations of seismic intensities and building losses were performed for a scenario earthquake. An Mw5.9 earthquake hit the Quetame district approximately 50 km southeast of Bogotá city and located in the Andes Mountain region in May 2008 and August 2023 (see Figure 11). In August 2023, an Mw6.0 earthquake also occurred in the active faults bounding the eastern range of the North Andes (see Figure 11). Much larger earthquakes have been anticipated in the active faults, including the Servita fault [41]. In this study, a scenario crustal earthquake with Mw7.0 corresponding to the Servita fault, an active fault located at the eastern range of North Andes close to Bogotá, as shown in Figure 11, was constructed based on active fault map information from a seismic microzonation project of Bogotá [42], as well as a methodology to estimate multi-wavelength finite fault slip distributions consistent with the fault geometry and seismic moment of the earthquake [43]. Using the slip model obtained, the ground motions and building losses for the scenario were estimated.

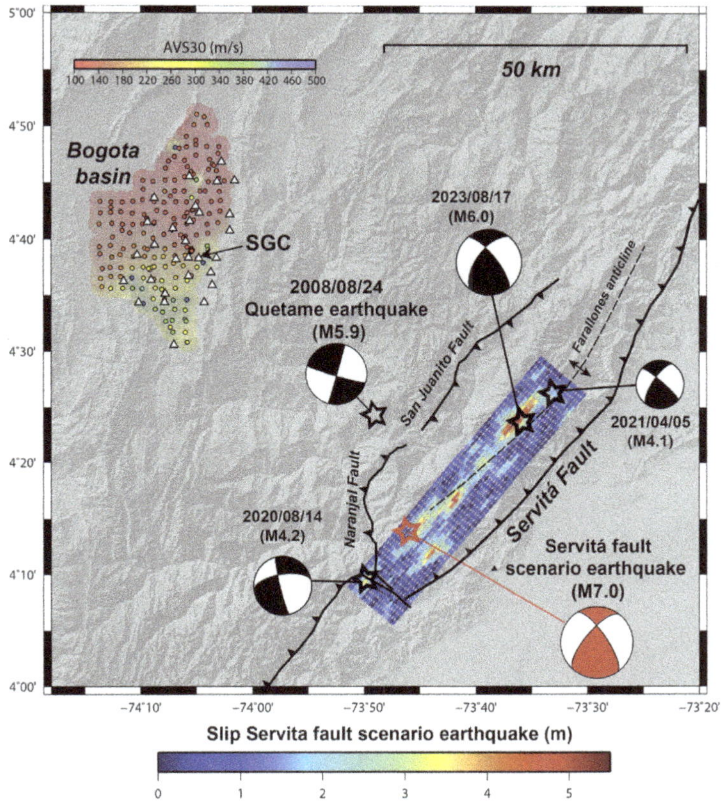

Figure 11. The slip model scenario of the Servita fault (Mw7.0) is shown near the southwest region of the map using blue to red color scale. The fault mechanism and epicenter of the Quetame earthquake (Mw5.8) on 24 May 2008, and of those of recent earthquakes near the Servita fault (2020, Mw4.2, 2021, Mw4.1, and 2023, Mw6.0) are displayed. The Vs30 map estimated within the Bogotá basin is shown by red to blue colored dots near the northwest corner of the map. White triangles show the strong motion stations of IDIGER and SGC in Bogotá. Trace and dip of the main active faults in the region are shown by black lines and triangles.

To perform the strong motion simulation, a hybrid simulation was used, which combines a deterministic simulation of ground motion at low frequencies (0.01–1 Hz), with a semi-stochastic simulation at high frequencies (1–30 Hz), based on a multi-scale finite fault rupture model [23,43,44]. Strong motion waveforms were calculated at all the sites of the strong motion stations by IDIGER for a bedrock outcrop condition defined as a site where Vs30 has a value of 500 m/s. This approximately corresponds to the geological condition for claystone–sandstone rock outcrops at the eastern hills of Bogotá. Distributions of strong motion indexes at the surface are obtained by multiplying amplification factors to the values of PGA, PGV, and acceleration response spectra of waveforms simulated at the bedrock by using the Vs30 map of Bogotá [23], and the empirical prediction equations for amplifications of strong motion indexes described in a previous section [24,25]. Since the shallow crustal earthquake is anticipated, surface wave amplifications were adopted to estimate surface ground motions for the scenario.

The case study presented here demonstrates the methodology's integration into simulated deterministic strong motion simulations. This integration is especially significant when prior recordings are unavailable in the seismic catalog of the region. It is worth emphasizing that the current methodology has been developed using seismic recordings whose peak accelerations are below 80 cm/s^2, as previously stated. However, its suitability for higher accelerations that induce nonlinear behavior necessitates further investigation. Consequently, it is imperative for future scientific research to prioritize the evaluation of these effects, enabling their full incorporation into the methodology.

Figure 12 shows the synthesized acceleration response spectra at the five stations. Since CJABO, CUSAQ, and CEING stations are located in the central part of Bogotá where the soil conditions are very soft, large amplitudes at approximately 2 s due to the surface wave site amplifications were found in the response spectra. Figure 13a–e shows the distribution of the predicted PGA, PGV, and Sa at periods of 0.35 s, 0.5 s, and 1.0 s. The PGAs and short-period spectral accelerations, such as 0.35 s and 0.5 s, show larger values in the southern part of the city because the epicentral distance is closer. The maximum PGA was estimated at 160 cm/s^2. On the contrary, the PGV and 1.0 s spectral accelerations show large values in the northern part since large amplifications are expected in longer period due to the very soft soil effect. The maximum PGV was estimated at 40 cm/s.

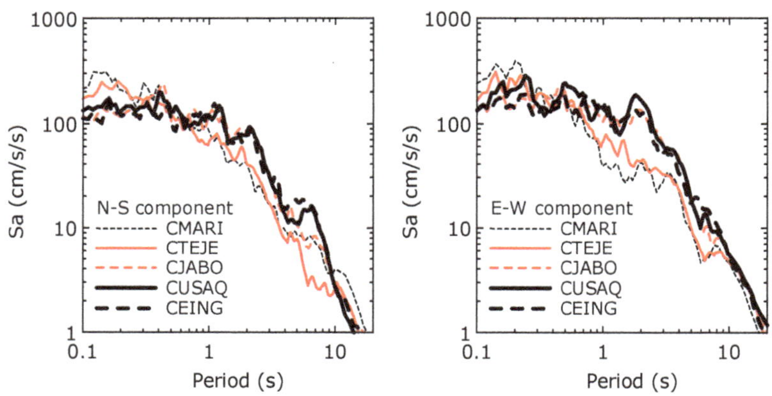

Figure 12. Synthesized acceleration response spectra of N-S and E-W components for the scenario earthquake at selected five stations.

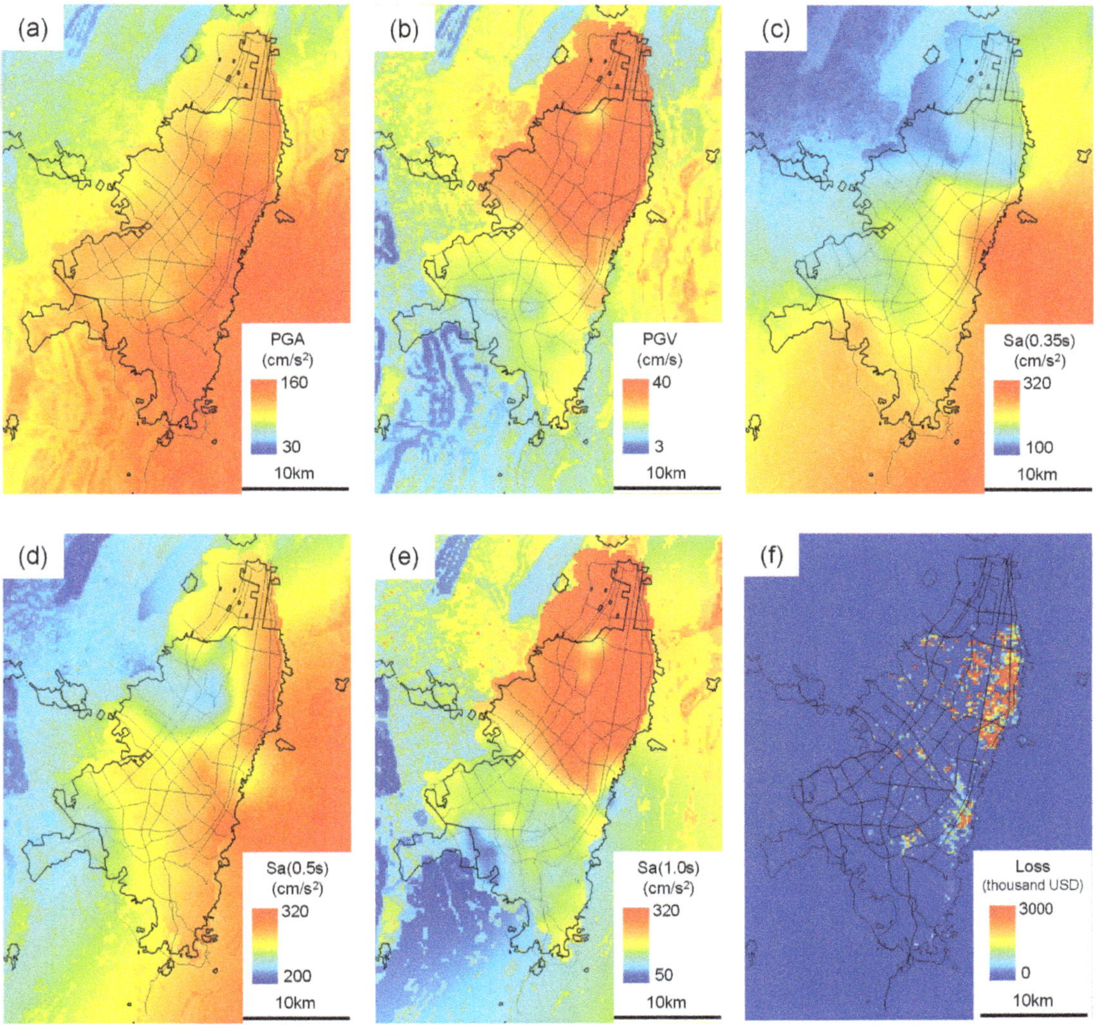

Figure 13. Results of seismic intensities and building losses for the scenario earthquake. (**a**) PGA, (**b**) PGV, (**c**) Spectral acceleration (Sa) at the period of 0.35 s, (**d**) Sa at the period of 0.5 s, (**e**) Sa at the period of 1.0 s, and (**f**) Building losses in thousand USD.

Figure 13f shows the distribution of the predicted building losses for the scenario. A great amount of building losses is predicted since the scale of the magnitude of the scenario is much larger than the 2019 event. Especially, heavy damage is concentrated in the northeastern part of the city because larger amplitudes of the seismic intensities were estimated due to the soft soil conditions and high-cost buildings, such as high-rise apartment houses and office buildings, are densely distributed in the area. Table 3 summarizes the predicted building loss and loss ratio for each building typology. The loss ratio in percentage is calculated from the ratio of the predicted loss to building replacement cost (i.e., using the MDR conditioned on the site spectral acceleration). The results show that the expected damages in low-rise unreinforced masonry buildings, such as ADOBE and MSC1_3, and mid- and high-rise buildings, such as PCRDMO6_12, PCRM_DMI6_12, PCRM_DMO6_12, and SC_20, are significant. The loss ratios of these typologies are expected to be more than 10%. Especially the loss ratio of high-rise buildings such as PCRM_DMO12_20 was higher than 50%. Significant losses in low-rise buildings were caused mainly by the low seismic capacities of the unreinforced masonries. That is why they are very sensitive to ground shaking. On the contrary, a large number of losses in high-rise buildings was estimated, although the seismic capacities of the building typologies are much higher than those of low-rise buildings. As shown in Figure 13f, the losses of the high-rise buildings were concentrated in the northeastern part of Bogotá. Strong shakings in the area were predicted especially in the longer period range (1–2 s) that affect high-rise buildings due to soft soil deposits and large amplification factors of approximately 1–2 s. These results indicate that not only low-rise unreinforced buildings, but also high-rise buildings would be vulnerable in Bogotá to large crustal earthquakes.

Table 3. Predicted building loss and loss ratio for each building typology in the scenario earthquake.

Structural Code	Number of Polygons (Section of Buildings)	Value (Million USD)	Loss (Million USD)	Loss Ratio (%)
ADOBE	19,193	1358	259	19.1
MSC1_3	1,186,283	69,319	601	0.9
MSC4_5	26,913	4517	69	1.5
MR1_3	206,205	32,196	0	0.0
MR4_5	15,645	3924	0	0.0
PCRDMI1_3	253	95	0	0.0
PCRDMO1_3	14,943	6956	0	0.0
PCRDMI4_5	64	65	0.3	0.5
PCRDMO4_5	5190	3077	0	0.0
PCRDMO6_12	3308	4161	517	12.4
PCRM_DMI4_5	10,299	3646	4	0.1
PCRM_DMO4_5	9904	5157	10	0.0
PCRM_DES4_5	1390	1172	0	0.0
PCRM_DMI6_12	588	585	97	16.6
PCRM_DMO6_12	4476	6899	904	13.1
PCRM_DES6_12	1365	2022	201	9.9
PCRM_DMO12_20	490	1415	732	51.7
PrFC4_5	12,553	2719	0	0.0
SC6_12	574	1279	105	8.2
SC_20	47	172	31	18.0
PRT_CER_MUR	11,058	4609	0	0.0
BOD_PEQ	42,023	7447	0	0.0
BOD_GRN	30,948	6795	0	0.0
Total	1,603,658	169,585	3520	2.1

The total amount of the predicted building losses is approximately 3.52 billion USD, corresponding to a total loss ratio of approximately 2%, as shown in Table 3. According to the report by the World Bank [45], the GDP (growth domestic product) of Colombia was about 300 billion USD in the past five years from 2016 to 2021. The predicted loss of the scenario corresponds to approximately 1% of the Colombian GDP. Moreover, attention should be paid to the fact that predicted losses cover only the direct building damage due to shaking, and the amount of losses would be larger by counting losses of infrastructures, such as roads and bridges, and indirect losses, such as business interruptions. Table 4 shows the comparison of the predicted losses with the actual economic losses of the recent worldwide M6-7 class earthquakes [46]. The table also shows the direct losses in Turkey caused by the Turkey-Syria earthquake (Mw 7.8) on 6 February 2023 [47]. It is difficult to directly compare our estimation with the actual losses because the distance from the source and the seismic capacities of the buildings are different from each other. However, the predicted losses are almost consistent with those in previous similar scale events. Attention must be paid to the fact that building damage and losses may significantly increase in the event of a much larger earthquake near the urban areas in the Bogotá region, such as the 2023 Turkey-Syria earthquake.

Table 4. Comparison of estimated loss by a scenario earthquake in this study with losses of the recent worldwide inland crustal earthquakes [46,47]. The losses of the 2023 Turkey and Syria earthquake represent the losses in Turkey estimated in [47]. The losses of the other earthquakes are referred from [46].

Earthquake	Date (YY/MM/DD)	M	Reported Loss (USD)
Haiti	2010/1/12	7.0	7.8 billion (4.3 billion direct)
Canterbury, New Zealand	2010/9/3	7.0	2.2 to 2.9 billion
Christchurch, New Zealand	2011/2/22	6.1	16.5 to 25 billion
Christchurch, New Zealand	2011/6/13	6.0	4.83 billion
Eastern Turkey	2011/10/23	7.1	500 million to 1.0 billion
Turkey and Syria	2023/2/6	7.8	18 billion (Direct loss of residential buildings) 9.7 billion (Direct loss of non-residential buildings) 6.4 billion (Direct loss of infrastructures) 34.2 billion (Direct loss in total)
Bogotá, Colombia (This study)	Scenario	7.0	3.52 billion (Predicted direct building losses due to shaking)

5. Conclusions

A quick estimation model of seismic intensities and ground-shaking-induced building losses for Bogotá, Colombia was developed. The model is based on the information of the strong motion observation network of Bogotá by IDIGER, the dense Vs30-based site amplifications, the building inventory, and vulnerability functions. The ground-shaking maps were developed from the observation records by a spatial interpolation technique and the Vs30 map. The economic building losses were estimated by the spectral accelerations at the period of typical building types, mean damage ratios of the vulnerability functions corresponding to the repair cost in percentage, and building replacement values in the building inventory. The shakemaps, including the distributions of PGAs, PGVs and spectral accelerations, and the building loss maps can be produced taking no longer than five minutes after receiving the seismic records from the observation sites.

The application of the system was demonstrated using the observed ground motion records during the Mesetas earthquake with Mw 6.0 in December 2019. It confirmed that larger amplitudes in the long period of approximately 1–2 s were recorded in the northeastern part of the Bogotá basin due to strong site amplifications of soft soils. In addition, the performance of the model was evaluated for a hypothetical earthquake scenario with the Mw 7.0 from the Servita fault at the eastern range of north Andes. The estimated PGAs and PGVs in Bogotá reached approximately 160 cm/s^2 and 40 cm/s, respectively. Furthermore, the distribution of the building damage and the total amount of the building losses were predicted. It was confirmed that the developed system allows us to quickly assess seismic intensities and shaking-induced building losses for earthquakes in real-time as well as for scenario earthquakes. The framework of the quick estimation model can be expanded to other regions and countries by implementing a seismic observation network, developing a site amplification model, and constructing the building exposure database with its corresponding fragility/vulnerability models.

It is acknowledged that the current model for the site amplifications will render better estimations on small-to-moderate earthquakes than for strong motions. However, the proposed estimation model in this study, compared to existing ones, represents a more realistic representation of the Bogotá basin given the use of dense geophysical explorations throughout the city. While better information is captured for stronger motions, future research could concentrate on utilizing additional sources of information to better represent the nonlinearities expected at the site-specific ground responses of the Bogotá basin. Likewise, future information on the impacts of strong motions in similar cities (i.e., with similar construction practices and design codes) could be used to improve the modeling inputs required for replacement cost and vulnerability functions definition. Such information contributes towards improving the accuracy of the seismic loss estimation tool. Finally, it is recognized that the proposed model outcome is portrayed as a deterministic value. Although it is found a valuable metric for communicating the expected earthquake impacts in Bogotá, future studies could include sources of uncertainty, such as the building's structural characteristics, the replacement costs, and the conditional damage ratio, among others, into the calculation procedure.

In the aftermath of disasters, national and local governments need to invest significant resources to address recovery needs. The developed model intends to help the city administrative offices and stakeholders in Bogotá to understand the significance of ground shakings and building damage immediately after an earthquake. The quick estimations of the impacts of an earthquake would lead to rapid and appropriate decision making for post-disaster responses, such as in identifying areas to be allocated resources for rescue and relief and in comprehensively estimating the budget needed for recovery and reconstruction. Furthermore, the model can be also used for considering pre-earthquake counter-measures by predicting ground shakings and building losses for anticipated future earthquakes.

Author Contributions: Conceptualization, Hiroyuki Miura, Masashi Matsuoka, Juan C. Reyes, Nelson Pulido, Andrea C. Riaño, Alvaro Hurtado and Raul Rincon; methodology, Hiroyuki Miura, Masashi Matsuoka, Juan C. Reyes, Nelson Pulido and Raul Rincon; software, Masashi Matsuoka and Mitsufumi Hashimoto; validation, Hiroyuki Miura, Masashi Matsuoka, Juan C. Reyes, Nelson Pulido, Andrea C. Riaño, Alvaro Hurtado and Raul Rincon; formal analysis, Hiroyuki Miura, Masashi Matsuoka, Juan C. Reyes, Nelson Pulido and Mitsufumi Hashimoto; investigation, Juan C. Reyes, Nelson Pulido, Andrea C. Riaño, Alvaro Hurtado, Raul Rincon and Helber García; resources, Juan C. Reyes, Nelson Pulido, Andrea C. Riaño, Alvaro Hurtado, Raul Rincon, Helber García and Carlos Lozano; data curation, Juan C. Reyes, Nelson Pulido, Andrea C. Riaño, Alvaro Hurtado, Raul Rincon, Helber García and Carlos Lozano; writing—original draft preparation, Hiroyuki Miura and Juan C. Reyes; writing—review and editing, Masashi Matsuoka, Juan C. Reyes, Nelson Pulido, Andrea C. Riaño, Alvaro Hurtado, Raul Rincon, Helber García and Carlos Lozano; visualization, Hiroyuki Miura, Nelson Pulido and Mitsufumi Hashimoto; supervision, Masashi Matsuoka and Juan C. Reyes; project administration, Masashi Matsuoka and Juan C. Reyes; funding acquisition, Masashi Matsuoka. All authors have read and agreed to the published version of the manuscript.

Funding: This study was supported in part by the Science and Technology Research Partnership for Sustainable Development (SATREPS) projects entitled "Application of State of the Art Technologies to Strengthen Research and Response to Seismic, Volcanic and Tsunami Events, and Enhance Risk Management in the republic of Colombia (Principal Investigator: Prof. Hiroyuki Kumagai)" and "The Project for Development of Integrated Expert System for Estimation and Observation of Damage Level of Infrastructure in Lima Metropolitan Area (Principal Investigator: Prof. Koichi Kusunoki)." This research was also supported by the Japan Society for the Promotion of Science (KAKENHI Grant numbers 22H01741).

Data Availability Statement: Earthquake source information estimated by the Red Sismológica Nacional de Colombia (RSNC) seismic network by SGC (Servicio Geológico Colombiano) were used in this study.

Acknowledgments: Earthquake source information estimated by the Red Sismológica Nacional de Colombia (RSNC) seismic network by SGC (Servicio Geológico Colombiano) were used in this study. This study was supported in part by the Science and Technology Research Partnership for Sustainable

Development (SATREPS) project entitled "Application of State of the Art Technologies to Strengthen Research and Response to Seismic, Volcanic and Tsunami Events, and Enhance Risk Management in the republic of Colombia (Principal Investigator: Hiroyuki Kumagai)".

Conflicts of Interest: Author Mitsufumi Hashimoto was employed by the company Kozo Keikaku Engineering Inc. The remaining authors declare that the research was conducted in the absence of any commercial or financial relationships that could be construed as a potential conflict of interest.

References

1. Wald, D.J.; Quitoriano, V.; Heaton, T.H.; Kanamori, H.; Scrivner, C.W.; Worden, C.B. TriNet "ShakeMaps": Rapid generation of peak ground motion and intensity maps for earthquakes in Southern California. *Earthq. Spectra* **1999**, *15*, 537–555. [CrossRef]
2. Wald, D.J.; Lin, K.W.; Quiotoriano, V. Quantifying and qualifying USGS ShakeMap uncertainty. In *USGS Open-File Report 2008-1238*; USGS: Reston, VA, USA, 2008; 26p. [CrossRef]
3. Worden, C.B.; Wald, D.J.; Allen, T.I.; Lin, K.; Garcia, D.; Cua, G. A revised ground-motion and intensity interpolation scheme for ShakeMap. *Bull. Seism. Soc. Am.* **2010**, *100*, 3083–3096. [CrossRef]
4. Midorikawa, S.; Abe, S. Real-time assessment of earthquake disaster in Yokohama based on dense strong-motion network. In Proceedings of the 12th World Conference on Earthquake Engineering, Auckland, New Zealand, 30 January–4 February 2000; Paper No. 1036. pp. 1–8.
5. Yamazaki, F. Seismic monitoring and early damage assessment systems in Japan. *Prog. Struct. Eng. Mater.* **2001**, *3*, 66–75. [CrossRef]
6. Matsuoka, M.; Yamamoto, N. Web-based quick estimation system of strong ground motion maps using engineering geomorphologic classification map and observed seismic records. In Proceedings of the 15th World Conference on Earthquake Engineering, Lisbon, Portugal, 24–28 September 2012. Paper No. 4016.
7. Zülfikar, A.C.; Fercan, N.O.Z.; Tunç, S.; Erdik, M. Real-time earthquake shake, damage and loss mapping for Istanbul metropolitan area. *Earth Planets Space* **2017**, *69*, 9. [CrossRef]
8. Michelini, A.; Faenza, L.; Lazano, G.; Lauciani, V.; Jozinovic, D.; Puglia, R.; Luzi, L. The new ShakeMap in Italy: Progress and advances in the last 10 yr. *Seismol. Res. Lett.* **2020**, *91*, 317–333. [CrossRef]
9. Guerin-Marthe, S.; Gehl, P.; Negulescu, G.; Auclair, S.; Fayjaloum, R. Rapid earthquake response: The state-of-the art and recommendations with a focus on European systems. *Int. J. Disaster Risk Reduct.* **2021**, *52*, 101958. [CrossRef]
10. Eguchi, R.T.; Goltz, J.D.; Seligson, A.; Flores, P.J.; Blais, N.C.; Heaton, T.H.; Bortugno, E. Real-time loss estimation as an emergency response decision support system: The early post-earthquake damage assessment tool (EPEDAT). *Earthq. Spectra* **1997**, *13*, 815–832. [CrossRef]
11. Kircher, C.A.; Nassar, A.A.; Kutsu, O.; Holmes, W.T. Development of building damage functions for earthquake loss estimation. *Earthq. Spectra* **1997**, *13*, 663–682. [CrossRef]
12. Kircher, C.A.; Reitherman, R.K.; Whitman, R.V.; Arnold, C. Estimation of earthquake losses to buildings. *Earthq. Spectra* **1997**, *13*, 703–720. [CrossRef]
13. Whitman, R.V.; Anagnos, T.; Kircher, C.A.; Lagorio, H.J.; Lawson, R.S.; Schneider, P. Development of a national earthquake loss estimation methodology. *Earthq. Spectra* **1997**, *13*, 643–661. [CrossRef]
14. Miura, H.; Midorikawa, S. Updating GIS building inventory data using high-resolution satellite images for earthquake damage assessment: Application to Metro Manila, Philippines. *Earthq. Spectra* **2006**, *22*, 151–168. [CrossRef]
15. Matsuoka, M.; Mito, S.; Midorikawa, S.; Miura, H.; Quiroz, L.G.; Maruyama, Y.; Estrada, M. Development of building inventory data and earthquake damage estimation in Lima, Peru for future earthquakes. *J. Disaster Res.* **2014**, *9*, 1032–1041. [CrossRef]
16. Yamin, L.E.; Hurtado, A.I.; Barbat, A.H.; Cardona, O.D. Seismic and wind vulnerability assessment for the GAR-13 global risk assessment. *Int. J. Disaster Risk Reduct.* **2014**, *10*, 452–460. [CrossRef]
17. Yamin, L.E.; Hurtado, A.; Rincon, R.; Dorado, J.F.; Reyes, J.C. Probabilistic seismic vulnerability assessment of buildings in terms of economic losses. *Eng. Struct.* **2017**, *138*, 308–323. [CrossRef]
18. Dimate, C.; Rivera, L.; Cisternas, A. Re-visiting large historical earthquakes in the Colombian Eastern Cordillera. *J. Seismol.* **2005**, *9*, 1–22. [CrossRef]
19. Cardona, O.D.; Ordaz, M.G.; Reinoso, E.; Yamin, L.E.; Barbat, A.H. CAPRA—Comprehensive approach to probabilistic risk assessment: International initiative for risk management effectiveness. In Proceedings of the 15th World Conference on Earthquake Engineering, Lisbon, Portugal, 24–28 September 2012. Paper No. 0276.
20. Cardona, O.D.; Bernal, G.A. Automatic system for post-earthquake evaluation of city damage in Bogotá. In Proceedings of the 16th European Conference on Earthquake Engineering, Thessaloniki, Greece, 18–21 June 2018. Paper No. 10916.
21. Servicio Geológico Colombiano. Available online: https://www.sgc.gov.co/sismos (accessed on 26 September 2023).
22. Pulido, N.; Senna, S.; Sekiguchi, T.; Yamanaka, H.; Eraso, J.; Perico, N.; Reyes, J.C.; García, H.; Pedraza, P.; Dimate, C. Estimation of velocity model of Bogotá basin (Colombia) based on microtremors array measurements. In Proceedings of the 16th World Conference on Earthquake Engineering, Santiago, Chile, 9–13 January 2017. Paper No. 4614.

23. Pulido, N.; García, H.; Montejo, J.; Senna, S. Long period ground motion simulations at the Bogotá basin, Colombia, based on a 3D velocity model of the basin from dense microtremors arrays measurements, gravity and geological data. In Proceedings of the 17th World Conference on Earthquake Engineering, Sendai, Japan, 13–18 September 2020. Paper No. C002830.
24. Miura, H.; Okamura, T.; Matsuoka, M.; Leal, M.; García, H.; Pulido, N. Empirical models for surface- and body-wave amplifications of response spectra in Bogotá Basin, Colombia. *Bull. Seism. Soc. Am.* **2019**, *109*, 987–1004. [CrossRef]
25. Miura, H.; Matsuoka, M.; Lozano, C.; García, H.; Pulido, N. Site amplification models for peak ground acceleration and velocity for the Bogotá Basin. In Proceedings of the 17th World Conference on Earthquake Engineering, Sendai, Japan, 13–18 September 2020. Paper No. C002297.
26. Ojeda, A.; Martinez, S.; Bermudez, M.; Atakan, K. The new accelerograph network from Santa Fe De Bogotá, Colombia and implications for microzonation. *Soil Dyn. Earthq. Eng.* **2022**, *22*, 791–797. [CrossRef]
27. Ojeda, A.; Havskov, J. Crustal structure and local seismicity in Colombia. *J. Seismol.* **2001**, *5*, 575–593. [CrossRef]
28. Pulido, N. Seismotectonics of the northern Andes (Colombia) and the development of seismic networks. *Bull. Int. Inst. Seismol. Earthq. Eng. Spec. Ed.* **2003**, 69–76. Available online: https://www.j-shis.bosai.go.jp/staff/nelson/papers/Pulido_IISEE_2003.pdf (accessed on 1 November 2023).
29. Cressie, N. The origins of Kriging. *Math. Geol.* **1990**, *22*, 239–252. [CrossRef]
30. IDECA (Infraestructura de Datos Espaciales de Bogotá). Bogotá's Open Data: The Spatial Data Infrastructure of Bogotá (in Spanish), Unidad Administrativa Especial de Catastro Distrital, Bogotá D.C., Colombia. Available online: https://www.ideca.gov.co (accessed on 1 February 2019).
31. Yamín, L.E.; Guesquiere, F.; Cardona, O.D.; Ordaz, M.G. Modelación Probabilista para la Gestión del Riesgo de Desastre, Banco Mundial, Universidad de Los Andes. 2013. Available online: http://repositorio.gestiondelriesgo.gov.co/handle/20.500.11762/19763 (accessed on 15 January 2019).
32. Uniandes (Universidad de los Andes). *Internal Report: Seismic Risk Studies of the Aburrá Valley and Development of the Methodology for the Post-Earthquake Buildings Assessment, as Part of the Metropolitan Plan of Seismic Risk of the Metropolitan Area of the Aburrá Valley*; Universidad de los Andes: Bogotá, Colombia, 2017. Archive: AMVA—Metropolitan Area of the Aburra Valley. (In Spanish)
33. Riaño, A.C.; Reyes, J.C.; Yamín, L.E.; Bielak, J.; Taborda, R.; Restrepo, D. Integration of 3D large-scale earthquake simulations into the assessment of the seismic risk of Bogotá, Colombia. *Earthq. Eng. Struct. Dyn.* **2021**, *50*, 155–176. [CrossRef]
34. DNP (Departamento Nacional de Planeacion, Colombia). Search System for Contracting Processes in the SECOP I, SECOP II, TVEC Platforms. Available online: https://consultaprocesos.colombiacompra.gov.co (accessed on 9 February 2023).
35. Turner & Townsend Inc. International Construction Market Survey. 2018. Available online: https://www.panganirealestate.com/images/international-construction-market-survey-2018.pdf (accessed on 9 February 2023).
36. Table of Average Prices per Square Meter per Quarter in Each Locality (2010–2019) Metrocuadrado: Price per Square Meter. Available online: https://www.metrocuadrado.com/noticias/herramientas/precios-m2 (accessed on 9 February 2023).
37. Fincaraiz. Apartments, Houses, Housing Projects and Real Estate News. Available online: https://www.fincaraiz.com.co/ (accessed on 9 February 2023).
38. UAECD (Unidad Administrativa Especial de Catastro Distrital). Resultados Censo Inmobiliario. 2019. Available online: https://www.catastrobogota.gov.co/recurso/resultados-censo-inmobiliario-2019 (accessed on 9 February 2023).
39. Poveda, E.; Pedraza, P.; Velandia, F.; Mayorga, E.; Plicka, V.; Gallovic, F.; Zahradnik, J. 2019 Mw6.0 Mesetas (Colombia) earthquake sequence insights from integrating seismic and morphostructural observations. *Earth Space Sci.* **2022**, *9*, e2022EA002465. [CrossRef]
40. Aguilar Suarez, A.L.; Stein, R.S. Largest Colombian crustal earthquake in 20 years strikes on Christmas eve. *Temblor* **2019**. [CrossRef]
41. Arcila Rivera, M.M.; García, J.; Montejo Espitia, J.S.; Eraso, J.F.; Valcárcel Torres, J.A.; Mora Cuevas, M.G.; Viganò, D.; Pagani, M.; Díaz Parra, F.J. (Eds.) *Modelo Nacional de Amenaza Sísmica Para Colombia*; Libros del Servicio Geológico Colombiano: Bogota DC, Colombia, 2020; Volume 43. [CrossRef]
42. INGEOMINAS (Instituto Nacional de Investigaciones Geológico Mineras). Microzonificación Sísmica de Santafé de Bogotá. 2017. Available online: https://www2.sgc.gov.co/biblioteca/Documents/Biblioteca/Microzonificacion%20sismica%20de%20santa%20fe%20de%20bogota.pdf (accessed on 9 February 2023).
43. Pulido, N.; Aguilar, Z.; Tavera, H.; Chlieh, M.; Calderon, D.; Sekiguchi, T.; Nakai, S.; Yamazaki, F. Scenario source models and strong motion for future mega-earthquakes: Application to Lima, Central Peru. *Bull. Seism. Soc. Am.* **2015**, *105*, 368–386. [CrossRef]
44. Pulido, N.; Dalguer, L.A. Estimation of the high-frequency radiation of the 2000 Tottori (Japan) earthquake based on a dynamic model of fault rupture: Application to the strong motion simulation. *Bull. Seism. Soc. Am.* **2009**, *99*, 2305–2322. [CrossRef]
45. The World Bank. GDP (current US$)—Colombia in World Bank National Accounts Data and OECD National Accounts Data File. Available online: https://data.worldbank.org/indicator/NY.GDP.MKTP.CD?locations=CO (accessed on 9 February 2023).

46. Jaiswal, K.; Wald, D.J. Estimating economic losses from earthquakes using an empirical approach. *Earthq. Spectra* **2013**, *29*, 309–324. [CrossRef]
47. The World Bank. Earthquake Damage in Türkiye Estimated to Exceed $34 Billion: World Bank Disaster Assessment Report. Available online: https://www.worldbank.org/en/news/press-release/2023/02/27/earthquake-damage-in-turkiye-estimated-to-exceed-34-billion-world-bank-disaster-assessment-report (accessed on 13 April 2023).

International Journal of
Geo-Information

MDPI

Article

Risk Assessment of Landslide Collapse Disasters along National Highways Based on Information Quantity and Random Forest Coupling Methods: A Case Study of the G331 National Highway

Zuoquan Nie [1], Qiuling Lang [1,*], Yichen Zhang [1], Jiquan Zhang [2], Yanan Chen [1] and Zengkai Pan [1]

1 School of Jilin Emergency Management, Changchun Institute of Technology, Changchun 130012, China
2 School of Environment, Northeast Normal University, Changchun 130117, China
* Correspondence: langqiuling@tom.com; Tel.: +86-15104417089

Abstract: Based on the data from two field surveys in 2015 and 2022, this paper calculates the weight of values using the entropy weight method and the variation coefficient method, and evaluates risk using the information quantity method. The information quantities of four levels of criteria (hazards, exposure, vulnerability, emergency responses, and capability of recovery) were extracted and inputted into a random forest model. After optimizing the hyperparameters of the random forest using GridSearchCV, the risk assessment was performed again. Finally, the accuracy of the two evaluation results was verified using an ROC curve, and the model with the higher AUC value was selected to create a risk map. Compared with previous studies, this paper considers the factors of emergency responses and recovery capability, which makes the risk assessment more comprehensive. Our findings show that the evaluation results based on the coupling model are more accurate than the evaluation results of the information method, as the coupling model had an AUC value of 0.9329. After considering the indices of emergency responses and capability of recovery, the risk level of the highest-risk area in the study area decreased.

Keywords: collapse; national highway; risk assessment; random forest; information quantity method

1. Introduction

The G331 national highway is located in Changbai Korean Autonomous County, Baishan City, Jilin Province, China (hereinafter referred to as Changbai County). It is close to the border between China and Korea; therefore, it is in an important strategic position and has a profound impact on China's geopolitical security [1]. The G331 national highway is a tourist route that leads to the Changbai Mountain scenic spot, and this attracts a large number of tourists and vehicles throughout the year. As other roads are far from the border, the special geographical location of the G331 national highway determines that its existence holds significant importance regarding national defense. The existence of this solution has resolved the geographical obstacles hindering the transportation of combat-ready materials by the Chinese army and improved the speed of responses to border emergencies.

The geological disasters along the G331 national highway in Changbai County are frequently influenced by human engineering activities. Among these activities, collapse disasters, also known as landslide disasters, are the most severe [2]. Collapse is a kind of geological phenomenon that the rock and soil mass on the steep slope is separated from the parent rock and accumulated at the foot of the slope under the action of gravity. Roads are an important type of infrastructure for economic development. The destruction of roads caused by collapses or landslides will affect the economic development of Changbai County. It also causes loss of life and property for tourists and residents [3]. The lack of information on the risks associated with collapse disasters and a reliable basis for decision-making hinders work on prevention and control in this study area. Prevention and control work is often carried out blindly and lacks focus, making it impossible to effectively utilize funds for disaster prevention and mitigation. The safety of local people's lives and property is difficult to guarantee, and meeting the needs of economic development is challenging.

In this paper, a coupling model is used to quantify the risk of collapse in the study area and create a collapse risk map. This provides a reliable foundation for the Changbai County government to develop a plan for preventing collapse disasters and implement an early warning system for such disasters. While minimizing economic investment in disaster prevention and mitigation, it can also achieve a better protection effect and reduce economic losses. The study area currently faces the problem of blind and non-targeted prevention and control strategies for its collapse disasters. Improving these strategies can enable defense work to proceed smoothly and effectively, and protect the safety of people's lives and property in order to meet the economic development needs of Changbai County.

Existing collapse risk research methods mainly include qualitative methods and quantitative methods [4]. The qualitative research method involves experts selecting the influencing factors of geological disasters based on their experience and evaluating the impact of these factors on the occurrence of geological disasters. This research method is often considered more subjective [5]. For example, the analytic hierarchy process [6], expert evaluation method [7], and fuzzy set integration method [8] are common qualitative research methods. The quantitative research method establishes a mathematical relationship between geological disasters and indicators using specific data, resulting in findings with a relatively high accuracy [9]. However, this research method is easily affected by the challenges of data collection, and the efficiency of data collection will impact the research progress [10]. Common quantitative research methods include the information quantity method [11], the logistic regression method [12], the frequency ratio method [13], the weight of evidence method [14], and so on. The information quantity method can identify the optimal combination of influencing factors that contribute significantly to the occurrence of collapse disasters in order to achieve an effective risk assessment of collapse disasters on roads [15–17]. With the rise of artificial intelligence, an increasing number of scholars have started utilizing machine learning models [18]. The most commonly used models are artificial neural networks [19], support vector machines [20], and random forest models [21]. Machine learning algorithms can be applied regardless of the dimension or type of data, making them widely applicable [22]. The random forest method combines the results of multiple classification trees. It effectively avoids the discontinuity of the predicted value in decision trees and reduces sensitivity to the training dataset so that the predicted value is smoother, thus preventing model overfitting and improving its stability [23–25].

Previous studies lack a uniform standard for assessing collapse disaster risks, and various scholars have different interpretations of the concept of a risk and what it entails. Gao [26] believes that a risk in this context refers to the potential loss of life, property, and economic activities resulting from geological disasters occurring in a specific area and at a specific time. Xiong [27] believes that the risk of geological disasters is a combination of geological and social attributes and is determined by sensitivity and vulnerability. Sui [28] believes that landslide risk is a measure used to describe the potential for slope damage to some extent; their research content includes hazard analyses, vulnerability analyses of the people affected by hazards, and risk zoning. Chang [29] considered that risk is the product of hazards, vulnerability, and exposure, which can be expressed as a multiplication of these three factors. On the basis of hazards, vulnerability, and exposure, this study also considers emergency responses and recovery capabilities to ensure that the evaluation results are more scientific and standardized. In this paper, the research area is the G331 national highway in Changbai County. The information quantity method, random forest method, variation coefficient method, entropy weight method, and other methods are primarily employed to create a collapse risk map along the G331 national highway. This map will assist future decision-makers and establishes a theoretical foundation for disaster prevention in the studied area.

2. Study Area and Data Sources

2.1. Study Area

The G331 national highway is an extremely important border road that connects China's northeast, China's northwest, and Inner Mongolia. It provides important assistance for the economy and trade of these three regions. The total length of this road is 9301 km. In this paper, the G331 national highway in Changbai County is studied (Figure 1), and the collapse of its back slope is identified by analyzing the elevation contours of the surrounding region. A buffer zone of 3.5 km is set on both sides of the G331 national highway. Our study area excludes the collapse of the back slope, and the remaining area in the buffer zone that belongs to the DPRK is also excluded from the study area of this paper. Changbai County is located in the southeastern part of Jilin Province in China, at the southern foot of Changbai Mountain, on the right bank of the upper reaches of the Yalu River. To the west is Linjiang, to the north is Fusong County, and to the southeast it is bounded by the Yalu River, with the Republic of Korea on the other side of the river. Its geographical coordinates are 127°12′20″ E–128°18′10″ E, 41°21′41″ N–41°58′02″ N. It is 82.9 km long from east to west and 30 km wide from north to south, with a total area of 2497.6 km^2. The surrounding region has a northern temperate continental humid monsoon climate. This climate is characterized by long, cold winters, short, rainy summers, and four distinct seasons.

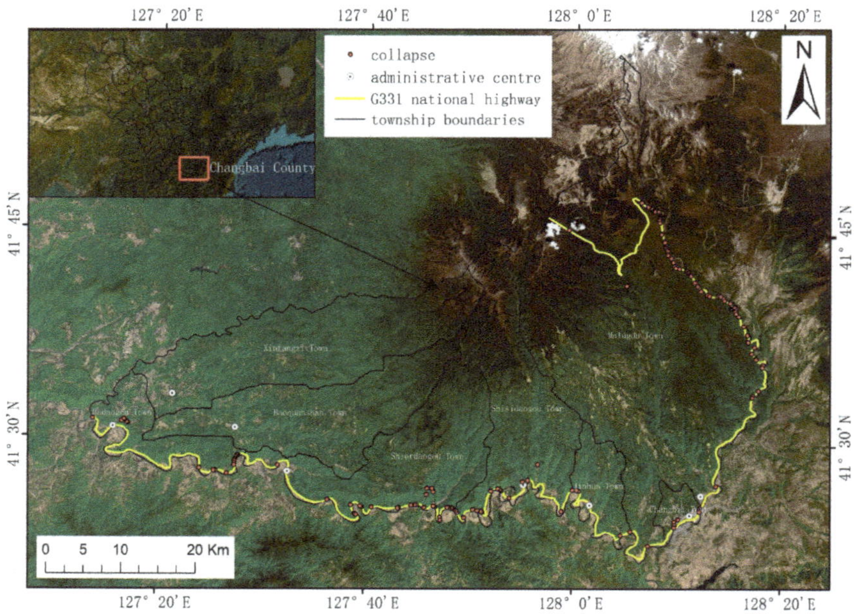

Figure 1. Geographical location of the study area.

2.2. Data Sources

In this paper, when consulting a large number of documents and considering the formation mechanism of collapse disasters and the related local conditions, eight indicators were used to assess the hazard. These indicators include the slope, aspect, curvature, NDVI, distance from the fault, lithology, mean annual precipitation, and distance from the road. For assessing exposure and vulnerability, factors such as data collectability, population density, road density, and building density were chosen as indicators. The age of the structure, highway classification, and building types were selected to indicate vulnerability. Through the official website of the Changbai County Government, one can fill out a form to apply for the provision of data and information on emergency responses and recovery capabilities, including GDP, educational status, and the number of medical staff. Table 1 shows the data source. Figure 2 depicts the workflow of the study.

Table 1. Data source.

Target Layer	Criterion Layer	Indicator Layer	Data Sources
Risk	Hazard	DEM	https://www.gscloud.cn (accessed on 24 May 2022)
		Slope, aspect, curvature	Extracted via DEM
		Mean annual precipitation	www.resdc.cn (accessed on 7 June 2022)
		NDVI	Using Landset8 satellite data produced via ENVI
		Lithology	http://dc.ngac.org.cn/Home (accessed on 19 June 2022)
		Distance from fault	http://dc.ngac.org.cn/Home (accessed on 19 June 2022)
		Distance from road	Changbai County Transportation Bureau
	Exposure	Population density	Changbai Yearbook
		Road density	Changbai County Transportation Bureau
		Building density	http://www.guihuayun.com/ (accessed on 24 May 2022)
	Vulnerability	Age of structure	Changbai Yearbook
		Road classification	Changbai County Transportation Bureau
		Building types	http://www.guihuayun.com/ (accessed on 24 May 2022)
	Emergency responses and recovery capability	GDP	www.resdc.cn (accessed on 7 June 2022)
		Educational status	Changbai County government official website
		Number of medical staff	(http://changbai.gov.cn/wzsy/) (accessed on 21 August 2022)

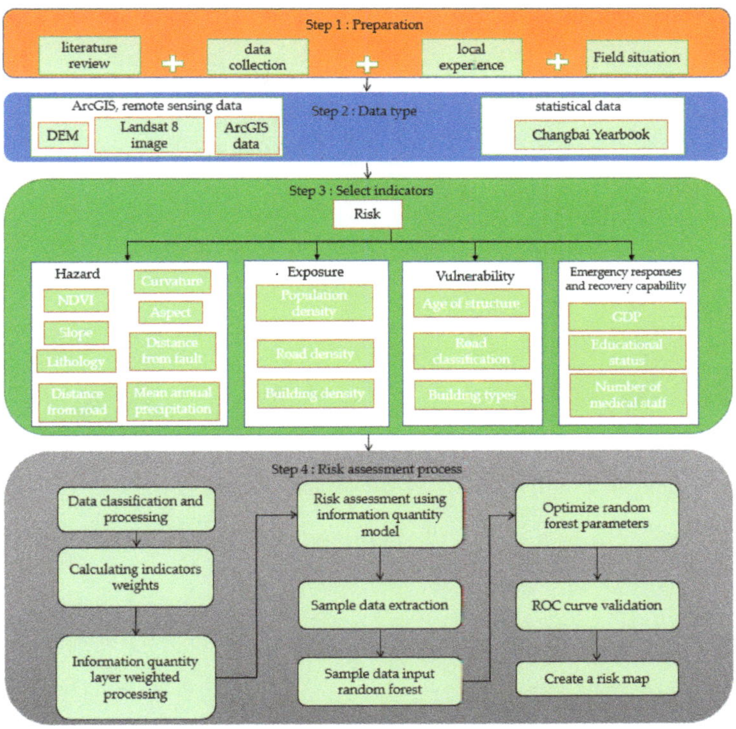

Figure 2. Workflow of the study.

3. Establishing an Evaluation Index System

In ArcGIS, after processing all of the collected data into a raster format, the raster data are resampled to a size of 30 m × 30 m for further use.

3.1. Hazard

The hazard level refers to the probability of a collapse disaster. Usually, the higher the hazard, the greater the risk. The slope measurement is a crucial factor that influences the stability of slopes, as collapses frequently happen on steeper slopes. These areas are more prone to collapse [30]. The aspect index reflects the light conditions and solar radiation of the slope. A sunny slope is more prone to collapse [31]. The curvature reflects the shape of the slope. The stability of a convex slope is low, and these are more prone to collapse [32]. The NDVI (Normalized Difference Vegetation Index) is used to characterize the soil and water conservation of regional geological disasters under varying vegetation conditions. Usually, an area with a smaller NDVI value is prone to collapse. The NDVI is measured according to Landsat 8 satellite data, which is produced via ENVI and imported into the ArcGIS platform [33]. The mean annual precipitation reflects the meteorological conditions of the study area. Areas with a higher mean annual precipitation often experience better vegetation development, which is not conducive to the occurrence of collapse [34]. The lithology and distance from the fault reflect how rock brittleness and geological conditions influence the occurrence and development of geological disasters in this area. The lithology is divided into three categories: soft rock, hard rock, and very hard rock. The hard rock area is more prone to collapse. The fault line data are analyzed in ArcGIS. The index data for the distance from the fault are calculated using the Euclidean distance tool. The closer an area is to the fault, the more susceptible it is to collapse [35]. The distance from the road reflects the impact of human engineering activities on a collapse. The closer an area is to the road, the more prone it is to collapse [36].

The spatial distribution of the hazard indicators is shown in Figure 3.

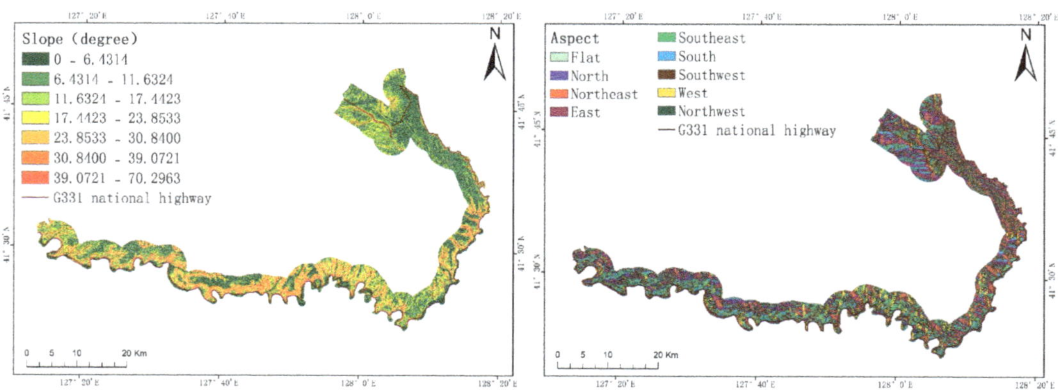

(a) (b)

Figure 3. *Cont.*

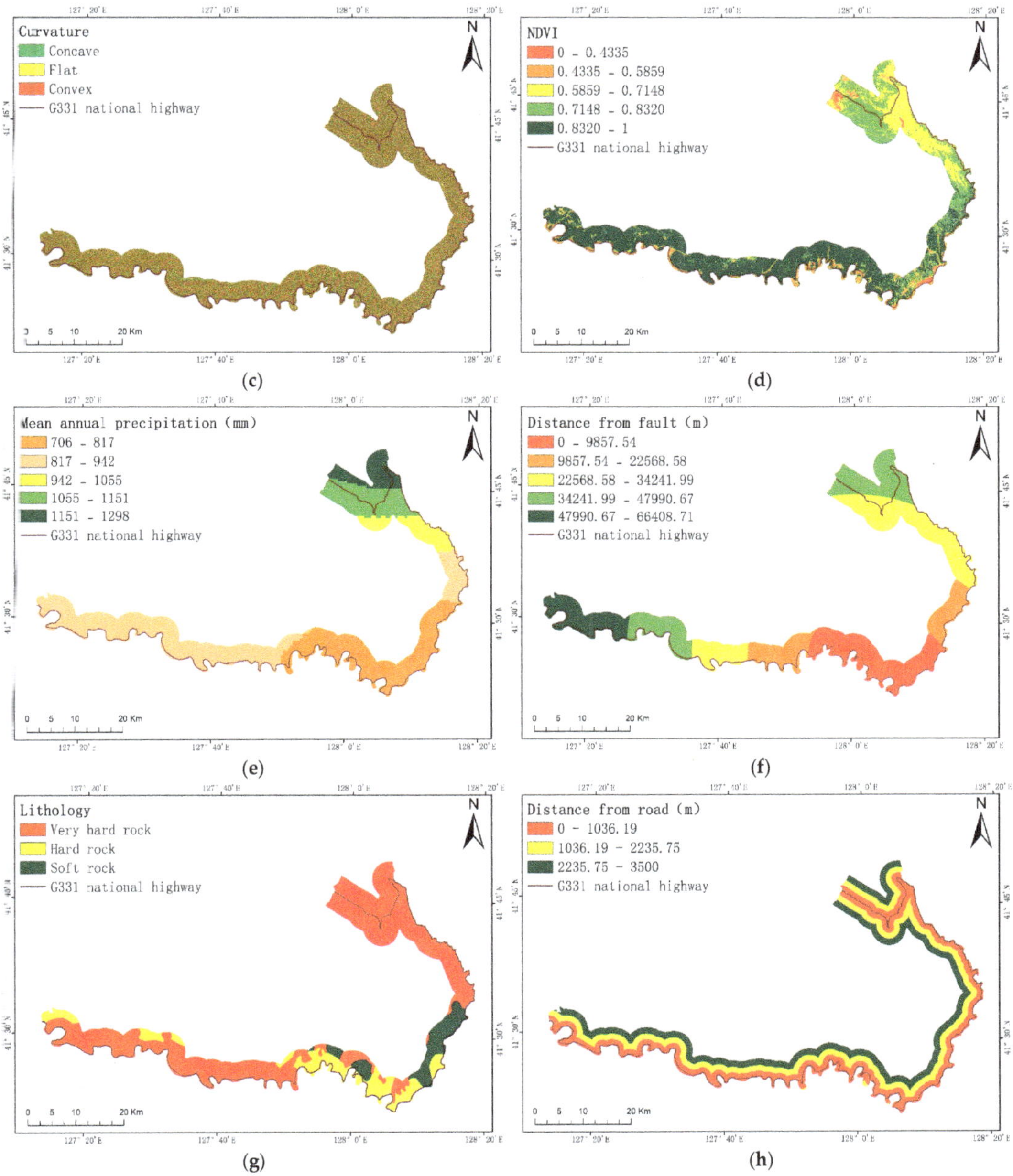

Figure 3. Spatial distribution of the hazard indicators. ((**a**) Slope, (**b**) aspect, (**c**) curvature, (**d**) NDVI, (**e**) mean annual precipitation, (**f**) distance from fault, (**g**) lithology, (**h**) distance from road).

3.2. Exposure

The exposure refers to the affected area exposed to the disaster. The greater the exposure, the higher the risk. The higher the population density, the greater the loss in a collapse disaster, and thus the greater the exposure [37]. Road infrastructure plays a crucial role in the economic development of urban areas. However, it is also susceptible to collapse disasters. This is the main part of our risk assessment along the studied road.

The higher the road density, the greater the exposure [38]. In this paper, the road density per square kilometer is calculated and then resampled to a 30 m grid data. On the basis of collecting point of interest (POI) data of Changbai County, ArcGIS was used to create a building density map of the study area. The grid size was 1 km, and the grid data needed to be resampled to 30 m for this study.

The spatial distribution of the exposure indicators is shown in Figure 4.

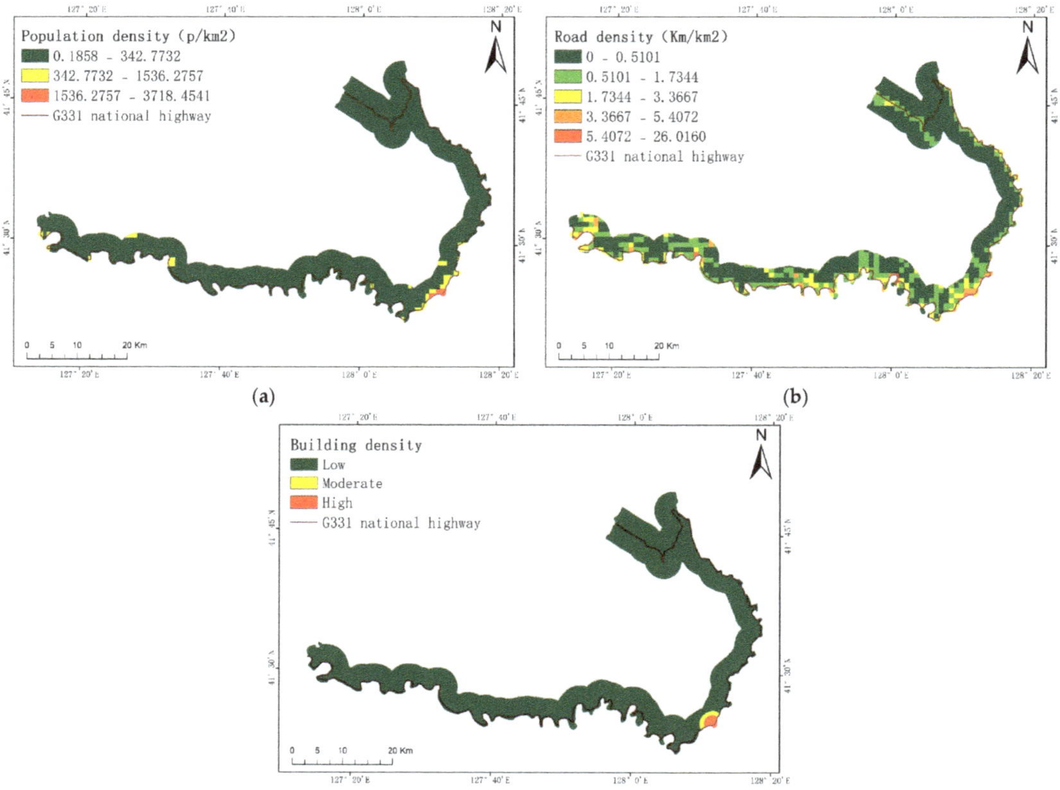

Figure 4. Spatial distribution of the exposure indices. ((**a**) Population density, (**b**) road density, (**c**) building density).

3.3. Vulnerability

Vulnerability refers to the extent of potential loss when the exposed area is affected by a disaster. The greater the vulnerability, the higher the risk. The percentage of people under the age of 14 and over the age of 60 was calculated for each township, and an age structure map was created using ArcGIS. People under the age of 14 and over the age of 60 are less capable of avoiding disasters compared to adults. As a result, they are more susceptible to injuries in collapse disasters and have a higher level of vulnerability. According to overall statistics on rural roads, it has been observed that the lower the road grade and technical standard, the more susceptible rural roads are to damage during disasters, compared to high-grade roads [39]. We measured the percentage of detached houses and bungalows in each township that are not surrounded by any other buildings. Based on the collection of POI data in Changbai County, this paper utilizes remote sensing images of the area to classify its buildings and create a map of building types in ArcGIS. Compared to multi-storey buildings, large hospitals, and shopping malls, houses and bungalows that stand alone and are not surrounded by other buildings are more prone to damage in collapse disasters [38].

The spatial distribution of the vulnerability indicators is shown in Figure 5.

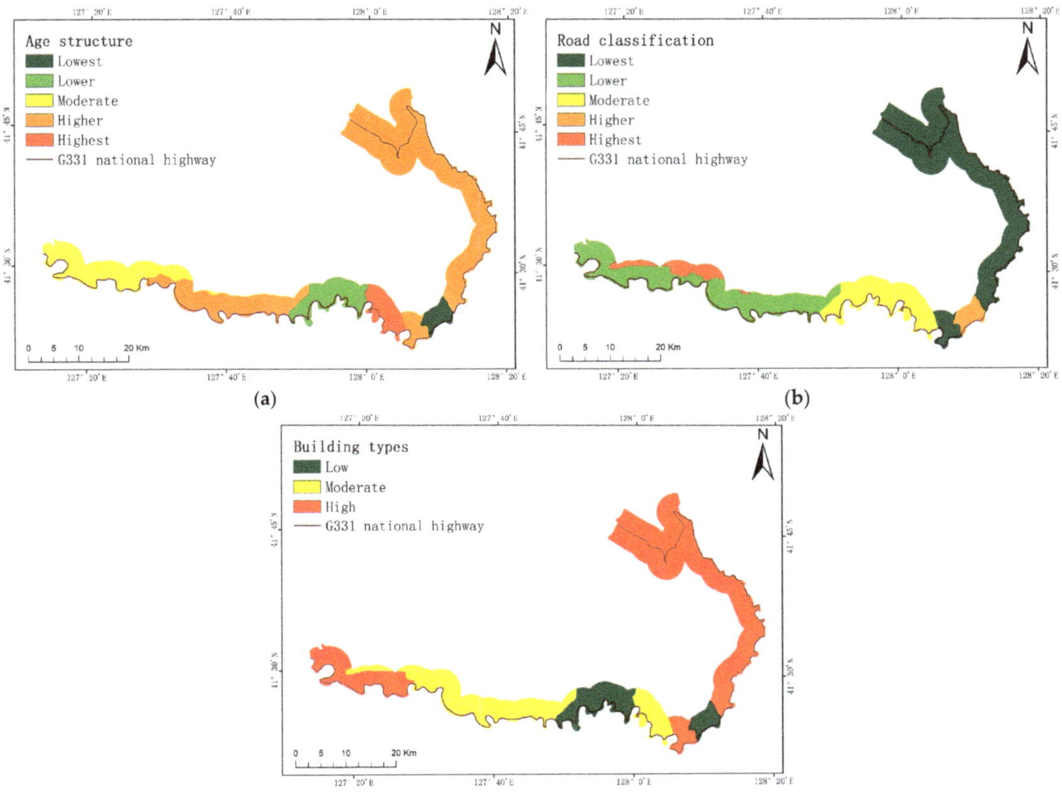

Figure 5. Spatial distribution of the vulnerability indices. ((**a**) Age of the structure, (**b**) road classification, (**c**) building types).

3.4. Emergency Responses and Recovery Capability

The emergency responses and recovery capability indices reflect the ability of a region to respond to disasters and recover from them afterwards. Usually, the higher the emergency response capability and recovery capability, the smaller the risk. The GDP value reflects the ability of a region to recover after a disaster. A region with a higher GDP can recover faster after being affected by a disaster. A higher level of emergency responses and recovery capability can also be correlated with higher levels of education [40], as education can improve disaster awareness and knowledge. People can better cope with disasters when they are prepared for them; therefore, well-educated individuals have greater disaster response capabilities. This paper calculates the average number of years of education completed among each township's population to assess the town's educational status. The higher the educational status, the higher the level of emergency responses and recovery capability [38]. Moreover, hospitals' rescue capabilities provide an important role in responding to disasters and post-disaster recovery. In areas with a higher number of medical staff, the ability to respond to disasters and recover from them is stronger, hence such areas have a higher level of emergency responses and recovery capability [40].

The spatial distribution of the emergency responses and recovery capability indicators is shown in Figure 6.

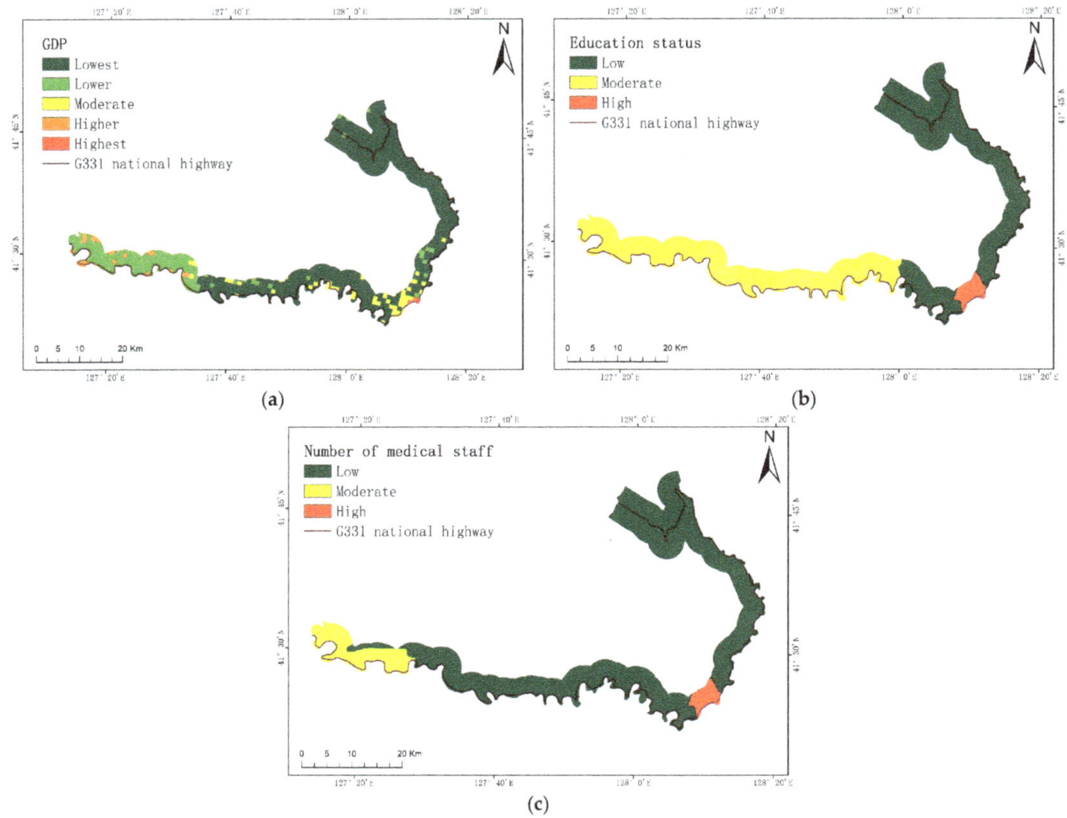

Figure 6. Spatial distribution of the emergency responses and recovery capability indices. ((**a**) GDP, (**b**) educational status, (**c**) number of medical staff).

4. Research Methods

4.1. Risk Assessment Model Construction

The risk of a collapse disaster is calculated according to a combination of the hazard, exposure, and vulnerability factors. In this paper, the entropy weight method was used to calculate the weights of each criterion layer, while the variation coefficient method is used to calculate the weights of each index layer. The risk index model for a collapse disaster is established as follows [40]:

$$Risk = H^{\omega H} \times E^{\omega E} \times V^{\omega V} \times (1 - R)^{\omega R} \tag{1}$$

$$Y = \sum_{i=1}^{n} \mu_i \times I(x_i) \tag{2}$$

where *Risk* refers to the risk index of a collapse disaster; *H*, *E*, *V*, and *R* correspond to the hazard, exposure, vulnerability, and emergency responses and recovery capability indices of collapse disasters, respectively; ωHe ωE, ωV, and ωR represent the weight value of each factor; x_i is the selected evaluation index; *e* represents the weight of the ith index; $I(x_i)$ represents the information quantity provided by the indicator layer x_i on the occurrence of collapse; and *Y* represents the *H*, *E*, *V*, and *R* criterion layers.

4.2. Entropy Weight Method

The entropy weight method was used to quantify the level of uncertainty of environmental factors [41]. The principle of this method is to calculate the weight of each index based on the information provided by the collected data. Because this calculation process does not take into account the influence of the factors, its objectivity is strong [42]. In this paper, the entropy weight method was used to calculate the weights of each criterion layer. First, the probability density was calculated based on a frequency ratio analysis, as shown in Equation (3), where P_{ij} is the FR value of each environmental factor, S_j represents the number of categories, and i and j represent the serial number and category of the environmental factors, respectively.

$$P_{ij} = \frac{P_{ij}}{\sum_{j=1}^{S_j} P_{ij}} \quad (3)$$

Secondly, the probability density P_{ij} was substituted into Equation (4) to obtain the entropy value H_j for each parameter. The calculation formula for the information coefficient I_j is as follows in Equation (5).

$$H_j = -\sum_{j=1}^{S_j} P_{ij} log_2 P_{ij} \quad (4)$$

$$I_j = \frac{log_2 S_j - H_j}{log_2 S_j}, I = (0,1) \quad (5)$$

Finally, the final weight value ω_j of each parameter was calculated by coupling the information coefficient I_j with the probability of collapse.

4.3. Variation Coefficient Method

The variation coefficient method was used to calculate the degree of change for each index based on statistical methods. It directly utilizes the information contained in each index to calculate the weight of the index, making it an objective weighting method [43]. This paper utilizes the variation coefficient method to calculate each indicator layer's weight.

Firstly, the mean value of each index was calculated. A_i represents the mean value of the ith index, and r_{ij} represents the jth value of the ith index, as shown in Equation (6):

$$e = \frac{1}{n} \sum_{j=1}^{n} r_{ij} \quad (6)$$

We calculated the coefficient of variation of each index as follows, where S_i represents the standard deviation of the ith index, and V_i represents the coefficient of variation of the ith index:

$$S_i = \sqrt{\frac{1}{n} \sum_{j=1}^{n} \left(r_{ij} - A_i\right)^2} \quad (7)$$

$$V_i = \frac{S_i}{A_i} \quad (8)$$

Finally, the weight of each index was obtained as follows, where e represents the weight of the ith index:

$$\mu_i = e \quad (9)$$

4.4. Information Quantity Method

After qualitatively classifying each evaluation index based on these data, the information quantity method can objectively reflect the relationship between each evaluation index and a collapse, avoiding the uncertainty caused by human subjective opinions [44].

In this study, the variation coefficient method was used to calculate the weight of each indicator layer in terms of evaluation index weight. The entropy weight method was used to calculate the weight of each criterion layer. Additionally, the information quantity of each index was weighted and superimposed to obtain the total information quantity.

The calculation formula for information quantity is shown in Equation (10):

$$I(x_i) = ln\frac{P(x_i|H)}{P(x_i)} \tag{10}$$

In this formula: x_i is the selected evaluation index, $I(x_i)$ is the information quantity provided by x_i at the collapse, $P(x_i|H)$ is the probability of x_i under the condition of collapse distribution, and $P(x_i)$ is the probability of x_i in the study area.

The above formula is a theoretical model, and is used in practical applications as follows (11):

$$I(x_i) = ln\frac{N_i/N}{S_i/S} \tag{11}$$

where S is the number of grids, N is the number of grids with collapses, S_i is the number of grids with evaluation index x_i, and N_i is the number of grids with collapses distributed in evaluation index x_i.

4.5. Random Forest

A random forest is an algorithm proposed by Leo Breiman in 2001. It is an algorithm that integrates multiple decision trees through an ensemble learning technique known as 'bagging' [45]. The concept of bagging [46] involves feeding the samples into multiple weak classifiers and then aggregating the classification results through voting to create a strong classifier. Random forests synthesize the results of all classification trees and regression trees, effectively avoiding the discontinuity of the predicted value in decision trees and their sensitivity to the training dataset. This makes the predicted value smoother, prevents the model from overfitting, and improves its stability [47].

The bootstrap sampling method is used to randomly extract N training samples from the dataset. These N training samples form a training set, and the training set for each tree is unique. When the node is split, the decision tree will select M feature factors from M features and use the best features among these M feature factors for node splitting. Each decision tree will then vote independently. Finally, the voting results from all decision trees are integrated, and the category with the highest number of votes is determined as the final output.

4.6. GridSearchCV

In machine learning models, there are certain hyperparameters that require manual selection. If these hyperparameters are not selected properly, a machine learning model may suffer from overfitting or underfitting [48]. When running the GridSearchCV, the selection interval and step length are specified in advance. The search will select hyperparameters based on this specified interval and step length. These hyperparameters will be incorporated into the model for training and, ultimately, identify the optimal performance hyperparameters [49]. This research utilized GridSearchCV to fine-tune the random forest model and select the optimal set of hyperparameters in order to enhance the accuracy of the model.

5. Research Results

5.1. Information Quantity Method

The weight of each criterion layer was calculated using the entropy weight method, while the weight of each index layer was calculated using the variation coefficient method. These calculation results are shown in Table 2. Simultaneously, the collapse disaster data from 2015 were used to calculate each evaluation index using the information quantity method. The corresponding information quantity values for each evaluation index are shown in Table 3.

Table 2. Weights of the indicator layers.

Target Layer	Criterion Layer	Criterion Layer Weights	Indicator Layer	Indicator Layer Weights
Risk	Hazard	0.2330	Slope (degree)	0.0665
			Aspect	0.1151
			Curvature	0.1155
			NDVI	0.1521
			Mean annual precipitation (mm)	0.1016
			Distance from fault (m)	0.1377
			Lithology	0.1664
			Distance from road (m)	0.1450
	Exposure	0.1221	Population density (p/km^2)	0.3526
			Road density (km/km^2)	0.2079
			Building density	0.4395
	Vulnerability	0.2729	Age of structure	0.4145
			Road classification	0.3422
			Building types	0.2433
	Emergency responses and recovery capability	0.3720	GDP	0.5461
			Educational status	0.2417
			Number of medical staff	0.2123

Table 3. Information quantities of the indicators.

Criterion Layer	Category	Information Quantity	Indicator Layer	Category	Information Quantity
Hazard	Low	−3.3451	Slope (degree)	0–6.4314	−1.8233
				6.4314–11.6324	−1.5189
				11.6324–17.4423	−0.6277
				17.4423–23.8533	−0.1079
				23.8533–30.8400	0.3197
				30.8400–39.0721	1.0671
				39.0721–70.2963	2.0259
	Medium	−2.7958	Aspect	Flat	0.0000
				North	0.2573
				Northeast	0.6518
				East	0.3086
				Southeast	0.1941
				South	−0.0200
				Southwest	−0.2762
				West	−2.0180
				Northwest	−1.0921
			Curvature	Concave	0.1300
				Flat	0.0000
				Convex	−0.0408
	Higher	−1.5567	NDVI	0–0.4335	−0.2555
				0.4335–0.5859	0.8270
				0.5859–0.7148	0.5974
				0.7148–0.8320	−0.0809
				0.8320–1	−0.7986
			Mean annual precipitation (mm)	706–817	−0.4274
				817–942	−0.0260
				942–1055	0.6101
				1055–1151	−0.0005
				1151–1298	0.4425
	Highest	1.2415	Distance from fault (m)	0–9857.54	−0.5391
				9857.54–22,568.58	0.2497
				22,568.58–34,241.99	0.2229
				34,241.99–47,990.67	0.1358
				47,990.67–66,408.71	−0.7098
			Lithology	Very hard rock	0.1582
				Hard rock	−0.2534
				Soft rock	−1.5518
			Distance from road (m)	0–1036.19	0.9099
				1036.19–2235.75	−2.6964
				2235.75–3500	−2.9131

Table 3. *Cont.*

Criterion Layer	Category	Information Quantity	Indicator Layer	Category	Information Quantity
Exposure	Low	−1.9625	Population density (p/km^2)	0.1858–342.7732 342.7732–1536.2757 1536.2757–3718.4541	−0.0185 0.4348 0.2052
	Medium	0.3414	Road density (km/km^2)	0–0.5101 0.5101–1.7344 1.7344–3.3667	−2.1271 0.3777 1.5542
	Higher	1.4598		3.3667–5.4072 5.4072–26.0160	1.0978 1.8249
	Highest	1.3088	Building density	Low Moderate High	−0.0134 −0.2681 1.1611
Vulnerability	Low	−0.9880	Age of structure	Lowest Lower Moderate Higher Highest	0.3363 0.1452 −0.3172 0.0856 −0.9880
	Medium	−0.3172	Road classification	Lowest Lower Moderate	0.0687 0.0917 −0.1332
	Higher	0.2076		Higher Highest	0.2410 0.0000
	Highest	0.0768	Building types	Low Moderate High	0.1961 −0.2068 0.0367
Emergency responses and recovery capability	Low	−1.4252	GDP	Lowest Lower	0.1511 −0.3131
	Medium	−0.9220		Moderate Higher Highest	−0.9223 −1.4275 0.0000
	Higher	−0.4696	Educational status	Low Moderate High	−0.0397 0.0166 0.3363
	Highest	0.1704	Number of medical staff	Low Moderate High	−0.0197 0.0251 0.3363

After calculating the value of each criterion layer using Formula (2), according to these values, hazard maps, exposure maps, and vulnerability maps were drawn in ArcGIS (Figure 7). Because the formation of the collapses in this area is mainly influenced by human engineering activities, the hazard level increases as one moves closer to the road (Figure 3h). The risk level in the town of Malugou is high, mainly due to the presence of hard rocks in this area and a low NDVI. The rocks in the towns of Badaogou and Shierdaogou are also hard, but the reason for their lower risk levels is that the NDVI values in these areas are higher (Figure 3d,g). The town of Changbai serves as the administrative center of Changbai County. It has the highest level of exposure, primarily due to the relatively large population density and building density of this town (Figure 4a,c). Other areas' higher levels of exposure are primarily attributed to the construction of rural roads for traffic purposes. In Malugou Town, there is a large proportion of houses and bungalows that stand alone and are not surrounded by other buildings (Figure 5c). In the areas around Shisidaogou and Changbai, the percentage of low-grade highways is higher (Figure 5b), so the towns of Malugou, Shisidaogou, and Changbai have the highest vulnerability levels. The region with the highest emergency response and recovery capability also has a higher GDP level (Figure 6a). The educational status and the number of medical staff in the town of Changbai are at the highest level, so the emergency responses and recovery capability in Changbai is also at a higher level (Figure 6b,c). The emergency responses and recovery capability levels are low in other areas.

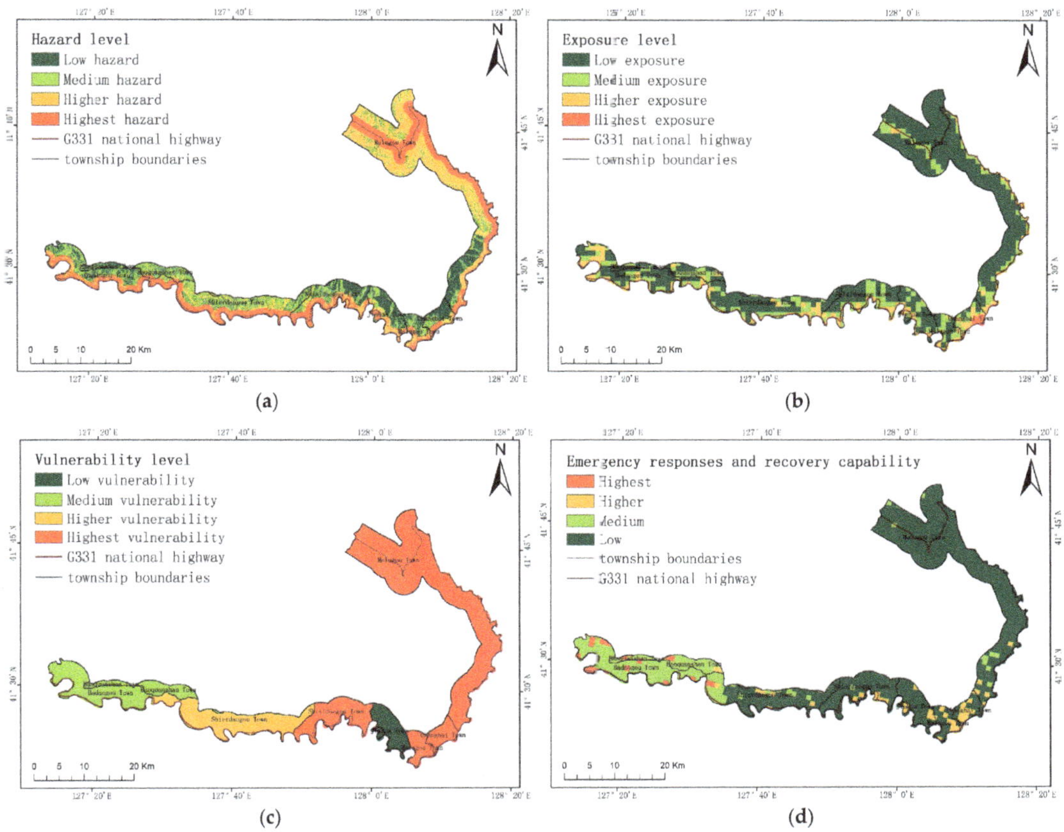

Figure 7. Drawing of each criterion layer. ((**a**) Hazard map, (**b**) exposure map, (**c**) vulnerability map, (**d**) emergency responses and recovery capability map).

After calculating the value of each target layer using Formula (1), the accuracy of this value was verified via an ROC curve using collapse data surveyed in 2022. In the ArcGIS system, a 100 m × 100 m grid is generated, and the grid points are generated simultaneously. The total amount of information quantity finally calculated is based on the values extracted from the grid points, taking into account whether there is a collapse in the grid. The sample data are then formed through spatial connection. The same number of non-collapse samples and collapse samples were randomly extracted and inputted into SPSS 26.0 software. Finally, the AUC value was observed at 0.776; the ROC curve is shown in Figure 8. Table 4 contains the curve's parameters.

Table 4. Area under the curve.

Area	Std. Error	Asymptotic Sig.	Asymptotic 95% Confidence Interval	
			Lower Bound	Upper Bound
0.776	0.02	0	0.736	0.815

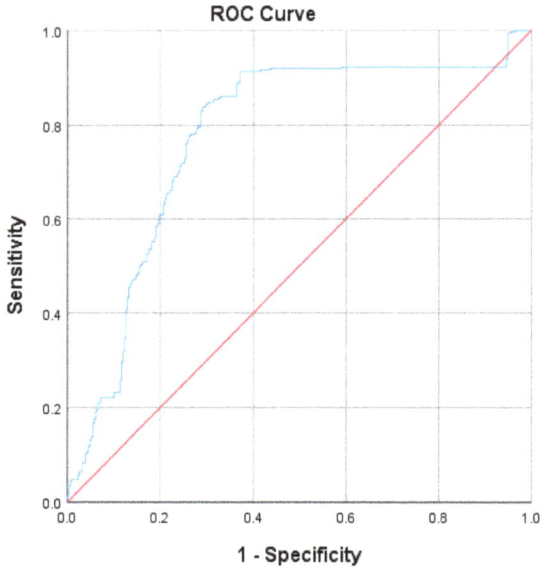

Figure 8. ROC curve for the information quantity method.

5.2. Random Forest

The AUC value for the random forest method is higher. A risk map was generated using a random forest algorithm, and the information quantity value of each criterion layer in the research area was extracted and assigned to the point. The collapse data surveyed in 2022 were selected, and an equal number of non-collapse instances were randomly selected in the study area to form the sample data. Using Python, the sample data were allocated in a ratio of 7:3 and inputted into the random forest. A total of 70% of the data was used for training the model, while the remaining 30% was used for testing the model. The model was tested using the ROC curve. Its AUC value is 0.9213 (Figure 9a). Subsequently, the GridSearchCV function from the Sklearn package was utilized to optimize the hyperparameters of the random forest model. The search range, step size, and optimized values from the GridSearchCV function are shown in Table 5. First, it optimized the n_estimators. The AUC value for this model is 0.9327 (Figure 9b). Finally, it optimized the min_samples_split and min_samples_leaf. The AUC value for this model is 0.9329 (Figure 9c).

Table 5. Hyperparameter optimization results.

Hyperparameters	Search Range and Step Size	Optimal Value of the Hyperparameters
n_estimators	(25, 500, 25)	75
min_samples_split	(1, 250, 1)	84
min_samples_leaf	(1, 110, 2)	25

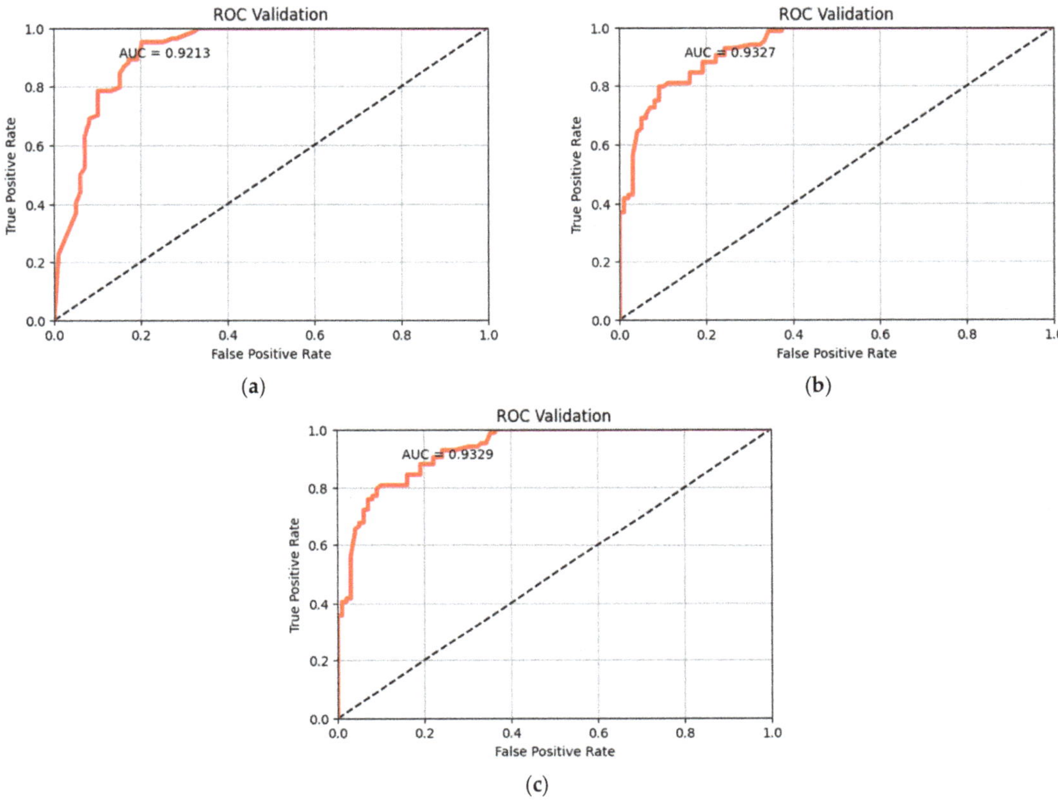

Figure 9. ROC curve for the random forest model. ((**a**) Initial model, (**b**) optimizing n_estimators, (**c**) final model).

5.3. Risk Map

Based on the 100 m × 100 m grid, the necessary data for predictions are extracted. The optimized random forest model is then selected to predict the entire research area, and the prediction probabilities are derived. In ArcGIS, the derived table is linked to the original grid attribute table, and the natural breakpoint method is used to reclassify the probability value into four categories. This process generates a risk map for the entire research area (Figure 10).

Table 6 presents the statistics concerning collapse distribution in each risk area based on the collapse data from 2015. The collapses are more evenly distributed in the areas with the higher and highest risk levels, accounting for 86.62%. When the highest-risk area comprises only 8.57% of the total area, the number of collapses reaches 70.06%. The collapses are highly concentrated in the higher-risk area and the highest-risk area, indicating that the collapse risk assessment aligns with the actual field situation.

Table 6. Collapse distribution.

Risk Level	Number of Grids	Proportion	Number of Collapses	Proportion
Low risk	400,325	51.25%	1	0.64%
Medium risk	190,432	24.38%	20	12.74%
Higher risk	123,452	15.80%	26	16.56%
Highest risk	66,974	8.57%	110	70.06%

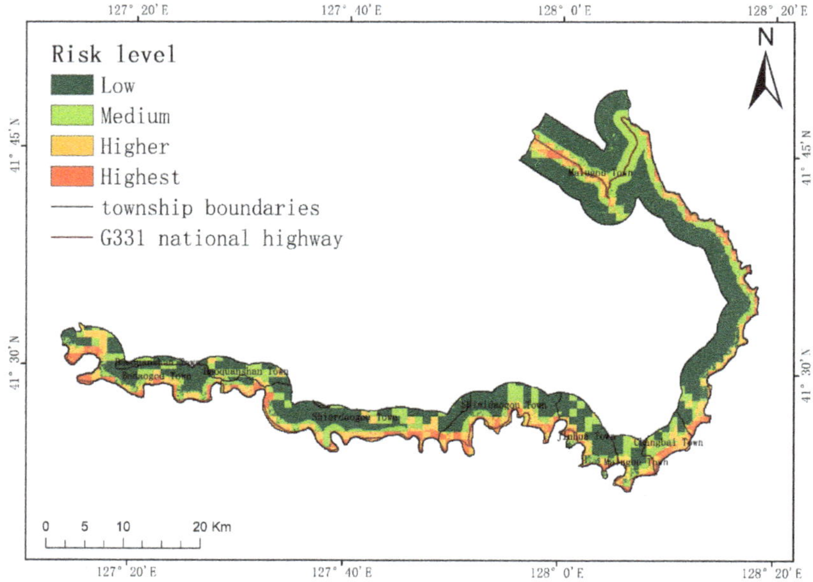

Figure 10. Risk map.

With the recent improvements on the economic level of Changbai County, the scale of various living, production, and engineering construction activities among residents has expanded unprecedentedly. Human activities have become a significant driving force be-hind changes in the geological environment and the occurrence of geological disasters in the surveyed area. The lack of foresight and disregard for scientific principles in human engineering activities has become the primary cause of geological disasters in this study area. This is due to a lack of coordination between engineering construction and the geo-logical environment. The human activities involved in the construction of the G331 national highway have resulted in frequent collapse disasters along the highway. Our risk map indicates that the highest-risk areas are predominantly situated on both sides of the G331 national highway. All townships exhibit a relatively high risk in proximity to both sides of the G331 national highway, whereas the risk is comparatively low in areas located far from the G331 national highway. However, the towns of Changbai, Badaogou, and Shierdaogou deviate slightly from this pattern. Changbai is the center of the county. From the population density map and building density map (Figure 4a,c), it can be observed that the majority of the buildings and population of this county are concentrated in this area. As a result, there are more medium-risk areas in this town compared to other towns. The slope along the G331 national highway through the town is gentle, and the lithology is soft (Figure 3a,g). Therefore, the risk along the G331 national highway in this town is higher still. Compared to other towns, the risk in the Badaogou Town and the north of Shierdaogou Townis higher. This is primarily due to the presence of rural roads that connect the villages in this region, as indicated by the road density map (Figure 4b). The increase in the amount of rural roads to hazard-affected areas leads to an increased risk in the northern parts of the two towns.

Figure 11 is a risk map that does not consider emergency responses and recovery capability. Table 7 similarly shows the area statistics for each risk level. It can be seen that the distribution of risk levels between these two is similar. However, after considering the emergency responses and recovery capability, a small number of low-risk areas were converted into medium-risk areas. The emergency response and recovery capability of these regions were at a low level. Also, a few of the highest-risk areas in the towns of Badaogou and Changbai were upgraded to higher-risk areas. The emergency responses and recovery capability of these transformed regions were at a high level (Figure 7d). On the whole, the areas with higher risk levels and the areas with the highest risk levels decreased. The proportion of higher-risk areas decreased by 2.65%, and the proportion of the highest-risk areas decreased by 2.01%.

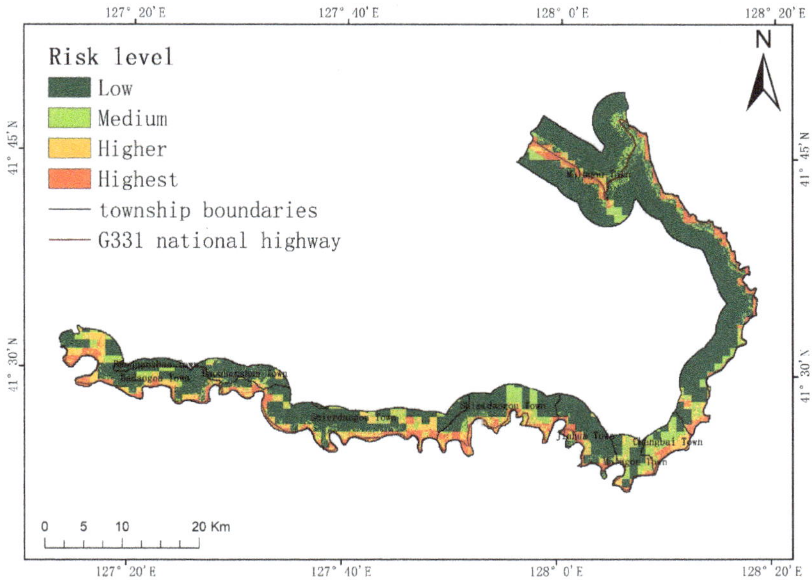

Figure 11. Risk map excluding emergency responses and recovery capability.

Table 7. Area statistics without considering emergency responses and recovery capability.

Risk Level	Number of Grids	Proportion
Low risk	483,874	61.92%
Medium risk	111,994	14.33%
Higher risk	102,814	13.16%
Highest risk	82,740	10.59%

6. Discussion

6.1. Comparison with Other Evaluation Methods

There are three main methods for assessing collapse risk. The first uses mathematical statistics. After collecting and processing data, this method establishes a probability model, analyzes the data using this model, calculates the probability distribution and risk value, and finally categorizes the risk level based on the risk value. Such methods have specific requirements regarding the data's quality and sources. The second method is the scene simulation method [50,51]. After collecting data, this method establishes a three-dimensional model and simulates the scene of a collapse using simulation technology. Such methods typically require corresponding software, which can often be costly and will result in increased research expenses. And, due to data processing and model accuracy, the simulation results may not meet expectations. Such methods are frequently employed in small-scale risk assessments. The third category is the index system method [52,53]. This method selects indicators by referring to the research of other scholars and then establishes an index system. It calculates and evaluates each layer of indicators, and then integrate them according to specific methods or standards to determine various levels of risk. This method has a certain level of subjectivity in selecting indices and calculating weights, which can impact the evaluation results.

Based on the index system method, this study integrates remote sensing data, statistical data, and raster data to minimize the subjectivity in selecting indices. The subjectivity of the weight calculation was reduced via qualitative weight calculation method. Finally, a random forest model was optimized using GridSearchCV for our risk assessment. By using the autonomous learning and self-adjusting characteristics of machine learning, the possible subjectivity in this research was further weakened. After minimizing the subjectivity, a risk map was obtained.

6.2. Comparison with Others' Studies

In previous studies, assessments of collapse risks have utilized various research methods, which vary depending on the research objectives and geographical locations. When conducting research on regional objects such as highways, railways, or historical relics, the index system and mathematical statistics methods are usually employed. When studying individual collapses, the scene simulation method is usually chosen. However, when selecting the index system method, various scholars have enhanced their approaches from different perspectives. In order to examine the relationship between landslides and the surrounding environment, Zhao utilized the spatial case-based reasoning method to assess the risk of landslides in Lushan County [39]. In order to eliminate the error of subjective weighting, Gao utilized a weight-based generalized objective function to evaluate vulnerability. This was combined with a Gaussian process classification model to assess hazard levels, and subsequently conduct a risk assessment [26]. Chang utilized the empirical model developed by Wischmeier to compute the intensity of rainfall for a hazard assessment. This approach allowed their study to examine the influence of rainfall on landslides. Subsequently, their risk assessment was conducted by considering three key factors: hazard, exposure, and vulnerability [29]. Sui studied a specific landslide at Shiyantan in Mayang County. They determined the threshold of landslide risk level in China using the ALARP method and discussed the acceptable range of risks to life and economic risk [30]. Karma Tempa evaluated the assets of the AH-48 highway using a system independently developed by Bhutan. He then assessed the risk of the AH-48 highway from three perspectives: asset value, threat value, and vulnerability value [54]. Zheng incorporated railway management into the evaluation system and assessed the risk of the Chengdu–Kunming railway using the analytic hierarchy process and triangular fuzzy number method [55]. In this study, the index system method was selected to collect remote sensing data, statistical data, and raster data. On the basis of hazard, exposure and vulnerability, emergency responses and recovery capability were also considered. The entropy weight method and the variation coefficient method were combined to eliminate subjective errors in our weight calculation process. Additionally, our random forest model's hyperparameters were adjusted using GridSearchCV to generate a risk map. The model eliminated subjective errors from two aspects: multi-source data and combined weights. This optimized machine learning model was used to generate the risk map, resulting in more accurate evaluation results. In this study, the more commonly used natural breakpoint method was selected as the threshold. In future research, the risk level threshold should be determined using a more scientific method, as the risk map will consequently be more accurate. Or, by utilizing a deep learning model, the results could be made more accurate.

6.3. The Social Impact of Our Research Results

The research results reported here are beneficial for developing geological disaster prevention plans and implementing geological disaster warning projects in the study area. When formulating plans for geological disaster prevention and control, different measures can be developed based on different levels of risk. In areas with higher elevation, collapses are often more likely to occur, and there are often more disaster-exposed people in these regions. Greater manpower and material resources are needed when formulating prevention and control measures. In urban planning, it is important to avoid constructing buildings in areas with the higher and highest risk levels. This is to prevent reductions in the vegetation coverage and to minimize the probability of collapses. At the same time, this can also reduce the loss of economic assets and personal safety when collapses occur in these regions. When conducting emergency evacuations, it is important to ensure that the layout of emergency material warehouses and emergency corridors in higher and highest risk areas is more rational. When a collapse occurs, emergency materials can reach the affected areas more quickly, and the evacuation of affected people will be expedited. Shelters should also be constructed in areas with a low level of risk. These results are also applicable in other countries. However, it is necessary to consider the local natural environment and human factors when selecting data indicators; the corresponding local data should be selected.

6.4. Limitations

The study area is located on the Chinese border. When collecting data, it was necessary to aggregate statistics into larger units in order to accurately represent the variations in values. Some of the data grid units provided by the relevant departments are large. As a result, it was not possible to fully cover the entire research area.

Collecting data on vulnerability, emergency responses and recovery capability indicators was too difficult. This led to the selection of township-level units for producing data maps of criterion layers. This further affected the accuracy of our vulnerability and emergency responses and recovery capability maps. If these indicators used smaller statistical units, for example, statistical units within a community, the quality of evaluation would be improved.

The random forest program runs automatically. Once it starts running, there are too few parts that can be debugged manually. This model is more suitable for solving classification problems. Its effectiveness in solving regression problems is not as high as that in solving classification problems.

7. Conclusions

In order to quantify the risks along the G331 national highway, this paper presents an inventory of data on collapses in Changbai County in 2015 and 2022. Seventeen risk assessment indices for road collapse disasters have been selected, establishing a spatial database with multi-source data. A risk map of collapse disasters along the road was created by combining weight calculations with the random forest algorithm and the information quantity method.

(1) The evaluation results obtained from the coupling model are more accurate compared to those obtained from the information quantity method, with an AUC value of 0.9329.
(2) According to the results of the risk assessment, the highest and higher risk areas are mainly concentrated within a 1 km radius of the G331 national highway. The towns of Badaogou, Shierdaogou, and the northern part of Malugou town are at a higher risk. The risk levels are lower in other regions.
(3) After considering the emergency response and recovery capability index, the risk levels in the higher- and highest-risk areas of the study area decreased. The proportion of higher risk areas decreased by 2.65%, and the proportion of the highest risk areas decreased by 2.01%.

In the future, additional research can be conducted using more comprehensive data. More detailed data on the natural environment and disaster-causing factor data should be collected to cover the study area located at the country's edge. Regarding the vulnerability data and the emergency responses and recovery capability index, statistics should be collected with smaller units to refine these two sections of the map. This will result in a more refined final risk map.

Author Contributions: Investigation, Zengkai Pan; Data curation, Zuoquan Nie, Yanan Chen and Zengkai Pan; Writing—original draft, Zuoquan Nie; Writing—review & editing, Qiuling Lang, Yichen Zhang and Jiquan Zhang; Visualization, Zuoquan Nie; Supervision, Qiuling Lang, Yichen Zhang and Jiquan Zhang; Funding acquisition, Qiuling Lang. All authors have read and agreed to the published version of the manuscript.

Funding: This research received no external funding.

Data Availability Statement: Data are contained within the article.

Conflicts of Interest: The authors declare no conflict of interest.

References

1. Nie, Z.; Lang, Q.; Zhang, Y. Hazard assessment of collapse along national highways based on information quantity model and geodetector. *J. Catastrophology* **2023**, *38*, 193–199.
2. Qian, L.; Zang, S. Differentiation Rule and Driving Mechanisms of Collapse Disasters in Changbai County. *Sustainability* **2022**, *14*, 2074. [CrossRef]
3. Li, Y.; Sheng, Y.; Chai, B.; Zhang, W.; Zhang, T.; Wang, J. Collapse susceptibility assessment using a support vector machine compared with back-propagation and radial basis function neural networks. *Geomat. Nat. Hazards Risk* **2020**, *11*, 510–534. [CrossRef]
4. Zhao, H.; Yao, L.; Mei, G.; Liu, T.; Ning, Y. A fuzzy comprehensive evaluation method based on AHP and entropy for a landslide susceptibility map. *Entropy* **2017**, *19*, 396. [CrossRef]
5. Panchal, S.; Shrivastava, A.K. Landslide hazard assessment using analytic hierarchy process (AHP): A case study of National Highway 5 in India. *Ain Shams Eng. J.* **2022**, *13*, 101626. [CrossRef]
6. Skilodimou, H.D.; Bathrellos, G.D.; Chousianitis, K.; Youssef, A.M.; Pradhan, B. Multi-hazard assessment modeling via multi-criteria analysis and GIS: A case study. *Environ. Earth. Sci.* **2019**, *78*, 47. [CrossRef]
7. Raghuvanshi, T.K.; Ibrahim, J.; Ayalew, D. Slope stability susceptibility evaluation parameter (SSEP) rating scheme—An approach for landslide hazard zonation. *J. Afr. Earth Sci.* **2014**, *99*, 595–612. [CrossRef]
8. Meten, M.; Bhandary, N.P.; Yatabe, R. Application of GIS-based fuzzy logic and rock engineering system (RES) approaches for landslide susceptibility mapping in Selelkula area of the Lower Jema River Gorge, Central Ethiopia. *Environ. Earth Sci.* **2015**, *74*, 3395–3416. [CrossRef]
9. Akgun, A.; Erkan, O. Landslide susceptibility mapping by geographical information system-based multivariate statistical and deterministic models: In an artificial reservoir area at Northern Turkey. *Arab. J. Geo. Sci.* **2016**, *9*, 165. [CrossRef]
10. Seong, Y.; Choi, C.-K.; Jung, Y. Assessment of Uncertainty in Grid-Based Rainfall-Runoff Model Based on Formal and Informal Likelihood Measures. *Water* **2022**, *14*, 2210. [CrossRef]
11. Alsabhan, A.H.; Singh, K.; Sharma, A.; Alam, S.; Pandey, D.D.; Rahman, S.A.S.; Khursheed, A.; Munshi, F.M. Landslide susceptibility assessment in the Himalayan range based along Kasauli—Parwanoo road corridor using weight of evidence, information value, and frequency ratio. *J. King Saud Univ. Sci.* **2022**, *34*, 101759. [CrossRef]
12. Park, S.; Choi, C.; Kim, B.; Kim, J. Landslide susceptibility mapping using frequency ratio, analytic hierarchy process, logistic regression, and artificial neural network methods at the Inje area, Korea. *Environ. Earth. Sci.* **2013**, *68*, 1443–1464. [CrossRef]
13. Gu, F.; Chen, J.; Sun, X.; Li, Y.; Zhang, Y.; Wang, Q. Comparison of Machine Learning and Traditional Statistical Methods in Debris Flow Susceptibility Assessment: A Case Study of Changping District, Beijing. *Water* **2023**, *15*, 705. [CrossRef]
14. Batar, A.K.; Watanabe, T. Landslide Susceptibility Mapping and Assessment Using Geospatial Platforms and Weights of Evidence (WoE) Method in the Indian Himalayan Region: Recent Developments, Gaps, and Future Directions. *ISPRS Int. J. Geo.-Inf.* **2021**, *10*, 114. [CrossRef]
15. Lin, J.; Chen, W.; Qi, X.; Hou, H. Risk assessment and its influencing factors analysis of geological hazards in typical mountain environment. *J. Clean. Prod.* **2021**, *309*, 127077. [CrossRef]
16. Wang, J.; Zhu, S.; Luo, X.; Chen, G.; Xu, Z.; Liu, X.; Li, Y. Refined micro-scale geological disaster susceptibility evaluation based on UAV tilt photography data and weighted certainty factor method in Mountainous Area. *Ecotoxicol. Environ. Saf.* **2020**, *189*, 110005. [CrossRef] [PubMed]
17. Lin, J.; Lin, M.; Chen, W.; Zhang, A.; Qi, X.; Hou, H. Ecological risks of geological disasters and the patterns of the urban agglomeration in the Fujian Delta region. *Ecol. Indic.* **2021**, *125*, 107475. [CrossRef]
18. Sun, D.; Shi, S.; Wen, H.; Xu, J.; Zhou, X.; Wu, J. A hybrid optimization method of factor screening predicated on GeoDetector and Random Forest for Landslide Susceptibility Mapping. *Geomorphology* **2021**, *379*, 107623. [CrossRef]
19. Su, Q.; Zhang, J.; Zhao, S.; Wang, L.; Liu, J.; Guo, J. Comparative Assessment of Three Nonlinear Approaches for Landslide Susceptibility Mapping in a Coal Mine Area. *ISPRS Int. J. Geo.-Inf.* **2017**, *6*, 228. [CrossRef]
20. Xu, S.; Zhang, M.; Ma, Y.; Liu, J.; Wang, Y.; Ma, X.; Chen, J. Multiclassification Method of Landslide Risk Assessment in Consideration of Disaster Levels: A Case Study of Xianyang City, Shaanxi Province. *ISPRS Int. J. Geo.-Inf.* **2021**, *10*, 646. [CrossRef]
21. Nurwatik, N.; Ummah, M.H.; Cahyono, A.B.; Darminto, M.R.; Hong, J.-H. A Comparison Study of Landslide Susceptibility Spatial Modeling Using Machine Learning. *ISPRS Int. J. Geo.-Inf.* **2022**, *11*, 602. [CrossRef]

22. Ðurić, U.; Marjanović, M.; Radić, Z.; Abolmasov, B. Machine learning based landslide assessment of the Belgrade metropolitan area: Pixel resolution effects and a cross-scaling concept. *Eng. Geol.* **2019**, *256*, 23–38. [CrossRef]
23. Kohestani, V.R.; Hassanlourad, M.; Ardakani, A. Evaluation of liquefaction potential based on CPT data using random forest. *Nat. Hazards* **2015**, *79*, 1079–1089. [CrossRef]
24. Wang, X.; Zhang, C.; Wang, C.; Liu, G.; Wang, H. GIS-based for prediction and prevention of environmental geological disaster susceptibility: From a perspective of sustainable development. *Ecotoxicol. Environ. Saf.* **2021**, *226*, 112881. [CrossRef] [PubMed]
25. Izquierdo-Horna, L.; Zevallos, J.; Yepez, Y. An integrated approach to seismic risk assessment using random forest and hierarchical analysis: Pisco, Peru. *Heliyon* **2022**, *8*, e10926. [CrossRef] [PubMed]
26. Gao, Z.; Ding, M.; Huang, T.; Liu, X.; Hao, Z.; Hu, X.; Chuanjie, X. Landslide risk assessment of high-mountain settlements using Gaussian process classification combined with improved weight-based generalized objective function. *Int. J. Disaster Risk Reduct.* **2022**, *67*, 102662. [CrossRef]
27. Xiong, J.; Sun, M.; Zhang, H.; Cheng, W.; Yang, Y.; Sun, M.; Cao, Y.; Wang, J. Application of the Levenburg–Marquardt back propagation neural network approach for landslide risk assessments. *Nat. Hazards Earth Syst. Sci.* **2019**, *9*, 629–653. [CrossRef]
28. Sui, H.; Hu, R.; Gao, W. Risk assessment of individual landslide based on the risk acceptable model: A case study of the Shiyantan landslide in Mayang County, China. *Hum. Ecol. Risk Assess. Int. J.* **2020**, *26*, 2500–2519. [CrossRef]
29. Chang, M.; Cui, P.; Dou, X.; Su, F. Quantitative risk assessment of landslides over the China-Pakistan economic corridor. *Int. J. Disaster Risk Reduct.* **2021**, *63*, 102441. [CrossRef]
30. Wu, Z.; Hu, M. Neotectonics, active tectonics and earthquake geology: Terminology, applications and advances. *J. Geodyn.* **2019**, *127*, 1–15. [CrossRef]
31. Ma, J.; Wang, X.; Yuan, G. Evaluation of Geological Hazard Susceptibility Based on the Regional Division Information Value Method. *ISPRS Int. J. Geo.-Inf.* **2023**, *12*, 17. [CrossRef]
32. Ohlmacher, G.C. Plan curvature and landslide probability in regions dominated by earth flows and earth slides. *Eng. Geol.* **2007**, *91*, 117–134. [CrossRef]
33. Mind'Je, R.; Li, L.; Amanambu, A.C.; Nahayo, L.; Nsengiyumva, J.B.; Gasirabo, A.; Mindje, M. Flood susceptibility modeling and hazard perception in Rwanda. *Int. J. Disaster Risk Reduct.* **2019**, *38*, 101211. [CrossRef]
34. Liu, K.; Wang, M.; Cao, Y.; Zhu, W.; Yang, G. Susceptibility of existing and planned Chinese railway system subjected to rainfall-induced multi-hazards. *Transp. Res. Part A Policy Pract.* **2018**, *117*, 214–226. [CrossRef]
35. Hadji, R.; Limani, Y.; Baghem, M.; Demdoum, A. Geologic, topographic and climatic controls in landslide hazard assessment using GIS modeling: A case study of Souk Ahras region, NE Algeria. *Quat. Int.* **2013**, *302*, 224–237. [CrossRef]
36. Siddique, T.; Pradhan, S.P.; Vishal, V.; Mondal, M.E.A.; Singh, T.N. Stability assessment of Himalayan road cut slopes along National Highway 58. *India Env. Earth. Sci.* **2017**, *76*, 759. [CrossRef]
37. Alam, R.; Quayyum, Z.; Moulds, S.; Radia, M.A.; Butler, A. Dhaka City Water logging Hazards: Area Identification and Vulnerability Assessment through GIS-Remote Sensing Techniques. *Res. Sq.* **2021**, *1*, 27. [CrossRef]
38. Duan, C.; Zhang, J.; Chen, Y.; Lang, Q.; Zhang, Y.; Wu, C.; Zhang, Z. Comprehensive Risk Assessment of Urban Waterlogging Disaster Based on MCDA-GIS Integration: The Case Study of Changchun, China. *Remote Sens.* **2022**, *14*, 3101. [CrossRef]
39. Zhao, Z.; Chen, J.; Xu, K. A spatial case-based reasoning method for regional landslide risk assessment. *Int. J. Appl. Earth Obs. Geoinf.* **2021**, *102*, 102381. [CrossRef]
40. Zhang, Y.; Lang, Q.; Chen, Y. Provincial geological disaster risk zoning method based on natural disaster risk assessmet framework: A case sthdy in Jilin Province. *Chin. J. Geol. Hazard Control* **2020**, *31*, 104–110. [CrossRef]
41. Zhan, J.; Norio, O.; Toda, Y. Integrative Natural Disaster Risk Management: Comprehensive Integration Model and China's Strategic Choice. *J. Nat. Disasters* **2006**, *1*, 29–37.
42. Azareh, A.; Rahmati, O.; Rafiei-Sardooi, E.; Sankey, J.B.; Lee, S. Modelling gully-erosion susceptibility in a semi-arid region, iran: Investigation of applicability of certainty factor and maximum entropy models. *Sci. Total Env.* **2019**, *655*, 684–696. [CrossRef] [PubMed]
43. Alamanis, N.; Dakoulas, P. Impact of Soil Properties' Spatial Correlation Lengths and Inclination on Permanent Slope Displacements Due to Earthquake Excitation. *Appl. Sci.* **2023**, *13*, 9868. [CrossRef]
44. Yang, X.; Dai, X.; Li, W.; Lu, H.; Liu, C. Socio-Ecological Vulnerability in Aba Prefecture, Western Sichuan Plateau: Evaluation, Driving Forces and Scenario Simulation. *ISPRS Int. J. Geo.-Inf.* **2022**, *11*, 524. [CrossRef]
45. Ba, Q.; Chen, Y.; Deng, S.; Wu, Q.; Yang, J. An Improved Information Value Model Based on Gray Clustering for Landslide Susceptibility Mapping. *ISPRS Int. J. Geo.-Inf.* **2017**, *6*, 18. [CrossRef]
46. Breiman, L. Random Forests. *Mach. Learn.* **2001**, *45*, 5–32. [CrossRef]
47. Miao, F.; Ruan, Q.; Wu, Y.; Qian, Z.; Kong, Z.; Qin, Z. Landslide Dynamic Susceptibility Mapping Base on Machine Learning and the PS-InSAR Coupling Model. *Remote Sens.* **2023**, *15*, 5427. [CrossRef]
48. Yeon, Y.-K.; Han, J.-G.; Ryu, K.H. Landslide susceptibility mapping in injae, korea, using a decision tree. *Eng. Geol.* **2010**, *116*, 274–283. [CrossRef]
49. Ahamad, G.N.; Shafiullah; Fatima, H.; Imdadullah; Zakariya, S.M. Influence of Optimal Hyperparameters on the Performance of Machine Learning Algorithms for Predicting Heart Disease. *Processes* **2023**, *11*, 734. [CrossRef]
50. Du, Y.; Lu, Y.; Xie, M. A new attempt for early warning of unstable rocks based on vibration parameters. *Bull. Eng. Geol. Env.* **2020**, *79*, 4363–4368. [CrossRef]

51. Du, Y.; Xie, M.; Jia, J. Stepped settlement: A possible mechanism for translational landslides. *CATENA* **2020**, *187*, 104365. [CrossRef]
52. Gargani, J. Influence of Relative Sea-Level Rise, Meteoric Water Infiltration and Rock Weathering on Giant Volcanic Landslides. *Geosciences* **2023**, *13*, 113. [CrossRef]
53. Mercurio, C.; Calderón-Cucunuba, L.P.; Argueta-Platero, A.A. Predicting Earthquake-Induced Landslides by Using a Stochastic Modeling Approach: A Case Study of the 2001 El Salvador Coseismic Landslides. *ISPRS Int. J. Geo.-Inf.* **2023**, *12*, 178. [CrossRef]
54. Li, W.; Fan, X.; Huang, F.; Chen, W.; Hong, H.; Huang, J.; Guo, Z. Uncertainties Analysis of Collapse Susceptibility Prediction Based on Remote Sensing and GIS: Influences of Different Data-Based Models and Connections between Collapses and Environmental Factors. *Remote Sens.* **2020**, *12*, 4134. [CrossRef]
55. Tempa, K.; Chettri, N.; Aryal, K.R.; Gautam, D. Geohazard vulnerability and condition assessment of the Asian highway AH-48 in Bhutan. *Geomat. Nat. Hazards Risk* **2021**, *12*, 2904–2930. [CrossRef]

geosciences

MDPI

Article

Comparison of the Piezocone Penetrometer (CPTU) and Flat Dilatometer (DMT) Methods for Landslide Characterisation

Kristijan Grabar [1], Jasmin Jug [2,*], Anja Bek [2] and Stjepan Strelec [2]

[1] SPP d.o.o., Koprivnička 47, 42000 Varaždin, Croatia; kristijan@spp.hr
[2] Faculty of Geotechnical Engineering, University of Zagreb, Hallerova aleja 7, 42000 Varaždin, Croatia; anja.bek@gfv.unizg.hr (A.B.); stjepan.strelec@gfv.unizg.hr (S.S.)
* Correspondence: jasmin.jug@gfv.unizg.hr; Tel.: +385-42-408916

Abstract: The increasing occurrence of landslides worldwide causes many human casualties and huge socio-economic losses. Therefore, the fastest and most accurate characterisation of landslides is important. The objective of this study is to compare how well the flat dilatometer (DMT) test and the piezocone penetration (CPTU) test can find the depth of a sliding zone. Inclinometers were used to measure horizontal changes in the soil to ensure the depth of the sliding zone was correct. The coincidence of the results of in situ static probes, and the displacements of the inclinometers is a sure confirmation of the depth of the sliding zone. In the example of Bedekovčina and Kravarsko landslides, in situ static probes were used to obtain values of input parameters on the sliding zone for parametric sensitivity analysis of parameters. Sensitivity analysis was performed by plotting the relationship between the above parameters and the vertical effective stress σ'_{v0} on the sliding zone. The sensitivity analysis of the parameters of 11 tested samples shows that for the parameters of the obtained DMT probe, a higher sensitivity of the parameters is obtained, closer to the values concerning the expected range, and a minor standard deviation. The parameter K_d obtained by dilatometer probing is the best indicator of the depth of the sliding zone. The literature value K_d = 1.8–2.0 on the sliding zone in this paper is extended to the range K_d = 1.8–2.5, and its detection sensitivity is influenced by over-consolidation in shallow soil layers. In general, the research results show that the dilatometer probe has an advantage over the piezocone penetrometer test for the needs of landslide characterisation.

Keywords: in situ static probes; landslide characterisation; sliding zone; inclinometer verification of results; parameter sensitivity analysis

1. Introduction

In 1991, Cruden defined a landslide as the movement of a rock mass, debris, or earth down a slope [1]. Landslides have been recorded for centuries, especially in Asia and Europe. The oldest recorded landslides occurred in central China in 1767 BC and were caused by an earthquake [2]. The investigation presented in this paper was conducted in northwestern Croatia, where landslides are also often present. Due to extreme weather conditions during the winter and spring of 2013, more than 900 shallow landslides were (re)activated in the mentioned area [3]. Landslides are spread over a wide area in Europe [4–6], most prevalent in mountainous, hilly, and coastal regions [7]. In the neighbouring countries of Slovenia and Italy, problems of landslides also occur [8–10]. Because of all of the above, landslide investigation is a very current, interesting, and frequent topic.

Saturation of the soil with water is considered the primary natural cause of landslide activation [11]. As the groundwater level rises in the slope, pore pressures increase, which reduces the effective stress and thus the shear strength of the soil that builds the slope [12]. Landslides cause human, animal, and economic losses. They destroy or damage residential and industrial buildings as well as agricultural and forest lands. In addition, they have a negative impact on the water quality in rivers and streams. Landslides are actually responsible for much more significant socioeconomic loss than is generally recognized and/or acknowledged. Much of the damage from landslides is not documented because they are wrongly included as part of reports of earthquakes [13], floods, volcanic eruptions, etc. However, the cost of damage from landslides can exceed all costs combined from the other listed geohazards [14].

Complex landslides are created by joining several sliding zones (sliding surfaces) [15]. Precisely, such complex landslides from the territory of the Republic of Croatia are the investigation areas in this paper. The locations of the landslide at the cemetery in Bedekovčina (Krapinsko-zagorska County) and the landslide in Kravarsko (Zagrebačka County) are shown in Figure 1. In both locations, the soil is extremely heterogeneous, and the in situ probes are used for detecting problematic layers prone to sliding [16]. This research also shows that in addition to lithological and geological investigations, it is very important to put comprehensive research into the context of hydrogeological conditions because they are the most common trigger for the initiation of the sliding process.

In situ probes have proven to be very effective in the detection of sliding zones, insight into the soil composition, geotechnical characteristics of the soil in the area of the manifested landslide, and the creation of geotechnical soil profiles [17–19]. The above was confirmed by Potvin et al. Based on the CPTU research, they profiled and detected the sliding zone in sensitive clay landslides in detail [20]. Totani et al. used 1997 an in situ dilatometer (DMT) for soil characterisation in slope stability [21].

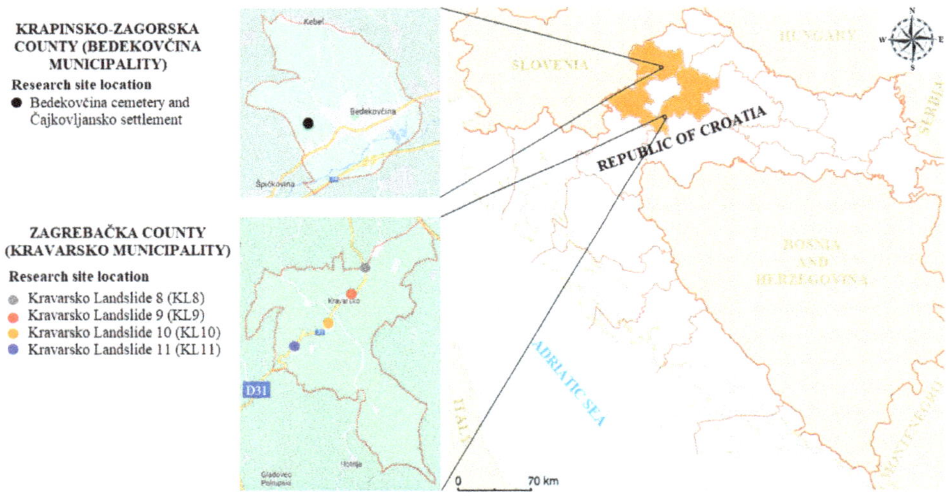

Figure 1. Locations of investigated landslides [22,23].

The novelty of this paper is the investigation of the sensitivity of an individual in-situ geotechnical static probe to the detection of sliding zones in a landslide, whereby the results of inclinometer observations are used for verification. The research sequence through three phases is shown in Figure 2. Based on the investigation results on the mentioned complex landslides, a parametric analysis of the soil inside the sliding area will be carried out by comparing the parameters of the probing results of two in situ methods, and it will be assessed which method gives better results for the characterisation of the landslide. The concept of landslide characterisation includes identifying the area affected by the landslide, as well as the geometry and determination of material properties. The purpose of characterisation is primarily to look at the problem from cause to effect. For this purpose, quantification of geometric relationships and material properties is required.

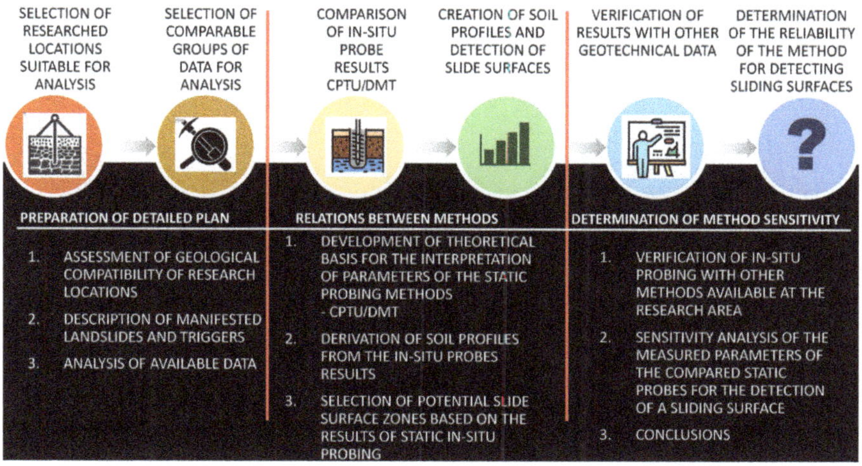

Figure 2. Investigation flow chart.

2. Investigation Area Description and Issue Presentation

2.1. Bedekovčina Landslide

Bedekovčina is a village and a municipality of the same name in Hrvatsko Zagorje (Krapinsko-zagorska County). The investigated area, the Bedekovčina cemetery, is located about 1200 m west of the inhabited area or about 1500 m northwest of the Krapina River (Figure 1). The area is morphologically hilly, and the gently sloping slopes belong to the southern slopes of the Ivančica mountain, at the transition to the Krapina river valley. The mentioned slopes are intersected by deeply incised gullies with steep sides and de-levelling that often exceeds 10 m [24].

In terms of lithology, the area of Bedekovčina is geologically built from Pliocene clastic deposits, sediments at the transition from the Pliocene to the older Quaternary (Pleistocene) and alluvial deposits in the valley of the Krapina river. The spatial arrangement of these deposits with corresponding lithological markings is shown on the overview geological profile in Figure 3. The description of the mentioned lithological members is as follows [25]:

- Lower Pliocene (Pl_1)—these deposits consist of clastites characterised by alternating sand, silt, clay, and sandy marl. Blue–grey marls are found in the superpositionally lowest parts of the sequence. They are overlain by well-stratified sandy marls with lenses of powder and clay and, locally, may also contain layers of coal. According to existing data, the thickness of these deposits is 100–450 m [25].
- Plio-Pleistocene (Pl,Q)—fluvial-type deposits at the transition between the Pliocene and the Quaternary are discordant to older Pliocene deposits. The contact with the older Pliocene deposits is tectonic, and in some cases, the boundary is erosional. Regarding lithology, there are dusty sands and sands with clay lenses. The location of Bedekovčina cemetery is located exactly in these deposits.
- Holocene (al)—alluvial deposits are present along the Krapina river. Among the geological materials, unbound clasts predominate, the granulometric composition of which varies from gravel, sand and dust to clay.

From a hydrogeological point of view, the described Pliocene and Plioquaternary clastic deposits are practically impermeable. This is evidenced by the deeply cut gullies and the existing erosion net. Surface runoff is dominant, while underground flow is limited and of local significance. However, in the described clastic deposits, there is a surface part that has been changed by external meteorological factors concerning the original soil. Due to this altered material, rainwater infiltration occurs, which contributes to the destabilisation of weathered deposits on the slope. Instability is also contributed by the fact that fine-grained deposits have a hard time releasing the infiltrated water [24,25].

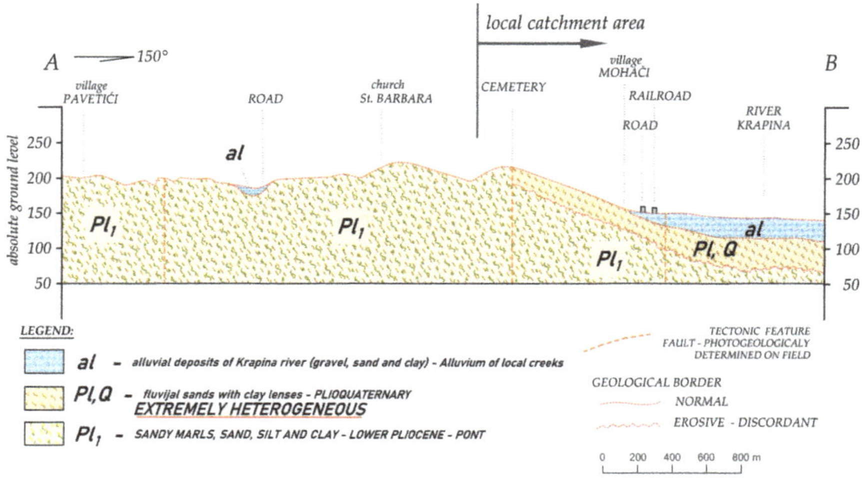

Figure 3. Geological profile of the wider area around the investigation location in Bedekovčina [20].

In the peripheral part of the investigated area, a morphological de-levelling was observed, which is the result of tectonic and erosion processes. The mentioned de-levelling is considered the limit of the surface local watershed since underground water inflows towards the cemetery from the north and west are impossible. Therefore, the groundwater that appears in the cemetery at a depth of 2–3 m originates from rainwater that has seeped into the surface erosion layer consisting of sand with clay lenses [24,25].

Figure 4 shows different types of clay that were sampled in the investigation area. Differences in sedimentation and composition are noticeable. Clays with high plasticity (first sample from left to right in Figure 4) are characterised by very low permeability, and when moisture is lost in the sample, shrinkage occurs (volume reduction). Clays with sand (fifth sample from left to right in Figure 4) have higher permeability than the previously described clay, while moisture loss does not result in shrinkage of the sample, or it is minimally expressed.

Figure 4. Different clay samples from the area of the Bedekovčina landslide.

2.2. Kravarsko Landslide

Kravarsko municipality is in the southern part of Zagrebačka County, and an important road passes through its area (State Road D31 Zagreb—Velika Gorica—Pokupsko—Glina). Numerous landslides were recorded, and four that are in the area of Kravarsko were geotechnically processed and analysed in this paper (Figure 1).

The broader research area is made up of middle Pliocene deposits (Pl_2), which are represented by clays and sands with interlayers of organic clay, peat, and lignite. Clays contain large amounts of dispersed organic substances (sometimes peat) and more or less silty and sandy components. Additionally, thinner and thicker layers of sand appear, which are mainly silty sandy and, in some places, bound with clay [26].

Specifically, extremely high groundwater levels in the piezometers were determined. The water levels reach almost to the surface, and some piezometers indicate artesian pressures (40 cm above the ground level). Recorded pressures are continuous. The structural-tectonic relations of the area are complex, so they significantly affect engineering-geological and hydrogeological conditions. The morphology of the area is, unfortunately, favourable for the formation of frequent landslides. Landslides in this area are not only present on the route of the state road, but soil sliding is a ubiquitous phenomenon of the area, conditioned by lithostratigraphic and hydrogeological conditions. The higher parts of the broken topography of the terrain are built by sands in which water is received by infiltration, and the disturbed continuity of the sandy layers in the lower parts of the terrain creates conditions for the formation of artesian aquifers. Sufosis (submergence) of surface streams and saturation of slopes with water that springs up at the foothills are frequent occurrences in Kravarsko [26,27].

3. Research Methodology

During in situ investigations, to characterise landslides at both investigation locations, the following geotechnical methods were used: the piezocone penetration test (CPTU), flat dilatometer test (DMT), and measurements of horizontal displacements with vertical inclinometers. The layout of the investigations carried out in the cemetery in Bedekovčina is shown in Figure 5. The investigation area is divided into two groups (Group 1 and Group 2).

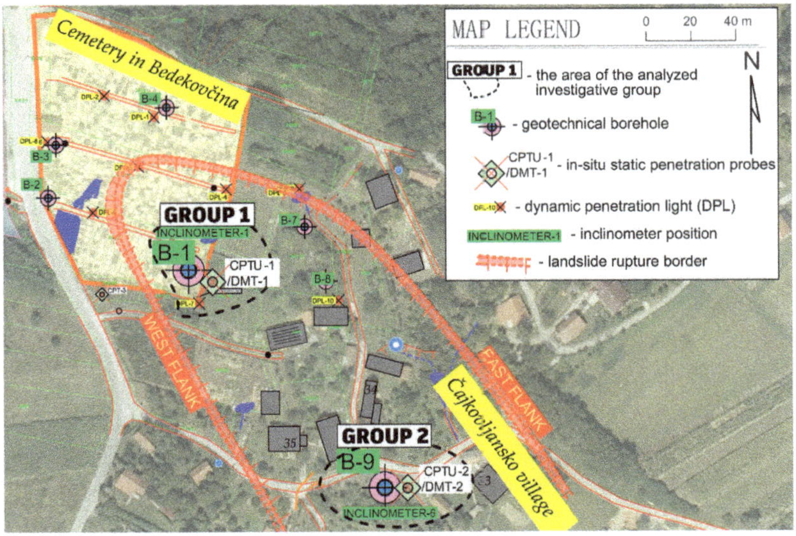

Figure 5. Plan view of in-situ investigations on the Bedekovčina landslide.

In a very narrow area of each group, probing was carried out with CPTU and DMT probes as well as an exploratory well in which an inclinometer was installed. In this way, the comparison of the selected parameters of the in situ probes for the detection of the sliding zone, as well as the subsequent verification with the results of the inclinometer observations, was enabled.

In the territory of Kravarsko municipality, four manifested landslides were analysed along the state road D31 (KL8, KL9, KL10, and KL11), so the investigation at this location was divided into four groups (from Group 3 to Group 6). The situations of in situ investigations for a particular landslide are shown in Figure 6.

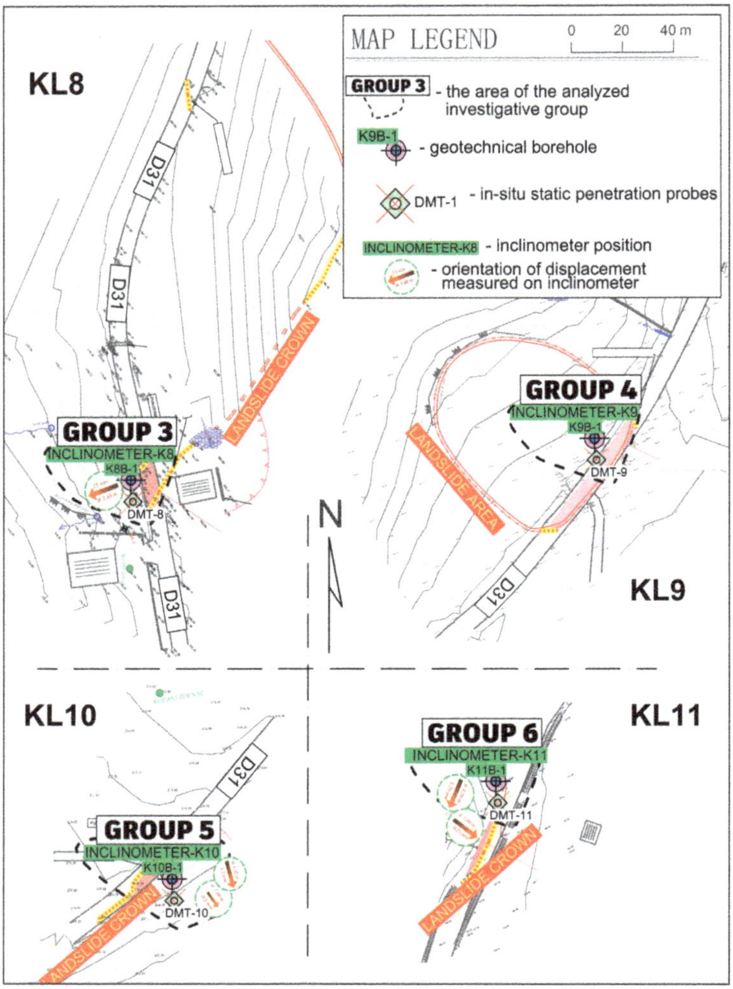

Figure 6. Situation plan of in-situ investigations on landslides in Kravarsko.

3.1. Penetration Theory of Piezocone Penetrometer

Testing with a CPTU is performed by pressing a measuring penetration piezocone (probe) at a constant velocity of 2 cm/s, using a hydraulic system and push penetration rods. The CPTU test is carried out according to HRN EN ISO 22476-1, "Geotechnical Investigation and Testing—Field Testing—Part 1: Electrical static penetration testing without pore pressure measurement and with pore pressure measurement" (ISO 22476-1:2012; EN ISO 22476:2012) [28]. All used equipment and investigation procedures are in accordance with the mentioned ISO norm. During indentation, the resistance to the probe tip (q_c), the friction on the probe jacket (f_s), the inclination of the accessory (i) and the pore pressure (u) on the filter above the probe tip are measured as standard [29,30].

Interpretation of CPTU results is carried out based on measured parameters and by applying conservative values of constants to ensure a lower and more conservative assessment of geotechnical parameters. Most of the correlations are semi-empirical, meaning they are supported by theoretical knowledge. Therefore, based on correlations, different geotechnical parameters are obtained [30,31]. To characterise landslides, undrained shear strength c_u, soil compressibility modulus M and over-consolidation ratio coefficient OCR were analysed in this paper.

Undrained shear strength c_u is the most commonly interpreted clay parameter in in-situ static cone penetration CPTU. Since the measured cone resistance q_c is influenced by the pore pressure u, Lunne et al. in 1985 proposed the calculation of the c_u based on the corrected cone resistance q_t [32], which is related to the net cone resistance q_{net}. The q_{net} is obtained as the difference between the corrected cone resistance q_t and the vertical stress σ_{v0}. The q_t presents the difference between the measured spike resistance q_c and the measured pore pressure u. Taking into account the previous statements, c_u in the CPTU test is defined by the following equation [33]:

$$c_u = \frac{q_c - u - \sigma_{v0}}{N_{kt}} = \frac{q_t - \sigma_{v0}}{N_{kt}} = \frac{q_{net}}{N_{kt}} \quad (1)$$

where is N_{kt} is the bearing factor of the net resistance at the tip (tip factor).

The range of the N_{kt} factor is in the range of 8 to 20 for different types of clay materials in different locations. If no additional data are available for the in situ sounding location, the initial value N_{kt} = 14 is adopted [34]. For interpretations of static cone penetration, and comparable to the results of undrained shear strength obtained with a flat dilatometer DMT, the cone factor N_{kt} = 18 was used at the research locations from this work.

Undrained shear strength c_u is defined as a parameter of total stresses measured at a constant sample volume. The shearing mode affects the measured amount of shear strength in soft to stiff clays [34]. This effect must be known to the researcher to be able to realistically interpret the research results. Thus, settings and test types affect the undrained shear strength measurement result. Different laboratory and field tests give different results for the same materials and depths. This is illustrated in Figure 7, published by Hight et al. for results from the Bothkennar geotechnical test site in Great Britain [35].

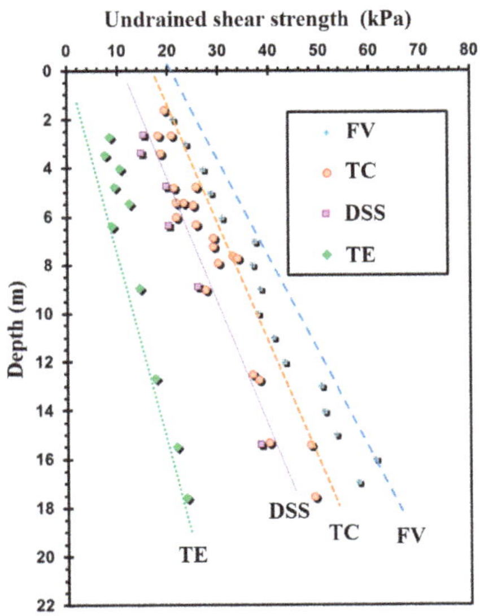

Figure 7. Groups of c_u profiles measured at the geotechnical test site of soft clays, Bothkennar in Great Britain. The tags are: FV—downhole wing probe; TC—triaxle compression (CKoUC); DSS—triaxle simple shear; TE—triaxle extension [35].

Compressibility modulus M (edometer modulus or modulus of one-dimensional compression of fine-grained soil) results from soil density due to mineralogical composition, microstructure, orientation of particles and water. The water component is responsible for the consistent state of the material, and due to the size of the mineral particles that make up clays and silts, it participates as a building element. It can overcome the mechanical stresses acting between the particles [36].

Estimating the drained consolidation parameter such as the one-dimensional modulus, i.e., the tangent oedometer compressibility modulus M can contain serious errors using the resistance parameter q_c in undrained conditions. Total undrained penetration stresses are difficult to correlate without relating them to measured pore pressures. Additionally, certain local experiences may correct the suggested correlations. The compressibility modulus M is calculated according to Equation (2) [37]:

$$M = \alpha_M \cdot q_c \tag{2}$$

where α_M is the cone factor and q_c is the probe tip resistance in MPa.

The over-consolidation ratio is defined as the ratio of the highest vertical stress in the previous period and the stress to which the soil is exposed at the observed moment. It is abbreviated as OCR (over-consolidation ratio). In relation to the over-consolidation coefficient, normally consolidated soils (OCR = 1) and over-consolidated (1 < OCR) are distinguished. In the static conical penetration CPTU test, the over-consolidation coefficient is determined from the resistance to the cone using the over-consolidation factor k_{OCR}. The expression for calculating the preconsolidation coefficient then reads [38]:

$$OCR = k_{OCR} \cdot \frac{q_{net}}{\sigma'_{v0}} = k_{OCR} \cdot \frac{q_t - \sigma_{v0}}{\sigma_{v0}} \tag{3}$$

where σ_{v0} is the vertical in situ stress in MPa; q_t is the corrected cone resistance in MPa; q_{net} represents the net cone resistance in MPa [38].

CPTU probing for this paper was carried out with a Pagani-type hydraulic indentation device (Figure 8). The interpretation of the measured data obtained by CPTU probing was carried out with the computer program CpeT-IT v.3.3.2.17 (2007), GeoLogismiki. Based on the CPTU data, the software performs interpretation by the type of soil behaviour (SBTn) and various geotechnical parameters, all following published empirical correlations according to Lunne et al. [39].

Figure 8. Self-propelled machine for hydraulic pressing of CPTU probe, Pagani TG 63-150.

3.2. *Penetration Theory of Flat Dilatometer*

Investigation with a flat dilatometer was developed in 1980 by the Italian Silvano Marchetti and is used to determine the strength and deformation properties of soil. The basis of the test is the stopping of the measuring steel probe (flat vane) at certain depths to measure specific pressures using a membrane built into the probe. In Croatia, testing is carried out according to HRN EN ISO 22476-11:2017, "Geotechnical Investigation and

Testing—Field Testing—Part 11: Flat Dilatometer Testing. The test suits clays, silts and soft rocks" [40].

By combining the measured dilatometer pressures (p_0—initial pressure at membrane displacement 0 mm and p_1- the pressure at which the membrane is displaced by 1.1 mm), the value of in situ pore water pressure u_0 and the effective value of vertical stresses σ_{v0}, other parameters can be determined, such as the value of the material index I_D, the horizontal stress index K_d and the dilatometer modulus E_D [40].

Horizontal stress index K_d is related to the in situ horizontal soil stress. In the case of a landslide, it indicates a disturbed layer corresponding to the sliding zone, where the value is $K_d \approx 2$ (1.8–2.0) [21,41,42]. Figure 9 shows an example of detecting a sliding zone by depth. At the depth at which the soil slide occurred, the value of the index K_d is 2, as shown in the profile in the lower right quarter of Figure 9.

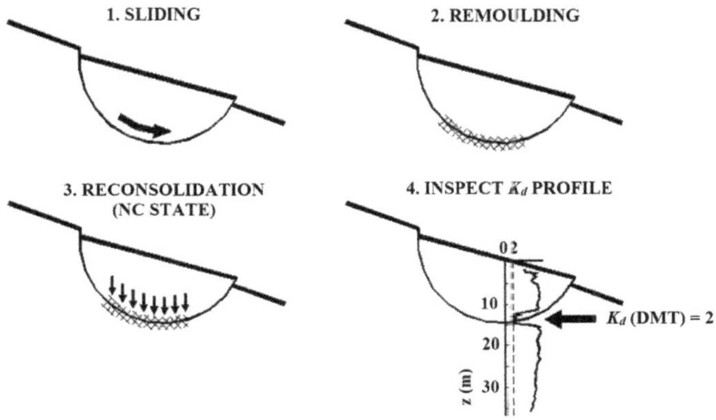

Figure 9. Detection of the sliding zone based on the index K_d [41].

The index K_d is obtained according to Equation (4) [41]:

$$K_d = \frac{(p_0 - u_0)}{\sigma'_{v0}} \tag{4}$$

where p_0 is the lifting pressure in kPa; u_0 represents the pore water pressure in kPa; σ'_{v0} is the vertical effective in-situ stress in kPa.

Over-consolidation Ratio *OCR* coefficient is highly dependent on the vertical effective stress because it shows the ratio of the stress to which the soil was exposed in the past to the stress to which the soil is exposed today. The correlation between the horizontal stress index K_d of the DMT and the over-consolidation coefficient *OCR* is of great importance. For normally consolidated materials, the value K_d = 2 corresponds to the value *OCR* = 1 [17,41,42]. The correlation mentioned above shows that the disturbed (squashed) material inside the sliding zone has lost all traces of ageing, structuring, or cementation. An increase in the K_d index indicates the presence of some form of ageing or cementation [40].

The original correlation for deriving the over-consolidation ratio *OCR* from the horizontal stress index K_d was proposed by Marchetti [43]:

$$OCR = (0.5 \cdot K_d)^{1.56} \tag{5}$$

The original correlation for determining c_u from DMT is the following [43]:

$$c_u = 0.22 \cdot \sigma_{v0} \, (0.5 \cdot K_d)^{1.56} \tag{6}$$

The compressibility modulus obtained from dilatometer sounding is determined according to the equation [41,43]:

$$M = R_M \cdot E_D \tag{7}$$

where R_M is a parameter that depends on the material index ranges I_D; E_D represents the dilatometer modulus in MPa.

In the area of the Bedekovčina landslide, two Marchetti-type DMT probes were carried out (positions shown in Figure 5). During the investigations in Kravarsko, one DMT probing was performed on each of the four landslides (Figure 6). The interpretation of the measured data obtained by DMT probing was carried out with the SDMT Elab program produced by Studio Prof. Marchetti.

3.3. Inclinometer Observations

A vertical inclinometer is a device for measuring the horizontal displacement of the normal to the axis of the pipe through which it passes. The device has a built-in sensor that works on the principle of measuring the deflection of the inclinometer tube from the vertical (normal). Inclinometers determine the size, velocity, direction and depth of sliding. This is important for understanding the causes, behaviour and rehabilitation of landslides. The results of inclinometer observations are presented as two diagrams, cumulative and incremental displacements. The diagram of cumulative displacements, analysed in this paper, shows the total (added) displacements measured at each measurement to the reference (zero measurement) [44].

To observe horizontal displacements during the phase of investigations and after the rehabilitation of the landslide in Bedekovčina, inclinometer tubes were installed in some exploration boreholes. Two inclinometers were analysed, and their positions are shown in Figure 5. In the Kravarsko municipality, four installed inclinometers were observed, one on each of the four analysed landslides. The positions of those inclinometers are shown in Figure 6. ABS inclinometer tubes manufactured by SISGEO S.rl., Milan, type model 0S131006000 ABS, with an outer diameter of 71.0 mm and an inner diameter of 60.0 mm were used at both investigation locations.

4. Prognostic Profiles Based on In Situ Geotechnical Probing Results

The following figures show results for Group 1 and Group 2 of the research area (the Bedekovčina landslide). Figure 10 shows the results of CPTU-1, and Figure 11 for CPTU-2. Figure 12 shows the overlay of the DMT-1 and DMT-2 investigation results for interpreted parameters. Figure 13 shows the raw data of DMT investigation for p_o, p_1 and interpreted data for E_d for the Kravarsko landslides (KL8, KL9 and KL10.) Figure 14 shows the interpreted parameters of DMT investigation for the Kravarsko landslides (KL8, KL9, KL10 and KL11).

Based on the results of CPTU and DMT probing, prognostic profiles of selected parameters for the sliding zone detection were created. Four parameters were chosen for analysis: undrained shear strength c_u, over-consolidation ratio OCR, compressibility modulus M and horizontal stress index K_d (only with DMT probing; CPTU probing does not provide K_d). During each analysis, potential sliding zones were identified.

As an example of the creation of prognostic profiles and the analysis of the obtained data, below is Figure 15, which shows the values of the analysed parameters by depth in the first group of the investigation area in Bedekovčina (DMT-1 and CPTU-1). Two potential sliding zones were detected at depths of 2.8 m and 5.6 m, and parameter values on the sliding zones were extracted.

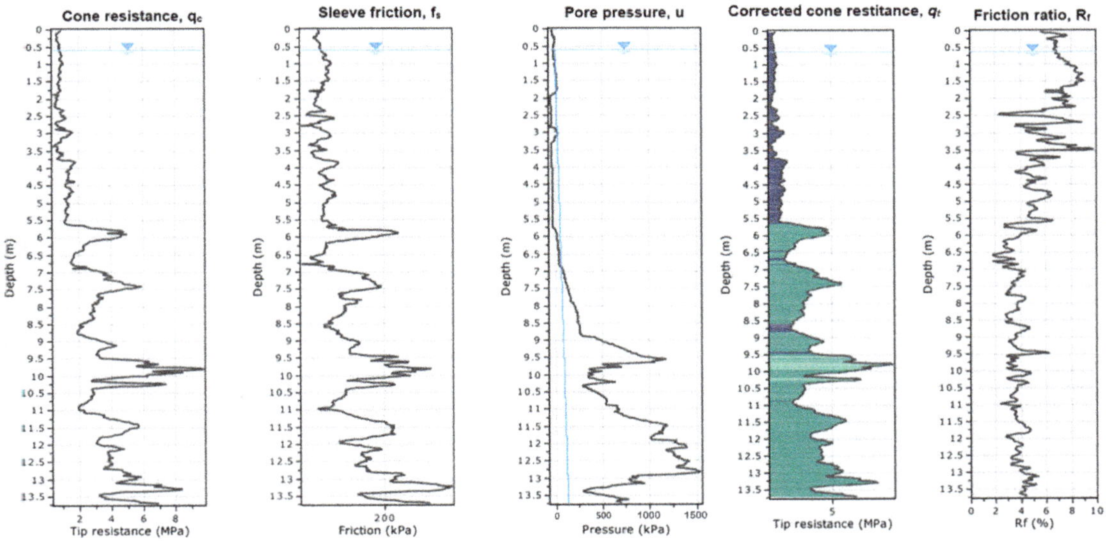

Figure 10. Results of DMT-1 investigation—Bedekovčina landslide.

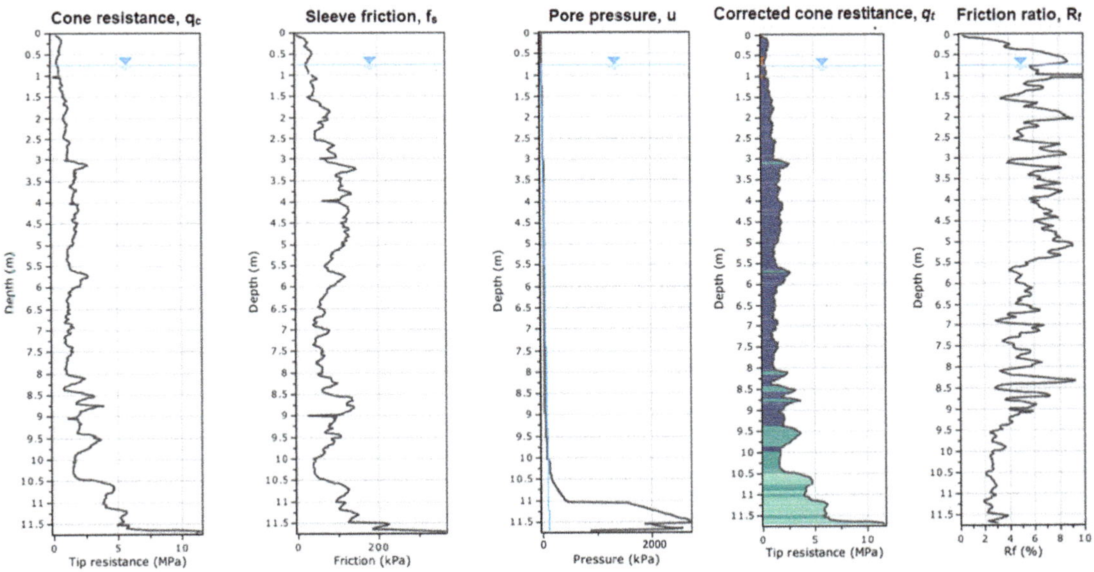

Figure 11. Results of for DMT-2 investigation—Bedekovčina landslide.

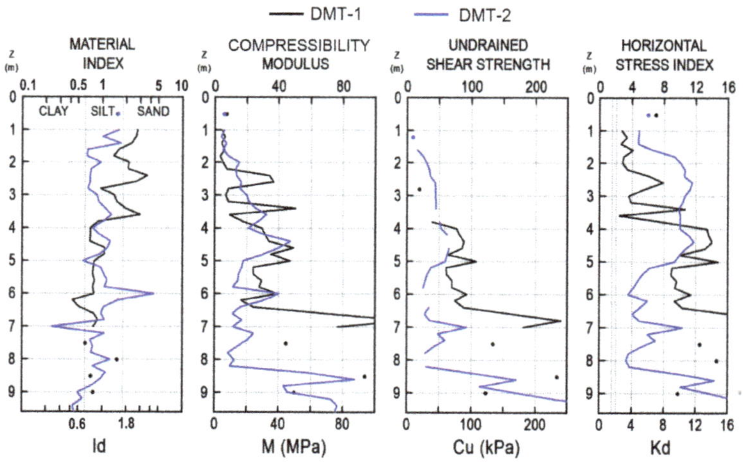

Figure 12. Overlay of research results obtained by DMT-1 and DMT-2 sounding—Bedekovčina landslide.

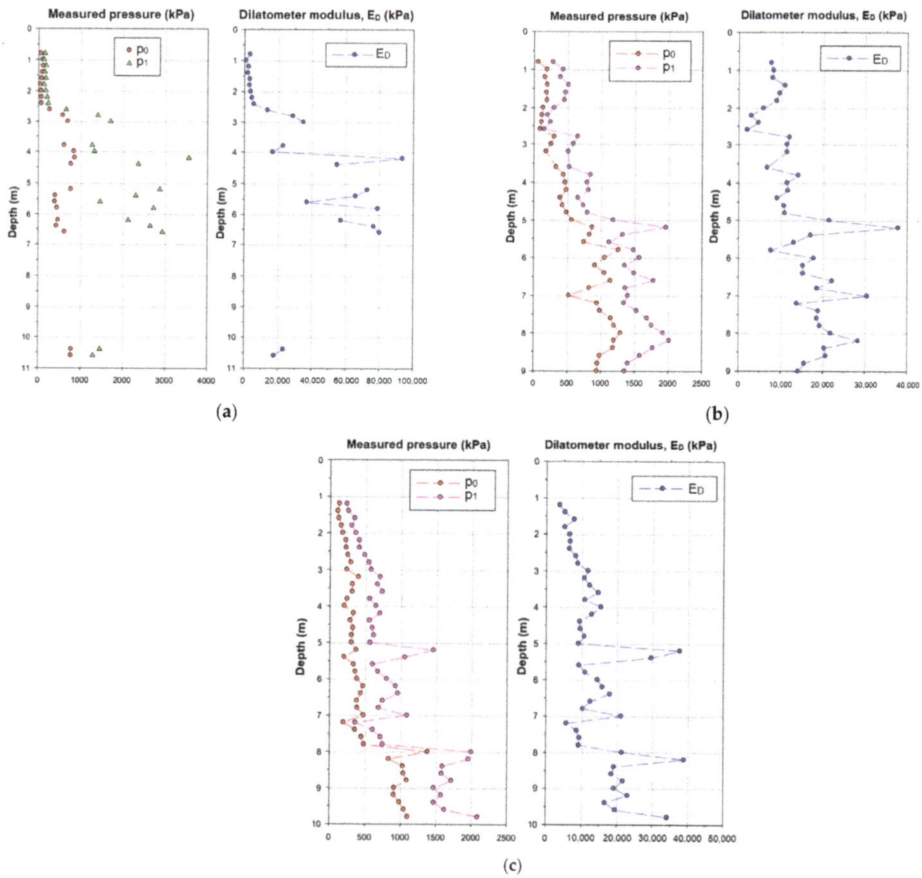

Figure 13. Results of DMT investigation for p_0, p_1, and interpreted data for E_D: (**a**) KL8; (**b**) KL9; (**c**) KL10.

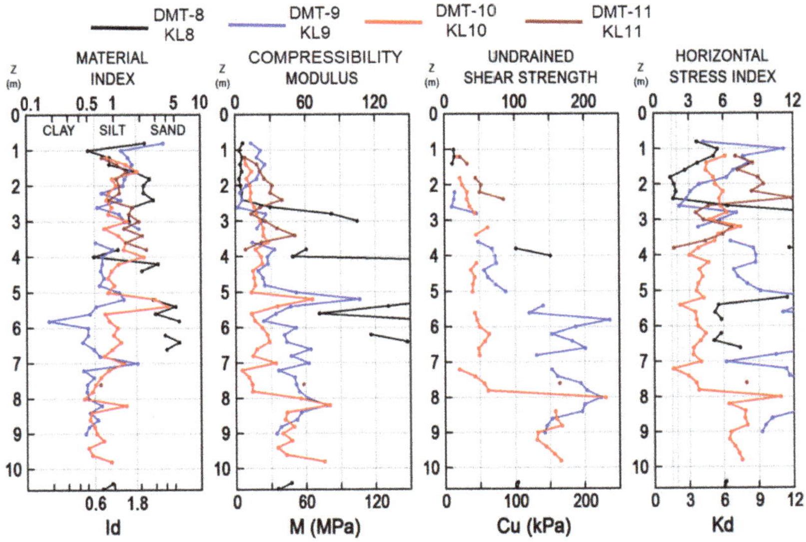

Figure 14. Results of DMT investigation on the Kravarsko landslides.

They are characterised by small values of the analysed parameters or their sudden drop, with the K_d parameter being approximately 2. Other results of the extracted parameters on the detected potential sliding zones for all studied research groups, including Group 1, are presented in Table 1.

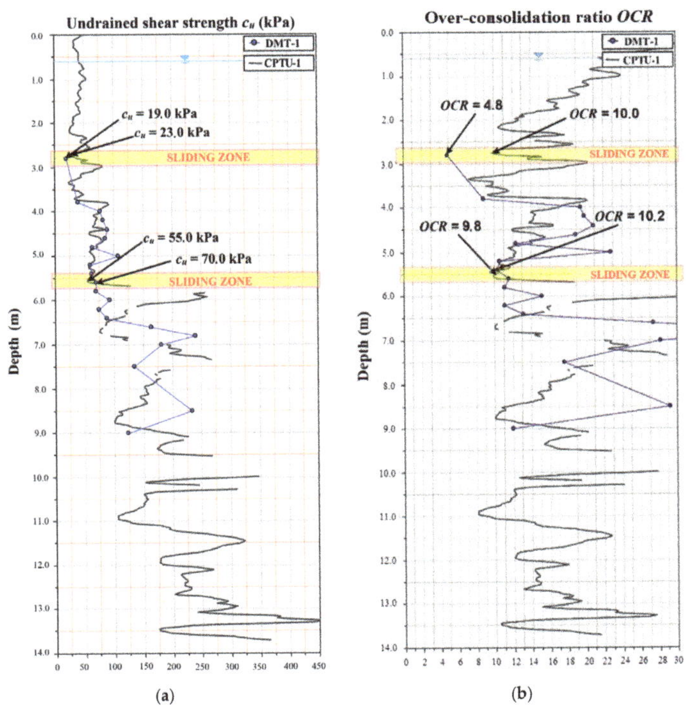

Figure 15. *Cont.*

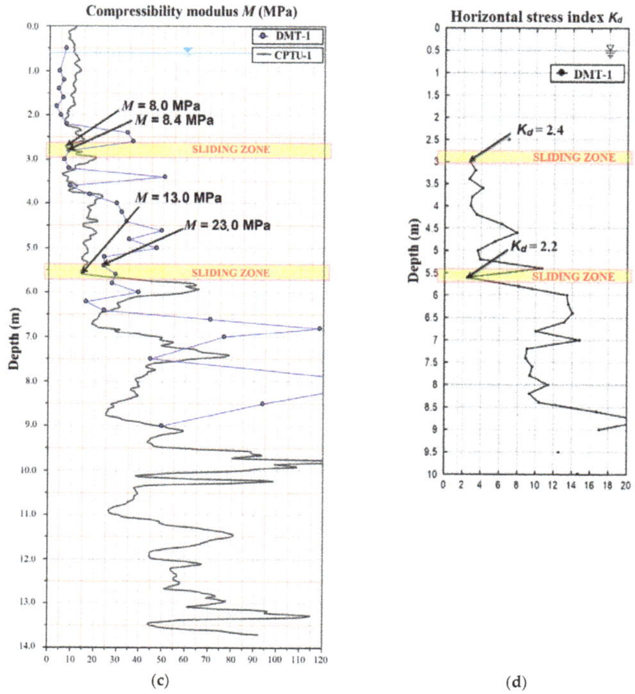

Figure 15. Prognostic profiles and the analysis of the obtained data by depth in the first group of the investigation area in Bedekovčina: (**a**) undrained shear strength c_u; (**b**) over-consolidation ratio *OCR*; (**c**) compressibility modulus *M*; (**d**) horizontal stress index K_d.

Table 1. Extracted parameters on the detected potential sliding zones for all analysed research groups.

Data Group	Location	Sliding Zone Depth (m)	GWL (m)	σ'_{v0} (kPa)	CPTU c_u (kPa)	DMT c_u (kPa)	CPTU *OCR* (-)	DMT *OCR* (-)	CPTU *M* (MPa)	DMT *M* (MPa)	DMT K_d (-)
Group 1	Bedekovčina	2.8 5.6	0.5 0.5	24.0 17.7	23.0 55.0	19.0 70.0	10.0 9.8	4.8 10.2	8.0 13.0	8.4 23.0	2.4 2.2
Group 2	Bedekovčina	6.0 8.2	0.5 0.5	17.7 16.7	47.0 26.0	25.0 28.0	7.5 3.5	3.1 2.5	16.5 8.0	11.4 9.3	3.5 3.7
Group 3	DC31, KL8	2.0	2.2	33.0	-	33.0	-	6.0	-	2.9	1.7
Group 4	DC31, KL9	2.2 3.2	2.3 2.3	37.0 45.0	- -	13.0 44.0	- -	2.0 7.1	- -	18.0 44.0	2.8 3.8
Group 5	DC31, KL10	3.0 5.4 7.2	6.8 6.8 6.8	50.0 92.0 120.0	- - -	50.0 40.0 20.0	- - -	7.0 2.4 0.8	- - -	18.0 13.4 4.9	4.2 2.2 1.6
Group 6	DC31, KL11	3.8	3.0	66.0	-	21.0	-	14.0	-	8.1	1.7

5. Verification of In Situ Probe Results with Displacements on Inclinometers

In order to verify the results of in situ probing, a comparison was made with the obtained values of horizontal displacements on inclinometers. Given that the K_d index is a good indicator of the sliding zone, diagrams were created for both investigation locations with overlaps of K_d values and horizontal displacements on the inclinometers. In Bedekovčina, in addition to DMT probing, CPTU probing was additionally performed, which does not provide a K_d index. Therefore, the values from the inclinometers were compared with the measured undrained shear strengths c_u.

Figure 16 shows selected examples of the comparison. Thus, Figure 16a shows the overlap of the depth profile of K_d index values with horizontal displacements for Group 1 data. Figure 16b shows the overlap of the same dana types but for data Group 3 (landslide

KL8 on road DC31 in Kravarsko). Figure 16c shows the overlap of undrained shear strength values obtained by CPTU probing in Bedekovčina and horizontal displacements obtained by inclinometers for Group 2 data.

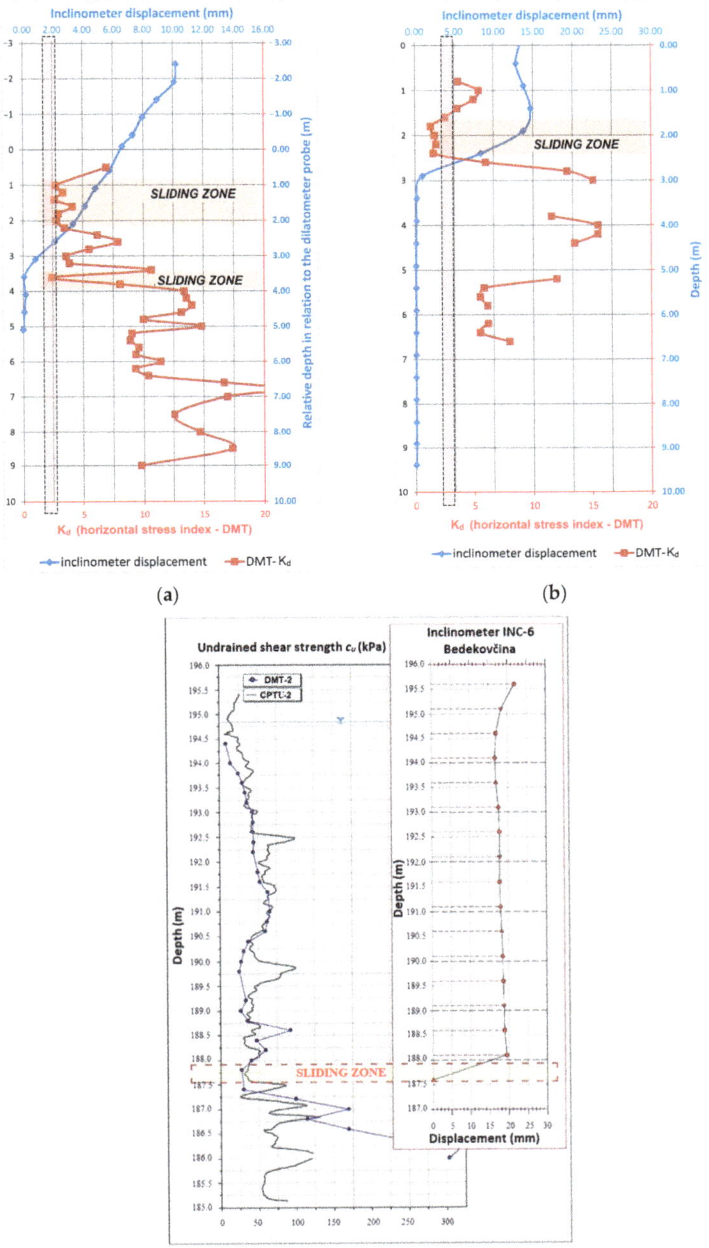

(c)

Figure 16. Comparison diagrams of in situ probing results with the obtained values of horizontal displacements on inclinometers: (**a**) Group 1 data K_d—horizontal displacements; (**b**) Group 3 data K_d—horizontal displacements; (**c**) Group 3 data c_u—horizontal displacements.

Figure 17 shows an example of the geotechnical soil profile, which refers to the KL10 landslide (for Group 5 data). The section shows the types of materials that build the slope on the KL10 landslide, as well as their properties and classification. The values and directions of movement on the inclinometers are also marked.

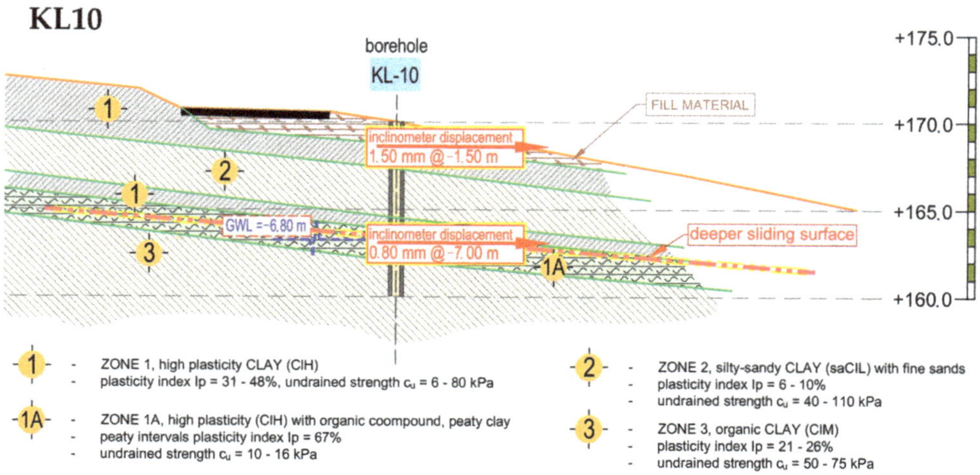

Figure 17. Geotechnical soil profile for KL10 landslide on road DC31 in Kravarsko.

6. Data Analysis

Diagrams for comparative analysis were obtained with the Microsoft Excel tool using the extracted data from Table 1. The sensitivity of a particular parameter is considered as the ability to accurately predict its state in relation to a given range (criterion). The goal of the procedure is to identify the fit of a specific parameter measured on the sliding zone in the expected ranges.

6.1. Analysis of the Horizontal Stress Index

From the diagram in Figure 18, it is evident that sliding zones can be classified in the expected range for vertical effective stresses $\sigma'_{vo} > 60$ kPa. When analysing shallow sliding zones or landslides where the groundwater level is high (conditions of low effective stresses $\sigma'_{vo} \leq 60$ kPa), the measured horizontal stress index K_d can be significantly higher than the literature range (K_d = 1.8–2.0) [21,39,40]. This finding is significant from interpreting the K_d index on shallow landslides.

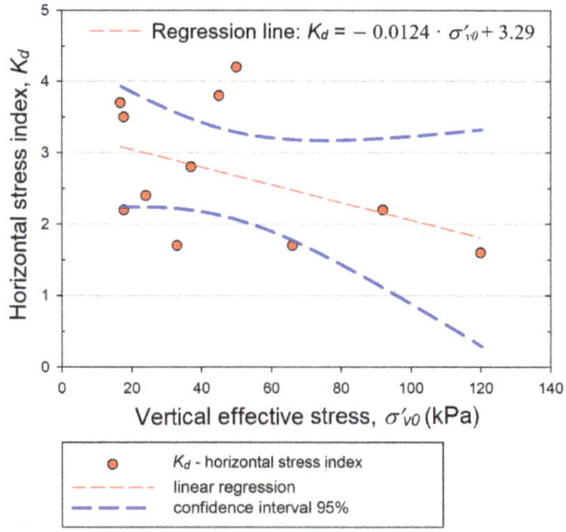

Figure 18. The relationship between the horizontal stress index K_d and the vertical effective stress σ'_{vo} on the detected sliding zones.

The fit of the results in the expected range or sensitivity is defined as the ratio of the number of samples within the criteria to the total number of samples. Data from the sensitivity analysis of the parameter K_d are presented in Table 2. Analysis of the results from this table confirmed the validity of the proposal made by Larsson et al. in 1989. They proposed a depth interval of 2.5 m as the boundary of the zone of influence on geotechnical parameters [45].

Table 2. Data analysis and parameter sensitivity for K_d.

Total Number of Samples, N	Mean Value of Data Set, $\bar{x}$	Sample Standard Deviation, "s"	Expected Range of Correct Result (Extended)	Parameter Sensitivity on the Total Data Sample	Sample Sensitivity, σ'_{vo} > 60 kPa
11	2.75	0.94	1.8–2.5	≅60%	100%

6.2. Analysis of the Undrained Shear Strength Parameter

The correlation diagram between undrained shear strength c_u and vertical effective stresses σ'_{vo} is shown in Figure 19. The data for the c_u parameter in the diagram are separated because they were collected in two ways: CPTU and DMT probing.

In geotechnical practice, it is considered that within the interval in which sliding occurs, the undrained shear strength should not exceed the value $c_u \leq 40$ kPa.

This limit can be regarded as relative because the result can be influenced by the residual strength of the surrounding soil above and below the sliding zone itself. The sensitivity analysis of the undrained shear strength for the DMT and CPTU probing is presented in Table 3.

Table 3. Data analysis and parameter sensitivity for c_u.

Total Number of Samples, N		Mean Value of Data Set, $\bar{x}$		Sample Standard Deviation, "s"		Expected Range of Correct Result		Sample Sensitivity	
DMT	CPTU	DMT	CPTU	DMT	CPTU	DMT	CPTU	DMT	CPTU
11.0	4.0	32.27	37.75	15.14	15.69	5.0–40.0	5.0–40.0	≅70%	≅50%

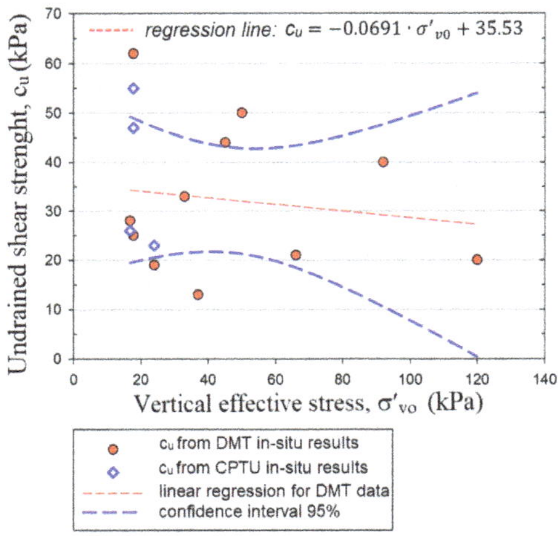

Figure 19. The relationship between the undrained shear strength c_u and vertical effective stresses σ'_{vo} on the detected sliding zones.

Comparing the two in situ methods concerning the undrained shear strength parameter measurement within the sliding zone, it can be concluded that the DMT penetration test is superior to the CPTU test.

6.3. Analysis of the Over-Consolidation Ratio

Figure 20 shows the relationship between the measured values of the over-consolidation ratio *OCR* on the sliding zones and the vertical effective stresses σ'_{vo} for two types of in situ static probes (DMT and CPTU). In geotechnical practice, it is common to consider soil over-consolidation conditions that are the result of surface soil erosion. For this reason, in most cases, shallower soil layers show a higher degree of over-consolidation, while in deeper layers, the influence of over-consolidation weakens. Based on experience, the criterion can be set that only slightly over-consolidated materials up to $OCR \leq 4$ occur within the sliding zone. The sensitivity analysis values of the *OCR* parameter are shown in Table 4.

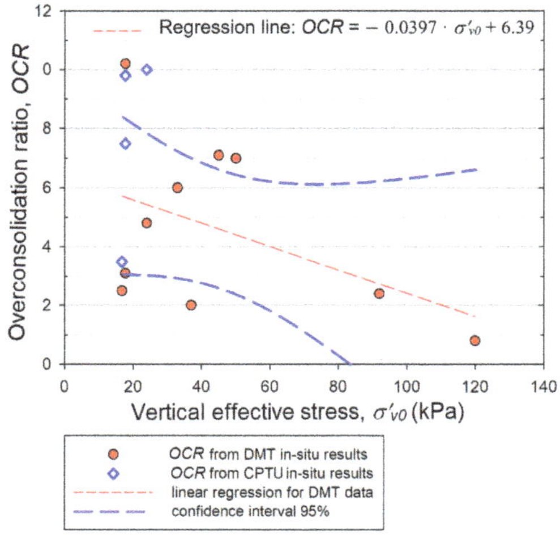

Figure 20. The relationship between the over-consolidation ratio *OCR* and vertical effective stresses σ'_{vo} on the detected sliding zones.

Table 4. Data analysis and parameter sensitivity for *OCR*.

Total Number of Samples, N		Mean Value of Data Set, $\bar{x}$		Sample Standard Deviation, "s"		Expected Range of Correct Result		Sample Sensitivity	
DMT	CPTU	DMT	CPTU	DMT	CPTU	DMT	CPTU	DMT	CPTU
11.0	4.0	4.59	7.70	2.95	3.02	1.0–4.0	1.0–4.0	≅45%	≅25%

Comparing the *OCR* values obtained by the DMT and CPTU probe on the considered samples, concerning sensitivity, the mean value of the set and the expected range of results, the DMT probe is again in advantage over the CPTU probe.

6.4. Analysis of the Compressibility Modulus

In this analysis, the basic assumption is that the soil structure has changed within the sliding zone, and consequently, the value of the compressibility modulus has changed. Based on the author's experience, the soil on the sliding zone is identified with soft compressible clay with a compressibility modulus $M < 6$ MPa. The correlation diagram between compressibility modulus M and vertical effective stresses σ'_{vo} is shown in Figure 21. Specific high values of the compressibility modulus were by no means to be expected, so it can be said that they are a surprise in this research.

The sensitivity analysis values of the *OCR* parameter from the DMT and CPTU probing are presented in Table 5. The obtained results indicate the fact that during sliding, the soil particles are oriented in the direction of movement. At the same time, this process does not negatively affect the compressibility modulus but even increases it. Thus, it can be concluded that the sliding process consolidates the soil with respect to the compressibility modulus by reorientating the soil particles in the sliding direction.

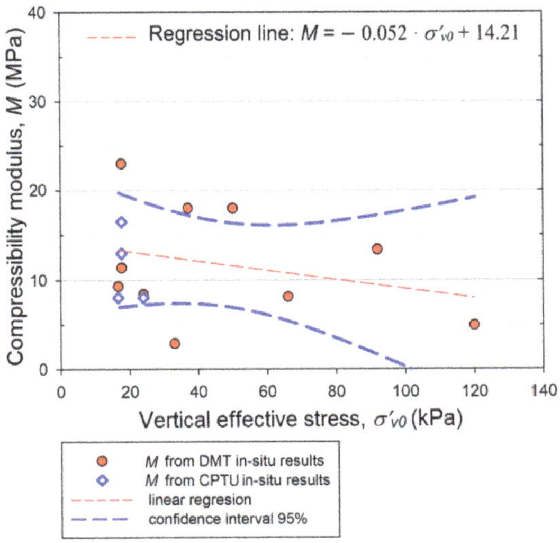

Figure 21. The relationship between the compressibility modulus M and vertical effective stresses σ'_{vo} on the detected sliding zones.

Table 5. Data analysis and parameter sensitivity for M.

Total Number of Samples, N		Mean Value of Data Set, $\bar{x}$		Sample Standard Deviation, "s"		Expected Range of Correct Result		Sample Sensitivity	
DMT	CPTU	DMT	CPTU	DMT	CPTU	DMT	CPTU	DMT	CPTU
11.0	4.0	11.74	11.37	6.35	4.20	1.0–8.0	1.0–8.0	≅35%	≅50%

7. Conclusions

Two methods, the piezocone penetration test (CPTU) and the flat dilatometer test (DMT), are most often used to investigate landslides. The main goal of this paper is to determine which of these methods is more suitable for determining the parameters of the sliding zone. For this to be possible, it was first necessary to conduct an in-situ investigation and collect data for analysis. Landslides from Bedekovčina and Kravarsko were selected for the investigation locations. DMT and CPTU probing were carried out at the relevant landslide locations, and vertical inclinometers were installed in the investigation boreholes. Later, the depths of potential sliding zones obtained by probing were verified by overlaying them with the results of horizontal displacements in inclinometers.

Four parameters (undrained shear strength c_u, over-consolidation ratio OCR, compressibility modulus M, and horizontal stress index K_d) were measured with the probes on the sliding zones, whose concordance with the expected range was assessed by means of sensitivity analysis. In the parametric analysis, relationship diagrams were created with regard to the effective vertical stress σ'_{vo} for each of the four parameters mentioned.

The characteristic scientific contributions of this research are as follows:

1. For the results of the DMT probe, a higher parameter sensitivity is generally obtained, closer to the expected range values, and the standard deviation (deviation from the arithmetic mean) is smaller.
2. The observed trend of the regression direction in the mentioned correlations has a negative direction coefficient for all considered parameters. From the above, it can be concluded that the detection of changes within the soil disturbed by sliding is more pronounced in the case of deeper sliding zones.
3. The measured parameters on deeper sliding zones fit more into the expected value ranges. This can be explained by the fact that almost every geotechnical investigated soil shows over-consolidated characteristics in the shallower layers. The described characteristic dominantly affects the results of static penetration at lower depths, i.e., at smaller amounts of vertical effective stresses.
4. The value of the effective vertical stress in sliding zones does not exceed 50 kPa, i.e., $\sigma'_{vo} \leq 50$ kPa. This means that the parameters measured by in situ penetration on the slip surface in the presented cases do not detect the slip surface within the expected parameter ranges. In the paper, the problematic case of shallow slips in the soil is particularly highlighted. Then, there are specific phenomena of high levels of groundwater or seepage water. It is shown that in such a case, the vertical effective stresses are extremely small.
5. The reorientation of soil particles caused by sliding primarily affects the lateral stresses in the soil. Regarding the stresses in the soil, the conducted research showed that in situ static penetrations do not detect soil over-consolidation in shallower layers.
6. It is confirmed that the horizontal stress index K_d is the parameter that best detects the sliding zones. However, the results show that the range of the expected value of the K_d index on the sliding zone can be extended. The range mentioned in the literature is K_d = 1.8–2.0. From the results of this paper, it is clear that the upper limit should definitely be raised. Figure 18 shows the linear regression of the results for K_d from Table 1, and it is concluded that the upper value should be increased by 0.5 to the range of K_d = 1.8–2.5. Another important conclusion can be drawn by looking at

Figure 18. As the test depth increases (the vertical effective stress), the deviation of the measured K_d values from the shown regression line decreases. This means that various surface disturbances on the landslide influence the measured K_d values at lower depths.

7. It was shown that the detection sensitivity of the K_d index is also influenced by overconsolidation in shallow soil layers. However, the results of this research show that caution should be exercised when establishing a relationship between K_d and OCR. A value of $K_d = 2$ corresponds to $OCR = 1$ only when testing normally consolidated clay. The surrounding soil significantly influences OCR, and in the zone of soil reorientation within the layer of the sliding surface, it reacts according to the type of material and the sliding process on the surface itself. Therefore, on landslides in which significant movements have already occurred, the clay in the sliding zone is completely disturbed (crushed), so the obtained values for OCR are much higher (OCR = 3–6) and can be unrealistically high ($OCR > 10$). From the above, it can be concluded that the measurement of the OCR coefficient by the DMT and CPTU methods cannot be a parameter for detecting a sliding zone.

In the end, it is pointed out once again that the penetration methods used for this research are of unequal sensitivity for the detection of the sliding zone, and it has been proven that the DMT dilatometer probe is superior to the CPTU piezocone penetrometer for the purposes of landslide characterisation.

Author Contributions: Conceptualization, K.G. and J.J.; methodology, K.G., A.B. and J.J.; software, K.G. and A.B.; validation, K.G., J.J., A.B. and S.S.; formal analysis, K.G. and A.B.; investigation, K.G. and A.B.; resources, S.S.; writing—original draft preparation, J.J. and A.B.; writing—review and editing, K.G. and S.S.; visualization, K.G. and A.B.; supervision, J.J. and S.S.; project administration, S.S.; funding acquisition, J.J. and S.S. All authors have read and agreed to the published version of the manuscript.

Funding: This research received no external funding.

Data Availability Statement: The data used in this research work are available upon request from the corresponding author.

Acknowledgments: Institutional projects GFV-IP-2024, Faculty of Geotechnical Engineering, University of Zagreb, Hallerova aleja 7, Varaždin (42000), Croatia.

Conflicts of Interest: The authors declare no conflicts of interest.

References

1. Cruden, D.M. A Simple Definition of a Landslide. *Bull. Int. Assoc. Eng. Geol.* **1991**, *43*, 27–29. [CrossRef]
2. Turner, A.K.; Schuster, R.L. *Landslides: Investigation and Mitigation (Report 247)*; National Academy Press: Washington, DC, USA, 1996.
3. Mihalić Arbanas, S.; Arbanas, Ž.; Krkač, M.; Bernat Gazibara, S.; Vivoda Prodan, M.; Đomlija, P.; Jagodnik, V.; Dugonjić Jovančević, S.; Sečanj, M.; Peranić, J. Landslide risk reduction in Croatia: Scientific research in the framework of the WCoE 2014–2017, IPL-173, IPL-184, ICL ABN. In *Advancing Culture of Living with Landslide, Proceedings of 4th World Landslide Forum, Ljubljana, Slovenija, 2017*; Sassa, K., Mikoš, M., Yin, Y., Eds.; Springer: Berlin/Heidelberg, Germany, 2017.
4. European Commision. Joint Research Centre. Lessons Learnt from Landslide Disasters in Europe. Available online: https://esdac.jrc.ec.europa.eu/ESDB_Archive/eusoils_docs/other/eur20558EN.pdf (accessed on 3 May 2023).
5. Van Den Eeckhaut, M.; Hervas, J.; Jaedicke, C.; Malet, J.P.; Montanarella, L.; Nadim, F. Statistical modelling of Europe-wide landslide susceptibility using limited landslide inventory data. *Landslides* **2012**, *9*, 357–369. [CrossRef]
6. Gunther, A.; Van den Eeckhaut, M.; Malet, J.P.; Reichenbach, P.; Hervas, J. Climatephysiographically differentiated pan-european landslide susceptibility assessment using spatial multi-criteria evaluation and transnational landslide information. *Geomorphology* **2014**, *224*, 69–85. [CrossRef]
7. Haque, U.; Blum, P.; da Silva, P.H.; Andersen, P.; Pilz, J.; Chalov, S.R.; Malet, J.P.; Auflič, M.J.; Andres, N.; Poyiadji, E.; et al. Fatal landslides in Europe. *Landslides* **2016**, *12*, 1545–1554. [CrossRef]
8. Jordanova, G.; Gariano, S.L.; Melillo, M.; Peruccacci, S.; Brunetti, M.T.; Jemec Auflič, M. Determination of Empirical Rainfall Thresholds for Shallow Landslides in Slovenia Using an Automatic Tool. *Water* **2020**, *12*, 1449. [CrossRef]
9. Corsini, A.; Borgatti, L. Mountain Landslides: Monitoring, Modeling, and Mitigation. *Geosciences* **2019**, *9*, 365. [CrossRef]
10. Tofani, V.; Bicocchi, G.; Rossi, G.; Segoni, S.; D'Ambrosio, M.; Casagli, N.; Catani, F. Soil characterization for shallow land-slides modeling: A case study in the Northern Apennines (Central Italy). *Landslides* **2017**, *14*, 755–770. [CrossRef]

11. Pilecka, E.; Szwarkowski, D.; Stanisz, J.; Blockus, M. Analysis of a Landslide on a Railway Track Using Laser Scanning and FEM Numerical Modelling. *Appl. Sci.* **2022**, *12*, 7574. [CrossRef]
12. Bogard, T.A.; Greco, R. Landslide hydrology: From hydrology to pore pressure. *WIREs Water* **2016**, *3*, 439–459. [CrossRef]
13. Schuster, R.L.; Highland, L.M. *Socioeconomic and Environmental Impacts of Landslides in the Western Hemisphere*; Open-File Report 01-276; U.S. Geological Survey: Denver, CO, USA, 2001. Available online: https://pubs.usgs.gov/of/2001/0276/report.pdf (accessed on 9 May 2023).
14. Sim, K.B.; Lee, M.L.; Wong, S.Y. A review of landslide acceptable risk and tolerable risk. *Geoenviron. Disasters* **2022**, *9*, 3. [CrossRef]
15. Miklin, L.; Podolszki, L.; Gulam, V.; Markotić, I. The Impact of Climate Changes on Slope Stability and Landslide Conditioning Factors: An Example from Kravarsko, Croatia. *Remote Sens.* **2022**, *14*, 1794. [CrossRef]
16. Grabar, K.; Strelec, S.; Špiranec, M.; Dodigović, F. CPT—DMT Correlations on Regional Soils from Croatia. *Sensors* **2022**, *22*, 934. [CrossRef]
17. Špiranec, M.; Grabar, K.; Strelec, S. Determination slip surface in situ by dilatometer (DMT) test ("Kd method"). In Proceedings of the 7th conference of Croatian Geotechnical Society with international participation, Varaždin, Croatia, 10–12 November 2016.
18. Strelec, S.; Mesec, J.; Grabar, K.; Jug, J. Implementation of in-situ and geophysical investigation methods (ERT and MASW) with the purpose to determine 2D profile of landslide. *Acta Montan. Slovaca* **2017**, *22*, 345–358.
19. Rocha, B.P.; Yacoub, J.D.; Fernandes, J.B.; Rodrigues, R.A.; Giacheti, H.L. Geotechnical site characterization by DMT and laboratory tests on an unsaturated tropical soil site for slope stability analysis. *Soils Rocks* **2023**, *46*, e2023005123. [CrossRef]
20. Potvin, J.J.; Woeller, D.; Sharp, J.; Take, W.A. Stratigraphic profiling, slip surface detection, and assessment of remolding in sensitive clay landslides using the CPT. *Can. Geotech. J.* **2021**, *59*, 1146–1160. [CrossRef]
21. Totani, G.; Marchetti, S.; Paola, M.; Calabrese, M. Use of in-situ flat dilatometer (DMT) for ground characterization in the stability analysis of slope. In Proceedings of the XIV International Conference on Soil Mechanics and Foundation Engineering, Hamburg, Germany, 6–12 September 1997; Balkema: Rotterdam, The Netherlands, 1997; Volume 1, pp. 607–610, ISBN 90-5410-891-6.
22. Google Maps. Municipality Bedekovčina and Kravarsko. Available online: https://www.google.com/maps (accessed on 9 May 2023).
23. Hrvatska.eu. Zemlja i Ljudi. Politički Ustroj. Županije. Available online: https://croatia.eu/index.php?view=article&id=30&lang=1 (accessed on 8 May 2023).
24. Hrvatski Geološki Institut; Zavod za Mineralne Sirovine. Rudarsko-Geološka Studija Krapinsko-Zagorske Županije. Zagreb. 2016. Available online: https://kzz.hr/sadrzaj/dokumenti/izvjesca-o-radu/2016/10.%20Rudarsko-geoloska-studija-KZZ.pdf (accessed on 3 May 2023).
25. Aničić, B.; Muriša, M. *Osnovna Geološka Karta SFRJ List Rogatec, M 1:100000, L 33–68 [Base Geological Map Page Rogatec]*; Geološki zavod: Ljubljana, Slovenia; i geološki zavod Zagreb: Zagreb, Croatia, 1985; pp. 1971–1981.
26. Pikija, M. *Osnovna Geološka Karta SFRJ List Sisak, M: 1: 100 000, L 33–93 [Base Geological Map Page Sisak]*; Geološki zavod Zagreb: Zagreb, Croatia, 1975–1976.
27. Juraić Mucko, J.; Mucko, I. ARHITEKTONSKI ATELIER DESET d.o.o. za Arhitekturu i Urbanizam Zagreb. Prostorni Plan Uređenja Općine Kravarsko. Plan br. 0301. Available online: https://www.kravarsko.hr/download/Obrazlozenje.pdf (accessed on 10 May 2023).
28. *HRN EN ISO 22476-1*; Geotehničko Istraživanje i Ispitivanje—Terensko Ispitivanje—1. Dio: Ispitivanje Električnim Statičkim Prodiranjem Bez Mjerenja Pornog Tlaka i s Mjerenjem Pornog Tlaka (-). Available online: https://www.iso.org/standard/57728.html (accessed on 23 November 2023).
29. Schneider, J.A.; Randolph, M.F.; Mayne, P.W.; Ramsey, N.R. Analysis of Factors Influencing Soil Classification Using Normalized Piezocone Tip Resistance and Pore Pressure Parameters. *J. Geotech. Geoenviron. Eng.* **2008**, *134*, 1569–1586. [CrossRef]
30. Robertson, P.K.; Cabal, K.L. *Guide to Cone Penetration Testing for Geotechnical Engineering*, 6th ed.; Gregg Drilling & Testing, Inc.: Signal Hill, CA, USA, 2015; pp. 38–71.
31. Mayne, P.W. Evaluating effective stress parameters and undrained shear strength of soft-firm clays from CPT and DMT. *Aust. Geomech. J.* **2016**, *51*, 27–55.
32. Lunne, T.; Christoffersen, H.P.; Tjelta, T.I. Engineering use of piezocone data in North Sea clays. In Proceedings of the 11th International Conference on Soil Mechanics and Foundation Engineering, San Francisco, CA, USA, 12–16 August 1985; Volume 2, pp. 907–912.
33. Mayne, P.W.; Peuchen, J. CPTu bearing factor Nkt for undrained strength evaluation in clays. In Proceedings of the Fourth International Symposium on Cone Penetration Testing (CPT'18), Delft, The Netherlands, 21–23 June 2018.
34. Hight, D.W.; Mcmillan, F.; Powell, J.J.M.; Jardine, R.J.; Allenou, C.P. Some characteristics of London Clay. In Proceedings of the 1st International Workshop on Characterisation and Engineering Properties of Natural Soils, Singapore, 29 November–1 December 2006; Tan, T.S., Phoon, K.K., Hight, K.W., Leroueil, S., Balkema, A.A., Eds.; Swets & Zeitlinge: Lisse, The Netherlands, 2023; pp. 851–907.
35. Bagheri, M.; Rezania, M. Geological and Geotechnical Characteristics of London Clay from the Isle of Sheppey. *Geotech. Geol. Eng.* **2021**, *39*, 1701–1713. [CrossRef]
36. Robertson, P.K. Interpretation of Cone Penetration Tests a Unified Approach. *Can. Geotech. J.* **2009**, *46*, 1337–1355. [CrossRef]
37. Mayne, P.W.; Kemper, J.B. Profiling OCR in Stiff Clays by CPT and SPT. *Geotech. Test. J.* **1998**, *11*, 139–147. [CrossRef]
38. Lunne, T.; Robertson, P.K.; Powell, J. Cone Penetration Testing in Geotechnical Practice. *Soil Mech. Found. Eng.* **1997**, *46*, 237. [CrossRef]

39. *HRN EN ISO 22476-11:2017*; Geotehničko Istraživanje i Ispitivanje—Terensko Ispitivanje—11. Dio: Ispitivanje Plosnatim Dilatometrom. 2017. Available online: https://www.iso.org/standard/66434.html (accessed on 23 November 2023).
40. Marchetti, S. The Flat Dilatometer: Design Applications. Keynote Lecture. In Proceedings of the 3rd International Geotechnical Engineering Conference, Colombo, Sri Lanka, 10–11 August 2020; Cairo University: Giza, Egypt, 2020; pp. 421–448.
41. Totani, G.; Marchetti, S.; Monaco, P.; Calabrese, M. *Use of the Flat Dilatometer Test (DMT) in Geotechnical Design*; In situ; Faculty of Engineering, University of L'Aquila: L'Aquila, Italy, 2001.
42. Löfroth, H. Undrained shear strength in clay slopes—Influence of Stress Conditions. A Model and Field Test Study. Ph.D. Thesis, Swedish Geotechnical Institute, Göteborg, Sweden, 2008. Available online: https://www.researchgate.net/publication/321528042 (accessed on 10 May 2023).
43. Marchetti, S. In Situ Tests by Flat Dilatometer. *J. Geotech. Eng. Div.* **1980**, *106*, 299–321. [CrossRef]
44. Vassallo, R.; Calcaterra, S.; D'Agostino, N.; De Rosa, J.; Di Maio, C.; Gambino, P. Long-Term Displacement Monitoring of Slow Earthflows by Inclinometers and GPS, and Wide Area Surveillance by COSMO-SkyMed Data. *Geosciences* **2020**, *10*, 171. [CrossRef]
45. Larson, W.E.; Blake, G.R.; Allmaras, R.R.; Voorhees, W.B.; Gupta, S.C. *Mechanics and Related Processes in Structured Agricultural Soils*; NATO Applied Science; Kluwer Academic Publishers: Dordrecht, The Netherlands, 1989; Volume 172, p. 273.

materials

MDPI

Review

Research Progress on the Geomechanical Properties of Block-in-Matrix Rocks

Songfeng Guo [1,2,3,*], Qianhui Wei [1,4], Shengwen Qi [1,2,3], Lei Xue [1,2,3], Bowen Zheng [1,2,3], Hongjian Wang [4], Jinxuan Li [1,5], Shuaihua Song [1,2,3], Ning Liang [1,2,3], Yu Zou [1,2,3] and Zhiquan Huang [4,6]

1 Key Laboratory of Shale Gas and Geoengineeering, Institute of Geology and Geophysics, Chinese Academy of Sciences, Beijing 100029, China; z20211020254@stu.ncwu.edu.cn (Q.W.); qishengwen@mail.iggcas.ac.cn (S.Q.); xuelei@mail.iggcas.ac.cn (L.X.); zhengbowen@mail.iggcas.ac.cn (B.Z.); lijinxuan0703@163.com (J.L.); shuaihuasong@mail.iggcas.ac.cn (S.S.); liangning@mail.iggcas.ac.cn (N.L.); zouyu@mail.iggcas.ac.cn (Y.Z.)

2 Institutions of Earth Science, Chinese Academy of Sciences, Beijing 100029, China

3 University of Chinese Academy of Sciences, Beijing 100049, China

4 College of Geosciences and Engineering, North China University of Water Resources and Electric Power, Zhengzhou 450046, China; wanghj@ncwu.edu.cn (H.W.); huangzhiquan_183@126.com (Z.H.)

5 School of Civil Engineering, Chang'an University, Xi'an 710061, China

6 Luoyang Institute of Science and Technology, Luoyang 471023, China

* Correspondence: guosongfeng@mail.iggcas.ac.cn; Tel.: +86-010-92998022

Abstract: The differences in geomechanical properties and the uncertainty in the spatial distribution of Bimrock pose significant challenges to the construction and disaster prediction of geotechnical engineering. To clarify the geomechanical characteristics of Bimrock, this paper summarizes the basic concepts and classification methods of Bimrock at home and abroad. It discusses the methods and characteristics of determining the geometric features of Bimrock blocks and explores the influencing factors and laws of failure modes and strength under different stress states of Bimrock. The study finds that the failure mode of Bimrock is mainly influenced by factors such as block proportion, degree of welding between blocks and matrix, strength ratio between blocks and matrix, and geometric properties of blocks. Among these factors, block proportion is the most significant, and the degree of welding is a controlling factor. However, due to the complexity of Bimrock structures, there is a lack of applicable methods and mechanical models for the evaluation of geomechanical characteristics of Bimrock in engineering practice. This article also explores the influence and research methods of the geological characteristics of Bimrock in slope and tunnel engineering and, finally, provides prospects for the future research trends relating to Bimrock.

Keywords: Bimrock; engineering geomechanics; rock mass strength; rock mass classification; failure modes

1. Introduction

Bimrock (mélange), initially defined as a rock mass with internal continuity that can be mapped at a scale of 1:24,000 or smaller, consists of rock fragments of different sizes, whether external or original, embedded in a matrix typically formed through shear processes such as fault rocks, volcanic mudstones, glacial deposits, and weathered rocks [1]. From an engineering perspective, it can be simplified as a block-in-matrix structure, referred to as Bimrock [2].

Bimrock has been discovered in over 60 countries [2], with the most in-depth studies conducted on the Franciscan mélange, covering approximately one-third of Northern California [3,4]. Other notable examples include the following: the Ahauser dam breccia in the northern part of the Rheno–Hercynian zone of the Variscan Mountains in Germany [5,6]; the Shale–Limestone Chaotic Complex found in the Santa Barbara open-pit mine in Italy [5–11]; and the Sille Agglomerate discovered in Koimesis Tes Panagias Church and Sekili hill settlement carved in Turkey [12,13]. China also has extensive occurrences of Bimrock, particularly in several large Bimrock belts on the Tibetan plateau, including the Luhuo–Daofu ophiolite Bimrock belt [14], the Ganzi–Litang accretionary Bimrock belt [15], the Jiali–Palong Zangbu ophiolite Bimrock belt [16], the Jinsha River accretionary complex Bimrock belt [16], the Bangong Lake–Nujiang subduction accretionary complex Bimrock belt [17], the Lancang River accretionary complex Bimrock belt [18], and the Yarlung Zangbo River ophiolite Bimrock belt [19].

Due to the prevalence of Bimrock belts in tectonically active regions, secondary disasters such as landslides are likely to occur. The complex lithology, structure, geological hydrology, and geomechanical characteristics of Bimrock pose significant challenges to engineering projects involving underground structures, tunnels [20–22], and slopes [23–25]. Notable incidents include the Bai Ge landslide in Changdu City, Tibet, and the Temigu landslide in Batang, which dammed the Jinsha River, leading to substantial downstream losses [26,27]. Therefore, it is very important to explore the physical and mechanical properties of Bimrock.

This paper analyzes indoor and outdoor mechanical tests, physical model tests, numerical simulation methods, and conclusions from both domestic and international research. It discusses the impact of factors such as the proportion of Bimrock masses, parameters related to the number and shape of blocks, and the strength ratio of Bimrock masses to matrix on their failure modes and strength. The study aims to derive general principles and explore the impact of Bimrock on engineering projects such as slopes and tunnels.

2. Definition and Classification of Bimrock

2.1. Definition of Bimrock

Raymond [3] classified mélange and several other aliases of mélange as "block-in-matrix" structures, which is one of the numerous geological terms used to represent fragmented rocks and Bimrocks, including friction carpets, wildflysch, broken formation, argille scagliose, olistostromes, mega-breccias, sedimentary chaos, mixed clays, etc., as listed by Laznicka [28] and Neuendorf et al. [29]. However, in engineering applications, the focus is more on the mechanical properties and structure of the block, and detailed classification based on rock evolution is not necessary. Therefore, Medley [2] created a neutral term "Bimrock" from Raymond's "block-in-matrix" [3] concept. Many geological processes can produce Bimrock, including sedimentary rocks (large conglomerate blocks, slope deposits, and glacial deposits); igneous rocks (clasts in volcanic ejecta, volcanic debris, lava, and xenoliths in intrusive igneous bodies); deep-seated rocks (xenoliths in intrusive igneous bodies); brittle fractured rocks in tectonic fault zones (fault breccias); rocks formed by structural ductile deformation in fold disturbance zones (mélange); rocks formed by bottom-up processes (some sheared serpentinites); and rocks formed by chemical and mechanical weathering processes (saprolite, weathered granite, and welded conglomerate).

Formally speaking, Bimrock is a type of rock mixture composed of blocks with engineering significance accompanying a weaker matrix. The engineering significance refers to the following three points.

Mechanical specifications of blocks and matrix. The key to considering a rock mass as Bimrock is the moderately strong ratio between blocks and matrix, as shown in Table 1. This ratio can be determined based on three factors: elastic modulus, internal friction angle, and uniaxial compressive strength, as described by Medley and Zekkos [30]. For cases below the lower limit shown in the table, the failure surface is more likely to cut through rock masses rather than around them.

Specification of the maximum and minimum sizes of blocks. Medley [2] defined Lc as the characteristic engineering size, similar to a reference object in a photograph for comparing the scale of images. Lc can be (1) a scale indicator of the study area, such as $\sqrt{A}$, where A is the area of the study; (2) the maximum size of blocks d_{max} measured or estimated in the study area; (3) the thickness of the failure zone below a landslide; (4) tunnel diameter; (5) foundation width or length; (6) the size of laboratory specimens. Medley [2] defined 0.05 Lc as the lower size limit to distinguish blocks from the matrix and 0.75 Lc as the upper limit for block size, d_{max}. Blocks larger than 0.75 Lc lead to a reduced proportion of matrix in local rock masses and can be considered as unfaulted rock masses composed of blocks, namely, jointed block rock masses. Blocks smaller than 0.05 Lc have negligible effects on the mechanical behavior of Bimrock and can be treated as part of the matrix. Therefore, at any scale, the threshold size between blocks and matrix is set as 0.05 Lc.

Table 1. The mechanics regulations of block and matrix.

Specification	Formula Expression	Reference
The ratio of block to matrix stiffness (E, Young's modulus)	$(E_{block}/E_{matrix}) > \sim 2.0$	Lindquist [31]
The ratio of internal friction angle between block and matrix	$(\tan \varnothing_{weakest\ block})/(\tan \varnothing_{matrix}) > 1.5 - 2.0$	Medley [2] Lindquist and Goodman [32]
The ratio of uniaxial compressive strength (UCS) between block and matrix	$UCS_{block}/UCS_{matrix} > 1.5$	Medley and Zekkos [30]

Specification of the volume ratio of blocks. The volume ratio of blocks should be greater than 25% and less than 75%. This criterion was proposed by Lindquist [31] and Lindquist and Goodman [32]: When the volume fraction of blocks is less than 25%, the overall mechanical behavior of the rock is similar to that of the matrix; if the proportion of blocks in the matrix exceeds 25%, the overall strength of the matrix will be enhanced at the characteristic engineering size of rock laboratory tests. When the volume ratio exceeds approximately 75%, the impact of the matrix on strength can be negligible, and it can be regarded as a jointed block rock mass.

However, recent research results indicate that, even if the strength contrast is not satisfied (e.g., Kahraman and Alber [5,6]; Kahraman et al. [33,34]) or the proportion of blocks does not fall between 25% and 75% (e.g., Delenne [35]; Afifipour and Moarefvand [36–38]), as long as there is a heterogeneous structure of matrix-embedded blocks, it can be defined as Bimrock, and its research methods and mechanical principles are still applicable.

2.2. Classification of Bimrock

The classification of Bimrock is currently based on the following three main methods: classification based on internal heterogeneity of blocks and the degree of disruption [39]; classification based on genesis [40]; and classification based on the degree of welding between blocks and matrix [41].

Esu [39] subdivided complex formations into three main groups (Figure 1), each with different internal heterogeneity and degrees of disruption. The first group ("A" group) includes coherent sedimentary (rock) units, ranging from well-layered and well-bedded sediments to sheared sediments. The second group ("B" group) includes sedimentary (rock) units with varying degrees of disruption, ranging from fractures in an orderly sediment (i.e., well layered; subgroup "B1") to chaotic rock units with blocked clasts (subgroup "B3"), where the blocks are embedded in a softer and sheared matrix. The last group ("C" group) includes highly heterogeneous sedimentary units consisting of weathered rock fragments embedded in clay matrix (e.g., residual soil and colluvium). The labels "residual soil and colluvium" for the "C" group indicate that these blocks are formed by weathering of parent rock and surface gravity transport (colluvium, landslides, etc.).

Festa [40] believes that the bim structure depends largely on the formation process and classifies mélanges into sedimentary mélanges, tectonic mélanges, and bottom-up mélanges [40,42–44]. Under the influence of crustal movements such as earthquakes, underground fluids entrain magma and erupt to the surface, forming bottom-up Bimrock [45]. Tectonic mélanges are mainly formed by shear processes and are characterized by an "ordered structure", i.e., scale-independent repetition, typically with elongated blocky bodies that have a lens-like or uniformly distributed aspect ratio. Sedimentary mélanges typically form in submarine debris flows, namely, landslides [46–48]. In contrast to tectonic mélanges, sedimentary mélanges have a highly disordered arrangement and different block shapes randomly distributed in a finer-grained matrix [49,50].

Altinsoy [41] adopted the Bimrock classification method proposed by Riedmüller et al. [51], categorizing Bimrock into welded Bimrock and unwelded Bimrock. For the former, the boundary strength between the matrix and blocks is considered approximately equal to the strength of the matrix, while, for the latter, the contact strength between the matrix and blocks is less than the strength of the matrix. Volcanic Bimrock formed under high-temperature and high-pressure conditions often belongs to welded Bimrock [52,53]. Figure 2 schematically illustrates the geomechanical characteristics of welded and unwelded Bimrock [54]. For unwelded Bimrock, the internal friction angle is greater than that of the matrix, and the cohesion is less than that of the matrix. Therefore, with increasing principal stress, the shear stress of Bimrock gradually changes from being less than that of the pure matrix to being greater than that of the pure matrix. The failure envelope of the mélange rocks changes with the increase in block proportion. For welded Bimrock, both cohesion and internal friction angle are greater than those of the pure matrix, so, under the same principal stress, the shear stress of Bimrock is always greater than that of the matrix. For unwelded Bimrock, the blocks can be easily separated from the matrix, but it is difficult to obtain undisturbed rock mass. For welded Bimrock, it is almost impossible to separate the block from the matrix without damaging the rock mass, but undisturbed samples can be obtained by means such as drilling and coring [54].

Based on these three simple classification methods of Bimrock, further refinements can be made. Nikolaidis and Saroglou [55] proposed a method based on six parameters (linear volume ratio of block bodies, strength of Bimrock, complexity of the matrix, complexity of Bimrock, classification of block bodies, and direction of block bodies) to characterize rock masses with bim structures, which can be directly assessed in the field. Napoli [56] considered rock type (water sensitivity), degree of anisotropy, internal disruption of complex formations, and block volume ratio (VBP) and proposed a more comprehensive classification method that allows for detailed classification based on laboratory mechanical tests and on-site photos.

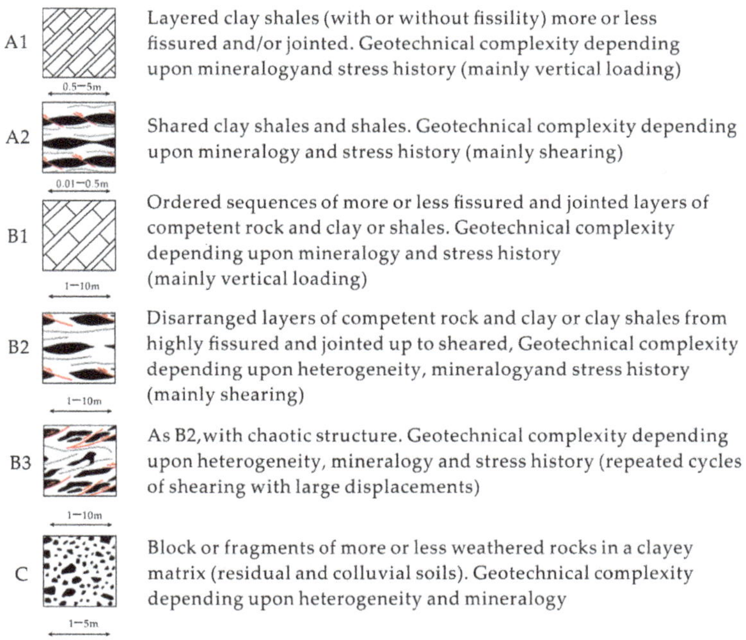

Figure 1. Classification of complex strata (modified from [56] (The red line shows the shearing action).

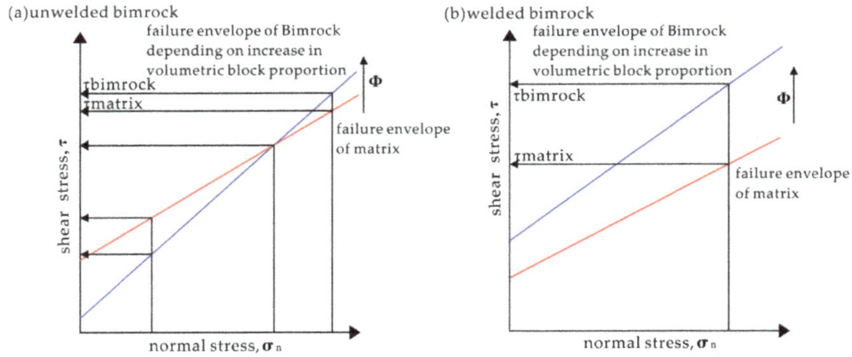

Figure 2. Schematic explanation of the mechanical behavior of a welded and unwelded Bimrock (modified from [54]).

3. Research on the Measurement Method and Uncertainty of Geometric Features of Blocks in Bimrock

Due to the unique structure of Bimrock with matrix-enclosed blocks, the content and shape parameters of block play a significant role in the mechanical properties of Bimrock. The effective determination of block content is crucial and can be expressed as volumetric block proportion (VBP), rock block proportion (RBP), or equivalent block proportion (EBP) [57]. Based on stereometric principles [58], the following four measurement methods can be used to determine the block content of Bimrock: 0D point sampling, 1D scan lines and drilling [59], 2D photo image analysis and mapping of windows [2,5,6,11–13,33,34], and 3D sieving analysis [9]. According to Delesse's principle [60], when the sample size is sufficiently large, all these measurements of different dimensions can yield the same result. Since 0D point sampling involves many uncertainties and is labor intensive, and 3D sieving analysis, while the most accurate method on a laboratory scale, is usually impossible due to the contact strength between strong block bodies and weaker matrix [61,62], the 1D and 2D methods are more widely used to determine the block volume ratio.

3.1. One-Dimensional Measurement

The block volume ratio is related to the linear ratio of blocks, the ratio of total chord length to drilling length, or the ratio of the length of block intersected by the line traversing the exposed rock outcrop (d_{mod}) to the total length of the outcrop [63]. When the rock is not obviously exposed and cannot be directly observed, one-dimensional measurement can be carried out by drilling and other methods, or, when the study area is large and it requires a lot of work to directly obtain the block proportion, one-dimensional measurement can be used for rough estimation first. However, estimating the three-dimensional block volume ratio using one-dimensional linear ratios introduces uncertainty because chord lengths and d_{mod} are generally shorter than the maximum size of block bodies (see Figure 3), and, thus, one-dimensional chord length distribution typically underestimates the actual three-dimensional block size distribution.

Although a material with strictly fractal characteristics should have a formation process that repeats itself, many scholars found that Bimrock, with a multiplicity of formative processes, also poses fractality properties. Sammis and Biegel [64] demonstrated that the size distribution of block bodies in Bimrock has self-similarity and can be described by a negative power law (fractal) distribution, where the absolute value of the negative slope of the frequency–size relationship plotted on logarithmic coordinates is called the fractal dimension (Equation (1)).

$$Y = CX^{-D} \tag{1}$$

Here, Y is the cumulative block frequency for each block size, X is the size of the block, C is a constant coefficient related to the maximum block size, and D is the fractal dimension [62,65,66].

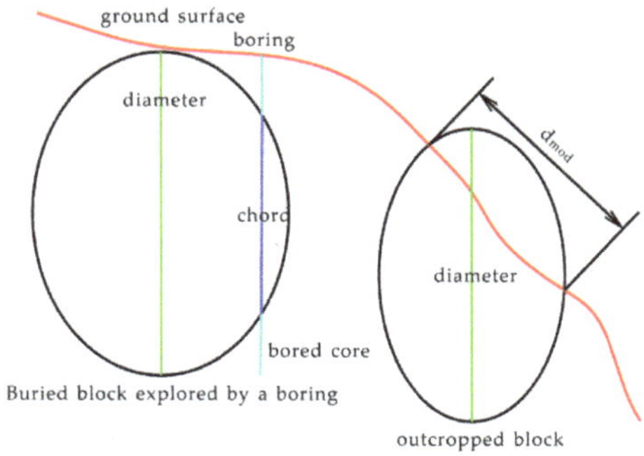

Figure 3. Relationship of d_{mod} or chord length to actual maximum size (modified from [30]).

Self-similarity means that, in study areas of different sizes, the size distribution of blocks has approximately the same fractal dimension [61]. Medley [2] measured d_{mod} for about 1900 blocks in the Franciscan mélange, covering an area range exceeding seven orders of magnitude, as shown in Figure 4. By transforming the block size distribution at different scales to a dimensionless form using $d_{mod}/\sqrt{A}$, and representing block frequencies in relative terms, it was found that the fractal dimension of block size distribution measured by one-dimensional methods ranged between 1.1 and 1.7, with an overall value of approximately 1.3. It was also observed that, when d_{mod} is $0.05\sqrt{A}$, the block frequency reaches its peak, and over 99% of block sizes are smaller than $0.75\sqrt{A}$. Based on the fractal characteristics of the Franciscan mélange, Lindquist [31] created physical models of different block arrangement directions and simulated drilling with scan lines. The results indicated that the peak of the chord length distribution is related to the high relative frequency of the smallest size block, once again confirming that the peak of the chord length distribution can reflect the size lower limit of the block relative to the matrix. Medley [67] demonstrated through physical modeling that there is almost no equivalence between chord length distribution and actual block sizes, with a pronounced tail effect. Small chord length distribution is greater than the actual block size, but this can be corrected and transformed through estimates of uncertainty factors determined by different scan line lengths and VBP.

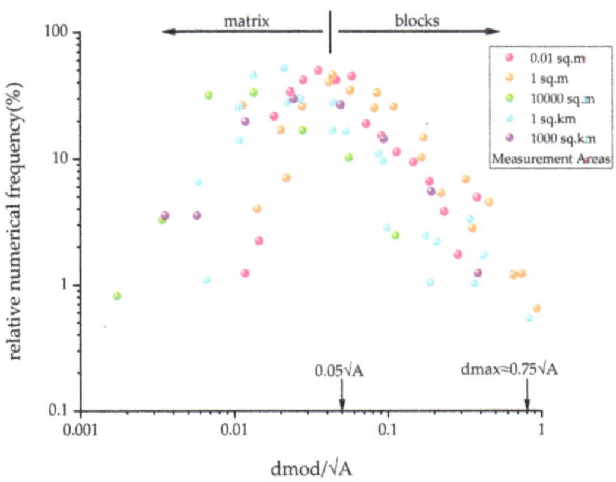

Figure 4. Histogram of logarithmic distribution of Franciscan mélange block size and relative frequency (modified from [30]).

However, the results of the above studies are empirical, and it is necessary to quantify the uncertainty of one-dimensional measurement estimates of three-dimensional block proportions. The magnitude of uncertainty depends primarily on the number and depth of boreholes [67], the VBP itself [68], and additional variables, including block size, shape, and distribution direction [69,70]. The introduction of the concept of the representative volume element (RVE) is the basis for studying the uncertainty of one-dimensional measurement [68,70]. As shown in Figure 5, for a given VBP, the RVE can be considered as a fixed-size circular block volume uniformly distributed in a rectangular matrix with an area ratio equal to the VBP. When a line of length L intersects a block with diameter D along one edge, n line segments of different lengths are randomly formed.

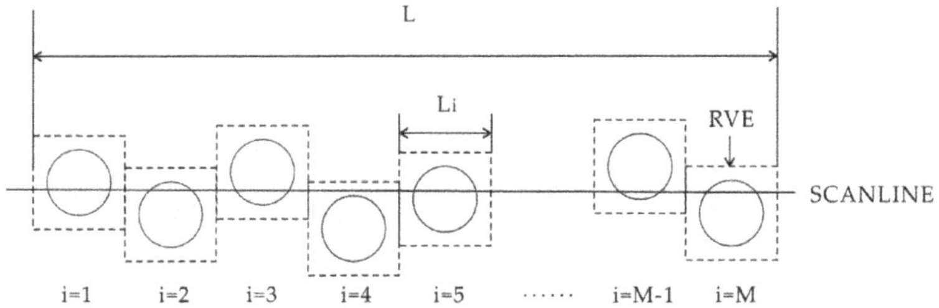

Figure 5. Repeated RVE array structure (modified from [70]).

Theoretical derivation of the estimation of VBP for random variables is carried out, and the results of the scan line are extended to the expected value of three-dimensional block estimates V_b. Simulations with different lengths of scan lines show that, as L/D increases, the histogram of estimates of VBP obtained from the RVE converges to a Gaussian distribution, and convergence is better with increasing VBP. Therefore, average block size and estimated VBP can be used to quantify uncertainty. Analytical equations for the mean and variance of estimates of VBP from scan lines and normalized coefficients of variation were derived.

Ramos-Cañón et al. [71] reconfirmed, based on 3D numerical simulations of PLB, that the variables with the greatest impact on the uncertainty of VBP determination are the number and depth of boreholes or one-dimensional scan lines and the VBP itself; the smaller the VBP, the greater the uncertainty. Lower bounds on uncertainty estimates were also quantified. Considering the anisotropy of the actual block distribution, Tien et al. [69] proposed the concept of an RVE containing an ellipse within a parallelogram. Analytical solutions were derived to quantify uncertainty in one-dimensional scan line estimates of volume fraction for anisotropic heterogeneous materials. The method includes block aspect ratio, orientation, diameter, and volume fraction and is further simplified with intercept length. Tien et al. [72] derived an analytical solution based on the concept of the RVE to quantify uncertainty in estimating the volume fraction (Vf) in inhomogeneous materials using a two-dimensional probe, showing that the uncertainty of estimates depends on block size, measured area, and Vf level. When the RVE method is transformed into the 2D domain, it produces uneven wall (edge) effects near the edges. To reduce the impact of wall effects, Tien et al. [68] trimmed four strips from the edges of an extended square (generated domain), but this method may lead to V_b dispersion. To address this, Lu et al. [70] adopted a periodic boundary method, as shown in Figure 6, each circle in figure represents a block. Where, if a block intersects one edge of the boundary, the external part of the block is mapped to its opposite side. This effectively avoids wall effects.

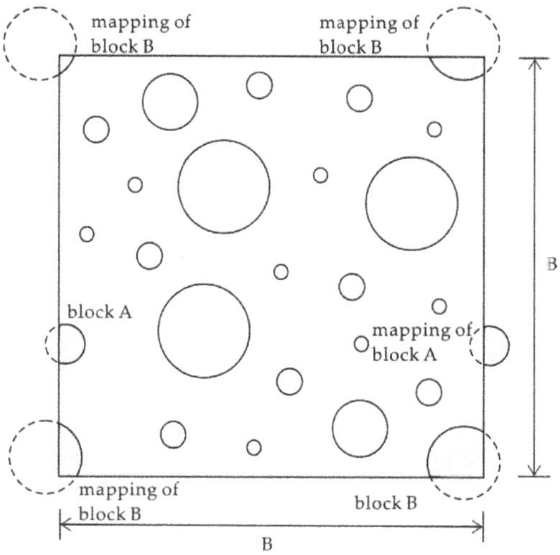

Figure 6. Conceptual diagram of periodic boundary (modified from [70]).

3.2. Two-Dimensional Measurement

Processing the two-dimensional images of Bimrock to estimate VBP requires satisfying the two following conditions to reduce uncertainty: the ratio of block length and axis being small enough and there being no obvious anisotropy. This is because, when the block distribution has obvious anisotropy, the outcrop block distribution will not reflect the internal situation of the mélange. The blocks have sufficient color contrast or clear boundaries with the matrix so that the blocks can be easily distinguished from the matrix, and the photos obtained for image processing can be more accurate in analyzing the geometric features. To determine whether there is anisotropy in the distribution of block bodies due to their random distribution, photos can be taken from multiple angles on the exposed surface of Bimrock, and the long and short axes of each block can be scanned to create a cumulative frequency distribution [52,53]. Since the distribution of blocks in Bimrock is random, it is generally considered that more than 70% of the aspect ratios of the long and short axes of blocks are less than 1.5, which can be regarded as approximately isotropic in both 2D and 3D [5,12,52].

When there is a significant contrast in color between the block and the matrix, scanning is typically performed on the circumferential surface of the core, and grayscale or RGB color photos with highlighted color tones are used for image processing [52]. Image processing methods include image classification and node counting methods. The overall goal of image classification programs is to use "supervised" or "unsupervised" classification to automatically categorize pixels into different themes [73]. Node counting methods involve overlaying a square grid on the photo and determining the percentage of each component by dividing the number of node intersections for each component by the total number of grid intersections [52]. For cores where the boundaries between the block and the matrix are not clearly defined due to severe weathering or the minerals themselves, or when the scanner accuracy cannot achieve the necessary precision due to small sizes, image processing may not automatically select blocks. In such cases, transparent film is wrapped around each core, and the boundaries are manually traced, after which the film is scanned with a scanner, and the block boundaries are redrawn for digitalization [5,12]. Digitalized images can be used not only to determine VBP but also to obtain parameters such as the number of blocks, roundness, perimeter, area, etc. [12].

According to the principles of stereovision, the uncertainty of two-dimensional measurements estimating VBP is similar to that of one-dimensional measurements and mainly comes from the sample size and the VBP itself [74]. In addition, Haneberg used Monte

Carlo computer simulations to generate ellipsoids with different eccentricities, orientations, and distributions (uniform and random) to simulate rock blocks [75]. The study explored the deviation introduced when inferring 3D block distribution from 2D projections and found that the error depends largely on block shape and orientation relative to the outcrop face. Sampling of outcrop photos can overestimate or underestimate block sizes and proportions, with errors reaching up to 80%. Napoli et al. developed Matlab code to generate 3D Bimrock models with given block size distributions and different VBP values. They assessed the deviation of 2D measurement values (ABP) from VBP as a function of the study area size and adjusted ABP measurement values to estimate VBP [24].

The resulting graph can be directly used or interpolated to analyze a series of ABP values (Figure 7) and sizes from outcrops, thereby obtaining estimates of VBP.

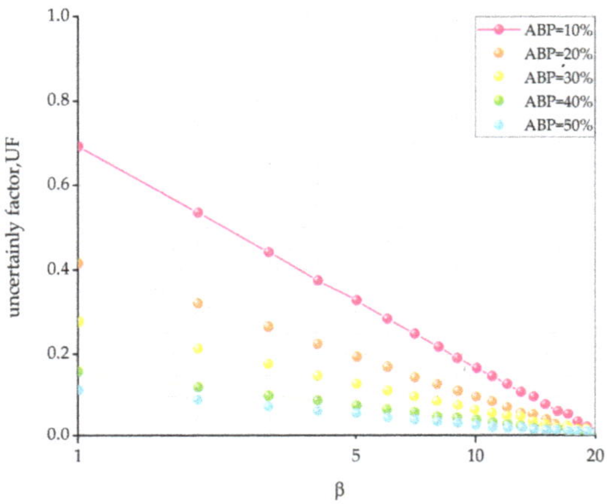

Figure 7. The uncertainty factor of the two-dimensional measurement VBP estimate is a linear fit on a semilog plot as a function of the two-dimensional measurement ABP (modified from [24]).

As different VBP values in Bimrock can also affect the overall density and seismic velocities, some researchers have attempted to estimate VBP values from density or velocity. If there is a significant density contrast between block bodies and the matrix, the overall density of the sample will be proportional to the VBP [31]. Kahraman et al. used regression analysis to study this and found that density is closely related to VBP and has statistical significance, making it suitable for estimating VBP [34]. However, VBP has no significant relationship with shear wave velocity and shows a weak linear relationship with compressional wave velocity. This is because acoustic wave velocities are not only related to VBP but also sensitive to block shape, size distribution, and the strength ratio of block bodies to the matrix. Mahdevari and Maarefvand conducted transverse and longitudinal wave tests on physically prepared samples and found that, for isotropic Bimrock with blocks, both transverse and longitudinal wave velocities increase with increasing VBP [76]. The estimated uncertainty depends on the maximum block size and fractal dimension. The transverse and longitudinal wave velocities decrease with increasing fractal dimension, but there is no significant trend between the variation of transverse and longitudinal wave velocities and the maximum block size, and the change in wave velocities with VBP is greater than with the fractal dimension.

4. Bimrock Failure Modes and Influencing Factors

4.1. Deformation and Failure Modes of Bimrock under Axial Compression

The typical failure mode of brittle homogeneous rocks in uniaxial compression tests is primarily axial splitting, where cracks propagate from top to bottom parallel or nearly parallel to the loading axis [77]. However, Bimrock, with its heterogeneous structure of matrix-embedded blocks, may exhibit interference with the original failure path. Afifipour and Moarefvand analyzed the failure process of Bimrock based on stress–strain curves (Figure 8) [38]. In the initial stage of loading, all curves show similar trends, with nonlinear convex portions in the stress–strain curves due to the specimen's porous structure and closure of voids and pre-existing cracks. This initial nonlinearity is more pronounced in samples with a higher rock block proportion (RBP).

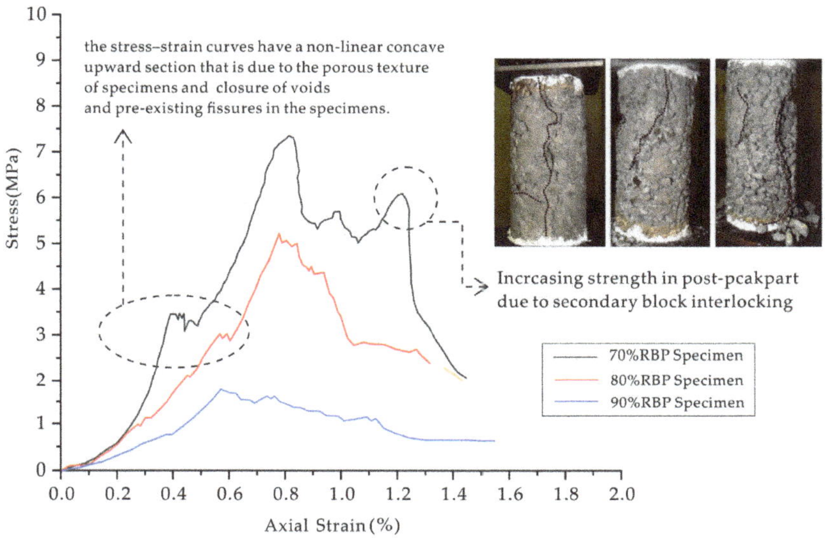

Figure 8. Typical uniaxial stress–strain curves of Bimrock model with different block mass ratios (RBP) (modified from [38]).

Subsequently, there is an approximately linear portion, followed by a nonlinear portion until the peak strength point. The initial yield strength is related to localized degradation of the bonding surface and matrix deformation, while the second yield strength is related to rock block interlocking or other interactions. Similar trends were observed in in situ shear tests conducted by Coli et al. [10] and two-dimensional numerical simulations by Xu et al. [78]. After reaching peak strength, increasing RBP in the sample transforms the deformation behavior from "rapid stress drop" to "slow stress drop with increasing strain", and post-peak behavior depends on the block proportion. In specimens with lower RBP, the post-peak region exhibits strain softening as the dominant trend, and an instantaneous hardening phase is also observed. This unique trend may be attributed to secondary interlocking caused by block displacement. The specific failure mode is mainly influenced by VBP, block geometric properties, block–matrix strength ratio, and the degree of bonding.

The strength contrast between blocks, the matrix, and their contact points is a primary factor influencing the failure mode of Bimrock. Kahraman et al. [33] found in the study of the Misis fault breccia with uneven matrix strength that, when the matrix strength is low, the failure surface follows along the block boundaries. When the matrix strength is high or when the interblock bonding strength is higher than the strength of individual blocks under loading conditions, block fracture may occur [79]. If the strength at the contact points is low, the failure surface still bypasses the blocks. Medley and Zekkos integrated three lower bounds of block–matrix strength ratio (refer to the first section) and suggested that, if the ratio is below the lower limit, the failure surface will pass through the blocks [30].

VBP, together with the degree of bonding, often determines the failure mode. Early studies have shown [31] that, when the VBP is between 25 and 75%, for unwelded Bimrock under uniaxial compression, due to the lowest strength at the contact point, the failure surface of the mixed rock extends around the block along the contact point, and the block strength has almost no effect on the failure mode Kahraman and Alber found that the

Ahauser dam breccia, a bonded mixture, still adheres to this rule, with a VBP ranging from 28.5% to 68.3% [5]. However, when the VBP is less than 28.5%, noticeable matrix failure is observed, and, when the VBP exceeds 68.3%, the failure surface penetrates the blocks. This may still be related to the specific low block strength of the Ahauser dam breccia. For columnar specimens of Bimrock with a high block proportion, as the block proportion increases, the failure mode under uniaxial compression gradually exhibits continuous zigzag failure surfaces, multiple localized shear failure surfaces, and detachment of blocks along the edges of the specimen [36]. The way of Bimrock strength changing with VBP is determined by the bond degree between blocks. When bonding, the trend of increasing or decreasing VBP depends on the curvature degree of the failure path and whether it passes through the blocks. If the block strength is less than the matrix, the strength changing trend with increasing VBP is more pronounced.

The failure mode is also related to the overall size and fractal dimension of the blocks [80]. Due to the larger surface area of coarse blocks, when the blocks stay on the failure surface, the surface may penetrate the blocks rather than extending along the periphery of the blocks. As the VBP decreases and the fractal dimension increases, the failure morphology of Bimrock transforms from block fracture to matrix and block fracture. In addition, at a low VBP and high fractal dimension, the failure mechanism is peeling; increasing VBP and decreasing fractal dimension transition to shear failure.

Xu et al. summarized the following failure modes under uniaxial compression of Bimrock with three schematic diagrams (see Figure 9, the red arrows indicate the destruction path) [78]:

1. Failure path around rock blocks. This type of failure occurs where the matrix is weaker than the rock blocks, and the block proportion is small, allowing enough relative space for the failure path to pass through. Examples include conglomerates with highly filled matrix, non-bonded, coarse-grained alluvial and colluvial deposits, mélanges [2,31], and agglomerates formed under high temperature and pressure conditions [52];
2. Branching along both sides of the block before merging. This situation also occurs where the matrix is weaker than the blocks but the block proportion is high and the relative positions or contacts between the blocks are small [36];
3. Penetration through rock blocks and matrix. This occurs when the mechanical contrast between the blocks and the matrix is small, or the blocks are weaker than the matrix [6].

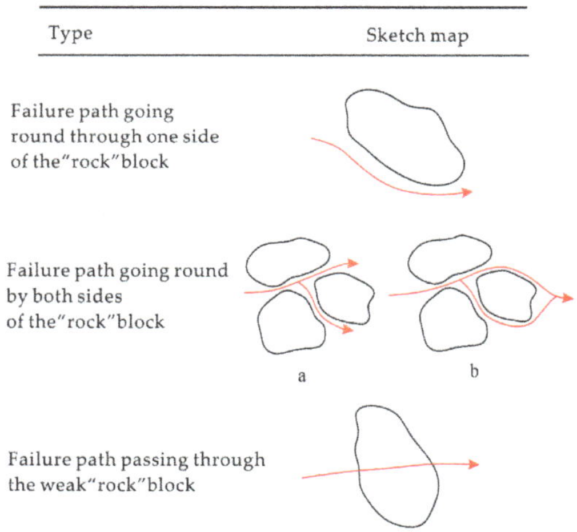

Figure 9. Schematic diagram of three possible failure modes of Bimrock (modified from [78]). (**a**) and (**b**) in the figure denote two failure paths going round two sides of block.

Considering the way the failure path passes through the block, Delenne defined the block damage value F^{pm} as a function of the product of the matrix volume fraction (ρ^{m}) and the degree of cementation between the block and matrix (σ^{pm*}), i.e., $F^{pm} \propto \sigma^{pm*}\rho^{m}$ [35]. Figure 10 shows the expansion states of the following three kinds of cracks in the Bimrocks, σ^{pm*}, normalized by the particle tensile strength σ^{p}:

1. Below the particle damage limit F^{pm}, cracks bypass the particles and propagate through the matrix, pores, or along the block–matrix contact surface due to high bulk content and weak cementation (Figure 10a);
2. If this limit is exceeded and $\rho^{m} < 0.2$, cracks will also penetrate from the solid bridge part into the block, thus concentrating stress strongly and causing surface wear of the block (Figure 10b);
3. Above this limit and where $\rho^{m} > 0.2$, cracks spread through the matrix and inside the block due to the uniform stress of the matrix, resulting in block rupture, and the proportion of damaged blocks increases with the increase in ρ^{m} and σ^{pm*} (Figure 10c).

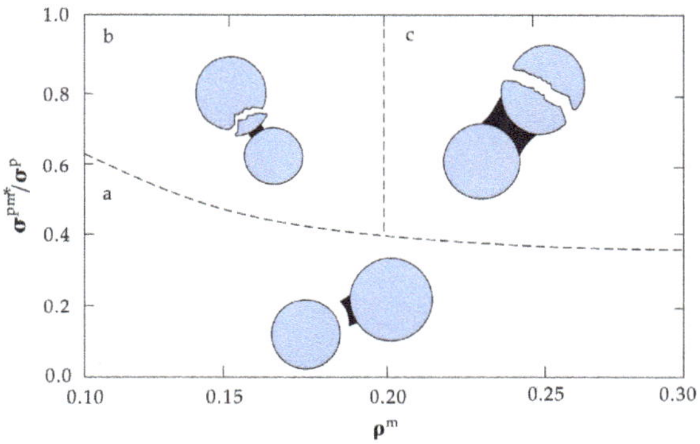

Figure 10. Schematic diagram of crack state changes with matrix volume fraction ρ^{m} and effective particle–matrix adhesion σ^{pm*} (modified from [35]). a denotes cracks sliding along block boundary, b denotes block surface fracturing and c denotes block fracturing extensively, which can be addressed in the main text in detail.

4.2. Deformation and Failure Characteristics of Bimrock under Shear Action

The mechanical behavior of Bimrock under shear action is also influenced by VBP. Avşar found that the shear failure mode is more influenced by normal stress than VBP [13]. Under all normal stresses, Bimrock undergoes strain softening after reaching the shear stress peak, and specimens with higher VBP values exhibit a smaller decrease in shear stress. Under low normal stresses, shear dilation is observed regardless of VBP values, and the dilation magnitude increases with increasing VBP. As normal stress increases, vertical displacement decreases, and primarily contraction behavior is observed. Additionally, under low normal stress, both the peak and residual strengths are significantly influenced by VBP.

5. Analysis of Factors Influencing the Mechanical Properties of Bimrock

Due to the unique characteristics of Bimrock, its mechanical properties are influenced by various factors, including the block proportion, the orientation of anisotropic block arrangement, block quantity parameters, block shape parameters, and the block–matrix strength ratio. The main areas of influence include the uniaxial compressive strength (UCS), deviatoric stress in triaxial tests, elastic modulus (E), mechanical parameters such as internal friction angle (φ) and cohesion (c), the trend of crack propagation under different deformation and failure conditions, and the relationship between acoustic wave velocity and various influencing factors. In recent years, scholars both domestically and internationally have expanded their analysis from early single-variable studies to establishing models with multiple combined parameters [12,13,34,52–54,81–83].

5.1. Block Proportion

Lindquist and Goodman indicated that when the VBP is <25%, the strength of the Bimrock is approximately equal to the strength of the matrix [32]. When the VBP is >25%, the Bimrock can be considered as a joint-filled rock mass, and the strength depends on the strength of the blocks. When the VBP is between 25% and 75%, the overall strength of the Bimrock is directly related to VBP and does not depend on the strength of the blocks. Therefore, block proportion is the most significant factor affecting the mechanical properties of Bimrock. In addition, the degree of weld is a determining factor in the trend of Bimrock with varying VBP [82].

In this paper, the data given in Table 2 are summarized, and the Bimrock is classified into welded and unwelded Bimrock for the discussion of different mechanical parameters. For ease of visual statistics, Figures 11–14 normalize the mechanical parameters shown on the vertical axis by the corresponding matrix parameters.

Table 2. Data source of research on physical and mechanical properties of mélange rocks.

Name	Block	Matrix	Welded or Unwelded	References
Hong Kong Colluvium	Coarse Gravel	Soil	No	[84]
Franciscan Mélange	Limestone, Volcanic Rock, Serpentinite, Flint, and Rare Limestone and Exotic Metamorphic Rocks	Shale, Mudstone, Siltstone, Serpentinite, or Sandstone	No	[2,61,63]
Model Experiment	Sand-Silicate Cement–Fly Ash Mixture	Bentonite–Silicate Cement Mixture	No	[31]
Ankara Agglomerates	Pink and Black Andesite Blocks	Tuff	Yes	[52–54,57]
Model Experiment	Tuff and Andesite	Calcined Gypsum, Bentonite, Cement, and Water	No	[85]
Ahauser Dam Breccia	Shale	Recrystallized Limestone	Yes	[5,6]
Misis Fault Breccia	Chalky Limestone	Fine-Grained Red Clay Rock with Iron-Rich Clay	Yes	[33,34]
Model Experiment	Lightweight Expanded Clay Aggregate (LECA) Beads	Sealant for Building Material Joints	Yes	[35]
The Shale–Limestone Chaotic Complex	Angular Limestone	Dark Gray Clay	No	[7–11]
Calaveras Dam Franciscan Mélange	Gray Sandstone, Serpentinite, Siliceous Schist, Greenstone (Altered Basalt), and Blueschist (Blueschist and Hornblende Schist)	Clay Shale	No	[86,87]
Model Experiment	River Aggregates	Portland Cement and Water Mixture	No	[36–38]
Model Experiment	Dolomite	Portland Cement, Silica Sand, and Water Mixture	Yes	[36–38]
Sille Agglomerate	Andesite	Tuff	Yes	[12,13]

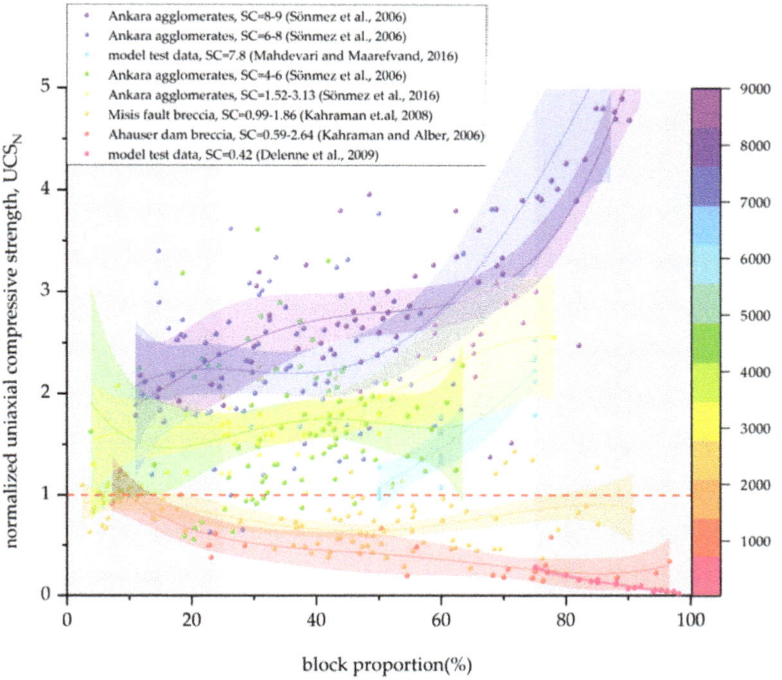

Figure 11. The relationship between the normalized uniaxial compressive strength and block proportion of welded Bimrock; SC is the block and matrix strength comparison. (The data in the figure are from [5,33,35,54,57,80]).

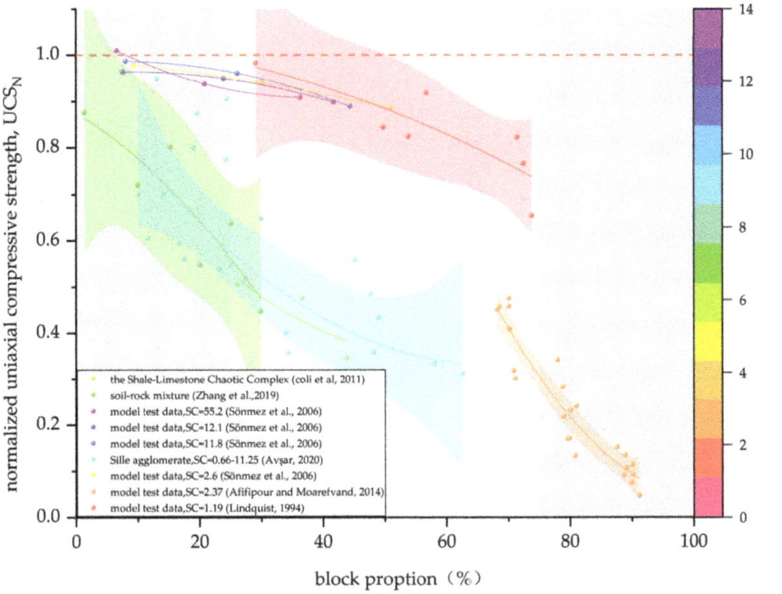

Figure 12. The relationship between the normalized uniaxial compressive strength and block proportion of unwelded Bimrock; SC is the block and matrix strength comparison. (The data in the figure are from [10,12,31,38,84,85]).

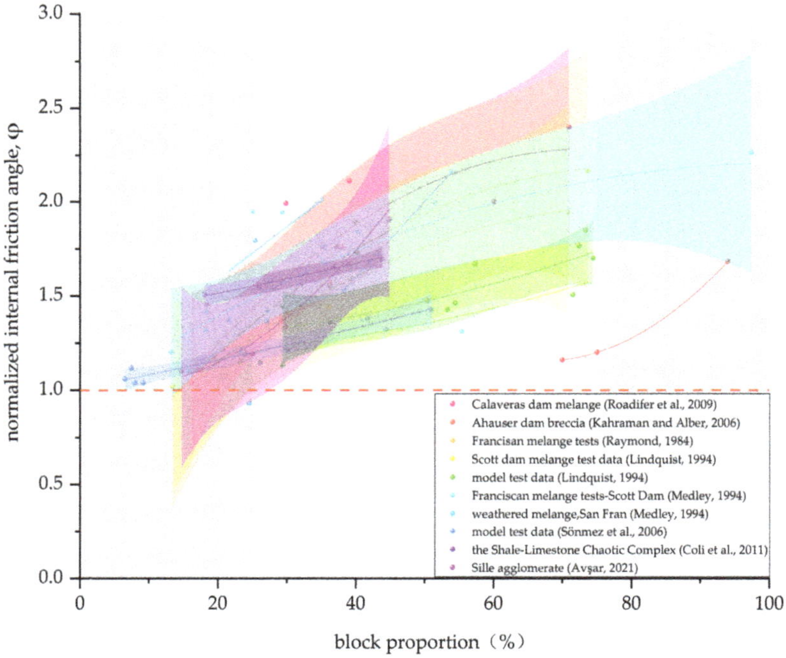

Figure 13. Correlation between normalized internal friction angle and block proportion. (The data in the figure are from [2,3,5,10,13,31,85,86]).

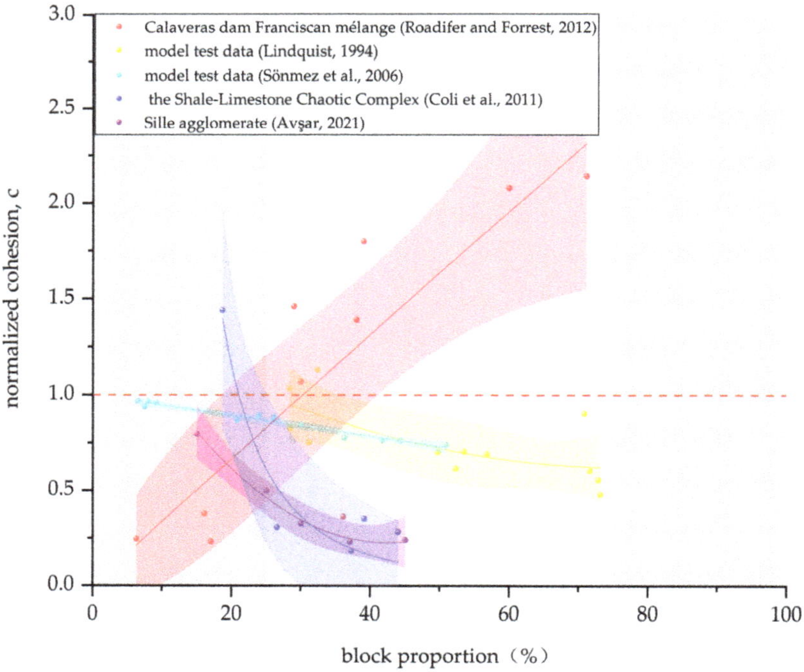

Figure 14. Correlation between normalized cohesion and block proportion. (The data in the figure are from [10,13,31,85,87]).

Figure 11 shows the trend of normalized uniaxial compressive strength (UCSN) of welded Bimrock with the change in block proportion. When the block proportion is >25%, UCSN generally increases with the increase in block proportion and is greater than 1. This indicates that the UCS of welded Bimrock is generally greater than that of pure matrix because the block strength is greater, and its presence increases the tortuosity of the failure surface, thereby increasing the overall strength of the Bimrock. Different from other welded Bimrocks, Kahraman and Alber found that the UCS of Ahauser dam breccia decreases with the increase in block proportion and is generally lower than that of the pure matrix [5]. The main reason is that the strength of the limestone matrix of Ahauser dam breccia is greater than that of the slate blocks. Similarly, for Misis fault breccia [33], there are two different trends when the VBP is below and above 25%, and there is no overall correlation, and the strength is lower than that of the pure matrix. This is because it only satisfies the bim structure, and, different from the conventional Bimrock block–matrix strength ratio, the matrix strength range is large, so there is a large difference in the overall strength of the Bimrock under the same VBP.

From Figure 11, it can also be seen that, when the block proportion is <25%, with the increase in block proportion, the overall UCS shows a downward trend or is approximately equal to the matrix strength. It can be considered that the presence of blocks destroys the heterogeneity of the matrix, causing stress concentration. Since the number is small, it is not enough to form a tortuous failure surface around the failure path. When the block proportion is >75%, UCSN still shows an increasing trend, and the increase is even greater. This is because, for samples with a high block proportion, even if the block strength is higher than the contact surface, the failure path is forced to pass through the blocks [57]. Delenne [35] found through model experiments that the uniaxial compressive strength under high block proportion has a good linear relationship with the product of the matrix volume fraction (ρm) and the block–matrix bond degree (σpm*). However, contrary to the trend of Sönmez et al. [57], this is because the LECA beads selected by Delenne [35] as the model experiment have a block strength smaller than the matrix.

Figure 12 shows the trend of normalized uniaxial compressive strength of unwelded Bimrock with the change in block proportion. In contrast to the trend in Figure 11, the strength is generally lower than that of the pure matrix, indicating that blocks have a weakening effect on the overall strength. This is because the contact surface between the blocks and the matrix is the weakest part of unwelded Bimrock, and the presence of blocks increases the contact surface area, making the failure surface easier to expand. Avşar observed that the failure surface of core samples mainly passes through the boundary between blocks and matrix, verifying that the boundary strength of Sille Agglomerate is the lowest, belonging to unwelded Bimrock, and UCS is significantly negatively correlated with VBP [12]. Therefore, changing the roughness and size distribution characteristics of the boundary length is crucial for predicting the UCS of unwelded Bimrock. The data from Afifipour and Moarefvand show that the strength of unwelded Bimrock with a high block proportion decreases more significantly with the increase in block proportion [36]. Medley proposed scale independence related to VBP [2].

Kahraman and Alber explained that the scale effect weakens with the increase in VBP [5]. Afifipour and Moarefvand also observed that, for Bimrock with a high block proportion, UCS values are independent of specimen size, and the difference between VBP and RBP is very small [36]. In addition, unwelded Bimrock shows the same decreasing trend regardless of the block proportion. This is because, at a high block proportion, unwelded Bimrock can be regarded as an extremely fragmented jointed rock mass [31]; therefore, the strength still decreases.

Block proportion also affects the shear properties of Bimrock. Figures 13 and 14 show the change in normalized internal friction angle and cohesion of Bimrock with block proportion, respectively. From Figure 13, it can be seen that, whether it is welded or unwelded Bimrock, the internal friction angle increases with the increase in block proportion because the block proportion changes the tortuosity of the failure surface of the Bimrock and is not affected by the strength of the contact surface [13,31].

Coli et al. [10] demonstrated through unconventional in situ direct shear tests that the internal friction angle has a good linear positive correlation with VBP ($R^2 = 0.85$); Avşar [13] also proved a similar linear relationship through direct shear tests, with a high correlation coefficient of 0.912. Early research [31] believed that, for unwelded Bimrock, when the block proportion is relatively small (0–10%), there is almost no increase in the internal friction angle relative to the matrix; when the block proportion is greater than 75%, the internal friction angle reaches the rest angle, and it can be assumed that there is no or very low interlocking contact between the blocks, and there is almost no increase with the increase in VBP [88]. However, in the studies of Medley [61], Coli et al. [10], and Avşar [13], it is shown that, when the block proportion is less than 30%, the internal friction angle also changes significantly. The data on Ahauser dam breccia from Kahraman and Alber [6] also show that the internal friction angle of Bimrock still changes significantly with VBP under a high block proportion (>75%). By comparing the effects of block proportion on uniaxial compressive strength and shear strength of unwelded Bimrock, it can be considered that the internal friction angle depends only on the matrix itself and the block proportion. For uniaxial compressive strength, the block direction and roundness also have a significant effect. Kalender et al. have given a good predictive model for this relationship [82].

Figure 14 shows that the cohesion changes with VBP are similar to the normalized uniaxial compressive strength. Welded Bimrock shows a positive correlation, and unwelded Bimrock shows a negative correlation. Also, the cohesion of welded Bimrock is greater than that of the pure matrix, while the opposite is true for unwelded Bimrock. Roadifer and Forrest [87] obtained a linear relationship between the cohesion of welded Bimrock and block proportion under two matrix strengths, indicating that the cohesion of welded Bimrock under weaker sheared shale matrix is smaller than that under a stronger matrix, and there seems to be a 25% threshold to distinguish between the two amplitude changes of cohesion. For unwelded Bimrock, the increase in the contact surface area between blocks and the matrix weakens the overall degree of welding, and Avşar [13] obtained a good inverse proportional linear relationship between cohesion and VBP ($R^2 = 0.857$) through direct shear tests. In the study of Coli et al. [10], cohesion suddenly decreases at a VBP around 20–25%, indicating that, when this threshold is reached, blocks begin to control the overall bonding strength of Bimrock.

From the influence of block proportion on the stress–strain curve of Bimrock, it can be seen that block proportion also affects the elastic modulus E of Bimrock. Since the blocks reduce the uniformity of stress transfer, the elastic modulus of Bimrock is usually smaller than that of the pure matrix. Figure 15 reflects the trend of elastic modulus with VBP. According to the test results of Kahraman and Alber [5], there is no obvious relationship between E and VBP, and it only has a good correlation when the VBP is between 20% and 70%, and other data points in other regions are more discrete. From the data of Sönmez et al. [54], it can be seen that the elastic modulus is weakly positively correlated with VBP. If introducing the block number Bc and the line roughness fractal dimension D, a combined parameter $CP = \frac{B_c\left(1-\left(\frac{VBP}{100}\right)\right)}{100(D-1)}$, it is found that elastic modulus and CP have a good correlation. Mahdevari and Maarefvand [76] found through reshaping specimens that, with the increase in VBP and the decrease in fractal dimension, the deformation modulus of Bimrock increases. The high correlation factor multivariate regression analysis shows that there is an indirect relationship between the deformation modulus and the fractal dimension and a direct relationship with the maximum block size and VBP.

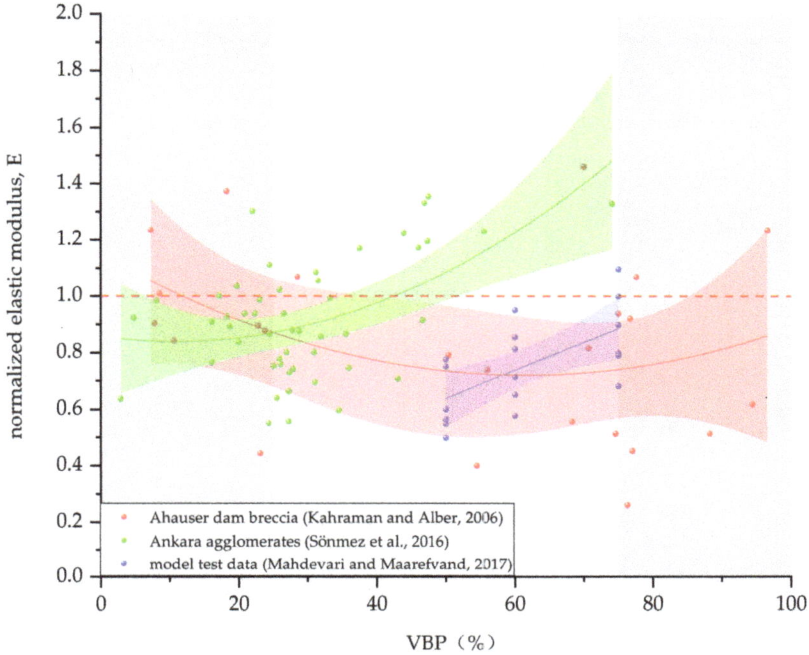

Figure 15. Relationship between the normalized elastic modulus E and the block proportion. (The data in the figure are from [5,54,76]).

5.2. Block Quantity Parameters

Block quantity parameters include the number of blocks contained in a single specimen (Number of Counted Blocks, denoted as BN, or Block Count, denoted as BC), and the block size fractal dimension (Fragmentation Fractal Dimension, denoted as DF or D).

The block size fractal dimension represents how many blocks of different sizes are present, with the formula shown in Equation (1). The block size fractal dimension of Bimrock is generally obtained using two-dimensional image processing methods. Its 3D fractal dimension is equal to the 2D fractal dimension plus one, usually between 2 and 3 [89], and sometimes exceeds 3 [90]. When the fractal dimension is greater than 3, Bimrock is mainly composed of fragmented blocks. If the fractal dimension is less than 3, larger blocks dominate. If the fractal dimension equals 3, blocks of all sizes have the same proportion, and the distribution function is based on the total weight's uniform distribution function [80].

The calculation process of the fractal dimensions (1D, 2D, and 3D) for the same shape is shown in Figure 16 [54]. Here, it is necessary to distinguish from the roughness fractal dimension (DR), which reflects the irregularity of the blocks.

Block quantity parameters are usually obtained using two-dimensional image processing methods. These two parameters are often considered in conjunction with VBP because, under the same VBP, a larger DF implies more fragmented blocks and hence a larger BN [12]. Under the same DF, a larger BN implies a larger VBP. From the data of Kahraman and Alber [5] (Figure 17), it can be seen that the more blocks there are, the smaller the UCS. This is because the block strength of Ahauser dam breccia is smaller than that of the matrix, and the possibility of the failure path passing through the blocks increases with more blocks. However, for Ankara agglomerate [54], its block strength is higher than that of the matrix, and UCS still shows a negative correlation with BN. The original paper [54] explains that BN increases the weakest structure of the block–matrix contact surface. However, Ankara agglomerate is a typical welded Bimrock, and, as shown in Figure 11, UCS increases with the increase in VBP. This also indicates that the contact surface strength is not the weakest. Therefore, this paper believes that the complexity of the failure mode of Bimrock needs to be considered. It can be considered that, under the same VBP and VBP < 75%, as BN increases, there are more fragmented blocks, and the failure path no longer follows the edge of the blocks but directly passes through the matrix. The impact of contact strength gradually decreases, so UCS decreases.

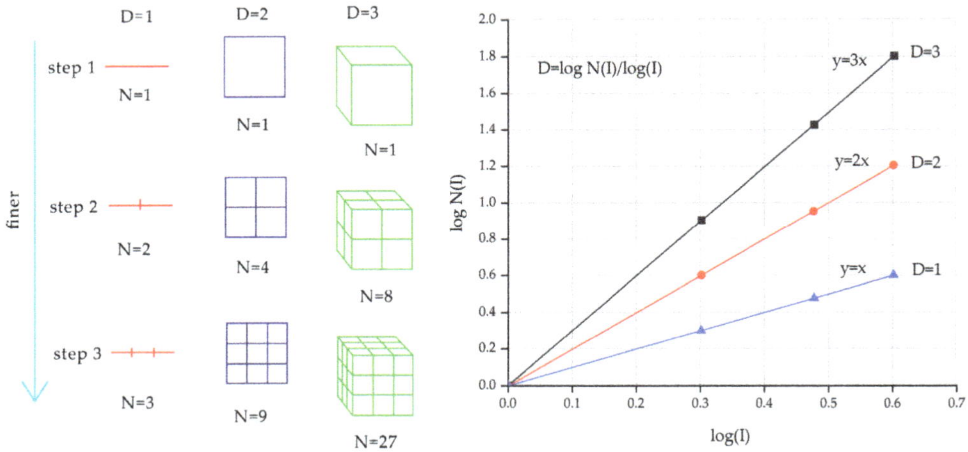

Figure 16. Determination of fractal dimensions in line (1D), square (2D), and cube (3D) (modified from [54]).

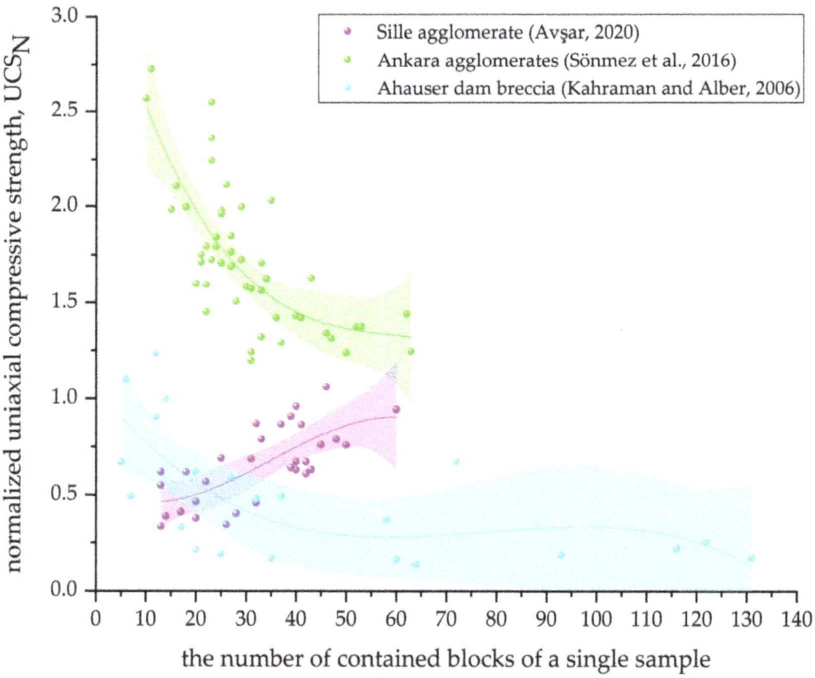

Figure 17. Relationship between the normalized uniaxial compressive strength and the number of contained blocks of a single sample. (The data in the figure are from [5,12,54]).

From the data of Mahdevari and Maarefvand [80], it can be seen that, when the block proportion is relatively high, and the fractal dimension increases from 2 to 2.5, UCS slightly increases. This is because, when the fractal dimension is small, the gaps between large blocks are filled with matrix and air. When the fractal dimension is high, small blocks can fill the gaps between large blocks, making the overall structure denser, thus increasing strength. Avşar's data [12] also reflect this pattern (Figure 18). However, when the fractal dimension increases from 2.5 to 3.5, the strength decreases. This is because, with the continuous increase in small blocks, shear failure has not yet spread to the interior of the specimen, and there is an overall transition from shear failure to peeling off along the surface of the specimen.

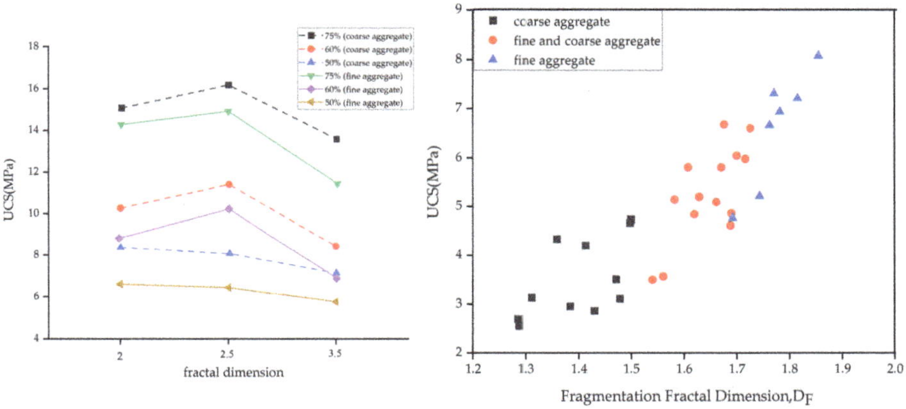

Figure 18. The relationship between the compressive strength of physical samples and the fractal dimension and the bulk thickness (modified from [12,80]).

5.3. Block–Matrix Strength Ratio

From Figure 11, it can be observed that, for welded Bimrock, there seems to be a trend where the greater the strength contrast (SC) between blocks and matrix, the larger the normalized strength of Bimrock. However, on the one hand, it is necessary to consider that, when the SC is small, the failure surface is more likely to pass through the blocks. When the SC is large, the failure path becomes more tortuous. Both scenarios can increase the overall strength. Therefore, for different Bimrock types, it is necessary to find the threshold of the SC or welding degree to determine the trend of the effect of SC on overall strength. As for non-welded Bimrock (Figure 12), the failure surface will generally bypass the blocks so SC has no effect on its strength [57].

6. Impact of Bimrock on Geological Engineering

Due to the uncertainty and variability of Bimrock, its engineering behavior is more complex than other rock-soil materials. It is necessary to consider not only block quantity and size parameters and block–matrix strength ratio, but also the arrangement of blocks relative to engineering features such as slope inclination and tunnel excavation direction, which can have significant effects on engineering projects.

6.1. Slope Engineering

Khorasani et al. first introduced the term "bimslope", defined as a Bimrock slope with matrix-embedded blocks [91]. Due to the difficulty in conducting in situ tests on Bimrock slopes, indoor model tests and numerical simulations are often used for analysis. Khorasani et al. [92,93] established an indoor physical model and conducted tests using a tilting table apparatus, measuring the face angle threshold of a bimslope during failure. Medley and Sanz [94] and Minuto and Morandi [95] developed a simple two-dimensional model simulating blocks with a horizontal rectangle, stating that a three-dimensional model is needed for Monte-Carlo-type random simulations to better approximate real-world conditions. Montoya-Araque et al. [96] and Napoli et al. [23,25] utilized the finite element method (FEM) and Limit Equilibrium Method (LEM) to simulate the uncertainty of actual block distributions, emphasizing the crucial role of random methods in Bimrock simulations.

The impact of VBP (Volume Block Proportion) on the safety factor of Bimrock slopes remains significant. Figure 19 illustrates the trend of the VBP's influence on the safety factor in various studies, with the safety factor normalized by dividing it by the safety factor of pure matrix. When block sizes and arrangements are randomly distributed, the presence of blocks increases the complexity of the failure surface, leading to an increase in the internal friction angle, regardless of the degree of welding. Khorasani et al. [92,93] found that the internal friction angle has a greater impact on the safety factor than cohesion in Bimrock slope stability. Therefore, the safety factor of Bimrock slopes is generally positively correlated with VBP, surpassing that of slopes composed solely of pure matrix [95]. Additionally, as the VBP increases, especially exceeding 20%, the arrangement of blocks becomes more complex, resulting in increased variability in the safety factor and greater uncertainty in slope stability analysis [92,93].

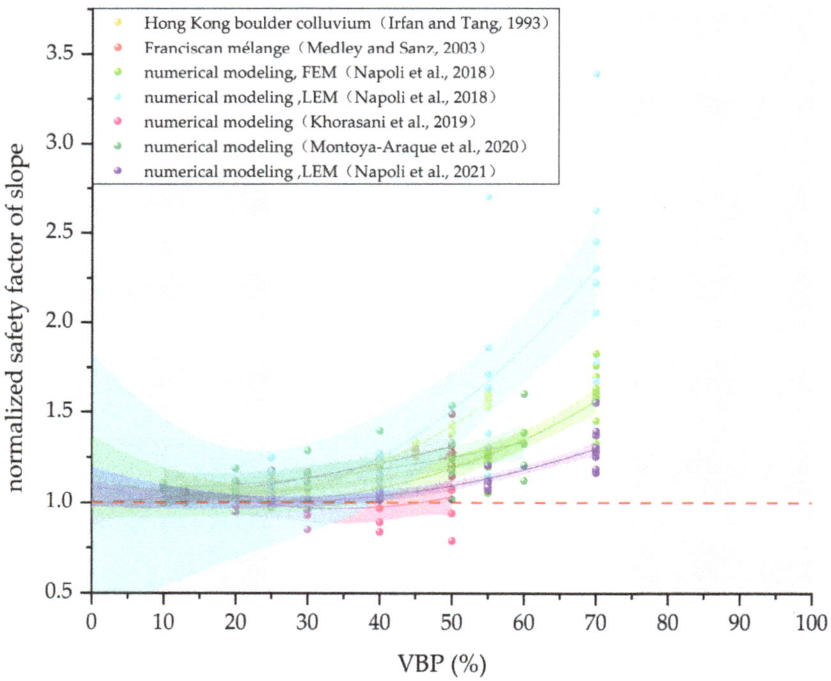

Figure 19. Relationship between normalized safety factor of slope and VBP. (The data in the figure are from [22,23,84,93,94,96]).

The degree of inclination of blocks relative to the slope angle plays a crucial role in the safety factor. If the block shapes exhibit anisotropy, and the angle between the block and the slope is relatively small (<15°), the blocks may not significantly affect slope stability [97]. Simultaneously, due to the minimal strength of the block–matrix contact surface, it might even reduce the slope safety factor [25,94]. As the angle between the block and slope increases, the impact of blocks on the safety factor also becomes more significant. Typically, when rock blocks are inclined at 0° relative to the horizontal plane, they tend to generate the most convoluted and irregular failure surface, resulting in the highest safety factor, particularly when the VBP exceeds 40% [25,97].

The degree of welding, represented by the strength of the block–matrix contact surface, also affects slope stability. To further investigate the impact of contact strength parameters, Khorasani et al. [97] defined the Interface Strength Ratio (*ISR*) with Equation (2).

$$ISR = \frac{C_1}{C_M} \cdot \frac{\tan(\varphi_1)}{\tan(\varphi_M)} \tag{2}$$

where C is cohesion, φ is the friction angle, and I (interface) and M (matrix) respectively refer to the contact surface and the matrix. The value of *ISR* varies between 0 and 1; for welded Bimrock, the *ISR* value is 1. Khorasani et al. demonstrated through the establishment of a finite element model that, under unchanged conditions, reducing *ISR* significantly decreases the stability of the bimslope [97]. Additionally, the influence of the friction angle between the contact surface and the matrix is more significant than cohesion [92,93].

In the areas where the failure surface of Bimrock slopes crosses at the engineering scale, the blocks are often unevenly distributed, leading to the presence of lean block zones and rich block zones [98]. This results in different regions of the same slope having varying internal friction angles, altering the failure path. Therefore, when modeling, the consideration of Monte Carlo random generation of blocks for the entire slope needs to account for actual working conditions.

6.2. Tunnel Engineering

The heterogeneity of Bimrock also poses significant challenges to tunnel engineering. One of the main construction issues is working under mixed face conditions that involve materials with different excavation characteristics. The pronounced spatial variations in rock mass stiffness and strength reduce the predictability, adversely affecting the project [20,99], especially when the block size approaches the excavation scale. Therefore, when tunneling in Bimrock formations, it is necessary to anticipate some complex scenarios such as: the necessity of using different excavation methods (such as drilling and blasting) in the same round; frequent stress and deformation variations in both rock mass and support, where the heterogeneity of structural Bimrock leads to stress concentration in blocks, potentially resulting in sudden brittle failure [100,101]; possibility of various failure mechanisms, including excessive over-excavation, face and crown collapse, brittle failure of hard blocks, and long-term creep of the matrix; and non-uniform distribution of groundwater, potentially leading to significant water influx, especially in fault zones [102].

On the other hand, the matrix during tunnel excavation may undergo significant, possibly time-dependent deformations, with the dominant rock mass behavior outside the excavation area influencing the response. Isolated blocks within the excavation area may contribute to the stability of the tunnel face but only temporarily affect the overall stability of the tunnel. Assessment of displacement monitoring data, such as displacement vector maps, time histories, deflections, and trend lines [103–106], indicates that displacement magnitudes generally increase with a decrease in the block volume ratio.

The position of relatively rigid blocks relative to the excavation perimeter plays a crucial role in the surrounding rock mass behavior [21]. Ignoring the presence of blocks can lead to severe overestimation of displacements and the extension and shape of plastic zones, with shear stress, displacement, and plastic zones strongly influenced by nearby blocks and their dimensions. Even when the VBP is equal to 25%, the presence of blocks may result in significant changes in rock mass strength. With an increase in VBP, this overestimation phenomenon becomes more pronounced [22]. However, lower VBP values (<25%) produce the opposite effect, attributed to block-deficient zones near the tunnel causing stress and shear strain concentration, thereby reducing rock mass stability.

7. Conclusions and Outlook

The unique structure of Bimrock, characterized by the inclusion of blocks within the matrix, along with significant differences in strength between the matrix and blocks, uneven distribution of blocks, and the anisotropy of blocks and matrix, has a profound impact on engineering construction and the prediction of induced disasters. Investigating the structure and strength characteristics of Bimrock and understanding the similarities and differences in various engineering applications are of great engineering and theoretical significance. Among these, the block volume percentage (VBP) is the most crucial factor influencing strength, but separating blocks from the matrix is often impractical. Internationally, one-dimensional scan lines and two-dimensional digital image processing methods are primarily used to obtain the VBP. However, the non-uniformity and anisotropy of block distribution in Bimrock, as well as the scale effect caused by small sample sizes, can introduce biases in VBP predictions. Therefore, corrections based on specific geological conditions are particularly important. In terms of strength analysis, since Bimrock is not a single rock mass and obtaining intact rock samples is challenging, strength determinations are typically based on uniaxial and triaxial compressive strength tests and shear strength tests on artificial Bimrock or a small number of intact samples. Predictive models are constructed or numerical simulations are performed to extend the findings to practical applications.

In specific engineering projects, slope engineering primarily considers the relative positions of block orientations to the slope surface and the distribution of blocks, while tunnel engineering focuses on variations in block distribution along the excavation direction and their relative proportions to the matrix. Both aspects can be used for short-term predictions of mechanical behavior through the development of numerical models.

Based on the analysis in this paper, it is believed that, although some research has been conducted on Bimrock, there is currently no available method or model for evaluating the engineering geological and mechanical properties of Bimrock for practical applications. Future research should address the following aspects.

Currently, predictive models for Bimrock primarily consider factors such as VBP, block shape, number, and block-to-matrix-strength ratio. The quantification of bonding degree is not taken into account in these models. The quantification of bonding degree is mostly limited to the establishment of numerical models for slopes. Therefore, the influence of welding degree should be studied.

The impact of roundness or block roughness on overall strength has been studied to some extent. Simultaneously, consideration should be given to the influence of block shape parameters on failure paths, expanding the scope to predict failure surfaces in engineering.

Presently, research on mechanical properties is confined to static responses in laboratory mechanical experiments. However, in real working conditions, dynamic loading conditions such as earthquakes, leading to stress disturbances, need to be considered.

Currently, whether in laboratory experiments or numerical simulations, the matrix is treated as a homogeneous rock mass. However, when the matrix is an anisotropic rock mass, such as laminated shale or slate, the relative positions of block arrangement and matrix anisotropy need to be considered.

When evaluating the mass quality of Bimrock, consideration should be given not only to the rock characteristics of Bimrock itself but also to the influence of structural features such as joints and faults. The feedback mechanism between the presence of blocks and the impact of structural features on rock mass quality needs further study.

Author Contributions: All authors contributed to the conception or design of the work. S.G. came up with idea of the problem. S.G. and S.Q. were responsible for the conception and design of this study. Q.W., S.G., H.W., B.Z., J.L., L.X. and S.S. organized and managed the data; S.G. and Q.W. performed the theoretical analysis and wrote the main manuscript; Q.W., S.G., J.L., L.X. and S.S. made the figure; S.G., S.Q., H.W., N.L., Y.Z. and Z.H. were responsible for the critical revision of the manuscript for important intellectual content and improved the language. All authors have read and agreed to the published version of the manuscript.

Funding: This research was funded by The Second Comprehensive Scientific Research Project on the Tibetan Plateau (no. 2019QZKK0904); the National Natural Science Foundation of China (nos. 41825018 and 42077266); and the Youth Innovation Promotion Association of Chinese Academy of Sciences (no. 2022062).

Conflicts of Interest: Author Songfeng Guo was employed by the Key Laboratory of Shale Gas and Geoengineeering, Institute of Geology and Geophysics, Chinese Academy of Sciences. The remaining authors declare that the research was conducted in the absence of any commercial or financial relationships that could be construed as potential conflicts of interest.

References

1. Greenly, E. *The Geology of Anglesey*; Geological Society of Great Britain Memoir; HMSO: London, UK, 1919; 1980p.
2. Medley, E.W. The Engineering Characterization of Mélanges and Similar Block-in-Matrix Rocks (Bimrocks). Ph.D. Thesis, University of California at Berkeley, Berkeley, CA, USA, 1994.
3. Raymond, L.A. (Ed.) Classification of mélanges. In *Melanges: Their Nature, Origin, and Significance*; Geological Society of America Special Paper; GeoScienceWorld: McLean, VA, USA, 1984; Volume 228, pp. 7–20.
4. Wakabayashi, J. The Franciscan complex, San Francisco Bay Area: A record of subduction complex processes. In *Geologic Field Trips in Northern California, Centennial Meeting of the Cordilleran Section of the Geological Society of America*; Wagner, D.L., Graham, S.A., Eds.; Division of Mines and Geology Special Publication: Sacramento, CA, USA, 1999; Volume 119, pp. 1–21.
5. Kahraman, S.; Alber, M. Estimating unconfined compressive strength and elastic modulus of a fault breccia mixture of weak blocks and strong matrix. *Int. J. Rock Mech. Min. Sci.* **2006**, *43*, 1277–1287. [CrossRef]
6. Kahraman, S.; Alber, M. Triaxial strength of a fault breccia of weak rocks in a strong matrix. *Bull. Eng. Geol. Environ.* **2008**, *67*, 435–441. [CrossRef]
7. Coli, N.; Berry, P.; Boldini, D.; Castellucci, P. Analysis of the Block Size Distribution in the Shale-Limestone Chaotic Complex (Tuscany, Italy). In Proceedings of the Symposium of American Rock Mechanics Association, San Francisco, CA, USA, 27 June–3 July 2008; 7p.
8. Coli, N.; Berry, P.; Boldini, D.; Bruno, R. In situ large size non conventional shear tests for the mechanical characterization of a bimrock in The Santa Barbara open pit mine (Italy). In Proceedings of the 43rd US Rock Mechanics Symposium and 4th US-Canada Rock Mechanics Symposium, Asheville, NC, USA, 28 June–1 July 2009. Paper No. ARMA 09-18.
9. Coli, N.; Berry, P.; Boldini, D.; Bruno, R. Investigation of block geometrical properties of the Shale–Limestone Chaotic Complex bimrock of the Santa Barbara open pit mine (Italy). In Proceedings of the Rock Engineering in Difficult Conditions, 3rd Canada–US Rock Mechanics Symposium, Toronto, CA, USA, 9–15 May 2009; 8p.
10. Coli, N.; Berry, P.; Boldini, D. In situ non-conventional shear tests for the mechanical characterisation of a bimrock. *Int. J. Rock Mech. Min. Sci.* **2011**, *48*, 95–102. [CrossRef]
11. Coli, N.; Berry, P.; Boldini, D.; Bruno, R. The contribution of geostatistics to the characterisation of some bimrock properties. *Eng. Geol.* **2012**, *137–138*, 53–63. [CrossRef]
12. Avşar, E. Contribution of fractal dimension theory into the uniaxial compressive strength prediction of a volcanic welded bimrock. *Bull. Eng. Geol. Environ.* **2020**, *79*, 3605–3619. [CrossRef]
13. Avşar, E. An experimental investigation of shear strength behavior of a welded bimrock by meso-scale direct shear tests. *Eng. Geol.* **2021**, *294*, 106321. [CrossRef]
14. Zhang, K.Z.; Li, Z.J.; Huang, C.; Xu, G. Discovery of Luhuo-Daofu ophiolite and its significance. *Sediment. Tethian Geol.* **2008**, *28*, 106–110.
15. Yan, S.T.; Zhu, L.D.; Wu, Q.S.; Li, H.; Dai, X.J. Sedientology, geochemistry and chronology of Jurassic strata in the Ganzi-Litang ophiolitic melange belt, Xinlong Area, Sichuan and its tectonic setting. *Acta Petrol. Sin.* **2022**, *38*, 1149–1168. [CrossRef]
16. Pan, G.T.; Ren, F.; Yin, F.G.; Wang, L.Q.; Wang, B.D.; Wang, D.B.; An, X.Y.; Tang, Y.; Liu, H.; Wang, Q.Y.; et al. Oceanic plate geology and key zone of Sichuan-Tibet Railway engineering geology. *Earth Sci.* **2020**, *45*, 2293–2304. [CrossRef]
17. Ren, F.; Yin, F.G.; Peng, Z.M.; Pan, G.T.; Wei, D. Zircon U-Pb age and Hf isotope characteristics of late Paleozoic diabase in the eastern part of the Bangong Hu-Nujiang subduction accretion complex and its tectonic significance. *Earth Sci.* **2022**, *29*, 164–179. [CrossRef]
18. Wang, F.; Liu, F.L.; Ji, L.; Liu, L.S. LA-ICP-MS U-Pb chronology of detrite zircons from shallow metamorphic rock series of Lancang Group in the Lancang River complex and its tectonic significance. *Acta Petrol. Sin.* **2017**, *33*, 2975–2985.
19. Zhang, X.Q.; Zhang, Z.L.; Wang, J.G.; Wang, S. A re-discussion on the formation and evolution of the Brahmaputra River junction zone. *Geol. China* **2020**, *47*, 944–970.
20. Button, E.A.; Schubert, W.; Riedmüller, G. Shallow tunneling in a tectonic melange: Rock mass characterization and data interpretation. In Proceedings of the NARMS-TAC 2002, Mining and Tunnelling Innovation and Opportunity, Toronto, ON, Canada, 7–10 July 2002; Hammah, R., Bawden, W., Curran, J., Telesnicki, M., Eds.; University of Toronto Press: Toronto, ON, USA, 2002; Volume 2, pp. 1125–1132.
21. Button, E.A.; Riedmüller, G.; Schubert, W.; Klima, K.; Medley, E. Tunnelling in tectonic melanges-accommodating the impacts of geomechanical complexities and anisotropic rock mass fabrics. *Bull. Eng. Geol. Environ.* **2004**, *63*, 109–117. [CrossRef]
22. Napoli, M.L.; Barbero, M.; Scavia, C. Tunneling in heterogeneous rock masses with a block-in-matrix fabric. *Int. J. Rock Mech. Min. Sci.* **2021**, *138*, 104655. [CrossRef]
23. Napoli, M.L.; Barbero, M.; Ravera, E.; Scavia, C. A stochastic approach to slope stability analysis in bimrocks. *Int. J. Rock Mech. Min. Sci.* **2018**, *101*, 41–49. [CrossRef]

24. Napoli, M.L.; Milan, L.; Barbero, M.; Scavia, C. Identifying uncertainty in estimates of bimrocks volumetric proportions from 2D measurements. *Eng. Geol.* **2020**, *278*, 105831. [CrossRef]
25. Napoli, M.L.; Barbero, M.; Scavia, C. Effects of block shape and inclination on the stability of melange bimrocks. *Bull. Eng. Geol. Environ.* **2021**, *80*, 7457–7466. [CrossRef]
26. Cao, P.; Li, Y.S.; Li, Z.L.; Ba, R.J.; Zhang, S.T. Geological structure characteristics and genetic mechanism of Baige landslide in Qamdu, Tibet. *Earth Sci.* **2021**, *46*, 3397–3409.
27. Chen, J.; Cui, Z.J.; Chen, R.C.; Zheng, X.X. Genesis and evolution of Temigu landslide barrier lake in the upper reaches of Jinsha River. *Earth Sci.* **2021**, *28*, 85–93. [CrossRef]
28. Laznicka, P. Breccias and Coarse Fragmentites—Petrology, Environments, Associations, Ores. In *Developments in Economic Geology*; Elsevier: Amsterdam, The Netherlands, 1988; Volume 25, 832p.
29. Neuendorf, K.K.E.; Mehl, J.P., Jr.; Jackson, J.A. *Glossary of Geology*, 5th ed.; American Geological Institute: Alexandria, VA, USA, 2005; 800p.
30. Medley, E.W.; Zekkos, D. Geopractitioner approaches to working with antisocial mélanges. Invited paper. In *Mélanges: Processes of Formation and Societal Significance*; Geological Society of America Special Paper; Wakabayashi, J., Dilek, Y., Eds.; GeoScienceWorld: McLean, VA, USA, 2011; Volume 480, pp. 261–277.
31. Lindquist, E.S. The Strength and Deformation Properties of Melange. Ph.D. Thesis, University of California at Berkeley, Berkeley, CA, USA, 1994.
32. Lindquist, E.S.; Goodman, R.E. The strength and deformation properties of a physical model melange. In Proceedings of the 1st North American Rock Mechanics Conference (NARMS), Austin, TX, USA, 1–3 June 1994; Nelson, P.P., Laubach, S.E., Eds.; Balkema: Rotterdam, The Netherlands, 1994.
33. Kahraman, S.; Alber, M.; Fener, M.; Gunaydin, O. Evaluating the geomechanical properties of Misis fault breccia (Turkey). *Int. J. Rock Mech. Min. Sci.* **2008**, *45*, 1469–1479. [CrossRef]
34. Kahraman, S.; Alber, M.; Fener, M.; Gunaydin, O. An assessment on the indirect determination of the volumetric block proportion of Misis fault breccia (Adana, Turkey). *Bull. Eng. Geol. Environ.* **2014**, *74*, 899–907. [CrossRef]
35. Delenne, J.Y.; Topin, V.; Radjai, F. Failure of cemented granular materials under simple compression: Experiments and numerical simulations. *Acta Mech.* **2009**, *205*, 9–21. [CrossRef]
36. Afifipour, M.; Moarefvand, P. Failure patterns of geomaterials with block-in-matrix texture: Experimental and numerical evaluation. *Arab. J. Geosci.* **2013**, *7*, 2781–2792. [CrossRef]
37. Afifipour, M.; Moarefvand, P. Experimental study of post-peak behavior of bimrocks with high rock block proportions. *J. Cent. South Univ.* **2014**, *21*, 761–767. [CrossRef]
38. Afifipour, M.; Moarefvand, P. Mechanical behavior of bimrocks having high rock block proportion. *Int. J. Rock Mech. Min. Sci.* **2014**, *65*, 40–48. [CrossRef]
39. Esu, F. *Behaviour of Slopes in Structurally Complex Formations*; General Report, Session IV, Proceedings of the International Symposium; The Geotechnics of Structurally Complex Formations: Capri, Italy, 1977; pp. 292–304.
40. Festa, A. Tectonic, sedimentary, and diapiric formation of the Messinian melange: Tertiary Piedmont Basin (northwestern Italy). In *Melanges Processes of Formation and Societal Significance*; Wakabayashi, J., Dilek, Y., Eds.; Geological Society of America: Boulder, CO, USA, 2011; Volume 480, pp. 215–232. [CrossRef]
41. Altinsoy, H. A Physical Based Model Investigation for Determination of Shear Strength of Block in Matrix Rocks. Master's Thesis, Geological Engineering Department, Hacettepe University, Ankara, Turkey, 2006.
42. Festa, A.; Dilek, Y.; Pini, G.A.; Codegone, G.; Ogata, K. Mechanisms and processes of stratal disruption and mixing in the development of mélanges and broken formations: Redefining and classifying mélanges. *Tectonophysics* **2012**, *568–569*, 7–24. [CrossRef]
43. Festa, A.; Dilek, Y.; Codegone, G.; Cavagna, S.; Pini, G.A. Structural anatomy of the Ligurian accretionary wedge (Monferrato, NW Italy), and evolution of superposed mélanges. *Bulletin* **2013**, *125*, 1580–1598. [CrossRef]
44. Festa, A.; Barbero, E.; Remitti, F.; Ogata, K.; Pini, G.A. Mélanges and chaotic rock units: Implications for exhumed subduction complexes and orogenic belts. *Geosyst. Geoenviron.* **2022**, *1*, 100030. [CrossRef]
45. Barber, T. Mud diapirism: The origin of mélanges: Cautionary tales from Indonesia. *J. Asian Earth Sci.* **2013**, *76*, 428–438. [CrossRef]
46. Ogata, K.; Festa, A.; Pini, G.A.; Pogačnik, Ž.; Lucente, C.C. Substrate deformation and incorporation in sedimentary mélanges (olistostromes): Examples from the northern Apennines (Italy) and northwestern Dinarides (Slovenia). *Gondwana Res.* **2019**, *74*, 105–129. [CrossRef]
47. Ogata, K.; Festa, A.; Pini, G.A.; Pogačnik, Ž.; Lucente, C.C. Submarine landslide deposits in orogenic belts: Olistostromes and sedimentary mélanges. In *Submarine Landslides: Subaqueous Mass Transport Deposits from Outcrop to Seismic Profiles*, 1st ed.; Geophysical Monograph 247; Ogata, K., Festa, A., Pini, G.A., Eds.; American Geophysical Union: Washington, DC, USA; John Wiley & Sons Inc.: Hoboken, NJ, USA, 2020; pp. 3–26. [CrossRef]
48. Pini, G.A.; Ogata, K.; Camerlenghi, A.; Festa, A.; Lucente, C.C.; Codegone, G. Sedimentary mélanges and fossil mass-transport complexes: A key for better understanding submarine mass movements? In *Submarine Mass Movements and their Consequences Advances in Natural and Technological Hazards Research*; Mosher, D.C., Shipp, R.C., Moscardelli, L., Chaytor, J.D., Baxter, C.D.P., Lee, H.J., Urgeles, R., Eds.; Springer Science Business Media: Berlin/Heidelberg, Germany, 2012; Volume 31, pp. 585–594.

49. Festa, A.; Ogata, K.; Pini, G.A.; Alonso, J.L. Origin and significance of olistostromes in the evolution of orogenic belts: A global synthesis. *Gondwana Res.* **2016**, *39*, 180–203. [CrossRef]
50. Festa, A.; Pini, G.A.; Ogata, K.; Dilek, Y. Diagnostic features and field-criteria in recognition of tectonic, sedimentary and diapiric mélanges in orogenic belts and exhumed subduction-accretion complexes. *Gondwana Res.* **2019**, *74*, 11–34. [CrossRef]
51. Riedmüller, G.; Brosch, F.J.; Klima, K.; Medley, E.W. Engineering geological characterization of brittle faults and classification of fault rocks. *Eng. Geol. Geomech. Tunneling* **2001**, *19*, 13–19.
52. Sönmez, H.; Gokceoglu, C.; Tuncay, E.; Medley, E.W.; Nefeslioglu, H.A. Relationship between volumetric block proportions and overall UCS of a volcanic bimrock: Felsbau. *J. Eng. Geol. Geomech. Tunn.* **2004**, *22*, 27–34.
53. Sönmez, H.; Tuncay, E.; Gokceoglu, C. Models to predict the uniaxial compressive strength and the modulus of elasticity for Ankara Agglomerate. *Int. J. Rock Mech. Min. Sci.* **2004**, *41*, 717–729. [CrossRef]
54. Sönmez, H.; Ercanoglu, M.; Kalender, A.; Dagdelenler, G.; Tunusluoglu, C. Predicting uniaxial compressive strength and deformation modulus of volcanic bimrock considering engineering dimension. *Int. J. Rock Mech. Min. Sci.* **2016**, *86*, 91–103. [CrossRef]
55. Nikolaidis, G.; Saroglou, C. Engineering geological characterization of block-in-matrix rocks. *Bull. Geol. Soc.* **2016**, *50*, 874–884.
56. Napoli, M.L.; Festa, A.; Barbero, M. Practical classification of geotechnically complex formations with block-in-matrix fabrics. *Eng. Geol.* **2022**, *301*, 106595. [CrossRef]
57. Sönmez, H.A.; Gokceoglu, C.; Medley, E.W.; Tuncay, E.; Nefeslioglu, H.A. Estimating the uniaxial compressive strength of a volcanic bimrock. *Int. J. Rock Mech. Min. Sci.* **2006**, *43*, 554–561. [CrossRef]
58. Russ, J.C.; Dehoff, R.T. *Practical Stereology*, 2nd ed.; Kluwer Academic/Plenum: New York, NY, USA, 2000.
59. Medley, E.W. Using stereologic methods to estimate the block volumetric proportion in melanges and similar block-in-matrix rocks (bimrocks). In Proceedings of the Congress of the International Association of Engineering Geologists, 7th, Lisbon, Portugal, 5–9 September 1994; A.A. Balkema: Rotterdam, The Netherlands, 1994; pp. 1031–1040.
60. Delesse, M.A. Procédé mécanique pour déterminer la composition des roches. *Comptes Rendues I'Acad. Des Sci.* **1847**, *25*, 544–545.
61. Medley, E.W. Estimating block size distributions of melanges and similar block-in-matrix rocks (bimrocks). In *North American Rock Mechanics Symposium*, 5th ed.; University of Toronto: Toronto, ON, Canada, 2002; pp. 599–606.
62. Goodman, R.E.; Ahlgren, C.S. Evaluating safety of concrete gravity dam on weak rock: Scott dam. *Geotech. Geoenviron. Eng.* **2000**, *126*, 429–442. [CrossRef]
63. Medley, E.W. Characterization of Franciscan melanges and other heterogeneous rock/soil mixtures. In *Engineering Geology Practice in Northern California*; Ferriz, H., Anderson, R., Eds.; Association of Engineering Geologists Special Publication 12, and California Geological Survey Bulletin; California Department of Conservation, Division of Mines and Geology: Sacramento, CA, USA, 2001; Volume 210, pp. 107–122.
64. Sammis, C.G.; Biegel, R.L. Fractals, fault gouge and friction. *J. Pure Appl. Geophys.* **1989**, *131*, 255–271. [CrossRef]
65. Archambault, G.; Daigneault, R.; Rouleau, A.; Tavchandjian, O. Mechanics of shear zones and fault belts development by anastomosing patterns of fractures at all scales. In *Mechanics of Jointed and Faulted Rock*; Rossmanith, H.P., Ed.; CRC Press: Boca Raton, FL, USA, 1990; pp. 197–204.
66. Boone, S.J.; Westland, J.; Busbridge, J.R.; Garrod, B. Prediction of boulder obstructions. In *Tunnels & Metropolises, Proceedings of the World Tunnelling Congress, Sao Paulo, Brazil, 25–30 April 1998*; Negro, A., Ferreira, A.A., Eds.; AA Balkema: Rotterdam, The Netherlands, 1998.
67. Medley, E.W. Uncertainty in estimates of block volumetric proportion in melange bimrocks. In Proceedings of the International Symposium of the International Association of Engineering Geologists, Athens, Greece, 23–27 June 1997; A.A. Balkema: Rotterdam, The Netherlands, 1997; pp. 267–272.
68. Tien, Y.M.; Lin, J.S.; Kou, M.C.; Lu, Y.C.; Chung, Y.J.; Wu, T.H.; Lee, D.H. Uncertainty in estimation of volumetric block proportion of bimrocks by using scanline Method. In Proceedings of the 44th U.S. Rock Mechanics/Geomechanics Symposium and 5th U.S.-Canada Rock Mechanics Symposium, Salt Lake City, UT, USA, 27–30 June 2010. Paper No. 10-158.
69. Tien, Y.M.; Lu, Y.C.; Wu, T.H.; Lin, J.S.; Lee, D.H. Quantify uncertainty in scanline estimates of volumetric fraction of anisotropic bimrocks. In Proceedings of the 45th U.S. Rock Mechanics/Geomechanics Symposium, San Francisco, CA, USA, 26–29 June 2011. Paper No. 11-345.
70. Lu, Y.C.; Tien, Y.M.; Juang, C.H.; Lin, J.S. Uncertainty of volume fraction in bimrock using the scan-line method and its application in the estimation of deformability parameters. *Bull. Eng. Geol. Environ.* **2019**, *79*, 1651–1668. [CrossRef]
71. Ramos-Cañón, A.M.; Castro-Malaver, L.C.; Padilla-Bello, N.V.; Vega-Posada, C.A. Incertidumbre en la determinación del Porcentaje Volumétrico de Bloques de BIMrocks/BIMsoil a partir de información unidimensional. *Rev. Bol. Geol.* **2020**, *42*, 69–80. [CrossRef]
72. Tien, Y.M.; Lu, Y.C.; Chang, H.H.; Chung, Y.C.; Lin, J.S.; Lee, D.H. Uncertainty of volumetric fraction estimates using 2-D measurements. In Proceedings of the 46th U.S. Rock Mechanics/Geomechanics Symposium, Chicago, IL, USA, 24–27 June 2012. Paper No. 12-494.
73. Lillesand, T.M.; Kiefer, R.W. *Remote Sensing and Image Interpretation*, 2nd ed.; Wiley: New York, NY, USA, 1987; 721p.
74. Sahagian, D.L.; Proussevitch, A.A. 3D particle size distributions from 2D observations: Stereology for natural applications. *Volcanol. Geotherm. Res.* **1998**, *84*, 173–196. [CrossRef]

75. Haneberg, W.C. Simulation of 3D block populations to characterize outcrop sampling bias in bimrocks: Felsbau Rock and Soil. *Eng. J. Eng. Geol. Geomech. Tunneling* **2004**, *22*, 19–26.
76. Mahdevari, S.; Moarefvand, P. Experimental investigation of fractal dimension effect on deformation modulus of an artificial bimrock. *Bull. Eng. Geol. Environ.* **2017**, *77*, 1729–1737. [CrossRef]
77. Fairhurst, C.; Cook, N.G.W. The phenomenon of rock splitting parallel to the direction of maximum compression in the neighborhood of a surface. In Proceedings of the 1st Congress of International Society of Rock Mechanics, Lisbon, Portugal, 25 September–1 October 1996.
78. Xu, W.J.; Yue, Z.Q.; Hu, R.L. Study on the mesostructure and mesomechanical characteristics of the soil–rock mixture using digital image processing based finite element method. *Int. J. Rock Mech. Min. Sci.* **2008**, *45*, 749–762. [CrossRef]
79. Buyukozturk, O.; Hearing, B. Crack propagation in concrete composites influenced by interface fracture parameters. *Int. J. Solids Struct.* **1998**, *35*, 4055–4066. [CrossRef]
80. Mahdevari, S.; Maarefvand, P. An investigation into the effects of block size distribution function on the strength of bimrocks based on large-scale laboratory tests. *Arab. J. Geosci.* **2016**, *9*, 509. [CrossRef]
81. Candan, G. A fuzzy triangular chart to predict the uniaxial compressive strength of the Ankara agglomerates from their petrographic composition. *Eng. Geol.* **2002**, *66*, 39–51. [CrossRef]
82. Kalender, A.; Sonmez, H.; Medley, E.W.; Tunusluoglu, C.; Kasapoglu, K.E. An approach to predicting the overall strengths of unwelded bimrocks and bimsoils. *Eng. Geol.* **2014**, *183*, 65–79. [CrossRef]
83. Mahdevari, S.; Maarefvand, P. Applying ultrasonic waves to evaluate the volumetric block proportion of bimrocks. *Arab. J. Geosci.* **2017**, *10*, 204. [CrossRef]
84. Irfan, T.Y.; Tang, K.Y. *Effect of the Coarse Fractions on the Shear Strength of Colluvium*; Geotechnical Engineering Office, Civil Engineering Dept.: Hong Kong, China, 1993.
85. Sönmez, H.; Altinsoy, H.; Gokceoglu, C.; Medley, E.W. Considerations in developing an empirical strength criterion for bimrocks. In Proceedings of the Asian Rock Mechanics Symposium, 4th, Singapore, 6–10 November 2006.
86. Roadifer, J.W.; Forrest, M.P.; Lindquist, E.S. Evaluation of shear strength of melange foundation at Calaveras Dam. In Proceedings of the U.S. Society for Dams, Annual Meeting and Conference, 29th, on "Managing Our Water Retention Systems", Nashville, TN, USA, 20–24 April 2009; pp. 507–521.
87. Roadifer, J.; Forrest, M.P. Characterization and treatment of mélange and sandstone foundation at Calaveras Dam. In Proceedings of the GeoCongress 2012 State of the Art and Practice in Geotechnical Engineering, Oakland, CA, USA, 25–29 March 2012; pp. 3362–3371.
88. Coskun, A. Development of an Empirical Approach to Overcome the Problems for Boundary Condition between Bimrocks and Jointed Rock Masses. Master's Thesis, Geological Engineering Department, Hacettepe University, Ankara, Turkey, 2010.
89. Turcotte, D.L. Fractals and fragmentation. *J. Geophys. Res.* **1986**, *91*, 1921–1926. [CrossRef]
90. Sheridan, M.F.; Zemtsov, A.N.; Macias, J.L. *A Detailed Study of Size Distribution of Large Volcanic Blocks at Ksudach, Kamchatka: Evidence for Fractal Fragmentation*; Western Pacific Geophysics Meeting; American Geophysical Union: Washington, DC, USA, 1992; p. 59.
91. Khorasani, E.; Amini, M.; Hossaini, M.F. The Influence assessment of blocks orientation on bimslopes stability using numerical modeling. In *Proceedings of the 4th ISRM Young Scholars Symposium on Rock Mechanics, Jeju, Republic of Korea, 10 May 2017*; International Society for Rock Mechanics and Rock Engineering: Lisbon, Portugal, 2017; pp. 269–272.
92. Khorasani, E.; Amini, M.; Hossaini, M.F.; Medley, E.W. Evaluating the effects of the inclinations of rock blocks on the stability of bimrock slopes. *Geomech. Eng.* **2019**, *17*, 281–287. [CrossRef]
93. Khorasani, E.; Amini, M.; Hossaini, M.F.; Medley, E.W. Statistical analysis of bimslope stability using physical and numerical models. *Eng. Geol.* **2019**, *254*, 13–24. [CrossRef]
94. Medley, E.W.; Rehermann, P.S. Characterization of bimrocks (rock/soil mixtures) with application to slope stability problems. In Proceedings of the Eurock 2004 and 53rd Geomechanics Colloquium, Salzburg, Austria, 7–9 October 2004.
95. Minuto, D.; Morandi, L. Geotechnical Characterization and Slope Stability of a Relict Landslide in Bimsoils (Blocks in Matrix Soils) in Dowtown Genoa, Italy. *Eng. Geol. Soc. Territ.* **2015**, *2*, 1083–1088. [CrossRef]
96. Montoya-Araque, E.A.; Suarez-Burgoa, L.O.; Medley, E.W. Application of the tortuous surface method to stochastic analysis of bimslope stability. *Bull. Eng. Geol. Environ.* **2020**, *79*, 5329–5340. [CrossRef]
97. Khorasani, E.; Amini, M.; Medley, E.W.; Tarigh Azali, S.; Hossaini, M.F. A General Solution for Estimating the Safety Factor of Bimslopes. *Rock Mech. Rock Eng.* **2022**, *55*, 7675–7693. [CrossRef]
98. Kim, C.; Snell, C.; Medley, E.W. Shear strength of Franciscan Complex melange as calculated from back-analysis of a landslide. In Proceedings of the 5th International Conference on Case Histories in Geotechnical Engineering, New York, NY, USA, 13–17 April 2004.
99. Püstow, H.; Riedmüller, G.; Schubert, W. Tunnelling in a tectonic melange of high structural complexity. *Felsbau* **2001**, *19*, 34–42.
100. Schubert, W.; Riedmüller, G. Tunnelling in fault zones—State of the art in investigation and construction. *Felsbau* **2000**, *18*, 715.
101. Schubert, W.; Riedmüller, G. Tunnelling in fault zones—Innovative approaches. In *Pacific Rocks 2000, Proceedings of the 4th North American Rock Mechanics Symposium, Seattle, Washington, DC, USA, 31 July–3 August 2000*; Balkema: Rotterdam, The Netherlands, 2000; pp. 113–124.

102. Moritz, B.A. Ductile Support Method for Tunnels in Squeezing Rock. Ph.D. Thesis, Graz University of Technology, Graz, Austria, 1999; Volume 5, 112p.
103. Schubert, W.; Budil, A. The importance of longitudinal deformation in tunnel excavation. In Proceedings of the 8th International Society for Rock Mechanics (ISRM), Tokyo, Japan, 25 September 1995; Balkema: Rotterdam, The Netherlands, 1995; pp. 1411–1415.
104. Vavrovsky, G.M.; Schubert, P. Advanced analyses of monitored displacements opens a new field to continuously understand and control the geotechnical behavior of tunnels. In *Proceedings of the 8th International Society for Rock Mechanics (ISRM), Tokyo, Japan, September 1995*; Balkema: Rotterdam, The Netherlands, 1995; pp. 1416–1420.
105. Steindorfer, A. Short Term Prediction of Rock Mass Behaviour in Tunnelling Using Advanced Analysis of Displacement Monitoring Data. Ph.D. Thesis, Graz University of Technology, Graz, Austria, 1997; Volume 1, 111p.
106. Schubert, W.; Steindorfer, A.; Button, E.A. Displacement monitoring in tunnels—An overview. *Felsbau* **2002**, *20*, 15–23.

International Journal of
Geo-Information

MDPI

Article

Modeling Shallow Landslide Runout Distance in Eocene Flysch Facies Using Empirical–Statistical Models (Western Black Sea Region of Türkiye)

Muge Pinar Komu [1], Hakan Ahmet Nefeslioglu [2] and Candan Gokceoglu [3,*]

1 Graduate School of Science and Engineering, Hacettepe University, Beytepe, Ankara 06800, Türkiye; mugekomu@gmail.com
2 Institute of Earth and Space Sciences, Eskisehir Technical University, Tepebasi, Eskisehir 26555, Türkiye; han@eskisehir.edu.tr
3 Department of Geological Engineering, Hacettepe University, Beytepe, Ankara 06800, Türkiye
* Correspondence: cgokce@hacettepe.edu.tr

Abstract: Uncertainties related to runout distances in shallow landslide analyses may not only affect lives but may also result in economic losses. Owing to the increase in shallow landslides, which are especially triggered by heavy rainfall, runout distances have been investigated to decipher whether applications of a functional runout distance are feasible. This paper aims to give insights into the modeling of the shallow landslide runout probability in Eocene flysch facies in the Western Black Sea region of Türkiye. There are two main stages in this study—which are dominated by empirical models, the detection of initiation points, and propagation—which help us to understand and visualize the possible runout distances in the study area. Shallow landslide initiation point determination using machine learning has a critical role in the ordered tasks in this study. Modified Holmgren and simplified friction-limited model (SFLM) parameters were applied to provide a good approximation of runout distances during the propagation stage using Flow-R software. The empirical model parameters suggested for debris flows and shallow landslides were investigated comparatively. The runout distance models had approximately the same performance depending on the debris flow and shallow landslide parameters. While the impacted total runout areas for the debris flow parameters were predicted to amount to approximately 146 km^2, the impacted total runout areas for the shallow landslide parameters were estimated to be about 101 km^2. Considering the inclusion of the RCP 4.5 and RCP 8.5 precipitation scenarios in the analyses, this also shows that the shallow landslide and debris flow runout distance impact areas will decrease. The investigation of runout distance analyses and the inclusion of the RCP scenarios in the runout analyses are highly intriguing for landslide researchers.

Keywords: shallow landslide; runout distance; empirical model; machine learning; Eocene flysch facies; RCP 4.5; RCP 8.5

1. Introduction

Considering the increase in extreme rainfalls in the last decade due to climate change, it is possible to claim that shallow landslides are like any other type of natural hazard and that researchers should increasingly focus on them [1]. Runout distance determination in shallow landslide analysis is crucial because hazard evaluations involving runout distance information constitute one of the bases of sustainable development for societies living in mountainous areas [2]. Runout methods are yielding extraordinary insights into landslide runout analysis. For many years, researchers have studied landslides using existing runout methods, which are empirical–statistical methods, physical methods, and numerical models. The empirical–statistical model was applied in studies by Poltnig et al. [3], Kaafarani et al. [4], Mergili et al. [5], Zhou et al. [6], Apriani et al. [7], Di Napoli et al. [8], and Guthrie and Befus [9] to inspect the runout of landslides. A novel terrain-matching targeted machine learning model, which depends on statistical methods, was used to estimate landslide runouts in a study by Ju et al. [10]. Additionally, a data-driven method for the estimation of shallow landslide runout was used in a study by Giarola et al. [11]. Observation of flow-like landslide runouts was accomplished using the physical method, which was applied as a flume test, in studies by Baselt et al. [12] and Gao et al. [13]. The numerical model, which has frequently been used to scrutinize flow-like landslide runouts in recent studies, was also used in studies by Gao et al. [13], Yang et al. [14], Chae et al. [15], Calista et al. [16], Abraham et al. [17], Dash et al. [18], Vicari et al. [19], Mikoš and Bezak [20], Oh et al. [21], Zhou et al. [22], Alene et al. [23], and La Porta et al. [24]. Investigations into the probabilities of failures were based on merged models in some studies. For instance, the combination of the numerical and physical models yielded enhanced comprehension of the failure mechanism in a study by Zhu et al. [25]. The physical method and numerical method were also combined in order to probe the runout distances of debris flows in studies by Clark [26], Melo et al. [27], and Gan and Zhang [28]. In addition, numerical models aiming to calculate the runout of flow-like landslides were successfully verified with a physical model (flume test) in a study by Gao et al. [13]. Numerical and empirical models were combined in order to investigate landslide runouts in a study by Peruzzetto et al. [29]. By considering data-driven approaches, runout analyses can be successfully prepared by applying empirical–statistical, physical, and numerical methods. The mentioned runout methods are often evaluated as extremely advantageous, but, in fact, they also have disadvantages. For example, based on results obtained during an experiment, the physical model is more suitable for in situ conditions than the numerical model [30]. However, it is demanding for physical models to not only fully reconstruct 3D topography, due to scaling effects [31], but also reflect the dynamic real conditions of field observations [32]. The numerical model's precision is contingent upon the parameters established within the model [25]. The numerical model, which is an expensive method, should be applied by experts who specifically possess experience in numerical analysis [33]. It is not suitable for rapid decision making because the preparation of all potential simulations is a time-consuming task [29]. In contrast, the empirical–statistical model enables quick assessments due to its lower computational times [34]. In fact, the empirical–statistical model can be used practically and realistically within a reasonable timeframe with a lower cost and minimal calculation demands. However, volume neglect and volume prediction based on past landslide events are challenging in complex geological conditions [34,35]. To decide on a suitable method for runout analysis, available data are important [36]. It may be claimed that the best runout model method is inconclusive, considering the advantages and disadvantages of both methods. Nevertheless, recent studies point out that although researchers have used various methods over the years to investigate methods of estimating runout distance that may contribute to landslide risk evaluations, empirical–statistical methods are still generally preferred because of computational costs [37]. The reach angle method was often considered as an empirical method by many researchers [3,4,38,39]. The multiple flow direction [40], Holmgren [41], modified Holmgren [42], cellular automata [43], and random walk [44,45] models provide opportunities to simulate possible runout paths. A combination of the empirical–statistical simplified friction-limited model (SFLM), which considers the reach angle, and flow direction algorithms, enables us to observe details such as feasible runout distances and paths using Flow-R software [42], as shown in recent studies [39,46–54]. Owing to the development of computer technology, recent studies on landslide runout distance prepared with software go well beyond the conventional techniques. DebrisFlow Predictor [9], DFLOWZ [55], and the Progressive Debris-Flow Routing and Inundation model [56] are alternative empirical model programs that currently give the ability to inspect not only debris flow propagations but also shallow landslide runout scenarios. Consequently, the main purpose of this study is to investigate the empirical models of the shallow landslide runout distance in Eocene flysch facies in the Western Black Sea region of Türkiye. Concentrated precipitations often contribute to the occurrence of shallow landslides [57]. Extreme rainfalls in the region containing Eocene flysch facies are on the rise, so residents have complained about high frequencies of shallow landslide occurrences. Considering shallow landslide occurrences and the spatial probability evaluation of the hazard, integration of landslide runout distance with a landslide susceptibility assessment seems to be indispensable [5,8,58,59]. Flow-R software, which was developed at the University of Lausanne [42], was used to empirically examine shallow landslide runout scenarios in this study. This software offers researchers the opportunity to try a variety of landslide runout simulations [39,46–54,60–63] to obtain runout distances by considering different parameter configurations. Table 1 provides a summary of previous Flow-R studies.

Table 1. Summary of studies that utilized Flow-R.

Authors	Research	Landslide Initiations	Runout Method	Flow Direction Algorithm
Horton et al. [42], Jiang et al. [51], Xu et al. [54], Park et al. [61], Sharma et al. [63]	Debris flow runout susceptibility map	Initiations detected by Flow-R	SFLM	Modified Holmgren
Pastorello et al. [46], McCoy [47], Paudel [48], Putra et al. [53]	Debris flow runout susceptibility map	User-defined initiations considering flow accumulation and slope [46], landslide susceptibility map [48], and Sentinel images [53]	SFLM	Modified Holmgren
Ali et al. [37], Bera et al. [62]	Debris flow runout susceptibility map	User-defined initiations based on remote sensing, slope angle distribution [37], and GPS and multi-temporal satellite images [61]	SFLM	Holmgren
Giano et al. [2]	Debris flow runout susceptibility map	Initiations detected by Flow-R	Perla	Modified Holmgren
Polat and Erik [60]	Debris flow runout susceptibility map	User-defined initiations considering landslide susceptibility map	Perla	Modified Holmgren
Charbel and El Hage Hassan [50]	Mudflow runout susceptibility map	Initiations detected by Flow-R	SFLM	Holmgren
Do et al. [49]	Landslide runout susceptibility map	Initiations detected by Flow-R	SFLM	Modified Holmgren
Liu et al. [52]	Landslide runout susceptibility map	User-defined initiations based on previous studies and D-InSAR technology	SFLM	Modified Holmgren

If the runout distance can be estimated, the resilience to shallow landslides will be stronger. Accordingly, this shallow landslide investigation focuses on the development of more accurate empirical models to obtain better estimations of shallow landslide runout distances. Using machine learning to detect landslide initiations is the critical stage of this study; however, accurate model parameter configuration also constitutes one of the most prominent stages in the algorithm provided by Flow-R. Modified Holmgren and SFLM parameters are used to estimate runout distances in the propagation stage in Flow-R. Considering the difficulty in distinguishing and the connectedness of shallow landslides and debris flow, the decision of whether a movement is a shallow landslide or debris flow depends on a researcher's judgment [64]. It is also impossible to estimate which types of landslides will occur (whether they are debris flows or shallow landslides) in Eocene flysch facies. Therefore, runout results are comparatively presented with respect to two different parameter configurations, which are shallow landslide and debris flow. Additionally, the question of whether the empirical–statistical method is suitable for our large study area is also discussed deeply in the later sections of this paper. Moreover, it is expected that the changes in precipitation trends caused by climate change will lead to changes in landslide occurrence in future years [65–69]. Thus, this study also aims to show that it is possible to analyze the near and distant future runout distances of shallow landslides considering the climate RCP scenarios' precipitation differences. The empirical model parameters suggested for debris flow and shallow landslides are also investigated comparatively by including RCP 4.5 and 8.5 monthly precipitation values.

Finally, the key research points of this paper can be listed as follows: (i) select the most suitable threshold values for detecting shallow landslide initiations for scenarios with and without RCP values, (ii) determine the appropriate propagation parameters, and (iii) estimate the probable maximum shallow landslide runout distances for scenarios with and without RCP values.

2. Materials

2.1. Study Area

The study area was the Eocene flysch facies located in the Western Black Sea region of Türkiye, covering an area of about 877 km^2 (Figure 1). Bartin and Zonguldak are the main cities located on the Eocene flysch facies. The research area was determined by considering the natural geological boundary consisting of Eocene-aged clastics and carbonates [70]. The Eocene flysch facies contain weak and weathered sandstone, claystone, and silt and have low topographic values. Their altitude varies between 4 and 1571 meters (m). The altitude in the northeast is low, while in the southwestern part it is relatively high. Rainfall is almost equally drastic at all meteorological stations in the study area. According to the Turkish State Meteorological Service [71], extreme rainfall of 252.8 mm was observed on 11 August 2021 in only three hours at the Bartin Ulus Ceyupler station, which is very close to the study area. Therefore, it is not surprising to posit that shallow landslides abound in this formation owing to heavy rainfall. In the recent past, this region was frequently exposed to heavy rains. For instance, Can et al. [72] stated that due to heavy rainfall in May 1998, shallow landslides were observed in Eocene flysch facies in this region. In addition, the Turkish State Meteorological Service's historical meteorological precipitation data indicate that 93.2 mm of daily rainfall occurred in Bartin on 21 May 1998 [73]. During this meteorological event, 91.8 mm of daily rainfall was also observed in Zonguldak. As mentioned, shallow landslides are frequently observed in Eocene flysch units due to heavy rainfalls [74] (Figure 2). Figure 2a clarifies the shallow landslides triggered after the May 1998 precipitations, while Figure 2b–d show examples of shallow landslide occurrences after heavy rainfall in November 2023. In total, 60 and 111 mm of daily precipitation were observed in the cities of Bartin and Zonguldak on 19 November 2023, respectively [75]. Figure 2 shows the types of landslides detected in the study area according to the Varnes (1978) classification frame [76]. The study area is predominantly covered by complex cultivation patterns and broad-leaved forests [77]. Mixed forests and land principally occupied by agriculture with significant areas of natural vegetation are also observable in the study area [77]. Shallow landslides are frequently observed in both mixed forests and broad-leaved forests in this study area, according to Corine Land Cover [77].

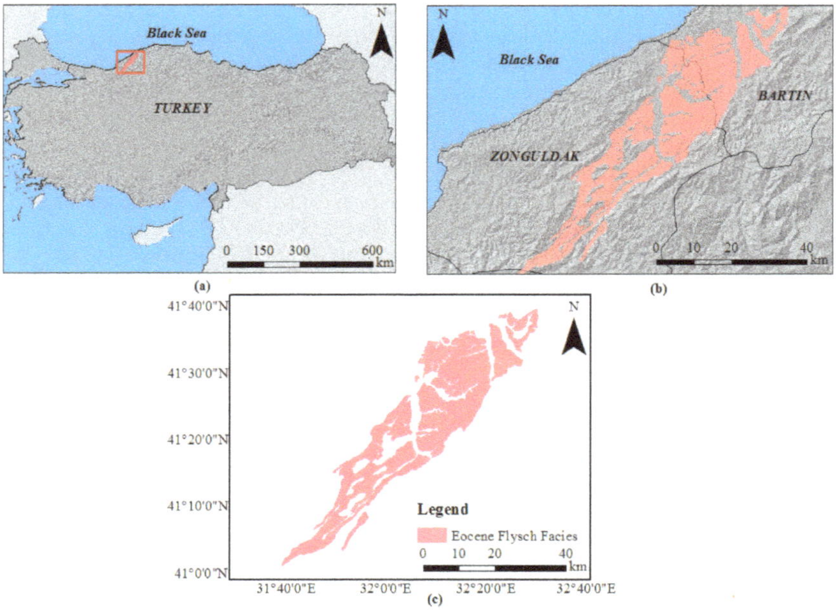

Figure 1. Map of the study area: (**a–c**) the boundaries of the Eocene flysch facies were modified based on Akbas et al. [70].

Figure 2. Some typical shallow landslides triggered in the study area (**a**) after the May 1998 rainfall (Google Earth image) and (**b–d**) after the November 2023 rainfall (field photographs).

Although shallow runout distances have not yet been investigated in Eocene flysch facies, some landslide analyses have already been conducted in the formation. Duman et al. [78] prepared a landslide inventory of Northwestern Anatolia involving Eocene flysch facies in the region. Akgun et al. [74] evaluated the general characteristics of the landslide sizes in the formation by considering magnitude and frequency relations. In their study, they emphasized that while rollover effects were noticed at 0.048 km^2, the fractal dimensions of the distributions were noticed as 1.41 [74] in the flysch facies.

Studying large areas makes parameter selection difficult. For this reason, the study area was divided based on the basin boundaries. ArcGIS software enabled the division of the study area with respect to the basin boundaries by considering the watersheds in the region. The Eocene flysch facies were divided into three sub-basins in which runout analyses could be performed (Figure 3). To perform appropriate parameter selection by decreasing the size of the investigation area, smaller sub-basins could also be considered. However, studying larger areas is important for researchers investigating shallow landslide runout distances. The areal distributions of the sub-basins are as follows: Egerci basin in the south is 269 km^2, Beycuma basin in the middle is 311 km^2, and Ihsanoglu basin in the north is 297 km^2.

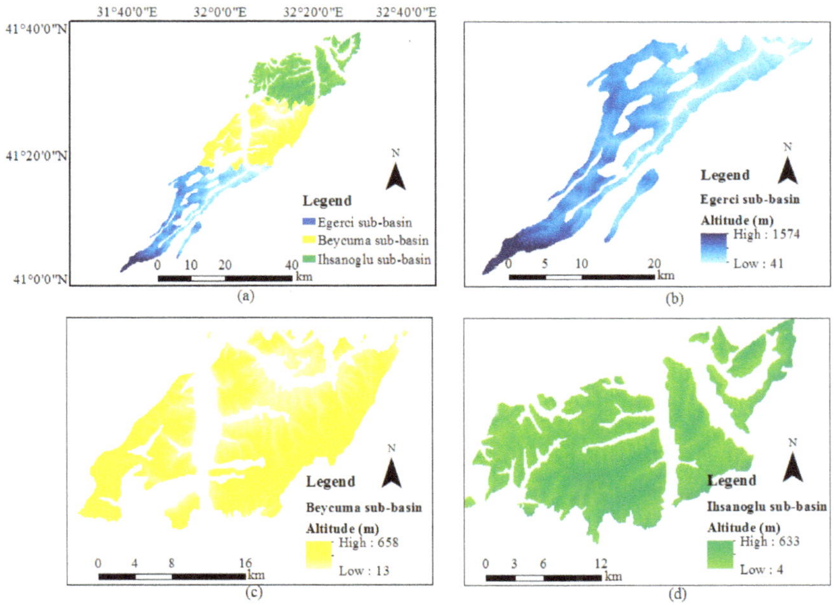

Figure 3. The sub-basins evaluated in the Eocene flysch facies: (**a**) general view of the Eocene flysch facies with sub-basins; (**b**) Egerci sub-basin in the south; (**c**) Beycuma sub-basin in the middle; and (**d**) Ihsanoglu sub-basin in the north.

2.2. Data Source and Pre-Processing

It should primarily be noted that the preparation of a susceptibility map was especially critical to determine the initiation points that were used during the runout analyses. Therefore, the following data were obtained to prepare a shallow landslide susceptibility map and to calculate runout distances in this study: First. Shuttle Radar Topography Mission Digital Elevation Model (SRTM DEM) data with a spatial resolution of ~30 m were obtained. They were converted to a 25 m resolution using the nearest resampling technique in ArcGIS. The coordinate system was set as WGS 1984 UTM zone 36N. Second, a DEM with a spatial resolution of 25 m was also implemented to obtain topographic features such as the slope gradient and slope aspect using the System for Automated Geoscientific Analyses (SAGA) GIS. In addition, the slope length factor, valley depth, topographic wetness index (TWI), plan curvature, profile curvature, convergence index, closed depressions, channel network base levels, and channel network distance were obtained in the SAGA GIS environment to identify the geomorphometric features of the study area. Third, the geological information was evaluated based on [38]. Subsequently, Corine Land Cover [77] was used to assess the land cover in the study area with a 100 m spatial resolution. Fifth, NDVI data were obtained from Sentinel-2 images with a 10 m spatial resolution.

$$\text{NDVI} = (\text{NIR} - \text{Red})/(\text{NIR} + \text{Red}) \quad (1)$$

Here, NDVI is the Normalized Difference Vegetation Index, NIR is the Near-Infrared Band (band 8), and Red is the Red Band (band 4) of a Sentinel-2 image. Finally, the spatial distributions of recent shallow landslides were mapped using Sentinel-2 images and Google Earth. Short field surveys were also carried out to perform ground checks of the mapped shallow landslides. Furthermore, available landslide inventories published in national databases and the recent literature [72,78–80] were also evaluated, particularly to identify shallow landslides during mapping using satellite images. When an available landslide inventory was examined considering historical satellite images from Google Earth, 262 shallow landslides were mapped. The landslides were geometrically classified in terms of failure depths, reach angles, and observed runout distances. The volume and area relations proposed by Hovius et al. [81] based on Jaboyedoff et al. [82] and the semi-ellipsoid approach were used to determine the landslide depths. The obtained values proved that the failures mapped in this study can be classified as shallow landslides since their depths are less than 5 m [83]. The descriptive statistics of the failure depths, reach angles, and observed runout distances are given in Table 2 with respect to the sub-basins. Figure 2b,d, which were taken after heavy rainfall in November 2023, provide good evidence that shallow landslides have continued to occur due to excessive rainfall in this region, and their observed runout distances are compatible with Table 2.

Table 2. The properties of the landslide inventories prepared for the sub-basins.

Sub-Basin	N	Statistics	Area (m^2)	Failure Depth (m)	Travel Angle (°)	Observed Runout Distance (m)
Egerci	111	Min.	21	0.2	1	7
		Max.	4116	3.2	49	122
		Mean	536	1.0	23	53
		Median	344	0.9	24	47
		Std. Deviation	572	0.5	12	29
Beycuma	15	Min.	23	0.2	4	7
		Max.	1843	2.1	28	83
		Mean	376	0.9	13	36
		Median	258	0.8	13	31
		Std. Deviation	438	0.5	8	20
Ihsanoglu	136	Min.	7	0.1	1	5
		Max.	1710	2.1	32	122
		Mean	140	0.5	11	25
		Median	65	0.4	10	17
		Std. Deviation	226	0.3	7	21

3. Methods

The method of this study was divided into four critical interconnected steps: (i) prepare a shallow landslide susceptibility map using the machine learning logistic regression (LR) method; (ii) determine shallow landslide initiations by considering the shallow landslide susceptibility map and the RCP scenarios' precipitation values; (iii) assess the shallow landslide runout distance methodology by taking into consideration detected shallow landslide initiations; and (iv) select suitable model parameters for runout distance assessment.

3.1. Preparation of Shallow Landslide Susceptibility Map

Considering the constant lithological information of the Eocene flysch facies, fifteen features, including the altitude, slope gradient, slope aspect, landslides, land use, NDVI, slope length factor, valley depth, topographic wetness index (TWI), plan curvature, slope profile curvature, convergence index, closed depressions, channel network base levels, and channel network distance, were used as the landslide conditioning factors for landslide susceptibility mapping (Table 3).

Table 3. Descriptive statistics of the topographic parameters with respect to the sub-basins.

Sub-Basin	Statistics	A	SG (°)	SA (°)	PLC	PRC	CI	CD	TWI	SLF	CNBL (m)	CND (m)	VD (m)
Egerci	Min.	41	0	0	−0.01	−0.01	−92	0	2	0	41	−49	−3
	Max.	1574	60	6.28	0.01	0.012	96	20	21	38	1382	449	549
	Mean	444	15	3.25	0	0	0	0	7	5	381	64	236
	Median	425	14	3.04	0	0	0	0	6	4	369	49	229
	Std. Deviation	259	8	1.91	0.002	0.002	8	1	2	3	233	60	133
Beycuma	Min.	13	0	0	−0.01	−0.01	−100	0	3	0	13	−27	0
	Max.	658	52	6.28	0.008	0.009	100	19	20	24	444	304	277
	Mean	170	11	3.27	0	0	0	0	7	3	129	41	140
	Median	158	9	3.33	0	0	0	0	7	2	118	32	146
	Std. Deviation	91	6	1.91	0.001	0.001	9	1	2	2	69	38	61
Ihsanoglu	Min.	4	0	0	−0.01	−0.02	−98	0	2	0	6	−39	0
	Max.	633	70	6.28	0.017	0.021	92	32	20	30	284	468	339
	Mean	121	11	3.35	0	0	0	0	7	3	79	42	117
	Median	104	10	3.43	0	0	0	0	6	2	69	32	117
	Std. Deviation	71	7	1.83	0.001	0.001	10	1	2	2	43	42	51

Abbreviations: altitude: A, slope gradient: SG, slope aspect: SA, plan curvature: PLC, profile curvature: PRC, convergence index: CI, closed depression: CD, topographic wetness index: TWI, slope length factor: SLF, channel network base level: CNBL, channel network distance: CND, valley depth: VD.

An open-source machine learning library in Python provided the opportunity to produce a landslide susceptibility map at this stage. The LR method is valuable, given its high prediction accuracy, in machine learning research, such as the preparation of landslide susceptibility maps [84–90]. For instance, Polykretis and Chalkias [91] prepared a landslide susceptibility map for shallow rotational and translational slides and the flow types of landslides using this method. The authors stated that the LR method is the better option. Nhu et al. [86] stated that the LR method was the second most successful and trustworthy model in their comparison of the production of shallow landslide susceptibility maps. Nwazelibe et al. [90] indicate that LR is the better method to produce landslide susceptibility maps. Landslides, which are shallow translational debris slides, slumps, and occasional debris topples, frequently occur in the Eocene-aged Ameki-Bende formation [90]. Taking these examples from the literature into account, in this study, LR was used to determine shallow landslide susceptibility.

$$P = 1/(1 + e^{-y}) = e/(1 + e^{y}) \quad (2)$$

Here, P is the landslide probability occurrence, which changes from 0 to 1, while e is the exponential constant [92]. In Equation (3), y is a dependent variable. When b_0 is a constant intercept from the model, b_n is the triggered landslide factor of a_n, which represents an independent variable.

$$y = b_0 + b_1a_1 + b_2a_2 + \cdots \ldots + b_na_n \tag{3}$$

The ratio of the training/testing samples was selected as 80:20. The data were divided into training (80%) and testing (20%) sets [91,93–97]. The "train_test_split" and "cross_val_score" functions of the sklearn model selection were used in the open-source Python libraries for training and testing the data.

Shallow landslide susceptibility maps obtained using the LR method for the sub-basins are given in Figure 4. There are three classes indicating low (0–0.4), moderate (0.4–0.7), and high (0.7–1) probabilities of shallow landslide occurrences.

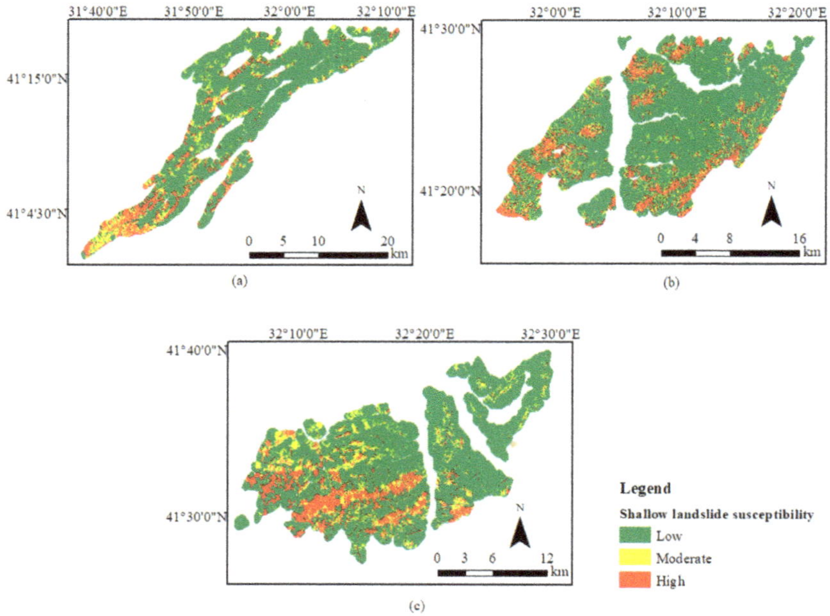

Figure 4. Shallow landslide susceptibility maps of Eocene flysch facies obtained using logistic regression for the sub-basins: (**a**) Egerci sub-basin in the south; (**b**) Beycuma sub-basin in the middle; and (**c**) Ihsanoglu sub-basin in the north.

3.2. Determination of Shallow Landslide Initiations

Possible shallow landslide initiation determination is a significant concern for landslide researchers. Even though it is difficult to make precise predictions about initiations, a susceptibility map can provide a basis to estimate possible shallow landslide initiations. Similar considerations were also given by Xu et al. [54]. Xu et al. [54] indicated that more susceptible areas on susceptibility maps have higher initiation abilities. Can et al. [72] marked that due to heavy rainfall in May 1998, shallow landslides were observed in very highly susceptible zones in the Eocene flysch facies in this region. In this study, highly susceptible areas, which are called shallow landslide initiations, were implemented for runout distance estimation using Flow-R software. In fact, the determination of the susceptibility threshold of each basin was key to identifying the shallow landslide initiations in this study because they were used as pre-defined initiations in the runout distance assessment stage. The QGIS environment was used to visualize the shallow landslide susceptibility maps and apply the necessary threshold value of susceptibility to determine the initiations (Figure 5).

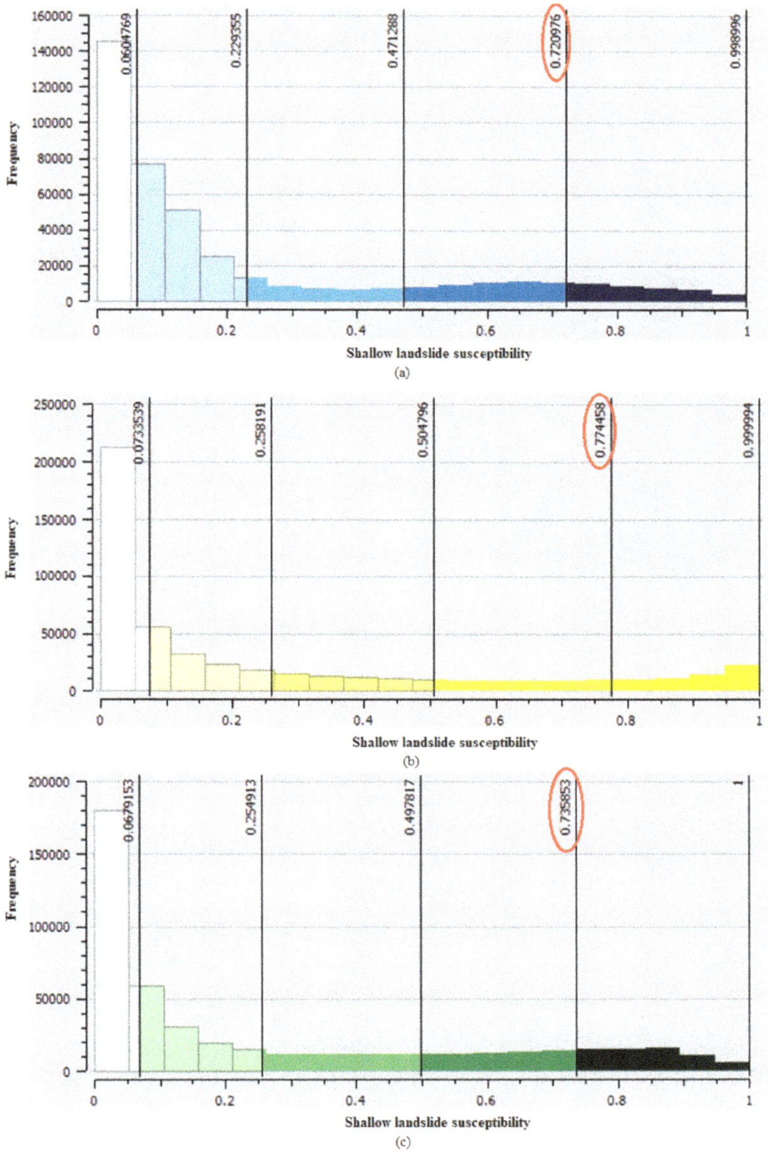

Figure 5. Shallow landslide initiation threshold selection: (**a**) Egerci sub-basin; (**b**) Beycuma sub-basin; and (**c**) Ihsanoglu sub-basin.

When shallow landslide initiations are identified according to a susceptibility threshold, it is a good idea to be on the safe side. The threshold value was selected according to the frequency histograms of the shallow landslide susceptibility values of the sub-basins (Figure 5). The threshold value for the shallow landslide initiations in the sub-basins was determined to be 0.70. Therefore, shallow landslide susceptibility grid cells with values equal to or higher than 0.70 and recent landslides mapped during the inventory stage were chosen as the shallow landslide initiations for the sub-basins (Figure 6). According to Corine Land Cover [77], the initiations in the Beycuma and Ihsanoglu basins are frequently abundant in broad-leaved forests, while those in the Egerci basin are abundant in broad-leaved forests and mixed forests.

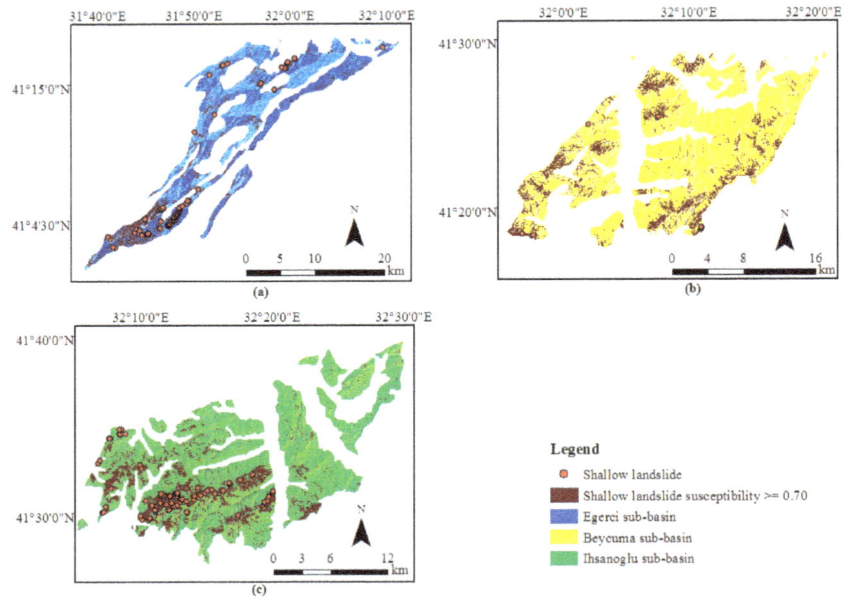

Figure 6. Shallow landslide initiations determined for the sub-basins: (**a**) Egerci sub-basin in the south; (**b**) Beycuma sub-basin in the middle; and (**c**) Ihsanoglu sub-basin in the north.

In this part of the study, Community Climate System Model (CCSM4) projections based on RCP 4.5 and RCP 8.5 monthly precipitation scenarios for both cities were also analyzed using an IDW heterogeneous environment method in ArcGIS based on data obtained from NCAR [98]. Geological hazards, such as shallow landslide initiation points, may exhibit a spatially heterogeneous distribution. IDW interpolation demonstrates greater success in heterogeneous environments [99]. The IDW method is considered the best method due to its reasonable processing timeframe and adequate precision for heterogeneous data [100]. The IDW method was also used to analyze the spatial distributions of the RCP scenarios' rainfalls in studies by Mohamed Yusof et al. [101] and Nasidi et al. [102]. Although 30 years is the most widely used reference period, 20 years has recently been used in some works [103]. Therefore, the time periods were classified as 2025–2044, 2044–2063, 2063–2082, and 2082–2100. In addition, 20-year monthly median values were used to analyze the general precipitation trend in order to apply the IDW method. Selected precipitation threshold values were utilized in order to detect critical shallow landslide initiations by considering the IDW results of the three sub-basins. Historical meteorological data for two stations, Bartin and Zonguldak, and RCP 4.5 and 8.5 scenario precipitation values for both cities were evaluated by separately using cross-correlation and choosing the reference period of 1995–2022. This step aimed to determine how extreme precipitations are represented in the obtained RCP data. Therefore, extreme precipitation values that resulted in shallow landslides in the study area were considered using this cross-correlation equation in order to reach predicted precipitation values. Considering these values, finding the median precipitation values was a critical stage. Both scenarios gave 81 mm monthly critical precipitation threshold values. Although the probability of this precipitation amount falling in one month or one day will be uncertain, it is possible to expect this amount of rainfall in one day by considering previous daily rainfall data and recent extreme daily rainfall data in this area. Hence, it was revealed that for the two scenarios the critical precipitation threshold of 81 mm would be applied in the analyses. Nevertheless, in this condition, shallow landslide initiations should supply both conditions together. These conditions are that the shallow landslide susceptibility is equal to or more than 0.70 and the RCP precipitation scenarios are equal to or more than 81 mm. In addition, while including the RCP precipitation scenarios' effects in the runout distance analyses, the impacts of the other triggering landslide parameters were neglected, and it is admitted that they might stay the same except for the RCP precipitations. Shallow landslide initiations, which were detected by considering the RCP precipitation scenarios with respect to divergent time periods, are shown in Figure 5. Despite the reduction in shallow landslide initiations, the Egerci sub-basin and the Beycuma sub-basin may be exposed to shallow landslides, while the Ihsanoglu sub-basin may not be affected by shallow landslides in the presented time range for both scenarios (Figures 7 and 8). Therefore, the Ihsanoglu sub-basin is not presented in Figures 7 and 8. Although there is no big difference between the results of the other three scenarios apart from the RCP 4.5: 2025–2044 scenarios in the Egerci sub-basin, differences can be observed in the northwestern part of the Egerci sub-basin in terms of shallow landslide initiations. The differences between in the presented RCP 4.5 and 8.5 scenarios are considerably larger in the period of 2025–2044 for both the Egerci and Beycuma sub-basins. In addition, the RCP 8.5 scenario only affects shallow landslide initiations in the period of 2025–2044 for both sub-basins. In the Beycuma sub-basin, shallow landslide initiations can be observed for the RCP 4.5 scenario in the periods of 2025–2044 and 2082–2100, while in the Egerci sub-basin shallow landslide initiations can be examined in the periods of 2025–2044, 2063–2082, and 2082–2100. All results show that shallow landslide initiations decrease if the RCP precipitation scenarios are considered; therefore, this study also has the same opinion about landslide susceptibility decrement depending on RCP scenarios as the previous studies by Ageenko et al. [65] and Park and Lee [104].

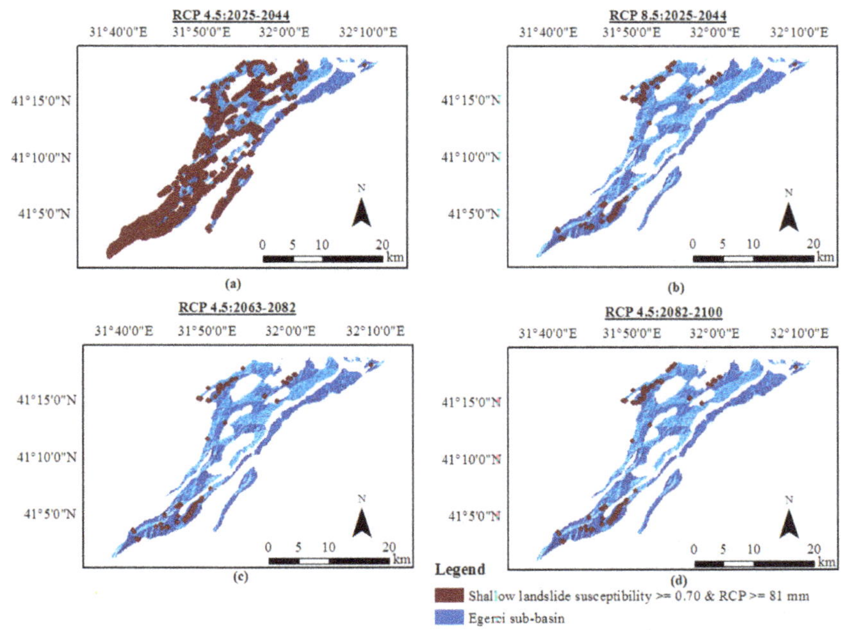

Figure 7. Shallow landslide initiations affected by RCP 4.5 and 8.5 scenarios determined for the Egerci sub-basin: (**a**) RCP 4.5: 2025–2044; (**b**) RCP 8.5: 2025–2044; (**c**) RCP 4.5: 2063–2082; and (**d**) RCP 4.5: 2082–2100.

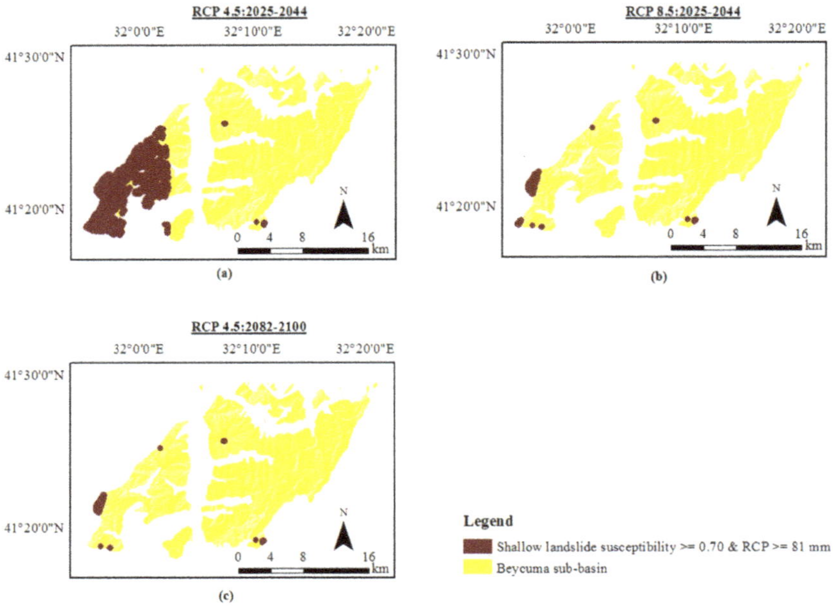

Figure 8. Shallow landslide initiations affected by RCP 4.5 and 8.5 scenarios determined for the Beycuma sub-basin: (**a**) RCP 4.5: 2025–2044; (**b**) RCP 8.5: 2025–2044; and (**c**) RCP 4.5: 2082–2100.

3.3. Shallow Landslide Runout Distance Assessment Methodology

The empirical–statistical methods were applied in this study to analyze the runout distances of shallow landslides. Models are generally prepared using the empirical method for large-scale studies [46]. It is important to choose a suitable method and algorithm to determine how to implement the software [42]. The flow direction algorithm and the simplified friction-limited model (SFLM) are two main stages that need to be applied to obtain probability of runout from the software during the stages of spreading and energy calculation. Essentially, the combination of the spreading algorithm and energy calculation gives the possible runout paths and distances. First, it is indispensable to choose the flow direction algorithm to be used in a study. Therefore, the question of which flow direction algorithm method is preferable for use in landslide runout studies should be pondered. Holmgren [41] added the "x" parameter to the multiple flow direction approach of this algorithm, allowing studies to be carried out in more detail. In Equation (1), while i and j represent flow directions, ρ_i^{fd} is the sensitivity ratio in the i direction, $\tan\beta_i$ is the gradient between the center cell and the cell in the i direction, $\tan\beta_j$ is the slope gradient between the center cell and the cell in the j direction, and x is the variable parameter that controls the variable deviation. As can be seen in Equation (4), x can take values between one and infinity.

$$\rho_i^{fd} = (\tan\beta_i)^x / (\sum^8{}_{(j=1)} (\tan\beta_j)^x) \ \forall \ \{(\tan\beta > 0@ \times \varepsilon[1;+\infty])\} \quad (4)$$

The modified Holmgren algorithm proposed by Horton et al. [42] is one of the most favorable flow direction algorithms because it includes the "dh" parameter, which varies the gradient values by varying the height of the central cell. It smooths DEM roughness. Since low-resolution data were used, the modified Holmgren algorithm [42] was used in this study. Putra et al. [53] stated that they also chose this algorithm because it is less sensitive to the DEM resolution. The SFLM is the second key part of the propagation stage.

In this stage, the energy should be sufficient when it travels to another cell. Energy loss is caused by friction [42,48].

$$E_{kin}{}^{i} = E_{kin}{}^{0} + \Delta E_{pot}{}^{i} - E_f{}^{i} \quad (5)$$

In Equation (5), $E_{kin}{}^{i}$, which is the kinetic energy (cell in direction i), is calculated using $\Delta E_{pot}{}^{i}$ and $E_f{}^{i}$. While $\Delta E_{pot}{}^{i}$ is the variability in the potential energy (cell in direction i), $E_f{}^{i}$ is energy loss due to friction (cell in direction i).

$$E_f{}^{i} = g\Delta x \tan\varphi \quad (6)$$

The $E_f{}^{i}$ calculation multipliers are g, Δx, and tan φ (Equation (6)). Δx represents the horizontal displacement increment, while tan φ is the energy line gradient. g is gravitational acceleration and is also used to calculate energy loss. The reach or travel angle is significant for the SFLM during the energy calculation. Past landslide runout travel angles are the key to understanding the region's landslide travel angle trends. Since the study area was divided into three sub-basins, different travel angles were determined for each sub-basin separately. The properties of the landslide inventory prepared in this study, involving the depth of the estimated failure surface, runout distances, and travel angles, were used to identify the parameter sets that achieved ideal model performance across the models of the sub-basins. The determination of the median travel angle for the sub-basins was important for the assessment of the probabilities of runout distances in larger areas.

The selection of an appropriate velocity and energy limit was also important for this study to remain within the acceptable energy levels and to avoid an endless energy loop during the analysis (Equation (7)).

$$V_i = \min\{\sqrt{(V_0{}^2 + 2g\Delta h - 2g\Delta x \tan\varphi)}, V_{max}\} \quad (7)$$

Here, V_0 and V_{max} indicate the initial speed and the maximum speed limit, respectively.

3.4. Selection of Model Parameters

Appropriate parameter selection considering the region is important during model preparation. In this study, Flow-R software, which is a distributed empirical model, was used to model possible runout distances for the sub-basins. The software enabled comparison of the runout distance results not only for debris flows but also for shallow landslides by considering proposed parameters. Flow-R software promoted the utilization of a range of parameters to enhance the accuracy of the research. Many simulation scenarios can be considered to investigate the effects of the selected parameters because the parameters can change more rapidly in actual conditions. The parameter x can become valuable for understanding the effect of the flow direction. The value of x is equal to 4 for debris flows, while it can be set as 22 to 26 for shallow landslides [105]. The value of x has also been set as 4 for debris flows in recent studies [37,47,51,61]. In this study, comparative analyses were performed depending on whether failures occurred as debris flows or shallow landslides. Therefore, considering the suggestions already given in the literature, the value of x was set as 4 or 25 for the debris flows and shallow landslides in the analyses, respectively. The value of dh is also a significant parameter that affects results by changing the reduced rate of roughness. The value of 1 was suggested for dh in recent studies [2,49,53]. Horton et al. [105] also proposed that dh can be equal to 1 for both debris flows and shallow landslides. Since slope morphology can change in various ways, travel angles can also change to various degrees. For instance, slope gradients are higher in the western part of the Eocene flysch facies. Travel angles were also evaluated separately for the three sub-basins by considering back analyses. Accordingly, median travel angle degrees were determined for the sub-basins separately as follows: 24° for the Egerci sub-basin, 13° for the Beycuma sub-basin, and 10° for the Ihsanoglu sub-basin. Horton et al. [105] suggested that travel angles of 8°–13° can be used for debris flows, while they may be 16° to 20° for shallow landslides. Different researchers considered different travel angles in the literature, e.g., 5° [38,49], 9° [51], 10° [47], 11° [42], 12° [37], and 13° [61]. Back analyses also show that if the other parameters are kept the same and only the travel angle is decreased, runout distances will increase. The last parameter that should be considered during analyses is velocity. The estimation of velocity is difficult. Back-calculated debris flow velocities were observed in the range of 5 to 15 m/s in a study by Prochaska et al. [106]. Similarly, different researchers considered different velocity values in the literature, e.g., 5 m/s [49], 10 m/s [37], 15 m/s [42,47,54], 16 m/s [107], 20 m/s [46], and 25 m/s [51]. By considering both back analyses and the local information obtained by residents having memorials related to the shallow landslides that occurred in May 1998 in this region, it was concluded that the velocity parameter could be set as 15 m/s for shallow landslides in the Eocene flysch facies. The same velocity value was also used for debris flow during the comparative analyses. This parameter selection was compatible with the suggestions that were given by Horton et al. [42]. Apart from the velocity parameter, energy limitation should also be considered to obtain logical runout distance results. Consequently, the DEM and pre-defined landslide initiations were imported into Flow-R. Files should be converted to ASCII format. It is also necessary to be aware that files should have the same spatial resolution and coordinate system. The model parameters implemented in Flow-R for shallow landslides and debris flow are summarized in Table 4 with respect to the different sub-basins.

Table 4. The model parameters implemented in Flow-R in this study.

Sub-Basin	Failure	Flow Direction Algorithm	x	dh (m)	Travel Angle (°)	Velocity (m/s)
Egerci	Debris flow	Modified Holmgren	4	1	24	15
	Shallow landslides	Modified Holmgren	25	1	24	15
Beycuma	Debris flow	Modified Holmgren	4	1	13	15
	Shallow landslides	Modified Holmgren	25	1	13	15
Ihsanoglu	Debris flow	Modified Holmgren	4	1	10	15
	Shallow landslides	Modified Holmgren	25	1	10	15

4. Results

4.1. Shallow Landslide Runout Distance Assessment

Effective determination of the runout distances of shallow landslides is significant to understand and assess the importance of the negative effects of shallow landslides. As runout research has been developing rapidly, it is difficult to agree on which methods will become the most practical in the future. This study has pointed out that the empirical–statistical method can determine the runout distances of shallow landslides. Runout distance models have approximately the same performance depending on the debris flow and shallow landslide parameters. Runout distances in the Eocene flysch facies can be estimated at a coefficient of determination level of 0.64 with debris flow parameters, while this value was obtained as 0.62 for the models developed using the shallow landslide parameters (Figure 9). Therefore, it is possible to bring the hazard regions up to date by integrating the resultant runout zones into landslide susceptibility maps. The results of the runout distance evaluations for the sub-basins are represented in Figures 10–12 with respect to the model parameters considered for debris flows and shallow landslides.

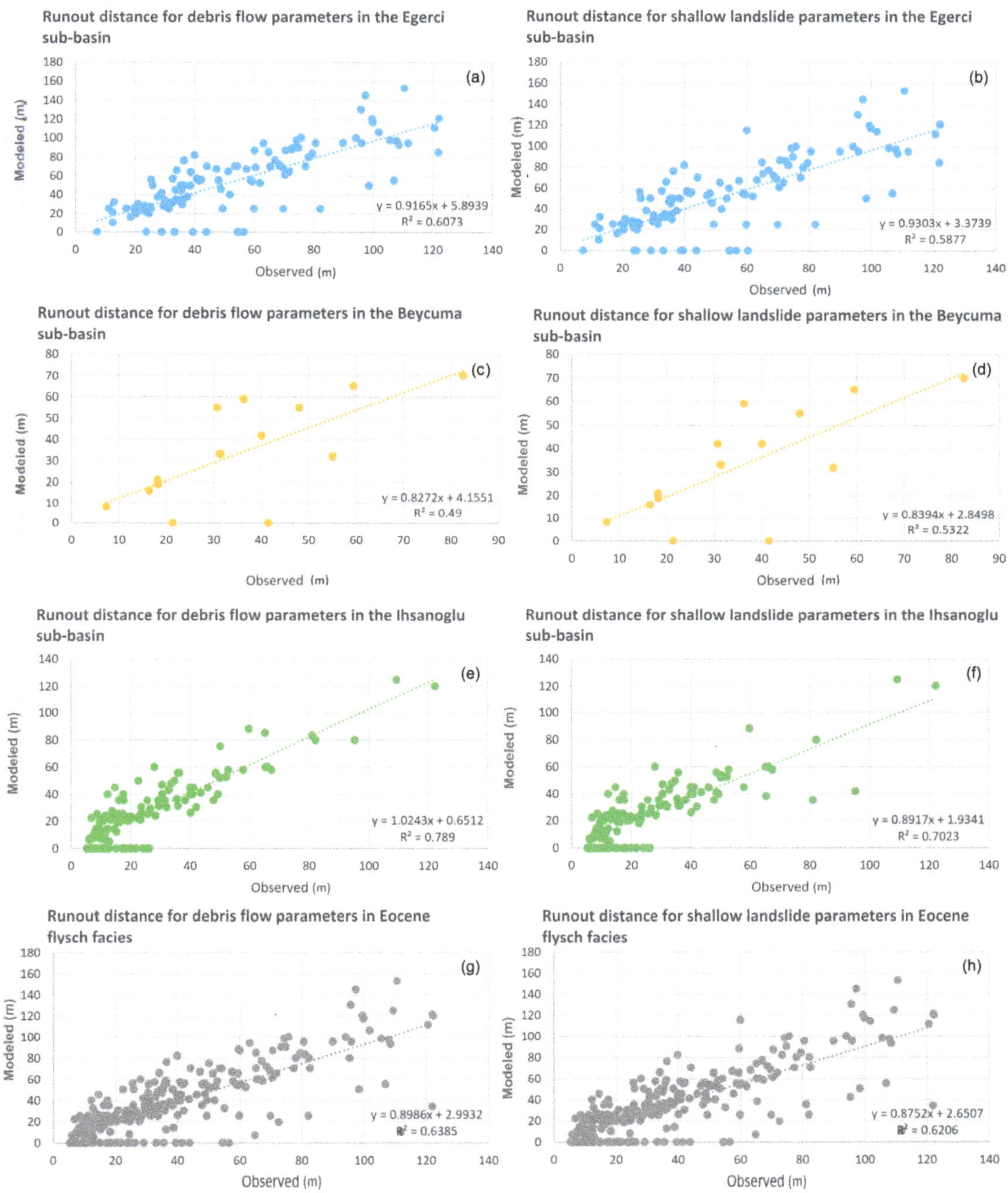

Figure 9. Cross-correlations of the observed and modeled runout distances obtained using the debris flow and shallow landslide model parameters (**a**,**b**) in the Egerci sub-basin; (**c**,**d**) in the Beycuma sub-basin; (**e**,**f**) in the Ihsanoglu sub-basin; and (**g**,**h**) in the Eocene flysch facies.

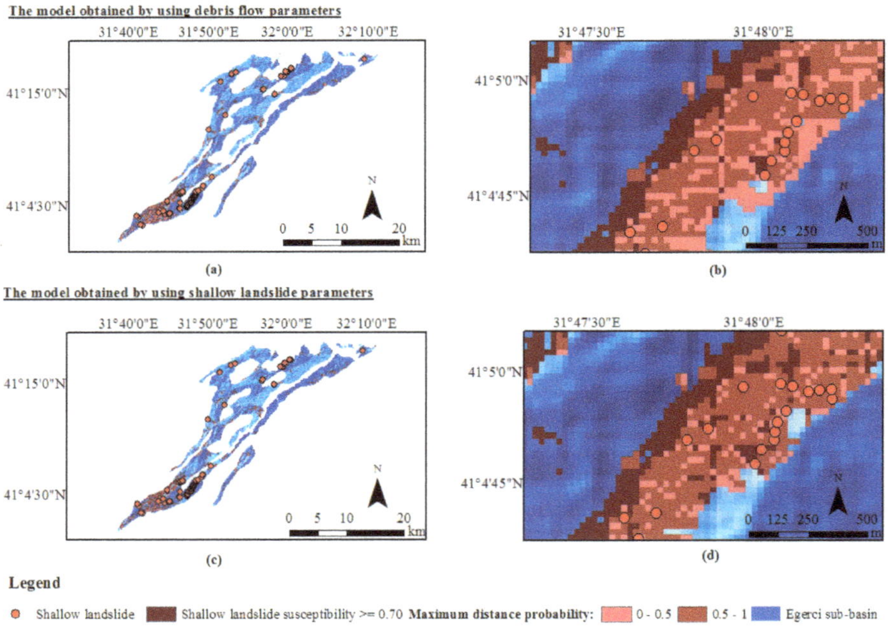

Figure 10. Probable runout distance models and close views of the spatial distributions of the results in the Egerci sub-basin (**a**,**b**) using the debris flow parameters and (**c**,**d**) using the shallow landslide parameters.

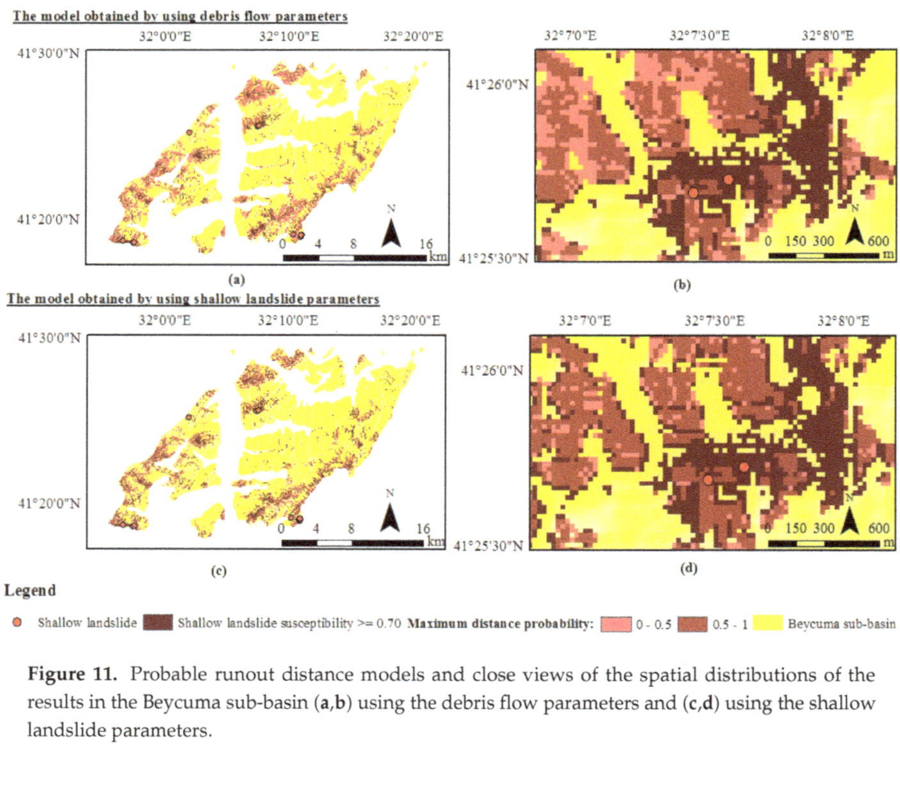

Figure 11. Probable runout distance models and close views of the spatial distributions of the results in the Beycuma sub-basin (**a**,**b**) using the debris flow parameters and (**c**,**d**) using the shallow landslide parameters.

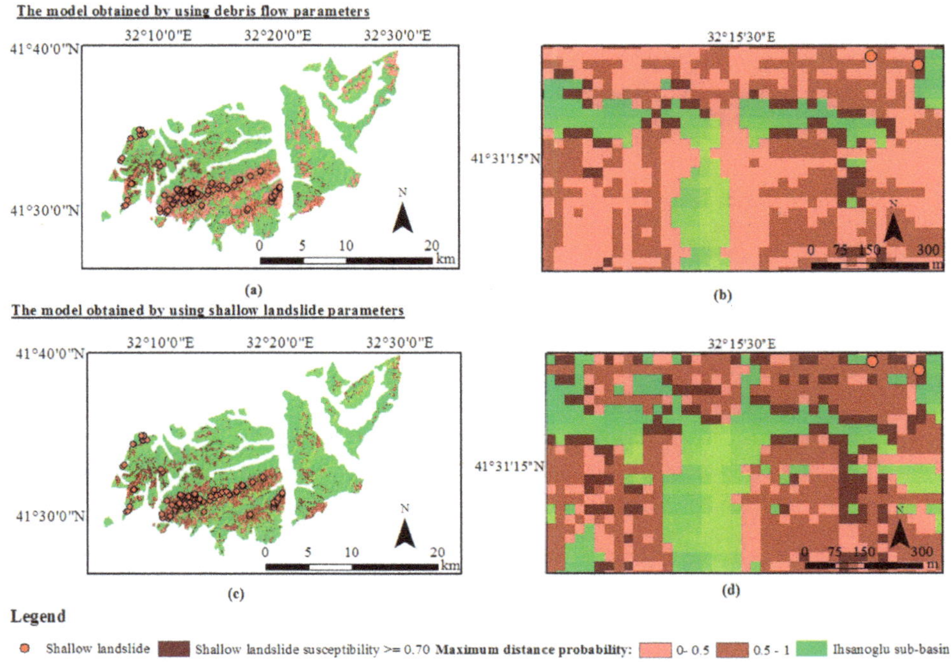

Figure 12. Probable runout distance models and close views of the spatial distributions of the results in the Ihsanoglu sub-basin (**a**,**b**) obtained using the debris flow parameters and (**c**,**d**) using the shallow landslide parameters.

The uncertainty about whether observed landslides are shallow landslides or debris flows continued in the statistical–empirical evaluations. Therefore, models were prepared for both parameter options to determine where shallow landslides or debris flows may reach because it was difficult to make decisions about the exact detection of whether landslides are shallow landslides or debris flows on a regional scale. It was possible to compute that the runout distances did not differ significantly when the simulations with both parameters were examined. It can also be stated that it was hard to distinguish differences between the debris flow analysis results and the shallow landslide analysis results. Changes in distances generally happened gradually in various areas for the debris flow analysis parameters. Moreover, the runout distances reached farther in the debris flow analyses. However, highly probable runout distances were not as common in the debris flow analyses as in the shallow landslide analyses. The highly probable range (0.5–1) indicates that a grid cell is involved in a runout zone with a probability value higher than 50%, while the low-probability range (0–0.5) means that a grid cell is included in a runout zone with a probability value lower than 50%. An assessment of the runout areas indicated that while the models with debris flow parameters impacted a total area of 146 km^2, the models with shallow landslide parameters impacted a total area of 101 km^2.

4.2. Shallow Landslide Runout Distance Assessment Considering RCP Scenarios

Including the RCP precipitation scenarios in the runout distance analyses in the study area adequately fulfilled the need for necessary precautions related to shallow landslides in the future. Simulations of possible runouts were carried out for shallow landslides and debris flow to investigate whether their parameters affect the runout distances in the Eocene flysch facies for detected landslide initiations, which were determined considering the RCP 4.5 and RCP 8.5 precipitation scenarios. Possible future runout areas that will be affected by debris flows or shallow landslides will decrease because of a reduction in shallow landslide initiations (Figures 13–15). It is also not surprising that due to a lack of shallow landslide initiations, shallow landslides will not occur in the Egerci, Beycuma, and Ihsanoglu sub-basins during some periods because shallow landslide initiations were factored into the estimation of the runout distances in this study. Although shallow landslide initiations will decrease in the Egerci sub-basin in the period of 2025–2044 in the RCP 4.5 scenario (Figure 14), it is possible to claim that there are no big differences between these runouts and the runouts shown earlier, which did not include the RCP precipitation scenarios. However, in the same period (2025–2044), there will be discernable declines in shallow landslide initiations and their runout distances in the Beycuma sub-basin. It is notable that dramatic declines in shallow landslide initiations will also be clearly visible in the Egerci and Beycuma sub-basins in the period of 2025–2044 for the RCP 8.5 scenario. In addition, Figures 13–15 give the opportunity to compare the effects of the RCP 4.5 and 8.5 precipitation scenarios for the Egerci sub-basin and the Beycuma sub-basin in the period of 2025–2044 in terms of the divergence of their runout distances. Compared with the initial analyses, there is a trend of decreases in shallow landslide initiations and their runout distances in the Egerci and Beycuma sub-basins. However, it is noteworthy that fluctuating changes in shallow landslide initiations and runout distances will be observed in both sub-basins. For instance, they will show an increasing trend after the period of 2044–2082 in the Beycuma sub-basin, despite the fact that they will not reach the levels of the initial analyses (Figure 13).

It should be underlined that all obtained runout model results depend on the RCP scenarios. If the RCP scenarios are changed or do not turn out as projected, results such as landslide initiations and their runout distances might change.

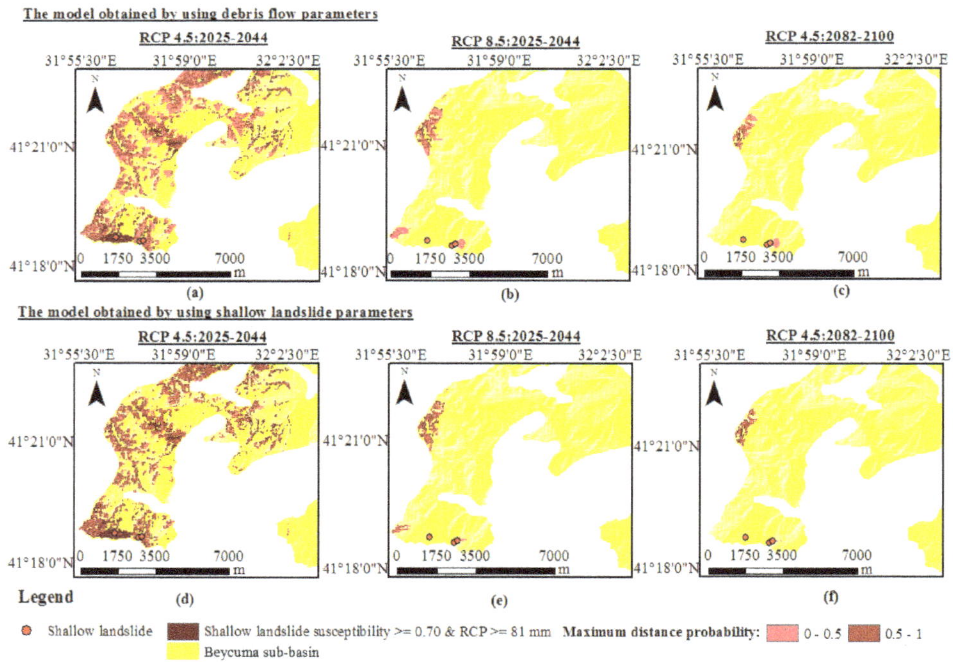

Figure 13. Probable runout distance models for the RCP 4.5: 2025–2044, RCP 8.5: 2025–2044, and RCP 4.5: 2082–2100 scenarios in the Beycuma sub-basin (**a–c**) using the debris flow parameters and (**d–f**) using the shallow landslide parameters.

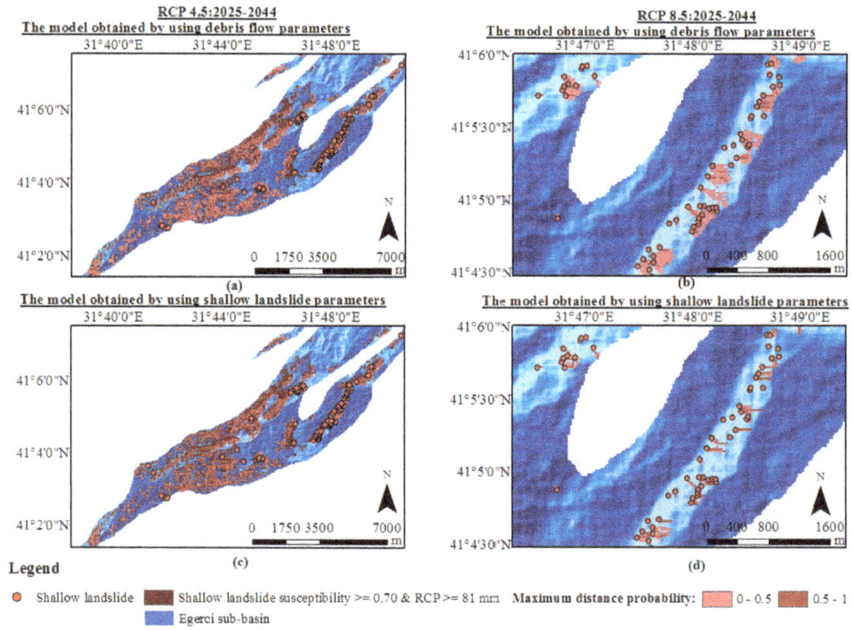

Figure 14. Probable runout distance models for the RCP 4.5: 2025–2044 and RCP 8.5: 2025–2044 scenarios in the Egerci sub-basin (**a**,**b**) using the debris flow parameters and (**c**,**d**) using the shallow landslide parameters.

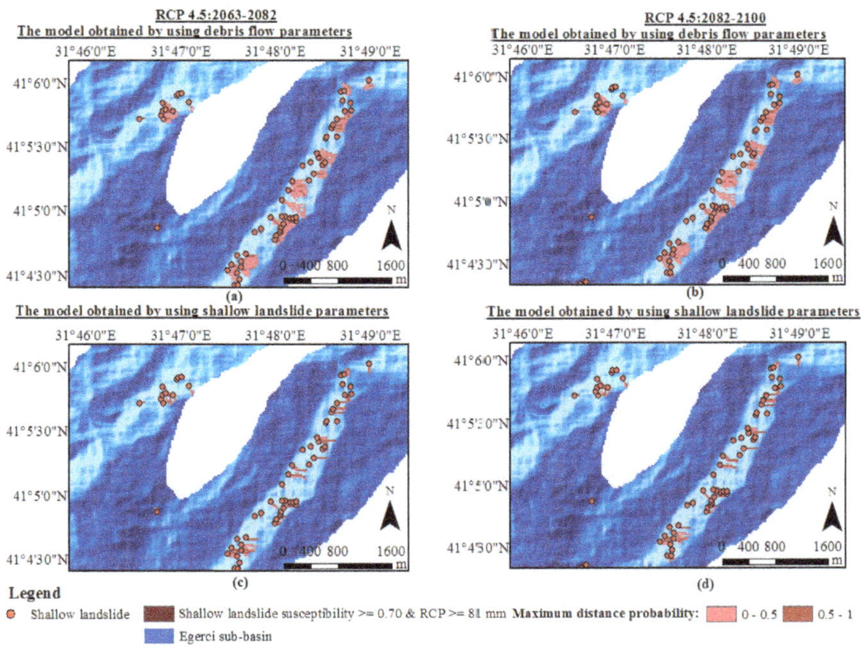

Figure 15. Probable runout distance models for the RCP 4.5: 2063–2082 and RCP 4.5: 2082–2100 scenarios in the Egerci sub-basin (**a**,**b**) using the debris flow parameters and (**c**,**d**) using the shallow landslide parameters.

5. Discussion

The runout distances of shallow landslides are important outputs that may facilitate improved resilience against shallow landslides. According to the results of empirical evaluations, runout model outputs indicate that shallow landslides can damage transportation networks and residential areas in this region. For example, Nie et al. [108] assessed the landslide risk along national highways based on information quantity and random forest coupling methods. Flow-R, a distributed empirical model, was implemented with great success not only in this study but also in many studies in the literature. Giano et al. [2], Polat and Erik [60], Charbel and El Hage Hassan [50], and Jiang et al. [51] discussed whether Flow-R and observed landslide runout results are compatible. The researchers concluded that Flow-R results are well matched with the observed results. Despite not considering rheological parameters, Flow-R model results are coherent [51,53]. Do et al. [49] stated that the Flow-R model may give exaggerated results. However, the results can be used to evaluate all possibilities. Putra et al. [53] also put forward that although Flow-R does not provide complete hazard mapping, it does enable the prediction of possible runout debris flow paths. In terms of analyzing shallow landslide runout distances in the three sub-basins, the selection of model parameters and data quality had significant roles. Despite employing a limited dataset, Flow-R was proven to attain accurate results, especially for the Ihsanoglu sub-basin, with an R-squared level of about 79%. Applying a machine learning algorithm was advantageous in this study during the preparation of a susceptibility map to obtain quite reliable shallow landslide initiations. Flow-R software can also be implemented to distinguish landslide initiations by applying a grid-based overlay approach [2,50,51,54]. However, the pre-determination of landslide initiations via machine learning is thought to have a positive effect, increasing the success of the analyses with respect to the accuracy of the runout distance probability. Shallow landslide initiations will decline in the future RCP scenarios for the sub-basins, which means the landslide runout distances will decrease [65,104]. If a more detailed examination is conducted on shallow landslide initiations, they are abundant in broad-leaved forests in the three sub-basins in the initial analyses. The literature review also showed that shallow landslide initiations are often observable in coniferous forests and broad-leaved forests [109,110]. However, climate change scenarios indicate that the susceptibility of broad-leaved forests will decrease [111,112]. If shallow landslide initiation detection analyses, including the RCP precipitation scenarios, are considered, shallow landslide initiations show a decreasing trend in broad-leaved forests, especially in the Beycuma sub-basin. The period between 2025 and 2044 will be the most affected period for both sub-basins in the RCP 4.5 scenario in terms of both shallow landslide initiations and their runout distances. These analysis results are critical for estimating the hazard rates of shallow landslides in the study area in terms of their contributions to disaster management strategies.

Limitations of this Study

Improving the spatial resolution of the DEM is important to obtain better results [50]. Horton et al. [42] stated that DEMs with a 10 m spatial resolution provide higher accuracies. However, a spatial resolution of 25 m can also be implemented successfully in analyses. The models in this study were also sensitive to DEM effects. Nevertheless, considering the size of the study area, the computational cost of higher resolutions could not be managed. Supercomputers may be a solution to this situation. Moreover, despite many attempts at changing the DEM resolution to analyze runout distances, propagations were not triggered for some initiations in datasets examined considering both 10 m and 25 m resolutions. The main reason for this peculiarity may be that the DEM stopped or was deflected by natural environment features and building features [53]. Furthermore, the selection of the velocity parameter also may have affected the results. Increasing runout distances should not be surprising in view of an increasing velocity. For instance, Horton et al. [105] suggest that the velocity limitation is 8 m/s for mudflows. Therefore, it has become clear that runout distances are affected by the maximum velocity limitation and should be considered cautiously during model parameter determinations. In this study, a back analysis was carried out just for this purpose. In addition, the landslide initiation process is primarily influenced by geological conditions [113]. Conducting further research that integrates lithological and geological investigations along with hydrogeological conditions would highlight the significance of these data in the initiation process of landslides [114]. Therefore, an additional focus on hydrogeological conditions may be required to improve the accuracy of probable shallow landslide initiations.

6. Conclusions

Runout distances have been ruled out in many studies. Considering the successful results, runout will be considered more in future studies because runout distance probabilities in susceptible areas enable us to make more appropriate decisions in hazardous regions. The goal of this study was to determine the shallow landslide runout distance probabilities in the Eocene flysch facies in the Western Black Sea region of Türkiye. This research offers evidence that the empirical–statistical method can successfully evaluate the runout distances in the study area. Despite the different empirical model parameters suggested for shallow landslides and debris flow, the models give nearly identical runout distance results in the Eocene flysch facies. Runout distances can be predicted at R-squared levels of 0.64 and 0.62 for the debris flow and shallow landslide models, respectively, in this study area. The total impacted runout areas were estimated to be approximately 146 km^2 and 101 km^2 for the debris flow and shallow landslide parameters, respectively. This study also reveals that machine learning algorithms can be remarkably fast and accurate in providing shallow landslide initiations in the sub-basins. The successful inclusion of the climate change scenarios in the runout distance analyses was ensured. The RCP 4.5 and RCP 8.5 scenarios also give the possible future runout distances in the Eocene flysch facies. The simulated runout results, including the RCP 4.5 and RCP 8.5 scenarios, prove that areas affected by shallow landslides or debris flows will decrease in the future due to the decrement in shallow landslide initiations. The analysis results also show that the time period of 2025–2044, in accordance with the RCP 4.5 scenario, persists as a challenging interval for the Egerci and Beycuma sub-basins due to shallow landslide occurrences and their runout distances. In conclusion, this study reveals that if runout analyses can be carried out successfully, they contribute to the preparation of effective disaster management policies to increase the resilience to landslides for societies living in mountainous areas. Using more advanced machine learning and statistical techniques to predict failure initiations for shallow landslides is recommended to improve the accuracy of the runout distance results in future work. It is foreseen that relations between the RCP precipitation scenarios and the runout distances of shallow landslides will increasingly seize attention in enhanced landslide studies in the future.

Author Contributions: Conceptualization, Muge Pinar Komu, Hakan Ahmet Nefeslioglu and Candan Gokceoglu; methodology, Muge Pinar Komu, Hakan Ahmet Nefeslioglu and Candan Gokceoglu; software, Muge Pinar Komu; writing—original draft preparation, Muge Pinar Komu; writing—review and editing, Candan Gokceoglu and Hakan Ahmet Nefeslioglu; supervision, Candan Gokceoglu and Hakan Ahmet Nefeslioglu. All authors have read and agreed to the published version of the manuscript.

Funding: This research received no external funding.

Data Availability Statement: Some or all of the data, models, or code that support the findings of this study are available from the corresponding author upon reasonable request.

Acknowledgments: This work is part of the Ph.D. thesis of Muge Pinar Komu, which is supervised by Candan Gokceoglu and Hakan Ahmet Nefeslioglu.

Conflicts of Interest: The authors declare no conflicts of interest.

References

1. Kritikos, T.; Davies, T. Assessment of Rainfall-Generated Shallow Landslide/Debris-Flow Susceptibility and Runout Using a GIS-Based Approach: Application to Western Southern Alps of New Zealand. *Landslides* **2015**, *12*, 1051–1075. [CrossRef]
2. Giano, S.I.; Pescatore, E.; Siervo, V. Morphometry and Debris-Flow Susceptibility Map in Mountain Drainage Basins of the Vallo Di Diano, Southern Italy. *Remote Sens.* **2021**, *13*, 3254. [CrossRef]
3. Poltnig, W.; Bak, R.; Berg, W.; Keršmanc, T. Runout-Modelling of Shallow Landslides in Carinthia (Austria). *Aust. J. Earth Sci.* **2016**, *109*, 59–67. [CrossRef]
4. Kaafarani, R.; Jaoude, G.A.; Wartman, J.; Tawk, M. Landslide Susceptibility Mapping Based on Triggering Factors Using a Multi-Modal Approach. In *MATEC Web Conference, Proceedings of the International Conference of Engineering Risk, Beirut, Lebanon, 3–5 April 2019*; Chehade, F.H., Ed.; EDP Sciences: Les Ulis Cedex, France, 2019; Volume 281, pp. 1–8. [CrossRef]
5. Mergili, M.; Schwarz, L.; Kociu, A. Combining Release and Runout in Statistical Landslide Susceptibility Modeling. *Landslides* **2019**, *16*, 2151–2165. [CrossRef]
6. Zhou, W.; Fang, J.; Tang, C.; Yang, G. Empirical Relationships for The Estimation of Debris Flow Runout Distances on Depositional Fans in The Wenchuan Earthquake Zone. *J. Hydrol.* **2019**, *577*, 123932. [CrossRef]
7. Apriani, D.W.; Crededi, C.; Khala, S. An Empirical-Statistical Model for Landslide Runout Distance Prediction in Indonesia. *Tahun* **2022**, *27*, 1. [CrossRef]
8. Di Napoli, M.; Martire, D.D.; Bausilio, G.; Calcaterra, D.; Confuorto, P.; Firpo, M.; Pepe, G.; Cevasco, A. Rainfall-Induced Shallow Landslide Detachment, Transit and Runout Susceptibility Mapping by Integrating Machine Learning Techniques and GIS-based Approaches. *Water* **2021**, *13*, 488. [CrossRef]
9. Guthrie, R.; Befus, A. DebrisFlow Predictor: An Agent-Based Runout Program for Shallow Landslides. *Nat. Hazards Earth Syst. Sci.* **2021**, *21*, 1029–1049. [CrossRef]
10. Ju, L.-Y.; Xiao, T.; He, J.; Wang, H.-J.; Zhang, L.-M. Predicting Landslide Runout Paths Using Terrain Matching-Targeted Machine Learning, Engineering. *Geology* **2022**, *311*, 106902. [CrossRef]
11. Giarola, A.; Meisina, C.; Tarolli, P.; Zucca, F.; Galve, J.P.; Bordoni, M. A Data-Driven Method for the Estimation of Shallow Landslide Runout. *CATENA* **2024**, *234*, 107573. [CrossRef]
12. Baselt, I.; Queiroz de Oliveira, G.; Fischer, J.-T.; Pudasaini, S.P. Evolution of Stony Debris Flows in Laboratory Experiments. *Geomorphology* **2021**, *372*, 107431. [CrossRef]
13. Gao, Y.; Li, B.; Zhang, H.; Wu, W.; Li, J.; Yin, Y. Numerical Modeling of Mixed Two-Phase in Long Runout Flow-Like Landslide Using LPF3D. *Landslides* **2023**, *21*, 641–660. [CrossRef]
14. Yang, L.; Wei, Y.; Wang, W.; Zhu, S. Numerical Runout Modeling Analysis of the Loess Landslide at Yining, Xinjiang, China. *Water* **2019**, *11*, 1324. [CrossRef]
15. Chae, B.-G.; Wu, Y.-H.; Liu, K.-F.; Choi, J.; Park, H.-J. Simulation of Debris-Flow Runout Near a Construction Site in Korea. *Appl. Sci.* **2020**, *10*, 6079. [CrossRef]
16. Calista, M.; Menna, V.; Mancinelli, V.; Sciarra, N.; Miccadei, E. Rockfall and Debris Flow Hazard Assessment in the SW Escarpment of Montagna del Morrone Ridge (Abruzzo, Central Italy). *Water* **2020**, *12*, 1206. [CrossRef]
17. Abraham, M.T.; Satyam, N.; Reddy, S.K.P.; Pradhan, B. Runout Modeling and Calibration of Friction Parameters of Kurichermala Debris Flow, India. *Landslides* **2021**, *18*, 737–754. [CrossRef]
18. Dash, R.K.; Kanungo, D.P.; Malet, J.P. Runout Modelling and Hazard Assessment of Tangni Debris Flow in Garhwal Himalayas, India. *Environ. Earth Sci.* **2021**, *80*, 338. [CrossRef]
19. Vicari, H.; Nordal, S.; Thakur, V. The Significance of Entrainment on Debris Flow Modelling: The Case of Hunnedalen, Norway. In *Challenges and Innovations in Geomechanics, IACMAG 2021*; Barla, M., Di Donna, A., Sterpi, D., Eds.; Lecture Notes in Civil Engineering; Springer: Cham, Switzerland, 2021; Volume 126, ISBN 978-3-030-64517-5. [CrossRef]
20. Mikoš, M.; Bezak, N. Debris Flow Modelling Using RAMMS Model in the Alpine Environment with Focus on the Model Parameters and Main Characteristics. *Front. Earth Sci.* **2021**, *8*, 605061. [CrossRef]
21. Oh, C.-H.; Choo, K.-S.; Go, C.-M.; Choi, J.-R.; Kim, B.-S. Forecasting of Debris Flow Using Machine Learning-Based Adjusted Rainfall Information and RAMMS Model. *Water* **2021**, *13*, 2360. [CrossRef]
22. Zhou, W.; Qiu, H.; Wang, L.; Pei, Y.; Tang, B.; Ma, S.; Yang, D.; Cao, M. Combining Rainfall-Induced Shallow Landslides and Subsequent Debris Flows for Hazard Chain Prediction. *CATENA* **2022**, *213*, 106199. [CrossRef]
23. Alene, G.H.; Vicari, H.; Irshad, S.; Perkis, A.; Bruland, O.; Thakur, V. Realistic Visualization Of Debris Flow Type Landslides Through Virtual Reality. *Landslides* **2023**, *20*, 13–23. [CrossRef]
24. La Porta, G.; Leonardi, A.; Pirulli, M.; Cafaro, F.; Castelli, F. Time-Resolved Triggering and Runout Analysis of Rainfall-Induced Shallow Landslides. *Acta Geotech.* **2023**. [CrossRef]
25. Zhu, C.; He, M.; Karakus, M.; Zhang, X.; Tao, Z. Numerical Simulations of the Failure Process of Anaclinal Slope Physical Model and Control Mechanism of Negative Poisson's Ratio Cable. *Bull. Eng. Geol. Environ.* **2021**, *80*, 3365–3380. [CrossRef]
26. Clark, B. Numerical Modelling of Debris Flow Hazards using Computational Fluid Dynamics. Master's Thesis, Norwegian University of Science and Technology, Trondheim, Norway, 2018.
27. Melo, R.; van Asch, T.; Zêzere, J.L. Debris Flow Run-Out Simulation and Analysis Using a Dynamic Model. *Nat. Hazards Earth Syst. Sci.* **2018**, *18*, 555–570. [CrossRef]

28. Gan, J.J.; Zhang, Y.X. Numerical Simulation of Debris Flow Runout Using Ramms: A Case Study of Luzhuang Gully in China. *CMES* **2019**, *121*, 981–1009. [CrossRef]
29. Peruzzetto, M.; Mangeney, A.; Grandjean, G.; Levy, C.; Thiery, Y.; Rohmer, J.; Lucas, A. Operational Estimation of Landslide Runout: Comparison of Empirical and Numerical Methods. *Geosciences* **2020**, *10*, 424. [CrossRef]
30. Zhu, C.; He, M.; Karakus, M.; Cui, X.; Tao, Z. Investigating Toppling Failure Mechanism of Anti-Dip Layered Slope due to Excavation by Physical Modelling. *Rock. Mech. Rock. Eng.* **2020**, *53*, 5029–5050. [CrossRef]
31. Wei, L.; Cheng, H.; Dai, Z. Propagation Modeling of Rainfall-Induced Landslides: A Case Study of the Shaziba Landslide in Enshi, China. *Water* **2023**, *15*, 424. [CrossRef]
32. Iverson, R.M. Scaling and Design of Landslide and Debris-Flow Experiments. *Geomorphology* **2015**, *244*, 9–20. [CrossRef]
33. Khalkhali, A.B.; Koochaksaraei, M.K. Evaluation of Limit Equilibrium and Finite Element Methods in Slope Stability Analysis Case Study of Zaremroud Landslide, Iran. *Comput. Eng. Phys. Model.* **2019**, *2*, 1–15.
34. Guo, J.; Cui, Y.; Xu, W.; Yin, Y.; Li, Y.; Jin, W. Numerical Investigation of the Landslide-Debris Flow Transformation Process Considering Topographic and Entrainment Effects: A Case Study. *Landslides* **2022**, *19*, 773–788. [CrossRef]
35. Milledge, D.G.; Densmore, A.L.; Bellugi, D.; Rosser, N.J.; Watt, J.; Li, G.; Oven, K.J. Simple Rules to Minimise Exposure to Coseismic Landslide Hazard. *Nat. Hazards Earth Syst. Sci.* **2019**, *19*, 837–856. [CrossRef]
36. Vadivel, S.; Sennimalai, C.S. Failure Mechanism of Long-Runout Landslide Triggered by Heavy Rainfall in Achanakkal, Nilgiris, India. *J. Geotech. Geoenviron. Eng.* **2019**, *145*, 04019047. [CrossRef]
37. Ali, S.; Haider, R.; Abbas, W. Empirical Assessment of Rockfall and Debris Flow Risk along the Karakoram Highway, Pakistan. *Nat. Hazards* **2021**, *106*, 2437–2460. [CrossRef]
38. Sturzenegger, M.; Holm, K.; Lau, C.; Jakob, M. Semi-Automated Regional Scale Debris-Flow and Debris-Flood Susceptibility Mapping Based on Digital Elevation Model Metrics and Flow-R Software. In Proceedings of the 7th International Conference on Debris-Flow Hazards Mitigation, Golden, CO, USA, 10–13 June 2019.
39. Pradhan, A.M.S. Quantification of Slope Unit Wise Vulnerability in Terms of Building Aggregation. *WRRDC Res. Lett.* **2021**, *7*, 1–2.
40. Quinn, P.; Beven, K.; Chevallier, P.; Planchon, O. The Prediction of Hillslope Flow Paths for Distributed Hydrological Modelling Using Digital Terrain Models. *Hydrol. Process* **1991**, *5*, 59–79. [CrossRef]
41. Holmgren, P. Multiple Flow Direction Algorithms for Runoff Modelling in Grid Based Elevation Models: An Empirical Evaluation. *Hydrol. Process* **1994**, *8*, 327–334. [CrossRef]
42. Horton, P.; Jaboyedoff, M.; Rudaz, B.; Zimmermann, M. Flow-R, a Model for Susceptibility Mapping of Debris Flows and other Gravitational Hazards at a Regional Scale. *Nat. Hazards* **2013**, *13*, 869–885. [CrossRef]
43. Guthrie, R.H.; Deadman, P.; Cabrera, R.; Evans, S.G. Exploring the Magnitude-Frequency Distribution: A Cellular Automata Model for Landslides. *Landslides* **2008**, *5*, 151–159. [CrossRef]
44. Gamma, P. dfwalk—Ein Murgang-Simulationsprogramm zur Gefahrenzonierung. In *Geographica Bernensia G66*; University of Bern: Bern, German, 2000.
45. Mergili, M.; Krenn, J.; Chu, H.-J. r.randomwalk v1, A Multi-Functional Conceptual Tool for Mass Movement Routing. *Geosci. Model. Dev.* **2015**, *8*, 4027–4043. [CrossRef]
46. Pastorello, R.; Michelini, T.; D'Agostino, V. On the Criteria to Create a Susceptibility Map to Debris Flow at a Regional Scale Using Flow-R. *J. Mt. Sci.* **2017**, *14*, 621–635. [CrossRef]
47. McCoy, K.M. Debris-Flow Susceptibility Mapping in Colorado using Flow-R: Calibration Techniques and Selected Examples. In Proceedings of the 7th International Conference on Debris-Flow Hazards Mitigation, Golden, CO, USA, 10–13 June 2019.
48. Paudel, B.P. GIS-Based Assessment of Debris Flow Susceptibility and Hazard in Mountainous Regions of Nepal. Ph.D. Thesis, Ottawa University, Ottawa, ON, Canada, 2019.
49. Do, H.M.; Yin, K.L.; Guo, Z.Z. A comparative study on the integrative ability of the analytical hierarchy process, weights of evidence and logistic regression methods with the Flow-R model for landslide susceptibility assessment. *Geomat. Nat. Haz. Risk* **2020**, *11*, 2449–2485. [CrossRef]
50. Charbel, L.; El Hage Hassan, H. Mudflow Modeling Using Flow-R Software: Case Study of Ras Baalbek Basin (Lebanon). *Geo-Eco-Trop* **2021**, *45*, 475–486.
51. Jiang, N.; Su, F.; Li, Y.; Guo, X.; Zhang, J.; Liu, X. Debris Flow Assessment in the Gaizi-Bulunkou Section of Karakoram Highway. *Front. Earth Sci.* **2021**, *9*, 660579. [CrossRef]
52. Liu, J.; Wu, Y.; Gao, X.; Zhang, X. A Simple Method of Mapping Landslides Runout Zones Considering Kinematic Uncertainties. *Remote Sens.* **2022**, *14*, 668. [CrossRef]
53. Putra, M.; Dinata, I.; Sadisun, I.; Sarah, D.; Aulia, A.; Sukristiyanti, S. Modeling of Individual Debris Flows Based on DEMNAS Using Flow-R: A Case Study in Sigi. Central Sulawesi. *RISET Geol. Dan Pertamb.* **2022**, *32*, 37–58. [CrossRef]
54. Xu, H.; Su, P.; Chen, Q.; Liu, F.; Zhou, Q.; Liu, L. Susceptibility Areas Identification and Risk Assessment of Debris Flow Using the Flow-R Model: A Case Study of Basu County of Tibet. *Geoenviron. Disasters* **2022**, *9*, 13. [CrossRef]
55. Berti, M.; Simoni, A. DFLOWZ: A Free Program to Evaluate the Area Potentially Inundated by a Debris Flow. *Comput. Geosci.* **2014**, *67*, 14–23. [CrossRef]
56. Gorr, A.N.; McGuire, L.A.; Youberg, A.M.; Rengers, F.K. A Progressive Flow-Routing Model for Rapid Assessment of Debris-Flow Inundation. *Landslides* **2022**, *19*, 2055–2073. [CrossRef]

57. Wang, X.; Yin, J.; Luo, M.; Ren, H.; Li, J.; Wang, L.; Li, D.; Li, G. Active High-Locality Landslides in Mao County: Early Identification and Deformational Rules. *J. Earth Sci.* **2023**, *34*, 1596–1615. [CrossRef]
58. Pradhan, A.M.S.; Kang, H.; Lee, S.; Kim, Y. Spatial Model Integration for Shallow Landslide Susceptibility and Its Runout Using a GIS-Based Approach in Yongin, Korea. *Geocarto Int.* **2017**, *32*, 420–441. [CrossRef]
59. Rahman, M.S.; Ahmed, B.; Di, L.P. Landslide Initiation and Runout Susceptibility Modeling in the Context of Hill Cutting and Rapid Urbanization: A Combined Approach of Weights of Evidence and Spatial Multi-Criteria. *J. Mt. Sci.* **2017**, *14*, 1919–1937. [CrossRef]
60. Polat, A.; Erik, D. Debris Flow Susceptibility and Propagation Assessment in West Koyulhisar, Turkey. *J. Mt. Sci.* **2020**, *17*, 2611–2623. [CrossRef]
61. Park, D.; Lee, S.; Nikhil, N.V.; Kang, S.; Park, J. Debris Flow Hazard Zonation by Probabilistic Analysis (Mt. Woomyeon, Seoul, Korea). *Int. J. Innov. Res. Sci. Eng. Technol.* **2013**, *2*, 2381–2390.
62. Bera, S.; Melo, R.; Guru, B. Assessment of Exposed Elements in a Changing Built Environment by Using an Integrated Model of Debris Flow Initiation and Runout (Kalimpong Region, Himalaya). *Bull. Eng. Geol. Environ.* **2021**, *80*, 7131–7152. [CrossRef]
63. Sharma, C.P.; Kumar, A.; Chahal, P.; Shukla, U.K.; Srivastava, P.; Jaiswal, M.K. Debris Flow Susceptibility Assessment of Leh Valley, Ladakh, Based on Concepts of Connectivity, Propagation and Evidence-Based Probability. *Nat. Hazards* **2023**, *115*, 1833–1859. [CrossRef]
64. Guo, J.; Wang, J.; Li, Y.; Yi, S. Discussions on the Transformation Conditions of Wangcang Landslide-Induced Debris Flow. *Landslides* **2021**, *18*, 1833–1843. [CrossRef]
65. Ageenko, A.; Hansen, L.C.; Lyng, K.L.; Bodum, L.; Arsanjani, J.J. Landslide Susceptibility Mapping Using Machine Learning: A Danish Case Study. *ISPRS Int. J. Geo-Inf.* **2022**, *11*, 324. [CrossRef]
66. Araújo, J.R.; Ramos, A.M.; Soares, P.M.M.; Melo, R.; Oliveira, S.C.; Trigo, R.M. Impact of Extreme Rainfall Events on Landslide Activity in Portugal Under Climate Change Scenarios. *Landslides* **2022**, *19*, 2279–2293. [CrossRef]
67. Nefros, C.; Tsagkas, D.S.; Kitsara, G.; Loupasakis, C.; Giannakopoulos, C. Landslide Susceptibility Mapping under the Climate Change Impact in the Chania Regional Unit, West Crete, Greece. *Land* **2023**, *12*, 154. [CrossRef]
68. Chen, Y.; Chen, E.; Zhang, J.; Zhu, J.; Xiao, Y.; Dai, Q. Investigation of Model Uncertainty in Rainfall-Induced Landslide Prediction under Changing Climate Conditions. *Land* **2023**, *12*, 1732. [CrossRef]
69. Guo, Z.; Ferrer, J.V.; Hürlimann, M.; Medina, V.; Puig-Polo, C.; Yin, K.; Huang, D. Shallow Landslide Susceptibility Assessment Under Future Climate and Land Cover Changes: A case Study from Southwest China. *Geosci. Front.* **2023**, *14*, 101542. [CrossRef]
70. Akbas, B.; Akdeniz, N.; Aksay, A.; Altun, İ.E.; Balcı, V.; Bilginer, E.; Bilgiç, T.; Duru, M.; Ercan, T.; Gedik, İ.; et al. 1:1250,000 olcekli Türkiye jeoloji haritasi. *Maden Tetk. Ve Aram. Genel Mudurlugu Yayin.* **2011**, 12p.
71. Turkish State Meteorological Service. Maximum Rainfall Intensity and Recurrence Analysis. Available online: https://mgm.gov.tr/veridegerlendirme/maksimum-yagislar.aspx (accessed on 23 December 2022).
72. Can, T.; Nefeslioglu, H.A.; Gokceoglu, C.; Sonmez, H.; Duman, T.Y. Susceptibility Assessments of Shallow Earthflows Triggered by Heavy Rainfall at Three Catchments by Logistic Regression Analyses. *Geomorphology* **2005**, *72*, 250–271. [CrossRef]
73. Turkish State Meteorological Service. Daily Historical Precipitation (mm=kg/m^2) OMGİ and MANUEL, 2023, Data Sheets provided by Turkish State Meteorological Service.
74. Akgun, A.; Gorum, T.; Nefeslioglu, H. Landslide Size Distribution Characteristics of Cretaceous and Eocene Flysch Assemblages in the Western Black Sea Region of Turkey. In *Understanding and Reducing Landslide Disaster Risk*; Guzzetti, F., Arbanas, S.M., Eds.; Springer: London, UK; Berlin/Heidelberg, Germany, 2021; Volume 2, pp. 299–303. [CrossRef]
75. Turkish State Meteorological Service. Daily Total Precipitation (mm). Available online: https://www.mgm.gov.tr/sondurum/toplam-yagis.aspx (accessed on 21 November 2023).
76. Varnes, D.J. Slope movement types and processes. In *Special Report 176: Landslides: Analysis and Control*; Schuster, R.L., Krizek, R.J., Eds.; Transportation and Road Research Board, National Academy of Science: Washington, DC, USA, 1978; pp. 11–33.
77. CORINE Land Cover Change 2012–2018 (Vector/Raster 100 m), Europe, 6-Yearly, CLC 2018 Version v2020 20u1. Available online: https://land.copernicus.eu/pan-european/corine-land-cover/lcc-2012-2018?tab=download (accessed on 16 August 2022).
78. Duman, T.; Emre, O.; Can, T.; Nefeslioglu, H.A.; Kecer, M.; Dogan, A.; Durmaz, S.; Ates, S. Türkiye Heyelan Envanteri Haritasi, 1/500,000 olcekli Zonguldak Paftasi. Maden Tetkik ve Arama Genel Mudurlugu (MTA). *Ozel Yayinlar Serisi* **2005**, *4a*, 10p.
79. Duman, T.Y.; Can, T.; Emre, O.; Kecer, M.; Dogan, A.; Ates, S.; Durmaz, S. Landslide Inventory of Northwestern Anatolia, Turkey. *J. Eng. Geol.* **2005**, *77*, 99–114. [CrossRef]
80. Duman, T.Y.; Can, T.; Emre, O. 1/1500000 olcekli Türkiye Heyelan Envanteri Haritasi. *Maden Tetkik ve Arama Genel Mudurlugu (MTA) Ozel Yayinlar Serisi* **2011**, *27*, 12p.
81. Hovius, N.; Stark, C.P.; Allen, P.A. Sediment Flux from a Mountain Belt Derived by Landslide Mapping. *Geology* **1997**, *25*, 231–234. [CrossRef]
82. Jaboyedoff, M.; Carrea, D.; Derron, M.H.; Thierry, O.; Penna, I.; Bejamin, R. A Review of Methods Used to Estimate Initial Landslide Failure Surface Depths and Volumes. *Eng. Geol.* **2020**, *267*, 105478. [CrossRef]
83. Zaruba, Q.; Mencl, V. *Landslides and Their Control*; Elsevier: Amsterdam, The Netherlands, 1969.
84. Nefeslioglu, H.A.; Gokceoglu, C. Probabilistic Risk Assessment in Medium Scale for Rainfall Induced Earthflows: Catakli Catchment Area (Cayeli, Rize, Turkey). *Math Probl. Eng.* **2011**, *2011*, 280431. [CrossRef]

85. Kalantar, B.; Pradhan, B.; Naghibi, S.A.; Motevalli, A.; Mansor, S. Assessment of the Effects of Training Data Selection on the Landslide Susceptibility Mapping: A Comparison between Support Vector Machine (SVM), Logistic Regression (LR) and Artificial Neural Networks (ANN). *Geomat. Nat. Haz. Risk* **2018**, *9*, 49–69. [CrossRef]
86. Nhu, V.-H.; Shirzadi, A.; Shahabi, H.; Singh, S.K.; Al-Ansari, N.; Clague, J.J.; Jaafari, A.; Chen, W.; Miraki, S.; Dou, J.; et al. Shallow Landslide Susceptibility Mapping: A Comparison between Logistic Model Tree, Logistic Regression, Naïve Bayes Tree, Artificial Neural Network, and Support Vector Machine Algorithms. *Int. J. Environ. Res. Public Health* **2020**, *17*, 2749. [CrossRef]
87. Crawford, M.M.; Dortch, J.M.; Koch, H.J.; Killen, A.A.; Zhu, J.; Zhu, Y.; Bryson, L.S.; Haneberg, W.C. Using Landslide-Inventory Mapping for a Combined Bagged-Trees and Logistic-Regression Approach to Determining Landslide Susceptibility in Eastern Kentucky, USA. *Q. J. Eng. Geol. Hydrogeol.* **2021**, *54*, qjegh2020-177. [CrossRef]
88. Crawford, M.M.; Dortch, J.M.; Koch, H.J.; Zhu, Y.; Haneberg, W.C.; Wang, Z.; Bryson, L.S. Landslide Risk Assessment in Eastern Kentucky. USA: Developing a Regional Scale, Limited Resource Approach. *Remote Sens.* **2022**, *14*, 6246. [CrossRef]
89. Dashbold, B.; Bryson, L.S.; Crawford, M.M. Landslide Hazard and Susceptibility Maps Derived from Satellite and Remote Sensing Data Using Limit Equilibrium Analysis and Machine Learning Model. *Nat. Hazards* **2022**, *116*, 235–265. [CrossRef]
90. Nwazelibe, V.; Unigwe, C.; Egbueri, J. Integration and Comparison of Algorithmic Weight of Evidence and Logistic Regression in Landslide Susceptibility Mapping of the Orumba North Erosion-Prone Region, Nigeria. *Model Earth Syst. Environ.* **2022**, *9*, 967–986. [CrossRef]
91. Polykretis, C.; Chalkias, C. Comparison and Evaluation of Landslide Susceptibility Maps Obtained from Weight of Evidence, Logistic Regression, and Artificial Neural Network Models. *Nat. Hazards* **2018**, *93*, 249–274. [CrossRef]
92. Menard, S. *Applied Logistic Regression Analysis*; Sage University Paper Series on Quantitative Applications in Social Sciences, Sage Publications No. 106; SAGE Publications: Thousand Oaks, CA, USA, 1995.
93. Ghasemian, B.; Shahabi, H.; Shirzadi, A.; Al-Ansari, N.; Jaafari, A.; Geertsema, M.; Melesse, A.M.; Singh, S.K.; Ahmad, A. Application of a Novel Hybrid Machine Learning Algorithm in Shallow Landslide Susceptibility Mapping in a Mountainous Area. *Front. Environ. Sci.* **2022**, *10*, 897254. [CrossRef]
94. Xie, W.; Li, X.; Jian, W.; Yang, Y.; Liu, H.; Robledo, L.F.; Nie, W. A Novel Hybrid Method for Landslide Susceptibility Mapping-Based GeoDetector and Machine Learning Cluster: A Case of Xiaojin County, China. *ISPRS Int. J. Geo-Inf.* **2021**, *10*, 93. [CrossRef]
95. Xie, W.; Nie, W.; Saffari, P.; Robledo, L.F.; Descote, P.-Y.; Jian, W. Landslide Hazard Assessment Based on Bayesian Optimization-Support Vector Machine in Nanping City, China. *Nat. Hazards* **2021**, *109*, 931–948. [CrossRef]
96. Karakaş, G.; Can, R.; Kocaman, S.; Nefeslioglu, H.A.; Gokceoglu, C. Landslide Susceptibility Mapping with Random Forest Model for Ordu, Turkey. *Int. Arch. Photogramm. Remote Sens. Spatial Inf. Sci.* **2020**, *XLIII-B3-2020*, 1229–1236. [CrossRef]
97. Ling, S.; Zhao, S.; Huang, J.; Zhang, X. Landslide Susceptibility Assessment Using Statistical and Machine Learning Techniques: A Case Study in The Upper Reaches of the Minjiang River, Southwestern China. *Front. Earth Sci.* **2022**, *10*, 986172. [CrossRef]
98. NCAR Geographic Information System Program. Climate Change Scenarios. Available online: https://gis.ucar.edu/inspector (accessed on 16 August 2022).
99. Sun, L.; Wei, Y.; Cai, H.; Yan, J.; Xiao, J. Improved Fast Adaptive IDW Interpolation Algorithm based on the Borehole Data Sample Characteristic and Its Application. *J. Phys. Conf. Ser.* **2019**, *1284*, 012074. [CrossRef]
100. Maleika, W. Inverse distance weighting method optimization in the process of digital terrain model creation based on data collected from a multibeam echosounder. *Appl. Geomat.* **2020**, *12*, 397–407. [CrossRef]
101. Mohamed Yusof, M.K.T.; A Rashid, A.S.; Abdul Khanan, M.F.; Abdul Rahman, M.Z.; Abdul Manan, W.A.; Kalatehjari, R.; Dehghanbanadaki, A. Assessing the Impact of RCP4.5 and RCP8.5 Scenarios on Landslide Susceptibility Mapping Using Support Vector Machine: A case study of Penang Island, Malaysia. *Phys. Chem. Earth Parts A/B/C* **2024**, *133*, 103496. [CrossRef]
102. Nasidi, N.M.; Wayayok, A.; Abdullah, A.F.; Kassim, M.S.M. Dynamics of Potential Precipitation under Climate Change Scenarios at Cameron Highlands, Malaysia. *SN Appl. Sci.* **2021**, *3*, 334. [CrossRef]
103. New Decade Brings Reference Period Change for Climate Data. Available online: https://climate.copernicus.eu/new-decade-brings-reference-period-change-climate-data (accessed on 9 October 2022).
104. Park, S.J.; Lee, D. Predicting Susceptibility to Landslides Under Climate Change Impacts in Metropolitan Areas of South Korea Using Machine Learning. *Geomat. Nat. Hazards Risk* **2021**, *12*, 2462–2476. [CrossRef]
105. Horton, P.; Oppikofer, T.; Michoud, C. Introduction to Flow R Online Tutorial © Terranum Ltd. Available online: https://www.terranum.ch/en/products/flow-r/ (accessed on 10 August 2022).
106. Prochaska, A.B.; Santi, P.M.; Higgins, J.D.; Cannon, S.H. A Study of Methods to Estimate Debris Flow Velocity. *Landslides* **2008**, *5*, 431–444. [CrossRef]
107. Nie, Y.; Li, X.; Xu, R. Dynamic Hazard Assessment of Debris Flow Based on TRIGRS and Flow-R Coupled Models. *Stoch. Environ. Res. Risk Assess* **2022**, *36*, 97–114. [CrossRef]
108. Nie, Z.; Lang, Q.; Zhang, Y.; Zhang, J.; Chen, Y.; Pan, Z. Risk Assessment of Landslide Collapse Disasters along National Highways Based on Information Quantity and Random Forest Coupling Methods: A Case Study of the G331 National Highway. *ISPRS Int. J. Geo-Inf.* **2023**, *12*, 493. [CrossRef]
109. Takaoka, S. A Landscape Level Study on Vegetation Richness of Ancient Landslide Areas. *Prog. Phys. Geogr.* **2023**, *48*, 1–15. [CrossRef]

110. Nishioka, M.; Inoue, H.; Ota, T.; Mizoue, N. Impact of Forest Type and Age on Shallow Landslide Susceptibility a Case Study from the 2017 Heavy Rainfall in Northern Kyushu. *Jpn. Socioecon. Plan. Manag.* **2023**, *28*, 389–396. [CrossRef]
111. Malla, R.; Neupane, P.R.; Köhl, M. Climate Change Impacts Vegetation Shift of Broad Leaved and Coniferous Forests. *Trees For. People* **2023**, *14*, 100457. [CrossRef]
112. Yu, J.; Berry, P.; Guillod, B.P.; Hickler, T. Climate Change Impacts on the Future of Forests in Great Britain. *Front. Environ. Sci.* **2021**, *9*, 640530. [CrossRef]
113. Du, Y.; Xie, M.; Jia, J. Stepped Settlement: A Possible Mechanism for Translational Landslides. *Catena* **2020**, *187*, 104365. [CrossRef]
114. Grabar, K.; Jug, J.; Bek, A.; Strelec, S. Comparison of the Piezocone Penetrometer (CPTU) and Flat Dilatometer (DMT) Methods for Landslide Characterisation. *Geosciences* **2024**, *14*, 64. [CrossRef]

applied sciences

MDPI

Article

Mechanics and Stability of Force Chain Arch in Excavated Granular Material

Meimei Wang *, Jianwei Zheng and Shanshan Xue

Deep Mining and Rock Burst Research Institute, Chinese Institute of Coal Science, Beijing 100013, China; zhengjanvy@163.com (J.Z.); mkzy_xsshan@126.com (S.X.)
* Correspondence: mmwang0302@hotmail.com

Abstract: Rock and soil masses in geotechnical engineering projects, such as tunnels, mines and slopes, undergo relative motion, exhibiting mechanical characteristics of solid–fluid transition under critical conditions. This work analyzes the characteristics of the solid–fluid transition interface and the mode of load transfer through biaxial compression particle flow photoelastic experiments on granular materials. The study documents that this interface forms an arch shape, marked by a force chain arch. The granular material exhibits two distinct states depending on its position: below the arch, the granular material is in a solid–fluid transitional state, with bearing capacity reduced, while above the arch, it is in a stable solid state, capable of bearing the overlying rock layer's load. The presence of the force chain arch alters the direction of the originally downward-transferring load, redirecting it along the trajectory of the arch. Analysis of the force and stability of the force chain arch revealed that the arch shape parameters and boundary loads control the instability of the arch. Changes in the overlying and lateral loads lead to different types of instability of the force chain arch. The findings of the study are crucial for underground engineering construction and for the prevention of geological disasters related to granular material.

Keywords: force chain arch; state of stress equilibrium; arch trajectory; critical transition; stability

1. Introduction

The excavation of geotechnical engineering projects—such as tunnel excavation, underground mining, foundation pit excavation and slopes—causes disturbances, which create free surfaces disrupting the original mechanical equilibrium of the rock and soil masses. The modified masses attain a state of self-stabilization through internal particle reorganization and stress redistribution [1–6]. The rock and soil masses below the free surface remain in a fluid state during this self-stabilization process. According to the natural equilibrium arch theory of Protodyakonov [7–9], unbraced excavated rock tunnels with joints experience collapse, eventually forming a temporarily stable equilibrium arch, known as the natural equilibrium arch. Terzaghi [10] proposed that under the assumption of equilibrium arch conditions, the rock and soil masses form a resistance to shearing along the vertical planes extending upwards from both sides of the arch foot, which can be used to estimate the pressure reduction above the trapdoor due to the arching of the rock and soil masses. The deformations generated by the Terzaghi principle in a porous medium, in the context of tunnel works in soft rock and granular terrain, have been resolved with the use of FEM [11]. Chen et al. [12] carried out a continuous loading experiment with a two-dimensional granular system and obtained the experimental data through digital image correlation (DIC). The experimental results indicated that the evolution of force chains was directly related to the number, geometric properties and permutation distribution of granules in direct contact with the external load. Li et al. [13] analyzed the stability of the arch structure via theoretical analysis and particle flow numerical simulation, and they found out that during the mining stage of the working face, the rock strata arch structure experiences a process of arch breaking, and the instability of the arch structure is the root cause of increasing surface subsidence damage. Handy [14] analyzed the soil arching effect behind retaining walls and provided predictions for the shape of the soil arch. Wang and Yen [15] conducted a comparable analysis for slopes. Kim et al. [16] detected a reduction in earth pressure due to the horizontal/vertical three-dimensional arching effect. They combined experimental and theoretical analyses to study lateral earth pressure on vertical circular shafts, quantifying the magnitude and distribution of lateral earth pressure measured by the three-dimensional arching effect. The obtained lateral earth pressure on vertical circular shafts considering the arching effect was 80% lower than that calculated by Rankine's theory. Shabanimashcool et al. [17] used a three-dimensional discontinuous numerical modeling method to investigate the interaction between rock bolts and rock masses and concluded that interlocking rock blocks constitute a stress concentration zone in the shape of an arch (i.e., a pressure arch) inside the rock mass. This pressure arch can resist the pull-out force of rock bolts and thus plays an important role in the rock mass's ability to withstand the forces from the rock anchors. Li et al. [18,19] employed numerical simulation methods to study the velocity distribution characteristics above the tunnel working face and arch apex, proposing a new permissible movement velocity field. They established an improved failure mechanism considering soil arching effects and internal plastic deformation of the soil through the application of spatial discretization technology. Trapdoor experimental studies have documented stress redistribution caused by soil arching. The results of numerous trapdoor experiments in geotechnical centrifuges have been reported in several studies, focusing on stress changes and redirection during the entire movement of the trapdoor [20–27]. The experiments confirmed that soil arches form during the movement of the trapdoor, with stress transferring from the yield zone to the non-yield zone.

In the present study, the photoelastic experimental method [28–30] is utilized, employing a biaxial compression bidirectional flow experimental apparatus for granular materials (model RLJW–2000) [31]. The method involves loading granular materials and progressively releasing particles to simulate the solid–fluid transition during deep mining processes, achieving visualization of the force chain arch. It is detected that the force chain arch acts as an interface during the transition from a solid to a fluid state in the granular system, delineating the quasi-solid stable zone from the quasi-liquid transitional zone. During this process, the force chain arch bears load transmitted from the upper and lateral directions, altering the original load transmission direction. In this study, the force acting on the force chain arch is analyzed, and its stability is discussed. The refined understanding of the force chain arch provides better insights into the formation, evolution and mechanisms of geological disasters and thus also contributes to the improvement of early warning models and monitoring systems [32,33].

2. Particle Flow Photoelastic Experiment Apparatus and Experimental Materials

The granular material samples were placed in a container, with circular polarizers arranged at the front and back. When illuminated with an LED light source on one side, high-resolution photographs (taken with a Canon 5D Mark III, 5760 × 3840 pixels, bought from Canon, Beijing, China) were captured from the other side, showcasing the samples at various experimental stages under both normal and polarized light conditions. The optical path during the experiment is illustrated in Figure 1a, and the set-up of the biaxial compression bidirectional flow experiment for the granular material along with its loading system is shown in Figure 1b.

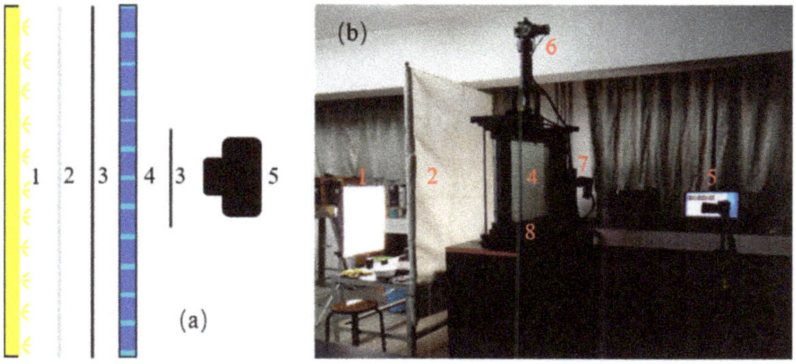

Figure 1. (**a**) Schematic of the photoelastic system; (**b**) Set-up of the experimental apparatus.

The components include the following: 1—Light source, 2—Light diffusion sheet, 3—Circular polarizers, 4—Photoelastic granular material (placed in a container made of a pair of transparent glass plates), 5—High-resolution camera, 6—Vertical servo loading cylinder, 7—Horizontal servo loading cylinder, 8—Eleven independent particle release ports (at the bottom).

The experimental apparatus achieves biaxial loading through two blades—one for normal and the other for tangential force—along with their corresponding servo loading cylinders. The entire frame of the apparatus is situated in a vertical structure, with the container holding the granular model located on the cabinet. The frame of the model can rotate around the vertical column, facilitating the insertion and removal of the particles, as illustrated in Figure 1b. The bottom and right sides of the particle container comprise grooves and are equipped with 11 steel plate baffles acting as trapdoors, respectively. When the load is constant, the opening of a trapdoor at the bottom allows particles within the container to flow out from the base. Similarly, opening of a lateral trapdoor induces the flow of particles from the side.

The materials used for the photoelastic experiments primarily include substances capable of producing birefringence, such as epoxy resin and polycarbonate. The photoelastic experimental material makes it possible to observe the path of force transformation under polarized light, which is called force chains. In the experiment, the granular materials were obtained by cutting Lexan 9030 polycarbonate sheets. This material provides excellent light transmission properties and is easy to process. The density of the polycarbonate is 1.2 g/cm^3, with an elastic modulus of 2.3 GPa and a Poisson's ratio of 0.36. The inherent bonding strength of polycarbonate is low, which results in minimal impact on its overall strength and stability. To prevent interfering effects related to crystallization due to a single particle size during the experiment, the polycarbonate sheets were processed into flat cylindrical particles of three sizes: Φ8 mm $\times$ 3 mm, Φ10 mm $\times$ 3 mm and Φ12 mm $\times$ 3 mm. The static and dynamic friction coefficients of the particles are 0.5 and 0.02, respectively. Pressure is the critical function in the study, whereas self-weight of the particles is negligible. Therefore, the Euler similarity criterion, which is based on pressure, was selected, resulting in the similarity ratios for the model, as summarized in Table 1.

Table 1. Comparison of model similarity ratio.

Actual Length	Model Length	Actual Pressure	Model Pressure	Length Ratio	Pressure Ratio	Density Ratio
100 m	760 mm	12.5 MPa	0.20 kN	130	140	1.1

3. Evolution of Force Chain Networks during the Solid–Fluid Transition Process

The granular material within the biaxial compression bidirectional flow experimental apparatus was subjected to loading, as shown in Figure 2a. The load includes a uniformly distributed load q_y at the top and q_x on the left side, with the bottom and right sides being fixed. During the experiment, the top and lateral loads are maintained at 200 N through servo control, equivalent to an actual rock layer geo-stress of 12.5 MPa. The geometric dimensions and loads of the released granular material are converted according to the similarity ratios (Table 1). Releasing particles from the bottom reliably simulates deep mining, and a force chain arch structure appears above the trapdoor, as depicted in Figure 2b. The green dots at the bottom indicates the trapdoor, and the red arrow represents the flowing direction.

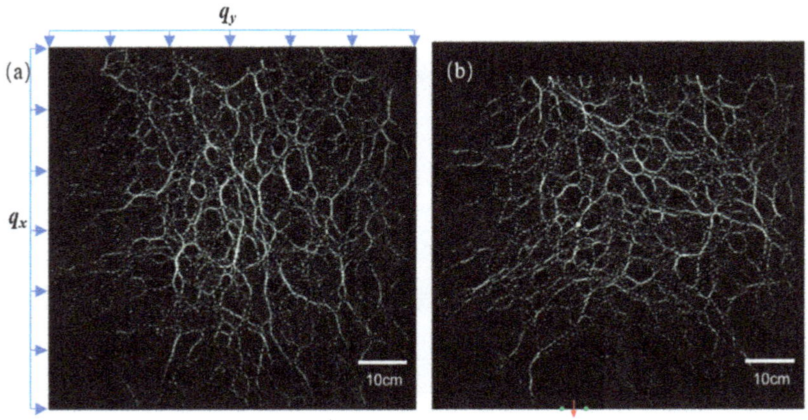

Figure 2. (**a**) Schematic of bidirectional loading on granular material; (**b**) Force chain network after particle release from the bottom.

Starting from the spatial distribution of photoelastic force chains, the spatial autocorrelation method commonly used in the field of geographical statistics was employed to determine the principal direction of the force chain arch structure [34]. Spatial autocorrelation [35] refers to the potential interdependencies among the observational data of a variable within the same distribution area. Its results are related to the relative size of the variable and are independent of its absolute value. In the photoelastic experiments, the intensity of light represents the relative magnitude of force. By calculating the local spatial autocorrelation coefficients of light intensity within a small window, the spatial distribution of local force chains is obtained. The characteristic direction of the distribution represents the development direction of local force chains, and the next window is determined. By repeating the calculation process above, starting from the bottom-left corner of the model, the arch trajectory is established via local iterative statistical method. The force arch identified using the autocorrelation algorithm after the release of particles from the bottom is shown in Figure 3a, where red points represent the centroid of the calculation window. Fitting these points, the trajectory of the force chain arch is shown in Figure 3b, following a quadratic function equation, $y = -0.0031x^2 + 2.51x - 235.66$, with R-square of 0.9925.

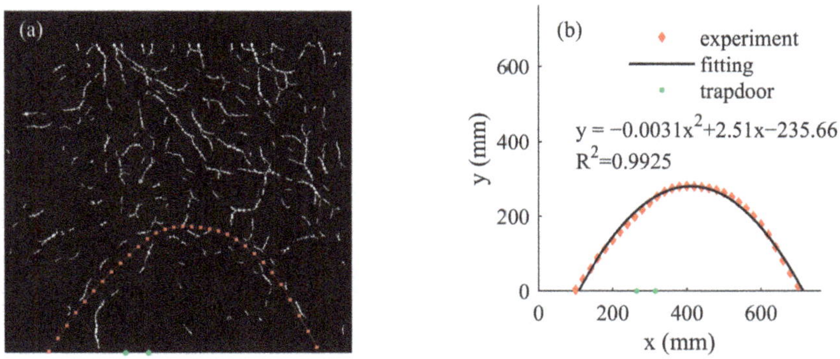

Figure 3. Force chain arch after particle release from the bottom: (**a**) Identified via autocorrelation algorithm; (**b**) Quadratic function fitting.

4. Analysis of the Forces and Stability of the Solid–Fluid Transition Interface in Bottom Particle Flow

4.1. Force Analysis of the Force Chain Arch Formed by Bottom Particle Flow

The force chain arch formed by releasing particles from the bottom under biaxial compression load is analyzed in the following section. The basic assumptions are as follows: the force chain arch ACB is an arch under plane strain, and its thickness has no impact on the arch; the overlying load q_y is a vertical uniformly distributed load, and the left lateral load q_x represents a horizontal uniformly distributed load, with the bottom and right sides being fixed, also with uniformly distributed loads q_y, q_x, respectively. As shown in Figure 4a, points A and B are the left and right arch foot, respectively; point C is the arch apex; and D is any point on the arch. The arch span is $2l$, and the arch height is h. First, the left half arch ADC is analyzed. The horizontal reaction force at the arch foot is F_H; the vertical reaction force is F_V; and the reaction force at the arch apex is F_m.

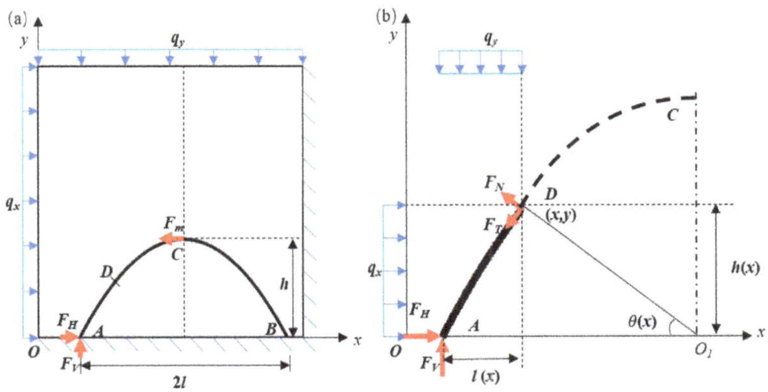

Figure 4. Bottom particle flow: (**a**) Schematic of the force chain arch equilibrium; (**b**) Schematic of the forces at any point on the force chain arch.

Based on the static equilibrium conditions at the arch apex (point C)

$$\begin{cases} \sum F_x = 0 \\ \sum F_y = 0 \\ \sum M_C = 0 \end{cases} \tag{1}$$

which leads to

$$\begin{cases} F_H + q_x h - F_m = 0 \\ F_V - q_y l = 0 \\ F_H h + q_y l \frac{l}{2} - F_V l + q_x h \frac{h}{2} = 0 \end{cases} \tag{2}$$

Solving the equations, the boundary conditions are obtained:

$$\begin{cases} F_H = -\frac{1}{2} q_x h + \frac{1}{2h} q_y l^2 \\ F_V = q_y l \\ F_m = \frac{1}{2} q_x h + \frac{1}{2h} q_y l^2 \end{cases} \tag{3}$$

Any point D on the arch, with coordinates (x, y), satisfies the interface trace equation. As shown in Figure 4b, the horizontal distance from point D to the left arch foot is $l(x)$; its height is $h(x) = y(x)$; and the angle between the central axis of the force chain arch and the x-axis, $\angle OO_1D = \theta(x)$. The normal and tangential forces are F_N and F_T, respectively, where $l(x)$, $h(x)$, $\theta(x)$ are functions uniquely determined by the abscissa x, with $q_x = q_y = 66.7$ kN/m being the known uniformly distributed load.

The equilibrium conditions for any point D(x,y) on the arch are calculated in the following equations:

$$\begin{cases} \sum F_x = 0 : F_H - F_T sin\theta(x) - F_N cos\theta(x) + q_x h(x) = 0 \\ \sum F_y = 0 : F_V + F_N sin\theta(x) - F_T cos\theta(x) - q_y l(x) = 0 \end{cases} \tag{4}$$

which leads to

$$\begin{bmatrix} cos\theta(x) & sin\theta(x) \\ sin\theta(x) & -cos\theta(x) \end{bmatrix} \begin{bmatrix} F_N \\ F_T \end{bmatrix} = \begin{bmatrix} q_x h(x) + F_H \\ q_y l(x) - F_V \end{bmatrix} \tag{5}$$

Therefore

$$\begin{aligned} \begin{bmatrix} F_N \\ F_T \end{bmatrix} &= \begin{bmatrix} cos\theta(x) & sin\theta(x) \\ sin\theta(x) & -cos\theta(x) \end{bmatrix}^{-1} \begin{bmatrix} q_x h(x) + F_H \\ q_y l(x) - F_V \end{bmatrix} \\ &= \begin{bmatrix} (q_y l(x) - F_V) sin\theta(x) + (q_x h(x) + F_H) cos\theta(x) \\ -(q_y l(x) - F_V) cos\theta(x) + (q_x h(x) + F_H) sin\theta(x) \end{bmatrix} \end{aligned} \tag{6}$$

Substituting Equation (3) to determine the force at any point D(x, y) on the arch yields

$$\begin{cases} F_N = q_y[l(x) - l] sin\theta(x) + \left[-\frac{1}{2} q_x h + \frac{1}{2h} q_y l^2 + q_x h(x)\right] cos\theta(x) \\ F_T = q_y[l - l(x)] cos\theta(x) + \left[-\frac{1}{2} q_x h + \frac{1}{2h} q_y l^2 + q_x h(x)\right] sin\theta(x) \end{cases} \tag{7}$$

Since x and θ have a one-to-one correspondence, the terms related to l and h in Equation (7) are represented as functions of θ to derive a more general conclusion. Thus, the force at any point D($x(\theta)$, $y(\theta)$) on the arch is dependent on θ, expressed as Equation (8). The normal and tangential forces at various points on the arch are shown in Figure 5a,b, respectively.

$$\begin{cases} F_N = q_y[l(\theta) - l] sin\theta + \left[-\frac{1}{2} q_x h + \frac{1}{2h} q_y l^2 + q_x h(\theta)\right] cos\theta \\ F_T = q_y[l - l(\theta)] cos\theta + \left[-\frac{1}{2} q_x h + \frac{1}{2h} q_y l^2 + q_x h(\theta)\right] sin\theta \end{cases} \tag{8}$$

The normal force F_N on the arch is zero at $\theta = \pi/2$, increases in absolute value toward the arch feet on both sides and attains a maximum value of approximately 1.5 kN at the arch feet. The tangential force F_T reaches the maximum of 20.75 kN at the arch feet and the arch apex, and the local minimum of 18.34 kN between the arch apex and the arch feet. Considering the stability of the arch shoulders, the relationship between the normal force and tangential force in equilibrium state is as follows:

$$F_N = F_T tan\alpha \tag{9}$$

where α is the natural rest angle of the rock and soil masses. Defining the maximum values of normal and tangential forces in the equilibrium state as their corresponding allowable stresses, the instability condition of the force chain arch is defined as

$$\begin{aligned} & F_N > \{F_N\} \\ & or\ F_T > \{F_T\} \\ & or\ F_N \geq F_T tan\alpha \end{aligned} \tag{10}$$

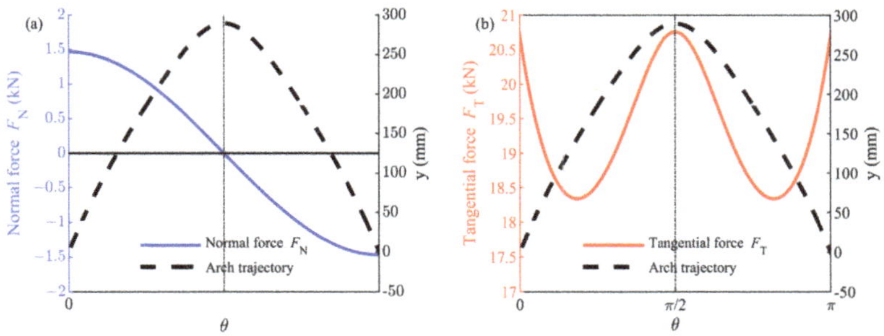

Figure 5. Under bottom particle flow, the forces at each point on the arch in equilibrium state: (**a**) Normal force F_N; (**b**) Tangential force F_T.

The values of $\{F_N\}$ and $\{F_T\}$ are related to the microscopic scale particle contact forces. Ideally, if the arch trace is composed of a series of particles, the normal force $\{F_N\}$ of the force chain arch is defined by the tangential contact force f_t between individual particles, while the tangential force $\{F_T\}$ of the force chain arch is provided by the normal contact force f_n between particles. According to the Hertz contact theory for static, elastic spherical particles, the normal force of particle contact is determined by the elastic modulus of the particles, the radius of the spheres and the amount of particle overlap. The criterion for particle sliding is generally expressed as

$$f_t \geq \mu f_n \tag{11}$$

where μ is the coefficient of translational friction of the granular material.

In actual complex rock and soil masses, the shear failure and cohesion c of the soil affect the equilibrium of the force chain arch. At the microscopic level, different models were proposed to describe particle contact collisions, including the linear viscoelastic model, the non-linear viscoelastic model and the hysteresis model considering plasticity, which include damping coefficients and elastic recovery coefficients. Due to the weak cementation properties of the particles in the experiment, some specific factors, such as cohesion, are negligible. However, the mechanical calculations of the force chain arch's equilibrium, which is based on visualization of the force chain in the photoelastic experiment, are still of referential significance.

4.2. Stability Analysis of the Force Chain Arch Formed by Bottom Particle Flow

The instability of the force chain arch can be categorized into two types:

1. Define arch span L = $2l$. Based on Equation (8), the force expressions at the arch feet are as follows: when the boundary loads q_y and q_x are constant, F_H is a linear function of the arch height h and quadratic function of the arch span L, while F_V is a linear function of the arch span L. If the particle system is disturbed, e.g., by further excavation, causing movement of the particles at the arch feet, the horizontal force F_H and vertical force F_V at the arch feet change, thereby breaking the force chain arch. Continuing downward transfer of the overlying load causes the originally unforced particles under the arch to possibly participate in the formation of a new force chain

network, or the particle stress may change. Thus, when the system attains a new equilibrium, the force chain network becomes updated.

2. When the particle system is subjected to a new stress disturbance, i.e., at least one of the boundary loads q_y or q_x increases, this causes a change in the forces at various points on the arch, and thus, the breaking of the stress arch and redistribution of the force chain network.

The following scenario considers the critical conditions for the instability of the force chain arch under changes in boundary loads q_y or q_x and arch span L:

1. Keeping the lateral load q_x constant and increasing or decreasing the overlying load q_y, the resulting normal and tangential forces are shown in Figure 6a,b, respectively, and the ratio of normal to tangential force is shown in Figure 6c.

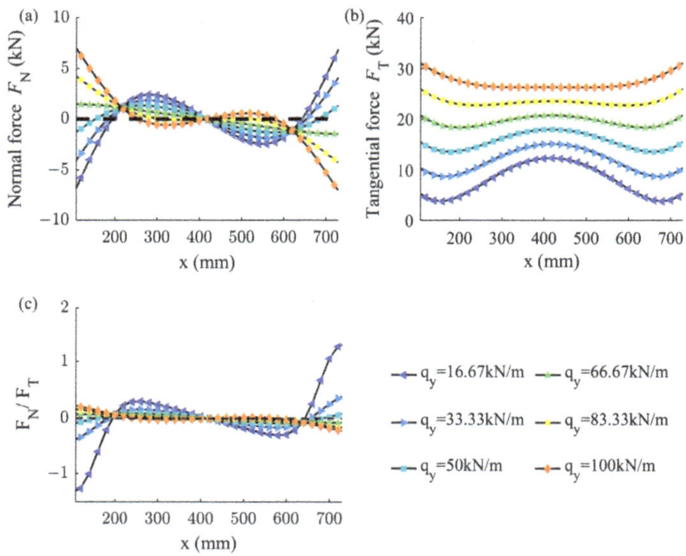

Figure 6. Bottom particle flow with the changes in forces at various points on the arch under variation in the overlying load q_y: (**a**) Normal force F_N; (**b**) Tangential force F_T; (**c**) The ratio of normal force to tangential force F_N/F_T.

When q_y decreases, the absolute value of normal force at each point increases and exceeds the value in the original equilibrium state (curve defined by green symbols). The normal force at the foot becomes the maximum value, and the critical point occurs at the arch foot. The tangential force decreases as a whole, and the value at the arch apex reaches the maximum. When q_y increases, the normal force at each point also rises and exceeds the maximum value in the original equilibrium state, with the arch feet being the critical points. The tangential force at each point increases along with q_y increases. For one particular curve, for example, q_y = 33.33 kN/m, the tangential force decreases from the arch foot to the arch apex, defining the arch foot as the critical section.

In summary, when the lateral load remains constant and the overlying load varies, the potential failure modes include normal buckling failure at the arch feet and tensile failure at the arch apex. When the overlying load increases, the potential failure modes include normal buckling failure and tensile failure at the arch feet.

2. Keeping the overlying load q_y constant and changing the lateral load q_x, the normal force, tangential force and their ratio are shown in Figure 7a–c.

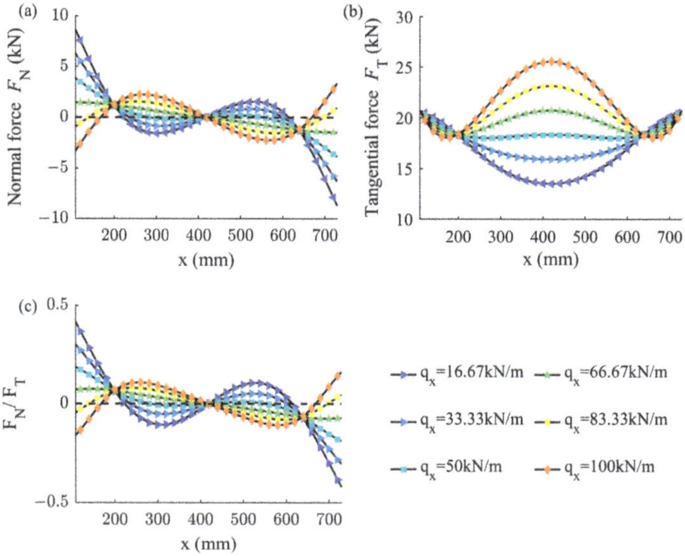

Figure 7. Bottom particle flow with the changes in forces at various points on the arch under variation in the lateral load q_x: (**a**) Normal force F_N; (**b**) Tangential force F_T; (**c**) The ratio of normal force to tangential force F_N/F_T.

When q_x decreases, the normal force at each point increases and exceeds the maximum value in the original equilibrium state (curve defined by green symbols), with the critical section located at the arch feet. The tangential force at each point decreases, remaining in a safe state, and the ratio of normal force to tangential force rises, with the critical section located at the arch feet. When q_x increases, the normal force and the ratio of normal force to tangential force also rise, exceeding the maximum value in the original equilibrium state, defining the arch feet as the critical section. The tangential force at each point increases and reaches the critical state at the arch apex.

In summary, when the overlying load remains unchanged and the lateral load decreases or increases moderately, the potential failure modes include normal buckling failure at the arch feet and tensile failure at the arch apex.

By considering both the load and the configuration of the arch, critical loads are identified under various combinations:

1. The tangential force at the arch apex and the normal force at the arch foot are calculated for different overlying loads q_y and arch spans L. The lateral loads q_x required to reach the allowable stress for each position are shown in Figure 8a. The red surface represents the boundary at which tangential tension leads to collapse of the arch apex; the green surface represents normal buckling at the arch foot; and the blue surface represents the boundary for tangential tension at the arch foot. Equation (6) demonstrates that both F_N and F_T are negatively correlated with q_x, indicating that the lateral load is an absolute adverse factor for the equilibrium of the arch. Therefore, the minimum q_x values from each surface (Equation (12)) constitute the equilibrium boundary of the arch, as shown in Figure 8b.

$$q_x \leq min\left\{\begin{array}{ll} q_x(F_T < \{F_T\}), & arch\ apex \\ max\left\{\begin{array}{l} q_x(F_T < \{F_T\}) \\ q_x(F_N < \{F_N\}) \\ q_x(F_N) < q_x(F_T) \end{array}\right\}, & arch\ foot \end{array}\right\} \tag{12}$$

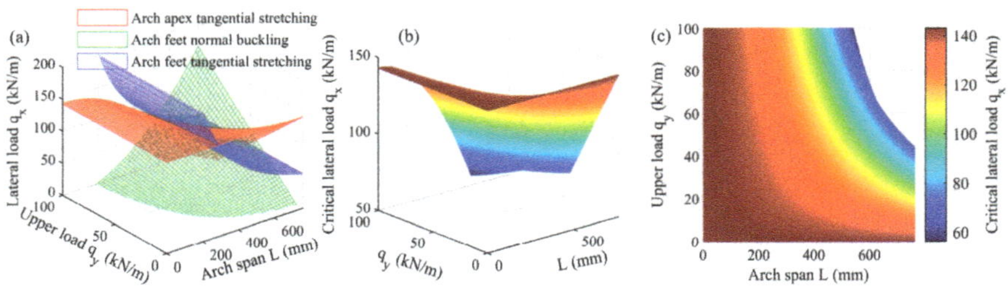

Figure 8. Bottom particle flow: (**a**) The critical lateral loads for failure at unfavorable sections of the arch; (**b**) The critical lateral loads for arch failure; (**c**) A color projection of the critical lateral loads for arch failure.

A projection of Figure 8b onto the q_x-L plane is shown in Figure 8c, indicating that the larger the arch span and the overlying load, the smaller the lateral load required to cause failure in the force chain arch. In other words, the value of the arch span and the overlying load need to be controlled to keep the system stable.

2. For different lateral loads q_x and arch spans L, calculations were conducted for the tangential force at the arch apex and the normal force at the arch feet to determine the overlying load q_y required for them to attain the allowable stress (Figure 9a). The red surface represents the boundary where tangential tension leads to collapse at the arch apex; the green surface represents normal buckling at the arch foot; and the blue surface illustrates the boundary for tangential tension at the arch feet.

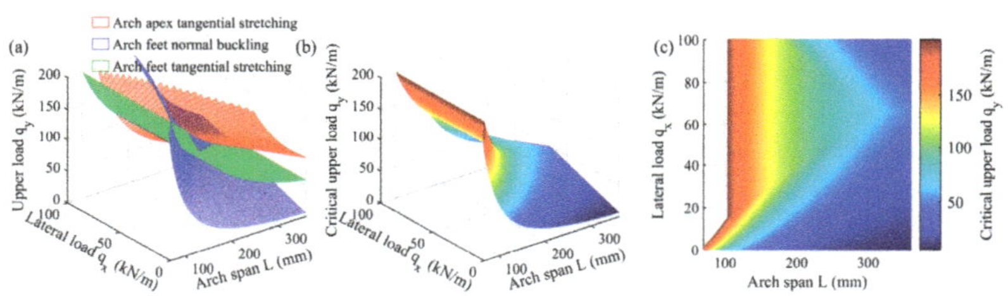

Figure 9. Bottom particle flow: (**a**) The critical overlying loads for failure at unfavorable sections on the arch; (**b**) The critical overlying loads for arch failure; (**c**) A color projection of the critical overlying loads for arch failure.

The relevant equation for the arch apex is

$$\begin{array}{l} F_T = -\frac{1}{2}q_x h + \frac{1}{2h}q_y l^2 + q_x h \leq \{F_T\} \\ q_y \leq \frac{h}{l^2}(2\{F_T\} - q_x h) \end{array} \tag{13}$$

and for the arch feet

$$\begin{array}{c} \left\{ \begin{array}{l} F_N = -\frac{1}{2}q_x h + \frac{1}{2h}q_y l^2 \leq \{F_N\} \\ F_T = q_y(l - l(\theta)) \leq \{F_T\} \end{array} \right. \\ \left\{ \begin{array}{l} q_y \leq \frac{h}{l^2}(2\{F_N\} + q_x h) \\ q_y \leq \frac{\{F_T\}}{(l - l(\theta))} \end{array} \right. \end{array} \tag{14}$$

Therefore, the area below the three surfaces constitutes the safety zone. The intersection of these zones provides the critical overlying load at which arch failure occurs, as shown in Figure 9b. The projection of Figure 9b onto the q_x-L plane is shown in Figure 9c.

The smaller the arch span L and the larger the lateral load q_x, the greater the critical overlying load q_y. With a larger q_x, for smaller arch spans, the greater the q_y required to maintain stability of the force chain arch. When the arch span L is very small, meaning the arch is relatively "flat", it is difficult to maintain balance due to the close proximity of the load application point to the arch trajectory. Considering the actual range of loads, it is proposed that for very narrow force chain arches under very large lateral loads, stability can no longer be maintained by applying an overlying load (as shown in the left part of Figure 9c, when $L < 200$ mm).

5. Analysis of the Forces and Stability of the Solid–Fluid Transition Interface in Lateral Particle Flow

5.1. Force Analysis of the Force Chain Arch Formed by Lateral Particle Flow

The force chain arch effect, resulting from the solid–fluid transition, can extend horizontally. Particles were released laterally [36] from the right edge of the model (indicated by the red arrow in Figure 10a, with green dots representing the release points) using the biaxial compression bidirectional flow experimental apparatus for granular materials. This visually demonstrated the evolution from localized slippage to landslides and other geological disasters in slope bodies. The trajectory of the force chain arch is determined using the spatial autocorrelation coefficient method, and then, it is fitted, as shown in Figure 10b,c.

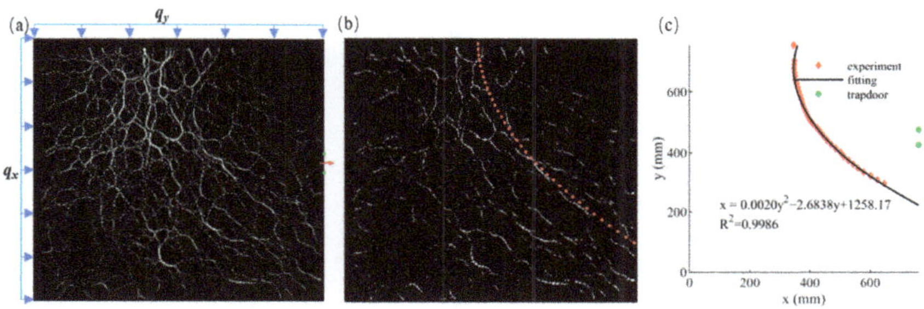

Figure 10. (**a**) Network of force chains during lateral flow of granular material; (**b**) Fitting of the force chain arch (lateral release of particles); (**c**) The fitting function.

Similarly, force analysis is conducted on the force chain arch formed by the lateral release of particles. The force chain arch ACB represents the form of a semi-arch and can be fitted with the equation of a parabola opening to the right, with the arch span denoted as h and the arch height as l. The maximum value for l is $L_0 = 760$ mm, which is the width of the model. The mechanical analysis is depicted in Figure 11a, where the horizontal reaction force at the arch foot near the release point is F_H; the vertical reaction force is F_V; and, due to the vertical load at the top, the stress is transmitted vertically downward as F_R. Based on the static equilibrium condition at the arch foot near the right-side release point, it follows that

$$\begin{cases} \sum F_x = q_x h - F_H = 0 \\ \sum F_y = F_V - q_y(L_0 - l) - F_R = 0 \\ \sum M = F_H h - q_y \frac{(L_0{}^2 - l^2)}{2} - F_V l - q_x h \frac{h}{2} = 0 \end{cases} \tag{15}$$

The solution of Equation (15) yields

$$\begin{cases} F_H = q_x h \\ F_V = \frac{h^2}{2l} q_x - \frac{(L_0{}^2 - l^2)}{2l} q_y \\ F_R = \frac{h^2}{2l} q_x - \frac{(L_0 - l)^2}{2l} q_y \end{cases} \tag{16}$$

As shown in Figure 11b, at any point C (x,y) on the arch, the normal force is F_N; the tangential force is F_T; and the angle between the line connecting C to the upper right corner and the horizontal direction is θ.

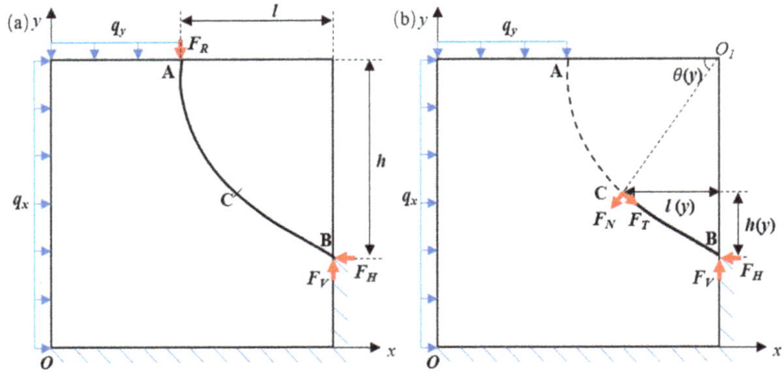

Figure 11. Lateral particle release: (**a**) Force chain arch equilibrium analysis diagram; (**b**) Force diagram at any point on the force chain arch.

The forces at any point C (x,y) on the arch fulfill the equilibrium condition:

$$\begin{aligned}\textstyle\sum F_x = F_T sin\theta(y) - F_N cos\theta(y) - F_H + q_x h(y) = 0\\ \textstyle\sum F_y = F_V - F_N sin\theta(y) - F_T cos\theta(y) - q_y l(y) = 0\end{aligned} \tag{17}$$

which leads to the following equation

$$\begin{bmatrix} cos\theta(y) & -sin\theta(y) \\ sin\theta(y) & cos\theta(y) \end{bmatrix}\begin{bmatrix} F_N \\ F_T \end{bmatrix} = \begin{bmatrix} q_x h(y) - F_H \\ -q_y l(y) + F_V \end{bmatrix} \tag{18}$$

and finally to Equation (19)

$$\begin{aligned}\begin{bmatrix} F_N \\ F_T \end{bmatrix} &= \begin{bmatrix} cos\theta(y) & -sin\theta(y) \\ sin\theta(y) & cos\theta(y) \end{bmatrix}^{-1}\begin{bmatrix} q_x h(y) - F_H \\ -q_y l(y) + F_V \end{bmatrix}\\ &= \begin{bmatrix} (q_y l(y) + F_V) sin\theta(y) + (q_x h(y) - F_H) cos\theta(y) \\ (-q_y l(y) + F_V) cos\theta(y) - (q_x h(y) - F_H) sin\theta(y) \end{bmatrix}\end{aligned} \tag{19}$$

Substituting Equation (16) provides the expressions for the normal and tangential forces at any point on the arch:

$$\begin{cases} F_N = \left[\frac{h^2}{2l} q_x - \frac{(L_0{}^2 - l^2)}{2l} q_y + q_y l(y)\right] sin\theta(y) + q_x[(h(y) - h] cos\theta(y) \\ F_T = \left[\frac{h^2}{2l} q_x - \frac{(L_0{}^2 - l^2)}{2l} q_y - q_y l(y)\right] cos\theta(y) - q_x[(h(y) + h] sin\theta(y) \end{cases} \tag{20}$$

As illustrated in Figure 12a,b, the normal force F_N and tangential force F_T on the arch both attain their maximum values on the far-right side (hereinafter referred to as the arch foot), with maximum absolute values of 16.91 kN and 15.05 kN, respectively. The normal force at the vertex located at the top of the model (hereinafter referred to as the arch apex) is zero.

Keeping the lateral load constant, the changes in normal forces at each point on the arch under different overlying loads are shown in Figure 13a, and the changes in tangential forces are shown in Figure 13b.

The data document that changes in the overlying load adversely affect both the normal force at the arch foot and tangential force at the arch apex, as shown in Figure 13a,b. The failure mode of the force chain arch may be due to yielding of the tangential force at the arch apex, causing a rightward collapse, or due to yielding of the normal force at the arch foot, inducing vertical movement of the arch foot.

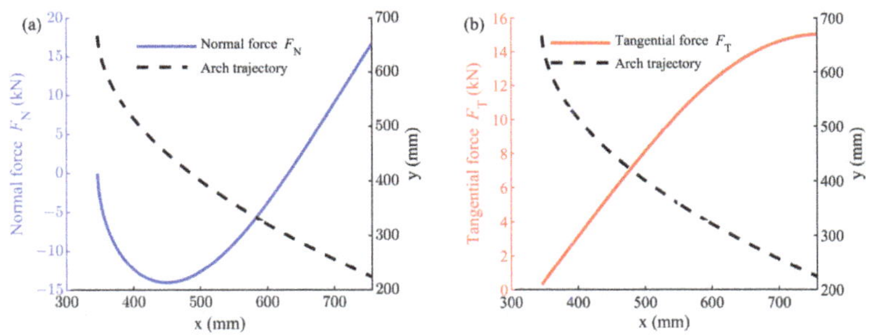

Figure 12. Lateral particle release: (**a**) Normal force F_N at each point on the arch in equilibrium; (**b**) Tangential force F_T at each point on the arch in equilibrium.

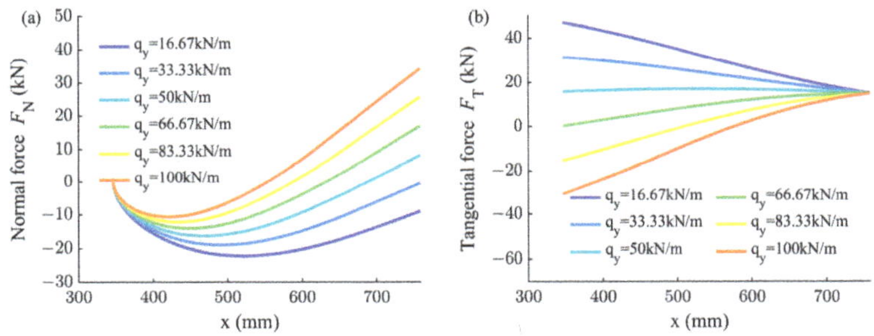

Figure 13. Lateral particle release under different overlying loads q_y. (**a**) Changes in normal force F_N at each point on the arch; (**b**) Changes in tangential force F_T at each point on the arch.

The variation in normal forces at each point on the arch and the variation in tangential forces for a constant overlying load and under the influence of different lateral loads are illustrated in Figure 14a,b, respectively. Changes in the lateral load are detrimental to the tangential force at the arch apex and both the normal and tangential forces at the arch foot. Under such conditions, in addition to yielding of the tangential force at the arch apex and the normal force at the arch foot, the arch may also fail due to yielding of the tangential force at the arch foot. Rightward sliding of the particles within the model is a consequence of this type of failure, indicative of a landslide tendency.

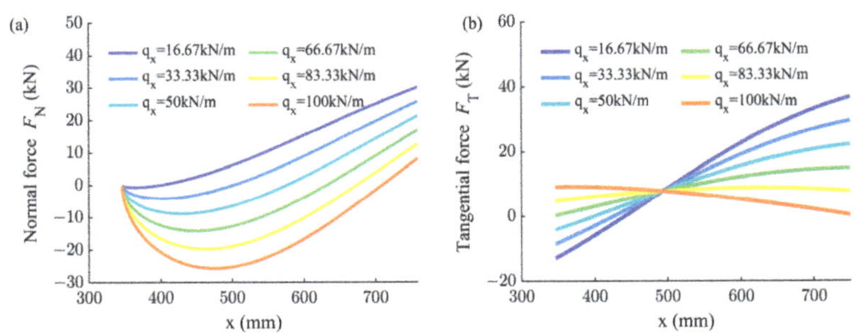

Figure 14. Lateral particle release under different lateral loads q_x. (**a**) Changes in normal force F_N at each point on the arch; (**b**) Changes in tangential force F_T at each point on the arch.

5.2. Stability Analysis of the Force Chain Arch Formed by Lateral Particle Flow

Subsequently, the tangential force is calculated at the arch apex, as well as the normal and tangential forces at the arch foot, under varying overlying loads q_y and arch spans h.

Here, h is called an arch span because the arch trajectory is half a parabola opening to the right. These calculations determined the lateral loads q_x required for these forces to attain their allowable stress. The red surface in Figure 15a represents the critical lateral load surface for tangential stretching, leading to the collapse at the arch apex, and the green surface marks the critical lateral load surface for normal buckling at the arch foot. The minimum value of q_x derived from these surfaces, according to Equation (15), constitutes the critical lateral load required for arch equilibrium, which must fulfill the following requirements:

$$q_x \leq \min\begin{Bmatrix} q_x(F_T < \{F_T\}), & arch\ apex \\ q_x(F_N < \{F_N\}), & arch\ foot \end{Bmatrix} \tag{21}$$

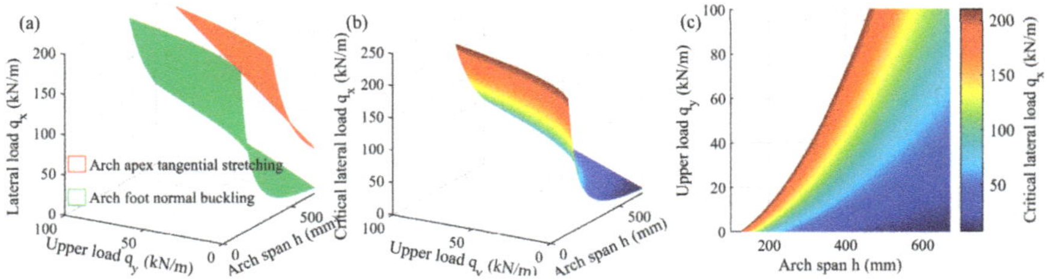

Figure 15. Lateral particle release: (**a**) The critical lateral loads for arch failure; (**b**) A color projection of the critical lateral loads for arch failure; (**c**) A color projection of the critical overlying loads for arch failure.

The critical lateral load surface for maintaining the balance of the force chain arch is presented in Figure 15b, with its projection onto the q_x-h plane shown in Figure 15c. The diagrams document that the smaller the arch span h and the larger the overlying load q_y, the greater the critical lateral load q_x. When the arch span h and the overlying load q_y are very small, as exhibited in the top-left corner in Figure 15c, maintaining balance is challenging due to the proximity of the load application point to the arch trajectory.

For different lateral loads q_x and arch spans h, considering the tangential force at the arch apex and the normal force at the arch foot when calculating the critical overlying load, the critical overlying load q_y required for these forces to attain their allowable stress is illustrated in Figure 16a, where the red surface represents the critical overlying load for tangential stretching, leading to the collapse at the arch apex; the green surface represents normal buckling at the arch foot; and the blue surface represents tangential stretching at the arch foot. The overlying load q_y, which fulfills the arch equilibrium, is located above the red surface and below the blue surface, thus constituting the critical overlying load surface for the arch equilibrium, as shown in Figure 16b, with its projection onto the q_x-h plane displayed in Figure 16c.

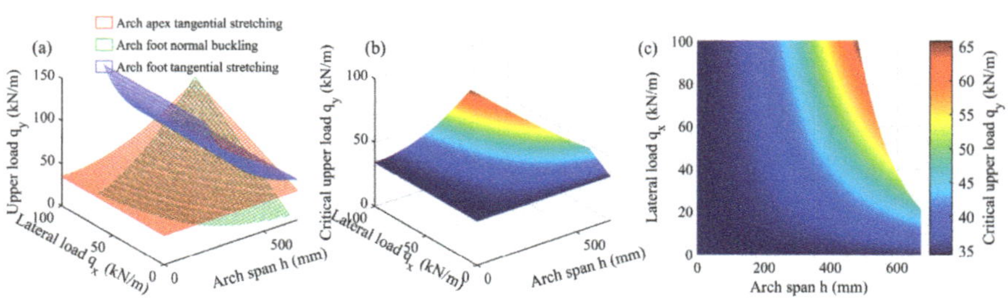

Figure 16. Lateral particle release: (**a**) The critical overlying loads for failure at unfavorable sections on the arch; (**b**) The critical overlying loads for arch failure; (**c**) A color projection of the critical overlying loads for arch failure.

The results document that the larger the arch span h, or the larger the lateral load q_x, the greater the critical overlying load q_y. When the arch span and lateral load are both large, the balance conditions could not be satisfied, indicating that the force chain arch is in a state of instability under such conditions. In other words, a combination of horizontal geo-stress and overburden is needed to keep the slope stable. For high and steep slopes, this is difficult to achieve, and there is a potential danger of slope instability.

6. Conclusions

This work presents visualized force chain arches via photoelastic experiments and their mechanical research, as well as stability. The main results are as follows.

1. Under biaxial compression conditions, the force chain arch formed within the granular medium via particle flow becomes the dividing line for solid–fluid transition. The particle assemblies above the force chain arch are in a stable solid state, still capable of bearing the overlying load, while those below the force chain arch are in a solid–fluid transitional state, experiencing less force. The force chain arch acts as an interface during the transition from solid to fluid in the granular system, separating the quasi-solid stable region from the solid–liquid transitional area. During this process, the force chain arch alters the original direction of load transfer, bearing most of the load transmitted from above and laterally.
2. The experimental model area is subdivided into grids by combining the results from indoor photoelastic experiments and employing autocorrelation algorithms to identify the main direction of local force chains. The trace line of the solid–fluid transition interface is obtained and fitted, resulting in an equation, which describes the parabolic shape of this interface.
3. A force and stability analysis is conducted on the fitted solid–fluid transition interface, i.e., the arch-shaped parabola. Under constant boundary load conditions on the solid–fluid transition interface caused by vertical particle flow, if the particle system is disturbed and the particles at the arch foot move, the horizontal force F_H and vertical force F_V at the arch foot change, leading to the disruption and rearrangement of the force chain arch. Under varying boundary load conditions, the arch apex and the arch foot are identified as critical sections.
4. When the trapdoor is located at the bottom of the model, changes in the overlying load or lateral load can lead to normal buckling and tensile stretching failure at the arch feet, as well as tensile stretching failure at the arch apex.
5. When the trapdoor is located on the right side of the model, changes in the overlying load adversely affect the normal force at the arch foot and the tangential force at the arch apex. Under different lateral loads, adverse effects on the tangential force at the arch apex and both the normal and tangential forces at the arch foot are observed.

7. Discussion

When the particles flow from the bottom of the model, the arch is broken due to apex collapse or foot shift toward the arch axis. When the grains flow from the right side of the model, the failure mode of the force chain arch may be rightward collapse due to yielding of the tangential force at the arch apex or vertical movement of the arch foot due to yielding of the normal force. In addition to yielding of the tangential force at the arch apex and normal force at the arch foot, the arch may fail due to yielding of the tangential force at the arch foot, leading to rightward particle movement within the model, indicative of a landslide tendency.

By analyzing the mechanical characteristics and stability of the force chain arch, we can evaluate the stress state of the loose rock and soil after excavation, predict the rock and soil mass stability in mining, slope or other engineering practices and finally prevent the potential geological disasters.

Author Contributions: Conceptualization, M.W.; Data curation, J.Z. and S.X.; Formal analysis, M.W.; Methodology, M.W.; Project administration, M.W.; Software, M.W.; Visualization, M.W.; Writing—original draft, M.W.; Writing—review and editing, J.Z. and S.X. All authors have read and agreed to the published version of the manuscript.

Funding: This work was funded by the Foundation of China Coal Technology and Engineering Group (CCTEG, Grant Number 2022-2-MS002, 2022-MS001).

Institutional Review Board Statement: Not applicable.

Informed Consent Statement: Not applicable.

Data Availability Statement: The original contributions presented in the study are included in the article, further inquiries can be directed to the corresponding author.

Conflicts of Interest: The authors declare that this study received funding from the Foundation of China Coal Technology and Engineering Group. The funder was not involved in the study design, collection, analysis, interpretation of data, the writing of this article or the decision to submit it for publication.

References

1. Al-Naddaf, M.; Han, J.; Xu, C.; Jawad, S. Experimental investigation of soil arching mobilization and degradation under localized surface loading. *J. Geotech. Geoenviron. Eng.* **2019**, *145*, 04019114. [CrossRef]
2. Bi, Z.; Gong, Q.; Guo, P.; Cheng, Q. Experimental study of the evolution of soil arching effect under cyclic loading based on trapdoor test and particle image velocimetry. *Can. Geotech. J.* **2019**, *57*, 903–920. [CrossRef]
3. Qiao, L.; Chen, L.; Dasgupta, G.; Li, Q.; Li, Y. Surface characterization and frictional energy dissipation characteristics of deep granite under high stress conditions. *Rock Mech. Rock Eng.* **2019**, *52*, 1577–1589. [CrossRef]
4. Chen, L.; Qiao, L.; Li, Q. Study on dynamic compaction characteristics of gravelly soils with crushing effect. *Soil Dyn. Earthq. Eng.* **2019**, *120*, 158–169. [CrossRef]
5. Xi, X.; Yin, Z.; Yang, S.; Li, C.Q. Using artificial neural network to predict the fracture properties of the interfacial transition zone of concrete at the meso-scale. *Eng. Fract. Mech.* **2021**, *242*, 107488. [CrossRef]
6. Cui, C. Seismic behavior and reinforcement mechanisms of earth dam and liquefiable foundation system by shaking table tests and numerical simulation. *Soil Dyn. Earthq. Eng.* **2023**, *173*, 108083. [CrossRef]
7. Huang, J.; Zhang, Y.; Ouyang, X.; Xu, G. Lagged settlement in sandy cobble strata and earth pressure on shield tunnel. *Math. Biosci. Eng.* **2019**, *16*, 6209–6230. [CrossRef]
8. Li, P.; Wang, F.; Fan, L.; Wang, H.; Ma, G. Analytical scrutiny of loosening pressure on deep twin-tunnels in rock formations. *Tunn. Undergr. Space Technol. Inc. Trenchless Technol. Res.* **2019**, *83*, 373–380. [CrossRef]
9. Li, P.; Zou, H.; Wang, F.; Xiong, H. An analytical mechanism of limit support pressure on cutting face for deep tunnels in the sand. *Comput. Geotech.* **2020**, *119*, 103372. [CrossRef]
10. Terzaghi, K. *Theoretical Soil Mechanics*; Wiley: New York, NY, USA, 1943. [CrossRef]
11. Rodríguez, C.A.; Rodríguez-Pérez, Á.M.; López, R.; Hernández-Torres, J.A.; Caparrós-Mancera, J.J. A Finite Element Method Integrated with Terzaghi's Principle to Estimate Settlement of a Building Due to Tunnel Construction. *Buildings* **2023**, *13*, 1343. [CrossRef]
12. Chen, F.; Liu, Y.; Wang, Y.; Gu, Y.; Yu, Y.; Sun, J. Deformation and Force chain of Two-Dimensional Granular Systems under Continuous Loading. *Materials* **2023**, *16*, 5441. [CrossRef]
13. Li, Q.; Zhang, Y.; Zhao, Y.; Zhu, Y.; Yan, Y. Study on Stability Discrimination Technology of Stope Arch Structure. *Sustainability* **2022**, *14*, 11082. [CrossRef]
14. Handy, R. The Arch in Soil Arching. *J. Geotech. Eng. ASCE* **1985**, *111*, 302–318. [CrossRef]
15. Wang, W.; Yen, B. Soil arching in slopes. *J. Geotech. Eng. Div.* **1974**, *100*, 61–78. [CrossRef]
16. Kim, K.; Lee, D.; Cho, J.; Jeong, S.; Lee, S. The effect of arching pressure on a vertical circular shaft. *Tunn. Undergr. Space Technol. Inc. Trenchless Technol. Res.* **2013**, *37*, 10–21. [CrossRef]
17. Shabanimashcool, M.; Bērziņš, A. A simplified approach to estimate anchoring capacity of blocky rock mass with pressure arch theory. *Rock Mech. Rock Eng.* **2023**, *56*, 7147–7175. [CrossRef]
18. Li, W.; Zhang, C. Face stability analysis for a shield tunnel in anisotropic sands. *Int. J. Geomech.* **2020**, *20*, 04020043. [CrossRef]
19. Li, W.; Zhang, C.; Zhang, D.; Ye, Z.; Tan, Z. Face stability of shield tunnels considering a kinematically admissible velocity field of soil arching. *J. Rock Mech. Geotech. Eng.* **2022**, *14*, 505–526. [CrossRef]
20. Chevalier, B.; Combe, G.; Villard, P. Experimental and numerical study of the response of granular layer in the trap-door problem. *AIP Conf. Proc.* **2009**, *1145*, 649–652. [CrossRef]
21. Chevalier, B.; Combe, G.; Villard, P. Experimental and discrete element modeling studies of the trapdoor problem: Influence of the macro-mechanical frictional parameters. *Acta Geotech. Int. J. Geoengin.* **2012**, *7*, 15–39. [CrossRef]

22. Costa, Y.; Zornberg, J.; Bueno, B.; Costa, C. Failure mechanisms in sand over a deep active trapdoor. *J. Geotech. Geoenviron. Eng.* **2009**, *135*, 1741–1753. [CrossRef]
23. Cui, C.; Wang, W.; Jin, F.; Huang, D. Discrete-element modeling of cemented granular material using mixed-mode cohesive zone model. *J. Mater. Civ. Eng.* **2020**, *32*, 04020031. [CrossRef]
24. Iglesia, G.; Einstein, H.; Whitman, R. Investigation of soil arching with centrifuge tests. *J. Geotech. Geoenviron. Eng.* **2014**, *140*, 04013005. [CrossRef]
25. Rui, R.; Van Tol, A.F.; Xia, Y.; Van Eekelen, S.; Hu, G. Investigation of soil-arching development in dense sand by 2D model tests. *Geotech. Test. J.* **2016**, *39*, 415–430. [CrossRef]
26. Van Eekelen, S.; Bezuijen, A.; Lodder, H.; Van Tol, A. Model experiments on piled embankments. Part II. *Geotext. Geomembr.* **2012**, *32*, 82–94. [CrossRef]
27. Van Eekelen, S.; Bezuijen, A.; Lodder, H.; Van Tol, A. An analytical model of arching in piled embankments. *Geotext. Geomembr.* **2013**, *39*, 78–102. [CrossRef]
28. Bi, D.; Zhang, J.; Chakraborty, B.; Behringer, R. Jamming by shear. *Nature* **2011**, *480*, 355. [CrossRef]
29. Behringer, R.; Chakraborty, B. The physics of jamming for granular materials: A review. *Rep. Prog. Phys.* **2019**, *82*, 12601. [CrossRef]
30. Wang, M.; Wang, D.; Socolar, J.; Zheng, H.; Behringer, R. Jamming by shear in a dilating granular system. *Granul. Matter* **2019**, *21*, 102. [CrossRef]
31. Wang, J.A.; Pang, W.D.; Liang, C.; Han, X. Photoelastic Experiment Apparatus of Aggregates with Bilateral Flowing. Chinese Patent CN204556449U, 12 August 2015.
32. He, M.; Zhao, J.; Deng, B.; Zhang, Z. Effect of layered joints on rockburst in deep tunnels. *Int. J. Coal Sci. Technol.* **2022**, *9*, 21. [CrossRef]
33. Wu, S.; Han, L.; Cheng, Z.; Zhang, X.; Cheng, H. Study on the limit equilibrium slice method considering characteristics of inter-slice normal forces distribution: The improved Spencer method. *Environ. Earth Sci.* **2019**, *78*, 1–18. [CrossRef]
34. Wang, M.; Wang, J.-A.; Li, F.; Liang, C.; Yang, L. Arching of force chains in excavated granular material. *Front. Earth Sci.* **2024**, *11*, 1322038. [CrossRef]
35. Zhang, J.; Zheng, Y.; Sun, J.; Qi, D. Flow prediction in spatiotemporal networks based on multitask deep learning. *IEEE Trans. Knowl. Data Eng.* **2020**, *32*, 468–478. [CrossRef]
36. Wang, J.; Yang, L.; Li, F.; Wang, C. Force chains in top coal caving mining. *Int. J. Rock Mech. Min. Sci.* **2020**, *127*, 104218. [CrossRef]

GeoHazards

MDPI

Article

Modelling and Validation of the Derna Dam Break Event

Alessandro Annunziato [1,*], Marzia Santini [2], Chiara Proietti [3], Ludovica de Girolamo [4], Valerio Lorini [2], Andrea Gerhardinger [3] and Michele Tucci [3]

1 Independent Researcher, 21020 Taino, VA, Italy
2 Joint Research Centre of the European Commission, 21027 Ispra, VA, Italy; marzia.santini@ec.europa.eu (M.S.); valerio.lorini@europarl.europa.eu (V.L.)
3 FINCONS SPA, 20871 Vimercate, MB, Italy; chiara.proietti@ext.ec.europa.eu (C.P.); andrea.gerhardinger@ext.ec.europa.eu (A.G.); michele.tucci@ext.ec.europa.eu (M.T.)
4 Seidor Italy SRL, 20129 Milano, MI, Italy; ludovica.de-girolamo@ext.ec.europa.eu
* Correspondence: alessandro.annunziato@gmail.com

Abstract: The catastrophic failure of two dams in Libya on 10 and 11 September 2023 resulted in the devastating flooding of the city of Derna, which is located downstream of the dams, causing more than 6000 fatalities and displacing thousands of residents. The failure was attributed to heavy rainfall from Storm Daniel, leading to the dams reaching full capacity and subsequently overflowing and failing. This paper presents an analysis of the dam break, including the modelling of flow discharge and the resulting flooding of Derna. For validation purposes, this study compares the modelled quantities with post-event satellite imagery from UNOSAT and Copernicus, local reports, and data collected from social media using AI detection. The findings provide valuable insights into the dynamics of the dam break and its initial parameters, as well as an assessment of the accuracy of the results. The analysis is performed using a rapid estimation technique developed by JRC to provide the international emergency community with a swift overview of the impact and damage assessment of potential or actual dam break events. The use of all available data shows a satisfactory comparison with the calculated quantities. The rapid modelling of dam break events and combined analysis of multiple data types are proven suitable for promptly assessing the expected dynamic of the event, as well as reconstructing the unknown initial conditions before the break. Incorporating sensitivity analyses provides an estimate of the uncertainties associated with the deduced values of the unknown parameters and their relative importance in the analysis.

Keywords: dam; dam break; flooding; modelling; early warning systems; digital elevation model; population density; remote sensing

1. Introduction

The objective of this paper is to analyse the large-scale dam break event that occurred in Libya in September 2023. This was accomplished by modelling the dam break, estimating the flood extent, establishing its timeline, and conducting a comparison of the model results with the available qualitative or quantitative post-event information.

The dam break methodology utilised in this analysis was developed by JRC and has been applied in the aftermath of recent major events or to assess potential imminent threats to infrastructure. Examples of these analyses include the failure of the dam in southern Lao People's Democratic Republic in 2018 or the potential, then actual dam break of the Nova Khavokva Dam in Ukraine in 2023. The added value of this rapid estimation approach is to provide the international emergency community with a swift overview of the damage assessment of potential or actual dam break events.

The Derna dam break provided an opportunity to validate the accuracy of the model's estimations of actual or potential damage. A comprehensive description of the methodology, including the dam break characteristics to be considered, required input quantities for the analysis, and the analytical solutions, was extensively outlined by Annunziato et al. [1].

2. Derna Dam

2.1. Storm Daniel and Dam Break Event

In September, the initial stages of the Mediterranean Cyclone (or Medicane), locally known as 'Storm Daniel', brought heavy rains to Greece, southern Bulgaria, and parts of Türkiye. The storm then remained slow-moving in the eastern Mediterranean for several days (Figure 1). On 10 and 11 September 2023, the storm's major rain bands caused extreme rainfall in northeastern Libya, with the city of Al-Bayda recording 414.1 mm in 24 h. The intense rainfall resulted in severe flooding in the region.

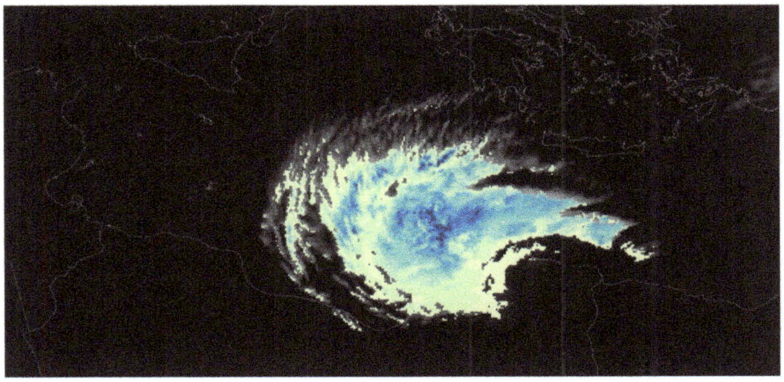

Figure 1. Satellite image showing Storm Daniel evolving in the Mediterranean Sea and affecting the northern areas of Libya (©EUMETSAT2023).

The most severe impacts were observed in the coastal city of Derna, located approximately 50 km east of Al-Bayda, where extensive flooding, exacerbated by the failure of two dams, led to the destruction of much of the city centre. As of 21 November 2023, there had been at least 4345 confirmed deaths and 8500 individuals reported as missing. Additionally, approximately 43,000 individuals were displaced across northeastern Libya due to the flooding caused by Storm Daniel [2,3] (Figure 2).

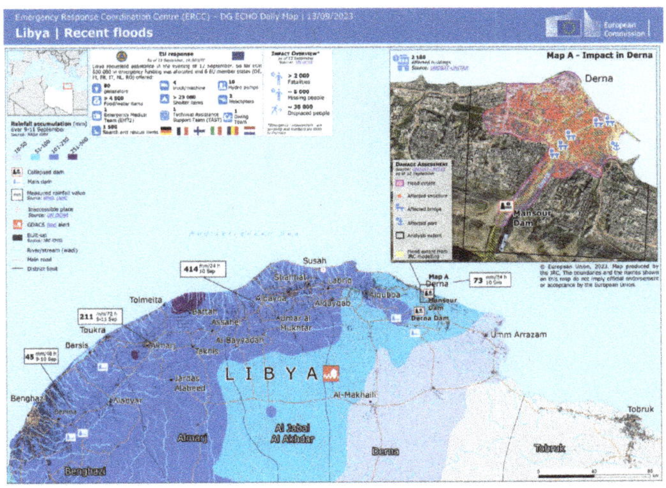

Figure 2. Situation map for the Libya floods (JRC and ECHO) [4].

Storm Daniel made landfall in Libya on 10 September, bringing strong winds and heavy rainfall. The following day, two dams located upstream of Derna collapsed under the intense pressure caused by the heavy rainfall, resulting in the release of 30 million cubic metres of water that ripped through the city (see Figure 3), home to approximately 100,000 inhabitants [2,3]. It is worth mentioning that Derna's vulnerability to flooding was well documented, as the city had previously experienced significant floods in 1942, 1959, 1968, and 1986 [5].

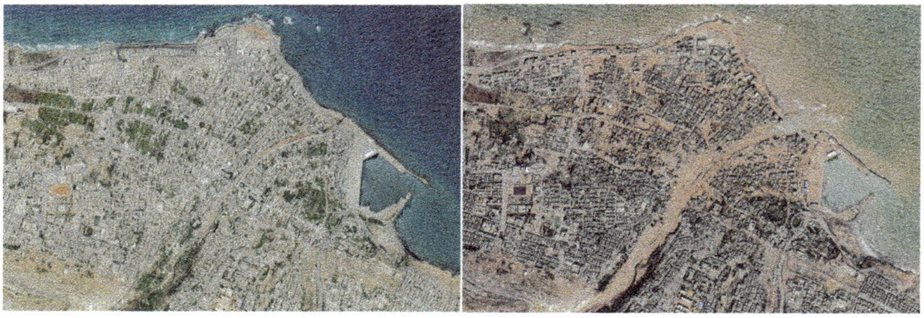

Figure 3. City of Derna before (left, 19 June 2023) and after (right, 13 September) the flood (©Google Earth).

The event involved two dams, as depicted in Figure 4. The Derna Dam, the larger of the two, is located approximately 12 km from the city, while the smaller Mansour Dam is situated just 1 km from the city.

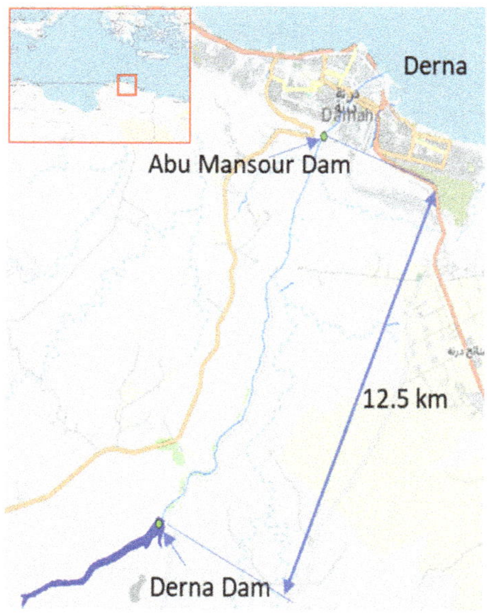

Figure 4. Geographic location of the dams with indication of approximate distance between the two (basemap from Open Street Map (OSM)).

Both dams are embankment dams with clay fill [6]. Historical imagery from Google Earth collected between 2009 and 2023 illustrates that the dams' reservoirs and the downstream river had been consistently at very low levels or completely empty. This is also confirmed by the adopted Shuttle Radar Topography Mission (SRTM) Digital Elevation Model (DEM) [7], which shows an empty riverbed upstream and downstream the dams. The sudden surge of water resulting from the intense precipitation associated with Storm

Daniel likely led to the overfilling of the Derna Dam and subsequent abrupt failure of the earth fill.

The dams were designed and constructed by the Hidroprojekt company of the former Yugoslavia Republic between 1973 and 1977. The construction of the dams aimed to facilitate irrigation of surrounding fields and to provide drinking water to Derna and nearby communities. The technical data of the dams, as per the company's website [6], are presented in Table 1; no further detail is available.

Table 1. Technical data of Derna and Mansour dams (Hidroprojekt) [6].

	Derna Dam	**Mansour Dam**
Dam height	75 m	45 m
Crest length	300 m	130 m
Foundation width	104 m	56 m
Embankment	735,000 m^3	104,000 m^3
Storage capacity	18 million m^3	1.5 million m^3

2.2. Previous Studies on Derna Dam

Limited studies on dam break risk assessments or specific analyses related to this event, whether published before or after the occurrence, have been identified, except for those mentioned below.

A comprehensive study conducted by Abdalla [8] focused on the potential failure of the Derna dam. In the study's abstract, the author reports the following: "*Piping failure was considered the most predictable one for Dema dam. For the current study it was shown that dam break causes a potential danger towards Dema City, since flooding occurs in the city for all hypothetical scenarios. It was recommended to predict the routing of the flow after failure due to lowering the retained water surface elevation in the reservoir and due to the use of levees and increasing walls heights in the city area*".

A relevant report by Ashoor [9] addressed potential floods in Derna caused by high precipitation. The model applied by Ashoor estimated the annual runoff volume for the study area for a forty-year period (1960–2000) to be 138.51 Mm^3. Additionally, it estimated the volume of floods based on historical events in October 1945 and late November 1986, which were associated with average precipitation of 145 mm and 64.14 mm, respectively. The 1945 rainfall resulted in a volume flood of 53.36 Mm^3, representing 40% of the annual runoff volume, while the November 1986 flood was estimated at 14.8 Mm^3, aligning well with the recorded flood in the basin. The findings underscore the high potential for flood risk in the study area, indicating the need for periodic maintenance of the dams in the Wadi Derna basin.

The European Commission Joint Research Centre (EC JRC) conducted a comprehensive analysis of the event, comparing the findings with satellite images, field reports, and social media [10].

A recent publication by Imran et al. [11] analysed the inundation of Derna following the destruction of the dam. While their results align with the findings of this study, a direct comparison with measured or observed quantities was not included.

Groenewege analysed the Derna Dam break event using Delft3D; also in this case, the author mentioned that it was used as a case study to try to understand the chain of events [12].

A further analysis of the Derna Dam break was presented by Prida [13] using the SFINCS model, taking into account different initial conditions to perform scenario analyses.

3. Methodology

3.1. Dam Break Analytical Studies vs. Rapid Analyses

The estimation of potential downstream consequences from dam-related events is an essential component of the risk assessment process implemented during the installation phase of a new dam. These calculations can be quite time-consuming and require the preparation of input decks containing all the relevant details of the dam under analysis.

However, in cases of specific ongoing or predicted natural or man-made events, there is also a need for rapid analyses that are able to provide information on areas at risk when original design analyses are unavailable or when different parameters are required. Akgun et al. [14] demonstrated the development of detailed calculations, highlighting differences in quality between 2D and 3D calculations, with 2D calculations often proving more practical in terms of computational time and yielding satisfactory results.

Fread used dam break modelling to analyse the consequences of floods caused by earth embankment dam breaks due to meteorological conditions [15].

Moreover, Yudianto et al. [16] emphasised the significance of dam break analyses in assessing inundation risks resulting from severe meteorological conditions. These analyses differ from those typically undertaken for dam certification during construction or authorization procedures, where analysts have detailed information and ample time to prepare and conduct case studies and risk assessments. In the context of emergency calculations, production time is a critical factor, and initial and boundary conditions might not be fully available. Nevertheless, the analyses must be completed within the shortest possible timeframe and with only the available data to inform decisions on evacuation and other preparedness and response measures.

This study highlights the ability of this approach to provide timely and high-quality information on the development and consequences of rapid-onset events, proving to be a valuable asset for emergency response and risk management.

3.2. Analytical Methodology

As per the methodology outlined by Annunziato et al. [1], the data were prepared to account for the water reservoir level upstream of the dam at its maximum height, which corresponds to the height of the crest of the dam. As per the SRTM DEM, which was used to perform the computations [7], the crest height is 215 m at its centre, while according to the Google Earth DEM, the value is 205 m. However, both sources suggest that the dam edges lie at a higher level (approximately 229 m) than its centre (either 215 or 205 m), which is not possible, as the crest of the dam is flat. Therefore, given the uncertainty around the actual absolute height of the dam crest, four different water levels were taken into consideration, namely 205, 215, 225, and 230 m (Figure 5). It is worth mentioning that during the event, it is unknown whether the overfill discharge pipe (Figure 6) was operational or not, either due to its inability to contain the excess water or because it was obstructed and could not perform its intended function. This means that the initial overflow could have been characterised by an even higher water level than the crest height.

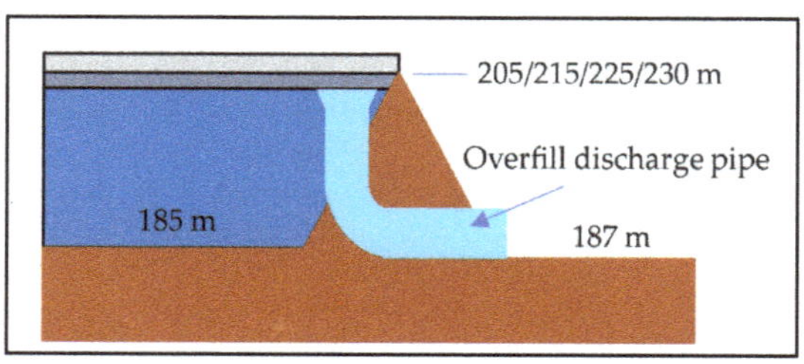

Figure 5. Dam geometry.

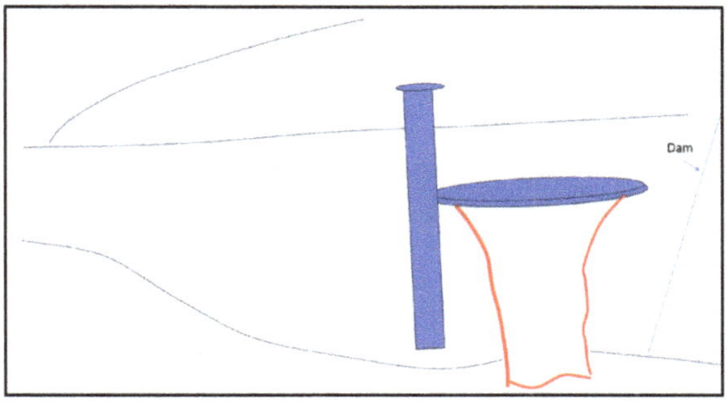

Figure 6. Drawing from image of Derna Dam overfill discharge pipe (image acquired from Erges) [17].

We should also note that the downstream Mansour Dam was not factored into the modelling, as it was reasonably assumed that it had already been destroyed or was immediately affected by the dam break flood. More details on this aspect are provided in Section 4.1.

3.3. Numerical Methods

In general, the analysis of a dam break through numerical modelling requires a significant amount of input data and output quantities. The applied methodology [1] aimed to establish a procedure that streamlines the preparation of the input deck and the derivation of the output quantities to enable a rapid analysis of a dam break event using the NAMIDANCE shallow-water model as the analytical tool [18,19]. This is crucial when, due to the urgency of understanding an ongoing or predicted event, a swift estimation of the potential impact is necessary. For the case of Derna, in the aftermath of the dam break event, very limited information was available to the international community to evaluate the consequences of the extensive flooding, while meteorological conditions did not allow for proper image acquisition and assessment.

NAMIDANCE has undergone validation through several benchmark problems and cross-code comparisons and has been applied to numerous tsunami events and modelling studies [19,20]. While NAMIDANCE has not been previously validated for dam break calculations, the tragic nature of this event presents an opportunity to assess the quality of the model for dam break simulations. The validation of the dam break scenario through the actual event is made possible by the availability of detailed satellite image analyses conducted by two independent expert teams from EU Copernicus EMS [21] and UNOSAT [22]. Furthermore, the JRC recently published a report that incorporates field reports and social media image analysis [10] and further enhances the quality of the validation exercise.

The calculations were performed using the Dam Break Platform [23], which allows the conditions to be set and subsequently performs the computations using NAMIDANCE. The DEM used for the Derna Dam calculations is the SRTM-30 m [7] resampled at 7.5 m. Each calculation was performed over a 24 h period to ensure that the dam reservoir was completely depleted of water and that the maximum flood extent was reached, even if, according to the calculations, this was reached in less than three hours. All processing was automated through the platform and is available for reference and consultation [23]. As the dynamics and extent of the dam break were unknown, an almost complete destruction of the dam was assumed in the modelling for the nominal case. Later on, this assumption was confirmed by the available post-event images, revealing extensive damage and significant sections destroyed (Figures 7 and 8).

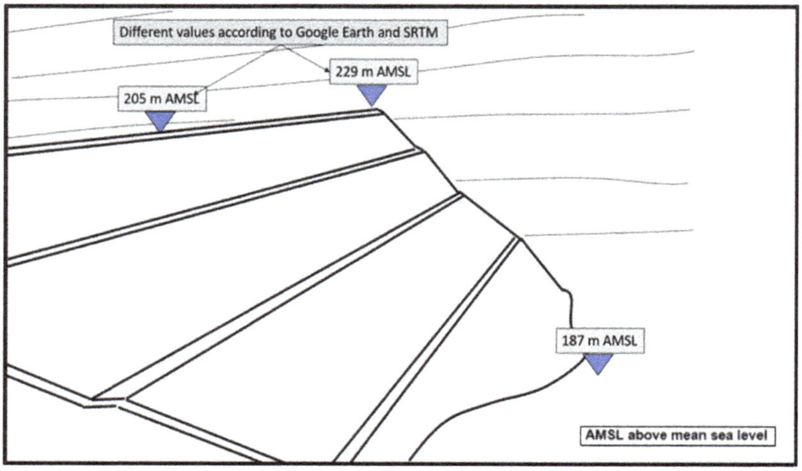

Figure 7. Drawing from image of the original form of the dam before the collapse (image acquired from DredgeWire) [24].

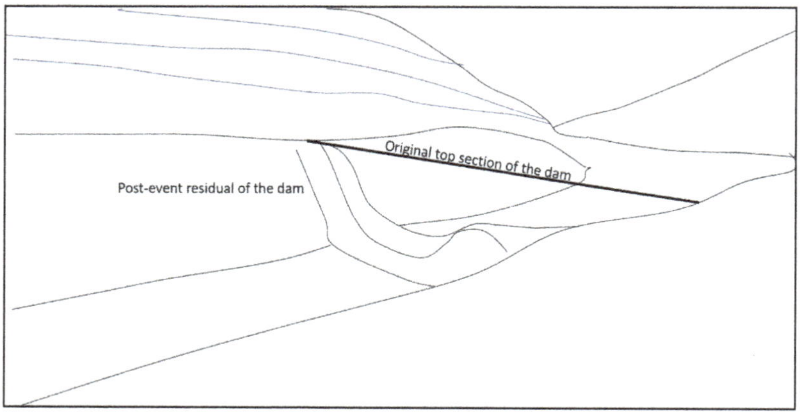

Figure 8. Drawing from image of the destroyed dam (image acquired from Eduvast) [25].

4. Derna Dam Break Scenario Modelling

4.1. Calculation of Nominal Case

The nominal case is performed considering a lake level height of 225 m above sea level, which is 10 m higher than the nominal elevation of the dam based on SRTM data. With these conditions, according to the scenario, the dam break would occur instantaneously, with a width of 260 m, which corresponds to 87% of the total crest width of 300 m. Subsequent sensitivity analyses, as discussed in the next section, showed that a width larger than 150 m does not substantially change the results because the narrow morphology of the upstream reservoirs limits the flow rate of the discharge. In the analysis, a Manning roughness coefficient of 0.015 $s/m^{1/3}$ was assumed to account for water flow resistance. The downstream river was assumed to be dry, in line with historical images from Google Earth. Nevertheless, the intense rainfall associated with the event could have caused an increase in the river's water level before the dam break, which will be the subject of a detailed analysis in the next sections.

The Mansour Dam was excluded from the analysis, as its break was assumed to be immediate. In fact, considering the initial modelled water flow rate of 49,720 m^3/s resulting from the break of the upstream dam, the maximum capacity of the Mansour reservoir of 1.5 Mm^3 would have been reached in less than 1 min. This assumption would have been valid if the reservoir were completely dry, but given the heavy rains of those days, this may not have been the case; therefore, the timeframe could have been even shorter. Consequently, the arrival time of the flood in the city of Derna would have been delayed by only one minute or even less.

The bounding box of the analysis is defined by the coordinates (22.528468, 32.6370721) to (22.696805, 32.77986), with a grid comprising 2425 × 2057 cells. A time step of 0.6 s was selected for the stability of the Courant number. The calculations were performed using a Windows server equipped with an NVIDIA GeForce RTX 3060 GPU card, with a typical runtime of approximately one hour and 25 min for a 24 h problem duration and about 40 min to process the results.

The calculated water flow from the dam break can be observed in the figure below, which shows water propagation every 10 min. In fact, Figure 9 indicates that under these conditions (reservoir water height at 225 m, break size of 260 m, Manning roughness coefficient set at 0.015 $s/m^{1/3}$, and empty downstream river), Derna would have been reached by water in less than 30 min, causing significant flooding in a large portion of the city.

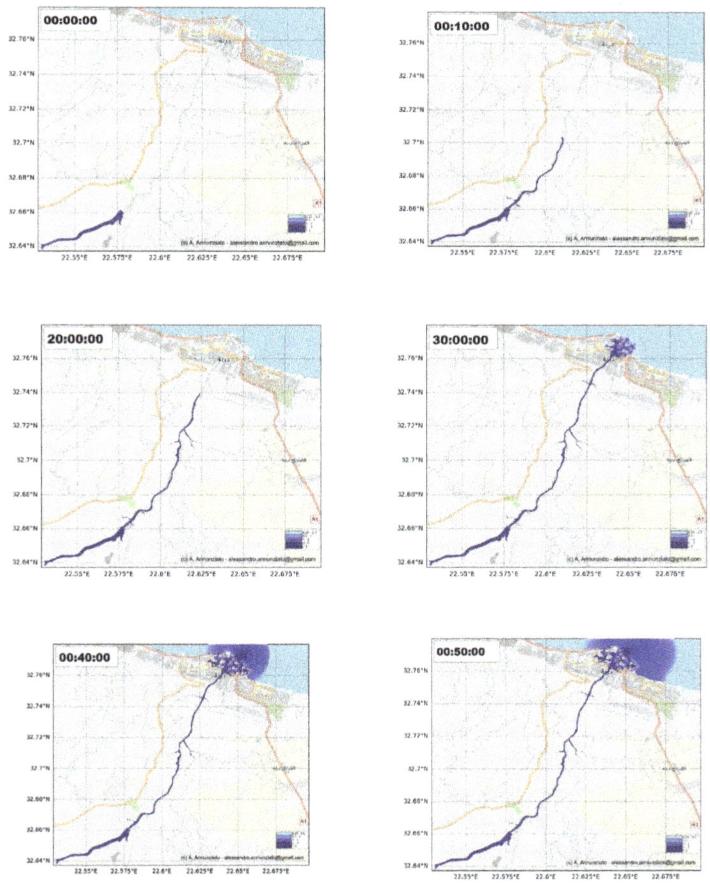

Figure 9. Calculated water propagation in the first 50 min following the Derna dam break, considering a reservoir water level of 225 m, a break size of 260 m, a Manning roughness coefficient set at 0.015 $s/m^{1/3}$, and an empty downstream river (basemap from OSM).

Table 2 presents the population affected by the event, categorised by water height. The combined population is approximately 111,000 based on the LandScan Global Populations 2014 model [26] and 71,000 according to the JRC Global Human Settlement Layer (GHSL) 2015 [27].

Table 2. Population affected by water height according to two population models (LandScan Global Populations 2014 and GHSL 2015).

Height Range (m)	LANDSCAN 2014	GHSL 2015
0.05–1.0	54,256	50,542
1.0–3.0	56,941	20,429
3.0–10	51	11
10–20	1	0
>20	0	0
Total	111,249	70,982

As already mentioned, the results discussed in this section can be accessed and consulted on the Dam Break Platform [23].

4.2. Sensitivity Analyses

Several details regarding the conditions at the time of the event remain unclear. Questions are still open regarding the rupture mode, including whether it was instantaneous or if it took some time to complete, as well as the size of the dam opening and the water levels in the reservoir before the break and in the downstream river.

It is assumed that prior to Storm Daniel, both the reservoir and the river were entirely empty and dry. Subsequently, as a result of the rainfall, the water level began to rise in the reservoir. It is unclear whether the water was released from the dam via the overfill discharge piping. If this was the case, the status and the level of the downstream river and that of the reservoir of the downstream dam would have been impacted, since they depend on the amount of water overflowing from the discharge pipe. However, the overfill pipe was probably inadequate or entirely ineffective in regulating water levels, ultimately causing the dam's destruction. However, the downstream river's level remains uncertain, ranging from completely dry to filled with water, depending on the amount of water overflowing from the overfill discharge pipe.

Given the uncertainties typical of these calculations, it was deemed necessary to conduct several sensitivity analyses to encompass the potential variations. Specifically, the following four parameters were considered: the initial water level of the upstream reservoir, the Manning roughness coefficient, the dam break width, and the level of the river downstream.

Regarding the reservoir's initial water level, we considered different hypotheses in addition to the nominal case described in the previous section, namely 205, 215, and 225 m above sea level. For the Manning roughness coefficient, we utilised the default value in the NAMIDANCE interface, 0.015 $s/m^{1/3}$, but we also examined lower and higher values (from 1/100 to 10 times) to verify the influence of this parameter on the results. In terms of the break size, while the nominal width of the dam is 300 m, our analysis of the SRTM indicates that the maximum break size could not exceed 260 m due to physical limitations imposed by the morphology of the valley itself, which, at its maximum height of 225 m, is approximately 260 m wide. As Figure 8 illustrates, at least 70–90% of the embankment appears to be missing as a result of the break, prompting the assumption of a nominal break width size of 260 m and a minimum size of 200 m, considering that in some sections, the break was not fully complete in height. For the level of the river downstream, three sets of arbitrary constant inlet flow values were imposed on the reservoir.

The calculation matrix is summarised in Table 3.

Table 3. Performed sensitivity analyses.

Lake Water Height (m)	Manning Roughness Coefficient (s/m$^{1/3}$)	Break Width (m)	Arbitrary Inlet Flows on the Reservoir Cells (m^2/s)
	Lake water level (m)		
205	0.015	260	
215	"	"	
225	"	"	
230	"	"	
	Manning Roughness Coefficient (s/m$^{1/3}$)		
225	0.00015	260	
"	0.0015	"	
"	0.015	"	
"	0.15	"	
	Break width (m)		
225	0.015	30	
"	"	75	
"	"	100	
"	"	150	
"	"	200	
"	"	260	
	Downstream River		
225	0.015	260	Dry river
"	"	"	0.0625
"	"	"	1.25
"	"	"	2.5

": This is the ditto mark, it indicates the same value as above.

4.2.1. Effect of the Lake Water Height

The initial water height of the lake was assumed to be between 205 and 230 m, considering the different SRTM data available for the area.

According to the calculation results shown in Table 4, as the water level rises in the upstream area, the initial peak flow also increases. For a reservoir water height of 225 m, the peak flow would reach a value of 49,720 m^3/s. Consequently, as illustrated in Figure 10, the arrival time would decrease from 83 to 20 min due to a significantly larger flow and depth in the river valley.

Table 4. Results of calculations with lake water height parameters.

	205 m	215 m	225 m	230 m
Discharged volume (million m^3)	4.3	20.14	48.2	61.7
Peak flow rate (m^3/s)	2493	27,357	49,720	65,003
Arrival time at Derna city entrance * (min)	83	26.0	21.0	20.0
Maximum flood wave height at the entrance of Derna * (m)	5.9	15.6	25.2	28.6

* Lat/Lon: 22.62784/32.74522.

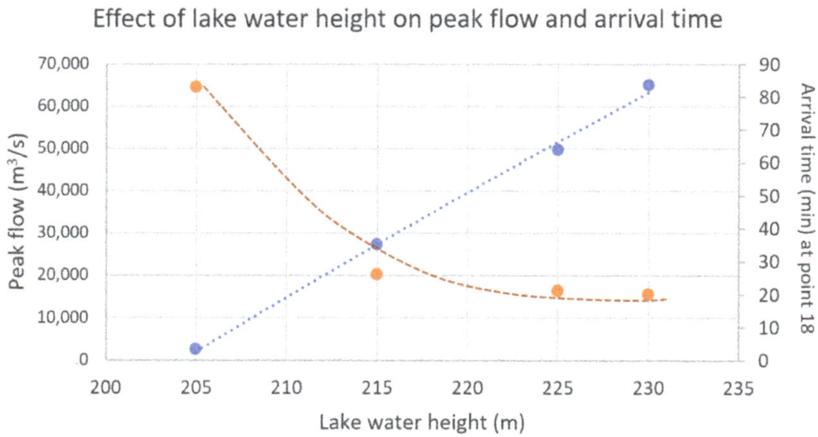

Figure 10. Effect of lake water height on peak flow (blue line) and arrival time (orange line) in Derna (22.62784/32.74522).

As per the model's results, the increase in the lake water level causes an increase of the initial peak flow, and as a result, the water level at the entrance of Derna increases notably from 5.2 to 28.6 m in the case of the highest lake water height (Figure 11). The resulting inundation extent in Derna would not change as much as the peak flow (Figure 12). Nevertheless, an increase in the inundation extent is noticeable with the increase in the lake elevation.

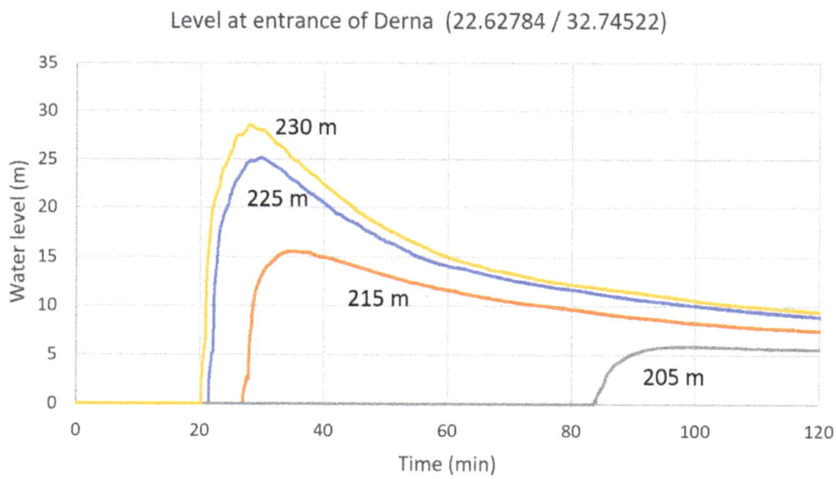

Figure 11. Flood water level at the entrance of Derna (22.62784/32.74522) with different lake water heights.

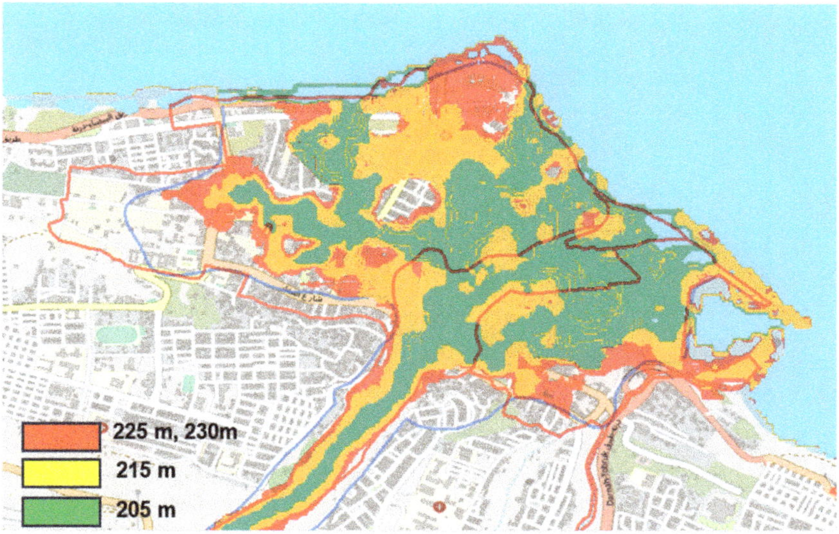

Figure 12. Effect of initial lake water height on the flood extent in Derna. The red line represents the flood extent assessed by Copernicus EMSR696 [21], and the blue line corresponds to UNOSAT FL20230912 as of 13 September 2023 [22] (basemap from OSM).

4.2.2. Impact of Manning Roughness Coefficient

All calculations involving different parameters for the Manning roughness coefficient were conducted using an initial water level of 225 m and a break width of 260 m.

As evident in Table 5, when the coefficient is increased from 0.0015 to 0.0015, then to 0.015 $s/m^{1/3}$, there are minimal changes in the peak flow, arrival time, and maximum flood wave height at the entrance of Derna, while a notable change is observed when the coefficient is increased by a factor of 10 to 0.15 $s/m^{1/3}$ (Figures 13–15).

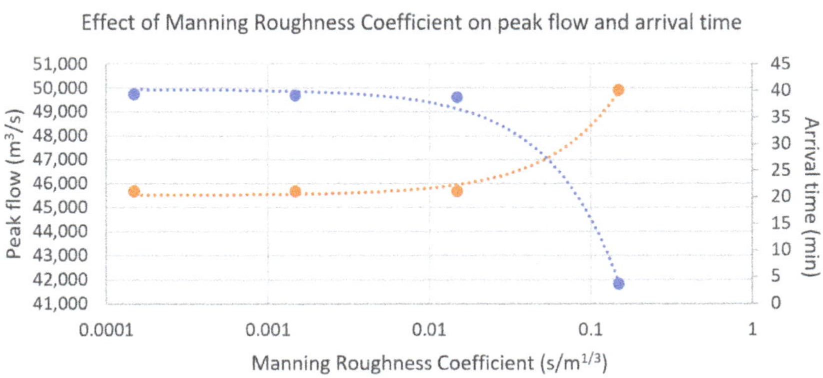

Figure 13. Effect of Manning roughness coefficient on the peak flow (blue line) and the arrival time (orange line) at the entrance of Derna (22.62784/32.74522).

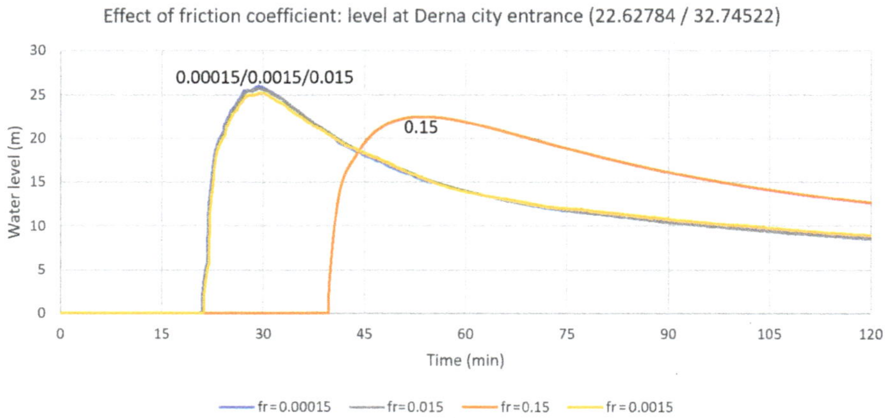

Figure 14. Flood wave height at the entrance of Derna (22.62784/32.74522) with different Manning roughness coefficients.

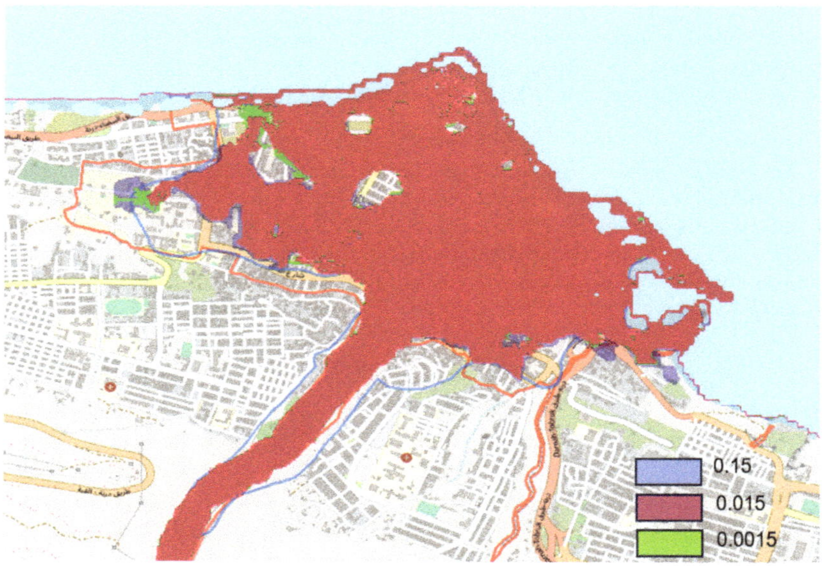

Figure 15. Effect of Manning roughness coefficient (s/m$^{1/3}$) on flooding in Derna. The red line represents the flood extent assessed by Copernicus EMSR696 [21], and the blue line corresponds to UNOSAT FL20230912 as of 13 September 2023 [22] (basemap from OSM).

Table 5. Results of calculations with Manning roughness coefficient parameters.

Manning roughness coefficient (s/m$^{1/3}$)	0.00015	0.0015	0.015	0.15
Discharged volume (million m^3)	49,887	49,852	49,720	41,175
Peak flow rate (m^3/s)	49.70	49.67	49.60	41.79
Arrival time (min)	21.0	21.0	21.0	40.0
Maximum flood wave height at the entrance of Derna * (m)	25.3	25.2	25.2	22.3

* Lat/Lon: 22.62784/32.74522.

4.2.3. Effect of Break Width

This section focuses on the impact of the break width, considering that all calculations were performed with an initial lake water height of 225 m and a Manning roughness coefficient of 0.015 $s/m^{1/3}$. The results are shown in Table 6.

Table 6. Results of calculations with different dam break width parameters.

	30 m	75 m	100 m	150 m	200 m	260 m
Discharged volume (million m^3)	48.2	48.2	48.2	48.2	48.2	48.2
Peak flow rate (m^3/s)	13,326	29,894	36,123	48,788	50,279	49,720
Arrival time (min)	31	23	22	21	21	21
Maximum flood wave height at the entrance of Derna * (m)	16.7	22	23.5	25	25	25.2

* Lat/Lon: 22.62784/32.74522.

It is notable that by increasing the parameter of the break width up to 200 m, the peak flow increases and the arrival time decreases (see Figure 16). Beyond this value, the changes become marginal. This is due to the morphology of the upstream reservoir, which, in some sections, is smaller than 200 m.

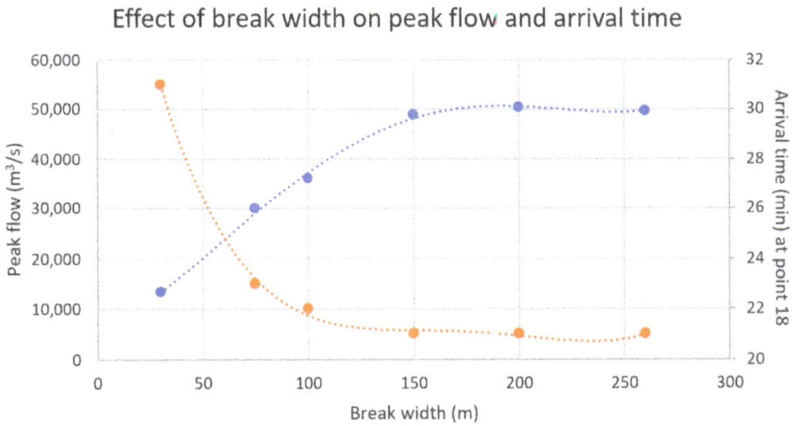

Figure 16. Effect of break width on the peak flow (blue line) and the arrival time (orange line) at the entrance of Derna (22.62784/32.74522).

4.2.4. Presence of River Flow Downstream of the Break

A sophisticated technique was employed to perform river flow calculations, as it is challenging to accurately consider the conditions during or just before an overflow event. The overflow was simulated by imposing a constant inlet flow in the upstream reservoir. As the overflow commenced, the flow began to spread across the river floodplain, with the depth of the river varying along its course, influenced by the outlet flow rate and the slope of the river in each location. In the literature, we found a similar method to take into account the initial river condition [28].

The overspilling was considered to last for 7 h, which is the time required for the water to reach the lower dam according to the calculation. The break could have occurred at any time during this period, although it typically occurs much sooner (see Figure 17). According to Froelich [29], all cases examined in his analysis displayed a break formation time of less than 1 h, with the sole exception of 8.5 h for the Oros Dam in Brazil.

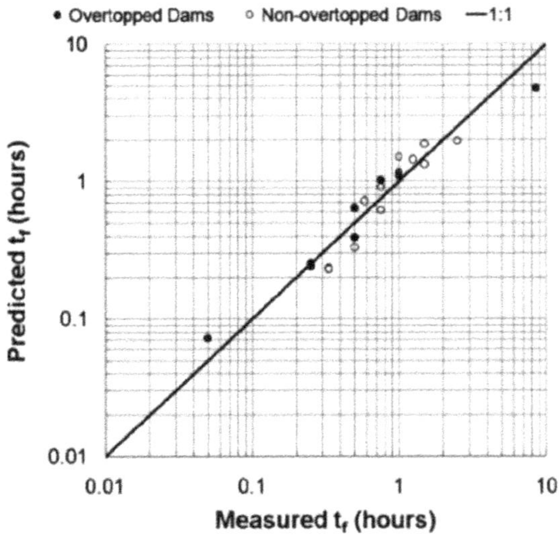

Figure 17. Measured and predicted break formation times, Froelich [29].

According to Froelich, the formation time for a potential dam break can be calculated using the following experimental formula:

$$T_f = 63.2\sqrt{\frac{V_w}{gH_b^2}}$$

where V_w represents the volume of water, and H_b represents the height of the dam. When applying this formula with the parameters specific to our case, that is, V_w = 18 Mm^3 and H_b = 75 m, the resulting time is calculated to be 1.1 h. This calculated time represents the expected duration required for a break to form based on the given water volume and dam height. However, the actual time of seven hours significantly exceeds this expected breaking time.

During the considered 7 h overspilling, three sets of arbitrary constant inlet flow values were imposed on the reservoir. When overtopping occurs, the water is discharged into the river plain, causing it to progressively fill up. The assumption is that the break would occur after the 7 h overspilling. The calculated water level along the river at the end of the 7 h was assumed as the initial condition for the river water level in the dam break calculation.

According to this scenario, once the break occurs, the travel time is significantly reduced compared to the dry case, as the wave velocity is higher due to the overall water height calculated as the sum of the river depth and the wave on top. Therefore, the velocity of the wave (v) is higher, as it is approximately v = $(g\ D)^{1/2}$, where D represents the overall water height and g is the gravity constant. By increasing the water level in the river, the arrival time decreases from 21 to 16 min, as shown in Figure 18.

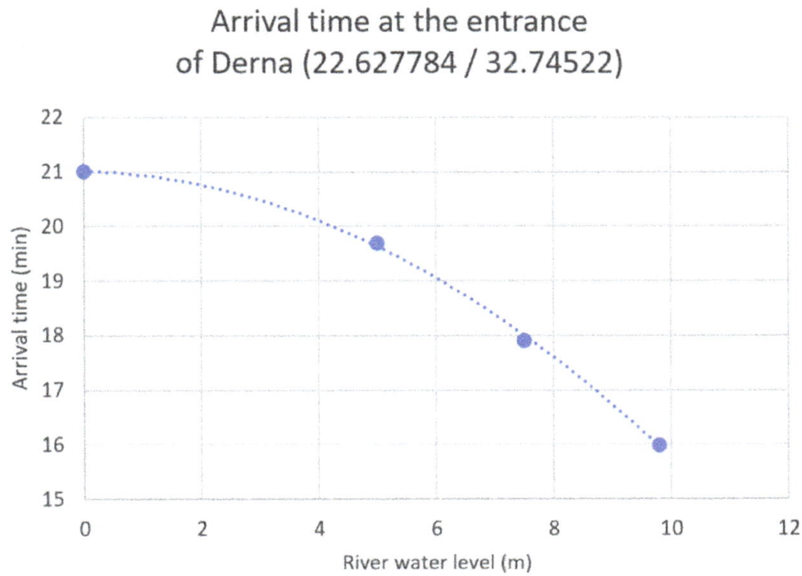

Figure 18. Arrival time as a function of initial water level in the river.

The modelled maximum water level at the entrance of Derna increases with the rise in the initial water level in the river, although not significantly, as shown in Figure 19.

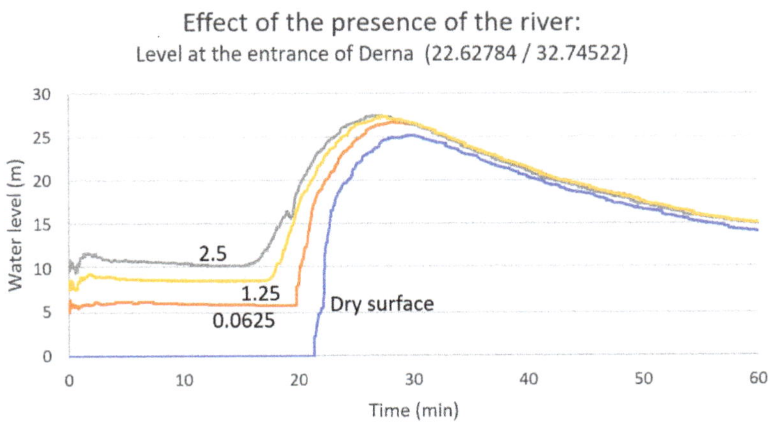

Figure 19. Effect of the presence of the river downstream of the dam when imposing 0.0625, 1.25, or 2.5 m^2/s as flux in each cell of the lake reservoir.

Therefore, the rising river level has a direct impact on the extent of downstream inundation, resulting in a more extensive area being affected, as illustrated in Figure 20.

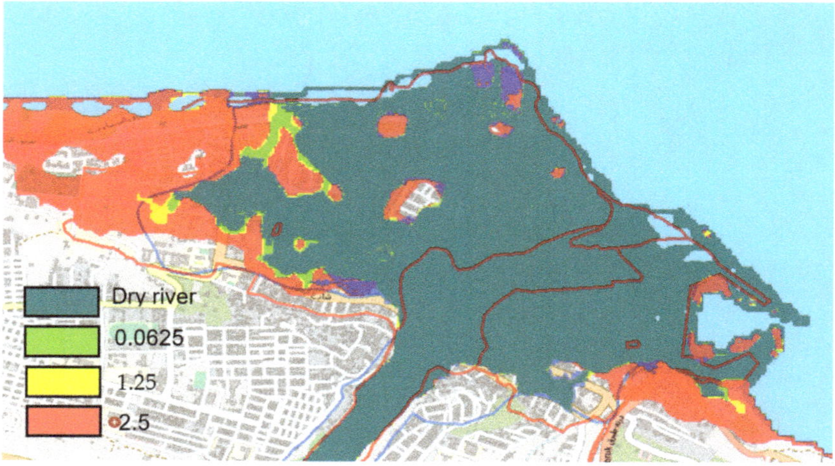

Figure 20. Effect of river flow on Derna inundation. The red line represents the flood extent assessed by Copernicus EMSR696 [21], and the blue line corresponds to UNOSAT FL20230912 as of 13 September 2023 [22] (basemap from OSM).

5. Model Validation

To validate the modelling results, we used a variety of available data sources, such as local reporting, post-event satellite mapping from Copernicus Emergency Management Service—Rapid Mapping, and information from social media. The integration of these diverse data sets enabled temporal and spatial validation.

5.1. Temporal Validation

As per the dam break scenario with the maximum reservoir water level height (230 m) and maximum break width (260 m), the water flow would take approximately 27 min to reach the nearest urban area of Derna (see Figure 9). It would reach the sea in around 40 min, and the maximum flood extent in Derna would be reached within 60 min.

The calculated flood travel time is illustrated in Figure 21, which indicates the arrival time across various time intervals. The water would arrive at the nearest urban area of Derna in approximately 25 min, and within another 10 min, most of the inundation would occur.

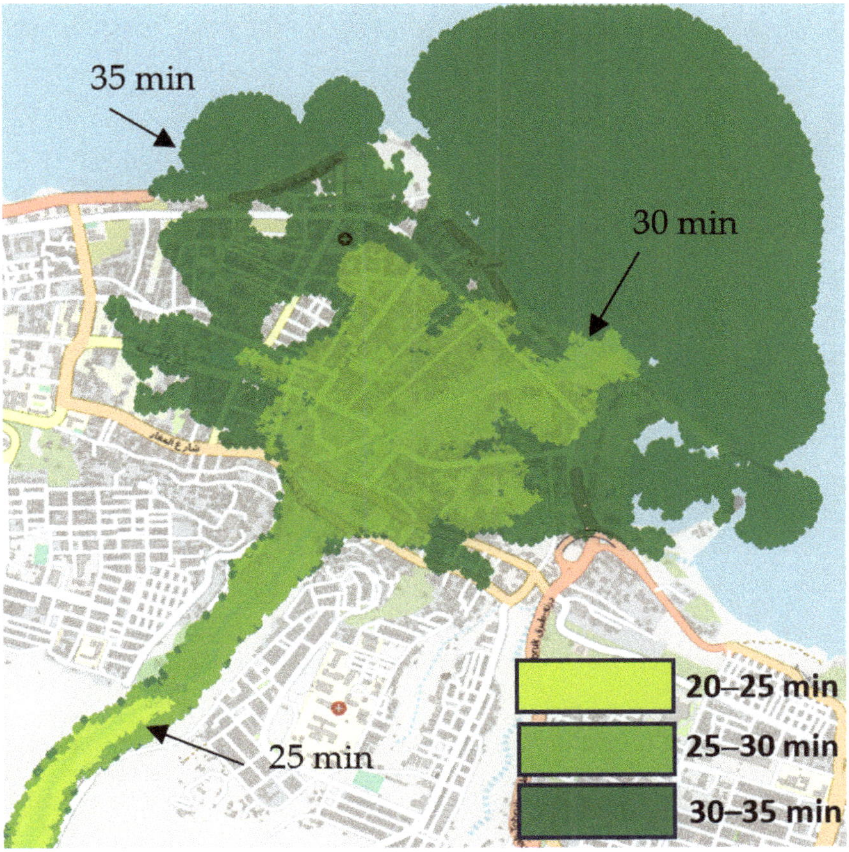

Figure 21. Calculated flood travel time (basemap from OSM).

These timeframes align closely with first-hand testimonies collected remotely by international experts, referring an approximate arrival time of around 30 min for the first arrival of the flow in the city (Figure 22).

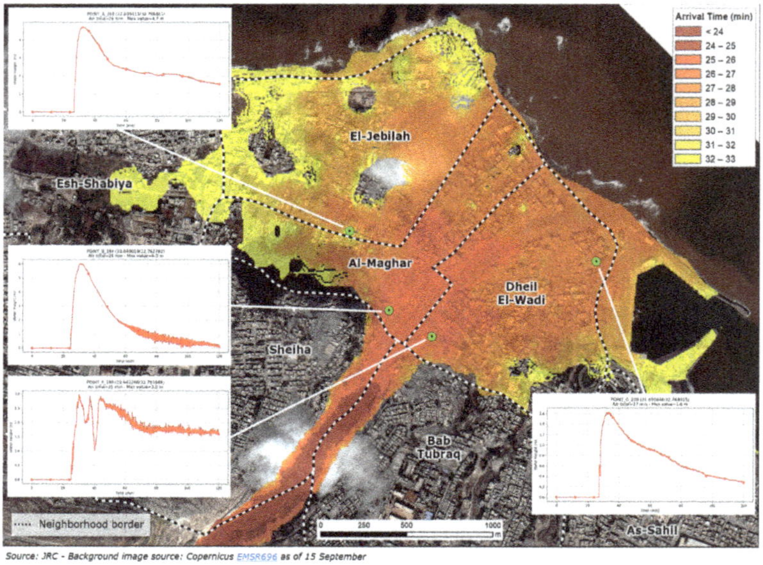

Figure 22. Flood arrival times in Derna (JRC and basemap from Copernicus).

Another aspect of the temporal validation process involves measuring the rate at which the flow depth increases over time. In this regard, a social media image collection and visual interpretation were conducted to determine the water depth in specific locations. For instance, data from security cameras in the Al-Maghar neighbourhood, which experienced the most significant flood extent and impact, indicated a water level rise of approximately two metres in less than one minute [30]. When comparing this observation with the charts in Figure 23, representing the simulation outcomes from the Al-Maghar neighbourhood, the order of magnitude is consistent.

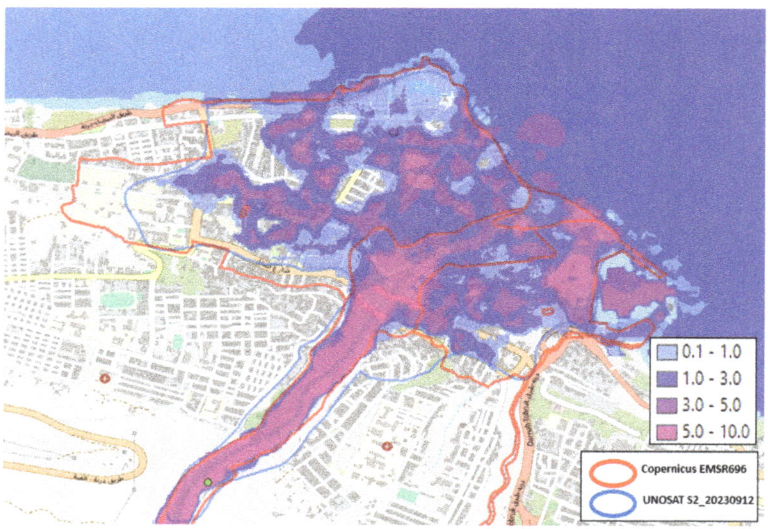

Figure 23. Maximum flow depth for the nominal case in the Al-Maghar neighbourhood. The computed data are compared with the flood extents identified by UNOSAT (blue line) [22] and Copernicus EMS (red line) [21] (basemap from OSM).

5.2. Spatial Validation

Figure 23 shows the modelled maximum flood extent, revealing that a substantial part of the Derna's city centre would have experienced flow depths exceeding 5 m. Moreover, the sea level in front of Derna would have experienced a rise, with depths ranging from 1 to 3 m. In particular, in the eastern coastal area, the water level would have increased by up to 1.15 m, suggesting potential coastal inundation in a significant portion of the eastern city. During the validation of the scenario with the actual event, the comparison of flood extents with Copernicus EMS [21] and UNOSAT [22] appeared to be largely accurate. However, some discrepancies were observed, particularly in western Derna, where some areas appeared unaffected in the calculation but were flooded according to satellite assessments.

These discrepancies are attributed to small variations in elevation in the SRTM digital elevation model [7], which would have obstructed water entry and flows into these areas (refer to the lower portion of Figure 24). It is unclear whether these are artefacts of the SRTM or actual elevated areas without local investigation. Moreover, the designation of these locations as 'flooded' or 'affected' by UNOSAT or Copernicus experts may indicate wet areas resulting from heavy precipitation rather than flooding due to the dam break. It is also important to note that the SRTM does not consider the presence of buildings, which can significantly alter water flow by creating channelling and higher-velocity paths.

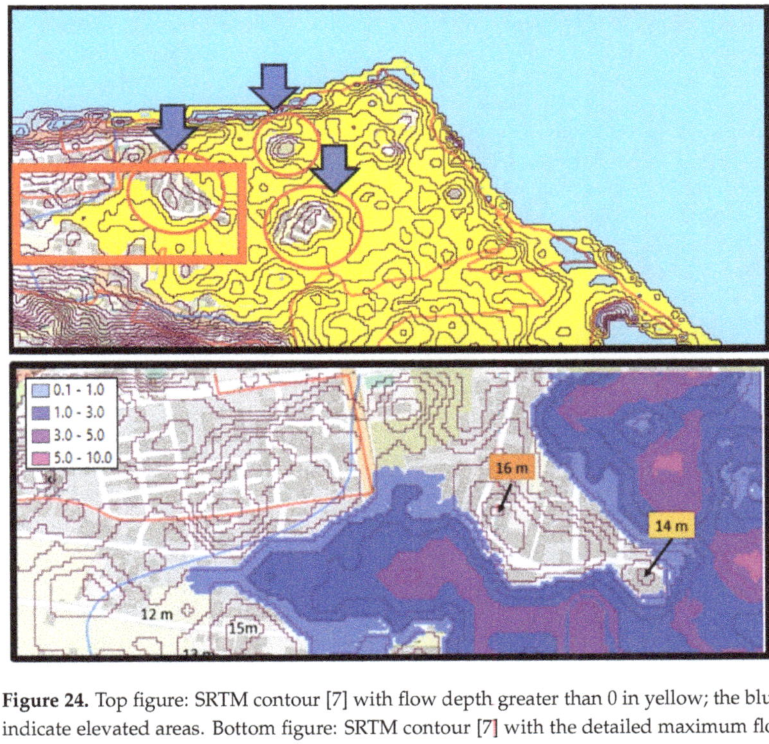

Figure 24. Top figure: SRTM contour [7] with flow depth greater than 0 in yellow; the blue arrows indicate elevated areas. Bottom figure: SRTM contour [7] with the detailed maximum flow depth (basemap from OSM).

The overall flooded area is detailed in Table 7. Both UNOSAT and Copernicus analyses estimated the overall area to be between 3.1 and 3.3 km^2, while the scenario indicates a slightly lower value of approximately 2.7 km^2.

Table 7. Inundated surface.

Mode	Surface (km^2)
UNOSAT	3.1
Copernicus	3.3
WL = 225 BW = 260	2.7

5.3. Flow Depth Validation

Ideally, field teams measuring flood heights relative to common reference points are deployed as early as possible to assess flow depths following sudden-onset and extensive flood events, as in the case of tsunami events. However, in this case, the assessment was conducted remotely by the international community.

To evaluate the accuracy of the flow depth in the model, a number of identifiable landmarks were selected and used to estimate the flow depth by visually analysing post-event images from social media referred to these points. Artificial intelligence techniques were utilised to extract flood images from social media, and text mining was applied to locate images containing information about water depth.

The visually interpreted water depth values were compared to the model-derived ones, revealing that the flood depth was generally underestimated. This discrepancy may be attributed to the same factors influencing the previously mentioned underestimation of flood extent, namely the height variations in the STRM [7] (refer to Figure 23). Additionally, it is important to note that the model does not account for the splash effects that can occur at the forefront of the flood wave. These splash effects might leave wet traces on building facades at higher levels with respect to the actual flood depth.

The results of the visual interpretation of the water height from the analysis of social media images compared to the modelled flood depths at selected locations are presented in Figures 25 and 26 and in Table 8. A list of image sources for selected locations can be found in Appendix A. The comparison of the visual interpretation of the images with the modelling results revealed an average underestimation of modelled flood depth of approximately 13% for the 230 m lake water level case and 17% with the 225 m case.

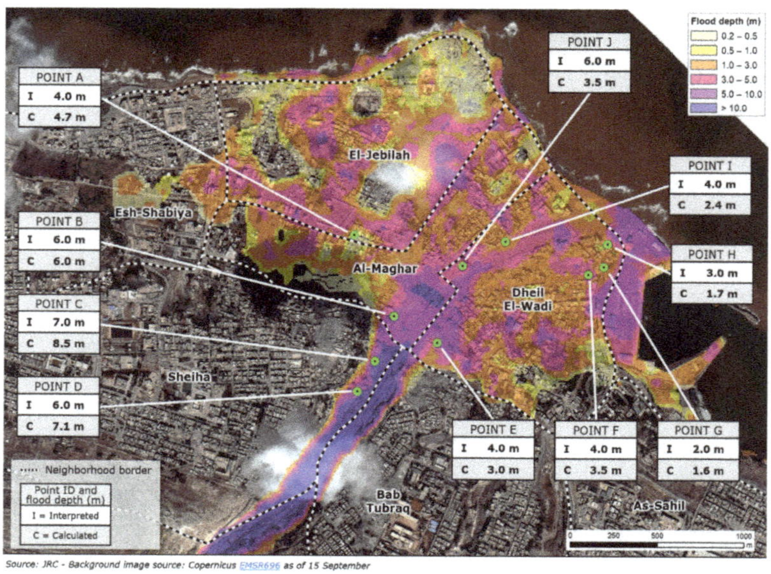

Figure 25. Comparison of the flood depth from modelling and from visual interpretation of images at selected locations (JRC and basemap from Copernicus).

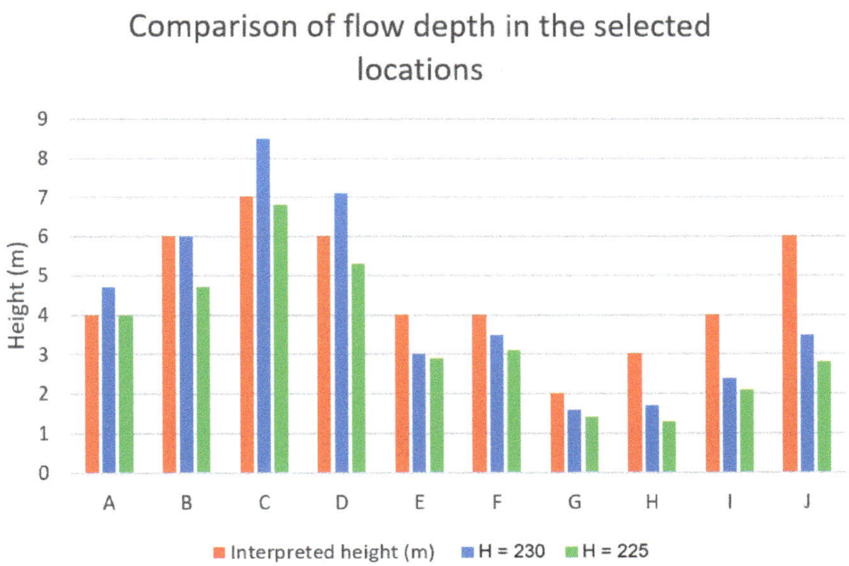

Figure 26. Comparison between interpreted and modelled flood depths at selected locations. Red bars: visual interpretation from social media; blue bars: model results with an assumed water height in the reservoir of 230 m; green bars: model results with an assumed water height in the reservoir of 225 m.

Table 8. List of selected locations and comparison of the flood depth values.

ID	Latitude	Longitude	Interpreted Flood Depth (m)	Calculated Height (230 m Case) (m)	Calculated Height (225 m Case) (m)	Delta for 230 m Case (%)	Delta for 225 m Case (%)
A	22.639111	32.766461	4.0	4.7	4	−18	15
B	22.640019	32.762782	6.0	6.0	4.7	0	22
C	22.639033	32.760849	7.0	8.5	6.8	−21	20
D	22.638136	32.759563	6.0	7.1	5.3	−18	25
E	22.642266	32.761646	4.0	3	2.9	25	3
F	22.65005	32.764593	4.0	3.5	3.1	13	11
G	22.650846	32.764915	2.0	1.6	1.4	20	13
H	22.651043	32.765885	3.0	1.7	1.3	43	24
I	22.645749	32.765969	4.0	2.4	2.1	40	13
J	22.643552	32.764949	6.0	3.5	2.8	42	20
		Average Difference (%)				13	17

6. Main Findings

The results of the validation exercise of the Derna dam break scenario conducted by integrating satellite mapping, media and social media images, and reports collected remotely from local testimonies are summarised in this section.

6.1. Temporal Assessment

Although the exact time of the dam break remains unknown, reports collected remotely from local testimonies suggest an approximate arrival time in Derna of 30 min. Therefore, the modelled arrival time of 27 min can be regarded as a highly accurate estimation.

The rate of flood height increase was assessed using data from security cameras [31] at key observation points, including the Al-Maghar neighbourhood, the Al Sahaba Mosque, and other easily identifiable landmarks. In the most heavily affected areas, such as Al-

Maghar and the Mosque, the observed rate of increase was approximately two metres per minute. This observed value corresponds with the modelled increase in flow depth extracted from the same observation points, confirming the consistency of the model's estimations.

6.2. Spatial Assessment

A comparison of flood extent data from Copernicus EMS [21] and UNOSAT [22] suggests that the spatial extent of the flood is well represented by the model's results. The modelled flood extent in Derna is approximately 57 hectares smaller when compared to post-event satellite mapping and assessments. This discrepancy is deemed acceptable, as the model does not account for factors such as rainfall contribution and resulting water accumulation in the smaller downstream dam. Moreover, the model only considers areas as flooded when the water depth exceeds 10 cm.

6.3. Flood Depth Assessment

The comparison between visual interpretation of social media information extracted through artificial intelligence with the modelled flood depth revealed an average underestimation of approximately 13% in the 230 m lake water level case and 17% in the 225 m case, confirming substantial consistency.

7. Conclusions

This study reveals that the flood arrival time in the city of Derna would have been between 15 and 30 min, making it extremely challenging to provide advance warning to the population, even if an active early warning system had been in place. Unfortunately, this was not the case, as an automatic alert system was not present at the time of the event. This underscores the importance of emergency preparedness measures in dam safety protocols.

Upon comparing the estimated inundation extent across different scenarios with the delineation by UNOSAT and Copernicus satellite products, it was found that the majority of the flooded area was correctly identified by the model, accounting for approximately 82 to 84% of the total actual flood extent. In only one calculation, characterised by a very high initial river flow, the inundation exceeded the delineation of the flooded area derived from satellite imagery. The comparison of the maximum water depth indicates that the difference between the computed and estimated height was less than 20%. This analysis demonstrates the added value of the combined use of high-resolution satellite images and social media information processing to determine a detailed flood extent and water height for the event.

This approach allowed us to address some of the significant uncertainties in the initial conditions of the Derna dam break event, including a plausible water height of the reservoir of between 225 and 230 m, indicating that the dam break probably occurred with the reservoir at its full capacity. The validation exercise conducted using satellite and social media-derived images also confirmed the accuracy of the temporally and spatially modelled flood.

Overall, the validation exercise demonstrated the feasibility and substantial reliability of a rapid assessment of the dam break dynamic through the selected modelling approach, despite the lack of precise local measurements. Acquiring detailed topography up to 1 m resolution, such as through drone flights equipped with LiDAR instrumentation, would allow for a more accurate assessment. In the case of the Casamicciola landslide in Ischia, Italy, Copernicus drone flights obtained a post-event digital terrain model with 5 cm resolution within two days [1].

The low volume of the Mansour Dam downstream of the Derna Dam should not have had a significant influence on the dynamic of the inundation in the city of Derna. While for other cases of cascading dam collapses, with reservoirs of similar dimensions, specialised techniques (e.g., accounting for geometry changes during the calculation) may be required, for this particular scenario, the downstream dam's reservoir volume was likely not a major factor in the flooding.

In conclusion, the rapid modelling of dam break events and combined analysis of multiple data types have proven suitable for promptly assessing the expected dynamic of the event, as well as reconstructing the unknown initial conditions before the break. Incorporating sensitivity analyses provides an estimate of the uncertainties associated with the deduced values of the unknown parameters and their relative importance in the analysis.

Author Contributions: Conceptualization, A.A., M.S. and C.P.; literature search, A.A., M.S. and C.P.; methodology, A.A., M.S. and C.P.; study design, A.A., M.S. and C.P.; data collection, A.A., M.S., C.P. and V.L.; data processing, A.A., M.S., C.P., A.G. and M.T.; data analysis, A.A., M.S., C.P. and V.L.; data interpretation, A.A., M.S. and C.P.; writing—original draft preparation, A.A., M.S., C.P. and L.d.G.; writing—review and editing, A.A., M.S., C.P. and L.d.G. All authors have read and agreed to the published version of the manuscript.

Funding: This research received no external funding.

Data Availability Statement: The calculations were performed using the Dam Break Platform. Results are available for reference and consultation at https://github.com/annunal/DamBreak/tree/main/DBP.

Acknowledgments: The authors acknowledge Andrey Zaytsev for his efforts in developing the numerical structure of NAMI DANCE and Bora Yalciner for his outstanding success in developing a GPU version of NAMI DANCE.

Conflicts of Interest: Authors Chiara Proietti and Andrea Gerhardinger were employed by the company FINCONS SPA. And Author Ludovica de Girolamo was employed by the company Seidor Italy SRL. The remaining authors declare that the research was conducted in the absence of any commercial or financial relationships that could be construed as a potential conflict of interest.

Appendix A

Table A1. List of image sources for selected locations.

ID	Image/Location Source	Latitude	Longitude	Interpreted Flood Height (m)
A	Source: X/@YeniSafakArabic	22.639111	32.766461	4.0
B	Source: El Pais	22.640019	32.762782	6.0
C	Source: X/@Alhamdhulillaah	22.639033	32.760849	7.0
D	Source: X/@AhmedMussa218	22.638136	32.759563	6.0
E	Source: X@ @lole06554	22.642266	32.761646	4.0
F	Source: Al Jazeera/Jamal Alkomaty/AP Photo	22.65005	32.764593	4.0
G	source: X/@magazine_wafaa	22.650846	32.764915	2.0
H	Source: Jamal Alkomaty/AP	22.651043	32.765885	3.0
I	Source: Al Jazeera/Jamal Alkomaty/AP Photo	22.645749	32.765969	4.0
J	Source: X/@WorldTimesWT	22.643552	32.764949	6.0

References

1. Annunziato, A.; Guney Dogan, G.; Yalciner, A.C. Modeling Dam Break Events Using Shallow Water Model. *Eng* **2023**, *4*, 1851–1870. [CrossRef]
2. World Meteorological Organization Report. Provisional State of the Global Climate 2023. Report. 2023. Available online: https://wmo.int/sites/default/files/2023-11/WMO%20Provisional%20State%20of%20the%20Global%20Climate%202023.pdf (accessed on 20 May 2024).
3. International Organization for Migration. Resilience, Recovery & Reconstruction Plan, Storm Daniel. Report. 2023. Available online: https://reliefweb.int/report/libya/iom-libya-resilience-recovery-reconstruction-plan-storm-daniel-september-2023-september-2024 (accessed on 20 May 2024).
4. ECHO Daily Map of 13 September 2023. Available online: https://erccportal.jrc.ec.europa.eu/ECHO-Products/Maps#/maps/4636 (accessed on 20 May 2024).
5. Pielke, R. Trends in Flooding in Africa. Available online: https://rogerpielkejr.substack.com/p/trends-in-flooding-in-africa (accessed on 20 May 2024).
6. Technical Data on Derna and Mansour Dams on Hidrotehnika Company Website. Available online: https://www.hidrotehnika.rs/en/libya/wadi-derna/ (accessed on 20 May 2024).
7. SRTM Topography Database (NASA) via Watkins, D. 30-Meter SRTM Tile Downloader Method. Available online: https://dwtkns.com/srtm30m/) (accessed on 20 May 2024).
8. Abdalla, M. Mathematical Model of Dam Break: Applied Study of Derna Dam. *MEJ—Mansoura Eng. J.* **2002**, *27*, 46–59. [CrossRef]
9. Ashoor, A.A.R. Estimation of the surface runoff depth of Wadi Derna Basin by integrating the geographic information systems and Soil Conservation Service (SCS-CN) model. *J. Pure Appl. Sci.* **2023**, *21*, 90–100. [CrossRef]
10. Santini, M.; Proietti, C.; Gerhardinger, A.; Tucci, M.; Barantiev, D.; Giustolisi, L.; Lorini, V. *Validation Analysis of the Dam Breach Calculations for Derna Dam, Libya*; JRC136254; Publications Office of the European Union: Luxembourg, 2023; under review.
11. Imran, J.; Nemnem, A.; Tanim, A.; Khan, S.; Nahian, A.; Goharian, E. Rising Waters, Falling Dams: Deciphering the Derna Flood Disaster. *Nat. Portf.* **2024**. *preprint version*. [CrossRef]
12. Groenewege, R. Rapid simulation and assessment of the Derna (Libya) dam failures. Conference Paper. 2024. Available online: https://www.researchgate.net/publication/378961823_Rapid_simulation_and_assessment_of_the_Derna_Libya_dam_failures (accessed on 20 May 2024).
13. Prida, A. Dam Break Simulation in Derna (Libya) Using HydroMT_SFINCS. Presented at Delft Software Days, November 2023. Available online: https://softwaredays.deltares.nl/-/dsd-int-2023-slides-dam-break-simulation-in-derna-libya-using-hydromt_sfincs?startDate=06-09-2023 (accessed on 20 May 2024).
14. Akgun, C.; Nas, S.S.; Uslu, A. 2D and 3D Numerical Simulation of Dam-Break Flooding: A Case Study of the Tuzluca Dam, Turkey. *Water* **2023**, *15*, 3622. [CrossRef]
15. Fread, D.L. Dam-Breach Floods. In *Hydrology of Disasters*; Singh, V.P., Ed.; Water Science and Technology Library; Springer: Dordrecht, The Netherlands, 1996; Volume 24. [CrossRef]
16. Yudianto, D.; Ginting, B.M.; Sanjaya, S.; Rusli, S.R.; Wicaksono, A. A Framework of Dam-Break Hazard Risk Mapping for a Data-Sparse Region in Indonesia. *ISPRS Int. J. Geo-Inf.* **2021**, *10*, 110. [CrossRef]
17. Erges. Available online: https://www.erges.com.tr/EN,532/derna-bu-mansur-dam.html (accessed on 20 May 2024).
18. Yalciner, B.; Zaytsev, A. Assessment of Efficiency and Performance of Tsunami Numerical Modeling with GPU. Abstract EGU2017–1246 Presented at the European Geoscience Union (EGU), April 2017. Available online: https://meetingorganizer.copernicus.org/EGU2017/EGU2017-1246.pdf (accessed on 20 May 2024).
19. Dogan, G.G.; Annunziato, A.; Hidayat, R.; Husrin, S.; Prasetya, G.; Kongko, W.; Zaytsev, A.; Pelinovsky, E.; Imamura, F.; Yalciner, A.C. Numerical simulations of December 22, 2018 Anak Krakatau tsunami and examination of possible submarine landslide scenarios. *Pure Appl. Geophys.* **2021**, *178*, 1–20. [CrossRef]
20. Velioglu Sogut, D.; Yalciner, A.C. Performance Comparison of NAMI DANCE and FLOW-3D® Models in Tsunami Propagation, Inundation and Currents using NTHMP Benchmark Problems. *Pure Appl. Geophys.* **2019**, *176*, 3115–3153. [CrossRef]
21. Copernicus Emergency Management System, Mapping Activation EMSR696. Available online: https://rapidmapping.emergency.copernicus.eu/EMSR696/download (accessed on 20 May 2024).
22. Flood Impact Assessment in Derna City Centre, East Province, Libya, UNOSAT Activation 3672. Available online: https://unosat.org/products/3672 (accessed on 20 May 2024).
23. Annunziato, A. The Dam Break Platform. Available online: https://github.com/annunal/DamBreak/tree/main/DBP (accessed on 20 May 2024).
24. DredgeWire. Available online: https://dredgewire.com/the-failed-dams-in-wadi-derna-in-libya/ (accessed on 20 May 2024).
25. Eduvast. Available online: https://www.eduvast.com/information/wadi-derna-dam-collapse-has-it-led-to-flooding-in-libya/ (accessed on 20 May 2024).
26. LandScan Global Data Set (Oak Ridge National Laboratory). 2014. Available online: https://landscan.ornl.gov/ (accessed on 20 May 2024).
27. EU Joint Research Centre Global Human Settlement Layer (GHSL). 2015. Available online: https://ghsl.jrc.ec.europa.eu/tools.php (accessed on 20 May 2024).

28. Kostecki, S.; Banasiak, R. The Catastrophe of the Niedów Dam—The Causes of the Dam's Breach, Its Development, and Consequences. *Water* **2021**, *13*, 3254. [CrossRef]
29. Froehlich, D.C. Embankment Dam Breach Parameters and Their Uncertainties. *J. Hydraul. Eng.* **2008**, *134*, 1708–1721. [CrossRef] [PubMed]
30. Security Camera Images from BBC. Available online: https://www.bbc.com/news/av/world-africa-66822697 (accessed on 20 May 2024).
31. Information Bulletin 164—The Copernicus Emergency Management Service Activities in the Aftermath of the Mudflow in Ischia (Italy). Available online: https://emergency.copernicus.eu/mapping/ems/information-bulletin-164-copernicus-emergency-management-service-activities-aftermath-mudflow (accessed on 20 May 2024).

GeoHazards

MDPI

Article

Combined Effect of the Microstructure and Mechanical Behavior of Lateritic Soils in the Instability of a Road Cut Slope in Rwanda

Roberto Valentino [1,*], Mattia Pizzati [1] and Jules Mizero [2]

[1] Department of Chemistry, Life Sciences and Environmental Sustainability, University of Parma, Parco Area delle Scienze 157/A, 43124 Parma, Italy; mattia.pizzati@unipr.it
[2] College of Science and Technology, University of Rwanda, Kigali P.O. Box 3900, Rwanda; mizerojules111@gmail.com
* Correspondence: roberto.valentino@unipr.it

Abstract: A very common hazard in Rwanda is represented by the instability of steep road cut slopes in lateritic soil. In its natural state, this material appears as a fine-grained weak and altered rock, generally in unsaturated conditions. Steep cut slopes made by this material could remain stable for a long time unless weathering weakens its mechanical behavior and heavy rainfall provokes a rapid landslide. This paper presents the results of an experimental investigation on the microstructural, petrophysical, and geotechnical properties of lateritic soil from a road cut slope located in Kabaya (Ngororero District—Rwanda), which was recently subjected to a landslide. The mechanical properties of the material are strictly related to the geological origin and history of the deposits, their formation environment, and weathering processes. These characteristics were revealed by peculiar microstructural features (micro-texture, porosity, and degree of alteration of original mineral paragenesis). The experimental investigations included identification and classification tests, direct shear tests on saturated samples, and swelling tests. This multidisciplinary approach provided insights into the relationship between geotechnical properties and the microstructural, petrophysical, and chemical characteristics of the altered rocks. This study showed how different levels of chemical alteration operated by weathering processes, in conjunction with brittle deformation related to the tectonic history, formed in the same site two shallow rock layers with similar macro-scale features and mechanical behaviors but markedly different microstructural and chemical properties. The innovative aspect of this research suggests an integrated multidisciplinary approach to considering microstructural aspects in addition to mechanical behavior in the slope stability analyses in lateritic soil. In particular, this study demonstrates the importance of such an approach since the failure mechanism is better explained if it is based on microstructural observations instead of considering the soil shear strength parameters only. This research helped to explain the formation of the landslide failure mechanism in a specific road cut slope, which could be assumed as representative of many other similar slopes subjected to landslides in Rwanda.

Keywords: lateritic soil; soil characterization; slope stability; microstructure; petrophysics

1. Introduction

Landslides are widespread natural phenomena that cause lots of damage to structures, infrastructures, human activities, and human life. Landslides have proven to be the most challenging socio-economic disasters along road corridors in hilly landscapes all over the world, which led to huge losses of life and property and environmental deterioration [1–4]. Some areas in the world are particularly affected by these phenomena. For example, mountainous Asian regions are impacted by extreme monsoonal rainfall that triggers many landslides, resulting in human impacts and damage to infrastructure every year [5]. Some mountainous areas in Sub-Saharan Africa, which include the Democratic

Republic of Congo (DRC), Uganda, and Rwanda, are particularly landslide prone and are representative of other mountainous regions in the tropics [6]. Moreover, these areas have many characteristics in common with tropical regions where landslides cause loss of lives and damage to infrastructure and, at the same time, landslide hazard is still poorly understood [7–11]. In particular, Rwanda extends over a mountainous region, which is characterized by very steep slopes. Mountain reliefs with heights of up to 4486 m above sea level mostly dominate the western and northern areas of Rwanda. Due to the steep topographic gradient, associated with the construction of extensive road cut slopes, landslides in Rwanda are very frequent phenomena [12–15].

Most failures, in Rwanda as in other countries, regard road cut slopes and are caused by the natural weathering of soft rocks, which appear rather stable in the short term after excavation but become unstable and may fail in the long term [16,17]. For some cut slopes, failures are due to extreme steepness, slope height, and inadequate estimation of soil resistance. In most cases, especially in Rwanda, slope failures are caused by heavy rainfalls associated with human interventions, such as excavations, the construction of roads and buildings, and logging [18]. Every year in Rwanda, landslides cause loss of human lives, damage to properties, and leave hundreds of people homeless due to house destruction. One of the most recent disasters caused by heavy rainfalls occurred during the night of 2–3 May 2023, resulting in significant damage and loss of lives in different parts of the western, northern, and southern provinces. The disasters claimed the lives of 131 people, injured several others, and damaged public infrastructures [19].

The soil that covers most of Rwanda's northern and western mountains derives from the weathering of the original bedrock. It is well known that weathering is one of the predisposing causes of landslides. Some researchers investigated the impact of weathering on slope stability in soft rocks, demonstrating that weathering produces both shear strength reduction in deeper layers and the disintegration of material on the surface of the slope [20].

Two dominant processes are recognized: chemical and physical weathering. Chemical weathering is the decomposition of the constituent minerals into more stable secondary mineral products, while physical weathering is related to the disaggregation of rocks without significant variation in mineralogical composition [20]. The properties of rocks and their behaviors during exposure to exogenic agents are mainly controlled by their mineralogical composition, pre-consolidation history, level of cementation and rock texture, and composition of binding material in their structure. Of course, water content plays a crucial role in the alteration of soft rocks [20]. Depending on the mineralogical composition of the rock and the chemical composition of water, different forms of chemical weathering can develop on the rock surface [20]. In combination with weathering, especially in road cut slopes, the removal of material during excavations produces a stress release that provokes the formation of internal weakness surfaces, which may be exploited during failure [20].

Although the effect of weathering on the shear strength reduction is widely studied in many regions characterized by the presence of tropical unsaturated soils, in Rwanda, there are very few detailed studies on both soils and soft rocks subjected to landslides at the moment.

In other similar environments in Africa, weathered soils are often named "lateritic soils". According to Oyelami and Van Rooy [21], the term "lateritic soil" is used to describe a highly weathered tropical or sub-tropical residual soil, which is generally well graded and coated with sesquioxide-rich concretions. Residual soils are formed directly or almost directly on the parent rock, making it possible to retain some of their characteristics [21,22]. These residual soils are generally produced by intense chemical weathering due to relatively high air temperature and high water content [23]. The geotechnical behavior of lateritic soils is mainly controlled by micro-textures, mineralogical compositions, and geochemical environmental conditions. Normally, alternating rainy and dry seasons favor the formation of thick lateritic soils. In fact, when the rainy season leaching of the parent rocks takes place during the dry season, capillary action transporting solutions of leached ions to the surface occurs. From the surface, the evaporation of water leaves the salts behind to be washed down during the following wet season. Thus, the whole zone could be progressively depleted of the more mobile elements, such as K, Na, and Ca [24]. Olanipekun [25] observed that the high proportion of Fe3þ oxides in laterites signifies a left-over accumulation as a result of the removal of alkalis and silica. Moreover, it seems that chemical weathering progresses more rapidly in warm than in cool climates, and provided there is good drainage, it is more prevalent in wet than dry climates [26]. This explains the more pronounced lateralization processes observed in the tropics compared to arid regions.

Kasthurba et al. [27] observed a downward softening of altered material as a result of an increase in the amount of clay-filled pores and a decrease in sesquioxide cementation. This explains why, in most cases, lateritic profiles are characterized by kaolinitization at lower levels, and the accumulation of sesquioxides occurs in the upper horizons.

Depending on the percentage of clay and the presence of expanding clay minerals, some lateritic soils change volume when exposed to humidity variations, while others are not affected [21]. Hence, some components, such as sand and gravel, are referred to as stable, while silt and clay are referred to as unstable. Stability in this sense is based on their ability to withstand variations in terms of moisture without a significant change in their properties [21].

Many studies have been carried out at a microscopic level to investigate the mechanical behavior of some lateritic soils in partially saturated conditions. Gao et al. [28] investigated the influences of drying–wetting cycles and dry density changes on the water retention characteristics of compacted lateritic soils. They used scanning electron microscopy and mercury intrusion porosimetry to explore the underlying microscopic mechanism. Kong et al. [29] evaluated the effects of multiple drying–wetting cycles on the soil–water characteristic curve (SWCC) of some undisturbed granite residual soils. They also applied advanced techniques to analyze the structure at the microscopic level to understand the possible microstructural variations of granite residual soils subject to a periodic drying and wetting process. Bai et al. [30] investigated the mechanical behavior of lateritic soil using the pressuremeter test, obtaining a pressuremeter curve that was analyzed in combination with the actual soil conditions. They determined the strength and deformation characteristics of lateritic soil under different free-iron oxide contents and different depths.

This paper presents the results of different investigations carried out on lateritic soil that was subjected to a landslide in early May 2018 in Kabaya, Ngororero district (Rwanda). The landslide is regarded as a road cut slope, which can be considered representative of many similar situations in the surrounding areas. The micro-investigations were carried out on two different soil samples, collected at different depths on site, and aimed at understanding which aspects of the microstructure could be considered directly related to the triggering mechanism of the landslide.

In particular, laboratory analyses included soil classification, which was carried out through both classical techniques and laser diffraction analysis, microstructural characterization connected with two-dimensional porosity estimation, mineralogical–chemical analysis via SEM, failure conditions through direct shear tests on saturated samples, and swelling tests.

The microstructural analysis showed how different levels of chemical alteration operated by weathering processes, which, in conjunction with the tectonic history, produced the formation of two shallow rock layers with similar macro-scale features and mechanical behavior but markedly different microstructural and chemical properties. Based on both microstructural and geotechnical characterization, this study demonstrated that the deepest layer was weaker than the more superficial one. This research helped to explain the formation of the landslide failure mechanism in correspondence with the weaker layer at the slope toe.

The main outcome of this research is emphasizing the importance of considering microstructural aspects in addition to mechanical behavior in the slope stability analyses in lateritic soil since the triggering mechanism of the landslide could be better explained based on microstructural observations than on shear strength parameters only.

2. Geological Origin and Site Description

The geology of Rwanda is dominantly composed of ancient metamorphic and igneous-intrusive rocks, followed by more recent lava flow emissions and subordinate sedimentary rocks [31]. The oldest rocks are located in the eastern Kivu lake district, at the southern border with Burundi, and in the central north-western region, where extensive exposures of Paleoproterozoic rocks (2.5 to 1.6 Ga) are present (Figure 1). These rocks are mainly composed of gneisses, schists, quartzites, and granites, which experienced severe alteration to sericite (alteration of alkali feldspar minerals to sericite) operated by weathering processes in conjunction with brittle deformation. Paleoproterozoic rocks are followed by the Mesoproterozoic series (1.6 to 1.0 Ga), which are composed of sedimentary rocks (sandstone and conglomerate) metamorphosed at high-temperature and low-pressure conditions (quartzite) (Figure 1). In the westernmost portion of Rwanda, dominated by Lake Kivu, there is extensive occurrence of lavas and associated pyroclastic rocks, the oldest of which can be dated back to the Cretaceous–Paleogene boundary (65 Ma). Volcanic rocks dated to the Cenozoic are found in the northwest and west of the country. Some of these volcanoes are highly alkaline and are the prosecutions of the Virunga volcanic area of southwestern Uganda. Tertiary and Quaternary clastic sediments fill parts of the Western Rift in the western part of the country, as well as other lakes in eastern Rwanda [32]. The Kabaya landslide, which is the focus of the present study, is located in the northwestern region, where rocks belonging to the Paleoproterozoic sequence crop out, at the boundary with Plio-Pleistocene age basaltic lava flows. In the study area, rocks are classified as gneisses, granites, and metamorphic schists (Figure 2) [31].

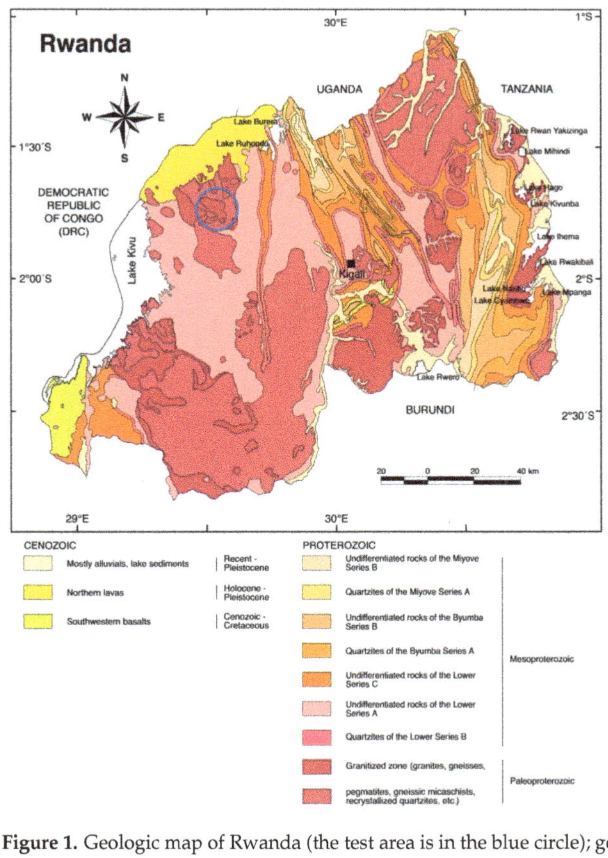

Figure 1. Geologic map of Rwanda (the test area is in the blue circle); geological atlas of Africa [32].

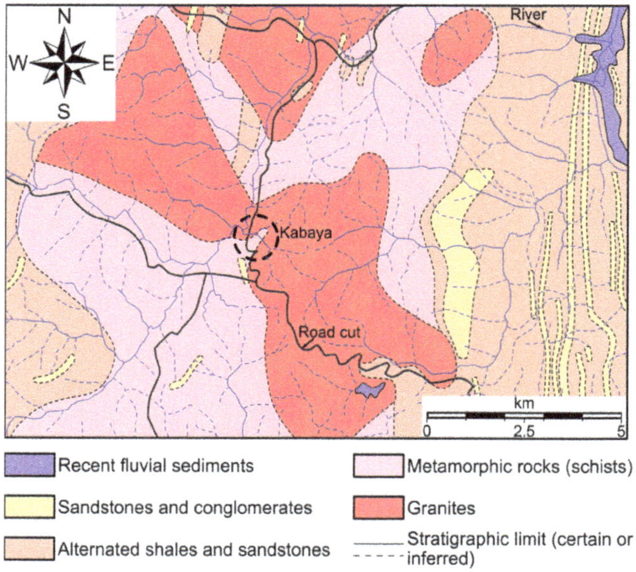

Figure 2. Detailed geological map of the study area. The dashed circle corresponds to the location of the Kabaya landslide [31].

Case Study in Ngororero District

The study site is located along National Road NR16 at Kabaya, Ngororero district, where a road cut slope was subjected to a landslide on 2 May 2018 (Figure 3). The landslide occurred during a wet period, characterized by a cumulative rainfall of 500 mm in one month. A daily rainfall of 17 mm was registered on the same day of the landslide occurrence (Figure 4).

The slope is close to the top of a hill (2300 m a.s.l.) in an urbanized area, where many houses, power lines, other roads, and facilities are located. The site elevation is comprised between 2260 and 2285 m a.s.l., the total height of the road cut is 25 m, and the mean slope angle is 37°. The thickness of the soil involved in the landslide was from 2 to 6 m. The slope was covered by trees and shrubs before the collapse. Immediately after the landslide occurred, the underlying soft rocks appeared completely disintegrated into heterometric material with sizes spanning from boulders to fine sand, which blocked the road drainage system and obstructed the roadway (Figure 3b). Figure 5 shows how the slope appeared on 3 December 2018, i.e., some months after the landslide occurred. The shallow bumpy soil had been removed from the scarp and exposed the soft bedrock of the slope. The bedrock appears composed of whitish (above) and reddish (below) soil layers with some thin yellow stripes (Figure 5). By comparing Figures 3 and 5, it is evident that the reddish soil layer was exposed at the toe of the road cut slope even before the collapse. Further rainfall runoff caused the formation of incised drainage pathways and gullies along the main slope (Figure 5). Figure 6 shows a sketch of the slope before failure, which has been determined based on a field survey. Figure 6 also reports the position and shape of the failure surface and the location of soil samples. It must be considered that the geometrical situation of the slope before failure was the result of a previous excavation of the natural slope due to the road construction. Therefore, it could be supposed that the original natural slope was composed of a shallow whitish soil layer and a deep reddish soil layer. After the road construction, the reddish layer appeared exposed at the slope toe, as reconstructed in Figure 6. For this reason, in the following, the whitish layer will be also called "shallow" and the reddish one will be assumed as "deep" by referring to the original "natural" situation. As will be explained, both whitish and reddish lateritic soils involved in the landslide were classified as sandy silt, derived by weathering of the parent bedrock. The lateritic soils in this site normally appear as cohesive soft rock in dry or partially saturated conditions, but they become completely de-structured when they are fully saturated.

Figure 3. Pictures of the studied landslide. (**a**) Detail of the slope along the road cut hosting the studied landslide. (**b**) Closure of the road due to the landslide.

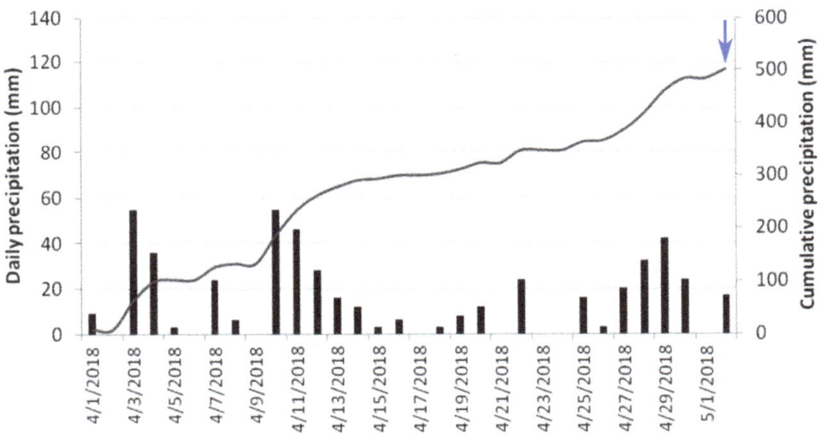

Figure 4. Kabaya–Ngororero study site: daily and cumulative precipitation from 1 April 2018 to the date of the event (2 May 2018) based on rainfall data from the Kabaya weather station. The blue arrow indicates the day when the landslide occurred.

Table 1. Soil classification.

Sample ID (Classification)	Depth (m)	Layer	USCS Classification
S4; S6; S7; S9; AR2	1.0	Whitish	Silty sand—SM
S1; S2	2.0	Whitish	Sandy silt—ML
S3; S5; S8; S10; AR1	2.0	Reddish	Sandy silt—ML

Figure 5. Scarp of the Kabaya road cut slope after the landslide occurred. The different labels refer to the several soil samples (see Table 1) at the position where they were collected along the slope. The white dashed line identifies the separation between the whitish (above) and reddish (below) soil layers.

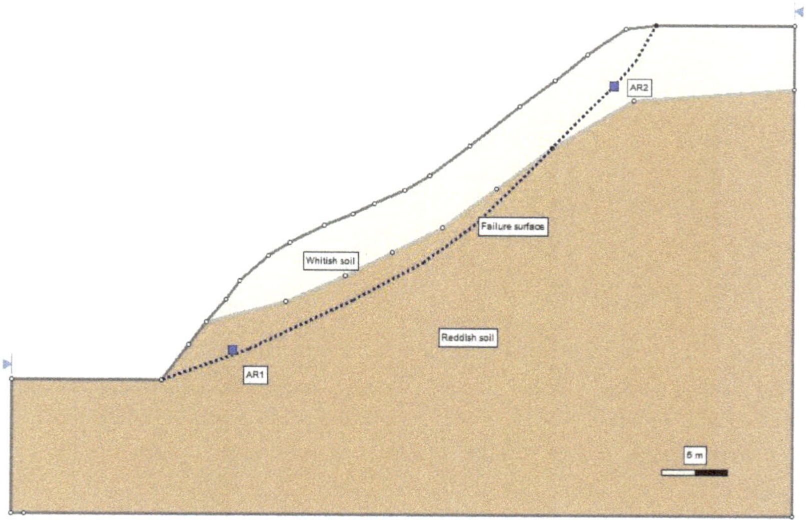

Figure 6. Sketch of the slope before failure. The blue dashed line represents the failure surface.

3. Materials and Methods

In this section, all the experimental procedures, tests, and methodologies adopted in classifying and characterizing the geological materials involved in the studied landslide process are described. Used procedures include sieving, sedimentation, laser optical grain size analyses, the definition of Atterberg limits, microstructural characterization, microporosity quantification, SEM-EDS chemical analyses, swelling tests, and direct shear tests. The results and observations deduced from the laboratory tests will be reported in a further section.

3.1. Classification of Materials

To obtain the geotechnical characterization of the soils involved in the landslide, a total of twelve soil samples were collected in different positions on the landslide scarp from the substrate that remained stable during the landslide (Figure 5). This material was considered representative of the soil that the landslide transported at the slope toe and that was lost. Ten of these samples (S1–S10) were completely disturbed, i.e., it was not possible to preserve their structure during in situ sampling, as the soil was suddenly disintegrated when subjected to the slightest pressure. Some samples were collected at 1 m of depth and were positioned at the top of the slope in the whitish layer. Instead, samples collected at 2 m were positioned in both the whitish (S1; S2) and reddish (S3; S5; S8; S10) layers, either in the center of the slope or near the slope toe (Figure 5). Two further samples, namely AR1 and AR2, were collected at 2 m and 1 m of depth, as representative of the reddish and the whitish layers, respectively. These two samples were almost undisturbed since they had the appearance of cohesive soft rocks in partially saturated conditions (Figure 7), and it was possible to collect them in "big" blocks. It should be noted that all the sampling points are located on the sliding surface for representativeness (Figures 5 and 6). Samples S1–S10 were analyzed to obtain the geotechnical classification according to the USCS system based on standard laboratory tests, such as sieving [33] and sedimentation [34]. Specifically, the sieve analysis was performed to obtain the size distribution of the fine particles (coarser than 0.075 mm). The sedimentation analysis was only used for four samples, namely S1, S4, S7, and S10, to determine the grain size distribution curve of materials passing through sieve No. 200 (0.075 mm). As will be discussed (Section 3.4), the grain size distribution curves of the AR1 and AR2 samples have been obtained with a different analytical procedure.

The Atterberg limits were only obtained according to standard tests [35] for samples AR1 and AR2.

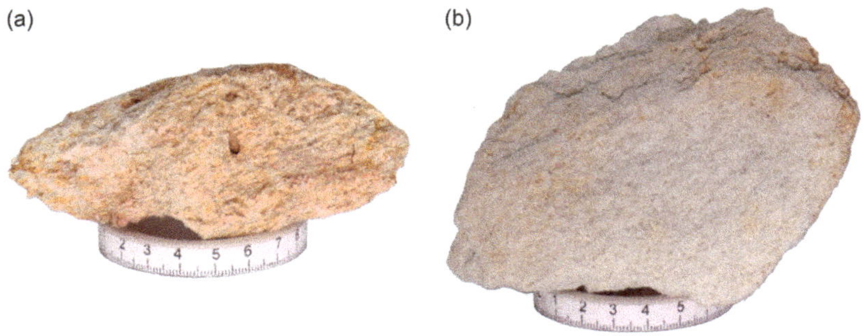

Figure 7. Pictures of two samples employed to characterize the microstructural, petrophysical, and mineralogical composition of the rocks involved in the studied landslide. The sample position is shown in Figure 5. (**a**) AR1 sample (reddish layer). (**b**) AR2 sample (whitish layer).

3.2. Microstructural Characterization

Microstructural and micro-textural characteristics of soft rocks involved in the studied landslide were evaluated with optical microscopy. AR1 and AR2 samples were impregnated with blue-dyed epoxy resin to emphasize both primary and secondary porosities of the rock and to harden the hand specimens. After resin impregnation, the samples were cut with a precision saw, and a total of 6 serial, 30 µm thick, thin sections were produced. Thin sections were polished with synthetic diamond powder to improve the transparency and facilitate mineral identification under an optical microscope. Rock-thin sections were scanned with a high-resolution Nikon SuperCoolScan 5000 scanner to describe the overall rock texture. Subsequently, thin sections were observed under a Zeiss Axioplan 2 optical, transmitted light, petrographic microscope, equipped with a Leica MC 170 HD high-resolution camera that is suitable to work on rock samples.

3.3. Mineralogical–Chemical Analysis via SEM

Detailed microstructural and chemical analyses were conducted on the AR1 and AR2 samples with a JEOL JSM 6400 Scanning Electron Microscope (SEM) attached to an Oxford-INCA X-ray Energy-Dispersive System (EDS) with an electric current of 15 kV and 240 nA. The diameter of the analytical EDS X-ray spots was 2 μm, much smaller than the crystal size of minerals composing the analyzed rocks (>1000 μm). The mineral composition was identified by comparing X-ray emission spectra with the provided datasets.

3.4. Laser Diffraction Granulometry Analysis

Grain size analysis through optical granulometry was performed on the AR1 and AR2 samples. Prior to the analysis, disaggregation of soft rocks was needed to reduce them to a granular, cohesionless material. Both rock specimens (sample mass of ~200 g) were inserted in tap water for ~one hour and manually stirred to facilitate disaggregation (see the procedure described by Pizzati et al. [36]). Disaggregated materials were inserted into an oven at a controlled temperature of 40 °C for three days to remove most of the moisture. After sample dehydration, a soft manual disaggregation was performed to remove all the remaining aggregates. The total mass of samples was split into smaller aliquots using Quantachrome Instruments macro- and micro-rifflers. The splitting procedure allowed us to gain smaller sample amounts (<1 g) while still preserving the original grain size classes of the total initial sample. All granulometric analyses were carried out with a Malvern Panalytical Mastersizer 3000 optical granulometer, operating at a grain size range from 10 nm to 3500 μm. The granulometer was equipped with a Hydro EV liquid-dispersion analysis modulus, and de-ionized water was used as a dispersant medium. The instrument calculated the grain size based on the Mie light diffraction theory [37], approximating non-spherical particles to perfect spheres with associated equivalent diameters. To extract reliable light scattering distributions, the Mie theory required the refractive and adsorption indexes of the analyzed material. In our samples, we adopted the optical parameters of crystalline quartz (refractive index of 1.54 and adsorption index of 0.1). Standard operating procedures were specifically set for both samples in order to perform reliable and reproducible analyses [38]. In particular, AR1 sample operating parameters included a sample mass of ~0.3 g, a stirrer speed of 2400 rpm, 50 measurements with a 10 s laser activation time, and ultrasonication at 10% for 10 min before the analysis. Conversely, for AR2, a sample mass of 0.7–0.8 g, a stirrer speed of 2200 rpm, 100 measurements with a 5 s laser activation time, and no ultrasonication were adopted.

3.5. Two-Dimensional Porosity Estimation

Porosity estimation was obtained from high-resolution petrographic photomicrographs taken under transmitted light microscopy on the AR1 and AR2 samples. A total of 75 photomicrographs were acquired to characterize the 2D porosity of different mineralogical phases composing the overall rock framework. All photomicrographs were acquired at 25× magnifications (picture area of 2352 × 1764 μm) and were imported into ImageJ version 1.53s, which is an open-source image analysis software [39]. Photomicrographs were calibrated and corrected to increase the image contrast-brightness and to remove background noise. Porosity was estimated by calculating the percentage of area composed of blue dye-filled voids with respect to the total photomicrograph area. Porosity calculation on the same mineralogical phase was performed on several photomicrographs to grant statistical robustness of data and to consider the variation of porosity distribution throughout the entire thin section.

3.6. Mechanical Characterization

To investigate the possible expansive behavior of the soil, the AR1 and AR2 samples have been subjected to swelling tests to determine the swelling pressure using the oedometer device and the constant volume method (Method C) based on [40].

The failure conditions of soils have been obtained by performing consolidated drained (CD) direct shear tests on the AR1 and AR2 samples based on British Standards [41]. The same laboratory tests have been recently performed by other researchers on lateritic soils from different geographical areas [42–44]. Direct shear tests were performed on saturated specimens, whose size was 60 mm × 60 mm × 30 mm, at a shear velocity of 0.01 mm/min. Three tests were performed on similar specimens at different normal stresses of 100, 200, and 300 kPa. The first shear test on each specimen allowed for obtaining the peak shear strength. The residual strength of the soil specimen was obtained by performing three cycles of direct shear tests on the same specimen (under the same normal stress). The shear box apparatus returned to its initial position every time it completed its travel. This process was repeated three times, i.e., until the residual shear stress remained constant at subsequent shearing stages. This procedure, although simplified, is known to be used as an alternative to the ring shear test. The objective of the shear tests was to study the shearing conditions in terms of effective stress and to determine the effective shear strength parameters of soils, related to the Mohr–Coulomb failure criterion, in both peak and residual conditions.

4. Results

4.1. Classification of the Lateritic Soil

The grain size distribution curves of ten samples (S1–S10) obtained through traditional tests and those obtained through laser diffraction granulometry analysis (AR1 and AR2) are all reported in Figure 8. It could be noticed that the percentage of fine particles (<0.075 mm) among all samples is between 20% (AR2) and 83% (AR1). The percentage of sand is between 17% (AR1) and 80% (AR2), while the percentage of gravel is nil. The percentage of clay is less than 6% for all samples (only 0.1% for AR2), except for AR1, with a percentage of 11%. The silt fraction is between a minimum of 19.9% (AR2) and a maximum of 72% (AR1). In particular, the laser granulometry analyses conducted on the AR1 and AR2 granular specimens show striking differences, with AR1 being finer with an average equivalent particle diameter of 53.4 μm, while AR2 has a mean diameter of 224 μm.

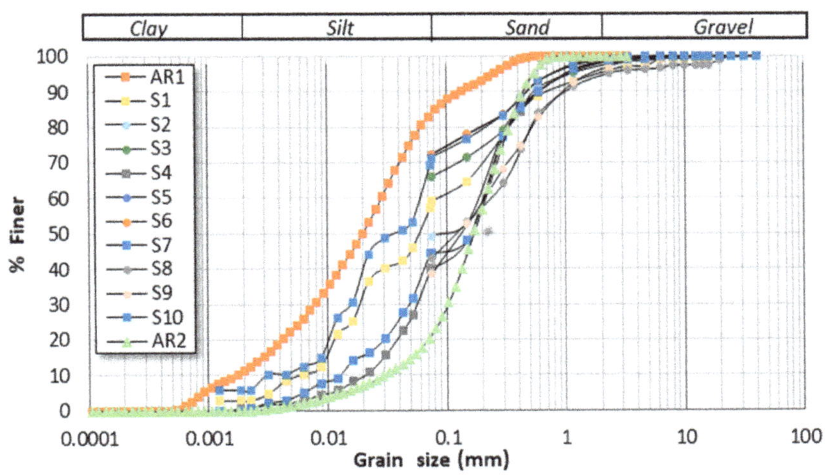

Figure 8. Grain size distribution curves of the soil samples.

With regard to the Atterberg limits, LL = 31% and LL = 29% were obtained for the AR1 and AR2 samples, respectively. It was impossible to perform the standard tests to obtain the plastic limit for all the analyzed samples so that the soils could be classified as non-plastic. Table 1 summarizes the results of soil classification based on grain size distribution. Due to the percentage of sand, silt, and clay, and also the lack of plasticity, the specimens were classified as silty sand (SM) or sandy silt (ML) according to USCS.

Based on the grain size distribution, the empirical Equation (1) proposed by Puckett et al. [45] was adopted to estimate the saturated hydraulic conductivity (K_s) of the soils as follows:

$$K_s = 4.36 \cdot 10^{-5} e^{-0.1975 \cdot C} \quad (1)$$

where C is the percentage of clay content. The obtained values of K_s are reported in Table 2. The hydraulic conductivity of the reddish layer is one order of magnitude higher than the whitish layer. These relatively high values are typical of residual soils in their intact state and are due to their distinctive formation process [46].

Table 2. Properties of the soil samples subjected to the swelling test.

Sample ID	Layer	Depth (m)	γ_n (kN/m^3)	γ_s (kN/m^3)	w_i (%)	w_f (%)	n (%)	e_0 -	Sr_i (%)	Sr_f (%)	K_s (m/s)
AR1	Reddish	2.0	18.9	26.5	15.8	23.2	38.5	0.63	68	100	4.9×10^{-6}
AR2	Whitish	1.0	19.5	26.5	13.9	20.2	35.5	0.55	68	100	4.3×10^{-5}

γ_n: unit weight of natural soil; γ_s: unit weight of saturated soil; w_i: initial gravimetric water content; w_i: final gravimetric water content; n: porosity; e_0: initial void ratio; Sr_i: initial degree of saturation; Sr_f: final degree of saturation; K_s: hydraulic conductivity.

4.2. Microstructural Features

The AR1 and AR2 samples display similar overall micro-textural characteristics but different mineral compositions. Both samples are composed of alternated layers of quartz and clay minerals (Figure 9). The areal distribution of this peculiar mineral association defines a schistosity at the scale of the entire thin section, which suggests a metamorphic origin of the analyzed samples. The AR2 sample displays metamorphic layers of quartz with a crystal size of <1 mm and an extensive occurrence of muscovite mica with characteristic needle–platy shape crystals (Figure 10a,b). Quartz crystals are characterized by 120° boundaries, which are typically documented in the case of dynamically recrystallized metamorphic rocks. Dark brown clay minerals only occur locally in between muscovite layers but are restricted to isolated areas. Iron oxides sparsely occur in tiny patches within muscovite-rich layers. Conversely, AR1 is dominantly composed of quartz, with a crystal size of up to 1–2 mm with more than half of the thin section area occupied by dark brown clay minerals with traces of iron oxides–hydroxides (Figure 10c,d). The overall porosity of both samples is high, and it is mainly of secondary origin, related to the brittle reactivation of crystal boundaries.

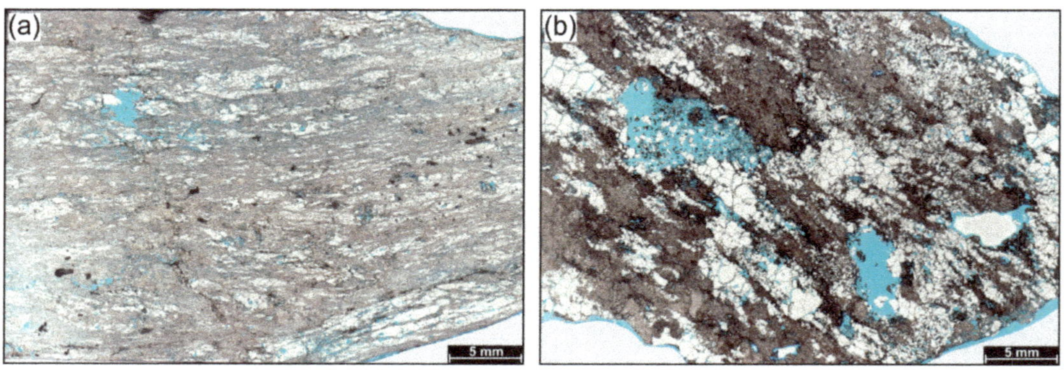

Figure 9. Thin section, high-resolution scan of the two representative samples. (**a**) AR2, least altered metamorphic schist. (**b**) AR1, most altered schist.

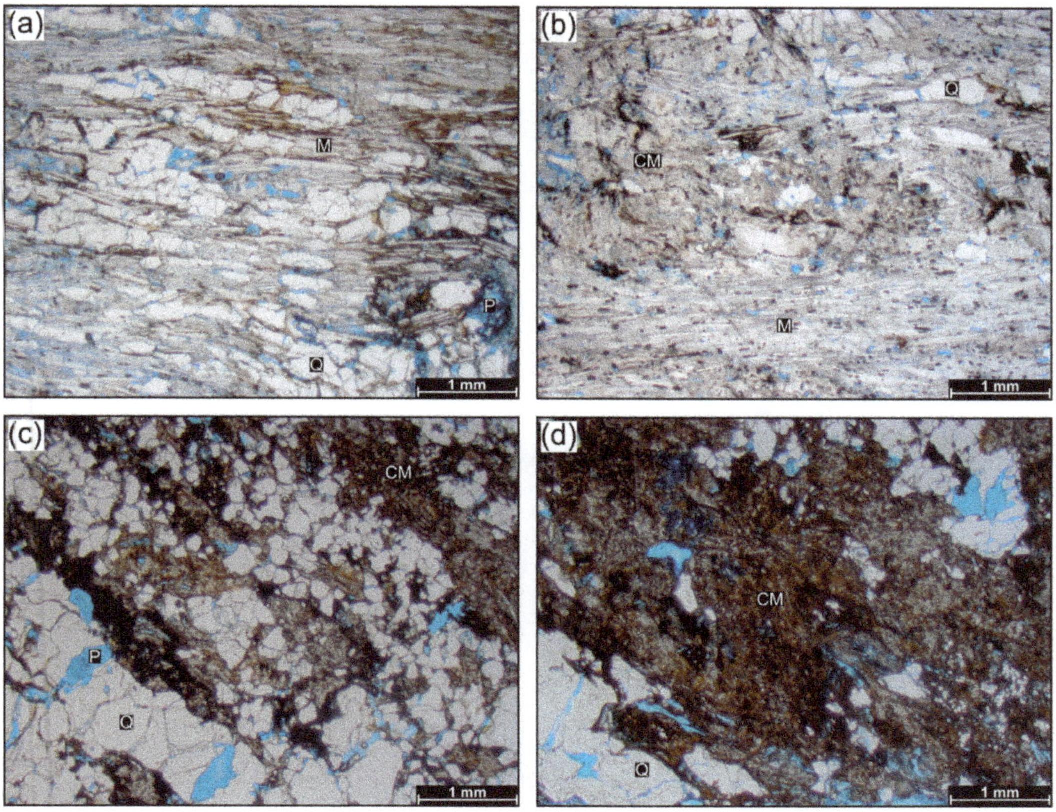

Figure 10. Micro-textural characteristics of the sampled soft rocks. (**a**) Metamorphic layering of quartz and muscovite mica in the AR2 sample. (**b**) Detail of muscovite layers with patchily distributed alteration clay minerals. (**c**) The AR1 sample with a sub-parallel layering of quartz and dark-brown clay minerals. (**d**) Detail of severe alteration with development of extensive area occupied by dark-brown clay minerals in the AR1 sample. Q, quartz; M, muscovite mica; CM, clay mineral; P, pore.

4.3. Mineralogical–Chemical Determination

EDS punctual X-ray analyses confirmed the presence of extensive layers of muscovite mica ($KAl_2(AlSi_3O_{10})(OH)_2$) in between quartz schistosity in the AR2 sample (Figure 11a). Conversely, in AR1, only small remaining fragments of muscovite crystals were found, and most of the clay minerals surrounding the quartz are composed of kaolinite, as confirmed by the emission spectrum ($Al_2(Si_2O_5)(OH)_4$) (Figure 11b). From the SEM observation in the AR1 and AR2 samples, it seems that kaolinite formed as the product of alteration of primary muscovite mica. This is also confirmed by the partial substitution of kaolinite at the expense of muscovite observed in discontinuous patches in the AR2 sample (Figure 11a).

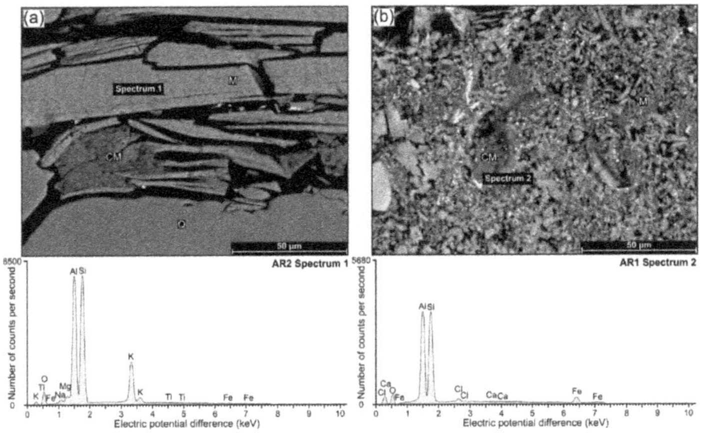

Figure 11. SEM-EDS X-ray analysis to determine the mineralogical composition of clay minerals. (**a**) Emission spectrum obtained from muscovite mica in the AR2 least-altered sample. (**b**) The emission spectrum gained from clay mineral identified to be kaolinite in the AR1 more-altered sample. Q, quartz; M, muscovite mica; CM, clay mineral; Al, aluminum; Ca, calcium; Cl, chlorine; Fe, iron; K, potassium; Mg, magnesium; Na, sodium; O, oxygen; Ti, titanium; Si, silicon.

4.4. Two-Dimensional Porosity

The AR1 sample, collected from the reddish deep layer, is characterized by a mean 2D porosity of quartz layers equal to 9.29%, while kaolinite composing most of the thin section area has a mean porosity of 5.01% (Figure 12). While porosity characterizing quartz is exclusively related to micro-fracturing along crystal boundaries, kaolinite displays both inter- and intra-crystalline micro-pores (Figure 12). The AR2 sample is characterized by a mean quartz porosity of 14.01%, while muscovite-dominated layers have an average porosity of 12.72%. Where present as isolated patches, kaolinite substituting muscovite shows a 2D porosity of 10.91% (Figure 12). Also, in this sample, quartz and muscovite porosity are related to the brittle reactivation of crystal boundaries, while kaolinite displays both inter- and intra-crystalline porosity. These findings on 2D porosity determined at a microscopic scale could be directly correlated to the values of the hydraulic permeability assessed by Equation (1) since the higher porosity of the whitish soil (AR2) corresponds to a higher value of hydraulic permeability.

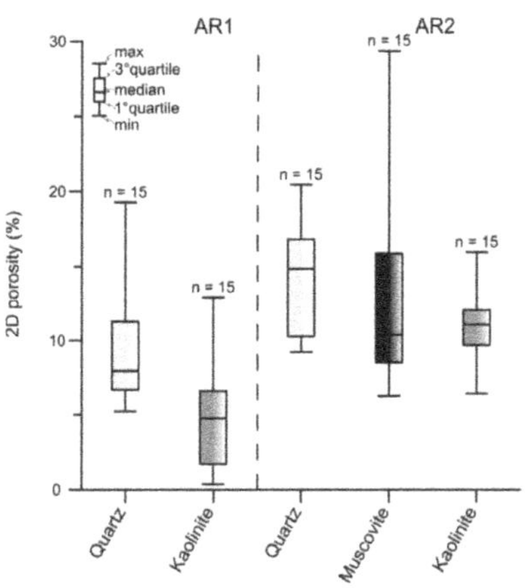

Figure 12. Box–whisker interquartile plot of 2D porosity gained via the image analysis technique on thin sections (n = the number of analyzed photomicrographs).

4.5. Mechanical Properties

4.5.1. Swelling Tests

The AR1 and AR2 samples have been subjected to oedometer tests to determine the swelling pressure. Table 2 reports the characteristics of the soil samples, including the values of gravimetric water content and saturation degree before and after the swelling test.

Figure 13 reports the experimental results of the swelling test. It is possible to notice how the maximum swelling pressure for the AR1 sample is 3.8 kPa and 2.5 kPa for the AR2 sample. Since the values of the swelling pressures are very low, it could be stated that both soil samples have very little swelling behavior, and this is due to the relatively low percentage of clay.

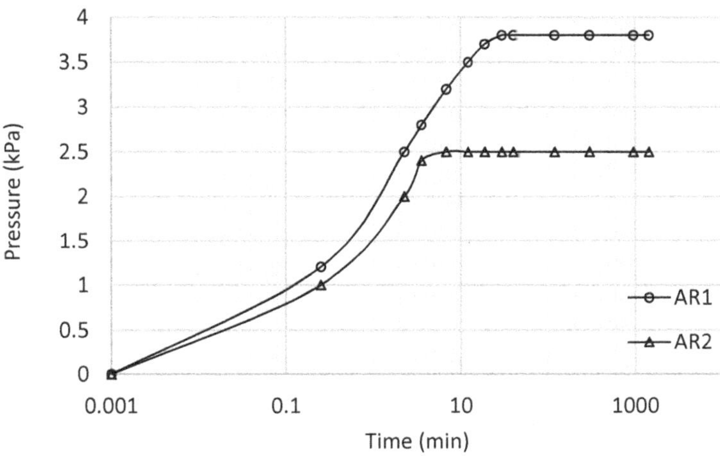

Figure 13. Experimental results of the swelling tests.

4.5.2. Direct Shear Tests

The mechanical characterization of soils involved in the landslide has been obtained by performing direct shear tests on two representative undisturbed samples, namely AR1 and AR2. The aim was to determine the Mohr–Coulomb shear strength parameters in both peak and residual conditions. Although the soil at the studied depths is mostly in unsaturated conditions, the partial saturation and the effect of suction have been neglected in this study, assuming that failure occurred when the soil was fully saturated due to rainfall infiltration, as will be shown in Section 4.5.3. Moreover, testing the saturated soil corresponded to the worst condition, and shear strength parameters could be interpreted in terms of effective stresses (drained conditions).

Figures 14 and 15 show the experimental results at both peak and residual stress states for the AR1 and AR2 samples, respectively. The corresponding values of failure shear stress were plotted on the Mohr plane (Figure 16), and the final shear strength parameters are reported in Table 3. It can be observed that, due to their type and the relatively low percentage of clay, soils are characterized by a low effective cohesion at peak conditions, i.e., 4.7 kPa and 2.3 kPa for AR2 and AR1, respectively. Moreover, at the peak stress state, AR2 (whitish) is characterized by a shear strength angle of 31°, which is higher than 29°, i.e., the value that has been obtained for AR1 (reddish). Instead, at residual conditions, both samples are characterized by nil effective cohesion, and the residual shear strength angle is equal to 28° for both soils (Figure 16). The results appear consistent with those obtained by other authors on similar materials [42–44].

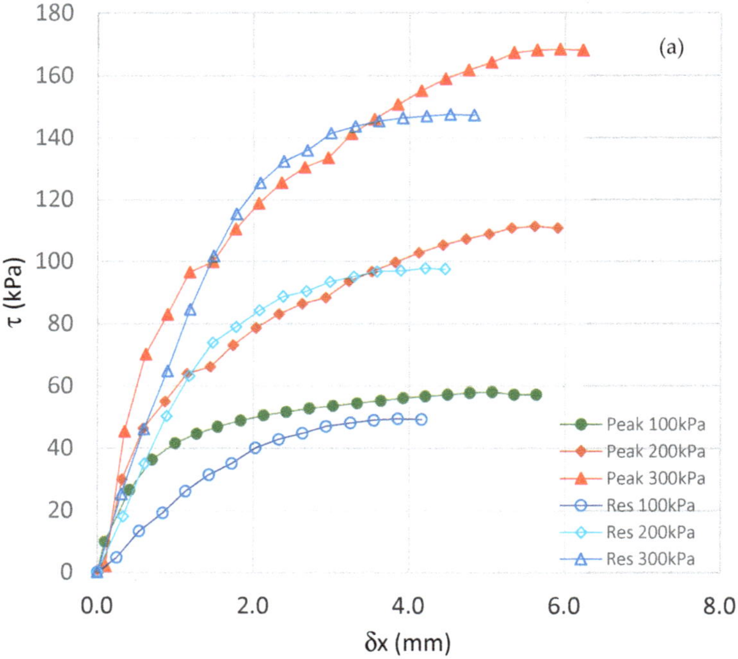

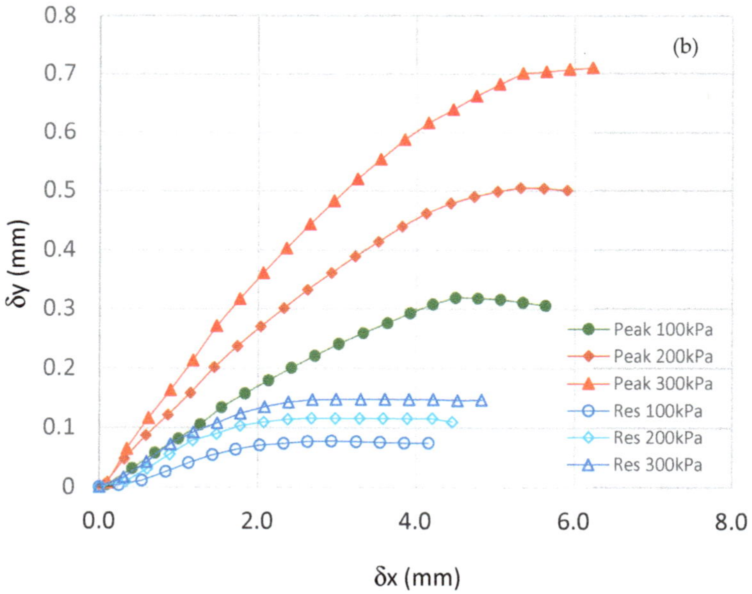

Figure 14. Experimental results of direct shear tests on the AR1 soil sample (reddish): (**a**) shear stress vs. horizontal displacement; (**b**) vertical displacement vs. horizontal displacement.

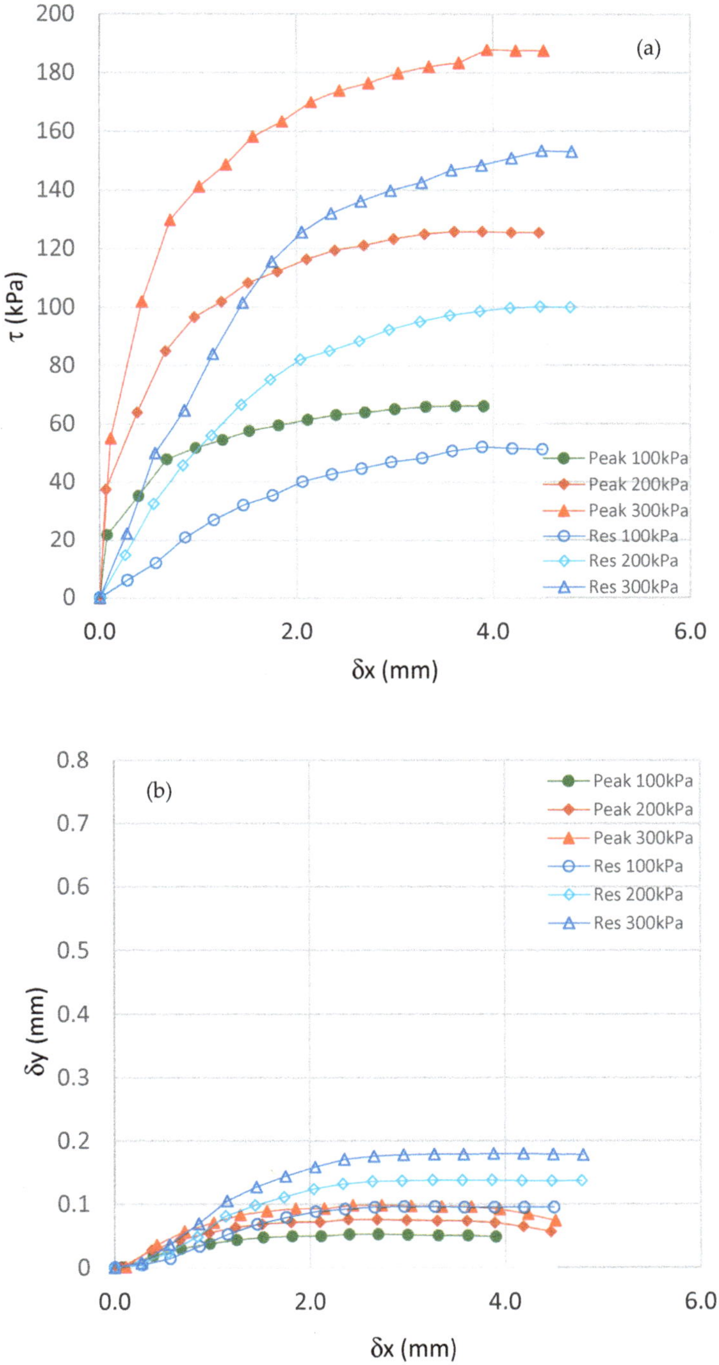

Figure 15. Experimental results of direct shear tests on the AR2 soil sample (whitish): (**a**) shear stress vs. horizontal displacement; (**b**) vertical displacement vs. horizontal displacement.

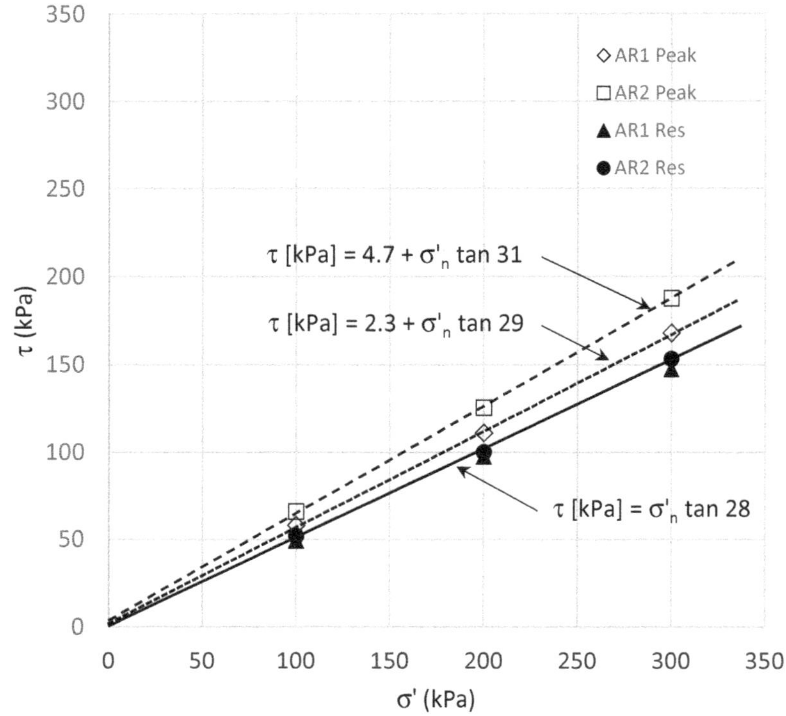

Figure 16. Plot of the peak and residual shear stress against normal stress and related Mohr–Coulomb failure criterion.

Table 3. Shear strength parameters.

Sample ID	Depth (m)	c_p' (kPa)	ϕ_p' (°)	c_r' (kPa)	ϕ_r' (°)
AR1	2.0	2.3	29	0	28
AR2	1.0	4.7	31	0	28

4.5.3. Slope Stability Analysis

A simplified slope stability analysis has been performed by the Morgenstern–Price limit equilibrium method using the code SLIDE-2 (Roscscience®, Toronto, ON, Canada). The shear strength input parameters are those corresponding to the peak conditions (c_p' and ϕ_p'). No water table has been assumed, as in the real situation, it is below the road level at a depth rather far from the failure mechanism. Figure 17 shows the results of the stability analysis. In particular, Figure 17 shows the failure surface with the minimum safety factor (0.8) and all the potential failure surfaces with a safety factor of less than one, i.e., the unstable condition. Although this is an oversimplified stability analysis, not least because it neglects the rain infiltration process, this result allows the following aspects to be highlighted:

- Before failure, the slope was stable for a long time thanks to the partially saturated conditions, which guaranteed an additional shear strength contribution beyond the peak effective shear strength condition;
- The slope became unstable when the soil became saturated due to rainfall, and the peak effective shear strength was attained;

- It is very likely that the first part of the slope to experience instability was the slope toe, consisting of reddish soil, which shows the lowest safety factor, and the whitish soil followed;
- The failure surfaces involving the top of the slope and the whitish soil are very similar to the field-observed situation (Figure 5).

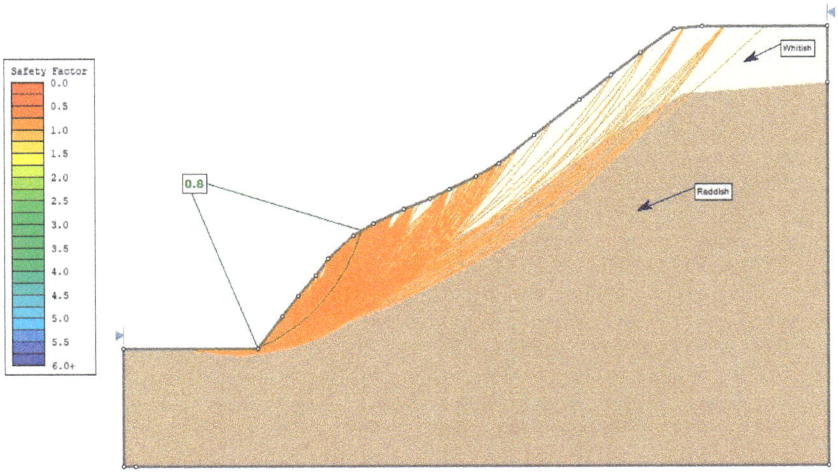

Figure 17. Results of the slope stability analysis carried out using SLIDE-2 (Roscscience®) (Morgenstern–Price method): only potential failure surfaces with a safety factor of less than one are shown, while the minimum safety factor is 0.8.

5. Discussion

The main objective of this research was to investigate the combined effects of microstructural features, mineral composition, and macroscopic mechanical behavior on the failure conditions of lateritic soil subjected to slope instability in the study site of Kabaya (Rwanda). Similar approaches of combining micro- and macro-tests on rocks involved in landslides have been recently used in other contexts to better understand the triggering conditions of some landslides [47–51]. The methodology used in this case study highlighted several aspects that can be considered useful in dealing with similar cases. They can be summarized as follows:

1. The grain size distribution curves obtained using both traditional methods (i.e., sieving and sedimentation) and innovative methods (laser diffraction granulometry) are different, depending on the method used. In fact, curves obtained using laser diffraction granulometry represent a lower and upper limit compared to all other curves obtained using traditional methods (Figure 8). This shows a higher degree of detail in the laser diffraction analysis, which could not have been achieved by traditional techniques, as already observed by [52].
2. The microstructural observations of two different samples (AR1 and AR2) helped to reconstruct the geological history of the slope and to shed light on the predisposing causes of the failure. In particular, the analyses at the thin section scale suggest a metamorphic origin of the two lateritic soils (reddish soil and whitish soil, respectively), which can also be classified as soft rocks. In particular, both samples can be described as medium-grade metamorphic schists, with original paragenesis mainly composed of quartz and muscovite mica organized in sub-parallel compositional layers (Figure 9). Kaolinite developed as an alteration product at the expense of muscovite mica, and it is patchily distributed in the whitish soil and is pervasive in the reddish soil, almost entirely substituting the muscovite layers. The two-dimensional porosity of both soft rocks is high, and it is mainly related to the brittle reactivation of crystal

boundaries and mineral substitution (Figure 12). It is likely that the original porosity of the unaltered pristine metamorphic rocks was rather low, as they form and deform at considerable depth and temperature conditions. The current high porosity and micro-texture could be related to the following aspects: (1) unloading undergone by the rocks following their exhumation and (2) alteration operated by the weathering processes. This is confirmed by the secondary porosity, which appears to be dominated by micro-fractures reactivating and exploiting original rock weaknesses, such as crystal boundaries (Figure 10a,c). The alteration by weathering was also responsible for the mineralogical change and substitution of muscovite to kaolinite (Figure 10a,b). This was facilitated by the lower stability of muscovite with respect to kaolinite under surficial temperature-pressure conditions. The alteration was likely eased by tropical temperatures and rainfall conditions experienced by the metamorphic rocks in the study site of Kabaya. Weathering processes were able to efficiently alter the pristine rocks due to the documented high porosity, facilitating the percolation of water from above. Quartz was not affected by chemical processes as it is more stable than phyllosilicates under surficial conditions. The chemical alteration was responsible also for physical changes in metamorphic rocks, with a significant shift in grain size and grain size distribution. Weathering produced a significant fining of grain size from 224 to 53.4 μm, with the sand-dominated initial sample becoming a silt-dominated one. It is possible to infer that the different degrees of alteration documented in the AR1 (reddish soil) and AR2 (whitish soil) samples are related to the different depths and positions with respect to the ground level in the original condition before the cut slope excavation. In particular, the whitish soil, which was positioned closer to the ground level, shows the most preserved mineralogical assemblage, while the reddish soil, which was the deeper layer before the construction of the cut slope, is the most altered. The difference in depth of the two soils, in the original condition before the cut slope excavation, could have played a significant role in driving the alteration, possibly due to the oscillation of a perched water table during wet and dry periods. In particular, the deeper reddish layer had experienced enduring water saturation conditions following heavy rainfall, while the whitish layer, closer to ground level, may have undergone unsaturated conditions for a longer period, with a consequent weaker alteration potential. This original situation was modified during the road cut slope excavation and weathering that affected the two layers in a different way, especially considering the exposure of the reddish soil at the slope toe after excavation. Although the geological history is rather far to be described in detail, it is clear that the combination of physical and chemical changes operated via weathering processes influenced the hydraulic and mechanical properties of the two soft rocks involved in the Kabaya landslide.

3. From a mechanical point of view, the reddish soil (AR1) shows a relatively higher swelling potential with respect to the whitish soil (AR2) (Figure 13) due to the relatively higher percentage of clay minerals. However, the swelling potential of both samples appears rather limited, and for this reason, it is unlikely that the slope failure mechanism was caused by the swelling behavior of the involved materials. With regard to the shearing behavior, it is worth considering that investigations by many authors in the past were focused on the effect of different water contents on the failure conditions of soft rocks involved in landslides (Fu et al., 2022 [50]; Falcão et al., 2023 [43]). In this research, only the worst condition, corresponding to the saturated material, has been considered. Moreover, only the saturated condition allowed us to interpret the shear strength parameters in terms of effective stresses. The obtained results show that the difference between peak and residual conditions for the reddish soil (AR1) is less pronounced than that observed for the whitish soil (AR2). As a consequence, the shear strength parameters of the reddish soil are weaker than those of the whitish soil for peak conditions; instead, shear strength parameters corresponding to residual conditions are the same for the two samples. It could be affirmed that the

weaker failure conditions of the reddish soil could be correlated to the higher level of weathering and degradation with respect to the whitish soil, as observed at the microstructural scale. This is consistent with the slope stability analysis. In fact, the simplified slope stability analysis (Section 4.5.3) showed that the failure surface with the lowest safety factor develops in the reddish layer at the slope toe, whose failure condition is represented by the weakest AR1 sample.

4. With regard to the triggering mechanism of the landslide that occurred on 2 May 2018, it can be assumed that the rainfall infiltrated through both the whitish and the reddish layers, which were already wet due to the antecedent rainfalls. Although the slope was characterized by a relatively high steepness, it was stable during dry periods thanks to the shear strength contribution given by the matric suction connected to the partial saturation condition. This contribution was completely lost after rainfall infiltration. Due to the different hydraulic conductivity of the whitish and reddish soils, the water infiltrated relatively easily through the whitish layer down to the reddish layer at the top of the slope and rather slowly in the reddish layer at the slope toe. As observed in similar phenomena, although in different geological contexts [53], the infiltrated rainfall created saturated zones, resulting in the loss of shear strength of the partially saturated soil until reaching the fully saturated condition, and the slope became unstable. As demonstrated by the simplified slope stability analysis (Section 4.5.3), the slope became unstable when the soil became saturated, and the peak effective shear strength was attained. The slope toe, made of reddish soil, was the first part of the slope to experience instability due to both higher steepness and lower peak shear strength with respect to the top part, and the whitish soil followed.

6. Conclusions

This paper intends to show the advantage of integrating routine geotechnical tests at a macroscopic scale with microstructural analyses in explaining the background and past causes of some failure phenomena, which could only be partially interpreted based on the shear strength parameters. In the presented case study, both whitish and reddish soil layers involved in the landslide are characterized by very similar shear strength parameters. In fact, the difference in peak shear strength angle is only 2° and the difference in peak effective cohesion is only 2.4 kPa, with the whitish soil stronger than the reddish one. Instead, the residual shear strength parameters are the same. Actually, microscopic analyses show many differences between the two materials. In particular, the difference in quartz crystal size is up to 2 mm, the difference in quartz porosity is 4.72% and the difference in kaolinite porosity is 7.71%. The whitish soil has a higher microstructural porosity than the whitish one. Correspondingly, the whitish soil is characterized by a hydraulic conductivity that is one order of magnitude higher than the reddish one. Therefore, it becomes easier to explain the predisposing causes and the triggering mechanism of the landslide, considering the weak characteristics of the reddish layer at the microscopic scale. Moreover, from the microstructural point of view, it was observed that kaolinite developed as an alteration product at the expense of muscovite mica, and it was observed patchily distributed in the whitish soil, while it was pervasive in the reddish soil. It should be considered that the whitish soil and the reddish soil were the shallow and the deep layers, respectively, of the slope in the original condition before the cut slope excavation. Therefore, the soil was more affected by weathering in depth than in the shallow part of the slope, in its original geometry. The higher degree of weakness of the deeper soil layer was only partially confirmed by the observed failure conditions through the direct shear tests. Moreover, although the maximum swelling pressure of the reddish soil was higher than the whitish one, with a difference of only 2.4 kPa, both soils showed very little swelling behavior.

The disadvantage of the combined approach is the need for different competencies and the use of microscopic techniques that cannot always be considered routine, given their level of sophistication. A further limit of this research is that all tests were performed without considering in detail the specific effect of water, except in the worst condition of full saturation. More in-depth comparative experiments to illustrate the specific role of water will be conducted in the future.

The following main outcomes, which are useful in practical applications in the considered context, can be drawn from this study:

1. The example given here demonstrates that the whitish soil (which was the shallowest layer in the original natural slope condition) is less altered and less poor in mechanical properties than the reddish (deeper) layer, although both of them may appear to have the consistency of a soft rock and are both subjected to weathering;
2. Assuming the same slope geometry, same water content, and same boundary conditions, when a road cut slope involves only the shallow, less altered soil, it could be more stable than a cut slope involving the deeper and weaker soil layer;
3. Although, from the point of view of classification (see, for example, the grain size curve and USCS classification) and mechanical characterization, the two materials may appear similar, with very close geotechnical parameters, and microstructural and petrophysical analyses demonstrate different characteristics in terms of the microstructure and mineralogical composition;
4. The selective chemical alteration of similar rocks at different depths may induce significant mineralogical and petrophysical changes, promoting the nucleation of shallow sliding surfaces.

In the specific case study of the Kabaya landslide, the microstructural features and mineralogical composition of the two involved soils revealed their geological history and helped to understand the predisposing factors of the triggering mechanism, which is more than what could be assessed only through standard geotechnical tests. However, further investigations will be performed on the mechanical behavior of the soil, even in partially saturated conditions, using triaxial tests, and on the experimental evaluation of the hydraulic conductivity of different soil layers, to better modeling the triggering mechanism of the landslide.

These considerations lead to the suggestion of a careful and detailed analysis of the materials involved in landslide phenomena, especially for laterites or soft rocks, whether they are superficial or deep, as minimal differences at the macroscopic scale may indicate more important differences at the microscopic scale, with risky consequences in slope stability analysis and the design of support structures.

Author Contributions: Conceptualization: R.V. and M.P.; material preparation, data collection, and analysis: R.V., M.P. and J.M.; supervision: R.V.; writing—original draft: R.V. and M.P.; writing—review and editing: R.V., M.P. and J.M. All authors have read and agreed to the published version of the manuscript.

Funding: This work has benefited from the equipment and framework of the COMP-HUB and COMP-R Initiatives and was funded by the "Departments of Excellence" program of the Italian Ministry for University and Research (MIUR, 2018–2022 and MUR, 2023–2027).

Data Availability Statement: The datasets generated and analyzed during the current study are not publicly available but are available from the corresponding author upon reasonable request.

Acknowledgments: Andrea Comelli (University of Parma) is kindly thanked for having prepared thin sections from difficult samples. The authors would also like to thank the two anonymous reviewers for their valuable comments.

Conflicts of Interest: The authors declare no conflicts of interest.

References

1. Froude, M.J.; Petley, D.N. Global fatal landslide occurrence from 2004 to 2016. *Nat. Hazards Earth Syst. Sci.* **2018**, *18*, 2161–2181. [CrossRef]
2. Guzzetti, F.; Gariano, S.L.; Peruccacci, S.; Brunetti, M.T.; Marchesini, I.; Rossi, M.; Melillo, M. Geographical landslide early warning systems. *Earth Sci. Rev.* **2020**, *200*, 102973. [CrossRef]
3. Du, Y.; Xie, M.; Jia, J. Stepped settlement: A possible mechanism for translational landslides. *Catena* **2020**, *187*, 104365. [CrossRef]

4. Cao, H.; Ma, G.; Liu, P.; Qin, X.; Wu, C.; Lu, J. Multi-Factor Analysis on the Stability of High Slopes in Open-Pit Mines. *Appl. Sci.* **2023**, *13*, 5940. [CrossRef]
5. Kirschbaum, D.; Kapnick, S.B.; Stanley, T.; Pascale, S. Changes in extreme precipitation and landslides over hogh mountains Asia. *Geophys. Res. Lett.* **2020**, *47*, 4. [CrossRef]
6. Dewitte, O.; Dille, A.; Depicker, A.; Kubwimana, D.; Maki Mateso, J.C.; Bibentyo, T.M.; Uwihirwe, J.; Monsieurs, E. Constraining landslide timing in a data-scarce context: From recent to very old processes in the tropical environment of the North Tanganyika-Kivu Rift region. *Landslides* **2021**, *18*, 161–177. [CrossRef]
7. Monsieurs, E.; Jacobs, L.; Michellier, C.; Basimike, J.; Bamulezi Ganza, G.; Kervyn, F.; Maki Mateso, J.C.; Mugaruka Bibentyo, T.; Kalikone Buzera, C.; Nahimana, L.; et al. Landslide inventory for hazard assessment in a data-poor context: A regional-scale approach in a tropical African environment. *Landslides* **2018**, *15*, 2195–2209. [CrossRef]
8. Nobile, A.; Dille, A.; Monsieurs, E.; Basimike, J.; Mugaruka Bibentyo, T.; d'Oreye, N.; Kervyn, F.; Dewitte, O. Multi-temporal DInSAR to characterise landslide ground deformations in a tropical urban environment: Focus on Bukavu (DR Congo). *Remote Sens.* **2018**, *10*, 626. [CrossRef]
9. Depicker, A.; Govers, G.; Jacobs, L.; Campforts, B.; Uwihirwe, J.; Dewitte, O. Interactions between deforestation, landscape rejuvenation, and shallow landslides in the North Tanganyika–Kivu rift region, Africa. *Earth. Surf. Dynam.* **2021**, *9*, 445–462. [CrossRef]
10. Deijns, A.A.J.; Dewitte, O.; Thiery, W.; d'Oreye, N.; Malet, J.P.; Kervyn, F. Timing landslide and flash flood events from SAR satellite: A 1 new method illustrated in African cloud-covered tropical environments. *Nat. Hazards Earth Syst. Sci.* **2022**, *22*, 3679–3700. [CrossRef]
11. Maki Mateso, J.C.; Bielders, C.; Monsieurs, E.; Depicker, A.; Smets, B.; Tambala, T.; Bagalwa Mateso, L.; Dewitte, O. Characteristics and causes of natural and human-induced landslides in a tropical mountainous region: The rift flank west of Lake Kivu (Democratic Republic of the Congo). *Nat. Hazards Earth Syst. Sci.* **2023**, *23*, 643–666. [CrossRef]
12. Valentino, R.; Sobio, Y.; Mizero, J.; Safari, J.; Nsengiyumva, F. Unstable road cut slopes and design of retaining structures in the Rwandan context. *Arab. J. Geosci.* **2021**, *14*, 1405. [CrossRef]
13. Nsengiyumva, J.B.; Luo, G.; Nahayo, L.; Huang, X.; Cai, P. Landslide Susceptibility Assessment Using Spatial Multi-Criteria Evaluation Model in Rwanda. *Int. J. Environ. Res. Public Health* **2018**, *15*, 243. [CrossRef] [PubMed]
14. Nsengiyumva, J.B.; Luo, G.; Amanambu, A.C.; Mind'je, R.; Habiyaremye, G.; Karamage, F.; Ochege, F.U.; Mupenzi, C. Comparing probabilistic and statistical methods in landslide susceptibility modeling in Rwanda/Centre-Eastern Africa. *Sci. Total Environ.* **2019**, *659*, 1457–1472. [CrossRef] [PubMed]
15. Nsengiyumva, J.B.; Valentino, R. Predicting landslide susceptibility and risks using GIS-based machine learning simulations, case of upper Nyabarongo catchment. *Geomat. Nat. Hazards Risk* **2020**, *11*, 1250–1277. [CrossRef]
16. Kumar, V.; Chandrasekaran, S.S. Analysis of failure of high slope subjected to rainfall infiltration at Peringavu in Kerala, India. *Eng. Fail. Anal.* **2022**, *138*, 106423. [CrossRef]
17. Du, Y.; Chen, C. Data Mining for Landslide Genetic Mechanism Analysis in the Yunnan Province of China. *Geotech. Geol. Eng.* **2022**, *40*, 5631–5642. [CrossRef]
18. Singh, R.; Umrao, R.K.; Singh, T.N. Stability evaluation of road-cut slopes in the Lesser Himalaya of Uttarakhand, India: Conventional and numerical approaches. *Bull. Eng. Geol. Environ.* **2014**, *73*, 845–857. [CrossRef]
19. Reuters. 2023. Available online: https://www.reuters.com/business/environment/rwanda-floods-kill-130-destroy-over-5000-houses-2023-05-04/ (accessed on 1 March 2024).
20. Miščević, P.; Vlastelica, G. Impact of weathering on slope stability in soft rock mass. *J. Rock Mech. Geotech. Eng.* **2014**, *6*, 240–250. [CrossRef]
21. Oyelami, C.; Van Rooy, J. A review of the use of lateritic soils in the construction/development of sustainable housing in Africa: A geological perspective. *J. Afr. Earth. Sci.* **2016**, *119*, 226–237. [CrossRef]
22. Kaze, R.C.; Naghizadeh, A.; Tchadjie, L.; Adesina, A.; Yankwa Djobo, J.N.; Deutou Nemaleu, J.; Kamseu, E.; Melo, U.C.; Tayeh, B.A. Lateritic soils based geopolymer materials: A review. *Constr. Build. Mater.* **2022**, *344*, 128157. [CrossRef]
23. Gonzalez de Vallejo, L.I.; Ferrer, M. *Geological Engineering*; CRC Press/Balkema: Leiden, The Netherlands, 2011.
24. Sinha, S.; Basu, A. Influence of parent lithology on field appearances and engineering properties of indurated laterites under same geographical conditions. *Eng. Geol.* **2021**, *295*, 106413. [CrossRef]
25. Olanipekun, E.O. Kinetics of leaching laterite. *Int. J. Min. Process.* **2000**, *60*, 9–14. [CrossRef]
26. Mitchell, J.K.; Soga, K. *Fundamentals of Soil Behavior*, 3rd ed.; John Wiley & Sons: Hoboken, NJ, USA, 2005.
27. Kasthurba, A.K.; Santhanam, M.; Mathews, M.S. Investigation of laterite stones for building purpose from Malabar region, Kerala state, SW India—Part 1: Field studies and profile characterisation. *Constr. Build. Mater.* **2007**, *21*, 73–82. [CrossRef]
28. Gao, Q.F.; Yu, H.C.; Zeng, L.; Huang, Y.X. Characterization of water retention behavior of cracked compacted lateritic soil exposed to wet-dry cycles. *Bull. Eng. Geol. Environ.* **2023**, *82*, 61. [CrossRef]
29. Kong, L.; Sayem, H.M.; Tian, H. Influence of drying-wetting cycles on soil-water characteristic curve of undisturbed granite residual soils and microstructure mechanism by nuclear magnetic resonance (NMR) spin-spin relaxation time (T2) relaxometry. *Can. Geotech. J.* **2018**, *55*, 208–216. [CrossRef]
30. Bai, W.; Zhang, W.; Kong, L.; Fan, H.; Guo, A.; Xu, G. Effect of soaking time of dithionite–citrate–bicarbonate solution on strength and deformation characteristics of lateritic soil. *J. Rock Mech. Geotech. Eng.* **2023**, *15*, 3039–3049. [CrossRef]

31. *Carte Lithologique du Rwanda*; Stanford Libraries: Stanford CA, USA, 1963.
32. Schlüter, T. *Geological Atlas of Africa*; Springer: Berlin/Heidelberg, Germany, 2006; p. 272. [CrossRef]
33. *ASTM D6913-04:2009-e1*; Standard Test Methods for Particle-Size Distribution (Gradation) of Soils Using Sieve Analysis. ASTM International: West Conshohocken, PA, USA, 2009.
34. *ASTM D7928-21e1*; Standard Test Method for Particle-Size Distribution (Gradation) of Fine-Grained Soils Using the Sedimentation (Hydrometer) Analysis. ASTM International: West Conshohocken, PA, USA, 2021.
35. *ASTM D4318:2010*; Standard Test Methods for Liquid Limit, Plastic Limit, and Plasticity Index of Soils. ASTM International: West Conshohocken, PA, USA, 2010.
36. Pizzati, M.; Balsamo, F.; Storti, F.; Mozafari, M.; Iacumin, P.; Tinterri, R.; Swennen, R. From axial parallel to orthogonal groundwater flow during fold amplification: Insights from carbonate concretion development during the growth of the Quattro Castella Anticline, Northern Apennines, Italy. *J. Geol. Soc.* **2018**, *175*, 806–819. [CrossRef]
37. Hergert, W.; Wriedt, T. *The Mie Theory—Basics and Applications*; Springer: Berlin/Heidelberg, Germany, 2012.
38. Storti, F.; Balsamo, F. Particle size distributions by laser diffraction: Sensitivity of granular matter strength to analytical operating procedures. *Solid Earth* **2010**, *1*, 25–48. [CrossRef]
39. Schneider, C.A.; Rasband, W.S.; Eliceiri, K.W. NIH Image to ImageJ: 25 years of image analysis. *Nat. Methods* **2012**, *9*, 671–675. [CrossRef]
40. *ASTM D4546-21*; Standard Test Methods for One-Dimensional Swell or Collapse of Soils. ASTM International: West Conshohocken, PA, USA, 2021.
41. *British Standards BS 1377-7:1990*; Methods of Test for Soils for Civil Engineering Purposes—Shear Strength Tests (Total Stress). British Standards Institution: London, UK, 1990.
42. Li, Z.; Li, J.; Han, M.; Liu, L. Investigating the Shear Strength Characteristics of Slip Zone Soil Based on In-situ Multiple Shear Tests. *KSCE J. Civ. Eng.* **2023**, *27*, 3793–3807. [CrossRef]
43. Falcão, P.R.; Baroni, M.; Masutti, G.C.; Rinaldo Pinheiro, J.B.; Fagundes, D.d.F. Assessment of the Impact of Inundation on the Strength of a Lateritic and Collapsible Soil. *Geotech. Geol. Eng.* **2023**, *41*, 4761–4773. [CrossRef]
44. Onwo, E.S.; Emeh, C.; Igwe, O. Effect of Geochemical Composition of Lateritic Soils on their Geotechnical Properties. *Indian Geotech. J.* **2022**, *52*, 877–894. [CrossRef]
45. Puckett, W.E.; Dane, J.H.; Hajek, B.F. Physical and mineralogical data to determine soil hydraulic properties. *Soil Sci. Soc. Am. J.* **1985**, *49*, 831–836. [CrossRef]
46. Shi, W.; Lehane, B.M.; Gover, A.; Terzaghi, S. Characterisation of a lateritic soil using laboratory and in-situ tests. In Proceedings of the Fifth International Conference on Geotechnical and Geophysical Site Characterization (ISC'5), Gold Coast, Australia, 5–9 September 2016; Australian Geomechanics Society: Sydney, Australia, 2016.
47. Sun, S.; Li, K.; Le, H.; Wei, G.; Liu, D.; Song, J.; Yu, Y. Study on the deterioration characteristics of greenschist under hydrochemical action and the disaster-causing mechanism in slope. *Bull. Eng. Geol. Environ.* **2022**, *81*, 315. [CrossRef]
48. Sardana, S.; Sinha, R.K.; Verma, A.K.; Singh, T.N. Investigations into the freeze–thaw-induced alteration in microstructure and deteriorative responses of physico-mechanical properties of Himalayan rock. *Bull. Eng. Geol. Environ.* **2022**, *81*, 269. [CrossRef]
49. Zhang, X.; Liu, X.; Chen, C.; Xu, Y.; Liu, H. Evolution of disintegration properties of granite residual soil with microstructure alteration due to wetting and drying cycles. *Bull. Eng. Geol. Environ.* **2022**, *81*, 93. [CrossRef]
50. Fu, X.; Du, Y.; Sheng, Q.; Zhang, Z.; Huang, J. Influences of water on the microstructure and mechanical behavior of the Xigeda formation. *Bull. Eng. Geol. Environ.* **2022**, *81*, 72. [CrossRef]
51. Cerri, R.I.; Rosolen, V.; Reis, F.A.G.V.; Pereira Filho, A.J.; Vemado, F.; do Carmo Giordano, L.; Marques Gabelini, B. The assessment of soil chemical, physical, and structural properties as landslide predisposing factors in the Serra do Mar mountain range (Caraguatatuba, Brazil). *Bull. Eng. Geol. Environ.* **2020**, *79*, 3307–3320. [CrossRef]
52. Bittelli, M.; Pellegrini, S.; Olmi, R.; Andrenelli, M.C.; Simonetti, G.; Borrelli, E.; Morari, F. Experimental evidence of laser diffraction accuracy for particle size analysis. *Geoderma* **2022**, *409*, 115627. [CrossRef]
53. Montrasio, L.; Valentino, R. Modelling rainfall-induced shallow landslides at different scales using SLIP—Part I, VI Italian Conference of Researchers in Geotechnical Engineering—Geotechnical Engineering in Multidisciplinary Research: From Microscale to Regional Scale, CNRIG2016. *Procedia Eng.* **2016**, *158*, 476–481. [CrossRef]

International Journal of *Geo-Information*

MDPI

Article

Monitoring and Cause Analysis of Land Subsidence along the Yangtze River Utilizing Time-Series InSAR

Yuanyuan Chen [1,*], Lin Guo [2], Jia Xu [3], Qiang Yang [1], Hao Wang [1] and Chenwei Zhu [1]

1 College of Civil Engineering, Nanjing Forestry University, Nanjing 210037, China
2 Key Laboratory of Mechanism, Prevention and Mitigation of Land Subsidence, Ministry of Education, Capital Normal University, Beijing 100048, China; 6382@cnu.edu.cn
3 School of Earth Sciences and Engineering, Hohai University, Nanjing 210098, China
* Correspondence: cheny@njfu.edu.cn

Abstract: Time-series monitoring of the land subsidence in the Yangtze River coastal area is crucial for maintaining river stability and early warning of disasters. This study employed PS-InSAR and SBAS-InSAR techniques to monitor the land subsidence along the Yangtze River in Nanjing, using a total of 42 Sentinel-1A images obtained between April 2015 and November 2021. The accuracy of both methods was compared and validated, while a comprehensive analysis was conducted to ascertain the spatial distribution characteristics and underlying causes of land subsidence. The maximum deviation between the two methods and six leveling point data did not exceed ±5 mm. Within the 5 km buffer zone on either side of the Yangtze River in Nanjing, four subsidence funnels were identified. Analysis of the factors contributing to land subsidence in this area indicates that underground engineering construction and operation, increasing ground building area, and geological condition all have certain correlations to the land subsidence. The results obtained through PS-InSAR and SBAS-InSAR technologies revealed a high degree of consistency in monitoring outcomes, and the latter method exhibited superior monitoring accuracy than the former one in this area. This study holds significant implications for guiding the scientific management of urban geohazards along the Yangtze River.

Keywords: land subsidence; PS-InSAR; SBAS-InSAR; time-series analysis; area along the Yangtze River

1. Introduction

The mainstream of the Yangtze River flows through the entire territory of Jiangsu, China. The geological conditions in the Nanjing area along the river are characterized by loose geology, poor bearing capacity, and shallow groundwater levels. Due to long-term river water accumulation, the area is prone to collapse and sedimentation, resulting in land subsidence. Additionally, with the rapid development of the economy in the Yangtze River Delta region, the demand for engineering construction is increasingly expanding. Various engineering activities, such as groundwater extraction, subway engineering, and large ground building construction, have accelerated the occurrence of regional ground subsidence [1,2]. Under the joint action of the above natural and human factors, the geological environment system along the Yangtze River Delta region has been damaged, leading to urban land subsidence disasters, which, in turn, cause serious hazards such as soil collapse, road facility damage, and building wall cracking, posing a certain threat to the security of residents. Therefore, it is crucial to strengthen the monitoring and analysis of surface deformation within a certain range along the Yangtze River using advanced techniques.

Traditional approaches for ground subsidence monitoring mainly include precision leveling, trigonometric leveling, and GNSS measurements [3]. Precision leveling requires less cost, a simple construction process, and can ensure measurement accuracy. However, it is limited in the number and scope of leveling points and has a long monitoring cycle, resulting in low temporal and spatial resolution [4]. Trigonometric leveling, as an indirect height measurement method, features minimal terrain limitations and fast measurement speed, but its accuracy is easily affected by atmospheric refraction and is difficult to calibrate [5,6]. While GNSS boasts all-weather operation, is fully automatic, and has high accuracy, the high monitoring cost makes it challenging to conduct large-scale ground subsidence monitoring [7,8]. Traditional manual measurement methods face various limitations such as measurement range, manpower consumption, timeliness, and cost. Synthetic aperture radar interferometry (InSAR), as a form of active microwave remote sensing technology enabling continuous, all-weather monitoring, can provide low-cost solutions for acquiring wide-ranging and high-precision deformation information [9–11]. Therefore, it serves as an important tool for large-scale ground deformation monitoring.

Conventional InSAR technology obtains DEM by calculating the phase difference between two SAR images [12], while D-InSAR technology utilizes DEM data for interferometric processing and filters out other phase information from the interferometric image, retaining only the deformation phase resulting from displacement [13,14]. This enables the extraction of precise ground deformation information at millimeter-level accuracy. However, the monitoring accuracy of D-InSAR is significantly hindered by various factors, including temporal and spatial decorrelation, atmospheric effects, and satellite orbit errors. Further, this method cannot monitor the development process of deformation [15,16]. To overcome the limitations and shortcomings of conventional InSAR techniques, researchers have proposed the Persistent Scatterer InSAR (PS-InSAR) [17,18] and Small Baseline Subsets InSAR (SBAS-InSAR) [19,20] approaches. These multi-temporal InSAR techniques effectively mitigate the impact of atmospheric effects and spatio-temporal decorrelation on the accuracy of surface deformation monitoring. Consequently, they have found widespread applications in various fields such as land subsidence monitoring, geological hazard assessment, and building deformation analysis [21–24].

This research acquired 42 Sentinel-1A images along the Yangtze River in Nanjing and applied PS-InSAR and SBAS-InSAR technologies to monitor and analyze the land subsidence in the area. The accuracy and reliability of the two technologies in land subsidence monitoring were evaluated through cross-validation of the continuous deformation field obtained from both techniques and comparison with external leveling data. Additionally, the causes of land subsidence in this area were analyzed from three perspectives: underground transportation engineering, building loads, and geological structure.

2. Research Area and Experimental Data

2.1. Research Area

Nanjing, situated in the lower reaches of the Yangtze River, serves as the capital of Jiangsu Province and holds strategic importance as a major intersection of the Belt and Road Initiative and the Yangtze River Economic Belt. The city is traversed by the Yangtze River, flowing from southwest to east, with its urban area encompassing significant water bodies such as the Qinhuai River, Qinhuai New River, and Xuanwu Lake. The region along both banks of the Yangtze River in Nanjing is profoundly influenced by the impact and sedimentation of river water, resulting in the formation of a broad floodplain. The geological structure in this area is characterized by soft soil, rendering it susceptible to collapse and sedimentation, and the bearing capacity is poor.

In terms of tectonic units, Nanjing is in the northern margin of the Yangtze ancient land, and the basement is mainly light metamorphic schist and metamorphic volcanic rocks. For the area along the Yangtze River in Nanjing, the overall geological conditions are not very ideal, the main characteristics are loose geology, poor bearing capacity, and shallowly buried groundwater levels. In recent years, Nanjing has experienced rapid urbanization, marked by the extensive construction of underground railways, intensive implementation of ground construction projects, and excessive exploitation of groundwater. Consequently, some areas along the Yangtze River in Nanjing have witnessed land subsidence, posing significant challenges not only to the economic development of the Yangtze River region, but also to the safety and well-being of urban residents. In light of these circumstances, this research focused on the land subsidence along the Yangtze River, especially the 5 km buffer zone on both sides of the Yangtze River in Nanjing as a typical area for land subsidence monitoring and analysis, as shown in Figure 1. A–F were the positions of six leveling points. The findings aim to provide valuable technical insights for urban management and protection.

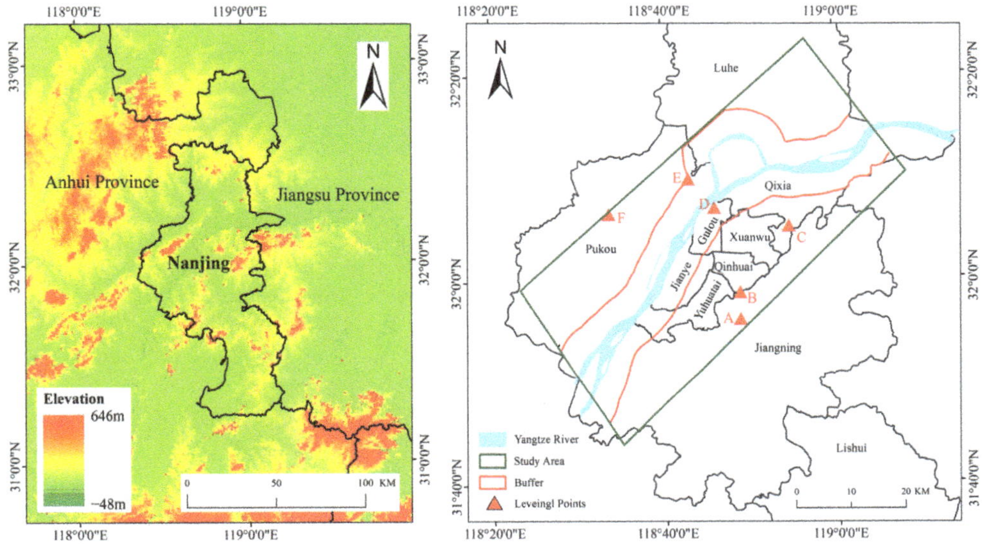

Figure 1. Study area.

2.2. Experimental Data

The Sentinel-1A satellite, operating in a solar synchronous orbit, was launched in April 2014. It is the inaugural environmental monitoring satellite deployed under the European Space Agency's Copernicus program and is equipped with a C-band SAR. Notably, it boasts high revisit frequency, wide coverage, and multiple polarimetric and working modes, which can meet the needs of time-series InSAR processing and analysis. For this study, approximately one image was selected every two months, culminating in a total of 42 Sentinel-1A images in Level-1 SLC format, which covered the areas along the Yangtze River in Nanjing from 8 April 2015 to 14 November 2021. The images were acquired in IW mode, with VV polarization, an incidence angle of 35.9°, and a spatial resolution of 5 m $\times$ 20 m (azimuthal direction $\times$ range direction). The Frame of the used data were 99 and 100, and the Path was 69.

Furthermore, this experiment utilized precision orbit data of the Sentinel-1A satellite, which was provided by the European Space Agency, to refine the orbit information. The reference digital elevation model (DEM) data utilized in this study were sourced from the SRTM elevation data with a resolution of 30 m. To validate the accuracy of the land subsidence information extracted by two methods, on-site leveling data spanning from 2015 to 2020 were employed. Figure 1 illustrates the spatial distribution of the six leveling points within the study area.

3. Technical Principles

3.1. PS-InSAR Technology

The Persistent Scatterer InSAR (PS-InSAR) technology, initially introduced by Ferretti in 2000, is designed to analyze objects within the study area that exhibit stable scattering characteristics over prolonged periods [17,18]. This involves identifying a set of ground target points with high coherence and stable reflection characteristics in image data, termed PS points, to extract phase changes. By doing this, this approach effectively avoids the problems of temporal and spatial decorrelation and reduces the influence of atmospheric effects on deformation information. As a result, PS-InSAR enables the acquisition of high-precision ground deformation information [25].

The fundamental principle of PS-InSAR technology is to select one SAR image as the master image, followed by the registration of N additional SAR images with the master image to generate N interference pairs. Subsequently, the differential processing of these pairs yields N interferograms, from which PS points with high coherence and stable reflection characteristics are extracted. Furthermore, the interference factors in phase information extraction are eliminated using orbital parameters, external DEM, and an established model. Ultimately, high-precision ground deformation information can be obtained. The interference phase of the PS point can be expressed as:

$$\varphi_{int} = \varphi_{def} + \varphi_{top} + \varphi_{fla} + \varphi_{noi} + \varphi_{atm} \quad (1)$$

Here, φ_{int} is the interference phase, φ_{def} is the deformation phase caused by the displacement of the target object during the satellite revisit cycle; φ_{fla} refers to the flat phase caused by interference on a flat surface caused by radar oblique range imaging, which can be eliminated with the assistance of precise orbital parameters; φ_{top} is the terrain phase caused by surface undulation, which can be eliminated with the assistance of DEM; φ_{atm} is the atmospheric delay phase caused by the influence of the ionosphere and troposphere on radar signals, resulting in signal propagation delay; and φ_{noi} is the phase error caused by radar thermal noise and inherent interference noise, which can be reduced by filtering.

With the assistance of precise orbit parameters and an external DEM, the interference image is flattened, and the terrain phase is eliminated. To express the relationship between the terrain residual increment and the interference phase of PS points, a model is established. Iterative calculations are performed using a linear model to estimate the surface deformation rate and the residual increment of terrain errors. Subsequently, after removing the residual deformation phase and terrain phase, further iterations are conducted to estimate the atmospheric delay phase error and noise error. This iterative process leads to the ultimate retrieval of the deformation rate and the residual elevation.

In this experiment, the selection of the master image was based on the principle of maximizing correlation among interference pairs. Ultimately, the SAR image obtained on 2 September 2019 was chosen as the master image. It was combined with the remaining 41 SAR images to form 41 interference pairs. As depicted in Figure 2, the spatial baseline exhibited a distribution primarily ranging -150~150 m, with the longest spatial baseline measuring 132.44 m and the shortest at -8.68 m. Subsequently, external DEM data were utilized to facilitate image registration and mitigate the impact of flat ground effects. The amplitude discrete index was employed to preliminarily select high-coherence target points. To obtain the initial deformation rate and residual elevation of the PS points, a linear inversion model was applied. Furthermore, high-pass filtering in space and low-pass filtering in time were employed to estimate the atmospheric delay phase. Geocoding of the PS points was accomplished using the reference DEM, with a coherence coefficient threshold of 0.7. After removing PS points with low coherence, a total of 976,871 effective PS points were obtained, enabling subsequent temporal and spatial analysis.

Due to the large coverage area of the acquired Sentinel-1A image, the processing was time-consuming. As a result, 42 images were cropped using geocoded vector data, and subsequent experimental processing was carried out based on these cropped images.

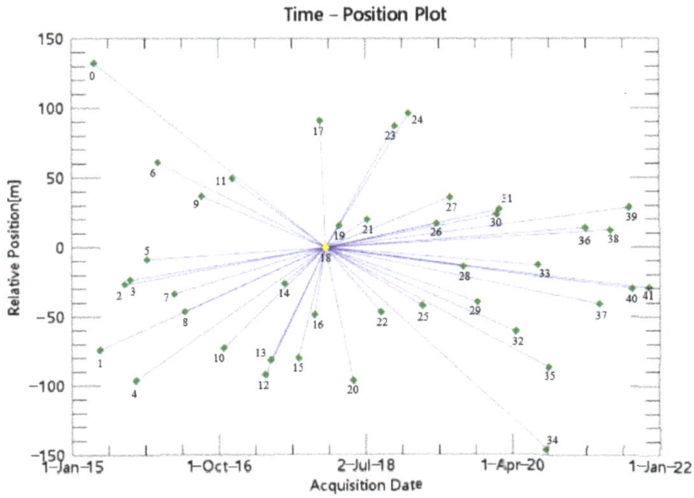

Figure 2. Image pairs and baselines generated by PS-InSAR.

3.2. SBAS-InSAR Technology

The Small Baseline Subset InSAR (SBAS-InSAR) technique was initially proposed by Berardino et al. in 2002 [19]. The fundamental concept of SBAS-InSAR technology involves the selection of M SAR images for flexible combination, ensuring that each image forms at least one interference pair with another image. By establishing thresholds for temporal and spatial baselines to exclude image data that do not meet the criteria, the adverse effects of spatiotemporal incoherence on monitoring results are mitigated. Subsequently, following the small baseline combination principle, differential interference processing is performed on all image pairs to derive N differential interferograms. The flat-earth phase and terrain phase in the interference phase are eliminated by removing the flat-earth effect and leveraging external DEM data. The interferometric phase of the interferogram generated by two SAR images can be expressed as the following formula,

$$\varphi = \varphi_1 - \varphi_2 = \frac{4\pi}{\lambda} \cdot (d_1 - d_2) + \Delta\varphi_{toperror} + \varphi_{atm} + \varphi_{noi} \tag{2}$$

Here, φ_1 and φ_2 are the interferometric phases obtained at the imaging time of two SAR images, respectively. $\Delta\varphi_{toperror}$ is the residual terrain phase error, and φ_{atm} is the atmospheric delay phase. φ_{noi} is the noise phase. In general, SBAS-InSAR technology operates under the assumption that surface deformation follows a linear pattern, and hence, a linear model was constructed to characterize the surface deformation. The residual terrain phase present in the interferometric phase was estimated and removed using the singular value decomposition method. Additionally, the atmospheric phase and nonlinear deformation phase were eliminated through spatial domain low-pass filtering and time domain high-pass filtering. Subsequently, the least-squares method was employed to calculate the average phase change rate and surface deformation for each given time.

In this experiment, the spatial baseline threshold was set to 48%, and the time baseline threshold was 180 days. The SAR image acquired on 7 September 2018 was selected as the master image. The generated image pairs and baselines are illustrated in Figure 3. Subsequently, the interference processing was assisted by external DEM data. The Goldstein filtering method and the Delaunay MCF unwrapping method were employed [26]. After evaluating the noise area in the unwrapping output and considering the phase information of the research area, the coherence threshold was finally set at 0.2. Following this, orbit refinement and re-phasing were conducted. The residual terrain error and atmospheric delay phase were estimated using two inversions, and finally, geocoding was performed.

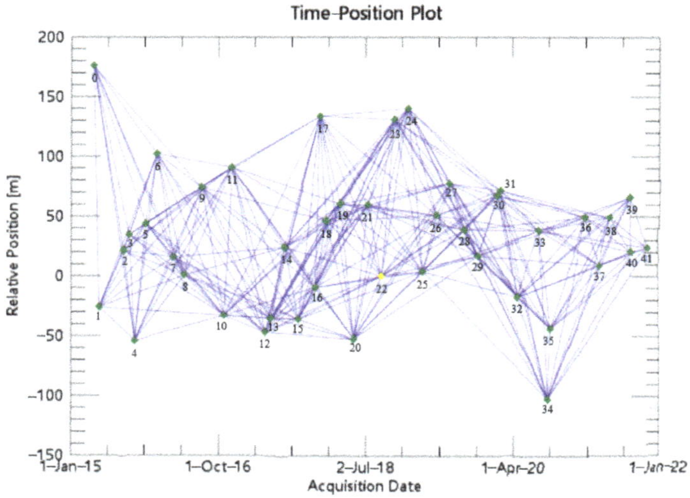

Figure 3. Image pairs and baselines generated by SBAS-InSAR.

4. Results Analysis

4.1. Comparative Validation of PS-InSAR and SBAS-InSAR Monitoring Results

In this study, PS-InSAR and SBAS-InSAR were employed to monitor the land subsidence using 42 Sentinel-1A SAR images. Figures 4 and 5 depict the distribution of Line-of-Sight (LOS) subsidence rates obtained from the two time-series InSAR methods, providing a clear and intuitive visualization of the land subsidence in the study area. The range of land LOS subsidence rates observed by PS-InSAR fell between −19.9 mm/a and 22.0 mm/a, while those derived from SBAS-InSAR were found to range from −12.1 mm/a to 17.7 mm/a.

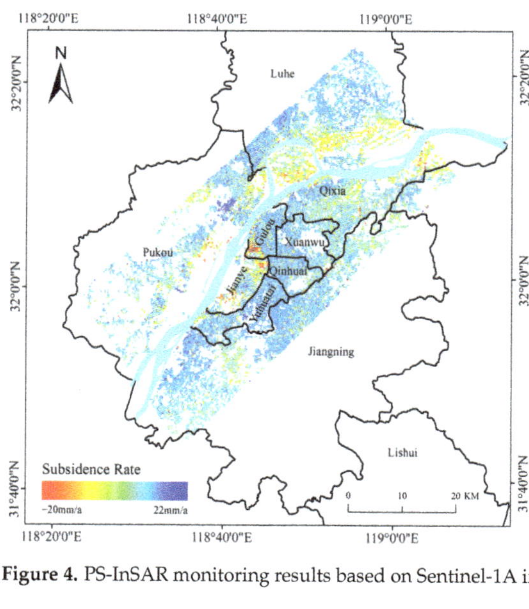

Figure 4. PS-InSAR monitoring results based on Sentinel-1A images.

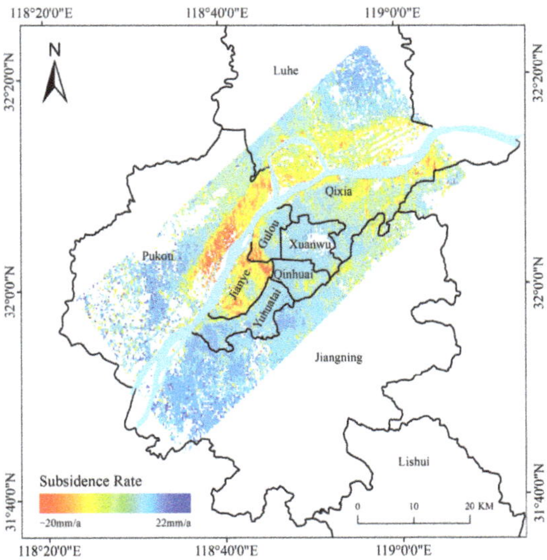

Figure 5. SBAS-InSAR monitoring results based on Sentinel-1A images.

To validate the reliability of the experimental findings in this paper, the results obtained from the two time-series analysis methods were subjected to linear fitting analysis. Linear fitting is a statistical modeling approach used to establish a linear relationship between variables [27]. The least-squares method is usually utilized to determine the most suitable fitting line [28]. The correlation coefficient, which assesses the degree of linear correlation between the two variables, was employed. The coefficient of determination R^2, ranging between 0 and 1, was utilized to gauge the goodness of fit of the model. In cases where variables exhibit linear correlation, the coefficient of determination represents the square of the correlation coefficient. The subsidence points sharing identical latitude and longitude were extracted from the outcomes of both PS-InSAR and SBAS-InSAR, facilitating a comparison of the linear relationship between the subsidence of these points observed by the two methods. As illustrated in Figure 6, the annual average land subsidence rates of 2462 subsidence points with the same latitude and longitude underwent linear fitting and analysis. It is evident that the points, with the PS-InSAR monitoring results plotted as vertical coordinates and the SBAS-InSAR monitoring results as horizontal coordinates, exhibit uniform distribution on either side of the fitting line. The coefficient of determination R^2 is 0.76, which indicates a high degree of consistency between the two processing outcomes.

4.2. Accuracy Verification of Monitoring Results Based on Leveling Data

This experiment utilized the leveling data from the region to validate the accuracy of these two methods. A total of six leveling points was chosen as reference points for testing the two time-series InSAR monitoring results. The locations of these reference leveling points are displayed in Figure 1. In order to ensure the correctness of the comparative analysis, the results of PS-InSAR and SBAS-InSAR were converted from LOS direction to vertical direction.

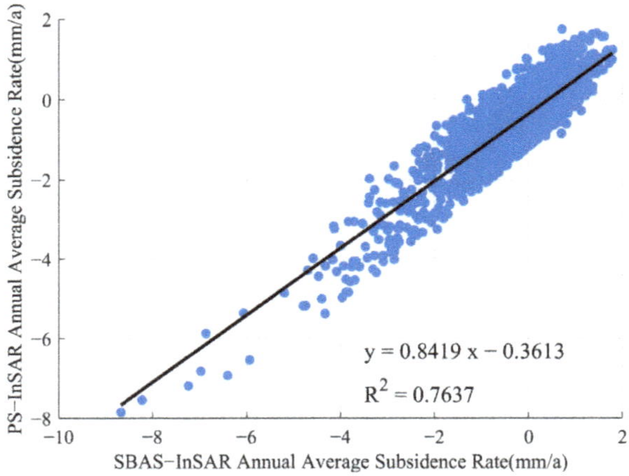

Figure 6. Comparison and validation of monitoring results between PS-InSAR and SBAS-InSAR.

To facilitate year-to-year comparisons, subsidence points with similar latitude and longitude to the same reference leveling points were extracted. Subsidence results from 2015 (from April to December) and 2016–2020 (from January to December) were selected for comparison with the leveling data in the corresponding time range. The errors between the leveling data and the deformation obtained by the two time-series InSAR techniques are shown in Figures 7 and 8. Figure 7 shows the comparison of deformation results of six points in three monitoring data at different times. Figure 8 shows the comparison of the deformation of each point obtained from the three monitoring data between 2015 and 2021. The maximum error between the PS-InSAR monitoring results and the leveling data was 4.54 mm, the minimum error was 0.02 mm, and the root mean square error was 2.12 mm. Meanwhile, the maximum error, minimum error, and root mean square between the SBAS-InSAR monitoring results and the leveling data were 4.15 mm, 0.14 mm, and 1.60 mm, respectively. These results suggest that both PS-InSAR and SBAS-InSAR are highly accurate methods for extracting ground deformation, as evidenced by the very small deviations between the monitoring results of the two methods and the leveling data. Notably, the maximum deviation did not exceed ±5 mm.

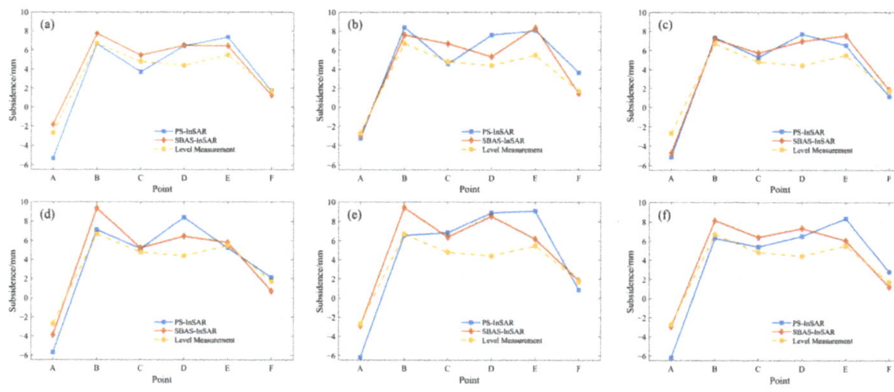

Figure 7. Accuracy comparisons between annual monitoring results from PS-InSAR and SBAS-InSAR and leveling data. (**a**) 2015.04–2015.12, (**b**) 2016.01–2016.12, (**c**) 2017.01–2017.12, (**d**) 2018.01–2018.12, (**e**) 2019.01–2019.12, (**f**) 2020.01–2020.12.

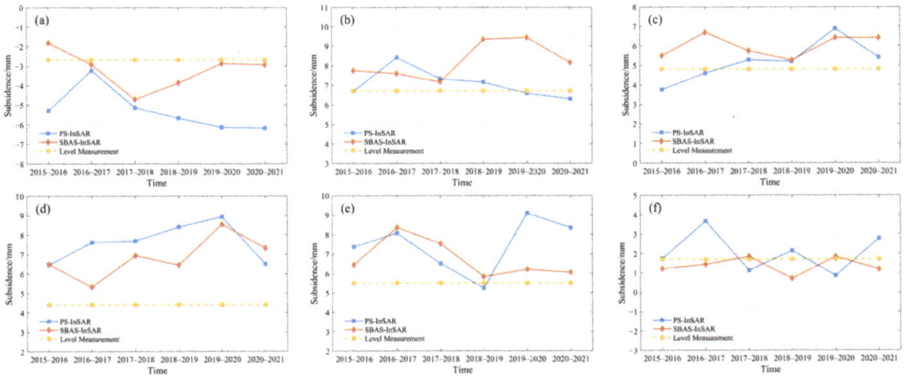

Figure 8. Accuracy comparison of PS-InSAR and SBAS-InSAR monitoring results with leveling data at each point from 2015 to 2021. (**a**) Point A, (**b**) point B, (**c**) point C, (**d**) point D, (**e**) point E, (**f**) point F.

To further illustrate the high correlation between the PS-InSAR and SBAS-InSAR monitoring results and the leveling data, the two monitoring results were subjected to linear fitting with the leveling data, and the fitting results are presented in Figure 9. The fitting results reveal an even distribution of points, with SBAS-InSAR and PS-InSAR monitoring results as the horizontal coordinates and leveling results as the vertical coordinates, on both sides of the fitting line. The coefficient of determination R2 between the PS-InSAR results and the leveling data was 0.8893, and that between the SBAS-InSAR results and the leveling data was 0.9381, indicating a high degree of correlation between the monitoring results of the two methods and the leveling data. These findings emphasize the reliability of land subsidence monitoring results based on PS-InSAR and SBAS-InSAR. Notably, in areas along the Yangtze River, especially where numerous high-rise buildings create strong dihedral scattering, the determination coefficient of SBAS-InSAR and leveling data exceeded that of PS-InSAR and leveling data, suggesting that SBAS-InSAR is more suitable for monitoring land subsidence in this area.

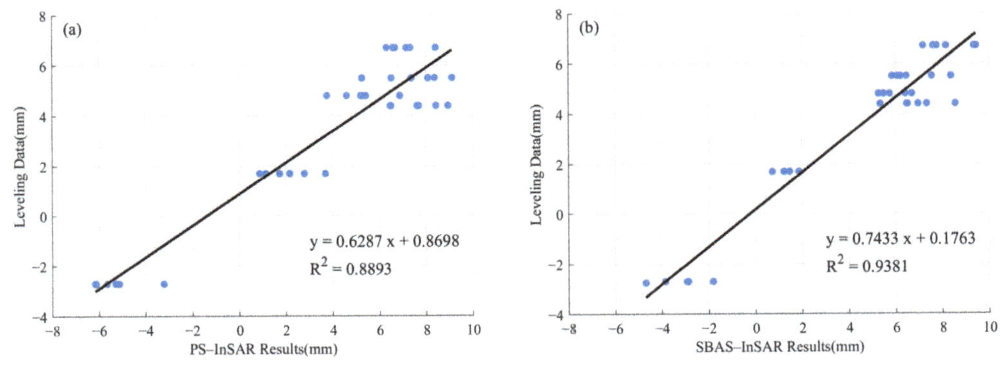

Figure 9. Linear fitting between leveling data and InSAR monitoring results. (**a**) PS-InSAR fitting results; (**b**) SBAS-InSAR fitting results.

4.3. Analysis of Spatiotemporal Variation Characteristics of Land Subsidence

A 5 km buffer zone was established on both sides of the Yangtze River, and the SBAS-InSAR results were cropped to obtain the average annual subsidence rate along the river. The spatial and temporal variation characteristics of land subsidence were analyzed in this buffer zone as the key area. Figure 10 displays the LOS subsidence and uplift areas, represented by red and blue colors, respectively. The ground deformation in the 5 km buffer showed stability from April 2015 to November 2021, with the annual average subsidence rate ranging from −12.5 mm/a to 17.2 mm/a. Within the buffer zone, there were four main subsidence areas, which were located at the area of the northeast of Jianye District and the southwest of Gulou District (S1), the eastern riverside area of Pukou District (S2), the Baguazhou area (S3), and the Longtan Street in the eastern part of Qixia District (S4). From 2015 to 2021, the spatial distribution of land subsidence in the buffer zone was relatively concentrated. Most areas in the northeast showed a slow sinking trend, with an average annual deformation rate ranging from −10 mm/a to 0 mm/a, while some areas in the southwest showed an upward trend, with an average annual deformation rate ranging from 0 mm/a to 10 mm/a.

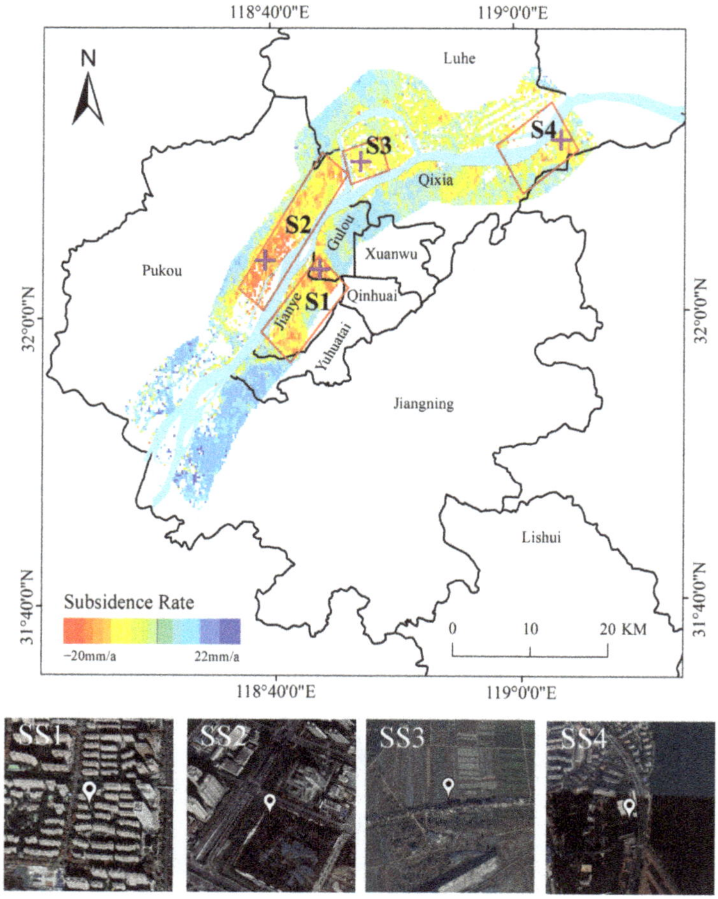

Figure 10. SBAS-InSAR monitoring results of the 5 km buffer zone along the Yangtze River from 2015 to 2021.

The distribution of the main subsidence areas in Figure 10 reveals a distinct pattern. It can be seen that the subsidence funnel S1 was gradually converging into a larger area. Notably, the southwest area of S1 near the Yuzui Wetland Park in Jianye District exhibited a maximum LOS subsidence rate of −10.4 mm/a, attributed to complex geological conditions. This area comprised an upper soil layer of alternating marshes and shoals of silty clay, and the lower part was the riverbed facies layer composed of medium to fine sand, making it susceptible to subsidence. The northeast area of S1, situated at the junction of the northeast of Jianye District and the southwest of Gulou District (now the Hexi development new area), experienced a maximum LOS deformation rate of −13.6 mm/a. In recent years, the area has been surrounded by high-rise buildings and developed rapidly. Metro Line 2 and Metro Line 10 passed through this area. Therefore, large-scale ground and underground construction was one of the main reasons for the formation of a subsidence funnel in this area. S2 was a super-large subsidence area composed of Jiangpu Street, Dingshan Street, Mount Taishan Street, and Yanjiang Street in Pukou District. This area boasted three metro lines and seven subway stations, with a maximum LOS subsidence rate of −13.5 mm/a. The subsidence rate of this region was directly influenced by its development planning. In June 2015, the Jiangbei New Area of Nanjing was established, formally commencing construction in 2017. The period from 2017 to 2022 witnessed rapid five-year growth in Nanjing Jiangbei New Area. The significant subsidence in this area primarily resulted from high-intensity underground engineering operations and alterations in surface loading. Area S3 was located in the southern part of Baguazhou, Qixia District, exhibiting a maximum LOS subsidence rate of −8.8 mm/a. Baguazhou, a large island in the Yangtze River, is marked by the division of the Yangtze River into its northern and southern branches. Despite the low building density and predominance of farmland, which minimizes the impact of human factors, the area's geological conditions are not stable due to its formation as a sandbar-type plain in the middle of the Yangtze River through alluvial action. Area S4, situated near Longtan Street along the river in the eastern part of Qixia District, exhibited a maximum LOS subsidence rate of −9.2 mm/a. This area falls within the Ningzhen folded belt and is characterized by soft soil. Given the presence of wharf companies, as well as multiple logistics and warehousing firms in the area, it is prone to subsidence.

To analyze the temporal characteristics of land subsidence in key areas along the Yangtze River in Nanjing, we selected four points with the highest LOS subsidence rates within the four identified subsidence funnels as representative feature points. Using these feature points, we generated a cumulative subsidence change map spanning the entire period. The spatial distribution of the four selected subsidence points is illustrated in Figure 10, while the SBAS-InSAR deformation curves for these four points are displayed in Figure 11. It is evident that between April 2015 and November 2021, all these four points exhibited distinct and varying patterns of uneven subsidence.

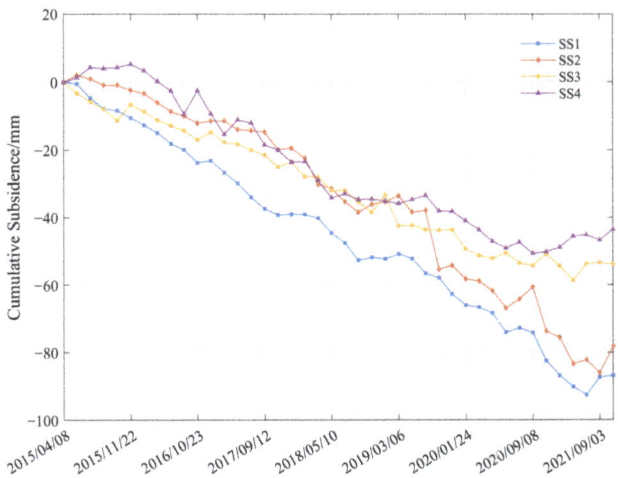

Figure 11. The time-series deformation trend of the feature points in the main subsidence area of SBAS-InSAR from 2015 to 2021.

As depicted in Figures 10 and 11, the feature point SS1 was located close to Jiangdong Street, Mochou Lake Street, and Binhu Street, with Metro Line 2 passing through. The LOS cumulative subsidence at SS1 amounted to −86.7 mm. This subsidence can be attributed to the excessive building density and load in this area, as analyzed earlier. The SS2 point, situated near Jiangpu Street in Pukou District, exhibited an overall gradual subsidence trend, with an LOS cumulative total subsidence of −78.1 mm. However, there were two rapid subsidence trends in 2019 and 2021, which can be closely linked to the development plan of the area. Point SS3 was situated in the Baguazhou area. While there was a slight rebound observed on the ground over a relatively short period, the overall trend of subsidence remained slow and stable. The LOS cumulative total subsidence at SS3 amounted to −53.8 mm. This trend can be attributed to the unstable geological conditions prevalent in the area. The LOS cumulative settlement at point SS4, situated in Longtan Street, Qixia District, has reached −43.5 mm, indicating a generally stable sinking trend, but with a noted rebound since September 2020.

5. Discussion of the Factors Related to Land Subsidence along the Yangtze River in Nanjing

5.1. The Influence of Underground Engineering on Land Subsidence

The construction of subway tracks is one of the key directions for the future development of cities. However, this infrastructure also impacts the soil structure, leading to urban land subsidence issues [29]. In this study, we aimed to investigate the influence of subway operation on urban land subsidence by analyzing the SBAS-InSAR subsidence monitoring results of three stations (Wanhuicheng Stantion, Linjing Station, and Lvboyuan Station) on Nanjing Metro Line 10 after they started operating in 2014. Figure 12 represents the subsidence distribution near the three subway stations during 2015 and 2021.

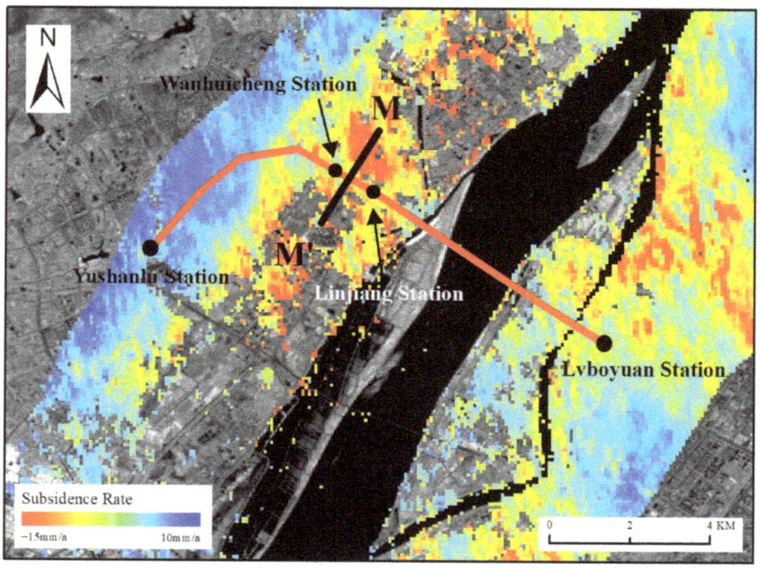

Figure 12. SBAS-InSAR LOS subsidence rate of some subway stations on Nanjing Metro Line 10.

As illustrated in Figure 12, the area has been undergoing low-speed subsidence since the three stations started operating in 2014. The primary subsidence-prone areas were concentrated near Wanhuicheng Station and Linjiang Station, both located in Pukou. Notably, the area between these two stations had been experiencing the most severe subsidence, and the area near Lvboyuan Station also exhibit a certain degree of subsidence, but it was smaller than that near Wanhuicheng and Linjiang Station.

To analyze the characteristics of the transverse ground deformation changes near Wanhuicheng and Lingjiang Stations, a transverse profile line MM′ across the subway line was selected in an area with serious subsidence. Subsidence rates of the points on the profile line were extracted for further analysis. Figure 13 shows the spatial distribution of subsidence rates on the transverse section line. The results indicated that the subsidence rates of the points on the MM′ profile line were unevenly distributed on both sides of the subway, with many small fluctuations. The LOS subsidence rates on the M side were significantly faster than that on the M′ side, and the LOS subsidence rates were in the range from −13.32 mm/a to −2.67 mm/a, which showed a general sinking trend. The different degrees of land subsidence around these subway stations indicate that there may be a certain relationship between subway station operation and ground subsidence. To prevent the occurrence of various situations, such as structural damage and safety hazards, regular monitoring and maintenance are necessary.

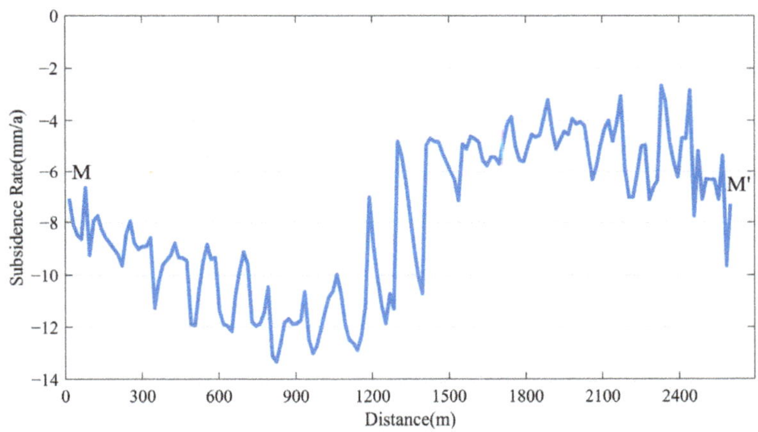

Figure 13. The LOS subsidence rates of the points on the MM′ profile line.

5.2. The Relationship between Ground Building Area and Land Subsidence

The acceleration of urbanization has increased a load of construction projects on the urban ground, which, in turn, poses a risk of land subsidence [30,31]. To investigate the impact of urban construction on land subsidence along the Yangtze River in Nanjing, the Landsat8 OLI images from 2016, 2019, and 2021 were used in this experiment to extract the ground buildings in the subsidence funnel areas. Based on the normalized building index (NDBI) obtained from Landsat 8, the ground buildings were extracted, and the areas were calculated [32]. The results found that the areas of ground building in the subsidence funnels along the Yangtze River have steadily increased from 2015 to 2021, as shown in Figure 14. By comparing the time-series deformation trend of the key feature points (Figure 11) extracted by SBAS-InSAR technology with the ground building areas, the impact of the building area changing on land subsidence was analyzed. Figure 14 shows that as the ground building areas continued to increase during 2016–2021, the cumulative subsidence of the feature points in the areas also increased. It can be inferred that the increase of ground building area brought by urban expansion is positively correlated with land subsidence to some extent.

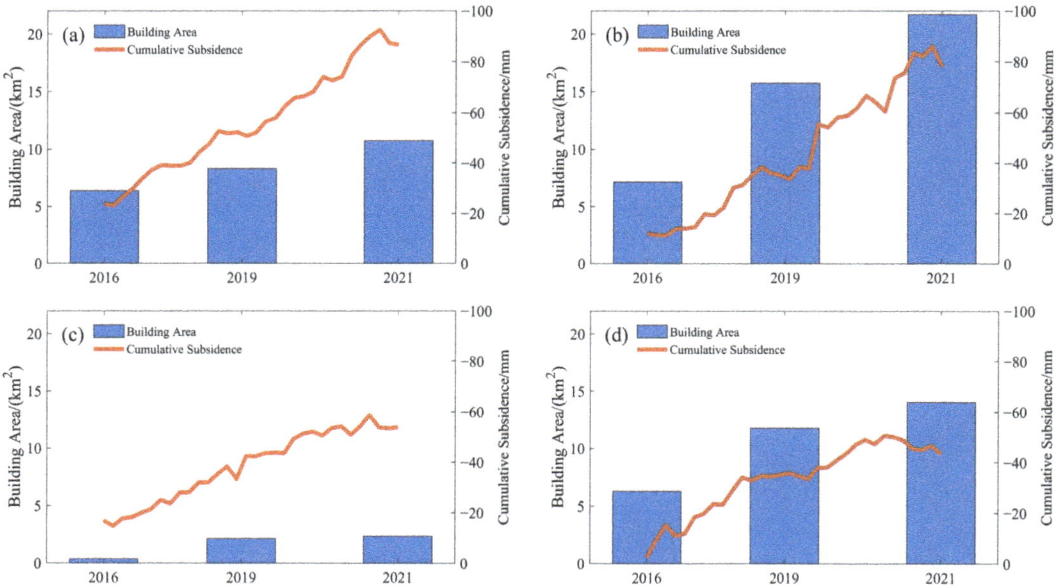

Figure 14. Relationship between LOS cumulative subsidence of the feature points and ground building area in subsidence funnels. (**a**) Area S1, (**b**) area S2, (**c**) area S3, (**d**) area S4.

5.3. The Influence of Geological Structure on Land Subsidence

The floodplain on both sides of the Yangtze River in Nanjing is broad. According to its topographic features, geomorphological units, and geotechnical combination, it can be classified as a floodplain soft soil area. This area primarily consists of Quaternary sedimentary soft soil layers that have been formed through the deposition and siltation process of the Yangtze River [33]. As a result, its geological composition is loose, resulting in poor bearing capacity. The structural characteristics of the soft soil layers in the Yangtze River floodplain are summarized in Table 1.

Table 1. Structural characteristics of the soft soil layer in the Yangtze floodplain.

Soil Layer	Characteristic
Surface recent and outdated fill	Mainly composed of building residue and cultivated soil, which is soft and has strong plasticity
Marsh facies with mucky and silty clay	Pure without sand, showing grayish-brown
Floodplain facies with muddy, silty clay and silt sand interbedding	With horizontal bedding and grey colors
Riverbed facies with medium-coarse gravel sand, coarse gravel sand mixed with gravel or boulders	Compactness increases with depth from loose to dense

According to Table 1, it can be seen that the floodplain of Nanjing along the Yangtze River predominantly consists of clay soil with high dispersion, strong hydrophilicity, a large void ratio, and high saturation. Owing to its thixotropic properties, this type of soil is susceptible to surface deformation when subjected to specific loads or high-intensity disturbances. Furthermore, excessive groundwater extraction and high-intensity engineering activities can disrupt the loose structure of the soft soil layer, rendering it more vulnerable to geological hazards such as subsidence, landslides, and soft foundations. The four primary subsidence areas identified in this study were all distributed within a 5 km radius along the Yangtze River. This led to the inference that the geological structure of this study area constitutes a significant factor contributing to land subsidence.

6. Conclusions

In this paper, two time-series technologies, PS-InSAR and SBAS-InSAR, were used to monitor the land subsidence in Nanjing along the Yangtze River using 42 Sentinel-1A images from April 2015 to November 2021. The following conclusions can be drawn:

(1) The ground deformation along the Yangtze River in Nanjing extracted by PS-InSAR and SBAS-InSAR technologies was consistent, and the two monitoring results had a high correlation with the leveling results. The maximum deviation of deformation didn't exceed ±5 mm, which was reliable in accuracy.

(2) The key areas within 5 km on both sides of the Yangtze River were analyzed. It was found that there were four subsidence funnels from 2015 to 2011, which were distributed in Jianye District and the southwest of Gulou District, the eastern riverside area of Pukou District, the Baguazhou area, and the Longtan Street in the eastern part of Qixia District, respectively. The subsidence in Jianye District was the most severe, and there was a trend of gradually connecting into a large subsidence funnel. The maximum LOS cumulative subsidence of Jianye district was −86.7 mm. The subsidence area along the river in Dingshan Street of Pukou District was the largest subsidence funnel in the study area. The maximum LOS cumulative subsidence reached −78.1 mm.

(3) The land subsidence along the Yangtze River has a certain correlation with the increasing of the ground building area, the construction and operation of some underground engineering such as subway stations, and the geological conditions of the region, but the contribution of each factor to the subsidence needs to be further analyzed in future studies.

Author Contributions: Conceptualization, Yuanyuan Chen, Lin Guo and Jia Xu; methodology, Yuanyuan Chen and Lin Guo; software, Qiang Yang, Hao Wang and Chenwei Zhu; validation, Yuanyuan Chen, Lin Guo, Jia Xu, Qiang Yang and Hao Wang; formal analysis, Jia Xu and Lin Guo; investigation, Yuanyuan Chen, Qiang Yang and Hao Wang; resources, Yuanyuan Chen; data curation, Hao Wang and Chenwei Zhu; writing—original draft preparation, Yuanyuan Chen and Lin Guo; writing—review and editing, Jia Xu, Qiang Yang and Chenwei Zhu; visualization, Hao Wang; supervision, Yuanyuan Chen; project administration, Yuanyuan Chen; funding acquisition, Lin Guo. All authors have read and agreed to the published version of the manuscript.

Funding: This research was funded by the Jiangsu Water Conservancy Science and Technology Project, grant number 2019001, the National Natural Science Foundation of China, grant number 41930109, and the Key Laboratory of Land satellite Remote sensing Application, Ministry of Natural Resources of the People's Republic of China, grant number KLSMNR-G202311.

Data Availability Statement: The data that support the findings of this study are available from the corresponding author upon reasonable requests.

Acknowledgments: The Sentinel-1A images were freely provided by the European Space Agency.

Conflicts of Interest: The authors declare no conflicts of interest.

References

1. Na, T.; Kawamura, Y.; Kang, S.S.; Utsuki, S. Hazard mapping of ground subsidence in east area of Sapporo using frequency ratio model and GIS. *Geomat. Nat. Hazards Risk* **2021**, *12*, 347–362. [CrossRef]
2. Bagheri-Gavkosh, M.; Hosseini, S.M.; Ataie-Ashtiani, B.; Sohani, Y.; Ebrahimian, H.; Morovat, F.; Ashrafi, S. Land subsidence: A global challenge. *Sci. Total Environ.* **2021**, *778*, 146193. [CrossRef] [PubMed]
3. Ding, X.L.; Liu, G.X.; Li, Z.L.; Li, Z.; Chen, Y.Q. Ground subsidence monitoring in Hong Kong with satellite SAR interferometry. *Photogramm. Eng. Remote Sens.* **2004**, *70*, 1151–1156. [CrossRef]

4. Yalçınkaya, M. Monitoring crustal movements in west Anatolia by precision leveling. *J. Surv. Eng.* **2003**, *129*, 44–49. [CrossRef]
5. Nestorović, Ž.; Delčev, S. Comparison of height differences obtained by trigonometric and spirit leveling method. *Geonauka* **2014**, *2*, 30–37. [CrossRef]
6. Yang, D.L.; Zou, J.; Shen, Y.W.; Zhu, H. Research and application of trigonometric leveling to replace precise leveling. *J. Surv. Eng.* **2021**, *147*, 04021012.
7. Xiang, Y.F.; Wang, H.; Chen, Y.Y.; Xing, Y. GNSS imaging of strain rate changes and vertical crustal motions over the Tibetan Plateau. *Remote Sens.* **2021**, *13*, 4937. [CrossRef]
8. Xiang, Y.F.; Yue, J.P.; Liu, G.; Chen, Y.Y. Characterizing the spatial patterns of vertical crustal deformations over the South American continent based on GNSS imaging. *Pure Appl. Geophys.* **2022**, *79*, 3569–3587. [CrossRef]
9. Zhu, J.J.; Li, Z.W.; Hu, W.J. Research progress and methods of InSAR for deformation monitoring. *Acta Geod. Cartogr. Sin.* **2017**, *46*, 1717–1733.
10. Wang, Y.; Feng, G.; Li, Z.; Luo, S.; Wang, H.; Xiong, Z.; Zhu, J.; Hu, J. A strategy for variable-scale InSAR deformation monitoring in a wide area: A case study in the Turpan–Hami Basin, China. *Remote Sens.* **2022**, *14*, 3832. [CrossRef]
11. Wang, Z.; Dai, H.; Yan, Y.; Ren, J.; Zhang, Y.; Liu, J. An InSAR deformation phase retrieval method combined with reference phase in mining areas. *Remote Sens.* **2023**, *15*, 4573. [CrossRef]
12. Feng, Y.; Zhou, Y.; Chen, Y.; Li, P.; Xi, M.; Tong, X. Automatic selection of permanent scatterers-based GCPs for refinement and reflattening in InSAR DEM generation. *Int. J. Digit. Earth* **2022**, *15*, 954–974. [CrossRef]
13. Fan, H.; Deng, K.; Ju, C.; Zhu, C.; Xue, J. Land subsidence monitoring by D-InSAR technique. *Min. Sci. Technol.* **2011**, *21*, 869–872. [CrossRef]
14. Fan, H.; Wei, G.; Yong, Q.; Xue, J.; Chen, B. A model for extracting large deformation mining subsidence using D-InSAR technique and probability integral method. *Trans. Nonferrous Met. Soc.* **2014**, *24*, 1242–1247. [CrossRef]
15. Chen, Y.; Yu, S.; Tao, Q.; Liu, G.; Wang, L.; Wang, F. Accuracy verification and correction of D-InSAR and SBAS-InSAR in monitoring mining surface subsidence. *Remote Sens.* **2021**, *13*, 4365. [CrossRef]
16. Wang, L.; Deng, K.; Zheng, M. Research on ground deformation monitoring method in mining areas using the probability integral model fusion D-InSAR, sub-band InSAR and offset-tracking. *Int. J. Appl. Earth Obs.* **2020**, *85*, 101981. [CrossRef]
17. Ferretti, A.; Prati, C.; Rocca, F. Analysis of permanent scatterers in SAR interferometry. In Proceedings of the IEEE 2000 International Geoscience and Remote Sensing Symposium, Honolulu, HI, USA, 24–28 July 2000; pp. 761–763.
18. Ferretti, A.; Prati, C.; Rocca, F. Permanent scatterers in SAR interferometry. *IEEE Trans. Geosci. Remote Sens.* **2001**, *39*, 8–20. [CrossRef]
19. Berardino, P.; Fornaro, G.; Lanari, R.; Sansosti, E. A new algorithm for surface deformation monitoring based on small baseline differential SAR interferograms. *IEEE Trans. Geosci. Remote Sens.* **2002**, *40*, 2375–2383. [CrossRef]
20. Boloorani, A.D.; Darvishi, M.; Weng, Q.; Liu, X. Post-war urban damage map** using InSAR: The case of Mosul City in Iraq. *ISPRS Int. J. Geo-Inf.* **2021**, *10*, 140. [CrossRef]
21. Joshi, M.; Kothyari, G.C.; Kotlia, B.S. Landslide detection in Kinnaur Valley, NW India using PS-InSAR technique. *Phys. Geogr.* **2024**, *45*, 160–174. [CrossRef]
22. Liu, J.; Li, T.; Ma, S.; Shan, Q.; Jiang, W. Study on LOS to Vertical Deformation Conversion Model on Embankment Slopes Using Multi-Satellite SAR Interferometry. *ISPRS Int. J. Geo-Inf.* **2024**, *13*, 58. [CrossRef]
23. Welikanna, D.R.; Jin, S.Y. Investigating ground deformation due to a series of collapse earthquakes by means of the PS-InSAR technique and Sentinel 1 data in Kandy, Sri Lanka. *J. Appl. Remote Sens.* **2023**, *17*, 014507. [CrossRef]
24. Zhang, P.; Guo, Z.; Guo, S.; Guo, S.; Xia, J. Land subsidence monitoring method in regions of variable radar reflection characteristics by integrating PS-InSAR and SBAS-InSAR techniques. *Remote Sens.* **2022**, *14*, 3265. [CrossRef]
25. Malik, K.; Kumar, D.; Perissin, D.; Pradhan, B. Estimation of ground subsidence of New Delhi, India using PS-InSAR technique and Multi-sensor Radar data. *Adv. Space Res.* **2022**, *69*, 1863–1882. [CrossRef]
26. Bahti, F.N.; Chung, C.C.; Lin, C.C. Parametric Test of the Sentinel 1A Persistent scatterer-and small baseline subset-interferogram Synthetic Aperture Radar processing using the stanford method for persistent scatterers for practical landslide monitoring. *Remote Sens.* **2023**, *15*, 4662. [CrossRef]
27. Cao, Y.; Li, P.; Hao, D.; Lian, Y.; Wang, Y.; Zhao, S. Analysis of the relationship between vegetation and radar interferometric coherence. *Sustainability* **2022**, *14*, 16471. [CrossRef]
28. Krystek, M.; Anton, M. A weighted total least-squares algorithm for fitting a straight line. *Meas. Sci. Technol.* **2007**, *8*, 3438. [CrossRef]
29. Xu, Y.; Ma, L.; Du, Y.; Shen, S. Analysis of urbanisation-induced land subsidence in Shanghai. *Nat. Hazards* **2012**, *63*, 1255–1267. [CrossRef]
30. Cui, Z.; Yang, J.; Yuan, L. Land subsidence caused by the interaction of high-rise buildings in soft soil areas. *Nat. Hazards* **2015**, *79*, 1199–1217. [CrossRef]
31. Yang, Q.; Ke, Y.; Zhang, D.; Chen, B.; Gong, H.; Lv, M.; Zhu, L.; Li, X. Multi-scale analysis of the relationship between land subsidence and buildings: A case study in an eastern Beijing urban area using the PS-InSAR technique. *Remote Sens.* **2018**, *10*, 1006. [CrossRef]

32. Karanam, H.K.; Neela, V.B. Study of normalized difference built-up (NDBI) index in automatically mapping urban areas from Landsat TN imagery. *Int. J. Eng. Sci. Math.* **2017**, *8*, 239–248.
33. Saito, Y.; Yang, Z.S.; Hori, K. The Huanghe (Yellow River) and Changjiang (Yangtze River) deltas: A review on their characteristics, evolution and sediment discharge during the Holocene. *Geomorphology* **2001**, *41*, 219–231. [CrossRef]

remote sensing

MDPI

Article

Performance Analysis of an Aerial Remote Sensing Platform Based on Real-Time Satellite Communication and Its Application in Natural Disaster Emergency Response

Xiangli He [1,2], Chong Xu [1,2,*], Shengquan Tang [3], Yuandong Huang [1,4], Wenwen Qi [1,2] and Zikang Xiao [1,2]

1 National Institute of Natural Hazards, Ministry of Emergency Management of China, Beijing 100085, China; xianglihe@ninhm.ac.cn (X.H.); huangyuandong23@mails.ucas.ac.cn (Y.H.); wenwenqi@ninhm.ac.cn (W.Q.); zikangxiao@ninhm.ac.cn (Z.X.)

2 Key Laboratory of Compound and Chained Natural Hazards Dynamics, Ministry of Emergency Management of China, Beijing 100085, China

3 Beijing Aerospace Institue for Metrology and Measurement Technology, Beijing 100076, China; htjlbaidu@163.com

4 School of Emergency Management Science and Engineering, University of Chinese Academy of Sciences, Beijing 100049, China

* Correspondence: chongxu@ninhm.ac.cn

Abstract: The frequency of natural disasters has increased recently, posing a huge threat to human society. Rapid, accurate, authentic, and comprehensive acquisition and transmission of disaster information are crucial in emergency response. In this paper, we propose a design scheme for an aerial remote sensing platform based on real-time satellite communication. This platform mainly includes a civilian heavy-duty unmanned aerial vehicle, ground observation system with the self-developed orthographic image stabilization device, wireless communication system with an airborne mobile communication device using Ku band, ground satellite information receiving station, and data processing and application analysis system. The image stabilization capability of the ground observation system and the communication capability of the wireless communication system were verified through ground and flight tests respectively. The results showed that the stability accuracy of the platform was better than the theoretical threshold, the system transmission rate was not less than 2 M bandwidth, the data packet loss rate was low, and the time delay was not more than 2 s. The images captured in the experiment were clear, with a resolution of less than 1cm and an overlap rate of more than 70%. These all results meet the emergency observation requirement, which indicates that the aerial remote sensing platform based on real-time satellite communication has great potential for application in natural disaster emergency response.

Keywords: natural disasters; aerial remote sensing; satellite communication; emergency response; data transmission

1. Introduction

Natural disasters such as earthquakes, landslides, floods, and forest and grassland fires have caused serious damage to human society and the ecological environment [1–3]. When natural disasters occur, the ground infrastructure and communication equipment are often destructed and damaged, which seriously restricts the timeliness of transmitting disaster area information to distant command centers and makes it difficult to achieve real-time information transmission [4,5]. Therefore, in the event of sudden major natural disasters, providing fast, accurate, and effective means of emergency information acquisition and transmission is crucial for real-time emergency monitoring and rapid response to emergency rescue [6–8].

So far, the methods of obtaining disaster area information mainly include ground survey (monitoring), satellite remote sensing, and aerial remote sensing [9–11]. When or after a disaster occurs, if relevant investigators are arranged to conduct a detailed survey in the disaster area, it will not only pose a great threat to their life safety but also be extremely costly and inefficient [12]. Although ground monitoring (such as ground remote sensing vehicles, ground monitoring stations, etc.) can obtain information chronically and quickly, it is difficult to achieve long-distance and large-area monitoring due to the limitations of ground conditions and environment [13]. Especially when roads in disaster areas are damaged, ground monitoring methods are basically ineffective and cannot complete the task. During the last two decades, there has been an increase in the use of remote sensing technology for large-scale and high-resolution mapping and monitoring of natural hazards [14]. Remote sensing has become an important tool for hazards risk assessment and rescue operations [15,16]. For example, satellite remote sensing uses optical sensors carried on satellites to observe the ground and transmit ground information to ground receiving stations in real time [9,17]. In terms of resolution, the Worldview-4 satellite launched by the United States on 11 November 2016 has a resolution of 0.31 m, which can obtain the highest resolution civilian optical remote sensing data currently available [18]. The successful launch of first sub-meter-level stereoscopic satellite, Gaofen-7, marks that the resolution of low-orbit remote sensing satellite can reach 0.65 m in China [19]. However, these satellites are not in geosynchronous orbit and have a long re-entry period, which makes the observation of the same location periodic and delayed. Such timeliness is not enough to meet the needs of post-disaster emergency response. In general, although satellite remote sensing can obtain information related to emergencies over a large area and with high resolution, it has the disadvantages of time limit, fixed orbit, long revisit period, and other inflexibility [20]. Even geosynchronous orbit satellites, such as Gaofen-5, can continuously observe the same location, but the information resolution is low, about 50 m [19]. If there is heavy fog or cloud cover, satellite remote sensing cannot meet the requirements of monitoring the details of the disaster area [13,21].

There has been a significant improvement in the spatial resolution of remote sensing technology in particular with the introduction of aerial remote sensing [22]. Low-cost aerial remote sensing is now indispensable for onsite rapid data collection to aid disaster management [23–25]. Aerial remote sensing is a method of collecting data on the ground topography and landforms by means of photography, video recording, and topographic mapping through photoelectric equipment on a flying platform [26]. Aerial remote sensing is very effective in obtaining relevant disaster information accurately truthfully and comprehensively [27]. In particular, through telephoto high-resolution photography, higher-resolution images can be obtained compared to satellite remote sensing, providing high-precision data support for the emergency response [21,28,29]. In addition, compared to floating platforms (near-space vehicles) and short-range unmanned aerial vehicles (UAV), long range manned (unmanned) aircraft have extremely fast flight speeds and can reach the disaster site within a few hundred kilometers around the airport, basically within two hours, which can meet the effect of rapid response. These indicate that aerial remote sensing technology is an effective supplementary means for ground observation and satellite remote sensing, having advantages such as high spatial resolution, high temporal resolution, high degree of automation, flexibility, and low cost [14,30,31]. Aerial remote sensing technology has also been successfully applied to multiple earthquake and geological disaster cases [27,31]. For example, after the Ms 8.0 Wenchuan earthquake in 2008 [32], the Ms 7.0 Jiuzhaigou earthquake in 2017 [33,34], and other disasters [35–38], aerial remote sensing was used to immediately investigate and obtain data such as high-resolution topography and real 3D models of the disaster area, which provided strong support for rapid and accurate assessment of the disaster situation and played an important role in emergency response and post-earthquake recovery and reconstruction [31].

However, although the image resolution obtained by aerial remote sensing is relatively high, due to the fact that airborne cameras are generally ordinary digital cameras with small image frames, a large number of images are obtained with significant image distortion [39]. Moreover, due to the large tilt angle, irregular tilt direction, and large amplitude of turbulence during shooting, there are differences in scale and rotation between adjacent images [40]. Chen et al. (2017) adopt a combination of oblique and vertical shooting methods to improve this problem [39], while others use image correction techniques on the images obtained by the aviation platform for detailed monitoring of natural hazards [14,41]. Additionally, the inertial stabilization platform is used to isolate the influence of aircraft carrier angular motion, vibration, etc. on the load visual axis, which is the basis and guarantee of high-precision aerial remote sensing imaging [42]. Regarding the aerial remote sensing platform, in addition to the continuous improvement of imaging technology, its data transmission stability and speed are also key performance factors affecting disaster rescue. The use of aerial remote sensing to obtain these observations is attractive, but often challenged by lack of suitable solutions to get live images back to decision makers [43]. In previous cases, the images were mostly read and interpreted after landing [31,34,35,44]. The aerial remote sensing platform may be beyond radio line of sight during flight, so its communication cannot rely solely on ground communication. Satellite communication is essential to ensure real-time acquisition of observation results from aerial remote sensing platforms [43]. At the beginning of aerial remote sensing applications, the bandwidth of satellite communication was relatively low and could not meet the real-time transmission requirements of high-definition remote sensing images. With the development of technology and the growth of application demand, the proportion of constellations used for satellite communication and earth observation is increasing, currently even reaching 68% [45]. In recent years, microsatellites are mainly used for earth observation and remote sensing, and large satellite constellations make many simultaneous and distributed measurements or observations possible, such as for disaster monitoring, and increase temporal resolution of collected data [45,46]. Moreover, the use of higher frequencies than the S-band (mainly for telemetry) and X-band (for data transmission), such as the Ku-, K-, and Ka-bands, has become more widely available, which allows higher data rates [45,47]. These are the latest technologies for large spacecraft and enable real-time transmission and acquisition of high-definition photos and videos required for disaster emergency response, but they are still emerging technologies in the field of small aerial remote sensing platform.

In this paper, we present a novel technical solution for an aerial remote sensing platform based on real-time satellite communication to achieve rapid arrival at disaster sites, high-quality acquisition of the latest disaster site information, and real-time transmission of on-site information. The innovation of this technical solution lies in the application of civilian heavy-duty unmanned aerial vehicles, self-developed orthographic image stabilization platform, and particularly airborne mobile communication devices that can achieve a transmission rate of no less than 2 Mbps in the Ku band. Additionally, as the actual performance of the aerial remote sensing platform potentially is influenced by multiple factors [14,42,47], we innovatively conducted double verification for the feasibility of this technical solution through both ground and flight tests. That is, we evaluate the performance of the proposed aerial remote sensing platform and its applicability for real-time emergency monitoring in disaster events by verifying the image stabilization capability of the image stabilization platform and the communication capability of the airborne satellite communication system in ground and flight tests.

2. Design Scheme of Aerial Remote Sensing Platform Based on Real-Time Satellite Communication

In terms of natural disaster monitoring, aerial remote sensing technology has the advantages of high imaging quality and flexibility. Combined with real-time satellite communication technology, it has the characteristics of real-time monitoring. This aerial remote sensing platform with real-time satellite communication capability mainly consists of aircraft platform, ground observation system, wireless communication system, ground satellite information receiving station, data processing and application analysis system, etc. (Figure 1). This design is based on aircraft as a platform, using remote sensing equipment installed in it to obtain raw data from natural disaster areas. Within the effective range, ground information can be effectively collected by adjusting the flight altitude to form images or other format data materials. The information is transmitted in real time to the geostationary satellite through the onboard satellite base station, and then sent to the designated ground fixed base station, so that global synchronous and uninterrupted data transmission can be achieved. Accordingly, the obtained on-site data of the disaster area can be transmitted in real time to the command center to formulate disaster response plans in the shortest possible time. Due to the real-time uplink and downlink communication capability of satellite transmission, the command center can also command aerial platforms to adjust their routes in a timely manner, conduct mobile investigations of disaster sites, and achieve real-time monitoring and emergency response in the event of natural disasters.

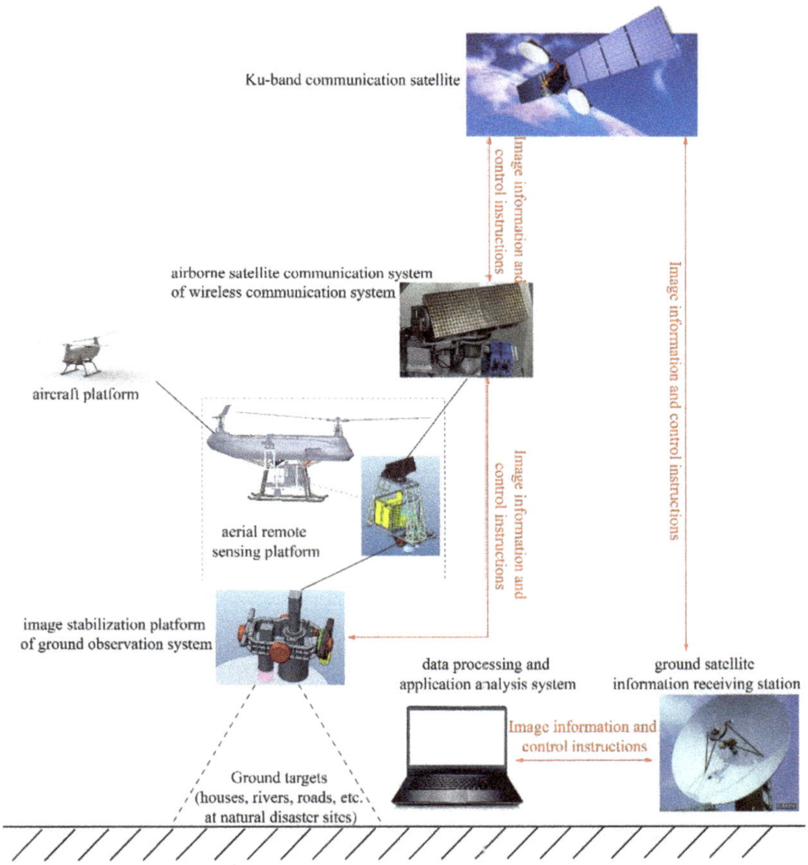

Figure 1. Composition diagram of the aerial remote sensing platform based on real-time satellite communication.

In the aerial remote sensing platform, the aircraft is the heavy-duty twin-rotor unmanned aerial vehicle, Camel 500, produced by Chongqing Tuohang Technology Co., Ltd., Chongqing, China This aircraft has a maximum takeoff weight of 500 kg and a load capacity of 50% (with the highest load ratio of unmanned aerial vehicles of the same class). It has a transportation capacity of flying with a load of 200 kg for 1 h, about 100 km, or a load of 150 kg for 3 h, about 300 km. This aircraft has a large loading space, multiple hanging points, rich delivery methods, and good adaptability to the center of gravity, which is much better than conventional helicopters. The ground satellite information receiving station and data processing and application analysis system are mature equipment, which will not be detailed in this paper.

The ground observation system is a combination of an orthographic image stabilization platform and a visible light camera loaded on it. The image stabilization platform is a mechanical image stabilization device. Its basic principle is to detect the shaking of the camera platform through inertial devices such as gyroscope sensors, and then adjust the servo system to isolate the external shaking of the photoelectric equipment, achieving the goal of stabilizing the image. Due to the equipment base fixed on the aircraft shaking with the turbulence of the aircraft, if there is no stabilization platform, the camera equipment will shake accordingly. In cases of significant shaking, this can seriously affect the image stitching effect and the shooting effect of photoelectric devices with longer exposure times. The airborne orthographic stabilization platform is generally installed at the belly position of the aircraft. Drill holes in the lower part of the camera and install a visible transparent cover. By controlling the self-stability of the frame in both pitch and roll directions, it ensures that the optical device lens loaded on the frame is always perpendicular to the ground plane during flight and perceives and collects data (mainly images) from ground targets. Optical systems have high resolution performance, so high-precision ground information can be obtained. Based on the spatial stability of the shooting of camera, multiple frames of images can be effectively concatenated to obtain ground remote sensing information for larger areas. The designed technical indicators of the image stabilization platform (Table 1) include external dimensions, working temperatures, total weight, effective load (camera and lens), operation range in both pitch and roll states, maximum angular velocity and acceleration in the search state, and stability accuracy within the swing range.

Table 1. Technical indicators of the orthographic stabilization platform.

Technical Indicators	States	Indicator Values
Operation range	Pitch	−10°–+10°
	Roll	−10°–+10°
Maximum angular velocity	Pitch	20°/s
	Roll	20°/s
Maximum angular acceleration	Pitch	$20°/s^2$
	Roll	$20°/s^2$
Stability accuracy		≤0.05°
Effective load (camera and lens)		≤8 kg
Maximum external dimensions		600 mm × 300 mm × 800 mm
Total platform weight (including load)		≤20 kg ± 20%
Working temperatures		−20 °C–+55 °C

The wireless communication system is composed of airborne mobile communication equipment. Through the mobile communication equipment, mobile carriers such as vehicles, ships, and airplanes can track satellites in real time during when they are moving, continuously transmitting multimedia information such as voice, data, and images, which can meet the needs of emergency communication for military and civilian under various mobile conditions. During the transmission process, the mobile communication equipment autonomously tracks satellites, fully utilizing the characteristics of large coverage area, strong anti-interference ability, and stable line of satellite communication. It can achieve communication functions of point-to-point, point-to- multipoint, and point-to-the main station of mobile satellites. The technical indicators of the mobile communication equipment in this design scheme are shown in Table 2.

Table 2. Technical indicators of the mobile communication equipment.

Technical Indicators	States	Indicator Values
Operating Frequency	Transmission	14 GHz–14.5 GHz
	Reception	12.250 GHz–12.750 GHz
Antenna Gain	Transmission	≥27 dBi
	Reception	≥27 dBi
Antenna main lobe beam width	Azimuth	≤2.5°
	Pitch	≤16.8°
First sidelobe		≤−11 dB
Polarization Type		Linear polarization
Polarization isolation (duplexer outlet)	Transmission	≥19 dB
	Reception	≥19 dB
Insertion loss of transmission branch (excluding radome)		≤1.5 dB
Port power capacity		≥200 W
Transmission and reception isolation		≥95 dB
Antenna motion range	Azimuth	±270° (continuous)
	Pitch	20°–80°
Polarization surface adjustment range		±90°
Maximum rotation speed of antenna	Azimuth	≥60°/s
	Pitch	≥60°/s
	Polarization	≥30°/s
Tracking method		Automatic tracking
Tracking Accuracy		≤0.3° (R.M.S)
Star capture time		≤10 s
Recapture time		≤6 s
Weight		≤27 kg
Reference source frequency stability		2×10^{-8}
Working temperature		−40 °C~60 °C
Storage temperature		−55 °C~70 °C
Working humidity		≤95% (+30 °C)
Standards for impact vibration		GJB367A [48]

3. Results and Analysis of the Ground Test

Due to the non-standard design of earth observation systems and wireless communication systems, the performance of products from different manufacturers varies greatly. Therefore, it is necessary to verify whether their performance meets the needs of real-time monitoring of natural disasters. The performance verification work is divided into two steps, namely the ground test and the flight test. The ground test involves simulating the flight environment through ground carriers to assess the performance of the orthographic stabilization platform and the transmission capability of the airborne mobile communication equipment.

3.1. *Performance Test of the Orthographic Stabilization Platform*

3.1.1. Analysis of Key Indicators

The purpose of the orthographic stabilization platform is to ensure that the optical axis of the visible light camera is always perpendicular to the earth's surface, ensuring clear imaging quality and accurate positioning of the target object. Therefore, its key indicator is stability accuracy.

In a stable state, the image stabilization platform should compensate for the swaying and undulating motion of the carrier body and maintain the spatial stability of the optical axis. Considering the typical indicators of the aircraft (flight altitude of 3000 m, speed less than 300 km/h, fuselage roll of ±10 °, cycle of 7 s), in order to meet the task of detecting ground targets and effectively concatenate the photos taken by the camera, the optical axis fluctuation must be limited to a certain range to ensure that the overlapping area of adjacent two frames of images is not less than 1/4. Considering the aircraft's speed and photography frequency (one frame of image per second), the aircraft's travel per second is 83.3 m. If the overlap area of the two frames of images is not less than 1/4, the coverage distance of each frame of image in the flight direction is 111 m. The corresponding field of view (FOV) angle and stability accuracy threshold can be expressed using the following formulas:

$$\alpha_{targ} = tg^{-1}(a/L) \tag{1}$$

$$\beta = \frac{1}{4}\alpha_{targ} \tag{2}$$

Among them, a is the coverage distance, L is the target distance, α_{targ} is the FOV angle, and β is the stable accuracy threshold. The stable accuracy calculation principle is shown in Figure 2. When the coverage distance is 111 m and the target distance is 3000 m, the FOV angle is 2.12°. Therefore, if the stable accuracy is less than 0.530°, the image stitching is effective.

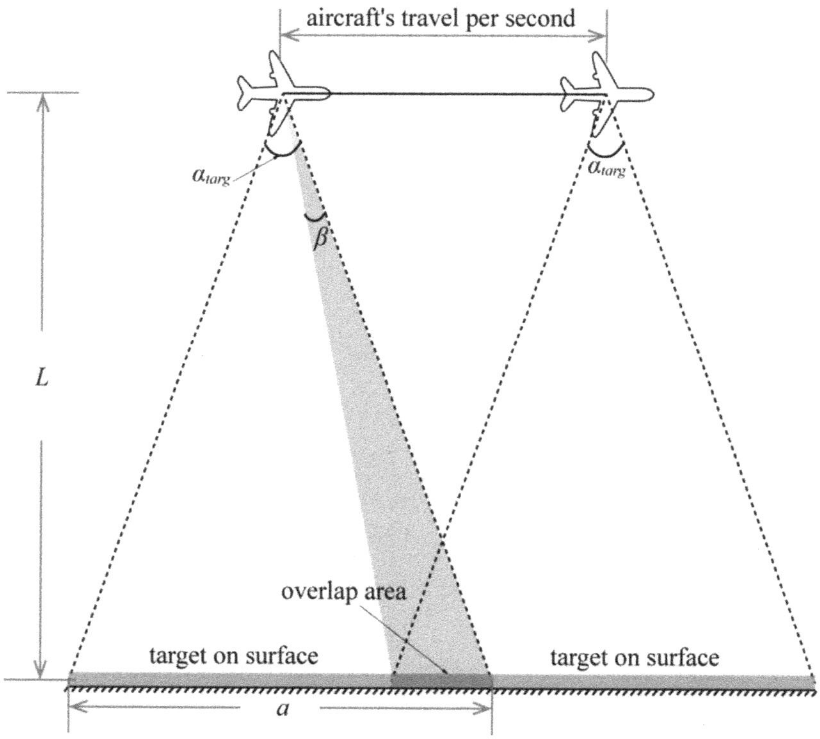

Figure 2. Schematic diagram of stable accuracy calculation principle. a is the coverage distance, L is the target distance, α_{targ} is the FOV angle, and β is the stable accuracy threshold.

3.1.2. Test Method and Results of the Stability Accuracy

The test of stability accuracy is achieved through optical path bending and an inclinometer scheme. The optical path turning test is designed with a two-axis manual swing table and a photoelectric bending device—that is, a 45° flat mirror is placed below the lens, the monitored target directly below it is converted into the target directly in front by bending the light path (Figure 3). The swing table is shaken with a swinging range and a swinging period, and the trajectory data of the target in front is recorded and imaged in the camera. Its peak is then taken as the stable accuracy.

On 8 September 2020, after setting up the optical path bending test conditions, different swing conditions were applied to the manual swing device. The test results are shown in Table 3. When the swinging period is 20 s and the amplitude is ±15 °, the stability accuracy is the best, only ±0.04°. When the swinging period is 15 s and the amplitude is ±15°, the stability accuracy is ±0.045°. When the swinging period is 7 s and the amplitude is ±10°, the stability accuracy is ±0.05°. All results are better than the calculated stability accuracy threshold of 0.535° mentioned above, which indicates that the stability accuracy test on ground under simulated airborne conditions meets the usage requirements.

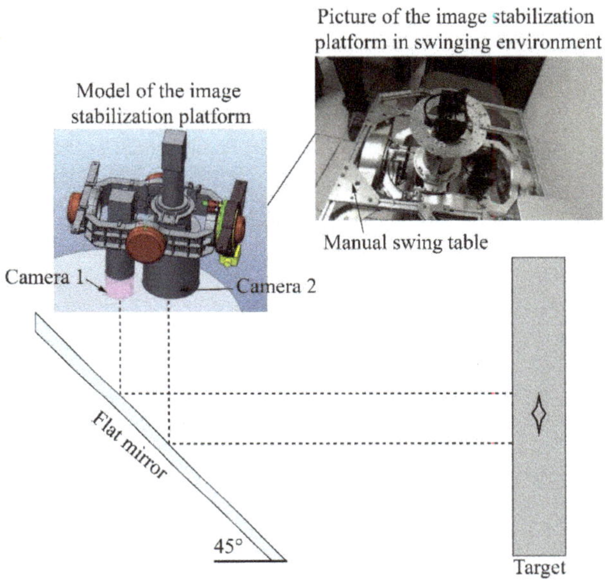

Figure 3. Layout diagram of the testing environment for the orthographic stabilization platform.

Table 3. Stability accuracy test results of the orthographic stabilization platform.

Swing Conditions		Stability Accuracy	
Swinging Period	**Amplitude**	**Optical Path Bending Test**	**Inclinometer Scheme**
20 s	±15°	±0.04°	±0.035°
15 s	±15°	±0.045°	±0.042°
7 s	±10°	±0.05°	±0.047°

The inclinometer scheme involves installing an inclinometer on the inner frame. When the swinging table shakes, the inclinometer data is collected and analyzed. This scheme is easy to operate. However, due to the poor performance of the inclinometer in dynamic environments, the inclinometer can only maintain accuracy in a stationary environment. When the equipment has high stability, the inclinometer on the stable surface is approximately in a stationary environment. At this time, the credibility of the test data is high, otherwise the data error is large. In general, the inclinometer scheme is a mutually influencing scheme. In the case of poor test results, the actual indicators of the equipment may not necessarily be worse than the test results. However, if the test results meet the requirements of the indicators, the actual indicators of the equipment must exceed the required indicators. On 8 September 2020, the load was equipped with an inclinometer and the manual swing device was subjected to the same swing conditions as the optical path bending test. The test results are shown in Table 3. When the swinging period is 20 s and the amplitude is ±15°, the stability accuracy is the best, only ±0.035°. When the swinging period is 15 s and the amplitude is ±15°, the stability accuracy is ±0.042°. When the swinging period is 7 s and the amplitude is ±10°, the stability accuracy is ±0.045°. All results are better than the results from optical path bending test, also indicating that its stability accuracy met the requirements of the usage conditions.

In the ground test, the swing environment of the orthographic stabilization platform fully compares the swing environment during the flight process, and a margin of over 50% is reserved for the motor capacity, which means that the stabilization platform can maintain stability accuracy in flight maneuvers that are more severe than the ground test environment. In addition, the components used in the orthographic stabilization platform are all wide temperature devices, and the machined parts have undergone surface anodizing or chrome plating treatment, which can adapt to external environments ranging from −30 °C to 65 °C. Moreover, the entire equipment was placed in a high and low temperature box for high and low temperature operation testing and assessment, and it was able to maintain normal working conditions, indicating that the equipment has the adaptability to the onboard environment.

3.2. Transmission Capability Test of the Airborne Communication System

3.2.1. Analysis of Key Indicators

The airborne communication system is used to achieve real-time communication between the flight carrier and the ground data processing center in the flight environment. The working link of the airborne communication system is shown in Figure 4, which operates in a bidirectional transmission mode of "aircraft–satellite–ground" and "ground–satellite–aircraft". In order to meet the data volume and speed requirements of real-time monitoring of natural disasters, airborne communication systems should have the following functions. First, the real-time transmission system has a bidirectional real-time transmission function. Second, the reliability of the servo system connecting to the satellite reaches over 98%, and the success rate of image transmission reaches over 80%. Third, the data transmission rate reaches 2 Mbps. Fourth, the data compression rate of the real-time transmission system can be selected between 2 and 128 times; fifth, the opening time of the real-time transmission system equipment shall not exceed 5 min (excluding aircraft takeoff time), and the data transmission delay time shall not exceed 5 s. Based on the above functions, the key indicators are transmission speed and transmission success rate. Only with high speed and success rate can real-time and data integrity requirements be met.

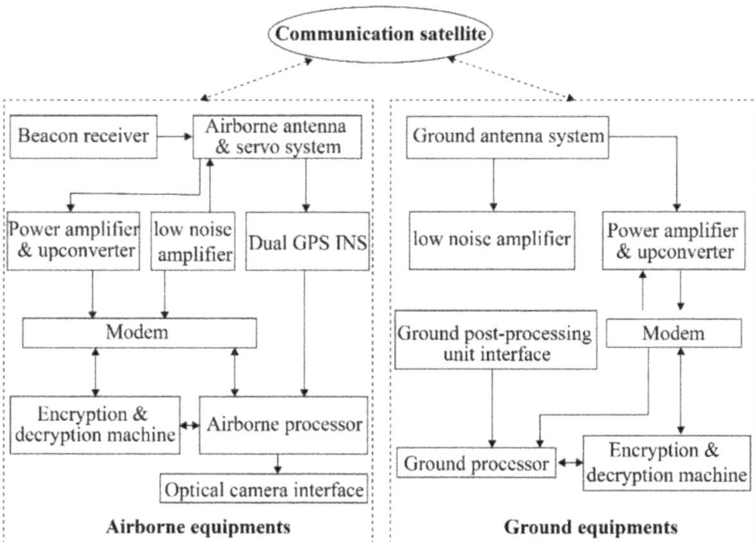

Figure 4. Schematic diagram of the composition of the "aircraft–satellite–ground" real-time transmission system.

3.2.2. Test Method and Results

In order to achieve test in the ground environment, a ground motion carrier is chosen for the communication system in the motion environment, approximating a bumpy flight state, and a ground simulation communication link is built. In this test, the manual pushing method is used in a suitable terrain, loading the communication antenna (Figure 5a). As shown in Figure 5a, the transmitting end is a flat antenna (with an equivalent diameter of 0.45 m) in the airborne mobile communication equipment. A parabolic antenna of 4.5 m is used to receive data in the ground station. The testing satellite is AsiaSat-5 communication satellite with a working frequency of 14,470 MHz. Static and dynamic tests are conducted at 256 Kbps, 512 Kbps, 1024 Kbps, and 2048 Kbps transmission bandwidths. The static test refers to the process where the onboard antenna is aligned with the satellite and no longer moves, while the dynamic test refers to the process of aligning the airborne antenna with the satellite and then moving the antenna base position along an 8-shaped trajectory on a site with a slight slope on the ground. During the testing process, the communication system will send the pre-stored image data to the processing center near the ground station via satellite. The processing center will then transmit the received images back to the communication system. Finally, the success rate of the first transmission is calculated.

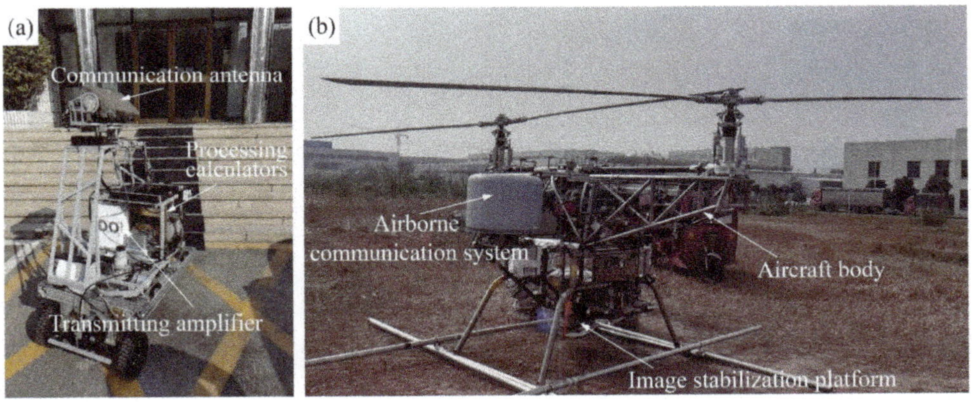

Figure 5. Photos of the testing systems. (**a**) The carrier of the mobile communication system in the ground test. (**b**) The system composition in the flight test.

On 20 September 2020, the communication capability test on ground was conducted in an open parking lot in Beijing with a slope of 20°. The road conditions were sandy and gravel, providing a certain shaking and bumpy environment for the test. The test results are shown in Table 4. As shown in the table, the transmission rate of the mobile communication equipment meets the requirement of 2 Mbps bandwidth. Under this bandwidth condition, the data packet loss rate is only 0.2% in the static environment, and data packet loss rate is 0.5% in the dynamic environment, both of which meet the requirements. Based on the use of image compression function (no less than 16 times), this indicates that 1080P video and high-resolution images of no less than 32 Mbps can be transmitted in real-time in the later stage.

In summary, the dynamic environment provided by the ground test conditions of the communication equipment is relatively harsh compared to the flight environment, with a pitch slope close to 20°, exceeding the typical aircraft sway amplitude of about 10°. At the same time, simulate the vibration of the aircraft fuselage through testing on sand and gravel road surfaces. Therefore, the experimental results in this environment have high reference value. In addition, the airborne communication system is driven using servo motors and leaves a margin of over 30% for motor capability, which means that the communication system can maintain tracking accuracy in flight maneuvers that are more severe than ground test environments. The components used in the communication system are also wide temperature devices, and the machined parts have undergone surface anodizing or chrome plating treatment, which can adapt to external environments ranging from −30 °C to 65 °C. The entire equipment was placed in a high and low temperature box for high and low temperature operation testing and assessment, and it can also maintain normal working conditions, indicating that the equipment can adapt to the airborne environment.

Table 4. Static and dynamic test data of the airborne communication system.

Transmission Bandwidths	Number of Sent Packets	Eb/No		Number of Received Packets		Number of Lost Packets		Packet Lost Rate (%)	
		Static	Dynamic	Static	Dynamic	Static	Dynamic	Static	Dynamic
256 Kbps	1000	11.0	11.1	999	997	1	3	0.1	0.3
512 Kbps	1000	10.1	9.8	1000	998	0	2	0	0.2
1024 Kbps	1000	7.2	7.1	1000	996	0	4	0	0.4
2048 Kbps	1000	4.3	4.2	998	995	2	10	0.2	0.5

4. Results and Analysis of the Flight Test

Flight test is an evaluation of the task completion ability of equipment with environmental adaptability working on a flight carrier after passing ground performance testing. The flight test was carried out on the premise of good performance of satellite communication equipment and orthographic stabilization platform by a heavy-duty twin rotor unmanned aerial vehicle-Camel 500, at the flight test site of unmanned aerial vehicle manufacturer Chongqing Camel Navigation Technology Co., Ltd. and on Nanpingba Island in Chongqing, China. The flight test site is relatively small, with a length of ~300 m and a width of ~300 m, so the test mainly focuses on hovering and small-scale flight for debugging of aircraft, platform equipment, and communication link. Nanpingba Island is located in the middle of the Yangtze River, with dense gullies on both sides. The narrow river channel—where the river flows rapidly and flood disasters, landslides, and mudslides are prone to occur—makes it a typical site for natural disasters. Therefore, it is used as the observation site for this flight test. Due to airspace control restrictions, this flight test site is limited to an island area of ~2000 m length and 1000 m width for turnaround flights to assess the reliability, stability, and real-time performance of the system.

4.1. Test Objects and Technical Indicators

The flight test also verifies whether the performance of ground observation systems and wireless communication systems meets the needs of real-time monitoring of natural disasters. In flight tests, the ground observation system suitable for natural disaster monitoring is a combination of orthographic stabilization platform and visible light cameras loaded on it, while the wireless communication system is composed of airborne communication equipment. The orthographic stabilization platform and airborne communication equipment are non-standard-design equipment, and their performance during the flight test to a certain extent reflects their ability to complete tasks. Therefore, this test mainly analyzes the performance of the orthographic stabilization platform and its optical load as well as the transmission capacity of the airborne communication equipment in the flight environment.

The flight system includes aircraft, communication equipment, and orthographic stabilization platform (Figure 5b). The technical indicators of the communication equipment and the orthographic stabilization platform in flight tests are the same as those of the ground test mentioned above. Due to the differences between the aircraft used in flight tests and the typical aircraft expected during ground testing, the key indicators of the stabilization platform vary due to changes in flight altitude, speed, and other parameters. Considering the index condition of Camel 500 (flight altitude of 200–500 m, speed less than 100 km/h, fuselage roll of ±10°, and cycle of 7 s) in flight tests, in order to meet the task of detecting ground targets and effectively splicing the photos taken by the camera, the optical axis must be limited to a certain range of fluctuations to ensure that the overlapping area of adjacent two frames of images is not less than 1/4. Considering the aircraft's speed and photography frequency of 1 s, the aircraft's travel per second is ~28 m. If the overlap area of the two frames of images is not less than 1/4, the coverage distance of each frame of images in the flight direction is ~37 m. When the target distance is 500 m, the corresponding FOV angle is calculated by Formula (1) to be 4.23°. According to Formula (2), if the stable accuracy is less than 1.05°, the effectiveness of image stitching can be ensured.

4.2. Flight Test Process

The working link of the airborne communication equipment is also shown in Figure 4, which is a bidirectional working mode. The communication link includes a mobile communication system, satellites, and satellite ground stations. The transmitting end is an airborne mobile communication antenna (equivalent diameter of 0.45 m) manufactured by China National Science and Technology Corporation and installed on the Camel 500 aircraft (Figure 5b). It can achieve a transmission rate of no less than 2 Mbps in the Ku band under propeller obstruction. The ground station uses a 4.5 m parabolic antenna for reception, located within the China Academy of Water Resources and Hydropower Sciences. The testing satellite is AsiaSat-5 communication satellite with a working frequency of 1419 MHz, for static and dynamic testing at a transmission bandwidth of 2048 Kbps.

The test work includes communication link debugging, aircraft installation matching, and quasi-static/small-scale hanging flight testing. On 16 January 2022, an uplink and downlink test was conducted on the ground station at the office area of the Chinese Academy of Water Resources and Hydropower Sciences, with a test bandwidth of 2 Mbps. The results show that its function is good. On 17 January 2022, a transmission and reception performance test was conducted at the site of Chongqing Tuohang Technology Co., Ltd. The test bandwidth was confirmed to be 2 Mbps and the transmission and reception were normal through self-transmit and self-receive. The installation and matching process of the aircraft took about 15 days. The Camel 500 aircraft was used as the main body and loaded under the fuselage. The total loaded weight is less than 150 kg. The installation-matching effect of the aircraft is shown in Figure 5b, with the communication equipment located on the right side of the aircraft and the stabilization platform located on the lower side of the aircraft.

The quasi-static hanging flight is conducted at the flight test site of Chongqing Tuohang Technology Co., Ltd. Through static hovering, slow turning, and small-scale flight, the imaging quality of the stabilization platform and the data transmission speed during the flight process are analyzed and the related parameters are adjusted in real time, providing technical preparation for subsequent typical field flight tests. Due to the lateral installation of the mobile communication system, when the flight direction is different, the space between the antenna and the satellite may be obstructed by metal propellers, resulting in a loss of some bandwidth. To test the impact of obstruction on communication bandwidth, the aircraft rotates 90° in the static process every 1 min to test the transmission effect. Small-scale flights are bounded by the airspace of the flight test site. During the flight, the aircraft tilts in both pitch and roll directions, with a tilt angle of no more than 10°, which tests the ability of the stabilization platform to maintain vertical shooting of the ground during maneuvering. During the testing process, the airborne communication system located in Chongqing sent the ground data of the flight test site captured by the industrial camera loaded on the image stabilization platform to the processing center near the Beijing ground station via satellite. The processing center stored the received images and performed image stitching and other tasks.

After completing flight tests at the test site and optimizing equipment performance, link connectivity, and bandwidth, a formal flight test was conducted on 18 January 2022 at a typical disaster site, Nanpingba Island. The communication bandwidth of the flight equipment was configured to be 2 Mbps, and the two flight heights were 250 m and 450 m, respectively. The flight is conducted within the airspace of Nanpingba Island, with a maneuvering speed of no more than 100 km/h. During the flight, the aircraft tilts to a certain extent in the pitch and roll directions, with a tilt angle of no more than 10°, to test the ability of the stabilization platform to maintain vertical shooting of the ground during maneuvering. During the testing process, the airborne communication system located in Chongqing sent the ground data of Nanpingba Island captured by the industrial camera loaded on the image stabilization platform to the processing center near the Beijing ground station via satellite. The processing center stored the received images and performed image stitching and other tasks.

4.3. Flight Test Results

The purpose of the flight test is to connect the real-time communication link of image information, assess the communication anti-interference ability and shooting vibration resistance of equipment in aircraft maneuvering state, and verify the engineering applicability of the system. Therefore, the main assessment includes imaging quality, image concatenability, and real-time data transmission.

In the flight test, the aircraft carried mobile communication equipment and image stabilization platform with a reliable working time of more than 2 h and conducted effective maneuvers at the flight test site. The results show that the image stabilization platform has good image stabilization ability and captured clear images in high-frequency fuselage vibration environments greater than 13 Hz. The communication link has a bandwidth of no less than 2 Mbps and a packet loss rate of no more than 1% in a 13 Hz propeller interference environment. In addition, the aircraft performed small-scale maneuvering flights, with periodic tilting vibrations applied in pitch and roll directions, and then the imaging quality was evaluated. The shooting object of the flight test site is weeds on the ground. Based on the actual size estimation of weeds, with clear leaf contours, the resolution of the image obtained in the flight altitude of 325.3 m is less than 1 cm (Figure 6), which meets the emergency observation requirements. Table 5 shows that the stability accuracy of the equipment under airborne conditions meets the usage requirements.

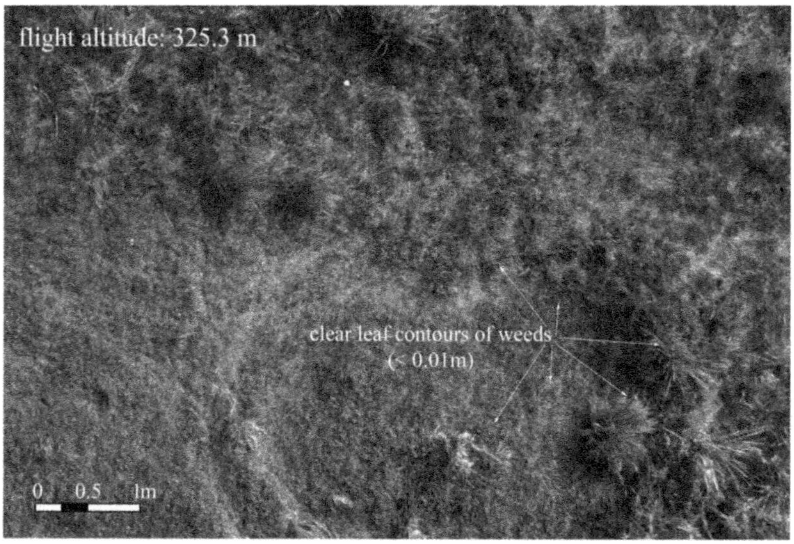

Figure 6. Ground weeds captured in the flight test site of Chongqing Tuohang Technology Co., Ltd.

During the flight test, satellite data transmission was carried out on the captured images of the image stabilization platform using the mobile Communication system. The aircraft hovered in four directions, with each direction separated by 90°, to analyze the impact of the relative positions of the antenna, fuselage, and satellite on communication quality. The ground receiving station located in Beijing received data and analyzed them. The results show that mobile communication equipment has a transmission rate of the bandwidth of 2 Mbps in 114 received packets, without data loss. Considering the use of image compression function (not less than 16 times), it can transmit 1080P real-time video and high-resolution images of not less than 32 Mbps in real-time in the later stage.

Table 5. Performance of image stabilization platform in low-speed small-scale maneuvering flights.

Swing Conditions			Image Quality	
Swinging Period	**Amplitude**	**Flight Height**	**Image Resolution**	**Overlap Rate**
20 s	±15°	300 m	<1 cm	>70%
15 s	±10°	300 m	<1 cm	>70%

The images collected during the Nanpingba Island flight test at altitudes of 250 m and 450 m were concatenated, and the results are shown in Figure 7a,d. Figure 7b,c are enlarged partial views of Figure 7a, from which the situation of farmland and small boats by the river can be clearly captured. Figure 7e is a frame from Figure 7d, from which the conditions of the houses and roads can be clearly seen. From the image quality and continuity of splicing, it can be seen that under the condition of 2 Mbps bandwidth, the packet loss rate is 0%, the transmission is very stable, the imaging and splicing effects are good, the ground buildings are clearly visible, and the landscapes on the ground can be effectively identified. These pieces of information are all of concern in disaster relief. After receiving image information, the Beijing ground station compared the timestamp of the image with the receiving time, with a time delay of no more than 2 s. The above data indicate that in the event of natural disasters, the aerial remote sensing combined with real-time satellite transmission can be used to emergency observation with mature and reliable engineering technology and application conditions.

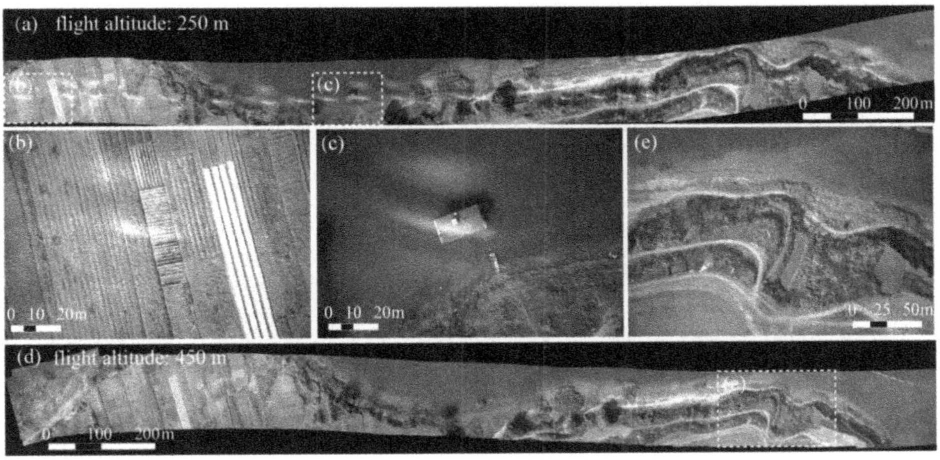

Figure 7. Image acquisition and stitching at flight altitudes of 250 m (**a**–**c**) and 450 m (**d**,**e**) during the Nanpingba Island flight test. (**b**,**c**) are enlarged partial views of (**a**), and (**e**) is an enlarged partial view of (**d**).

5. Discussion

In the event of a major natural disaster, providing fast, accurate, and effective means of emergency information collection and transmission is crucial for real-time emergency monitoring and rapid response for emergency rescue [6–8]. Aerial remote sensing has gradually become an irreplaceable technical means in modern spatial data acquisition due to its advantages of high spatial and temporal resolution, large scale, and fast mapping over a large area and has been widely applied in many fields [31]. Aerial remote sensing platforms can be divided into two categories based on whether they are piloted: manned and unmanned. In the Wenchuan earthquake in 2008 and the Ya'an earthquake in 2013, manned aircraft were involved in disaster assessment and post-disaster dynamic remote sensing monitoring [31]. However, the cost of this method is high, and it is not flexible enough. With the continuous advancement of science and technology and the further expansion of market demand, unmanned aerial vehicle remote sensing systems dominate in both quantity and application fields [49]. Unmanned aerial vehicles can be further classified into military and civilian, with many different sizes and models. The original aerial remote sensing platforms relied heavily on military aircraft or used small unmanned aerial vehicles. The purchase, maintenance, and operation costs of military aircraft are very high, and flight usually requires strict airspace management and approval, resulting in poor operational flexibility [50]. Small unmanned aerial vehicles have limited battery life and are suitable for short-term tasks [51]. They cannot conduct long-term monitoring and have low payload capacity. They can only carry lighter sensors and camera equipment, which limits the types and accuracy of data collection. Their flight altitude and range are limited, making it difficult to cover large areas. In the solution presented in this paper, we use the Camel 500 produced by Chongqing Tuohang Technology Co., Ltd., which is a civilian heavy-duty twin rotor unmanned aerial vehicle. The cost of using civilian heavy-duty unmanned aerial vehicles for aerial remote sensing platforms is lower than that of military aircraft. They can fly at lower altitudes, are suitable for a wider range of mission requirements, and can operate in areas that are not strictly restricted airspace. Moreover, their operational risks are relatively low, especially when used in densely populated and hazardous areas, making them safer. They also have a long battery life, suitable for long-term monitoring and data acquisition tasks, equipped with high-precision sensors and camera equipment that can obtain high-resolution and high-precision remote sensing data. In summary, civilian heavy-duty drones have significant advantages in terms of cost, flexibility, safety, and data quality [50,52,53].

In the operation of aerial remote sensing systems, although high-quality sensors are used, the image quality obtained is not ideal. The main factor limiting high-resolution imaging is often not caused by electronic or optical systems [54]. The aircraft carrier is difficult to maintain stability due to the influence of gusts, turbulence, etc., which in turn makes it difficult to maintain the stability of the imaging payload visual axis, resulting in degradation of imaging quality [42]. The inertial stabilization platform is used to isolate the influence of aircraft carrier angular motion, vibration, etc., on the load visual axis and is the foundation and guarantee of high-precision aerial remote sensing imaging [55,56]. The orthographic image stabilization platform in our solution is independently developed and designed. Based on the calculation principle (Figure 2), stability accuracy, that is, angular position error less than 0.53°, can meet the requirements of image overlay and stitching. In the ground test, the stability accuracy of the stabilization platform is less than 0.05° (Table 3), far better than the principle's threshold. Zhou et al. (2019) adopt parameter optimized adaptive composite control combining single neuron with proportional-integral-derivative (PID) to significantly improve the stability accuracy of a stabilization platform, with an angular position error of 0.2904° under dynamic base conditions, which was 39.9% higher than traditional PID control [42]. Nevertheless, the angular position error of the stable platform in our solution is smaller. Additionally, in flight testing, the stabilization platform demonstrated excellent image stabilization capability in high-frequency fuselage vibration environments greater than 13 Hz. The images captured during the test are clear, with a resolution of less than 1 cm and an overlap rate of over 70%. This resolution is much higher than that of satellite remote sensing platforms. For example, the resolution of Worldview-4 satellite can only reach 0.31 m, which is the highest resolution of civilian optical remote sensing data currently available [18]. Generally, the ultra-high-resolution data acquired using aerial remote sensing platforms can reach up to centimeter levels [26]. Bagheri (2017) developed a high-resolution aerial remote-sensing system for precision agriculture and the spatial resolution of the imagery also just was 3.6–95 mm [57]. It can be seen that the image resolution obtained in our testing is comparable to or even better than the current level of aerial remote sensing platforms.

With the continuous improvement of emergency management systems, the requirements for post disaster emergency response to natural disasters such as major earthquakes and geological hazards are becoming increasingly high. After a major disaster occurs, it is urgent not only to quickly obtain information about the disaster area but also to transmit it in real time to decision-making departments to determine emergency rescue plans [8]. Therefore, the speed and accuracy of data transmission are crucial for aerial remote sensing platforms to perform emergency response tasks. These emerging applications and the associate need for transmitting at higher data rate or performing more complex tasks, have raised the need for larger bandwidths and higher-frequency bands [45]. In recent years, the use of higher frequencies than the common VHF/UHF bands such as S-band and X-band has become more widely. Downlinks on S-band would be expected to be able to implement data rates from 100 Kbps to 1 Mbps [45]. In our solution, higher frequency bands such as Ku are used. The testing results show the transmission rate of the mobile communication system meets the requirement of 2 Mbps bandwidth. Under these bandwidth conditions, the packet loss rate in static environments is only 0.2%, while in dynamic environments it is 0.5% (Table 4), with a latency of no more than 2 s. Zulkifley et al. (2021) proposed that for payload communication, drones need to send back real-time telemetry data, images, or videos under remote monitoring, which results in drones requiring approximately 4 Mbps of data throughput under certain settings to achieve 1080P quality streaming video resolution [58]. In our solution, we consider the use of image compression function (not less than 16×), which can transmit 1080P real-time video and high-resolution images of not less than 32 M in real time in the later stage. From this, it can be seen that our solution can fully meet emergency response needs.

So far, disaster management based on aerial remote sensing has been commonly used for building damage assessment after earthquakes. In this assessment, due to its ability to obtain on-site data and process it as information on the area, quantity, rate, and type of damage, rescue teams will have a better understanding of the safe path and potential damage caused by secondary impacts [26]. In addition, it is also used for landslide dynamic monitoring [14,35] and change detection of coastal facilities to help assess the vulnerability caused by natural disasters such as tornadoes [59,60]. However, these applications do not require high real-time performance, but for natural disaster emergency response, fast information acquisition and transmission are required. With the rise of low-cost and high-payload commercial drones in recent years, as well as the maturity of airborne mobile communication equipment technology in the Ku- and KA-bands [45], aerial remote sensing platforms are more maturely applied to natural disaster emergency rescue. For example, during the rainstorm disaster in Henan Province, China on 20–21 July 2021, the "pterosaur-2H" emergency relief UAV produced by Aviation Industry Corporation of China, Ltd., Beijing, China operated for about 8 h in the disaster area. The aerial camera, optoelectronic equipment, and synthetic aperture radar were used to take pictures and monitor the disaster area, and relevant information was sent back to the command center in real time to realize efficient and accurate command of emergency rescue operations (https://mp.weixin.qq.com/s/pk0JGhBuyyrR9a-y1hr65w, accessed on 25 July 2021). According to the test results, it is believed that the aerial remote sensing platform based on real-time satellite communication proposed in this paper will be successfully applied to natural disaster rescue in the future.

6. Conclusions and Prospect

In order to respond to precise and rapid disaster emergency needs, we have proposed a technical solution for emergency monitoring based on an aerial remote sensing platform and verified its feasibility through testing. The conclusions are as follows:

(1) In this study, we designed an aerial remote sensing platform based on real-time satellite communication, which mainly includes aircraft, ground observation system, wire-

less communication system, ground satellite information receiving station, and data processing and application analysis system. Among them, the aircraft platform is the heavy-duty twin rotor unmanned aerial vehicle, Camel 500, produced by Chongqing Tuohang Technology Co., Ltd. The ground observation system is a combination of an orthographic stabilization platform and a visible light camera loaded on it. The wireless communication system is composed of a mobile communication system using the Ku-band.

(2) In the ground test, under different swinging conditions, the stability accuracies of the orthographic stabilization platform are less than 0.05°, which is better than the theoretical threshold of stability accuracy. The transmission rate of the mobile communication system meets the requirement of 2 Mbps bandwidth. Under these bandwidth conditions, the data packet loss rates are only 0.2% in the static environment and 0.5% in the dynamic environment, both of which meet the requirements.

(3) In the flight test, the orthographic stabilization platform showed good image stabilization ability in high-frequency fuselage vibration environments greater than 13 Hz. The images captured in the test are clear, with a resolution of less than 1 cm, an overlap rate of more than 70%, a transmission rate bandwidth of no less than 2 Mbps, a packet loss rate of no more than 1%, and a time delay of no more than 2 s, all of which meet the emergency observation requirements.

(4) The research results indicate that aerial remote sensing platforms based on real-time satellite communication have enormous potential in natural disaster emergency response. They can not only provide high-quality remote sensing data support but also achieve fast and stable communication with ground command centers. However, further research is needed on the performance of image stabilization platforms in complex flight environments. In the future, as the stability and reliability of satellite communication technology continue to improve and the equipment and technology of aerial remote sensing platforms continue to innovate and improve, the platform will adapt to the needs of different types of natural disaster prevention, reduction, and relief.

Author Contributions: Conceptualization, X.H. and C.X.; methodology, S.T.; validation, X.H., Y.H. and W.Q.; formal analysis, Z.X.; investigation, X.H. and Y.H.; data curation, X.H. and C.X.; writing—original draft preparation, X.H.; writing—review and editing, X.H., C.X., S.T., Y.H., W.Q. and Z.X.; visualization, X.H. and S.T.; funding acquisition, X.H. and C.X. All authors have read and agreed to the published version of the manuscript.

Funding: This research was funded by the National Natural Science Foundation of China (grant number 42077259) and the Foundation of National Institute of Natural Hazards, Ministry of Emergency Management of China (grant number 2023-JBKY-57; ZDJ2020-10).

Data Availability Statement: The data and materials are available and transparent. The original contributions presented in the study are included in the article. Further inquiries can be directed to the corresponding author.

Acknowledgments: We appreciate editors and reviewers for their helpful and professional comments on the manuscript.

Conflicts of Interest: The authors declare no conflicts of interest.

References

1. Ali, R.; Kuriqi, A.; Kisi, O. Human–environment natural disasters interconnection in China: A review. *Climate* **2020**, *8*, 48. [CrossRef]
2. Mata-Lima, H.; Alvino-Borba, A.; Pinheiro, A.; Mata-Lima, A.; Almeida, J.A. Impacts of natural disasters on environmental and socio-economic systems: What makes the difference? *Ambiente Soc.* **2013**, *16*, 45–64. [CrossRef]
3. Xu, C.; Xu, X.; Yao, X.; Dai, F. Three (nearly) complete inventories of landslides triggered by the May 12, 2008 Wenchuan Mw 7.9 earthquake of China and their spatial distribution statistical analysis. *Landslides* **2014**, *11*, 441–461. [CrossRef]
4. Ma, X.; Chen, W.; He, S.; Gao, A.; Zhu, R.; Liu, A. Assumption of application research of field communication system for seismic emergency based on BeiDou Navigation Satellite System. *Prog. Earthq. Sci.* **2020**, *50*, 1–7.
5. Ming, H. Discussion on Emergency Communication Support Plan for Disaster Site Rescue. *Smart City Appl.* **2023**, *6*, 28–30. [CrossRef]

6. Damaševičius, R.; Bacanin, N.; Misra, S. From sensors to safety: Internet of Emergency Services (IoES) for emergency response and disaster management. *J. Sens. Actuator Netw.* **2023**, *12*, 41. [CrossRef]
7. Zhou, L.; Wu, X.; Xu, Z.; Fujita, H. Emergency decision making for natural disasters: An overview. *Int. J. Disaster Risk Reduct.* **2018**, *27*, 567–576. [CrossRef]
8. Chen, K.; Fang, C.; Li, R. Research on airborne remote sensing emergency flight organization strategy in natural disaster. *Remote Sens. Land Resour.* **2018**, *30*, 55–59. [CrossRef]
9. Van Westen, C. Remote sensing for natural disaster management. *Int. Arch. Photogramm. Remote Sens.* **2000**, *33*, 1609–1617.
10. Bello, O.M.; Aina, Y.A. Satellite remote sensing as a tool in disaster management and sustainable development: Towards a synergistic approach. *Procedia-Soc. Behav. Sci.* **2014**, *120*, 365–373. [CrossRef]
11. Dong, X.; Deng, B.; Yuan, F.; Fu, X.; Zhang, W.; Ju, Y.; Ren, X. Application of Aerial Remote Sensing in Geological Hazards: Current Situation and Prospects. *Geomat. Inf. Sci. Wuhan Univ.* **2023**, *48*, 1897–1913. [CrossRef]
12. Wang, S.; Tang, C.; He, J.; Zhang, W.; Fang, Q.; Chen, X. Use of unmanned aerial vehicle for precise investigation of geological hazard in strong seismic zone. *J. Eng. Geol.* **2016**, *24*, 713–719. [CrossRef]
13. Li, Y.; Xu, L.; Chen, Y.; Deng, Z. Intergrated Space-Air-Train-Ground muti-source techniques for early detection of subgrade disasters and service status of railway subgrade. *Geomat. Inf. Sci. Wuhan Univ.* **2024**, *49*, 1392–1406. [CrossRef]
14. Lucieer, A.; de Jong, S.M.; Turner, D. Mapping landslide displacements using Structure from Motion (SfM) and image correlation of multi-temporal UAV photography. *Prog. Phys. Geogr.* **2014**, *38*, 97–116. [CrossRef]
15. Dong, L.; Shan, J. A comprehensive review of earthquake-induced building damage detection with remote sensing techniques. *ISPRS J. Photogramm. Remote Sens.* **2013**, *84*, 85–99. [CrossRef]
16. Joyce, K.E.; Belliss, S.E.; Samsonov, S.V.; McNeill, S.J.; Glassey, P.J. A review of the status of satellite remote sensing and image processing techniques for mapping natural hazards and disasters. *Prog. Phys. Geogr.* **2009**, *33*, 183–207. [CrossRef]
17. Kaku, K. Satellite remote sensing for disaster management support: A holistic and staged approach based on case studies in Sentinel Asia. *Int. J. Disaster Risk Reduct.* **2019**, *33*, 417–432. [CrossRef]
18. Sefercik, U.G.; Alkan, M.; Jacobsen, K.; Atalay, C.; Buyuksalih, G. Quality analysis of Worldview-4 DSMs generated by least squares matching and semiglobal matching. *J. Appl. Remote Sens.* **2021**, *15*, 034515. [CrossRef]
19. Chen, L.; Letu, H.; Fan, M.; Shang, H.; Tao, J.; Wu, L.; Zhang, Y.; Yu, C.; Gu, J.; Zhang, N. An introduction to the Chinese high-resolution Earth observation system: Gaofen-1~7 civilian satellites. *J. Remote Sens.* **2022**, 9769536. [CrossRef]
20. Lu, M.; Liao, X.; Yue, H.; Huang, S.; Xu, C.; Lu, H.; Bai, Y. Determining the distribution of unmanned aerial vehicles airports for the emergency monitoring of floods in China. *J. Geo-Inf. Sci.* **2019**, *21*, 854–864. [CrossRef]
21. He, J.; Tang, C.; Wang, S.; Zhang, K. Application of UAV images to geological disaster investigation. *Eng. Surv. Mapp.* **2017**, *26*, 40–45. [CrossRef]
22. Niethammer, U.; James, M.; Rothmund, S.; Travelletti, J.; Joswig, M. UAV-based remote sensing of the Super-Sauze landslide: Evaluation and results. *Eng. Geol.* **2012**, *128*, 2–11. [CrossRef]
23. Bendea, H.; Boccardo, P.; Dequal, S.; Giulio Tonolo, F.; Marenchino, D.; Piras, M. Low cost UAV for post-disaster assessment. *Int. Arch. Photogramm. Remote Sens. Spat. Inf. Sci.* **2008**, *37*, 1373–1379.
24. Boccardo, P.; Chiabrando, F.; Dutto, F.; Giulio Tonolo, F.; Lingua, A. UAV deployment exercise for mapping purposes: Evaluation of emergency response applications. *Sensors* **2015**, *15*, 15717–15737. [CrossRef] [PubMed]
25. Xu, Z.; Yang, J.; Peng, C.; Wu, Y.; Jiang, X.; Li, R.; Zheng, Y.; Gao, Y.; Liu, S.; Tian, B. Development of an UAS for post-earthquake disaster surveying and its application in Ms7. 0 Lushan Earthquake, Sichuan, China. *Comput. Geosci.* **2014**, *68*, 22–30. [CrossRef]
26. Yao, H.; Qin, R.; Chen, X. Unmanned aerial vehicle for remote sensing applications—A review. *Remote Sens.* **2019**, *11*, 1443. [CrossRef]
27. Fan, X.; Xu, Q.; Alonso-Rodriguez, A.; Subramanian, S.S.; Li, W.; Zheng, G.; Dong, X.; Huang, R. Successive landsliding and damming of the Jinsha River in eastern Tibet, China: Prime investigation, early warning, and emergency response. *Landslides* **2019**, *16*, 1003–1020. [CrossRef]
28. Chen, G.; Qiang, X.; Peng, S.; Sun, T.; Bai, J.; Ye, Z. Application research of UAV photogrammetry technology in the emergency rescue of Baige landslide. *J. Catastrophology* **2020**, *35*, 203–210. [CrossRef]
29. Zheng, S.; Li, Z. Geohazard Monitoring Based on Tilt Photography. *Bull. Surv. Mapp.* **2018**, 88–92. [CrossRef]
30. Lei, T.; Zhang, Y.; Wang, X.; Li, L.; Pang, Z.; Zhang, X.; Kan, G. The application of unmanned aerial vehicle remote sensing for monitoring secondary geological disasters after earthquakes. In Proceedings of the Ninth International Conference on Digital Image Processing (ICDIP 2017), Hong Kong, China, 19–22 May 2017. [CrossRef]
31. Xu, Q.; Guo, C.; Dong, X. Application status and prospect of aerial remote sensing technology for geohazards. *Acta Geod. Et Cartogr. Sin.* **2022**, *51*, 2020–2033. [CrossRef]
32. Zhang, Z.; Guo, D.; Ke, T.; Wang, J. The use of aerial photogrammetry in the fast response for China earthquake rescue. *J. Remote Sens.* **2008**, *12*, 852–857.
33. Li, Q.; Zhang, W.; Yi, Y. An information extraction method of in 2017 Jiuzhaigou earthquake-induced landslide: A case study of the Jiuzhaigou earthquake in 2017. *J. Univ. Chin. Acad. Sci.* **2020**, *37*, 93–102. [CrossRef]

34. Liang, R.; Dai, K.; Shi, X.; Guo, B.; Dong, X.; Liang, F.; Tomás, R.; Wen, N.; Fan, X. Automated mapping of ms 7.0 jiuzhaigou earthquake (China) post-disaster landslides based on high-resolution UAV imagery. *Remote Sens.* **2021**, *13*, 1330. [CrossRef]
35. Turner, D.; Lucieer, A.; De Jong, S. Time series analysis of landslide dynamics using an Unmanned Aerial Vehicle (UAV). *Remote Sens.* **2015**, *7*, 1736–1757. [CrossRef]
36. Fan, X.; Xu, Q.; Scaringi, G.; Dai, L.; Li, W.; Dong, X.; Zhu, X.; Pei, X.; Dai, K.; Havenith, H.-B. Failure mechanism and kinematics of the deadly June 24th 2017 Xinmo landslide, Maoxian, Sichuan, China. *Landslides* **2017**, *14*, 2129–2146. [CrossRef]
37. Fan, X.; Zhan, W.; Dong, X.; van Westen, C.; Xu, Q.; Dai, L.; Yang, Q.; Huang, R.; Havenith, H.-B. Analyzing successive landslide dam formation by different triggering mechanisms: The case of the Tangjiawan landslide, Sichuan, China. *Eng. Geol.* **2018**, *243*, 128–144. [CrossRef]
38. Ma, S.; Wei, J.; Xu, C.; Shao, X.; Xu, S.; Chai, S.; Cui, Y. UAV survey and numerical modeling of loess landslides: An example from Zaoling, southern Shanxi Province, China. *Nat. Hazards* **2020**, *104*, 1125–1140. [CrossRef]
39. Chen, T.; Hu, Z.; Wei, L.; Hu, S. Data processing and landslide information extraction based on UAV remote sensing. *J. Geo-Inf. Sci.* **2017**, *19*, 692–701.
40. Jin, W.; Ge, H.; Du, H.; Xu, X. A review on unmanned aerial vehicle remote sensing and its application. *Remote Sens. Inf.* **2009**, *1*, 88–92. [CrossRef]
41. Travelletti, J.; Delacourt, C.; Allemand, P.; Malet, J.-P.; Schmittbuhl, J.; Toussaint, R.; Bastard, M. Correlation of multi-temporal ground-based optical images for landslide monitoring: Application, potential and limitations. *ISPRS J. Photogramm. Remote Sens.* **2012**, *70*, 39–55. [CrossRef]
42. Zhou, X.; Shi, Y. Single neuron/PID adaptive compound control and parameter optimization for the inertially stabilized platform. *Chin. J. Sci. Instrum.* **2019**, *40*, 189–196. [CrossRef]
43. Skinnemoen, H. UAV & satellite communications live mission-critical visual data. In Proceedings of the 2014 IEEE International Conference on Aerospace Electronics and Remote Sensing Technology, Yogyakarta, Indonesia, 13–14 November 2014. [CrossRef]
44. Deng, F.; Dou, A.; Wu, W.; Chen, Z.; Yuan, X. Rapid Investigation of Disaster Situation in Extreme Disaster Area ofJiu-zhaigou Earthquake in Sichuan Based on UAV Remote Sensing. *J. Catastrophology* **2018**, *33*, 210–215. [CrossRef]
45. Burleigh, S.C.; De Cola, T.; Morosi, S.; Jayousi, S.; Cianca, E.; Fuchs, C. From Connectivity to advanced internet services: A comprehensive review of small satellites communications and networks. *Wirel. Commun. Mob. Comput.* **2019**, *2019*, 6243505. [CrossRef]
46. Sandau, R.; Roeser, H.-P.; Valenzuela, A. *Small Satellite Missions for Earth Observation: New Developments and Trends*; Springer: Berlin/Heidelberg, Germany, 2010; Available online: https://link.springer.com/book/10.1007/978-3-642-03501-2 (accessed on 2 July 2024).
47. Cui, X.; Xi, X.; Zhang, Y. Design of high dynamic Ku/Ka dual-antenna civil aviation system. *Comput. Meas. Control* **2019**, *27*, 240–244. [CrossRef]
48. *GJB 367A-2001*; General Specifications for Military Communication Equipment. General Equipment Department of the Chinese People's Liberation Army, Chinese National Standards: Beijing, China, 2001.
49. Liao, X.; Xiao, Q.; Zhang, H. UAV remote sensing: Popularization and expand application development trend. *J. Remote Sens.* **2019**, *23*, 1046–1052. [CrossRef]
50. Watts, A.C.; Ambrosia, V.G.; Hinkley, E.A. Unmanned aircraft systems in remote sensing and scientific research: Classification and considerations of use. *Remote Sens.* **2012**, *4*, 1671–1692. [CrossRef]
51. Zhang, J.; Liu, F.; Wang, J. Review of the light-weighted and small UAV system for aerial photography and remote sensing. *Natl. Remote Sens. Bull.* **2021**, *25*, 708–724. [CrossRef]
52. Yin, W.; Wang, Y. Research on autonomous capability assessment technology for large civilian UAVs. In Proceedings of the 2021 7th International Conference on Mechanical Engineering and Automation Science (ICMEAS), Seoul, Republic of Korea, 28–30 October 2021. [CrossRef]
53. Li, D.; Li, M. Research Advance and Application Prospect of Unmanned Aerial Vehicle Remote Sensing System. *Geomat. Inf. Sci. Wuhan Univ.* **2014**, *39*, 505–513. [CrossRef]
54. Qian, Y.; Liang, W.; Gao, X. Design of Optoelectronic Platform for stable image on Airborne Reconnaissance. *Acta Photonica Sin.* **2009**, *38*, 2108–2111.
55. Zhou, X.; Gao, H.; Zhao, B.; Zhao, L. A GA-based parameters tuning method for an ADRC controller of ISP for aerial remote sensing applications. *ISA Trans.* **2018**, *81*, 318–328. [CrossRef] [PubMed]
56. Wang, H.G.; Williams, T.C. Strategic inertial navigation systems-high-accuracy inertially stabilized platforms for hostile environments. *IEEE Control Syst. Mag.* **2008**, *28*, 65–85. [CrossRef]
57. Bagheri, N. Development of a high-resolution aerial remote-sensing system for precision agriculture. *Int. J. Remote Sens.* **2017**, *38*, 2053–2065. [CrossRef]
58. Zulkifley, M.A.; Behjati, M.; Nordin, R.; Zakaria, M.S. Mobile network performance and technical feasibility of LTE-powered unmanned aerial vehicle. *Sensors* **2021**, *21*, 2848. [CrossRef] [PubMed]

59. Sturdivant, E.J.; Lentz, E.E.; Thieler, E.R.; Farris, A.S.; Weber, K.M.; Remsen, D.P.; Miner, S.; Henderson, R.E. UAS-SfM for coastal research: Geomorphic feature extraction and land cover classification from high-resolution elevation and optical imagery. *Remote Sens.* **2017**, *9*, 1020. [CrossRef]
60. Barlow, J.; Gilham, J.; Ibarra Cofrã, I. Kinematic analysis of sea cliff stability using UAV photogrammetry. *Int. J. Remote Sens.* **2017**, *38*, 2464–2479. [CrossRef]

applied sciences

MDPI

Article

The Influence of the Key Characteristics of Overburden Rock Structure on the Development Height of Water-Conducting Fracture in Yushenfu Coal Mine Area, China

Shijie Song [1,2,3,*], Hao Ruan [1,2,3], Jiangbo Wei [1,2,3], Ruilin Niu [1,2,3], Xing Cheng [1,2,3] and Baodeng Chen [1,2,3]

1 College of Geology and Environment, Xi'an University of Science and Technology, Xi'an 710054, China; 22209226108@stu.xust.edu.cn (H.R.); weijiangbo@xust.edu.cn (J.W.); 22209226109@stu.xust.edu.cn (R.N.); 22209226107@stu.xust.edu.cn (X.C.); 19408010319@stu.xust.edu.cn (B.C.)

2 Research Institute of Coal Green Mining Geology, Xi'an University of Science and Technology, Xi'an 710054, China

3 Key Laboratory of Geological Guarantee for Coal Green Development of Shaanxi Province, Xi'an 710054, China

* Correspondence: songshijie@xust.edu.cn

Abstract: The destruction of shallow aquifers by water-conducting fractures of overlying strata caused by underground coal mining is the most representative form of mining-induced damage in the Yushenfu mining area. It has become an important factor restricting the green mining of coal in the Yushenfu mining area and even the ecological protection and high-quality development of the middle reaches of the Yellow River. As the key scientific problem of water-preserved coal mining, the scientific understanding of the development law and main influencing factors of water-conducting fractures in overlying strata has attracted great attention. Taking the geological occurrence characteristics of the main coal seam in Yushenfu mining area as the prototype, 24 different types of numerical models are constructed with the key characteristics of the overburden structure, such as the number of layers of sandstone in the overburden (sand layer coefficient) and the thickness ratio of sandstone and mudstone in the overburden (sand–mud ratio), as the main variables. By means of numerical simulation experiment and theoretical calculation, combined with field measurement and comparison, the influence of the key characteristics of overburden structure on the development height of water-conducting fracture is studied and revealed. It is proposed that the effective area for the study area to achieve water-preserved coal mining by using the height-limited mining method must conform to the coal seam overburden structure characteristics of "sand–mud ratio 6:4 and sand layer coefficient less than 70%" and "sand–mud ratio 8:2 and sand layer coefficient less than 80%". The results not only enrich and deepen the research on the influence of geological factors and the law of controlling the development of water-flowing fractures in overlying strata, but also provide theoretical support for the precise protection of groundwater resources in the Yushenfu mining area in the middle reaches of the Yellow River.

Keywords: overburden rock structure; water-conducting fractures; sand–mud ratio; sand layer coefficient; water-preserved mining; theory of broken rock strata

1. Introduction

The energy endowment pattern of "lack of oil, less gas, and relatively rich coal" determines the ballast status of coal resources in China's energy security [1,2]. In 2022, China's total raw coal production will reach 4.56 billion tons, an increase of 9% [3]. Driven by China's "dual-carbon goal" strategy, the energy status of coal resources has gradually changed from the main energy source to the main energy source [4], and the shift of coal production centers to the west is becoming more and more obvious. The Yellow River Basin, especially the middle reaches, has become an important continuation of the westward shift of China's coal production center strategy due to its rich coal resources and superior mining conditions. At present, 9 of the 14 large coal bases planned and constructed in China are distributed in the Yellow River Basin, of which 6 are densely distributed in the middle reaches of the Yellow River [5]. By 2022, coal production in the Yellow River Basin will have reached 3.59 billion tons [6], accounting for 78.7% of the total coal production in the country. Among them, coal production in the middle reaches of the Yellow River has exceeded 1.77 billion tons, accounting for about 40% of the total coal production in the country [7]. However, the ecological environment in the middle reaches of the Yellow River is fragile, soil erosion is serious, and water resources are scarce [8]. The coal mining area in the middle reaches of the Yellow River, represented by the Yushenfu mining area, is also characterized by the coal–water spatial relationship of "upper and lower water coal" [9] and the ecological response characteristics of "underground shallow aquifer supporting the surface ecosystem" [10]. Coal mining often causes serious damage to groundwater resources and further deteriorates the regional ecological environment. According to the latest survey data [11,12], after large-scale mining in the Yushenfu mining area, the shallow groundwater level in the whole area decreased by 5–20 m, the number of springs decreased by 84%, and the flow of springs decreased by 70% compared with that before mining. The resulting surface river cutoff, vegetation degradation, soil desertification, and other phenomena are common. In October 2021, the "Outline of the Yellow River Basin Ecological Protection and High-quality Development Plan" issued by the Central Committee of the Communist Party of China and the State Council clearly requires that the whole Yellow River Basin should unswervingly implement the "four principles" of determining the city by water, determining the land by water, determining the people by water, and determining the production by water. The middle reaches of the Yellow River highlight the work of soil and water conservation and water resource protection [13]. Therefore, the contradiction between the large-scale mining of coal resources in the middle reaches of the Yellow River and the protection of water resources has become more acute under the constraints of national strategies.

Scientific understanding of the development law, degree, and main influencing factors of water-flowing fracture in overlying strata is the key to solving the sharp contradiction between coal resource exploitation and groundwater resource protection in the Yushenfu mining area. It has attracted the attention of relevant scholars at home and abroad and has achieved a number of valuable results. Because mining activities are the direct cause of the development of water-flowing fractures in the overlying strata of coal seams, the influence of mining factors on the development of water-flowing fractures in overlying strata of coal seams is first concerned. The relevant research results mainly focus on mining height, working face geometry, mining methods, and roof management methods. Wu Jianhong and colleagues [14] established two models of multivariate nonlinear regression and GA-BP neural network based on field-measured data and analyzed that mining thickness and inclined length of working face are the main controlling factors affecting the development height of water-conducting fractures. Palchik [15] revealed that the development height of water-flowing fracture is related to the thickness of coal seam mining through the combination of theoretical analysis and experiment. Singh and colleagues [16] used stratum monitoring technology to monitor the study area and obtained that the longwall mining method can effectively control the development height of a water-flowing fracture; Wang H. et al. [17] used the borehole television method to study the temporal and spatial evolution law of the water-conducting fractures generated during the mining process from the coal seam to the surface and obtained the multiple relationship between the development height of the water-conducting fractures and the mining height. Yin H. and colleagues [18] analyzed the influence of mining thickness and working width on the development height of water-flowing fracture and constructed a three-dimensional structure model of the development height of water-flowing fracture to visualize the spatial distribution of water-flowing fracture. Li Zhenhua and others [19] obtained that the development height of water-conducting fractures is proportional to the mining thickness by means of field measurement, indoor simulation, and theoretical analysis. With the development of a large number of monitoring works on the development of water-conducting fractures in coal seam overburden and the in-depth analysis of the measured data of the maximum development height, it is found that the development degree of water-conducting fractures in coal seam overburden will be completely different under the same mining factors and different geological factors. Therefore, the study of the influence of geological factors on the development of water-conducting fractures in overlying strata of coal seams has gradually become a research hotspot. At present, the relevant research results mainly focus on mining depth, faults, key strata, loose strata, the physical and mechanical properties of overlying strata, and so on. Ju Jinfeng and colleagues [20] believed that the mining depth is also the main controlling factor of the development height of water-conducting fractures; especially in the shallow coal seam mining area, it is not appropriate to use the "mining height multiple" to predict the height of water-conducting fractures. Huang Bingxiang and colleagues [21] used a similar material simulation test method to study the influence of small faults on the development height of mining-induced fractures in overlying strata and obtained that normal faults would increase the development height of water-conducting fractures, while reverse faults had little effect on the development height of water-conducting fractures. Wang F. et al. [22] established the mechanical model of the arch structure in the loose layer and deduced that the arch structure of the loose clay layer can effectively block the development of the water-conducting fracture zone and the formation mechanism of the arch structure. Wang Shuangming and colleagues [23] used numerical simulation software to study the influence of thick sandstone on the development characteristics and laws of water-conducting fractures and concluded that the position of thick sandstone is the main factor affecting the development height of water-conducting fractures. Based on the key stratum theory, Xu Jialin and colleagues [24,25] analyzed the influence of different key stratum positions on the development height of water-flowing fracture and determined the relationship between the critical height of the key stratum fracture and the mining height. Wang Xiaozhen and colleagues [26] believed that the development height of water-flowing fracture was affected by the key stratum structure and mining height at the same time, and under the control of the key stratum structure, the development height of water-flowing fracture showed a step-type mutation characteristic with the mining height. Wang Xu and colleagues [27] concluded through comparative analysis, multiple linear regression, multiple nonlinear regression, BP neural network, and other methods that the physical and mechanical properties of overlying rock have a great influence on the development height of water-flowing fracture and the greater the proportion of hard rock, the greater the development height of water-flowing fracture, and the overall logarithmic relationship. However, coal is a sedimentary mineral, and the layered structure of its overlying strata is not only a very significant and important geological factor but also a carrier for the upward propagation of underground mining activity effects [28]. Therefore, the shape structure and key characteristics of the overlying strata control the whole process of the development of the water-flowing fracture in the overlying strata of the coal seam and also determine the maximum development height of the water-flowing fracture. However, the current research results mainly focus on mining depth, faults, key strata, loose strata, the physical and mechanical properties of overlying strata, and so on. There are few reports on the influence of overburden structure on the development law of water-conducting fractures at home and abroad.

In view of this, this paper takes the Yushenfu mining area in the middle reaches of the Yellow River in China as the research area and takes the 2^{-2} main coal seams and overburden strata in the area as the geological prototype. The numerical simulation test and the theoretical calculation of rock fragmentation are used to study and reveal the influence of the key characteristics of the overburden structure on the development height of water-conducting fractures. It not only has important scientific value for enriching and deepening the research on the influence of geological factors and controlling the development of water-conducting fractures in mining overburdens but also has important practical significance for supporting the protection of groundwater resources in the Yushenfu mining area in the middle reaches of the Yellow River.

2. Overview of the Study Area

2.1. Characteristics of Natural Geographical Environment

The Yushenfu mining area is located in the transition zone between the Loess Plateau and the Mu Us Desert in northern Shaanxi, China (see Figure 1). The terrain is high in the northwest and low in the southeast. The landform in the area can be divided into two types: aeolian sand beach and loess gully. It belongs to the semi-arid continental monsoon climate in the temperate zone. The maximum temperature is 35.9 °C, the minimum temperature is −8.5 °C, the average temperature is 8.1 °C, the average annual rainfall is 402.7 mm, and the evaporation is 1753.8–1978.7 mm. The soil in this area is mainly loessial soil and aeolian sandy soil. The soil structure is loose, the corrosion resistance is poor, the soil erosion and desertification are serious, and the surface vegetation is scarce. The vegetation coverage in the study area is only about 45%, and it is dominated by artificial herbs, and the ecological environment is very fragile [29].

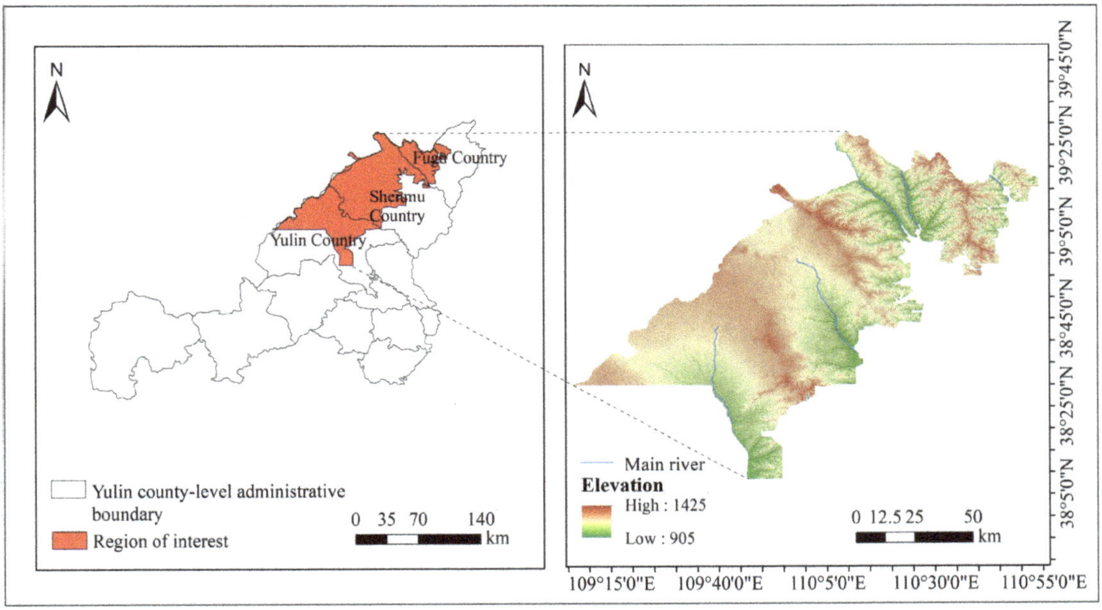

Figure 1. Location of study area.

2.2. Characteristics of Strata and Coal Seams

The main strata in the study area from old to new are the Upper Triassic Yongping Formation (T_3y), the Lower Jurassic Fuxian Formation (J_1f), the Middle Jurassic Yan'an Formation (J_2y), the Zhiluo Formation (J_2z), and the Anding Formation (J_2a). Lower Cretaceous Luohe Formation (K_1l), Neogene, and Quaternary. Yan'an Formation is the coal-bearing strata in the study area, generally containing 3–5 layers of mineable coal seams, of which 2^{-2} coal seams are the main coal seams. The buried depth is between 180 and 316.36 m, the average buried depth is 300 m, the thickness is between 0.26 and 12.16 m, the average thickness is 6.5 m, the dip angle is less than 3°, and it is nearly horizontal. A large number of boreholes in the study area revealed that: First of all, the macroscopic layered structure of 2^{-2} coal seam overburden is most widely distributed in the type of "sand layer–red soil layer–bedrock layer". Secondly, the overlying bedrock of 2^{-2} coal seam is mainly composed of sandstone and a small amount of mudstone interbedded structure, and there are two distinct sections of upper and lower characteristics, that is, the upper characteristic section is "sandy mudstone–fine sandstone" interbedded structure; the lower characteristic section is an "siltstone–fine grained sandstone" interbedded structure; the sand–mud ratio of the overlying bedrock of the 2^{-2} coal seam is generally 60~80%, while the sandstone is dominant. The number of sandstone layers is at least five layers, up to 35 layers, and the thickness is generally between 3 and 35 m. There are four layers of sandy mudstone, and the thickness is generally between 10 and 30 m. Thirdly, sandy clay or laterite is developed on the overlying bedrock of 2^{-2} coal seam, with an average thickness of 20 m, and is covered by modern aeolian sand with an average thickness of 30 m. The physical and hydraulic parameters of various rock and soil layers are shown in Table 1. There are aquifers in the upper and lower parts of the study area (see Figure 2), and the roof and floor of the coal seam are fine-grained sandstone and siltstone aquifuge. Under natural conditions, the aquifuge separates the coal seam from the aquifer, but there is a huge head difference under the condition of coal seam mining. The aquifuge will become a weak permeable layer or even a partially permeable layer, so that the groundwater in the aquifer will flow to the mine.

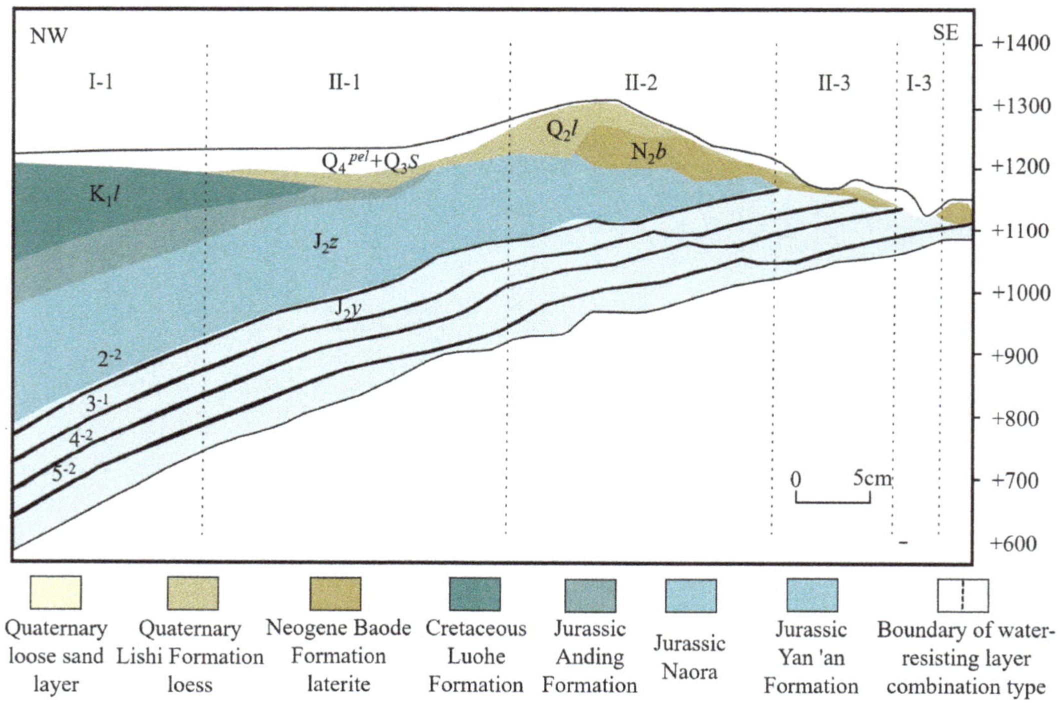

Figure 2. Hydrogeological profile of the study area.

Table 1. Physical and mechanical parameters of various rock and soil layers used in the numerical model.

Lithology	Saturated Uniaxial Compressive Strength/MPa	Elastic Modulus/MPa	Tensile Strength/MPa	Weight/KN/m^3	Internal Friction Angle/°	Poisson's Ratio	Cohesion/MPa	Porosity/%	Natural Moisture Content/Mpa	Permeability Coefficient
Sand–soil layer	0.72	23	0.03	17.2	37.0	0.31	0.07	0.38–0.47	11.9~17.3	4.58×10^{-4}
Red soil layer	0.85	69	0.20	18.6	37.2	0.30	0.08	0.42	17.4~18.7	3.27×10^{-6}
Sandy mudstone	30.4	2400	3.13	25.1	38.0	0.24	1.58	2.70	0.47	3.21×10^{-8}
Fine-grained sandstone	36.1	6100	2.50	26.1	41.7	0.20	5.33	0.8–10.2	0.10–2.03	2.13×10^{-7}
Silt sandstone	34.9	6300	2.45	25.5	37.7	0.21	6.06	0.4–9.34	0.22–2.26	4.32×10^{-6}
2^{-2} coal seam	22.1	5100	0.24	13.6	38.5	0.22	5.11	-	-	-
Bottom plate	80.7	35,000	1.86	26.8	43.0	0.29	48.22	-	-	-

3. Research Methods and Process

3.1. Numerical Simulation Test (FLAC^{3D}6.0)

The perspective and core purpose of this paper is to reveal the influence of the layered structure of coal seam overburden on the development of water-flowing fractures in the process of large deformation of mining overburden. FLAC^{3D}6.0 is a three-dimensional fast Lagrangian analysis numerical software of continuous medium mechanics developed by the ITASCA company in the United States (Minneapolis, MN, USA). It is a numerical analysis method based on a three-dimensional explicit finite difference. FLAC^{3D}6.0 software of the finite difference method has outstanding advantages in simulating large deformation of rock strata and calculating the efficiency of a three-dimensional numerical simulation test. It has good convenience in model modification and real-time visualization. In addition, the deformation and damage of overlying strata in the process of coal seam mining are mainly caused by the renewal and change of overlying strata stress caused by mining. After the excavation of underground coal seams, the stress of surrounding rock in goaf is redistributed and renewed, which leads to the pressure relief deformation of overlying strata. When the deformation reaches the limit state, damage will occur. The mining failure of overlying rock is mainly caused by the concentration of local shear stress and tensile stress of rock mass. The Mohr–Coulomb yield criterion in FLAC^{3D} considers both shear yield and tensile yield of materials, which can better simulate the shear and tensile failure characteristics of rock and soil. Based on the above factors, this paper uses the finite difference method FLAC^{3D} software to carry out the numerical simulation study on the development of an overburden mining water-flowing fractured zone.

3.1.1. Numerical Model Construction

Taking the buried depth of 2^{-2} coal seam in the study area, the macroscopic layered structure of the overburden, the thickness of the rock and soil layer, and the coal seam as the geological prototype, the numerical model is constructed in FLAC^{3D} software. Firstly, the geometric size parameters of the numerical model are designed. The total thickness of the model is 315 m, of which the thickness of the floor is 10 m, the thickness of the 2^{-2} coal seam is 5 m, the total thickness of the bedrock is 250 m, and the total thickness of the loose layer is 50 m (including the thickness of the red soil layer is 20 m, and the thickness of the sand layer is 30 m). The width of the model is 300 m and the length is 1000 m. It is used to simulate the width and annual advance distance of the 2^{-2} coal seam mining face in the study area. Secondly, according to the characteristics of the number of strata, lithology, and interbedded structure of the overlying bedrock section of the 2^{-2} coal seam in the study area, the bedrock layered structure is designed as the upper and lower sections. The upper section is the "sandy mudstone–fine sandstone" interbedded structure, which is fixedly divided into 4 groups of cycles, and the lower section is the "siltstone–fine sandstone" interbedded structure. The number of cycles is set according to the geological prototype and research needs. Thirdly, the constraints of each boundary of the numerical model are set, specifically: the left and right boundaries are set as single constraint boundaries, and $u = 0,\ v \neq 0,\ w \neq 0$ is taken. The front and rear boundaries are set as single constraint boundaries, taking $u \neq 0,\ v = 0,\ w \neq 0$. The bottom boundary is set as the fully constrained boundary, taking $u = 0,\ v = 0,\ w = 0$. The upper boundary is set as a free boundary and is not constrained, as shown in Figure 3.

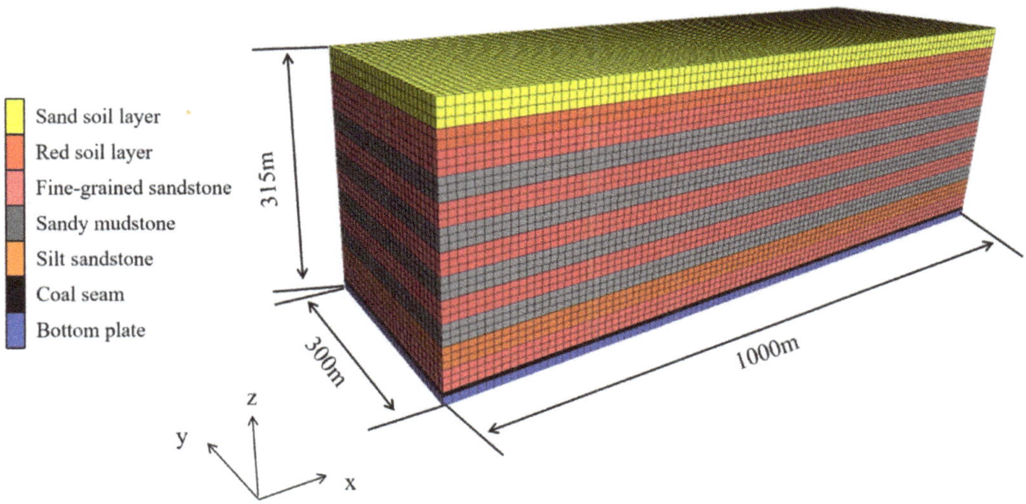

Figure 3. Three-dimensional geological model diagram.

3.1.2. Model Type Design

According to the research results of Song Shijie [28,30], the sand layer coefficient (defined as the ratio of the total number of sandstone layers to the total number of strata in the coal seam overburden) and the sand–mud ratio (defined as the ratio of the total thickness of sandstone to the total thickness of mudstone in the coal seam overburden) are the key characteristics of the coal seam overburden structure in the study area. Because this paper is to study the influence of the relationship between the number and thickness ratio of sandstone and mudstone in the overlying bedrock section on the water-flowing fracture rather than the morphological changes of the rock layer and the slope, the model is simplified according to the study area. The number of sandstone layers in the bedrock section of the numerical model is designed as 6 layers, 8 layers, 10 layers, 20 layers, 30 layers, and 40 layers. The number of sandy mudstone layers is designed as 4 layers, and the sand–mud ratios are designed as 6:4 and 8:2. The single layer thickness of fine sandstone and siltstone is treated as equal thickness, and the mining thickness is designed as 3 m and 5 m. After permutation and combination of the above variables, 24 different types of numerical models are obtained, as shown in Table 2.

Table 2. Types of numerical models and characteristics of overlying strata structure.

Numbering	Deep Mining/m	Height Mining/m	Loose Layer Characteristics		Bedrock Layered Structure Characteristics						
			Thickness of Sand Layer/m	Red Soil Layer Thickness/m	Total Number of Rock Strata/Layer	Sand–Mud Ratio	Sandstone Layer Coefficient/%	Sandstone		Sandy Mudstone	
								Number of Layers/Layer	Thickness /m	Number of Layers/Layer	Thickness/m
M1	300	5	30	20	10	6:4	60.0	6	25.00	4	25.00
M2	300	5	30	20	12	6:4	66.7	8	18.75	4	25.00
M3	300	5	30	20	14	6:4	71.4	10	15.00	4	25.00
M4	300	5	30	20	24	6:4	83.3	20	7.50	4	25.00
M5	300	5	30	20	34	6:4	88.2	30	5.00	4	25.00
M6	300	5	30	20	44	6:4	90.9	40	3.75	4	25.00
M7	300	3	30	20	10	6:4	60.0	6	25.00	4	25.00
M8	300	3	30	20	12	6:4	66.7	8	18.75	4	25.00
M9	300	3	30	20	14	6:4	71.4	10	15.00	4	25.00
M10	300	3	30	20	24	6:4	83.3	20	7.50	4	25.00
M11	300	3	30	20	34	6:4	88.2	30	5.00	4	25.00
M12	300	3	30	20	44	6:4	90.9	40	3.75	4	25.00
M13	300	5	30	20	10	8:2	60.0	6	33.33	4	12.50
M14	300	5	30	20	12	8:2	66.7	8	25.00	4	12.50
M15	300	5	30	20	14	8:2	71.4	10	20.00	4	12.50
M16	300	5	30	20	24	8:2	83.3	20	10.00	4	12.50
M17	300	5	30	20	34	8:2	88.2	30	6.67	4	12.50
M18	300	5	30	20	44	8:2	90.9	40	5.00	4	12.50
M19	300	3	30	20	10	8:2	60.0	6	33.33	4	12.50
M20	300	3	30	20	12	8:2	66.7	8	25.00	4	12.50
M21	300	3	30	20	14	8:2	71.4	10	20.00	4	12.50
M22	300	3	30	20	24	8:2	83.3	20	10.00	4	12.50
M23	300	3	30	20	34	8:2	88.2	30	6.67	4	12.50
M24	300	3	30	20	44	8:2	90.9	40	5.00	4	12.50

3.1.3. Extraction of Test Results

The excavation test of each numerical model is carried out step by step. The plastic zone method and the maximum principal stress method are used to identify and extract the development height of the water-conducting fracture of the overburden rock produced by each step of excavation. Until the full mining is reached, the maximum development height of the overburden water-flowing fracture is obtained.

3.2. Theoretical Calculation of Rock Expansion

3.2.1. Calculation Principle

The overlying strata of the coal seam will undergo a series of movements and deformations such as bending, cracking, breaking, and caving as the underground coal mining activities continue, and the volume will increase, which is called the swelling characteristics of the rock strata [31–33]. When the full mining is achieved, the volume of the underground mining space formed by the coal seam mining is basically equal to the sum of the volume of the rock mass in the overburden caving zone, the volume of the rock mass in the fracture zone, and the subsidence volume of the bending zone. On the main section of the subsidence basin, this law is expressed as (1) [34]:

$$M = \Delta h_k + \Delta h_l + \Delta h_d \quad (1)$$

where M—mining height, m; Δh_k—the vertical expansion of the rock mass in the overburden caving zone after compaction, m; Δh_l—the vertical expansion of the rock mass in the overburden fracture zone after compaction, m; and Δh_d—the subsidence of the bending zone, m.

Considering that the size of Δh_k is directly related to the height of the overlying rock fracture, the calculation method of the maximum development height of the overlying rock water-flowing fracture based on the theory of rock fragmentation and expansion proposed by Wei Jiangbo [35] can be used to calculate the maximum development height of the overlying rock water-flowing fracture of the coal seam in 24 numerical models, as shown in Equation (2).

$$H_l = \sum_{i=1}^{k} h_i K_{ci} - (1-q)M \quad (2)$$

where q—coefficient of sinking; M—height mining, m; Δh_l—thickness of overlying i-th strata of coal seam, m; K_{ci}—residual bulking coefficient after mining failure of the i-th overlying strata of coal seam; and H_l—the maximum development height of water-conducting fracture in overburden rock, m.

3.2.2. Calculation Process

The main calculation steps are as follows:

The first step: Determine the subsidence coefficient q. According to a large number of survey and observation results [28,30], the coal mining subsidence in the study area is generally between 0.6 and 0.8, so the q of the study area is 0.7.

The second step: Determine the position of the holding layer. Based on the discriminant theory and method of the key stratum [24,25], it is considered that when the load (i.e., the maximum vertical stress) borne by the hard rock stratum k is less than or equal to its ultimate shear strength, the rock stratum plays a controlling role in the upper rock stratum, which is the bearing stratum. The diagram of the position of the bearing layer with the change in sand–mud ratio and sand layer coefficient is shown in Figure 4.

The third step: Determine the height of the caving zone. According to a large number of measured caving ratio ranges in the study area (generally 4–10) [36], the height of the caving zone is determined.

The fourth step: Determine the Kc of each rock stratum between the coal seam roof and the overlying bearing stratum. According to Wei Jiangbo [35], based on the theory of rock bulking, the residual bulking coefficients of the bottom of the caving zone, the top of the caving zone, and the bearing layer are taken as 1.05, 1.02, and 1.002, respectively [37], and the residual bulking coefficients of each rock layer between the bearing layer and the top of the caving zone are taken according to Formula (3) [38].

$$Kc(H) = a - \eta \ ln(H + b) \tag{3}$$

where $K_C(H)$—the residual expansion coefficient of rock mass after mining failure and compaction of overburden rock with high H from the coal seam; H—the distance between each rock stratum of the overlying rock and the roof of the coal seam; η—the attenuation coefficient of the residual expansion coefficient of the rock stratum is taken according to Table 3; and a, b—all of them are the coefficients related to the occurrence characteristics of the stressed strata, and the values are taken according to Table 3.

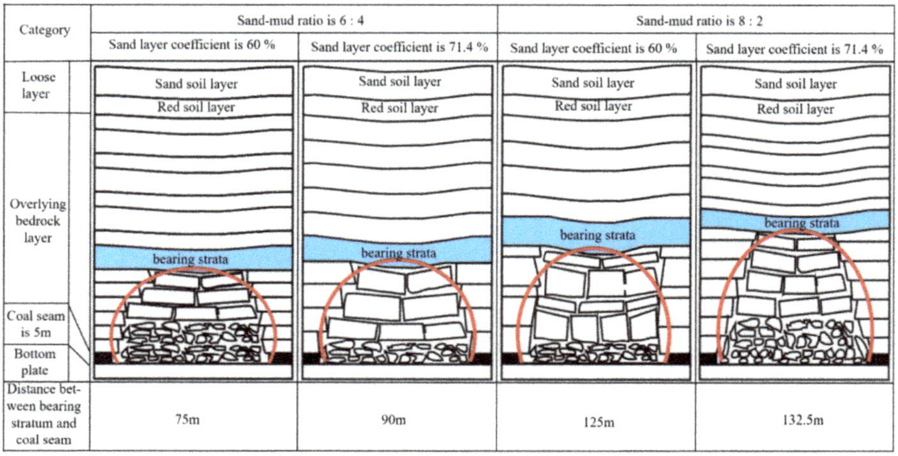

Figure 4. Relationship between bearing stratum and water-flowing fracture zone. Note: The range of the red line is the "stress arch" in the stress arch theory.

Table 3. Position of bearing layer and correlation coefficient table.

Numbering	a	η	b	Numbering	a	η	b
M1	1.071~1.109	0.016~0.024	3.692~11.836	M13	1.057~1.073	0.011~0.015	1.788~4.705
M2	1.071~1.111	0.016~0.024	3.772~12.208	M14	1.054~1.067	0.010~0.013	1.420~3.616
M3	1.064~1.090	0.014~0.019	2.800~8.132	M15	1.056~1.071	0.011~0.014	1.659~4.315
M4	1.057~1.074	0.012~0.015	1.886~5.006	M16	1.053~1.065	0.010~0.012	1.312~3.309
M5	1.056~1.071	0.011~0.014	1.700~4.438	M17	1.052~1.063	0.010~0.012	1.190~2.973
M6	1.055~1.070	0.011~0.014	1.620~4.199	M18	1.051~1.062	0.010~0.012	1.137~2.829
M7	1.088~1.097	0.021~0.021	5.989~8.984	M19	1.063~1.091	0.014~0.020	2.611~8.016
M8	1.066~1.098	0.015~0.022	2.906~9.251	M20	1.055~1.071	0.011~0.014	1.561~4.266
M9	1.060~1.082	0.013~0.017	2.147~6.252	M21	1.052~1.066	0.010~0.013	1.250~3.312
M10	1.054~1.069	0.011~0.014	1.428~3.852	M22	1.050~1.061	0.009~0.012	0.977~2.521
M11	1.053~1.066	0.010~0.013	1.282~3.408	M23	1.048~1.059	0.009~0.011	0.882~2.256
M12	1.052~1.065	0.010~0.013	1.219~3.221	M24	1.048~1.058	0.009~0.011	0.841~2.143

4. Results and Analysis

4.1. Results

The maximum development height of water-flowing fracture in coal seam overburden of 24 numerical models is obtained by numerical simulation test method and rock fragmentation theory calculation method, respectively, and the results are shown in Table 4.

Table 4. The maximum development height of water-flowing fracture in 24 numerical models of coal seam overburden.

Numbering	Numerical Simulation Test Results		Theoretical Results		Results	Numbering	Numerical Simulation Test Results		Theoretical Results		Results
	Maximum Height/m	Crack Production Ratio	Maximum Height/m	Crack Production Ratio	Average Relative Error/%		Maximum Height/m	Crack Production Ratio	Maximum Height/m	Crack Production Ratio	Average Relative Error/%
M1	78.5	15.7	82.7	16.5		M13	138.0	27.6	135.5	27.1	
M2	97.6	19.5	81.6	16.3		M14	148.3	29.6	158.7	31.7	
M3	99.1	19.8	97.7	19.5	3.1	M15	150.0	30.0	139.4	27.9	1.9
M4	117.5	23.5	128.1	25.6		M16	157.2	31.4	164.7	32.9	
M5	128.4	25.7	138.4	27.7		M17	170.5	34.1	176.8	35.4	
M6	133.2	26.6	147.9	29.6		M18	177.9	35.6	182.9	36.6	
M7	48.0	16.0	57.1	19.0		M19	64.6	21.5	90.6	30.2	
M8	74.1	24.7	81.5	27.2		M20	98.8	32.9	112.6	37.5	
M9	99.0	33.0	97.6	32.5	10.6	M21	118.1	39.4	142.1	47.4	17.0
M10	118.0	39.3	127.8	42.6		M22	157.9	52.6	164.7	54.9	
M11	129.0	43.0	138.4	46.1		M23	171.3	57.1	176.9	59.0	
M12	134.2	44.7	147.5	49.2		M24	178.2	59.4	183.1	61.0	

According to Table 4 the relative error rate of the maximum height of water-flowing fracture calculated by the two methods used in this paper is 1.9~17.0%, and the average absolute error rate is only 8.15%. It can be seen that the accuracy of the two results is high.

4.2. Analysis

4.2.1. The Influence of Sand Layer Coefficient on the Maximum Development Height of Water-Conducting Fracture

According to Table 4, when the mining height is 5 m and 3 m, the corresponding relationship curves between the sand layer coefficient of coal seam overburden and the maximum development height and fracture–mining ratio (N-MH-5m and N-MH-3m) of water-conducting fracture based on a numerical simulation test and the maximum development height and fracture–mining ratio (T-MH-5m and T-MH-5m) of water-conducting fracture based on the theory of rock fragmentation and expansion are drawn, as shown in Figures 5 and 6.

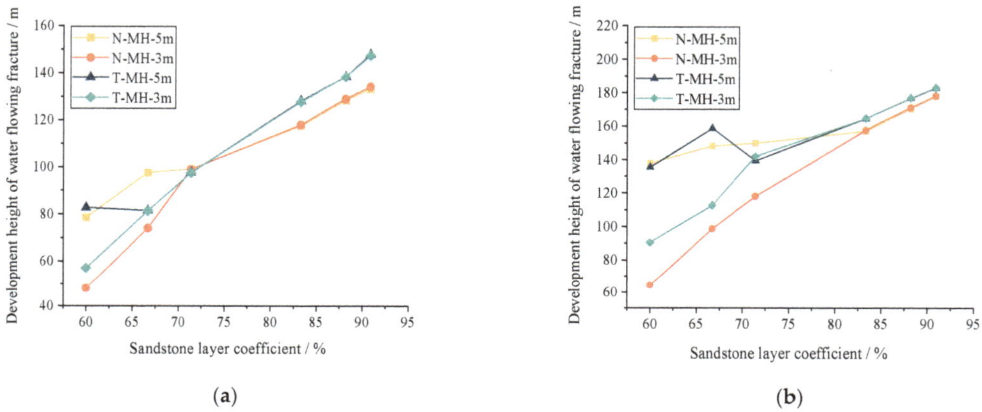

Figure 5. The corresponding relationship curve between sand layer coefficient and the maximum development height of water-conducting fracture. (**a**) The sand–mud ratio is 6:4. (**b**) The sand–mud ratio is 8:2.

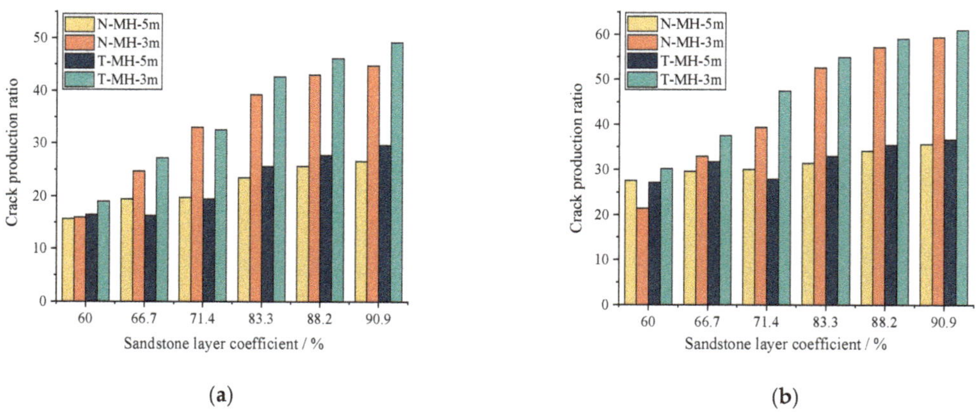

Figure 6. Columnar diagram of the corresponding relationship between sand layer coefficient and fracture-mining ratio. (**a**) The sand–mud ratio is 6:4. (**b**) The sand–mud ratio is 8:2.

The following can be seen from Table 4, Figures 5 and 6:

(1) Under the condition of a given sand–mud ratio, the maximum development height of a water-flowing fracture increases with the increase in the sand layer coefficient. That is to say, under the same buried depth, the more sandstone layers in the coal seam overburden, the greater the maximum development height of the water-flowing fracture. The numerical simulation results show that when the "sand–mud ratio is 6:4 + mining height is 5 m", the maximum development height of water-flowing fracture presents a three-stage change process of "rapid increase–stable–slow increase" with the change in sand layer coefficient. That is, when the sand layer coefficient is between 60~66.7%, the maximum development height of water-flowing fracture increases rapidly with the increase in sand layer coefficient, from 78.5 m to 97.6 m, with an average increase of 24.3%. When the sand layer coefficient is between 66.7% and 71.4%, the maximum development height of water-conducting fractures does not change much. When the sand layer coefficient is between 71.4% and 90.9%, the maximum development height of water-flowing fracture increases slowly with the increase in sand layer coefficient, from 99.1 m to 133.2 m, with an average increase of 10.5%. When the "sand–mud ratio is 6:4 + mining height is 3 m", the maximum development height of water-conducting fracture shows a two-stage change process of "rapid increase–slow increase" with the change in sand layer coefficient; that is, when the sand layer coefficient is between 60% and 71.4%, the maximum development height of water-conducting fracture increases rapidly with the increase in sand layer coefficient, from 48.0 m to 99.0 m, with an average increase of 43.99%. When the sand layer coefficient is between 71.4% and 90.9%, the maximum development height of a water-flowing fracture increases slowly with the increase in the sand layer coefficient, from 99.0 m to 134.2 m, with an average increase of 10.85%. Under the two conditions of "sand–mud ratio of 6:4 + mining height of 5 m" and "sand–mud ratio of 6:4 + mining height of 3 m", the distance between the bearing stratum and the coal seam increases from 75 m and 50 m to 135 m with the increase in sand layer coefficient, and the trend of the maximum development height of water-conducting fracture with the change in sand layer coefficient is consistent with the numerical simulation test results, which can be confirmed by each other. In particular, the two methods are highly consistent in identifying the maximum development height of water-conducting fractures when the "sand–mud ratio is 6:4 + mining height is 5 m", and the average error rate is only 3.1%.

The numerical simulation results show that when the "sand–mud ratio is 8:2 + mining height is 5 m", the maximum development height of water-flowing fracture increases slowly and evenly with the change in sand layer coefficient; that is, when the sand layer coefficient increases from 60% to 90.9%, the maximum development height of water-flowing fracture increases from 138 m to 177.9 m, with an average increase of 5.4%. When the "sand–mud ratio is 8:2 + mining height is 3 m", the maximum development height of water-flowing fracture increases rapidly and evenly with the change in sand layer coefficient; that is, when the sand layer coefficient increases from 60% to 90.9%, the maximum development height of water-flowing fracture increases from 64.6 m to 178.2 m, with an average increase of 23.74%. Under the two conditions of "sand–mud ratio of 8:2 + mining height of 5 m" and "sand–mud ratio of 8:2 + mining height of 3 m", the distance between the bearing stratum and the coal seam increases from 120 m and 79.2 m to 180 m with the increase in sand layer coefficient, and the trend of the maximum development height of water-conducting fracture with the change in sand layer coefficient is consistent with the numerical simulation test results, which can be confirmed by each other. In particular, the two methods are highly consistent in identifying the maximum development height of water-conducting fractures when the sand–mud ratio is 8:2 + mining height is 5 m, and the average error rate is only 1.9%.

It can be seen that the results of the theoretical calculation of rock fragmentation and the results of the numerical simulation test tend to be more consistent when the mining height and sand–mud ratio are large and tend to be more discrete when the mining height and sand–mud ratio are small. Based on the results of the two methods, combining the results of the two methods, the quantitative relationship between the maximum development height of water-conducting fractures and sand layer coefficient, sand–mud ratio, and mining thickness is shown in Equation (4).

$$\begin{aligned} H_l &= -11.76M^2 - 64.78x^2 + 0.07y^2 + 101.35M + 319.50x - 5.61y - 251.04 \\ R^2 &= 0.818 \end{aligned} \tag{4}$$

where H_l—the maximum development height of water-conducting fracture, m; x—sandstone layer coefficient, %; y—sand–mud ratio; and M—height mining, m.

(2) Under the condition of the same sand–mud ratio and the sand layer coefficient greater than 67%, the increase in the sand layer coefficient will eliminate the difference of the positive effect of different mining heights on the maximum development height of the overburden water-flowing fracture.
The numerical simulation test results show that when the sand–mud ratio is 6:4 and the sand layer coefficient exceeds 71.4%, the maximum development height of the water-conducting fracture produced by 5 m and 3 m mining height is very small, only within 0.6 m. When the sand–mud ratio is 8:2 and the sand layer coefficient exceeds 83.3%, the maximum development height of water-conducting fractures produced by 5 m and 3 m mining heights is very small, only within 0.8 m. It can be seen that when the sand–mud ratio is 6:4 and the sand layer coefficient exceeds 72% or the sand–mud ratio is 8:2 and the sand layer coefficient exceeds 83%, the influence of mining height changes on the maximum development height of water-flowing fracture tends to disappear. The theoretical calculation results of rock fragmentation and expansion show that when the sand–mud ratio is 6:4 and the sand layer coefficient exceeds 66.7%, the maximum development height of water-conducting fractures produced by 5 m and 3 m mining heights is very small, only within 0.4 m; when the sand–mud ratio is 8:2 and the sand layer coefficient exceeds 83.3%, the maximum development height of water-conducting fractures produced by 5 m and 3 m mining heights is very small, only within 0.2 m. The results of the two methods are highly consistent, which shows that: First, under the same buried depth, the increase in the number of sandstone layers weakens the anti-interference ability of the overlying rock mass to the small mining height, so that when the sand layer coefficient exceeds 70~80%, the maximum development height of the water-flowing fracture caused by the small mining height is equivalent to the large mining height. Secondly, the overburden structure of coal seam in the study area should meet the conditions of "sand–mud ratio 6:4 and sand layer coefficient less than 70%" and "sand–mud ratio 8:2 and sand layer coefficient less than 80%" so as to achieve the purpose of water-preserved coal mining through the method of height-limited mining.

(3) Under the condition of a fixed sand–mud ratio, the fracture–mining ratio will increase with the increase in the sand layer coefficient, and the increase in mining height will significantly reduce the increase in the fracture–mining ratio.
Both the numerical simulation test results and the theoretical calculation results of rock fragmentation and expansion show that when the "sand–mud ratio is 6:4 + mining height is 5 m or 3 m", the fracture–mining ratio increases uniformly with the increase in sand layer coefficient. When the "sand–mud ratio is 8:2 + mining height is 5 m", the fracture–mining ratio increases with the increase in sand layer coefficient, and the increase is not obvious. When the "sand–mud ratio is 8:2 + mining height is 3 m", the fracture–mining ratio increases significantly with the increase in sand layer coefficient. When "the sand–mud ratio is 6:4 or 8:2 + mining height is 3 m", the increase in sand layer coefficient leads to an average increase of about 24%. When "the ratio of sand to mud is 6:4 or 8:2 + mining height is 5 m", the increase in sand

layer coefficient leads to an average increase of about 8%, and the former is 3 times that of the latter.

4.2.2. The Influence of Sand–Mud Ratio on the Maximum Development Height of Water-Conducting Fracture

According to Table 4, when the mining height is 5 m and 3 m, the corresponding relationship curves between the sand–mud ratio of coal seam overburden and the maximum development height and fracture–mining ratio of water-conducting fracture based on numerical simulation tests (N-SMR-6:4 and N-SMR-8:2) and the maximum development height and fracture–mining ratio of water-conducting fracture based on the theory of rock fragmentation and expansion (T-SMR-6:4 and T-SMR-8:2) are drawn, as shown in Figures 7 and 8.

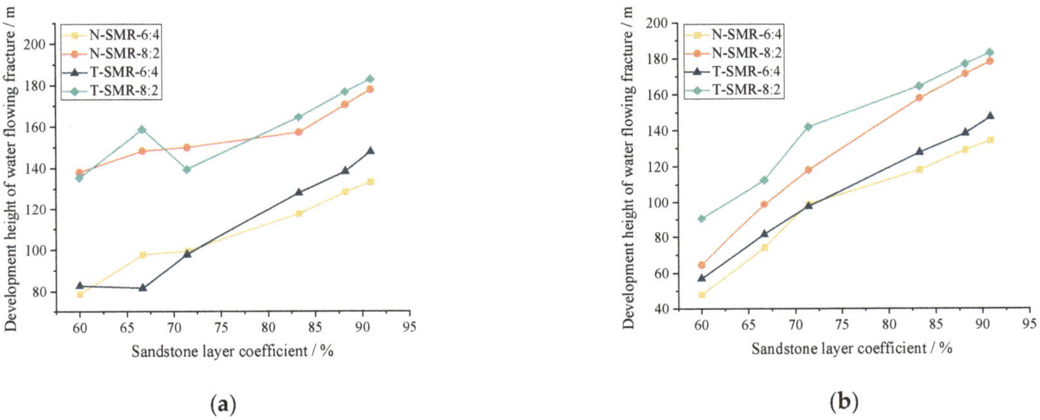

Figure 7. The corresponding relationship curve between sand–mud ratio and the maximum development height of water-conducting fracture. (**a**) The mining height is 5 m. (**b**) The mining height is 3 m.

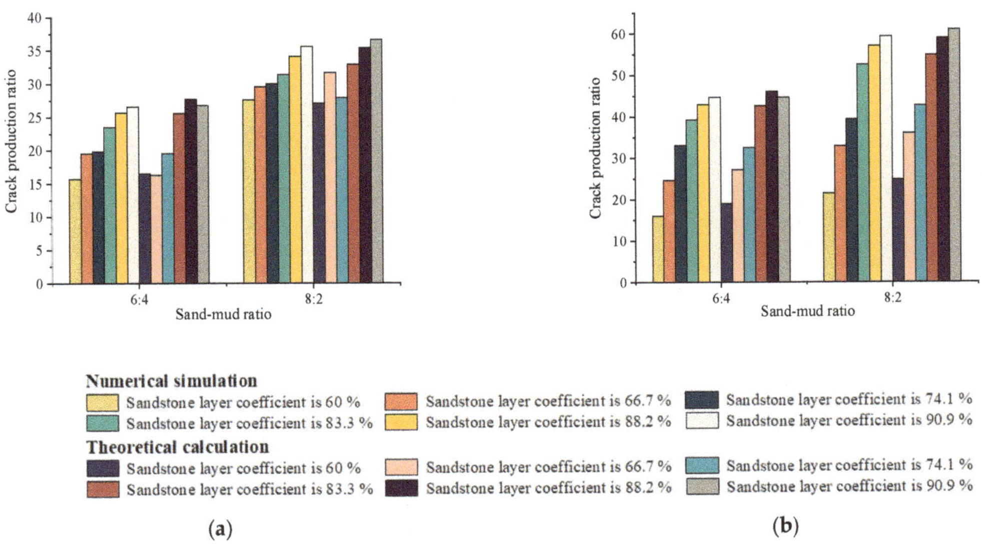

Figure 8. Columnar diagram of the corresponding relationship between sand–mud ratio and fracture-mining ratio. (**a**) The mining height is 5 m. (**b**) The mining height is 3 m.

The following can be seen from Table 4 and Figures 7 and 8:

(1) Under the condition of a fixed sand layer coefficient, the maximum development height of a water-conducting fracture increases with the increase in the sand–mud ratio. That is to say, under the same buried depth, the greater the total thickness of sandstone in the coal seam overburden is, the greater the maximum development height of the water-conducting fracture.
The results of the numerical simulation test show that under the conditions of "any sand layer coefficient + mining height 3 m" and "any sand layer coefficient + mining height 5 m", with the sand–mud ratio increasing from 6:4 to 8:2, the maximum development height of water-flowing fracture increases by 16.6~44 m and 39.7~59.5 m, respectively, with an average increase of 31.1 m and 47.9 m; that is, for every 1% increase in the proportion of total thickness of sandstone in bedrock, the maximum development height of water-flowing fracture increases by 1.56 m and 2.40 m. The theoretical calculation results of rock layer expansion show that under the conditions "arbitrary sand layer coefficient + mining height 3 m" and "arbitrary sand layer coefficient + mining height 5 m", as the sand–mud ratio increases from 6:4 to 8:2, the distance between the bearing layer and the coal seam increases by 40~76 m and 29.2~45 m, respectively, and the maximum development height of the water-conducting fracture increases by 31.1~44.5 m and 35.0~77.1 m, respectively, with an average increase of 36.7 m and 46.9 m. That is, for every 1% increase in the proportion of the total thickness of sandstone in the bedrock, the maximum development height of the water-conducting fracture increases by 1.84 m and 2.35 m, respectively. This is highly consistent with the results of the numerical simulation. The results of the two methods show that the increase in the sand–mud ratio of the overlying strata of the coal seam will amplify the positive effect of the increase in the mining height on the maximum development height of the water-conducting fracture of the overlying strata.

(2) Large mining height will amplify the positive effect of the increase in sand–mud ratio on the maximum development height of water-conducting fractures.
The results of the numerical simulation test show that the maximum development height of water-flowing fracture in overlying strata of coal seam is 48~134.2 m under the condition of "sand–mud ratio of 6:4 + mining height of 3 m", and the maximum development height of water-flowing fracture in overlying strata of coal seam is 64.6~178.2 m under the condition of "sand–mud ratio of 8:2 + mining height of 3 m", which is 31.1 m higher than that of the former. Under the condition of "sand–mud ratio of 6:4 + mining height of 5 m", the maximum development height of water-flowing fracture in overlying strata of coal seam is 78.5~133.2 m. Under the condition of "sand–mud ratio of 8:2 + mining height of 5 m", the maximum development height of water-flowing fracture in overlying strata of coal seam is 138~177.9 m, which is 47.9 m higher than that of the former. The theoretical calculation results of rock fragmentation and expansion show that the maximum development height of water-conducting fractures in overlying strata of coal seam is 57.1~147.5 m under the condition of "sand–mud ratio of 6:4 + mining height of 3 m", and the maximum development height of water-conducting fractures in overlying strata of coal seam is 90.6~183.1 m under the condition of "sand–mud ratio of 8:2 + mining height of 3 m", which is 36.7 m higher than the former on average. Under the condition of "sand–mud ratio of 6:4 + mining height of 5 m", the maximum development height of water-conducting fractures in overlying strata of coal seam is 82.7~147.9 m. Under the condition of "sand–mud ratio of 8:2 + mining height of 5 m", the maximum development height of water-conducting fractures in overlying strata of coal seam is 135.5~182.9 m, which is 46.9 m higher than the former on average. It can be seen that the results obtained by the two methods are very close, which together show that after the mining height increases from 3 m to 5 m, the increase in the sand–mud ratio

makes the average increase in the maximum development height of water-conducting fractures by about 51.0%.

(3) Under the condition of any sand layer coefficient, the fracture–mining ratio will increase with the increase in sand–mud ratio, and the increase in mining height will have a certain inhibitory effect on the increase in fracture–mining ratio.
The results of the numerical simulation test show that the ratio of fracture to production is 16~44.7 under the condition of "sand–mud ratio of 6:4 + mining height of 3 m", and the ratio of fracture to production is 21.5~59.4 under the condition of "sand–mud ratio of 8:2 + mining height of 3 m", which is 10.37 higher than that of the former. Under the condition of "sand–mud ratio of 6:4 + mining height of 5 m", the crack–production ratio is 15.7~26.6, and under the condition of "sand–mud ratio of 8:2 + mining height of 5 m", the crack–production ratio is 27.6~35.6; the latter is 9.58 higher than the former on average. The theoretical calculation results of rock fragmentation expansion show that under the condition of "sand–mud ratio of 6:4 + mining height of 3 m", the fracture–mining ratio is 19~49.2, and under the condition of "sand–mud ratio of 8:2 + mining height of 3 m", the fracture–mining ratio is 30.2~61, which is an average increase of 12.23 compared with the former. Under the condition of "sand–mud ratio of 6:4 + mining height of 5 m", the fracture–mining ratio is 16.5~29.6. Under the condition of "sand–mud ratio of 8:2 + mining height of 5 m", the fracture–mining ratio is 27.1~36.6. The latter is 9.38 higher than the former on average. It can be seen that the results obtained by the two methods are very close, and together they show that after the mining height increases from 3 m to 5 m, the increase in the sand–mud ratio makes the fracture–mining ratio decrease by about 15.0% on average.

5. Example Verification and Mechanism Analysis

5.1. Example Verification

This paper collected the measured data of the 'three-zone hole' overburden structure and the observation hole of the water-conducting fracture development in the typical working face of the Yushenfu mining area and compared with the results obtained by the prediction method of the maximum height prediction model of the water-conducting fracture development established in this paper (see Table 5).

From Table 5, based on the prediction model established in this paper, the relative error rate between the calculated maximum height of water-conducting fracture development and the measured value is −5.41–11.99%, and the average absolute error rate is only 7.17%. The accuracy of the prediction model is relatively high. Therefore, the constructed prediction model can provide a scientific reference for the study of the development of an overburden water-conducting fracture zone in the Yushenfu mining area and the guarantee of coal mine safety production.

Table 5. Comparison of the maximum development height of overburden water conduction fractures in typical working faces in Yushenfu mining area.

Name	Drilling Number	Thickness/m	The Ratio of Sandstone Thickness to Total Thickness/%	Sand Layer Coefficient/%	Theoretical Calculation Results/m	Measured Results/m	The Relative Error Rate Between Theoretical Calculation and Measured Results Is/%
Jinjitan coal mine	ZK1	5.4	92.67	87.17	122.88	109.72	11.99
Yu-shu-wan mine	ZK2	5	98.57	92.86	149.21	137.30	8.67
	ZK3	5	98.40	96.00	128.80	117.80	9.34
Hang lai bay coal mine	ZK4	4.49	95.74	91.30	113.88	114.38	−0.44
	ZK5	4.49	97.38	89.29	88.79	93.87	−5.41

5.2. Mechanism Analysis

(1) As the most important mining factor affecting and controlling the development of water-conducting fractures in the overlying strata of coal seams, mining height has been recognized by scholars at home and abroad [14–16]. The goaf formed by mining height is actually the space of movement, deformation, and fracture of the coal seam overburden. The size of the mining height determines the size and size of this "space", which also profoundly affects the degree of movement and deformation of the overlying strata of the coal seam. Generally speaking, the mining height has a significant positive effect on the development of water-conducting fractures in overlying strata; that is, the larger the mining height is, the larger the size of the goaf is, the larger the space for the movement and deformation of the overlying strata is, and the higher the development height of water-conducting fractures in overlying strata is [18,19]. Under the same mining depth and overburden structure conditions, the maximum development height of overburden water-conducting fractures caused by 5 m mining height is greater than that of 3 m mining height, and the former is 1.1~1.2 times of the latter on average, which not only verifies the positive effect of mining height on the development of overburden water-conducting fractures but also is consistent with the research results of Wang et al. [26] and Xun et al. [39].

(2) As a typical geological occurrence factor of coal resources in Yushenfu mining area, the number of sandstone layers (sand layer coefficient) in overlying strata will also have an important impact on the development of water-flowing fractures in overlying strata of coal seams. Under the condition of equal burial depth, the larger the sand layer coefficient is, the more the number of sandstone layers in the overlying strata of the coal seam is, and the smaller the thickness of the single-layer sandstone is. This will produce the following two effects: First, the increase in the sand layer coefficient will not only lead to the increase in the number of sandstone layers in the overlying strata of the coal seam but also lead to the increase in the number of layers between the sandstones (the main structural plane or weak surface in the rock mass), which makes the overall stability and anti-interference ability of the overlying strata of the coal seam worse [28,30] and produces a higher degree of failure response under the influence of the same mining activities. In addition, the latest research results show that [40] when the sand layer coefficient is large, that is, the number of sandstone layers is large, and the sandstone of each layer will lead to significant differences in the size, shape, and arrangement of internal particles due to differences in deposition time and environment. The particles between the sandstone layers are difficult to deform synergistically and produce relative displacement when subjected to external forces, so that the cracks are more likely to expand rapidly along the particle boundary, resulting in a greater development height, as shown in Figure 9a;

when the sand layer coefficient is small, that is, the number of sandstone layers is small, the thickness of each layer of sandstone is large, and the internal particles are relatively uniform and dense, so that under the same external force, it is easier to block and limit the generation and expansion of cracks, which in turn restricts the development of cracks in the rock layer, as shown in Figure 9b. Second, the increase in the sand layer coefficient will lead to the decrease in the thickness of the stressed rock layer (such as sandstone) in the overburden of the coal seam, which will lead to the deterioration of the mechanical properties of the stressed rock layer [41]. Under the same load and mining activities, it is more likely to crack, break, and eventually lose the ability to support the load of the overlying rock mass [42] and the ability to block the development of the underlying mining cracks [22]. In this case, the overlying strata of the coal seam will seek a new balance upward in the process of overall movement and deformation, that is, to determine the new stressed rock strata and to provide a larger space for the development of water-conducting fractures while completing the upward movement of the stressed rock strata in the entire overlying strata. The coupling superposition of the above two effects may be one of the main reasons for the larger sand layer coefficient of coal seam overburden and the larger maximum development height of water-conducting fractures.

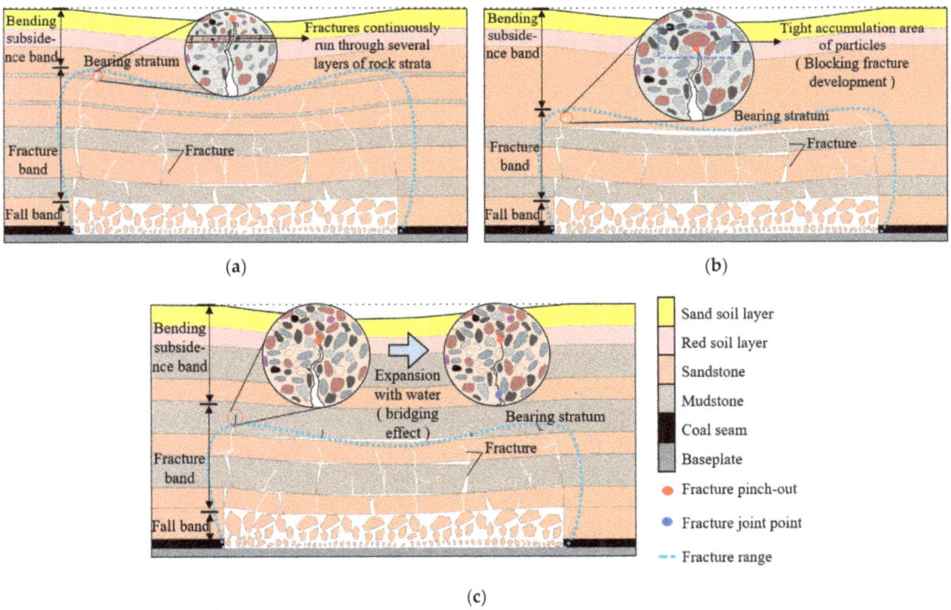

Figure 9. Relationship between the key characteristics of the layered structure of overburden rock and the development height of water conduction fractures. (**a**) When the sand layer coefficient is larger. (**b**) When the sand layer coefficient is small. (**c**) When the sand–mud ratio is small.

(3) The thickness ratio of sandstone and mudstone in overlying strata (sand–mud ratio), as another typical geological occurrence factor of coal resources in Yushenfu mining area, will also have an important impact on the development of water-conducting fractures in overlying strata. Under the condition of equal burial depth, the greater the sand–mud ratio, the greater the total thickness of the sandstone (i.e., hard rock) in the overlying rock of the coal seam, the stronger the overall rigidity of the overlying rock. According to the Griffith energy release rate criterion [43,44], according to the energy point of view of linear elastic fracture mechanics, the condition of fracture is that when the deformation energy released by crack propagation is equal to or

greater than the energy required for crack propagation, the crack will expand unstably. The effect of sandstone accumulating the load potential energy of the overlying rock mass is stronger [45], and the energy released during the fracture is larger and more concentrated, which makes the development rate and degree of mining-induced fractures in the overlying rock significantly improved. This will also produce the following two effects: First, under the influence of the same mining activities, the greater the sand–mud ratio of the overlying strata of the coal seam, the greater the total thickness of the sandstone and the thickness of the single layer, the more energy accumulated in the mining process, and the stronger the damage to the overlying strata after the fracture. The result is mainly manifested in the rapid development and expansion of the water-conducting fractures of the overlying strata in a short period of time, which can simultaneously cut through multiple overlying strata and even directly reach the surface. The large-area "overhang" in some mines in the Yushenfu mining area is the best example [46]. Second, the increase in sand–mud ratio will lead to the decrease in the total thickness of mudstone (soft rock) in the overlying strata of coal seam, which not only weakens the inhibitory effect of mudstone on the development of water-conducting fractures [47], but also indirectly causes the position of the stressed rock strata in the overlying strata of coal seam to move up [48] and finally promotes the development of water-conducting fractures (see Figure 3). In addition, the majority of studies have shown that once the mudstone is broken and cracked, once it encounters water, it will show a very obvious phenomenon of breaking fracture closure or even pinch-out [37,49]. The decrease in the sand–mud ratio leads to an increase in the total thickness of the mudstone in the overlying rock of the coal seam, which greatly strengthens the above-mentioned effect, thus reducing the development height of the water-conducting fracture of the overlying rock (see Figure 9c). The coupling superposition of the above two effects may be one of the main reasons for the larger sand–mud ratio of overlying strata and the larger maximum development height of water-conducting fractures.

6. Conclusions

(1) The maximum development height of the water-flowing fracture in the overlying strata of the Yushenfu Coal Mine will increase with the increase in the sand layer coefficient; that is, under the same buried depth, the more sandstone layers in the overlying strata of the coal seam, the greater the maximum development height of the water-flowing fracture. This effect will show different change processes under different sand–mud ratio and mining thickness combination conditions. When the sand layer coefficient is greater than 67%, the increase in the number of sand layers in the overlying rock will eliminate the difference in the positive effect of different mining thicknesses on the maximum development height of the overlying rock water-flowing fracture.

(2) The maximum development height of water-conducting fractures in the overlying strata of the study area will increase with the increase in sand–mud ratio; that is, under the same buried depth, the greater the total thickness of sandstone in the overlying strata of the coal seam, the greater the maximum development height of water-conducting fractures. The two methods together show that when the sand–mud ratio increases by 1%, the maximum development height of the water-conducting fracture increases by 1.56–2.40 m in the range of 60–90.9%; that is, the increase in the sand–mud ratio of the overlying strata of the coal seam will amplify the positive effect of the increase in the mining thickness on the maximum development height of the water-conducting fracture of the overlying strata.

(3) Based on sand layer coefficient, sand–mud ratio, and mining thickness, the prediction model of maximum development height of a water-conducting fracture is constructed. It is verified by an example that the average absolute error rate between the calculated

results of the model and the measured values is 7.17%, which is in line with the actual situation.

(4) Based on the numerical simulation test and the theoretical calculation results of rock fragmentation, it is proposed that the effective area of water-preserved coal mining in the study area can be achieved by using the height-limited mining method and must conform to the coal seam overburden structure characteristics of 'sand–mud ratio 6:4 and sand layer coefficient less than 70%' and 'sand–mud ratio 8:2 and sand layer coefficient less than 80%'.

Author Contributions: S.S.: conceptualization, methodology, and writing—review and editing. H.R.: methodology, software drawing, data curation, writing—original draft, and experiment. J.W.: investigation, software drawing, and experiment. X.C.: investigation and writing—review and editing. R.N.: investigation and writing—review and editing; B.C.: writing—review and editing. All the authors reviewed the manuscript. All authors have read and agreed to the published version of the manuscript.

Funding: This paper is supported by the National Natural Science Foundation of China (41402308), the Key Research and Development Program of Shaanxi Province (2023-YBSF-458), and the Key Laboratory of Geological Guarantee for Coal Green Development of Shaanxi Province (DZBZ2022Z-03).

Data Availability Statement: The original contributions presented in the study are included in the article, further inquiries can be directed to the corresponding author.

Conflicts of Interest: The authors declare that they have no known competing financial interests or personal relationships that could have appeared to influence the work reported in this paper.

References

1. Wang, G.; Ren, S.; Pang, Y.; Qu, S.; Zeng, D. Development achievements of China's coal industry during the 13thFive-Year Plan period and implementation path of "dual carbon" target. *Coal Sci. Technol.* **2021**, *49*, 1–8. [CrossRef]
2. Wu, Q.; Tu, K.; Zeng, Y.; Liu, S. Discussion on the main problems and countermeasures for building an upgrade version of main energy (coal) industry in China. *J. China Coal Soc.* **2019**, *44*, 1625–1636. [CrossRef]
3. *Statistical Bulletin of National Economic and Social Development of the People's Republic of China in 2022*; National Bureau of Statistics: Beijing, China, 2023; pp. 12–29.
4. Wang, S.; Shen, Y.; Song, S.; Liu, L.; Gu, L.; Wei, J. Change of coal energy status and green and low-carbon development under the "dual carbon" goal. *J. China Coal Soc.* **2023**, *48*, 2599–2612. [CrossRef]
5. Peng, S.; Bi, Y. Strategic consideration and core technology about environmental ecological restoration in coal mine areas in the Yellow River basin of China. *J. China Coal Soc.* **2020**, *45*, 1211–1221. [CrossRef]
6. Wu, Q. One line and one chess boost the high-quality development of the Yellow River Basin. *Sci. Technol. Ind. China* **2023**, *4*, 1–2. [CrossRef]
7. Zhang, H. Preliminary study on ecological mining mode in the coal-rich area of the Yellow River Basin. *Coal Sci. Technol.* **2021**, *49*, 233–242. [CrossRef]
8. Fan, L. Scientific connotation of water-preserved mining. *J. China Coal Soc.* **2017**, *42*, 27–35. [CrossRef]
9. Bian, Z.; Yu, H.; Lei, S.; Yin, D.; Zhu, G.; Mu, S.; Yang, D. Strategic consideration of exploitation on coal resources and its ecological restoration in the Yellow River Basin, China. *J. China Coal Soc.* **2021**, *46*, 1378–1391. [CrossRef]
10. Qian, M.; Xu, J. Concept and Technical Framework of Sustainable Mining. *J. China Univ. Min. Technol. (Soc. Sci.)* **2011**, *13*, 1–7+23. [CrossRef]
11. Fan, L.; Ma, W.; Chang, B.; Sun, K.; Miao, Y.; Lu, B.; Tian, S.; Yang, L. Hydrochemical characteristics and formation mechanism of groundwater in Yushenfu Mining Area. *Coal Sci. Technol.* **2023**, *51*, 383–394. [CrossRef]
12. Wang, S.; Fan, L.; Huang, Q.; Shen, T. Northern Shaanxi coal based on ecological water level protection. *Min. Saf. Environ. Prot.* **2010**, *37*, 81–83. [CrossRef]
13. Xie, X.; Hou, E.; Wang, S.; Zhao, B.; Long, T.; Feng, D.; Hou, P.; Zhang, Q. Study on Water Loss Model and Prediction Technology of Aquifer Induced by Coal Mining in Yushenfu Mining Area in the Middle Reaches of the Yellow River. *Coal Sci. Technol.* **2024**, *51*, 197–207. [CrossRef]
14. Wu, J.; Pan, J.; Gao, J.; Yan, Y.; Ma, H. Research on prediction of the height of water-conducting fracture zone in Huanglong Jurassic Coalfield. *Coal Sci. Technol.* **2023**, *51* (Suppl. S1), 231–241. [CrossRef]
15. Palchik, V. Localization of mining-induced horizontal fractures along rock layer interfaces in overburden: Field measurements and prediction. *Environ. Geol.* **2005**, *48*, 68–80. [CrossRef]
16. Singh, R.N.; Jakeman, M. Strata monitoring investigations around longwall panels beneath the cataract reservoir. *Mine Water Environ.* **2001**, *20*, 55–64. [CrossRef]

17. Wang, H.; Zhang, D.; Wang, X.; Zhang, W. Visual exploration of the spatiotemporal evolution law of overburden failure and mining-induced fractures: A case study of the Wangjialing coal mine in China. *Minerals* **2017**, *7*, 35. [CrossRef]
18. Yin, H.; Dong, F.; Zhang, Y.; Cheng, W.; Zhai, P.; Ren, X.; Liu, Z.; Zhai, Y.; Li, X. Height prediction and 3D visualization of mining-induced water-conducting fracture zone in western ordos basin based on a multi-factor regression analysis. *Energies* **2022**, *15*, 3850. [CrossRef]
19. Li, Z.; Li, S.; Du, F.; Wang, W.; Li, J.; Jiao, Y.; Fan, X. Research on the development law of karst caves on water conducting fractures under the influence of mining in Southwest Karst Mining Areas. *Coal Sci. Technol.* **2023**, *51*, 106–117. [CrossRef]
20. Ju, J.; Ma, X.; Zhao, F.; Liu, Y.; Wang, Y.; Liu, L.; Xu, J. Development and zoning characteristics of water-conducted fractures in Dongsheng Coalfield. *Coal Sci. Technol.* **2022**, *50*, 202–212. [CrossRef]
21. Huang, B.; Liu, C.; Xu, J. Effect of little fault in working face on water conducted fissure height. *J. China Coal Soc.* **2009**, *34*, 1316–1321. [CrossRef]
22. Wang, F.; Xu, J.; Xie, J. Effects of arch structure in unconsolidated layers on fracture and failure of overlying strata. *Int. J. Rock Mech. Min. Sci.* **2019**, *114*, 141–152. [CrossRef]
23. Wang, S.; Wei, J.; Song, S.; Wang, S.; Sun, T. Influence of thick sandstone on development of overburden mining fissures in northern Shaanxi coal mining area of Yellow River Basin and suggestions on water-preserved coal mining. *Coal Geol. Explor.* **2022**, *50*, 2. [CrossRef]
24. Xu, J.; Wang, X.; Liu, W.; Wang, Z. Effects of primary key stratum location on height of water flowing fracture zone. *Chin. J. Rock Mech. Eng.* **2009**, *28*, 380–385. [CrossRef]
25. Xu, J.; Zhu, W.; Wang, X. New method to predict the height of fractured water-conductingzone by location of key strata. *J. China Coal Soc.* **2012**, *37*, 762–769. [CrossRef]
26. Wang, X.; Xu, J.; Han, H.; Ju, J.; Xing, Y. Stepped development characteristic of water flowing fracture heightwith variation of mining thickness. *J. China Coal Soc.* **2019**, *44*, 3740–3749. [CrossRef]
27. Wang, X.; Yin, S.; Xu, B.; Cao, M.; Zhang, R.; Tang, Z.; Huang, W.; Li, W. Study on height optimization prediction model of overburden water-conducting fracture zone under fully mechanized mining. *Coal Sci. Technol.* **2023**, *51* (Suppl. S1), 284–297. [CrossRef]
28. Song, S.; Wang, S.; Zhao, X.; Shen, T. Stratification transfer method of the mining subsidence based on thecharacteristics of layered structure in coal overburden. *J. China Coal Soc.* **2018**, *43* (Suppl. S1), 87–95. [CrossRef]
29. Du, H.; Cao, Y.; Nie, W.; Song, S. Evolution of soil properties under artificial and natural revegetation in loess gully coal mining subsidence area. *J. China Coal Soc.* **2021**, *46*, 1641–1649. [CrossRef]
30. Song, S. Study on the Stratification Transfer Prediction Method of the Mining Subsidence Based on the Key Geological and Mining Factors. Ph.D. Thesis, University of Science and Technology, Xi'an, China, 2014.
31. Alejano, L.R.; Alonso, E. Considerations of the dilatancy angle in rocks and rock masses. *Int. J. Rock Mech. Min. Sci.* **2005**, *42*, 481–507. [CrossRef]
32. Liang, B.; Wang, B.; Jiang, L.; Li, G.; Li, C. Broken expand properties of caving rock in shallow buried goaf. *J. China Univ. Min. Technol.* **2016**, *45*, 475–482. [CrossRef]
33. Wang, H.; Wang, Q.; Zhao, Z.; Feng, Q. Study on bearing deformation characteristics and lateral pressure distribution law of caved gangue in gob. *Coal Sci. Technol.* **2023**, *51*, 20–29. [CrossRef]
34. Wang, L.; Wang, Z.; Huang, J.; Zhou, D. Prediction on the Height of Water-Flowing Fractured Zone for Shallow Seam Covered with Thin Bedrock and Thick Windblown Sands. *J. Min. Saf. Eng.* **2012**, *29*, 607–612.
35. Wei, J. Study on Control Mechanism of Overlying Thick Sandstone on Mining Fractures in Yushenfu Mining Area. Ph.D. Thesis, Xi'an University of Science and Technology, Xi'an, China, 2023.
36. *Northern Shaanxi Coal Mining Area Mining Damage Investigation Report*; Shaanxi Provincial Institute of Geological Survey: Xi'an, China, 2016.
37. He, M.; Gai, Q.; Gao, Y.; Zhang, X. Research on the mechanism and control of bulking equilibrium of non-pillar mining with automatic entry formation for hard roof. *J. China Univ. Min. Technol.* **2023**, *52*, 831–844+930. [CrossRef]
38. Zhang, G.; Tao, G.; Cao, Z.; Zhao, Y.; Xu, Z.; Li, Y.; Zuo, H. Study on failure characteristics of stope overlying rockconsidering strain hardening characteristics of caved rock mass. *Coal Sci. Technol.* **2022**, *50*, 46–52. [CrossRef]
39. Xun, B.; Lyu, Y.; Yao, X. Comparison of prediction models for the development height ofwater-conducting fractured zone. *Coal Sci. Technol.* **2023**, *51*, 190–200. [CrossRef]
40. He, B.; Wang, J.; Zhang, Y.; Yan, S.; Chen, T. Microscopic failure of yellow sandstone with different-sized grains and mineral composition. *J. Cent. South Univ.* **2023**, *30*, 2035–2047. [CrossRef]
41. Liu, W.; Shen, J. Experimental study of propagation mode of crack in real rock specimens with a single crack. *Chin. J. Rock Mech. Eng.* **2016**, *35*, 1182–1189. [CrossRef]
42. Li, Y.; Yang, C. Influence of geometric characteristics of pre-existing cracks on mixed mode fractures under compression-shear loading. *Chin. J. Rock Mech. Eng.* **2006**, *25*, 462–466. [CrossRef]
43. Baouche, R.; Sen, S.; Radwan, A.E.; Abd El Aal, A. In Situ Stress Determination Based on Acoustic Image Logs and Borehole Measurements in the In-Adaoui and Bourarhat Hydrocarbon Fields, Eastern Algeria. *Energies* **2023**, *16*, 4079. [CrossRef]
44. Lazarus, V. Perturbation approaches of a planar crack in linear elastic fracture mechanics: A review. *J. Mech. Phys. Solids* **2011**, *59*, 121–144. [CrossRef]

45. Gupta, V.; Duarte, C.A.; Babuška, I.; Banerjee, U. A stable and optimally convergent generalized FEM (SGFEM) for linear elastic fracture mechanics. *Comput. Methods Appl. Mech. Eng.* **2013**, *266*, 23–39. [CrossRef]
46. Lin, Z.; Li, Y.; Wu, J.; Yang, Y. Influence of hydraulic fracturing on the behavior of mine pressure at the end of mining. *J. Min. Saf. Eng.* **2023**, *40*, 714–721. [CrossRef]
47. Han, H.; Shi, C.; Xu, J.; Liu, Y.; Wang, X.; Zuo, W.; Agus, P.S.; Shi, Z. Mining-induced stress field prediction model based on the "slabbeam" structure of overburden key strata. *J. Min. Saf. Eng.* **2024**. 1–10. Available online: http://kns.cnki.net/kcms/detail/32.1760.TD.20231103.1030.002.html (accessed on 15 August 2024).
48. Feng, G.; Yu, F.; Dai, Z.; Chen, S.; Li, J. Experimental study on time effect characteristics of red mudstone swelling in Central Sichuan. *Chin. J. Rock Mech. Eng.* **2022**, *41* (Suppl. S1), 2780–2790. [CrossRef]
49. Ju, J.; Li, Q.; Xu, J.; Yang, L. Water-CO_2-rock interaction experiments to reveal the self-healing effect of fractured mining damaged rock. *J. China Coal Soc.* **2019**, *44*, 3700–3709. [CrossRef]

applied sciences

MDPI

Article

Experimental Investigation on Unloading-Induced Sliding Behavior of Dry Sands Subjected to Constant Shear Force

Wengang Dang [1,2,3], Kang Tao [4], Jinyang Fu [3,5,*] and Bangbiao Wu [1]

1 State Key Laboratory of Hydraulic Engineering Simulation and Safety, Tianjin University, Tianjin 300072, China; dangwg@mail.sysu.edu.cn (W.D.); bbwu@tju.edu.cn (B.W.)
2 State Key Laboratory for Tunnel Engineering, School of Civil Engineering, Sun Yat-sen University, Zhuhai 519086, China
3 National Engineering Research Center of High-Speed Railway Construction Technology, Changsha 410075, China
4 Institut für Geotechnik, TU Bergakademie Freiberg, 09599 Freiberg, Germany
5 School of Civil Engineering, Central South University, Changsha 410075, China
* Correspondence: jy.fu@csu.edu.cn

Abstract: Infilled joints or faults are often subjected to long-term stable shear forces, and nature surface processes of normal unloading can change the frictional balance. Therefore, it is essential to study the sliding behavior of such granular materials under such unloading conditions, since they are usually the filling matter. We conducted two groups of normal unloading direct shear tests considering two variables: unloading rate and the magnitude of constant shear force. Dry sands may slide discontinuously during normal unloading, and the slip velocity does not increase uniformly with unloading time. Due to horizontal particle interlacing and normal relaxation, there will be sliding velocity fluctuations and even temporary intermissions. At the stage of sliding acceleration, the normal force decreases with a higher unloading rate and increases with a larger shear force at the same sliding velocity. The normal forces obtained from the tests are less than those calculated by Coulomb's theory in the conventional constant-rate shear test. Under the same unloading rate, the range of apparent friction coefficient variation is narrower under larger shear forces. This study has revealed the movement patterns of natural granular layers and is of enlightening significance in the prevention of corresponding geohazards.

Keywords: dry sands; direct shear test; unloading normal force; shear force

1. Introduction

Granular matter is widely distributed in the earth's lithosphere and surface. Under long-term complex geological action, faults and joints on the earth are usually filled by grained gouges (quartz, clay, and phyllosilicates, etc.). The convex bodies on the ordinary fault surfaces will break under shear and compression. This can cause the fracture zone of the fault to be filled by small-sized grains, forming a fault gouge with a thickness of several meters [1–3]. When the gouge reaches a certain thickness, the shear strength of the structural surface will depend entirely on the characteristics of the filler [2,3], making the particle-filled structure more prone to damage under critical loading conditions.

The granular materials acting as gouges in joints or faults are often subjected to long-term stable shear forces. Human activities (e.g., reservoir drainage or excavation works), earth's surface processes (river cutting or glacial ablation, etc.) or tectonic movements [4–7] can change the stress state of these earth masses in the disturbed areas. Mostly, a pre-existing fault zone subjected to normal and shear stresses was initially in a frictional equilibrium state. The normal stress was larger than the shear stress. The reduced normal stress applied to the fault zone causes the deformation of the fault gouges in the fault segment. Specifically, if excavating a tunnel near a fault zone, the normal stress on the fault will be reduced [5]. Also, for example, when an earthquake wave from a distance is transmitted to a sheared fault, the normal stress applied will also drop sharply. Consequently, intensive unloading and stress redistributions can lead to the fast sliding of earth, causing landslides, ground surface cracking, and earthquakes [4,6,7]. Many former studies have focused on baldly contacted rock joints [8–10] or a small block sample with a thin layer of gouge [11–16] under the conditions of normal turbulence [8–16]. The effects of the characteristics of fault gouge, including humidity, mineral compositions, particle radii, and shapes, on the slip patterns have been studied mechanically and microscopically [17–20]. In particular, tests involving velocity stepping and slide-hold-slide have shown that the frictional weakening phenomenon is strongly related to instability. However, the mechanical behavior of granular layers in the thick-filled faults in nature have not attracted enough academic attention.

The shear strength of dry granular matters mainly depends on the normal force, shear rate, and the physical properties of the grains [21–23]. Wu et al. [24] investigated the particle crushing of isotropically consolidated silica sand specimens. Jiang et al. [25] studied the fundamental role of particle size and shear speed in the frictional instabilities of locally sheared granular materials. Wu and Zhao [26] investigated the dynamic triggering of frictional slip on simulated granular gouges. Shear experiments of geomaterials with complex stress routines were widely performed, and some of them also considered hydraulic effects [27–32].

Because of the un-prescribed sliding velocity, the shear properties of granular materials under normal unloading conditions can be quite different from the conventional experiments. However, related topics have not been explored before. Therefore, in this study, we intend to reveal the mechanical patterns of dry sands under invariable shear forces during the normal unloading process. This work is useful for better comprehending the movement patterns of natural granular layers and the prevention of corresponding geohazards.

2. Experimental Setup

2.1. Apparatus and Sample Preparation

We use the DJZ-500 shear box device (Figure 1a) to perform the unloading shear tests under constant shear forces. The shear displacement is measured by a horizontal LVDT, which is attached to the upper part of the shear box. Normal and shear loads are measured by the load cells with a sampling rate of 50 times per second. The granular material used in the experiments is directly placed into the shear box with a length × width × height = 400 mm × 200 mm × 100 mm (Figure 1a). Steel balls are placed between the upper and lower shear boxes to avoid frictional contact. At the same time, PVC film is laid on the inner wall of the shear box to prevent sand leakage.

The gouge layers are sometimes considered to have no cohesion [20]. Therefore, it is reasonable to use dry sands to study the behavior of gouge layers. We use natural sand collected from the coastal areas of southeast China as the experimental granular material, as shown in Figure 1b. The gradation diagram of the sand is shown in Figure 1c, which indicates that it is poorly graded. For each test, we layer 23.0 kg of sand into the shear box and fully compact it to reduce the porosity as much as possible. Since the height of the specimen is kept at 100 mm, the initial density of sand in the shear box is 2875 kg/m^3, i.e., the initial porosity of the granular sample is a constant, and the porosity distribution in the experimental samples is uniform. Typical particle sizes and other detailed properties of the sand are reported in Table 1. The relevant literature [33] confirms that under the authorized pressure in this study, no obvious particle breakage occurs. Therefore, we hold the precondition that the properties of particulate matter remain unchanged during the experiment.

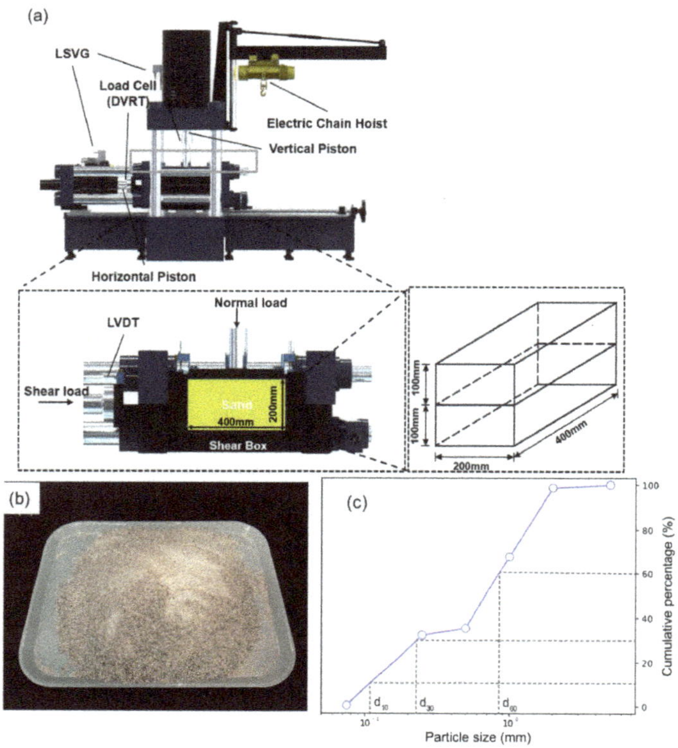

Figure 1. Testing apparatus and experimental materials: (**a**) the DJZ-500 shear box device; (**b**) dry sands used in tests; and (**c**) grain grading curve of the sand samples.

Table 1. Grain properties of granular matter used in experiments.

d_{10} (mm)	d_{10} (mm)	d_{10} (mm)	Uniformity Coefficient (C_u)	Coefficient of Curvature (C_c)
0.11	0.22	0.84	7.64	0.524

2.2. Experimental Scheme

As illustrated in Figure 2a, the granular matter in the shear box is subjected to both vertical force (normal force F_N) and horizontal force (shear force F_S). Consequently, the shear plane in the middle of the granular specimen is subjected to a normal stress σ and a stable shear stress τ. In the experiment, the sliding accelerations are minimal, so the inertia force on the shear box and specimen can be ignored, i.e., the force on the horizontal loading cell is equal to the shear force. During the unloading process, the shear force is kept constant by the horizontal servo system, and the applied normal force F_N decreases linearly with a constant rate a:

$$F_N = F_{N0} - at \tag{1}$$

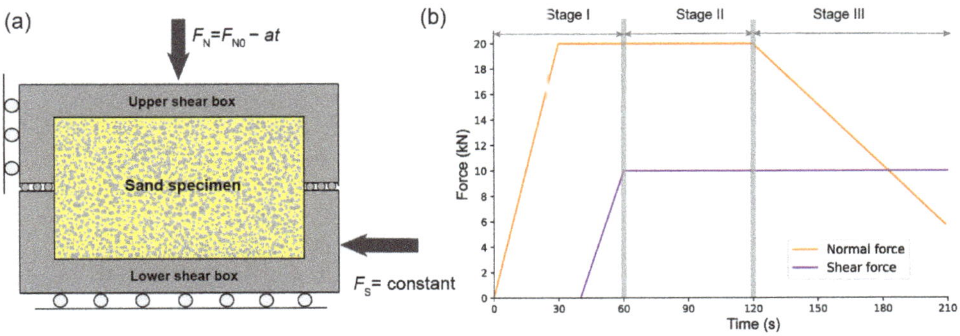

Figure 2. Experimental configuration: (**a**) setup of direct shear test under unloading normal force and constant shear force; (**b**) normal and shear force application scheme during three-stage loading process.

According to the loading scheme illustrated in Figure 2b, the test procedure includes three stages:

- Stage I: Load the vertical force to 20 kN at a loading rate of 40 kN/min. Then, maintain this normal force for 30 s. After that, load the shear force to the predetermined value within 20 s by activating the horizontal piston.
- Stage II: Maintain the shear force and the normal force at their respective values for 60 s, where the stress state of the whole system remains unchanged.
- Stage III: Unload the normal force linearly at a predetermined rate. The gradient of the descent line in Figure 2b equals the unloading rate. When the normal force descends to a certain level, the shear box begins to slide.

We conducted two groups of unloading direct shear tests (Group A and Group B) at room temperature, considering two variables: unloading rate and magnitude of shear force. The parameters used in the tests are shown in Table 2. Referring to a similar experiment studying rock specimens [30], the constant shear forces used in the tests are less than the ultimate shear strength under the given normal force and greater than the shear strength under zero normal force.

Table 2. Experimental scheme under a series of unloading rate and shear force conditions. Group A: varying unloading rate, keeping constant shear force of 10 kN. Group B: varying shear force, keeping a constant unloading rate of 0.24 kN/s.

Tests	No.	Unloading Rate (kN/s)	Shear Force (kN)	Initial Shear Force (kN)	Maximum Sliding Velocity (mm/s)
Group A	A1	0.08	10	20	3.7
	A2	0.16	10	20	
	A3	0.32	10	20	
	A4	0.4	10	20	
	A5	0.48	10	20	
	A6	0.56	10	20	
	A7	0.72	10	20	
	A8	0.8	10	20	
Group B	B1	0.24	8	20	3.7
	B2	0.24	9	20	
	B3	0.24	11	20	
	B4	0.24	12	20	
	B5	0.24	14	20	

3. Results and Analysis

3.1. Displacement and Velocity During Sliding Process

Unlike unloading slip behavior of rock joints [28,30], the mechanical behavior of the tested dry sands during the normal unloading process is more complex. The displacement of the shear box does not change for one minute during Stage II, where the normal and shear forces remain constant. When the unloading begins, the corresponding frictional sliding does not start immediately. We define the corresponding moment when the shear box begins to move as 'start time', t_i, and the corresponding moment when the shear box reaches the maximum velocity as the 'end time', t_j. The values of t_i and t_j recorded in the tests are shown in Table 3.

Table 3. The timing that sliding begins ('start time') and ends ('end time').

Tests	No.	Start Time (s)	End Time (s)
Group A	A1	152.2	221.2
	A2	137.4	194.6
	A3	127.5	160.3
	A4	125.6	151.8
	A5	125.1	147.4
	A6	125.1	143.3
	A7	123.1	139.0
	A8	123.3	139.6
Group B	B1	134.0	191.1
	B2	134.9	183.5
	B3	132.9	172.0
	B4	131.2	168.2
	B5	127.6	136.4

The variation curves of normal displacement and sliding displacement versus experimental time are drawn in Figure 3, based on the original data recorded by the shear device for each test (Group A and Group B). Figure 4 illustrates the results of sliding velocity as a function of elapsed time for each test, from the beginning moment of normal unloading (i.e., the slopes of sliding displacement vs. time curve). As can be seen from the lower part of Figure 3, the sliding displacement does not increase uniformly over time. More related details can be seen in Figure 4. During the early seconds of normal unloading, the sliding velocity fluctuates (except for test B5), and in some tests the velocity value drops to zero, i.e., the sliding deceleration or intermission occur absolutely in the experiment, but the shear box accelerates persistently in the last seconds of normal unloading. Also, the slope change in sliding displacement has a strong correlation with the normal displacement.

For our test apparatus, the maximum moving velocity of the horizontal piston is around 3.7 mm/s. Therefore, when the sliding velocity reaches 3.7 mm/s, the test will automatically be stopped, so the final vertical ordinate values of the curves in Figure 4 are the same. Note that even at the fastest sliding rate, the acceleration of the shear box is still minimal, and the unbalanced force of the shear test system (acceleration times total mass) is negligible. This is consistent with the anticipation in Section 2.2 that the shear force of the granular matter is the force on the horizontal loading cell, i.e., the constant value that we set.

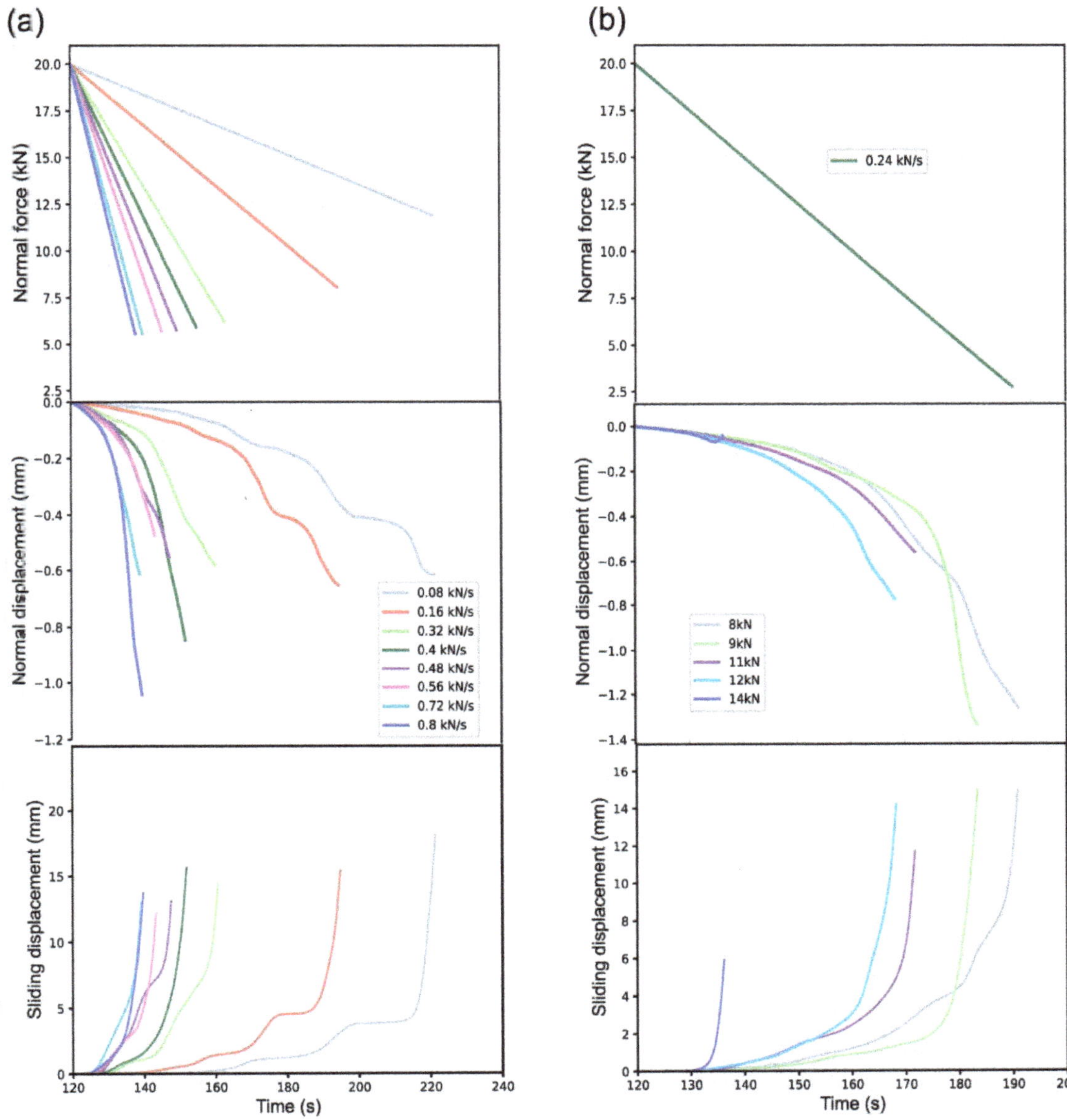

Figure 3. Normal force variation as function of elapsed time and experimental results of normal and sliding displacement versus time for each test. (**a**) Normal force, normal displacement, and sliding displacement versus time for different unloading rates (Group A). (**b**) Normal force, normal displacement, and sliding displacement versus time for different shear force (Group B).

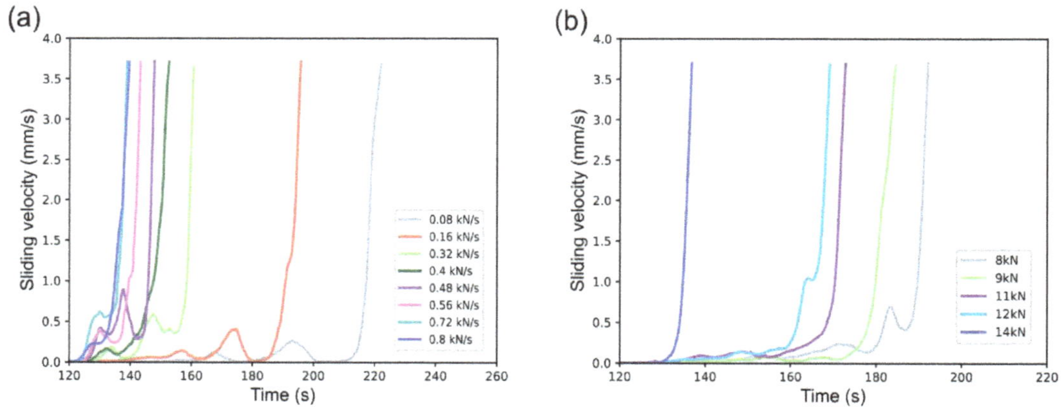

Figure 4. Sliding velocity versus time for each test since the beginning moment of normal unloading. (**a**) Sliding velocity versus time for different unloading rates. (**b**) Sliding velocity versus time for different shear force.

3.2. Shear Strength and Sliding Deceleration/Intermission

The conventional direct shear tests (direct shear at a constant shear displacement rate with a constant normal force) are widely used to measure the frictional criteria of granular matters. In Coulomb's theory, as the peak shear strength increases linearly with larger normal stress, there are two frictional criteria: frictional angle (φ) and cohesion. Since the cohesion of the granular matter equals zero, the relationship between the ultimate shear stress and normal stress can be described as Equation (2). In Equation (2), τ_m reflects the maximum shear capacity of the granular material, and the ratio of shear force to normal force (F_s/F_N) is friction coefficient (μ), which equals the tangent of friction angle when the cohesion is zero (Equation (3)).

$$\tau_m = \sigma \tan \varphi \quad (2)$$

$$\mu = \tan \varphi = 0.6897 \quad (3)$$

To study the frictional criterion (friction angle) of the granular matter used in this study, we additionally perform a series of conventional direct shear tests (selected constant shear rate = 0.0833 mm/s) (Figure 5a,b). The obtained friction coefficient and friction angle are 0.6897 and 34.6°, respectively, as shown in Figure 5b. As illustrated in Figure 5a, at the normal force of 30 kN, the shear resistance of the experimental granular material keeps increasing with a wide shear displacement range (from 0 to 25 mm), before reaching the potential peak strength τ_m in Equation (2).

The experimental results (Figure 5a) show that the shear resistance of the sands increases with higher normal load and larger shear displacement in a wide range. As the horizontal loading is being applied, the system is in a delicate state of equilibrium, which has a strong relationship with the accumulated sliding displacement. When the normal unloading begins, the upper restraint on the material is gradually relaxed and the balance tends to be broken, so the shear box moves forward in an attempt to find a new balance.

From a mesoscopic point of view (Figure 5c), there are two kinds of effects in the sliding process: particle interlacing and normal relaxation. If the unloading effect is not taken into consideration (when the unloading rate is low) after horizontal displacement (Δd), the sheared particles are embedded. Then, the system coordination number (average contact number for a single particle) increases. This leads to shear shrinkage and enhances shear strength. On the other hand, normal unloading makes the compressed particles bounce up and reduces the contact force (the darker balls in Figure 5c) between the particles, and the shear resistance may disappear. Within a certain unloading range, the effect of particle slip embedding may be greater than that of unloading, resulting in sliding velocity fluctuation and even sliding pause. Nevertheless, if the unloading rate or shear force is relatively high, the unloading effect will play a dominant role. In this case, there will be no obvious sliding deceleration/intermission phenomenon. The corresponding change in the normal displacement confirms our conjecture. Meanwhile, at the accelerating stage of sliding, the granular matter needs to slide rapidly to achieve the required shear force.

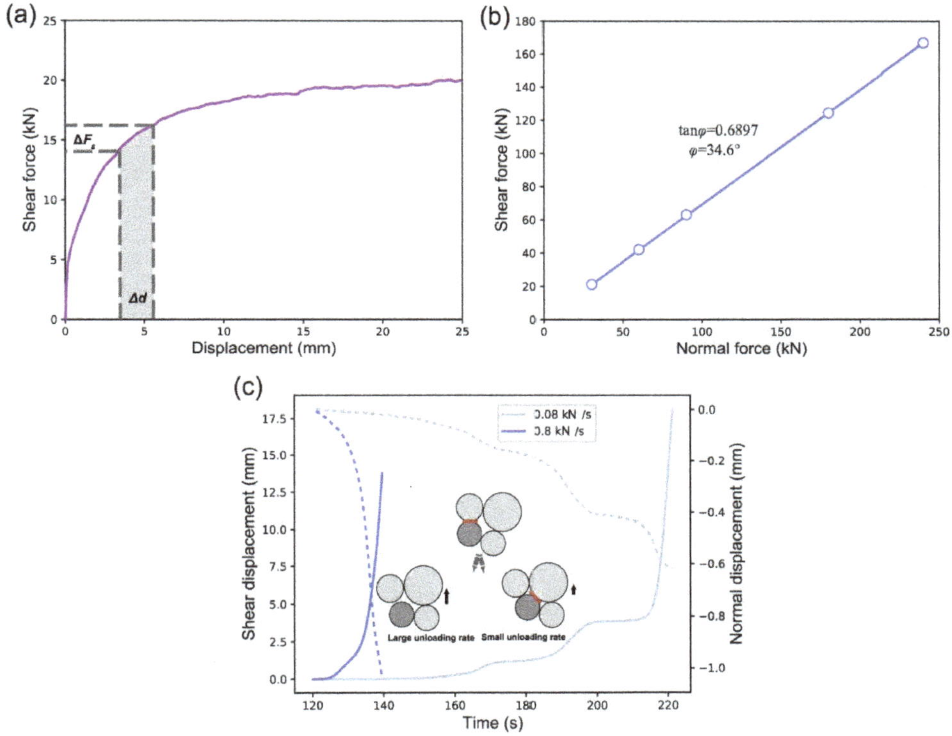

Figure 5. (**a**) Relationship between shear displacement and shear force under F_N = 30 kN in a conventional direct shear test. Within the displacement range, the shear strength of the granular material keeps increasing with larger shear displacement. (**b**) The peak shear force (shear strength) for different normal force in 0.0833 mm/s shear-displacement-controlled conventional direct shear test. (**c**) Shear displacement (solid lines) and normal displacement (dash lines) versus time for low (0.08 kN/s) and high (0.8 kN/s) unloading rates extracted from Figure 3a and illustration to explain sliding deceleration/intermission.

3.3. *Effects on the Normal Force of Unloading*

Values of instantaneous normal forces corresponding to the slip velocity of 3.7 mm/s, 2.5 mm/s, and 1.5 mm/s during the final fast sliding stage in test Group A and Group B are shown in Figure 6. As can be seen, the corresponding normal force has a strong relationship with both normal unloading rate and shear force with different sliding velocities. With a higher normal unloading rate, the corresponding normal force during fast sliding decreases. The change in corresponding normal force is more drastic when the unloading rate is too low or too high. When the unloading rate is 0.3~0.7 mm/s, the normal force is less sensitive to the change in the unloading rate. For the first several tests of Group A, due to the low normal unloading rates, the decrease in the corresponding normal force is insignificant. So, the three curves in Figure 6a are very close at the beginning. Thanks to the effects of a high unloading rate, the interval between these three curves enlarges gradually. In Figure 6b, the corresponding normal force during fast sliding increases with larger shear force. To better demonstrate the influence of normal unloading rates and constant shear forces on the corresponding normal force at different velocities, we fit the results above to synthesize 3D surfaces in Figure 7. When both shear force and unloading rate are in their respective middle ranges, the variation in normal force is not very conspicuous.

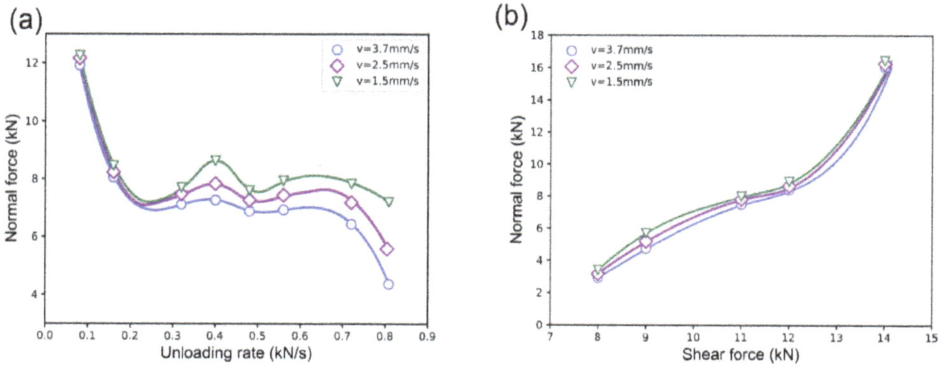

Figure 6. Variation in normal force at different sliding velocities. (**a**) For different unloading rate (Group A) and (**b**) for different shear force (Group B).

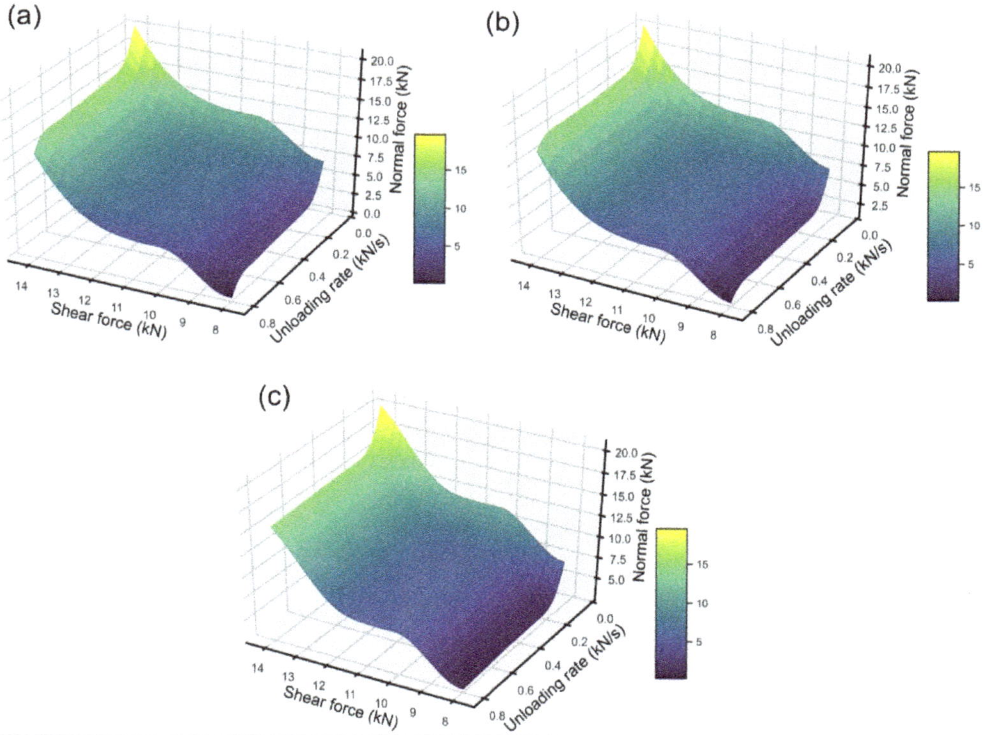

Figure 7. Normal force versus unloading rate and shear force at sliding velocity of (**a**) 3.7 mm/s, (**b**) 2.5 mm/s, and (**c**) 1.5 mm/s.

In the process of normal unloading, the sheared sand is in an unstable and fast motion state, where the normal constraint is constantly relaxed. Therefore, its shear strength characteristics cannot be explained by the Coulomb theory obtained from conventional low-rate direct shear tests. In Figure 6a, the normal forces are much lower than the theoretical value calculated by the Coulomb theory (10/0.6897 kN ≈ 14.50 kN). In addition, as shown in Figure 6b, the curves are not straight lines, and the corresponding normal force is always less than the theoretical value given by Equation (2).

3.4. Variation in Apparent Friction Coefficient

The ratio of shear force to normal force is defined as the apparent friction coefficient in the unloading period. Different from conventional direct shear tests, the temporal variation in the apparent friction coefficient in unloading shear tests is artificially designed. According to Equation (4), since F_s is a constant and F_N linearly decreases with unloading time, the friction coefficient will increase inversely and reach the maximum value when the normal unloading stops. As illustrated in Figure 8a,b, at the maximum sliding velocity of 3.7 mm/s, the corresponding friction coefficient increased with the increase in unloading rate, and decreased with a larger applied shear force.

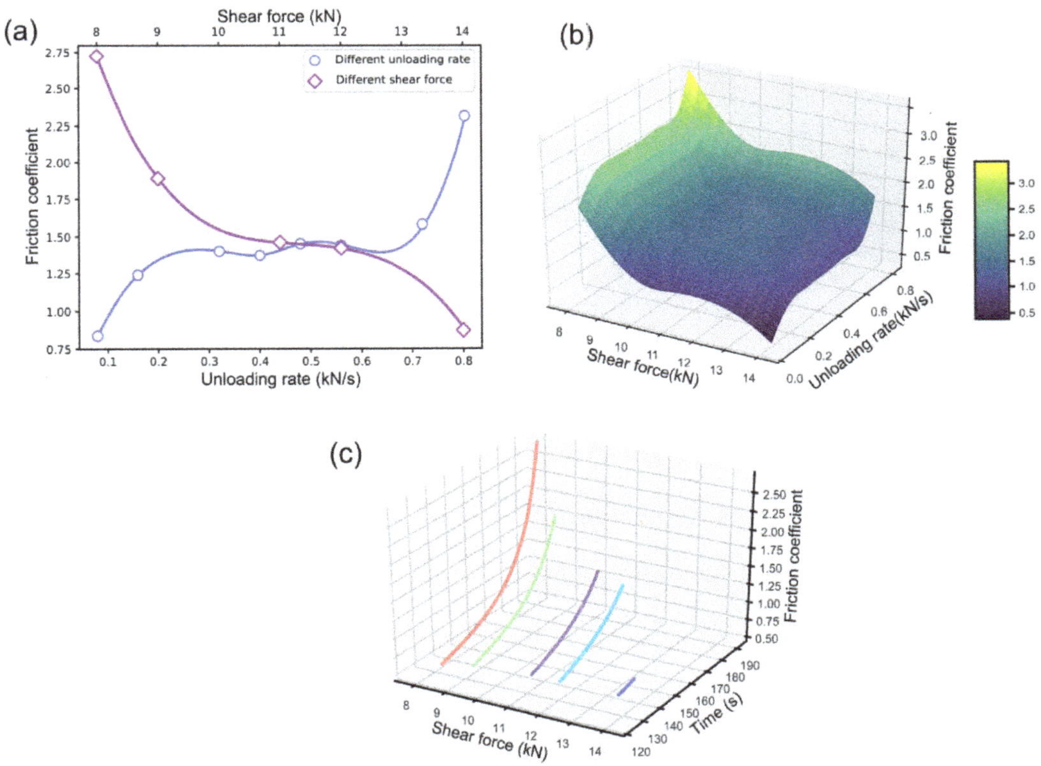

Figure 8. (**a**) Apparent friction coefficient when the sliding velocity reaches 3.7 mm/s (i.e., ultimate value in the test) for different unloading rates and different shear force. (**b**) Apparent friction coefficient versus unloading rate and shear force at sliding velocity of 3.7 mm/s. (**c**) The variation in friction coefficient versus normal unloading time for different shear force (the recorded data in Group B). The different colors represent different single tests.

In test Group B (with no intermission during the sliding), since the unloading rate is a constant (=0.24 kN/s), the apparent friction coefficient is a function of time and shear force, as formulated in Equation (5). Its corresponding surface in the space is an inclined concave surface. Obviously, after the same normal unloading time, the smaller the shear force, the smaller the friction coefficient. Because the sliding friction coefficient exists only during the stage of continuous slip, the range of friction coefficient variation is limited. This is reflected by the inverse proportional function curves from t_i to t_j in Figure 8c. The points beyond these lines are the values that the coefficient of friction has not reached in our tests. It can be seen that the greater the shear force, the narrower the range of the friction coefficient will be. The initial value of the friction coefficient is lower than that of the conventional direct shear test, and it changes according to the inverse proportional function. Also, the ultimate value of the friction coefficient is significantly greater than the tan φ value in Equation (3), which matches the result in Figure 6b when the normal force is fairly low.

$$\mu = \frac{F_S}{F_N} = \frac{F_s}{F_{N0} - at} \tag{4}$$

$$\mu = \mu(F_S, t) = \frac{F_s}{20 - 0.24t}, t \in (t_i, t_j) \tag{5}$$

4. Conclusions

We performed sliding tests on dry sands which were subjected to constant shear loads, and the normal loads were linearly unloaded. Normal force, shear force, and shear displacement were recorded during the unloading process. The main conclusions drawn were as follows:

The granular sample discontinuously slides during the unloading of normal force, and the slip velocity does not increase uniformly with the elapsed time. Due to the increase in inner material friction strength caused by the accumulated shear displacement, the sliding velocity of dry sands fluctuates, and temporary sliding intermission occurs. With different normal unloading rates and different constant shear forces applied, the dry sands begin to slide under almost the same corresponding normal force. The corresponding normal force decreases with larger unloading rates and increases with larger shear force at different sliding velocities in the final fast sliding stage. These normal force values are less than the values given by Coulomb's theory based on conventional displacement-controlled shear tests. The temporal variation in the apparent friction coefficient under the same normal unloading rate satisfies the inverse proportional function relation during the sliding, and the greater the constant shear force, the variation range of friction coefficient is narrower. Under a specified sliding velocity (the maximum experimental velocity), the apparent friction coefficient (i.e., the maximum friction coefficient) decreases with a higher normal unloading rate, and increases with larger shear force.

Regarding the limitations of this study, the sliding intermission phenomenon is not analyzed quantitatively, and the hypothesis in this paper can be verified by FEM and DEM numerical simulations in the future. Further investigations considering more environmental factors (e.g., temperatures and fluids [28,34]) are expected to better reflect the conditions in situ.

Author Contributions: Conceptualization, W.D., K.T., J.F. and B.W.; methodology, K.T.; software, K.T.; data curation, W.D.; writing—original draft preparation, K.T.; writing—review and editing, W.D. and J.F.; visualization, K.T.; supervision, W.D.; project administration, J.F. and B.W.; funding acquisition, J.F. and B.W. All authors have read and agreed to the published version of the manuscript.

Funding: This research is financially supported by the National Natural Science Foundation of China (No. 52474122), the Open Foundation of National Engineering Laboratory for High-Speed Railway Construction (HSR202105), the open research fund of the State Key Laboratory of Hydraulic

Engineering Simulation and Safety, Tianjin University (No. HESS-2317), and the Natural Science Foundation of Guangdong Province of China (2022A1515240009).

Institutional Review Board Statement: Not applicable.

Informed Consent Statement: Not applicable.

Data Availability Statement: The original contributions presented in this study are included in the article. Further inquiries can be directed to the corresponding author.

Conflicts of Interest: The authors declare no conflicts of interest.

References

1. Barton, N. *A Review of the Shear Strength of Filled Discontinuities in Rock*; Norwegian Geotechnical Institute Publication; Norwegian Geotechnical Institute: Oslo, Norway, 1974; p. 105. [CrossRef]
2. Li, Z.; Li, J.; Han, M.; Liu, L. Investigating the Shear Strength Characteristics of Slip Zone Soil Based on In-Situ Multiple Shear Tests. *KSCE J. Civ. Eng.* **2023**, *27*, 3793–3807. [CrossRef]
3. Huang, X.; Qi, S.; Xia, K.; Shi, X. Particle crushing of a filled fracture during compression and its effect on stress wave propagation. *J. Geophys. Res. Solid Earth* **2018**, *123*, 5559–5587. [CrossRef]
4. Song, K.; Wang, F.; Yi, Q.; Lu, S. Landslide deformation behavior influenced by water level fluctuations of the Three Gorges Reservoir (China). *Eng. Geol.* **2018**, *247*, 58–68. [CrossRef]
5. Wu, W.; Zhao, Z.; Duan, K. Unloading-induced instability of a simulated granular fault and implications for excavation-induced seismicity. *Tunn. Undergr. Space Technol.* **2017**, *63*, 154–161. [CrossRef]
6. Brideau, M.A.; Sturzenegger, M.; Stead, D.; Jaboyedoff, M.; Lawrence, M.; Roberts, N.J.; Ward, B.C.; Millard, T.H.; Clague, J.J. Stability analysis of the 2007 Chehalis lake landslide based on long-range terrestrial photogrammetry and airborne LiDAR data. *Landslides* **2012**, *9*, 75–91. [CrossRef]
7. Xia, K.; Chen, C.; Yang, K.; Zhang, H.; Pang, H. A case study on the characteristics of footwall ground deformation and movement and their mechanisms. *Nat. Hazards* **2020**, *104*, 1039–1077. [CrossRef]
8. Dang, W.; Konietzky, H.; Li, X. Frictional responses of concrete-to-concrete bedding planes under complex loading conditions. *Geomech. Eng.* **2019**, *17*, 253–259. [CrossRef]
9. Tao, K.; Dang, W.; Liao, X.; Li, X. Experimental study on the slip evolution of planar fractures subjected to cyclic normal stress. *Int. J. Coal Sci. Technol.* **2023**, *10*, 67. [CrossRef]
10. Kilgore, B.; Lozos, J.; Beeler, N.; Oglesby, D. Laboratory observations of fault strength in response to changes in normal stress. *J. Appl. Mech.-Trans. ASME* **2012**, *17*, 253–259. [CrossRef]
11. Boettcher, M.S.; Marone, C. Effects of normal stress variation on the strength and stability of creeping faults. *J. Geophys. Res. Solid Earth* **2004**, *109*, B03406. [CrossRef]
12. Perfettini, H.; Schmittbuhl, J. Periodic loading on a creeping fault: Implications for tides. *Geophys. Res. Lett.* **2001**, *28*, 435–438. [CrossRef]
13. Lockner, D.A.; Beeler, N.M. Premonitory slip and tidal triggering of earthquakes. *J. Geophys. Res. Solid Earth* **1999**, *104*, 20133–20151. [CrossRef]
14. Linker, M.F.; Dieterich, J.H. Effects of variable normal stress on rock friction: Observations and constitutive equations. *J. Geophys. Res. Solid Earth* **1992**, *97*, 4923–4940. [CrossRef]
15. Tao, K.; Dang, W. Frictional behavior of quartz gouge during slide-hold-slide considering normal stress oscillation. *Int. J. Coal Sci. Technol.* **2023**, *10*, 34. [CrossRef]
16. Tao, K.; Dang, W.; Li, Y. Frictional sliding of infilled planar granite fracture under oscillating normal stress. *Int. J. Min. Sci. Technol.* **2023**, *33*, 687–701. [CrossRef]
17. Ikari, M.J.; Carpenter, B.M.; Marone, C. A microphysical interpretation of rate-and state-dependent friction for fault gouge. *Geochem. Geophys. Geosyst.* **2016**, *17*, 1660–1677. [CrossRef]
18. Ikari, M.J.; Niemeijer, A.R.; Marone, C. The role of fault zone fabric and lithification state on frictional strength, constitutive behavior, and deformation microstructure. *J. Geophys. Res. Solid Earth* **2011**, *116*, B08404. [CrossRef]
19. Proctor, B.P.; Mitchell, T.M.; Hirth, G.; Goldsby, D.; Zorzi, F. Dynamic weakening of serpentinite gouges and bare surfaces at seismic slip rates. *J. Geophys. Res. Solid Earth* **2014**, *119*, 8107–8131. [CrossRef]
20. Carpenter, B.M.; Ikari, M.J.; Marone, C. Laboratory observations of time-dependent frictional strengthening and stress relaxation in natural and synthetic fault gouges. *J. Geophys. Res. Solid Earth* **2016**, *121*, 1183–1201. [CrossRef]
21. Alwalan, M.; Almajed, A.; Lemboye, K.; Almuaim, A. A direct shear characteristics of enzymatically cemented sands. *KSCE J. Civ. Eng.* **2023**, *27*, 1512–1525. [CrossRef]

22. Liu, S.; Mao, H.; Wang, Y.; Weng, L. Experimental study on crushable coarse granular materials during monotonic simple shear tests. *Geomech. Eng.* **2018**, *15*, 687–694. [CrossRef]
23. Samanta, M.; Punetha, P.; Sharma, M. Effect of roughness on interface shear behavior of sand with steel and concrete surface. *Geomech. Eng.* **2018**, *14*, 387–398. [CrossRef]
24. Wu, Y.; Hyodo, M.; Aramaki, N. Undrained cyclic shear characteristics and crushing behaviour of silica sand. *Geomech. Eng.* **2018**, *14*, 1–8. [CrossRef]
25. Jiang, Y.; Wang, G.; Kamai, T.; McSaveney, M.J. Effect of particle size and shear speed on frictional instability in sheared granular materials during large shear displacement. *Eng. Geol.* **2016**, *210*, 93–102. [CrossRef]
26. Wu, W.; Zhao, J. A dynamic-induced direct-shear model for dynamic triggering of frictional slip on simulated granular gouges. *Exp. Mech.* **2014**, *54*, 605–613. [CrossRef]
27. Wang, J.J.; Liu, M.; Jian, F.; Chai, H. Mechanical behaviors of a sandstone and mudstone under loading and unloading conditions. *Environ. Earth Sci.* **2019**, *78*, 30. [CrossRef]
28. Ji, Y.; Wu, W.; Zhao, Z. Unloading-induced rock fracture activation and maximum seismic moment prediction. *Eng. Geol.* **2019**, *262*, 105352. [CrossRef]
29. Duan, K.; Ji, Y.; Wu, W.; Kwok, C.Y. Unloading-induced failure of brittle rock and implications for excavation-induced strain burst. *Tunn. Undergr. Space Technol.* **2019**, *84*, 495–506. [CrossRef]
30. Zhu, T.; Huang, D. Experimental investigation of the shear mechanical behavior of sandstone under unloading normal stress. *Int. J. Rock Mech. Min. Sci.* **2019**, *114*, 186–194. [CrossRef]
31. Yin, S.; Liu, P.; Yan, P.; Li, X. Analysis of undrained shear characteristics and structural damage of granite residual soil. *KSCE J. Civ. Eng.* **2023**, *27*, 3753–3764. [CrossRef]
32. Sonnekus, M.C.H.; Smith, J.V. Comparing shear strength dispersion characteristics of the triaxial and direct shear methods for undisturbed dense sand. *KSCE J. Civ. Eng.* **2022**, *26*, 1560–1568. [CrossRef]
33. Lu, Z.; Yao, A.; Su, A.; Ren, X.; Liu, Q.; Dong, S. Re-recognizing the impact of particle shape on physical and mechanical properties of sandy soils: A numerical study. *Eng. Geol.* **2019**, *253*, 36–46. [CrossRef]
34. Li, Y.; Dang, J.; Zhang, S.; Shen, B.; Zheng, D.; Zhang, H.; Hou, J. Shear mechanical properties and surface damage characteristics of granite fractures treated by real-time high temperature. *J. Cent. South Univ.* **2023**, *30*, 2004–2017. [CrossRef]

International Journal of
Geo-Information

MDPI

Article

Evaluating Shallow Landslide Prediction Mapping by Using Two Different GIS-Based Models: 4SLIDE and SHALSTAB

Federico Valerio Moresi [1,*], Mauro Maesano [1], Marco di Cristofaro [1], Giuseppe Scarascia Mugnozza [2] and Elena Brunori [1]

[1] Department of Innovation in Biological, Agro-Food and Forest Systems, University of Tuscia, 01100 Viterbo, Italy; m.maesano@unitus.it (M.M.); marco.dicristofaro@unitus.it (M.d.C.); brunori@unitus.it (E.B.)
[2] European Forest Institute, Biocities, 00189 Roma, Italy; giuseppe.scarascia@efi.int
* Correspondence: f.v.moresi@unitus.it; Tel.: +39-076-135-7395

Abstract: Landslides affecting soil layers up to 1–2 m deep pose a significant hazard in mountainous and hilly regions, particularly in the Mediterranean, where intense precipitation is increasing. Identifying landslide-prone areas is crucial for risk assessment and mitigation, as landslides can severely impact land surfaces, infrastructure, and inhabited areas. Forest cover and management play a fundamental role in stabilizing soil and reducing landslide susceptibility. This study focuses on landslide forecasting models, which integrate geological, climatic, and topographic data to predict landslide probability and severity. Specifically, we compare the predictive accuracy of the 4SLIDE model with the established SHALSTAB model in a forested mountain catchment within Sila National Park, Southern Italy, using GIS-based analysis. The results demonstrate that both models effectively identify high-risk areas, with ROC analysis confirming the superior predictive capability of the 4SLIDE model. Our findings underscore the critical importance of early landslide identification, supporting timely interventions and the implementation of forest engineering and Civil Protection measures to mitigate the impact of landslides on communities and infrastructure.

Keywords: soil stability; land surface transformations; risk prediction; 4SLIDE; SHALSTAB; landslide; Mediterranean region

1. Introduction

Landslide susceptibility indicates the likelihood that a particular area will experience landslides or soil slippage. This susceptibility is shaped by various factors, including geological, environmental, and human-related influences [1]. Landslide susceptibility depends on factors such as topography (steep slopes), soil type (clayey/silty), and intense rainfall, which saturates the soil. These factors increase the risk of landslides and mudflows [2–4]. Human activities (like construction or deforestation) and past landslides increase regional vulnerability [1,3]. Risk assessment, through geotechnical and hydrological analyses, allows the identification of critical areas and the planning of interventions (engineering structures, soil stabilization, or monitoring) [5,6]. Natural disasters such as landslides, cause fatalities, but preventive measures and rapid responses can reduce them. Landslide hazard refers to the potential for slope failure, while risk quantifies the resulting economic and human losses [5]. Landslides are the result of a combination of natural factors (like heavy rainfall or floods) and human factors (like land use changes or deforestation) [7,8]. The interaction between rain, surface runoff, and subsurface runoff destabilizes slopes by eroding materials and modifying the regolith. Deforestation, hill cutting, and inadequate rural land planning, which represent forms of poor land management, are primary causes that exacerbate triggering factors and increase landslide risk [9,10]. According to the United Nations Office for Disaster Risk Reduction (UNDRR), landslides accounted for approximately 5% of all natural disaster-related deaths between 1998 and 2017, with over 18,000 fatalities globally [11]. Natural disasters in 2016 caused 8733 deaths and $154 billion in damages, highlighting the devastating socio-economic impact of these events [11]. Italy is among the most vulnerable European countries in this regard, with 620,808 landslides affecting 23,700 km^2, or 7.9% of its national territory [12–14]. Vegetation cover plays a crucial role in reducing hydrogeological risks, preventing runoff and surface erosion, and increasing soil shear resistance through root systems [5,15]. One of the key approaches for developing risk reduction strategies is the creation of a landslide susceptibility map (LSM). An LSM offers a spatial distribution of potential landslides, thus playing an essential role in mitigating landslide risk [16]. Models estimate risks and generate probability maps under various forest management and rainfall scenarios. Models for predicting landslide susceptibility are crucial tools for managing and mitigating natural risks [5]. These models help identify areas potentially prone to landslides, allowing planners and decision-makers to take preventive measures and reducing the impact of such events on people, infrastructure, and the environment [17]. Various methods for generating landslide susceptibility maps have been demonstrated over the past decades [5,15,18–20]. Landslide susceptibility mapping models offer several benefits, such as risk prevention and mitigation by identifying at-risk areas for planning interventions, emergency management by locating hazardous areas during critical events, land planning by supporting urban development and environmental conservation, and environmental monitoring by assessing the impact of climatic and environmental changes. Many researchers have explored different approaches for creating LSMs. In addition to crisis management, LSMs are crucial for identifying areas at risk of landslides, as well as for managing and reducing such risks [21]. These maps can be generated using an appropriate model, incorporating landslide data and a set of independent variables [22]. There are three main groups of landslide susceptibility methods: innovative, deterministic, and statistical [23]. Innovative models rely on expert judgment to assign weights to each factor. Therefore, this type of model has a higher potential for errors [24]. Deterministic models are developed based on mathematical relationships. These models are grounded in physical laws, requiring the calculation of the relationship between resisting forces and the drivers of mass movements [25].

4SLIDE is a physical model based on a deterministic approach, used to simulate slope stability and predict susceptibility to shallow landslides [19]. The 4SLIDE model integrates geological, topographical, and hydrological parameters to estimate a Factor of Safety (FS) index, indicating the stability of a location (0 represents the most unstable state, while 1.5 represents the most stable state). SHALSTAB (Shallow Landsliding Stability Model) is another widely used physical-deterministic model for predicting shallow landslides. It relies on analyzing slope stability alongside a hydrological model to assess soil saturation levels from rainfall and the potential for shallow landslides [26]. Both SHALSTAB and 4SLIDE are powerful tools in landslide risk management, each suited to specific contexts (Figure 1). SHALSTAB is ideal for large-scale analyses, especially in high-rainfall regions, where it predicts risk effectively over broad areas. In contrast, 4SLIDE is tailored for detailed site-specific analysis, especially in locations where soil characteristics are crucial to understanding risk. Both models utilize numerical geotechnical parameters distributed spatially within a raster-based GIS framework. However, their integration enables a more comprehensive and detailed understanding of the dynamics of landslide susceptibility. Comparing landslide susceptibility models not only helps assess the robustness and reliability of predictions but also identifies differences and similarities between the two approaches, thereby enhancing the accuracy of risk estimates. This dual analysis allows for the combination of the strengths of both models, enabling the identification of the most appropriate one for specific scenarios and, in some cases, suggesting their integrated use.

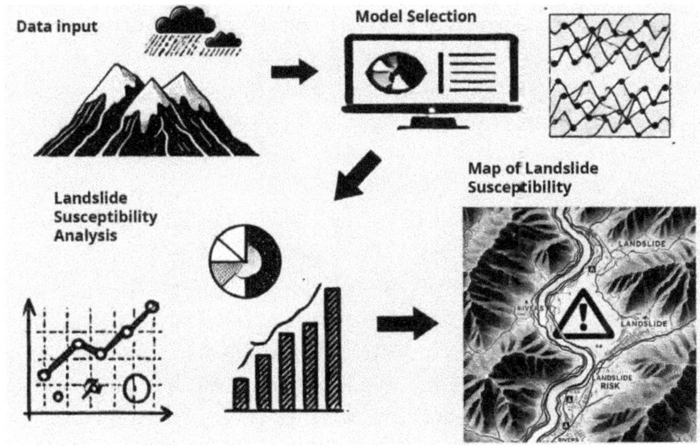

Figure 1. Theoretical flow chart illustrating the functioning of the two models.

This study compares the 4SLIDE model with another landslide susceptibility mapping model, SHALSTAB, highlighting the innovation of an in-depth comparative analysis based on the combined use of two distinct modeling approaches. For a comparative analysis of the different parameter percentages in landslide-prone areas, such as susceptibility assessments provided by models like 4SLIDE or SHALSTAB, it is useful to examine the distribution of various risk categories (e.g., low, medium, and high) over a specific map or region. This approach may include comparing the percentages of different risk levels or analyzing variations in a key parameter, such as terrain slope or soil saturation, within distinct risk zones. The novelty of the study lies in the direct comparison between the two models, which represent complementary but distinct methodologies. Focusing on detailed geotechnical analysis, 4SLIDE considers physical parameters of the terrain, such as slope stability and sliding phenomena, with high local resolution. On the other hand, SHALSTAB adopts a more generalized hydrological approach, based on hydrological thresholds and a broader scale of analysis, suitable for estimating risk areas in regional contexts. This process represents a methodological innovation that not only supports decision-making in risk management and landslide prevention planning but also makes a significant contribution to improving the practical applicability and reliability of existing models, ensuring more effective and sustainable land management. The comparison of the two models was conducted in data from the Bonis catchment, an experimental forested area in Calabria, southern Italy, which serves as an ideal research site for studying hydrology, meteorology, and forestry in a Mediterranean mountain environment. This catchment features varied topography, including steep slopes, diverse soil types, and vegetation. Thus, it reflects the complexities of landslide-prone regions. Equipped with an advanced monitoring network, the Bonis catchment tracks geotechnical and hydrological parameters, such as surface runoff, groundwater levels, soil moisture, and precipitation, providing high-resolution data essential for model calibration and validation [27]. The catchment's small size and consistent land use minimize external influences, offering a controlled environment for focused analysis of landslide susceptibility factors like soil properties, slope stability, and hydrology. Overall, the Bonis catchment is a representative site for evaluating landslide susceptibility models in Mediterranean mountain regions, making it an ideal location for this comparative study [27].

2. Materials and Methods

2.1. The 4SLIDE Model

4SLIDE is a GIS-implemented toolbox for ArcGIS© 10.3 (and later versions), a comprehensive geographic information system developed by ESRI as closed-source software. In contrast, the 4SLIDE model is an open-source .tbx file that can be easily imported as an external toolbox within ArcMap [19]. This model integrates the infinite slope stability equation [28] with the hydrological model "TOPMODEL" [29] to compute the Factor of Safety Index, assessing landslide risk at the catchment scale. The model also incorporates the soil reinforcement effect via a vegetation root strength model, following the principles established by Wu [30]. It focuses on the analysis of landslide bodies with well-defined sliding surfaces, often circular or planar in shape. The software provides a detailed analysis of internal soil forces and the interactions between the different geological layers present in the slope [19,28]. In particular, 4SLIDE allows an in-depth analysis of the effect of the water table on slope stability. This includes evaluating the contribution of pore water pressures and drainage conditions, providing a more accurate representation of the site's hydrogeological conditions [29]. The Factor of Safety (FS) is determined by the following formula:

$$FS = \frac{c_r + c_s + \left(\rho_s * g * z * cos^2\theta - \rho_w * g * hcos^2\theta\right) * tan\varnothing}{\rho_s * g * z * sin\theta * cos\theta} \tag{1}$$

The wet soil density, denoted as ρ_s (kg/m^3), represents the soil weight, including its moisture content. Another important parameter is the soil depth, indicated by z and measured in meters (m). The slope angle of the terrain is given by θ and is measured in degrees. Next, we have the water density, ρ_w, also expressed in kilograms per cubic meter (kg/m^3), which represents how heavy the water is per unit of volume. The water level, denoted by h, is measured in meters (m). Soil and plant roots enhance slope stability by providing additional cohesion. The cohesion due to the roots is denoted by c_r, while the cohesion provided by the soil is denoted by c_s. Both cohesive forces are measured in Newtons per square meter (N/m^2). Finally, gravity, denoted as g and measured in meters per second squared (m/s^2), represents the acceleration due to the gravitational force acting on all these elements. The use of wet soil density in the equation reflects its similarity to field conditions during landslide initiation. FS values have specific meanings: an FS of 1 indicates a critical equilibrium state, suggesting imminent failure, an FS below 1 signals slope collapse, and an FS above 1.5 indicates stability [5,7]. The model's output is a "soil stability map", a raster representation of the Factor of Safety (FS). This map categorizes the terrain into risk classes, each associated with specific FS values, as detailed in Table 1.

Table 1. Classes of Factor of Safety (FS) values on the map for 4SLIDE and SHALSTAB.

Map Color Code	Color Codes (rgb)	Predicted Stability Zone	Susceptibility Ranking	Factor of Safety Index
	(215, 25, 28)	very unstable	high	0 < (FS) ≤ 0.5
	(253, 174, 97)	unstable	moderate	0.5 < (FS) ≤ 1
	(255, 255, 192)	slightly unstable	low	1 < (FS) ≤ 1.3
	(166, 217, 106)	slightly stable	very low	1.3 < (FS) ≤ 1.50
	(26, 150, 65)	stable	stable	(FS) > 1.50

2.2. SHALSTAB (Shallow Landslide Stability Model)

SHALSTAB (Shallow Landslide Stability Model) is a computational tool designed to assess slope stability and predict susceptibility to shallow landslides, especially those triggered by rainfall [26]. SHALSTAB is a hydro-geomechanical model for large-scale slope stability analysis. The model combines a simplified infinite slope stability analysis with a water balance approach. It uses the ratio of height of the groundwater table to soil depth to estimate landslide susceptibility. It is particularly suitable for hydrogeological risk mapping on a regional or watershed scale. This model is valuable for evaluating the potential for soil slippage in hilly or mountainous regions, where heavy rainfall or surface runoff can saturate the soil, increasing the risk of shallow landslides [26]. Grounded in geotechnical analysis, SHALSTAB considers parameters such as slope steepness, soil characteristics, soil saturation levels, and prevailing weather conditions. It serves as a proactive tool to identify areas at risk of shallow landslides, aiding in decision-making processes related to planning and risk mitigation. It is essential to recognize that SHALSTAB is one of several models available for assessing slope stability, and its efficacy depends on the accuracy of data input and calibration for specific site conditions [26]. Engineers and geologists use models like SHALSTAB in conjunction with on-site information and data to evaluate and prevent shallow landslides. The SHALSTAB model integrates infinite slope and steady-state hydrological models, operating within the ArcMap GIS (Geographic Information System) [31]. Specifically designed for shallow landslides, SHALSTAB couples slope stability analysis based on the Mohr–Coulomb failure criterion [32] with a hydrological steady-state model [33]. Assumptions include runoff occurring at the soil–rock interface, a groundwater table parallel to the soil surface, shallow subsurface flow, and the absence of deep drainage and bedrock flow. The model calculates the factor of safety (FS) for each location to estimate stability. FS values are computed by the formula:

$$FS = \frac{(c_r + c_s) + (\rho_s - h * \rho_w) * zcos^2\theta * tan\varnothing}{\rho_s * z * sin\theta * cos\theta} \tag{2}$$

The symbols in Equation (2) refer to the same ones as in Equation (1). The application of SHALSTAB in a GIS context facilitates mapping susceptibility to shallow landslides geographically [31]. This process is beneficial for land planning, risk management, and disaster preparedness. Integrating SHALSTAB within a GIS framework enhances land management efficiency, offering a proactive approach to preventing shallow landslides and addressing associated risks (Table 1).

2.3. Case Study

The research was conducted within the Bonis Basin, an experimental catchment in Southern Italy, located at coordinates 39°29′1.13″ N; 16°31′41.35″ E, straddling the municipalities of Acri and Longobucco (Figure 2) [34,35]. The Bonis watershed, located in the Sila National Park in Calabria, is a mountainous environment characterized by a combination of a Mediterranean mountain climate, shallow soils, and mixed forest cover.

The catchment covers an area of 1385 km^2 and is mostly forested, with a dense network of streams. The Bonis stream features steep slopes with gradients ranging from 35% to 40%, along with small waterfalls and isolated pools [36]. The complex morphology of the catchment, with steep sections, enhances surface runoff and increases soil erosion risk during heavy rainfall. This configuration promotes a rapid hydrological response, with an immediate increase in water flow toward the outlet. The catchment's elevation ranges between 975 m and 1258 m above sea level. This variation contributes to the presence of a variety of local microclimates that influence vegetation distribution and hydrogeological processes. Higher areas tend to be more humid and cooler, while lower areas exhibit relatively drier and warmer climatic conditions [36]. From a geological perspective, the area is composed of four main complexes from different periods. The oldest is composed of acidic igneous and metamorphic rocks, followed by calcareous sediments dating back to the Mesozoic. Then, we find Pliocene sediments, which were deposited in more recent times. Finally, Quaternary sediments represent the youngest phase of this geological sequence [37]. The basin is mainly characterized by acidic intrusive plutonic rocks. The soils in the area are primarily composed of coarse-textured colluvium, which is derived from the weathering and erosion of volcanic rocks. These soils are classified as ultic haploxeralfs according to soil taxonomy, a classification that indicates they are well-drained, acidic, and typically found in regions with a warm climate. The volcanic material in the soil contributes to its coarse texture, which influences water retention and nutrient availability [35]. The soil thickness averages approximately 100 cm and demonstrates high permeability, allowing water to pass through easily. This characteristic contributes to the rapid drainage of rainfall and influences the overall hydrological behavior of the area. The water table responds to rainfall, fluctuating with rain intensity. The region receives an average annual precipitation of 1250 mm (Scheme 1) and experiences an average air temperature of 8.8 °C, which is indicative of a typical Mediterranean mountain climate [36]. This climate is characterized by cool and wet winters, where precipitation is concentrated, and mild to warm summers, marked by moderate temperatures and relatively low rainfall. These climatic conditions support a diverse range of vegetation and influence the overall ecosystem dynamics of the area.

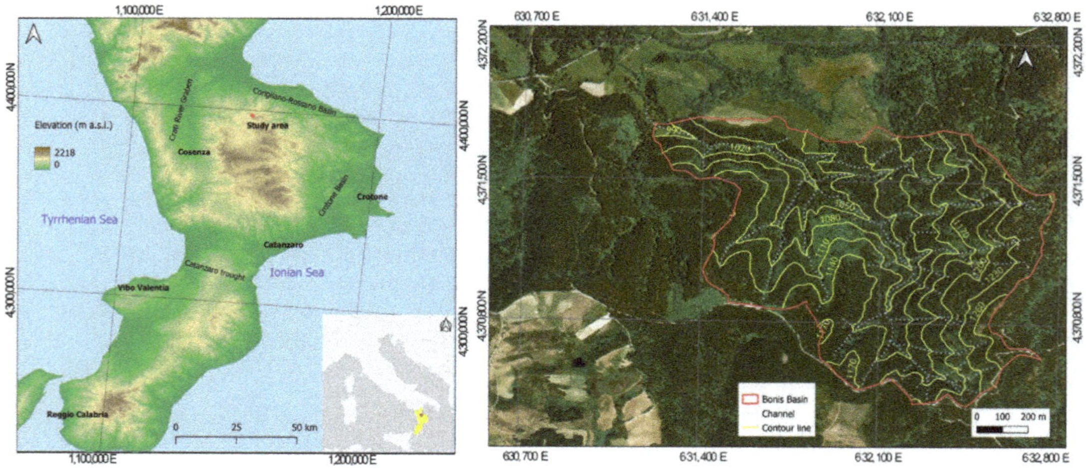

Figure 2. Study area, the Bonis watershed (base map by Google Maps).

The seasonal variation between wet and dry periods plays a significant role in shaping the hydrological and biological characteristics of the landscape. For this study, critical soil properties were analyzed to assess their influence on slope stability and potential landslide risks. These included cohesive strength, which measures the soil's ability to resist internal shear stress; internal friction angle, indicating the soil's resistance to sliding between particles; unconfined compressive strength, which assesses the soil's ability to endure compression without lateral support; unit weight, reflecting the soil's density and the weight it exerts on underlying layers; and hydraulic conductivity, which gauges how easily water can pass through the soil, affecting its saturation and potential for erosion. This comprehensive analysis aimed to provide insights into the soil's behavior under different environmental conditions. These properties were correlated through granulometric analyses performed on numerous soil samples collected within the study area [38]. This detailed examination of soil properties is crucial for understanding the geological and hydrological characteristics of the catchment, providing valuable insights for slope stability assessments and landslide risk evaluations. The landscape is characterized by extensive Calabrian pine plantations (*Pinus laricio* Poiret), along with mixed stands of chestnut (*Castanea sativa* Mill.) and riparian alder forests (*Alnus glutinosa* (L.) Gaertn) [33]. These plant species contribute to soil stabilization and affect the region's hydrological dynamics.

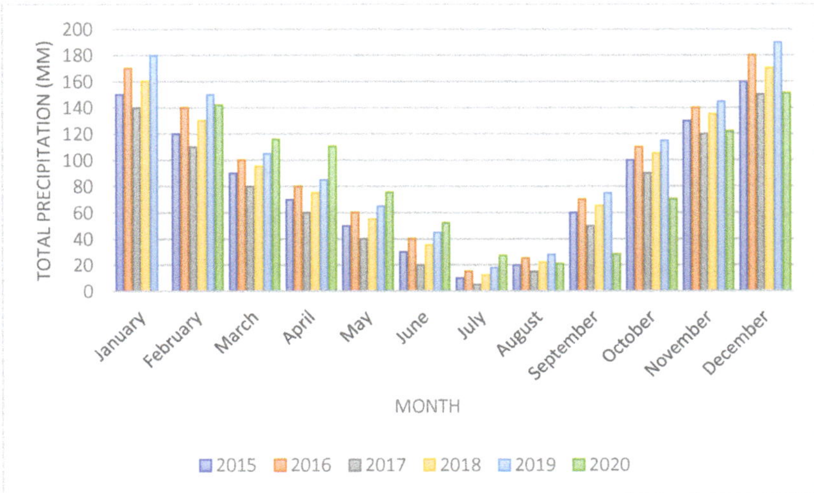

Scheme 1. Climatic graph representing the monthly (2020) distribution of precipitation in the Bonis watershed (https://www.cfd.calabria.it/ (accessed on 10 February 2025)).

2.4. Input Parameters

The parameters used for the geotechnical classification of soils are crucial for predicting landslide risk, as they directly affect soil stability. The soil properties considered in this study include cohesion, internal friction angle, unconfined compressive strength, unit weight, and saturated hydraulic conductivity, as shown in Table 2. These soil parameters were derived from 135 soil samples collected within the study area [15,34]. The soil sampling strategy was designed not only to ensure an adequate number of samples but also to achieve a spatially representative distribution across the study area. This approach aimed to capture variations in soil properties associated with different lithological and geomorphological settings, thus enhancing the robustness of the geotechnical characterization. To obtain a continuous spatial representation of soil properties, the collected data were interpolated using the moving average interpolation method [39]. This method was selected due to its ability to smooth local variations while preserving regional trends, reducing the influence of outliers and measurement noise. Additionally, the moving average approach ensures that interpolated values remain within the observed data range, preventing the generation of extreme values that might occur with other interpolation techniques, such as kriging. This method is particularly effective in areas where sample density is relatively uniform, as it provides a balanced estimation of soil property distributions across the study area [39].

Table 2. Geotechnical properties of the study area.

Lithology	Unit Weight of Soil (kN/m^3)	Unit Weight of *Saturated Soil* (kN/m^3)	Soil Transmissivity m^2/h	Cohesion (kPa)	Internal Friction Angle (rad)
Silt	1.66	2.06	0.502	12.74	0.42
Sand and Silt	1.48	1.88	0.680	8.16	0.52
Sand	1.34	1.69	0.172	0	0.61
Clay and sand	1.53	1.92	0.072	13.72	0.37
Clay	1.36	1.85	0.072	20.30	0.34

2.5. Model Calibration

The application of 4SLIDE and SHALSTAB models in the Bonis Basin involved the use of various input parameters to generate the safety factor map. The primary input was a high-resolution Digital Terrain Model (DTM) (pixel size 1*1 m), derived from a LiDAR survey conducted as part of the ALForLab project (https://www.cnr.it/en/news/8071 (accessed on 10 February 2025)). Root cohesion values for tree and shrub species present in the area were obtained from Moresi et al. [15], while geotechnical data were sourced from Bellotti [38] and reported in Table 2. Meteorological data were gathered from three stations within the study area, i.e., the basin outlet at 975 m a.s.l., Petrarella at 1258 m a.s.l., and Don Bruno at 1157 m a.s.l. [40]. Given the small size of the study area, rainfall distribution was considered sufficiently uniform, allowing for the application of a single average rainfall value of 50 mm/h, spatialized across the entire basin. This value was selected based on the least conservative climatic scenario derived from the pluviometric data analysis. Additionally, a geomorphological survey was performed to create a database of landslides (Figure 3) in the study area. Additionally, a geomorphological survey was conducted to create a database of landslides (Figure 3) present in the study area. After completing the geomorphological survey, 50 random sampling points were selected using GIS software (ArcMap 10.3) to obtain ground truth points. Each sampling point was associated with a sample area with a diameter of 20 m, which was carefully examined (Figure 4). The 50 sampling points were selected using a stratified approach to ensure a balanced distribution across the Bonis Basin based on slope, land use, and lithology. The sample included both areas with past landslides (confirmed by historical data and field surveys) and stable zones for a balanced comparison. Additionally, the points were chosen to be accessible for on-site verification. These details have been incorporated into the manuscript to enhance clarity and reproducibility. The resulting database included detailed information on the location, size, and characteristics of landslides, as well as data on areas where no landslides were observed. Furthermore, a Digital Terrain Model (DTM) was generated using LiDAR point clouds, which provided detailed data on ground elevation with a ground resolution of 1 pixel per square meter. The LiDAR dataset had a point cloud density ranging from 5 to 20 points per square meter, depending on the vegetation coverage, and was calibrated using ground targets surveyed with RTK GPS, ensuring centimetric accuracy. To improve the identification of specific landslide characteristics, the "Hillshade" visualization technique was applied to the DTM. This method uses simulated lighting effects to highlight terrain variations, making landslide features such as scarps, flow paths, and deposition more visible (Figure 4).

Figure 3. Five examples (**a–e**) of landslide phenomena present in the study site, which were detected during the geomorphological survey.

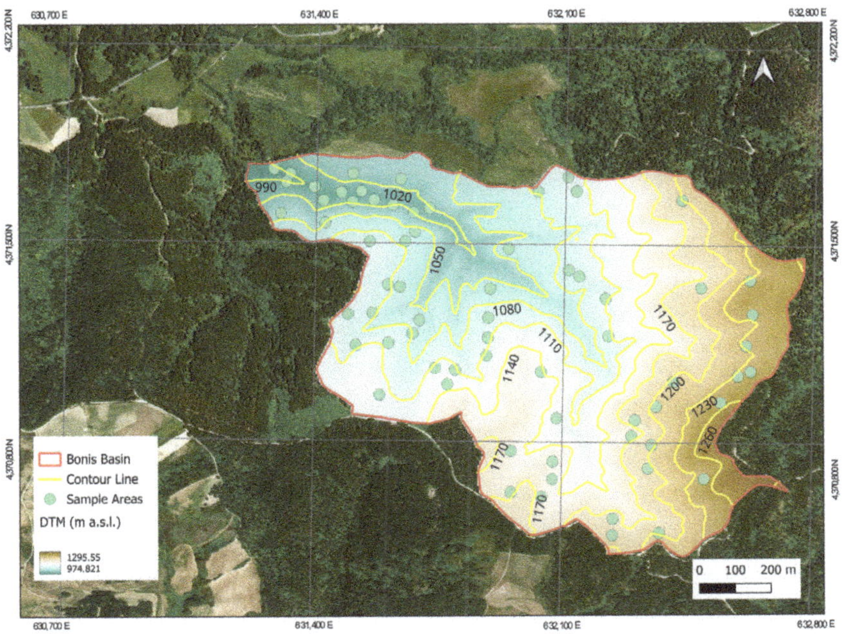

Figure 4. The 50 sampling plots (20 m diameter), randomly distributed, are shown in blue, using the digital elevation model as the topographic map.

2.6. Model Validation

Both ROC-AUC analysis and Cohen's Kappa were employed in tandem to evaluate the predictive performance and consistency of the 4SLIDE and SHALSTAB models. The accuracy of the landslide prediction maps produced by the 4SLIDE and SHALSTAB models was assessed using the Receiver Operating Characteristic (ROC) curve analysis [40,41]. The ROC curve analysis is a widely accepted technique for evaluating the performance of predictive models in binary classification tasks, where the goal is to correctly classify pixels into landslide or non-landslide categories. For this purpose, the method involves constructing a curve based on a confusion matrix, used in binary classification tasks. The confusion matrix considers four possible outcomes: true positive (TP), true negative (TN), false positive (FP), and false negative (FN). These outcomes are determined by comparing the model's predicted results with the ground truth data collected from field surveys (GTS) and the DTM-LiDAR survey. The ROC curve plots the true positive rate (sensitivity) against the false positive rate (1-specificity), providing a graphical representation of the model's performance. The area under the ROC curve (AUC), which ranges from 0.5 to 1.0, is used as an index to assess the model's accuracy. A higher AUC value indicates better predictive performance, with 1.0 representing perfect accuracy and 0.5 suggesting performance no better than random chance. This method offers a robust means of evaluating and comparing the predictive capabilities of the landslide models [40,41]. To further investigate the models' performance, a sensitivity analysis was conducted to assess the influence of slope on the predictive accuracy of 4SLIDE and SHALSTAB [42]. This analysis examined how the models respond to varying slope conditions by categorizing the terrain into three slope ranges and evaluating the corresponding differences in predictive performance. By analyzing the models' stability across different slope gradients, we aimed to determine whether specific terrain conditions significantly impact their reliability in landslide prediction. This assessment provides valuable insights into the strengths and limitations of each model when applied to diverse topographic settings. Additionally, to further assess the agreement between the two models, Cohen's Kappa (κ) coefficient [43,44] was used as a complementary statistical validation method. The Kappa coefficient not only provides a measure of agreement but also corrects for random chance, offering a more nuanced evaluation of model reliability. Unlike the ROC curve, which measures the overall predictive performance, Cohen's Kappa quantifies the level of agreement between the 4SLIDE and SHALSTAB classifications while correcting for agreement occurring by chance. This method allows for a direct comparison of the models' outputs by evaluating the consistency in landslide susceptibility classification across different areas. A κ value close to 1.0 indicates strong agreement, while values near 0 suggest poor agreement beyond chance. The analysis for comparing the ROC curves was performed using R software (Version 4.3.3) (www.r-project.org (accessed on 10 February 2025)), employing a non-parametric method to calculate the Area Under the Curve (AUC). Meanwhile, Cohen's Kappa analysis was conducted by constructing a binary confusion matrix where each pixel was classified as either landslide (0) or non-landslide (1) in both models. This classification allowed for a more detailed evaluation of inter-model agreement, complementing the predictive accuracy assessment provided by the ROC-AUC analysis. The ROC and Kappa analyses were performed using the database from the geomorphological survey and the DTM_LiDAR. The data collected from both methods—Ground Truth Survey (GTS) and LiDAR—were overlaid to accurately assess the presence or absence of landslides in each area, as described by Moresi [18,19]. Based on this comparison, each pixel in the dataset was classified with a value of 0 (presence of a landslide) or 1 (absence of landslide), providing a reliable ground truth for the validation process. This classification process helped create a reliable ground truth for the analysis and further validation of the results of the two models. By incorporating both ROC-AUC analysis and Cohen's Kappa, this study ensures a comprehensive evaluation of the models, considering both their predictive performance and classification consistency. This dual approach enhances the reliability and robustness of the landslide susceptibility assessment, providing valuable insights for risk management and land planning.

3. Results

The results from the two methodologies used to assess landslide susceptibility led to the creation of a susceptibility index, highlighting areas that are at risk. Each pixel on the generated maps corresponds to a Factor of Safety (FS) value, which categorizes the landslide susceptibility of various regions. The analysis highlighted a clear correlation between higher susceptibility areas and steep slope regions, particularly those with established river networks, underscoring the critical role that slope plays in evaluating landslide risk. The maps produced by both models allow for the classification of five classes based on the FS values. For the 4SLIDE model, the analysis revealed that 15% of the study area was classified as "high susceptibility to landslides". The Factor of Safety (FS) value ranged between 0.0 and 0.5, indicating a high probability of landslide occurrence. In contrast, 36% of the area showed medium susceptibility, with FS values between 0.5 and 1.3, suggesting a moderate risk of landslides. Additionally, 20% of the study area was classified as having "low to very low" susceptibility to shallow landslides, with FS values between 1.3 and 1.5, indicating a relatively stable condition but still subject to a certain degree of potential instability. This indicates a relatively stable condition, although some instability potential remains (Figure 5).

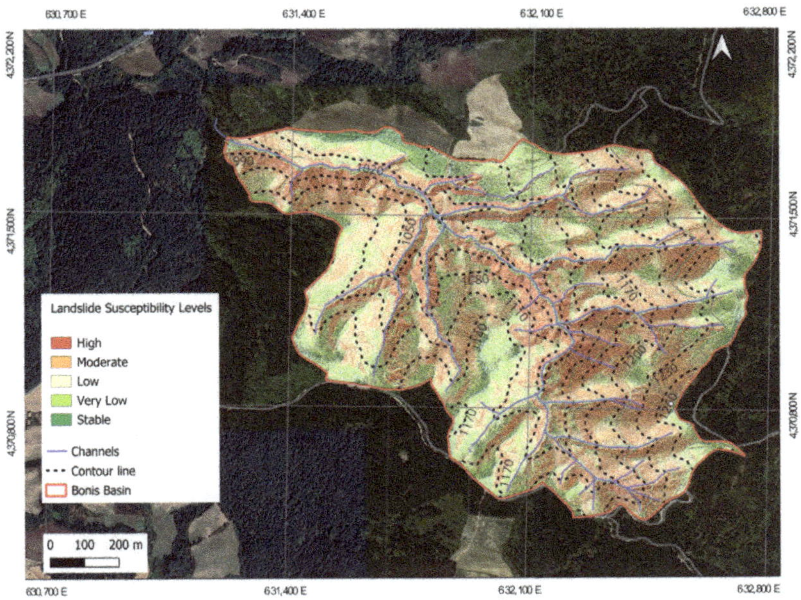

Figure 5. Landslide susceptibility map based on the 4SLIDE model.

In comparison, the SHALSTAB model (Figure 6) also identified 15% of the area as having high instability (0.0 < FS < 0.5). However, a larger portion of the area (49%) was classified as medium susceptibility (0.5 < FS < 1.3), while only 6% was classified as having low stability (1.3 < FS < 1.5).

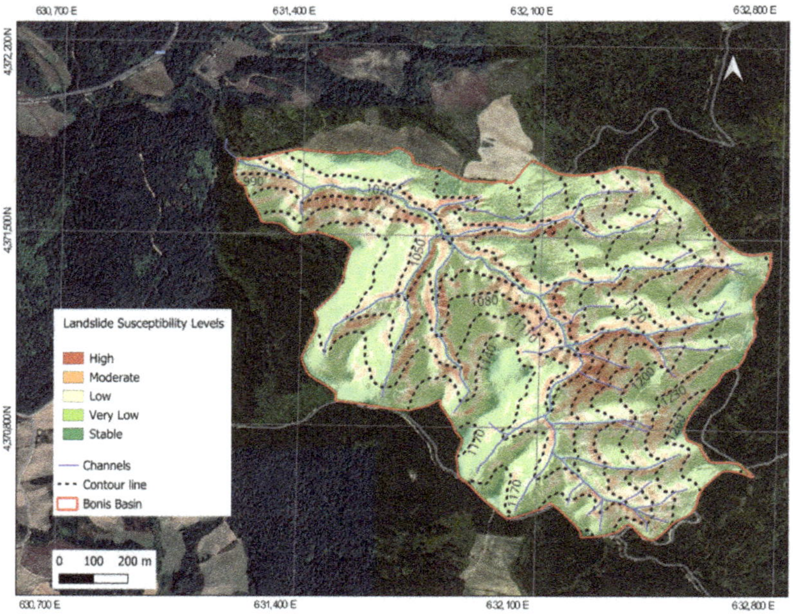

Figure 6. Landslide susceptibility map based on the SHALSTAB model.

Both models consistently identified the most vulnerable areas along steep slopes and unconsolidated geological materials, such as colluvium, particularly those in proximity to watercourses. In contrast, the more stable regions were typically located further away, upslope from these watercourses. The sensitivity analysis conducted across different slope categories further confirmed that both models exhibit varying predictive performances depending on terrain inclination. In particular, the models demonstrated higher sensitivity on steeper slopes (35–90°), where landslides were more frequent, while performance decreased on more gradual slopes (0–15°), where stable conditions prevailed. The analysis of the factor of safety (FS) across different slope classes provided further insights into the behavior of the two models. For low slopes (0–15°), both models indicated stable conditions, but 4SLIDE produced a slightly higher average FS (1.81) compared to SHALSTAB (1.67), suggesting that it considers the terrain marginally more stable. In the medium slope range (15–35°), SHALSTAB exhibited a slightly higher minimum FS value (1.03) compared to 4SLIDE (0.95), implying that it tends to classify certain areas as more stable. However, the overall average FS remained similar between the two models, with 4SLIDE at 1.74 and SHALSTAB at 1.68. The most significant differences were observed in the high slope category (35–90°), where landslides are more likely to occur. Here, SHALSTAB showed a higher minimum FS (0.97) compared to 4SLIDE (0.23), suggesting a more conservative approach to instability classification, whereas the average FS remained relatively close (1.60 for 4SLIDE and 1.65 for SHALSTAB). In evaluating the models' performance, the Area Under the Curve (AUC) method was used. This methodology allows for the verification of the reliability of the two susceptibility models for landslide risk through two parameters: sensitivity and specificity. Sensitivity measures the models' ability to correctly identify true positive cases (actual landslides), while specificity measures the models' ability to correctly identify true negative cases (stable areas). These two indicators are crucial in ROC analysis for understanding a model's ability to accurately predict susceptible or non-susceptible areas for landslides, minimizing the number of false positives and false negatives [37,38]. The 4SLIDE model provided AUC values of 0.76 and 0.70 (Figure 7a)when compared with ground truth data (GTS) and the Digital Terrain Model (DTM) derived from LiDAR, respectively. Similarly, the SHALSTAB model produced AUC results of 0.73 and 0.69 for the same comparisons(Figure 7b). These values indicate that both models demonstrate strong predictive capabilities in identifying areas susceptible to shallow landslides, with slightly better performance for the 4SLIDE model. The sensitivity analysis also revealed that while both models effectively captured landslide-prone areas on steep slopes, discrepancies were observed in lower slope categories (0–15° and 15–35°), where SHALSTAB tended to classify a slightly larger proportion of stable areas as susceptible compared to 4SLIDE. This difference may be attributed to variations in the models' underlying assumptions regarding hydrological and geomorphological processes influencing slope stability. To further assess the agreement between the two models, Cohen's Kappa (κ) coefficient was calculated. The results indicate a strong agreement between the 4SLIDE model and the ground truth survey (GTS) data, with a Cohen's Kappa value of κ = 0.77, while the SHALSTAB model showed a slightly lower but still good agreement, with κ = 0.72. When compared with the DTM-LiDAR data, both models exhibited moderate agreement, with κ = 0.65 for 4SLIDE and κ = 0.66 for SHALSTAB. These values suggest that while both models perform well, 4SLIDE demonstrates slightly stronger agreement with the ground truth data (GTS) than SHALSTAB. Both models showed moderate agreement when validated with the DTM-LiDAR data.

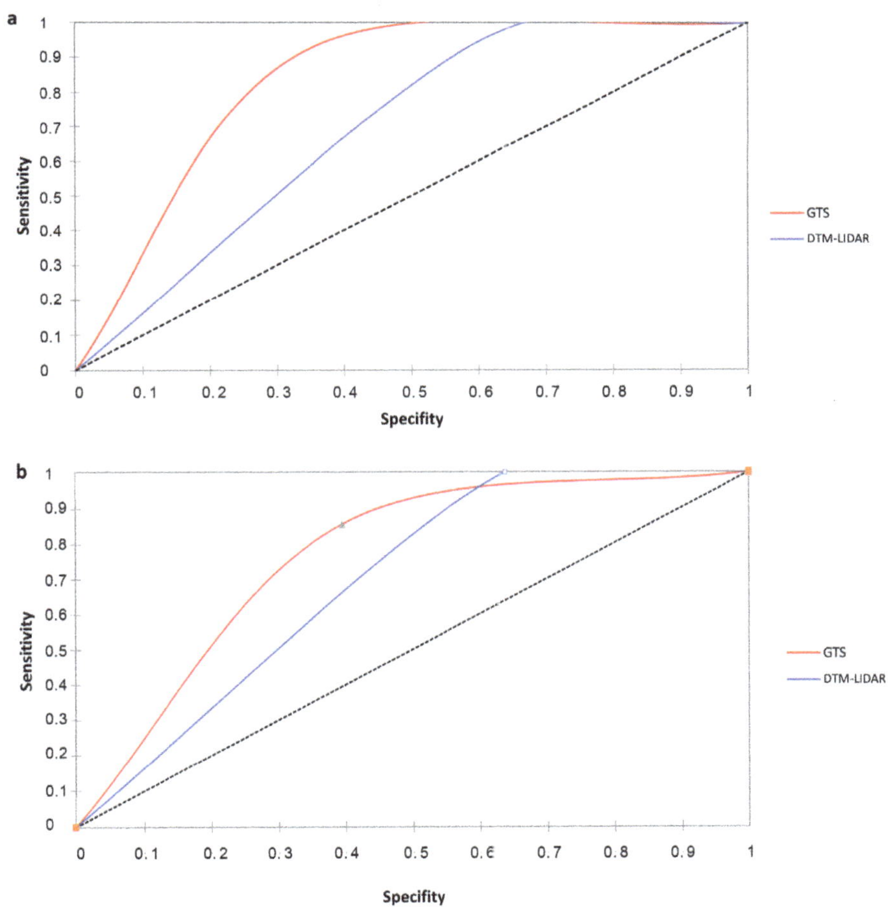

Figure 7. (**a**) ROC curve for the landslide susceptibility map generated by the 4SLIDE model. (**b**) ROC curve for the landslide susceptibility map generated by the model SHALSTAB.

Overall, the results show a strong correlation between the generated susceptibility maps and the landslide location recorded through field data. Both models effectively identified risk areas, although, like other predictive models, they rely on comprehensive input data to ensure high reliability. The sensitivity analysis further highlighted that the models' predictive performance varies depending on slope gradients, with a notable decline in accuracy on gentle slopes. This study demonstrates the utility and accuracy of these methodologies in landslide susceptibility assessment, paving the way for better-informed risk management and mitigation strategies in susceptible regions.

4. Discussion

Landslides are catastrophic events that cause significant damage, ranking among the natural disasters with the highest economic impact [9,10,13,14]. This has led to a growing interest in predictive studies aimed at mitigating risks and ensuring the safety of people and property. Over the past 30 years, many authors have proposed predictive models to identify areas prone to landslides, providing valuable tools for preventive actions [45–48]. Identifying these areas is essential for effective regional management and the implementation of mitigation strategies. In this context, the 4SLIDE model is particularly useful, as it generates landslide susceptibility maps for both forested and non-forested regions, highlighting the impact of vegetation on soil stability. The model accounts for various factors, including root cohesion (which stabilizes soil through vegetation), soil properties (such as texture, cohesion, and permeability), and terrain morphology (including slope, elevation, and curvature). By integrating these parameters, 4SLIDE offers a comprehensive approach to assessing landslide risk, helping decision-makers prioritize areas for intervention and improve land-use planning in landslide-prone regions [17].

Numerous models exist for such analyses, each with its own strengths and weaknesses in terms of ease of use, graphical interface, input parameter determination, and, in many cases, economic cost. Both 4SLIDE and SHALSTAB analyze landslide susceptibility at the catchment scale, employing steady-state hydrological processes and the infinite slope model [3,5,7]. Both models rely on numerical parameters and raster GIS data to process terrain data.

4SLIDE has primarily been used in mountainous basins and some urban and peri-urban contexts. In contrast, SHALSTAB, due to its longer history, has been widely used in various contexts by many researchers and field experts [45–48]. Given the similarities between the two models, a comparison was carried out to further evaluate 4SLIDE's performance. Both models utilize terrain slope, lithology, soil moisture, and vegetation data to generate susceptibility maps [19,26]. These maps can be further combined with information on landslide-triggering conditions, such as rainfall or earthquakes, to produce probabilistic hazard maps assessing risks to critical infrastructure (or other vulnerable elements).

The models were tested in the Bonis Forest catchment, an area where data could be effectively collected and spatially represented through raster maps. These maps provided detailed information on soil properties and vegetation characteristics, which were essential for model analysis. By integrating these data layers, the models processed relevant environmental factors, such as soil texture, root cohesion, and land morphology, in order to predict landslide susceptibility. This integration enabled a comprehensive understanding of the area's vulnerability, facilitating accurate model assessments and comparisons [15,19,26,31].

Both models identified unstable areas near watercourses, mainly due to steep slopes and reduced vegetation cover in these regions. The combination of steep terrain and limited vegetation reduces natural reinforcement, making these areas more susceptible to landslides. These findings emphasize the critical role of both slope steepness and vegetation in maintaining slope stability. In contrast, the most stable areas were found on gentler slopes with denser vegetation, which contributed to maintaining soil integrity and reducing the likelihood of landslides.

Importantly, both models highlighted vulnerable areas along steep slopes and unconsolidated geological materials, such as colluvium, particularly near watercourses. Stable regions were typically found further upslope. The sensitivity analysis revealed that both models performed better on steeper slopes (35–90°), where landslides were more frequent, and less effectively on gentler slopes (0–15°), where stable conditions prevailed. This variation highlights the critical interaction between slope steepness and rainfall intensity as key risk factors for landslide susceptibility. On steeper slopes, the effects of heavy rainfall tend to exacerbate instability, further increasing susceptibility to landslides, particularly in areas with insufficient vegetation.

In terms of the factor of safety (FS), both models indicated stability on low slopes (0–15°), but 4SLIDE showed a higher average Fs (1.81) compared to SHALSTAB (1.67), suggesting that it considered the terrain to be more stable. On medium slopes (15–35°), SHALSTAB had a slightly higher minimum FS (1.03) than 4SLIDE (0.95), although their average FS values were similar (4SLIDE: 1.74, SHALSTAB: 1.68). For high slopes (35–90°), SHALSTAB had a higher minimum FS (0.97) compared to 4SLIDE (0.23), indicating a more conservative approach to instability classification. However, the average FS values remained close (1.60 for 4SLIDE and 1.65 for SHALSTAB).

The Area Under the Curve (AUC) analysis revealed strong predictive capabilities for both models. The 4SLIDE model achieved AUC values of 0.76 and 0.70 when compared with ground truth data (GTS) and LiDAR-derived Digital Terrain Models (DTM), respectively. SHALSTAB showed AUC values of 0.73 and 0.69 for the same comparisons, indicating a slightly better performance by 4SLIDE. The sensitivity analysis highlighted discrepancies on lower slopes (0–15° and 15–35°), where SHALSTAB tended to classify more stable areas as susceptible compared to 4SLIDE. This may be due to differences in the models' assumptions about hydrological and geomorphological processes.

Cohen's Kappa coefficient confirmed strong agreement between both models and GTS data, with 4SLIDE showing $\kappa = 0.77$ and SHALSTAB showing $\kappa = 0.72$. Both models demonstrated moderate agreement with the DTM-LiDAR data (4SLIDE: $\kappa = 0.65$, SHALSTAB: $\kappa = 0.66$), suggesting a better accuracy with ground truth data. In conclusion, both models were well-aligned with the locations of landslides and effectively identified high-risk areas. However, their performance varied with slope gradients, showing better accuracy on steeper slopes and some discrepancies on gentler slopes. These results underscore the utility of both models for landslide susceptibility mapping and risk management in vulnerable regions. These results suggest that both models provide more accurate predictions when validated against direct field observations (GTS), which offer a highly detailed and site-specific assessment of landslide occurrences. These values indicate a good level of agreement, further validating the reliability of both models in classifying landslide-prone areas. Moreover, the consistency between the model predictions and the ground truth data, as confirmed by both AUC and Cohen's Kappa analyses, further supports the accuracy and effectiveness of both models in landslide risk assessment. The DTM generated from LiDAR data, which closely mirrors the field observations, demonstrates the utility of remote sensing in landslide susceptibility mapping.

Both simulations confirmed the key areas at risk for shallow landslides, demonstrating that 4SLIDE can be effectively compared with well-established models like SHALSTAB. However, 4SLIDE still has limitations, particularly due to its reliance on proprietary software, such as ArcMap. To enhance its accessibility, the next steps will focus on adapting 4SLIDE for use with open-source GIS software, such as QGIS or GRASS.

This adaptation will involve modifying the model to integrate seamlessly into these platforms, ensuring broader accessibility for users who prefer open-source tools. By integrating 4SLIDE into QGIS or GRASS, users will be able to conduct landslide susceptibility analyses directly within these widely used free GIS environments. This will require developing custom plugins or scripts, incorporating key data layers (such as terrain morphology, soil properties, and vegetation), and optimizing model outputs (e.g., landslide risk maps and susceptibility assessments) for better visualization and analysis.

Beyond improving accessibility, this transition to open-source GIS will enhance the model's role in regional planning and risk mitigation strategies. The integration of 4SLIDE into open-source GIS platforms will make it more adaptable to a wide range of geographic contexts and land-use planning needs. Specifically, it could be used to i. support decision-making in land-use planning by ensuring that high-risk areas are identified and considered in development projects, ii. improve early warning systems by integrating susceptibility maps into disaster response frameworks, iii. guide sustainable land management strategies, including afforestation, slope stabilization, and zoning regulations to reduce landslide hazards, and iv. facilitate participatory planning, enabling local governments and communities to assess and visualize risk levels more effectively.

This transition to open-source GIS represents a significant advancement, making the model more flexible, cost-effective, and accessible to researchers, policymakers, and land managers worldwide. By bridging the gap between scientific modeling and practical application, 4SLIDE has the potential to become a powerful tool for landslide risk assessment, contributing to safer and more resilient land planning.

Despite these benefits, it is crucial to acknowledge the limitations of both models. Their reliance on simplified assumptions, such as steady-state hydrological processes and general terrain data, may lead to inaccuracies in complex and dynamic environments. Additionally, the models' performance varies depending on the availability and quality of input data, particularly soil properties, vegetation cover, and rainfall data. These limitations should be considered when applying the models to real-world scenarios, and further improvements in data collection, model calibration, and computational methods are needed to enhance their accuracy and applicability in diverse environments.

5. Conclusions

Natural disasters over the past three decades have had a significant socio-economic and environmental impact, highlighting the need for effective disaster risk management. While advancements in early warning systems have helped reduce casualties, mitigation and planning remain essential to minimize future impacts of such events. The development of detailed and innovative risk reduction measures, such as updated hydrogeological risk maps through numerical modeling and spatially distributed parameterization, is crucial for landslide susceptibility assessments. This study evaluated and compared the performance of the open-source 4SLIDE model with the well-established SHALSTAB model, both used for mapping shallow landslides. While both models are open-source, 4SLIDE offers a more comprehensive approach, integrating terrain morphology, soil properties, and vegetation characteristics. SHALSTAB, on the other hand, uses a simplified slope stability analysis. The comparison assessed their predictive accuracy by comparing the outputs of both models with field data and recorded landslides. The results showed that 4SLIDE performs as reliably as SHALSTAB, with strong potential for application in more complex landscapes. Both models were calibrated using the same input data—geological, topographic, and land-use information—ensuring a direct and consistent comparison. Both models predicted landslide distribution with a moderate level of accuracy, and validation through two distinct evaluation methods further confirmed the results. Given the similar outcomes, 4SLIDE was found to be a valid alternative to SHALSTAB for landslide susceptibility mapping. The open-source nature of 4SLIDE also enhances its accessibility, making it a valuable tool for researchers, planners, and decision-makers, with integration into different GIS platforms allowing broad application. This study contributes to GIS-based landslide modeling by offering a more detailed approach than simplified models like SHALSTAB. 4SLIDE's ability to consider multiple environmental variables provides a more comprehensive risk assessment, making it useful for both disaster prevention and land management. This research also sets a structured methodology for comparing landslide susceptibility models using quantitative validation techniques, strengthening the reliability of predictive models in geomorphological studies.

However, this study acknowledges several limitations. The accuracy of 4SLIDE depends on the quality and resolution of input data, such as DEM, soil maps, and hydrological parameters. Furthermore, factors like extreme weather events and human-induced land changes can introduce uncertainties that are not fully accounted for by the model. Future research should explore the integration of machine learning techniques to improve the model's adaptability to different geographic contexts and conditions. In conclusion, this study reinforces the importance of GIS-based landslide modeling in disaster risk management. The findings demonstrate that 4SLIDE is an effective and accessible tool for landslide susceptibility assessment, supporting its adoption in real-world applications to enhance risk mitigation strategies and land-use planning.

Author Contributions: Conceptualization, Federico Valerio Moresi and Mauro Maesano; methodology, Federico Valerio Moresi and Mauro Maesano; software, Federico Valerio Moresi; validation, Federico Valerio Moresi, Marco di Cristofaro, and Elena Brunori; formal analysis, Federico Valerio Moresi, Mauro Maesano, and Elena Brunori; investigation, Federico Valerio Moresi and Mauro Maesano; resources, Giuseppe Scarascia Mugnozza; data curation, Federico Valerio Moresi, Mauro Maesano, and Elena Brunori; writing—original draft preparation, Federico Valerio Moresi, Mauro Maesano, Marco di Cristofaro, Elena Brunori, and Giuseppe Scarascia Mugnozza; writing—review and editing, Federico Valerio Moresi, Mauro Maesano, Marco di Cristofaro, Elena Brunori, and Giuseppe Scarascia Mugnozza; visualization, Federico Valerio Moresi, Mauro Maesano, Marco di Cristofaro, Elena Brunori, and Giuseppe Scarascia Mugnozza; supervision, Giuseppe Scarascia Mugnozza; project administration, Giuseppe Scarascia Mugnozza; funding acquisition, Giuseppe Scarascia Mugnozza. All authors have read and agreed to the published version of the manuscript.

Funding: This research was supported by the PON-MISE "INLEGNO" (F/2000 03/01-03/X45), scientific coordinator prof. Giuseppe Scarascia Mugnozza, and by POR FESR Lazio: Gruppi di ricerca 2014–2020 "BIOEDILCARBON" (A0375-2020-36712).

Data Availability Statement: Data available on request due to restrictions (e.g., privacy, legal, or ethical reasons).

Conflicts of Interest: The authors declare no conflicts of interest.

References

1. Cruden, D. A simple definition of a landslide. *Bull. Eng. Geol. Environ.* **1991**, *43*, 27.
2. Guo, J.; Yi, S.; Yin, Y.; Cui, Y.; Qin, M.; Li, T.; Wang, C. The effect of topography on landslide kinematics: A case study of the Jichang town landslide in Guizhou, China. *Landslides* **2020**, *17*, 959–973.
3. Varnes, D.J. *Landslide Hazard Zonation: A Review of Principles and Practice*; No. 3; United Nations: New York, NY, USA, 1984.
4. Ray, R.L.; Jacobs, J.M. Relationships among remotely sensed soil moisture, precipitation and landslide events. *Nat. Hazards* **2007**, *43*, 211–222.
5. Highland, L.M.; Bobrowsky, P. *The Landslide Handbook-A Guide to Understanding Landslides*; No. 1325; US Geological Survey: Reston, VA, USA, 2008.
6. Winter, M.G. Landslide hazards and risks to road users, road infrastructure and socio-economic activity. In Proceedings of the XVII European Conference on Soil Mechanics and Geotechnical Engineering, Reykjavík, Iceland, 1–6 September 2019.

7. Sidle, R.; Ochiai, H. Processes, prediction, and land use. In *Water Resources Monograph*; American Geophysical Union: Washington, DC, USA, 2006; p. 525.
8. Du, Y.; Xie, M.; Jia, J. Stepped settlement: A possible mechanism for translational landslides. *Catena* **2020**, *187*, 104365.
9. Trigila, A.; Iadanza, C. Assetto del territorio e difesa del suolo. *Riv. Giuridica Del Mezzog.* **2018**, *32*, 1053–1066.
10. Trigila, A.; Iadanza, C.; Esposito, C.; Scarascia-Mugnozza, G. Comparison of Logistic Regression and Random Forests techniques for shallow landslide susceptibility assessment in Giampilieri (NE Sicily, Italy). *Geomorphology* **2015**, *249*, 119–136.
11. UN Office for Disaster Risk Reduction. *The Human Cost of Disasters: An Overview of the Last 20 Years (2000–2019)*; UN Office for Disaster Risk Reduction: Geneva, Switzerland, 2020.
12. Herrera, G.; Mateos, R.M.; García-Davalillo, J.C.; Grandjean, G.; Poyiadji, E.; Maftei, R.; Filipciuc, T.-C.; Auflič, M.J.; Jež, J.; Podolszki, L.; et al. Landslide databases in the Geological Surveys of Europe. *Landslides* **2018**, *15*, 359–379.
13. Peruccacci, S.; Gariano, S.L.; Melillo, M.; Solimano, M.; Guzzetti, F.; Brunetti, M.T. The ITAlian rainfall-induced LandslIdes CAtalogue, an extensive and accurate spatio-temporal catalogue of rainfall-induced landslides in Italy. *Earth Syst. Sci. Data* **2023**, *15*, 2863–2877. [CrossRef]
14. Trigila, A.; Iadanza, C.; Lastoria, B.; Bussettini, M.; Barbano, A. *Dissesto Idrogeologico in Italia: Pericolosità e Indicatori di Rischio—Edizione 2021*; ISPRA. Available online: https://www.isprambiente.gov.it/it/pubblicazioni/rapporti/dissesto-idrogeologico-in-italia-pericolosita-e-indicatori-di-rischio-edizione-2021 (accessed on 1 February 2025).
15. Moresi, F.V.; Maesano, M.; Matteucci, G.; Romagnoli, M.; Sidle, R.C.; Mugnozza, G.S. Root biomechanical traits in a montane Mediterranean forest watershed: Variations with species diversity and soil depth. *Forests* **2019**, *10*, 341. [CrossRef]
16. Nohani, E.; Moharrami, M.; Sharafi, S.; Khosravi, K.; Pradhan, B.; Pham, B.T.; Lee, S.; Melesse, A.M. Landslide Susceptibility Mapping Using Different GIS-Based Bivariate Models. *Water* **2019**, *11*, 1402. [CrossRef]
17. Pal, S.C.; Chowdhuri, I. GIS-based spatial prediction of landslide susceptibility using frequency ratio model of Lachung River basin, North Sikkim, India. *SN Appl. Sci.* **2019**, *1*, 416.
18. Chae, B.G.; Lee, J.H.; Park, H.J.; Choi, J. A method for predicting the factor of safety of an infinite slope based on the depth ratio of the wetting front induced by rainfall infiltration. *Nat. Hazards Earth Syst. Sci.* **2015**, *15*, 1835–1849.
19. Moresi, F.V.; Maesano, M.; Collalti, A.; Sidle, R.C.; Matteucci, G.; Mugnozza, G.S. Mapping landslide prediction through a GIS-based model: A case study in a catchment in southern Italy. *Geosciences* **2020**, *10*, 309. [CrossRef]
20. Reichenbach, P.; Rossi, M.; Malamud, B.D.; Mihir, M.; Guzzetti, F. A review of statistically-based landslide susceptibility models. *Earth Sci. Rev.* **2018**, *180*, 60–91.
21. Holec, J.; Bednarik, M.; Šabo, M.; Minár, J.; Yilmaz, I.; Marschalko, M. A small-scale landslide susceptibility assessment for the territory of Western Carpathians. *Nat. Hazards* **2013**, *69*, 1081–1107.
22. Kanungo, D.P.; Arora, M.K.; Sarkar, S.; Gupta, R.P. Landslide susceptibility zonation (LSZ) mapping: A review. *J. South Asia Disaster Stud.* **2009**, *2*, 81–105.
23. Chen, W.; Li, X.; Wang, Y.; Liu, S. Landslide susceptibility mapping using LiDAR and DMC data: A case study in the Three Gorges area, China. *Environ. Earth Sci.* **2013**, *70*, 673–685.
24. Hjort, J.; Luoto, M. 2.6 Statistical methods for geomorphic distribution modeling. *Treatise Geomorphol.* **2013**, *2*, 59–73.
25. Caloiero, T.; Biondo, C.; Callegari, G.; Collalti, A.; Froio, R.; Maesano, M.; Matteucci, G.; Pellicone, G.; Veltri, A. Results of a long-term study on an experimental watershed in southern Italy. *Forum Geogr.* **2016**, *15*, 55–65.
26. Tsangaratos, P.; Ilia, I.; Rozos, D. Case event system for landslide susceptibility analysis. In *Landslide Science and Practice: Landslide Inventory and Susceptibility and Hazard Zoning*; Springer: Berlin/Heidelberg, Germany, 2013; Volume 1, pp. 585–593.
27. Dietrich, W.E.; Montgomery, D.R. *SHALSTAB: A Digital Terrain Model for Mapping Shallow Landslide Potential*; University of California: Berkeley, CA, USA, 1998.
28. Montgomery, D.R.; Dietrich, W.E. A physically based model for the topographic control on shallow landsliding. *Water Resour. Res.* **1994**, *30*, 1153–1171.
29. Beven, K.; Freer, J. A dynamic topmodel. *Hydrol. Process.* **2001**, *15*, 1993–2011.
30. Wu, T.H.; McKinnell, W.P., III; Swanston, D.N. Strength of tree roots and landslides on Prince of Wales Island, Alaska. *Can. Geotech. J.* **1979**, *16*, 19–33.
31. Dietrich, W.E.; Bellugi, D.; Real de Asua, R. Validation of the shallow landslide model, SHALSTAB, for forest management. In *Land Use and Watersheds: Human Influence on Hydrology and Geomorphology in Urban and Forest Areas*; American Geophysical Union: Washington, DC, USA, 2001; Volume 2, pp. 195–227.
32. O'loughlin, E.M. Prediction of surface saturation zones in natural catchments by topographic analysis. *Water Resour. Res.* **1986**, *22*, 794–804.
33. Skempton, A.W.; DeLory, F.A. Stability of natural slopes in London clay. In *Selected Papers on Soil Mechanics*; Thomas Telford Publishing: London, UK, 1984; pp. 70–73.

34. Collalti, A.; Biondo, C.; Buttafuoco, G.; Maesano, M.; Caloiero, T.; Lucà, F.; Pellicone, G.; Ricca, N.; Salvati, R.; Veltri, A.; et al. Simulation, calibration and validation protocols for the model 3D-CMCC-CNR-FEM: A case study in the Bonis' watershed (Calabria, Italy). *Forest* **2017**, *14*, 247–256.
35. Raab, G.; Halpern, D.; Scarciglia, F.; Raimondi, S.; Norton, K.; Pettke, T.; Hermann, J.; Portes, R.d.C.; Sanchez, A.M.A.; Egli, M. Linking tephrochronology and soil characteristics in the Sila and Nebrodi mountains, Italy. *Catena* **2017**, *158*, 266–285.
36. Callegari, G.; Ferrari, E.; Garfì, G.; Iovino, F.; Veltri, A. Impact of thinning on the water balance of a catchment in a Mediterranean environment. *For. Chron.* **2003**, *79*, 301–306.
37. Bonardi, G.; Cavazza, W.; Perrone, V.; Rossi, S. Calabria-Peloritani terrane and northern Ionian sea. In *Anatomy of an Orogen: The Apennines and Adjacent Mediterranean Basins*; Springer: Dordrecht, The Netherlands, 2021; pp. 287–306.
38. Bellotti, R.; Selleri, G. Correlazione tra le caratteristiche geotecniche di alcuni terreni di fondazione e confronto tra i risultati ottenibili con l'applicazione di diversi metodi di calcolo del carico ammissibile. *Riv. Ital. Geotec.* **1969**, *2*, 95–103.
39. Worobec, R.B. *7 Moving Average Interpolation. Toxicity Assessment Alternatives: Methods, Issues, Opportunities;* Springer: Berlin/Heidelberg, Germany, 1999; p. 71.
40. Ravazzani, G.; Caloiero, T.; Feki, M.; Pellicone, G. Impact of infiltration process modeling on runoff simulations: The Bonis River Basin. *Proceedings* **2018**, *2*, 638. [CrossRef]
41. Fawcett, T. An introduction to ROC analysis. *Pattern Recognit. Lett.* **2006**, *27*, 861–874.9.
42. Morris, M.D. Factorial sampling plans for preliminary computational experiments. *Tech-Nometrics* **1991**, *33*, 161–174.
43. Sun, S. Meta-analysis of Cohen's kappa. *Health Serv. Outcomes Res. Methodol.* **2011**, *11*, 145–163.
44. Więckowska, B.; Kubiak, K.B.; Jóźwiak, P.; Moryson, W.; Stawińska-Witoszyńska, B. Cohen's kappa coefficient as a measure to assess classification improvement following the addition of a new marker to a regression model. *Int. J. Environ. Res. Public Health* **2022**, *19*, 10213. [CrossRef] [PubMed]
45. Dikshit, A.; Pradhan, B.; Alamri, A.M. Pathways and challenges of the application of artificial intelligence to geohazards modelling. *Gondwana Res.* **2021**, *100*, 290–301.
46. Stanley, T.; Kirschbaum, D.B. A heuristic approach to global landslide susceptibility mapping. *Nat. Hazards* **2017**, *87*, 145–164.
47. Dietrich, W.E.; Reiss, R.; Hsu, M.L.; Montgomery, D.R. A process-based model for colluvial soil depth and shallow landsliding using digital elevation data. *Hydrol. Process.* **1995**, *9*, 383–400.
48. Toth, C.K.; Mora, O.E.; Lenzano, M.G.; Grejner-Brzezinska, D.A.; Beach, K. Landslide hazard detection from LiDAR data. In Proceedings of the ASPRS 2013 Annual Conference, Baltimore, MD, USA, 26–28 March 2013; pp. 24–28.

applied sciences

MDPI

Article

Dual-Indicator Micro-Electro-Mechanical System Monitoring Method for Rock Instability Early Warning

Chen Chen [1], Mowen Xie [1,*], Yan Du [2] and Xiaoyong Zhang [3]

1 School of Resources and Safety Engineering, University of Science and Technology Beijing, Beijing 100083, China; chenchen_powerful@163.com
2 School of Future Cities, University of Science and Technology Beijing, Beijing 100083, China; mutulei@163.com
3 Beijing Water Science and Technology Institute, Beijing 100048, China
* Correspondence: mowenxie@126.com

Abstract: Traditional displacement monitoring struggles to provide early warnings for sudden rock collapses. This study proposes a Micro-Electro-Mechanical System (MEMS) sensor-based monitoring method using dual dynamic indicators. By analyzing sensor mechanisms through vibration dynamics theory and establishing theoretical models via moment equilibrium equations, we derived a quantitative correlation between natural frequency (NF) and safety factor, identifying a 4:3 scaling coefficient specific to toppling-type unstable rocks. An innovative stability criterion algorithm integrating NF and the root mean square velocity amplitude ratio (RMS-VAR) was developed, revealing that RMS-VAR detects stability degradation three times faster than tilt measurements. Laboratory tests confirmed MEMS sensors' reliability in monitoring NF, amplitude ratio, and tilt angles, demonstrating that sensor deployment strategies and rock geometry jointly determine model accuracy. This quantitative approach offers a novel solution for monitoring sudden geological hazards, combining timeliness with cost-effectiveness.

Keywords: MEMS sensors; rock stability; natural frequency; RMS velocity amplitude ratio; dual dynamic indicators

1. Introduction

As a major energy-producing country, China has gained significant economic benefits from the development of cascade power stations and mining operations. Nevertheless, geohazards, including slope collapse disasters, continue to impose substantial socioeconomic impacts [1]. Due to complex geological conditions, frequent seismic activities, and engineering construction disturbances, rock mass internal stress imbalance frequently leads to collapse accidents [2]. The prevention of unstable rock mass collapse poses a major challenge in engineering geological disaster mitigation, given its sudden occurrence and spatially dispersed nature. Consequently, real-time stability monitoring and the early warning of collapsing unstable rock masses remain critical practical issues in geotechnical engineering.

For rock masses or blocks exhibiting potential instability on natural slopes, the deterioration of their primary structural planes plays a decisive role in stability control. Experimental validations and practical applications have demonstrated the effectiveness of monitoring methods utilizing vibration characteristics to assess structural plane degradation. The NF of an unstable rock mass, as its intrinsic dynamic parameter, is uniquely determined by the mass matrix and stiffness matrix. Iannucci et al. [3] employed passive seismic techniques to analyze the quantitative relationship between the NF and the stability of toppling rock masses. Bottelin et al. [4] employed ambient vibration response to verify stability variations in rock mass reinforcement effectiveness. Mercerat et al. [5] conducted a dynamic characteristic analysis of coastal slopes using ambient vibration, delineating deep structural features that were superficially undetectable. Additional studies have explored unstable rock mass vibration responses through vibration mechanics and dynamic mathematical analyses [6]. Chen Hongkai et al. [7] employed dynamics theory to analyze energy transitions during rock mass sliding, establishing theoretical formulas for elastic–impact velocity and acceleration. Internationally, seismic monitoring equipment has been predominantly employed for large unstable rock mass vibration studies [8]. Burjánek et al. [9] identified directional amplification effects correlated with structural plane orientation and slope deformation through environmental vibration data analysis. Glueer et al. [10] utilized passive seismic measurements to map subsurface structural plane distributions (30–90 m depth). Hausler et al. [11,12] performed modal analyses using 1 h environmental vibration datasets, enabling the spatial mapping of undetectable subsurface fractures via frequency-dependent vibration pattern correlations. Iannucci et al. [3] observed linear polarization in unstable rock mass particle trajectories through spectral ratio comparisons with the parent rock. These investigations confirm that vibration-derived dynamic indicators offer innovative approaches for collapse monitoring. However, traditional seismic equipment and single-index methodologies prove inadequate for ensuring monitoring accuracy and early-warning reliability. MEMS sensing, as a miniaturized technology integrating micromechanical structures, sensors, actuators, and electronic circuits, provides a reliable method for monitoring data acquisition in unstable rock masses. Compared to conventional vibration sensing, MEMS technology provides distinct advantages including miniaturization, low cost, low power consumption, and wide frequency response—essential features for complex engineering scenarios. Wu Di et al. [13] utilized MEMS sensors to conduct monitoring verification on highway slope displacement. Praveen et al. [14] implemented a real-time monitoring methodology for low-speed machinery using IoT technology and MEMS sensors. The convergence of MEMS sensing with wireless transmission and big data analytics has emerged as a novel research frontier in geotechnical disaster monitoring, though applications specifically targeting unstable rock mass collapse remain limited.

Unstable rock mass collapse prediction constitutes a multivariate challenge influenced by numerous factors [15–17], compounded by harsh environmental conditions including construction disturbances, blasting activities, poor network connectivity, and power supply limitations [18]. In order to achieve the early warning of unstable rock mass collapse and solve the problem of the real-time evaluation of unstable rock mass stability, this study investigates MEMS sensor operational principles, establishes vibration mechanics-based quantitative relationships between NF/RMS-VAR and rock stability, and proposes a dual dynamic index algorithm. Furthermore, we discuss real-time stability evaluation models and critical implementation considerations for MEMS-based monitoring methods. The developed MEMS sensing methodology provides enhanced temporal warning capabilities and novel non-engineering control strategies for preventing sudden geotechnical disasters.

2. Fundamental Theory and Methodology

2.1. MEMS Sensing Mechanisms

The stability of unstable rock masses gradually decreases with the deterioration of their main structural planes [19]. Under normal fretting conditions, the vibration characteristics of unstable rock masses differ from those of the bedrock to some extent, and these differences become more pronounced with increasing degrees of structural deterioration. MEMS sensing provides a novel technology for monitoring unstable rock mass collapse, owing to its high sensitivity, low power consumption, and compact dimensions. As shown in Figure 1a, vibration signals can be acquired by affixing MEMS sensors to the unstable rock mass and bedrock using gypsum bonding or bolting methods, combined with wired or wireless transmission systems. Figure 1b schematically represents the MEMS sensor as a cubic enclosure containing a miniature metallic sphere. When subjected to gravitational forces and micro-vibrations from unstable rock masses, this sphere undergoes inertial displacement, thereby inducing electrical signals of varying magnitudes through wall impacts. The system ultimately outputs triaxial acceleration, velocity, and displacement measurements.

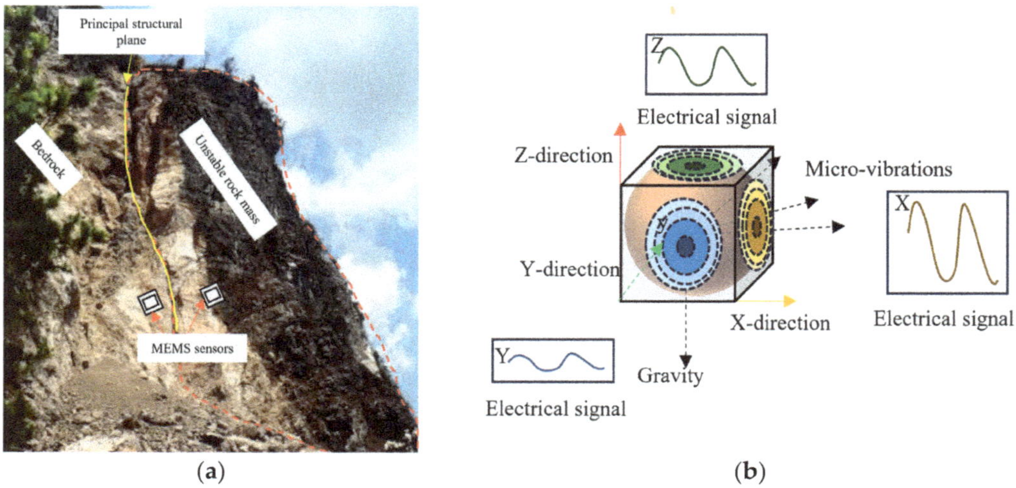

Figure 1. Working principle of MEMS sensor. (**a**) Application method of MEMS sensors. (**b**) MEMS sensing mechanisms.

MEMS sensors acquire vibration data from rock masses through direct contact measurements. As illustrated in Equation (1), a conventional MEMS accelerometer continuously measures triaxial acceleration components, including gravitational acceleration, through its embedded micro-electro-mechanical structures.

$$\begin{cases} \vec{a}(t) = \frac{\vec{F}(t)}{m} \\ \left|\vec{a}(t)\right| = \sqrt{a_x^2(t) + a_y^2(t) + a_z^2(t)} \end{cases} \tag{1}$$

where $\vec{F}(t)$ represents the impact force acting on a rock mass of mass m. $\vec{a}(t)$ represents the triaxial acceleration measured by MEMS sensors, $\vec{a}(t) = (a_x(t), a_y(t), a_z(t))$.

2.2. *Dynamic Interdependence: NF Versus Safety Factor*

According to vibration mechanics theory, unstable rock masses exhibit characteristic vibration under external excitation, which inherently reflect their mechanical properties [20]. To establish quantitative correlations between rock mass stability and vibrational signatures, we systematically analyzed dominant stability control factors through mechanical characterization. Figure 2 illustrates the force equilibrium of typical toppling-prone rock masses, where two critical moments coexist: (1) toppling moment induced by gravitational forces; (2) resisting moment is generated through structural plane bonding.

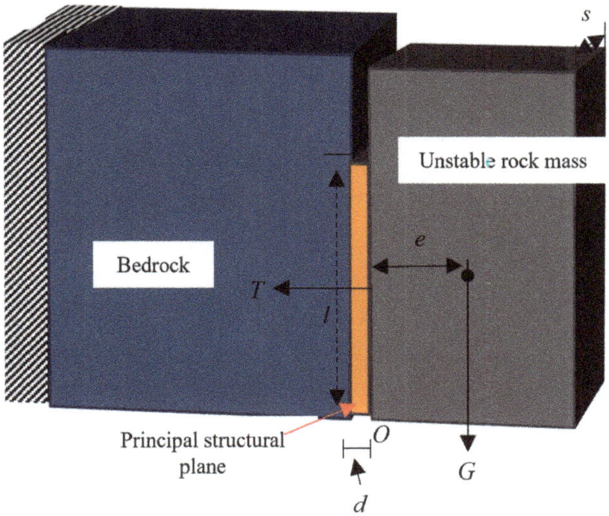

Figure 2. Schematic diagram of overthrown unstable rock mass.

Where l denotes the bonding length of the principal structural plane. When l decreases below critical thresholds, the resisting moment becomes insufficient to counteract gravitational toppling forces, ultimately triggering collapse. The instability failure characteristics of this unstable rock mass are primarily manifested as a rotational failure mode around a specific pivot point, caused by unbalanced moments resulting from either load increase or cohesion decrease. This demonstrates that structural plane bonding length constitutes the pivotal control parameter for rock mass stability. According to limit equilibrium principles, the safety factor K for toppling-prone rock mass is defined as the ratio of the resisting moment to the toppling moment, as expressed in Equation (2):

$$K = {(F \times l/2)}\Big/{G \times e} = {\delta_s T l^2 s}\Big/{2Ge} \tag{2}$$

where F is the resisting force; e is the distance between the center of gravity of rock mass and the structural plane; δ_s is the tensile strength of rock mass; T is the thickness of rock mass; and s is the width of rock mass. G represents the gravitational force of the unstable rock mass.

The square of the bonding length l^2 in unstable rock mass structural planes exhibits a positive correlation with the safety factor K. When l diminishes to the critical stability threshold l_c, the safety factor reaches the unity ($K = 1$) [2]. Rock mass failure initiates when $K < 1$, indicating a loss of structural equilibrium. This early-warning methodology employs MEMS sensors to track vibrational frequency modulations that correlate with the reduction in the bonding length of the principal structural plane. The NF constitutes an intrinsic parameter of rock mass vibrational dynamics. As shown in Figure 3, the unstable rock mass can be simplified into a single-degrees-of-freedom (SDOF) spring–mass oscillator. The motion balance shown in Equation (3) is established according to Newton's Second Law, and NF expression is shown in Equation (4) [20].

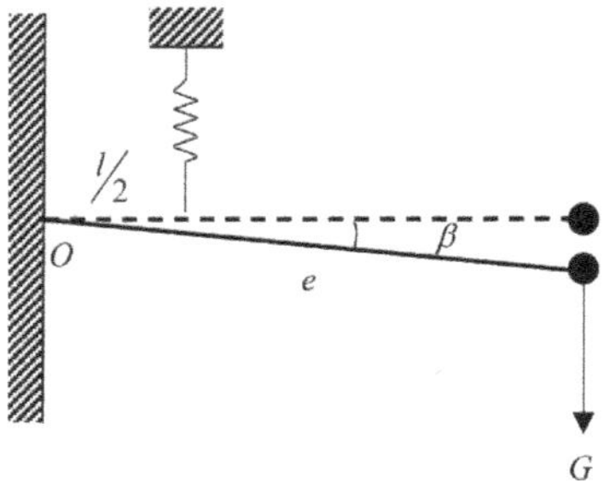

Figure 3. SDOF spring–mass oscillator of an unstable rock.

Where the stiffness decreases proportionally with bonding length reduction. This stiffness degradation induces measurable frequency attenuation, enabling the quantitative assessment of structural integrity through frequency-domain analysis.

$$me\ddot{\beta}(t) + \left({}^{l^3 ETs}\!\Big/_{3d} \right) \beta(t) = 0 \quad (3)$$

$$f_0 = (1/2\pi)\sqrt{(ETl^3 sg)/(3dGe^2)} \quad (4)$$

where $\beta(t)$ is the change in tilt angle in the vibration process; m is the quality of unstable rock mass; and f_0 is the NF of unstable rock mass.

According to Equations (2) and (4), the mathematical expression of NF and safety factor is obtained, as shown in Equation (5):

$$K = \sqrt[3]{18\pi^4 eTd^2 \delta_s f_0^4 / GE^2 g^2} \quad (5)$$

Equation (5) established a bijective relationship between the NF of unstable rock masses and their safety. The NF of toppling-type unstable rocks exhibits a quantitative proportional relationship of 4:3 with the safety factor. Under the condition that other parameters of the rock mass remain constant, a decrease in the NF induces a monotonic decline in the safety factor. A critical NF necessarily exists at which the safety factor equals 1. Consequently, the real-time monitoring of NF variations enables instability early warning for rock masses, leveraging this deterministic frequency–safety correlation. The quantitative relationship between the NF of unstable rock mass and the safety factor can be calibrated through laboratory experiments or finite elements simulation, and the safety factor of unstable rock mass can be determined by the NF value of the field measurable index and the multi-stage early-warning threshold of unstable rock mass can be set.

2.3. Quantification Metrics for MEMS-Acquired Stability Indicators

Through MEMS sensing, the vibration time history curve of bedrock is obtained as the vibration input of the spring–mass model, and the input curve of unstable rock mass is serves as the response output. The frequency response function (FRF) $H(f)$ representing the vibration response of unstable rock mass under ambient vibration can be obtained, and its calculation formula is shown in Equation (6). Under the excitation of a sinusoidal signal, the theoretical magnification factor of the spring–mass model system $D(s)$ is shown in Equation (7). The NF and damping ratio of unstable rock mass can be obtained by analyzing the relationship between the theoretical amplitudes and the actual amplitudes under the multi-frequency excitation [21].

$$H(f) = P_{xy}(f) \Big/ P_x(f) \tag{6}$$

$$D(s) = 1/\sqrt{(1-s^2)^2 + (2\xi s)^2} \tag{7}$$

where $P_{xy}(f)$ is the cross-correlation power spectral density function between bedrock and unstable rock mass; $P_x(f)$ is the matrix autocorrelation power spectral density function; and f is the excitation frequency of the input unstable rock mass under constant fretting.

When the frequency ratio reaches the unity ($f = f_0$), the amplification factor $D(s)$ attains its maximum value. This indicates that the peak value in the actual obtained FRF is the magnification of the response amplitude caused by the resonance of the unstable rock mass, and the horizontal coordinate frequency corresponding to the peak value is the NF of the unstable rock mass.

The amplitude–amplification factor $D(s)$ is the ratio of the vibration amplitude of unstable rock mass to that of bedrock under different excitation frequencies. As the NF f_0 of unstable rock mass decreases, the amplitude–amplification factor of the system under the same excitation frequency increases. Therefore, the ratio of the vibration amplitude of unstable rock mass to that of bedrock also increases. In order to reduce the error caused by the random interference of the MEMS sensor in the acquisition process of the vibration amplitude of the unstable rock mass and bedrock, the ratio of mean square velocity to the amplitude of the unstable rock mass and bedrock is adopted as the monitoring index of unstable rock mass collapse. The equation is as follows:

$$\eta = \sqrt{\int_{t_1}^{t_2} (1/(t_2 - t_1))y^2(t) \Big/ \int_{t_1}^{t_2} (1/(t_2 - t_1))x^2(t)} \tag{8}$$

where $x(t)$ is the velocity time history curve of the bedrock; and $y(t)$ is the velocity time history curve of unstable rock mass.

The stability state of an unstable rock mass evolves dynamically. MEMS-derived NF and RMS-VAR parameters effectively characterize the progressive destabilization of unstable rock masses. Using MEMS sensing technology, triaxial vibration time–history curves of unstable rock masses and bedrock are collected in real time, with a principal component analysis (PCA) employed to identify dominant vibration directions. Through time-domain window optimization, the FRF of unstable rock masses is calculated, and their natural frequencies are determined through fitting of amplitude amplification factors. The RMS-VAR between unstable rock masses and bedrock is derived by converting detrended acceleration curves to velocity curves. Furthermore, in engineering practice, unstable rock masses are typically formed through prolonged geological processes and often exhibit precursor phenomena such as localized rockfall events and minor deformations prior to collapse [22,23]. MEMS sensors are utilized to monitor instantaneous impact acceleration and subtle tilt angles caused by surrounding rockfalls, providing critical data support for early-warning systems of rock mass collapse.

3. Experimental Validation and Performance Evaluation

3.1. Experimental Setup and Modeling

To validate the applicability of MEMS sensing technology for unstable rock mass collapse monitoring, we developed a MEMS-based simulation monitoring system (as shown in Figure 4). The system comprises three key components: (1) artificially fabricated rock blocks with material ratios matching those of natural bedrock and parent rock formations; (2) integrated sensing instrumentation including MEMS accelerometers and a laser Doppler vibrometer (LDV) system; and (3) hydraulic stabilization jacks for bedrock anchorage.

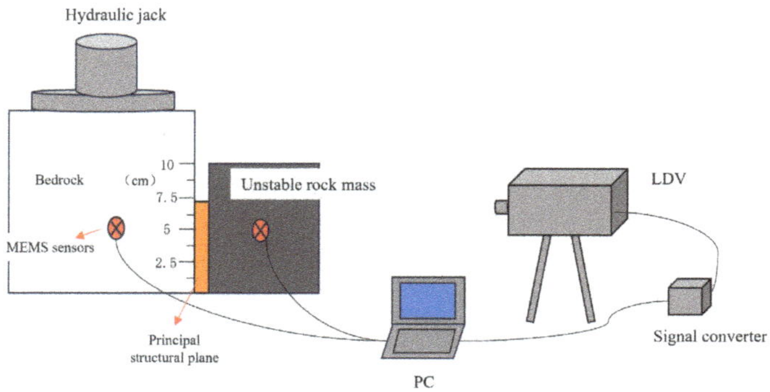

Figure 4. Simulated unstable rock collapse monitoring system.

3.1.1. Rockfall Model

In this experiment, the unstable rock mass and bedrock were simulated using proportionally equivalent materials. The material composition consisted of quartz and barite powder as aggregates, gypsum as the cementitious binder, with a retarding agent and glycerol incorporated as admixtures. The optimized mass ratio for the synthetic rock matrix was formulated as: Quartz Sand:Barite Powder:Gypsum:Retarder:Water:Glycerol = 50:30:8:0.02:10:1.5. As detailed in Table 1, the fabricated specimens exhibited comparable physical properties to natural geological formations. Figure 5 illustrates the geometric configuration, where the protrusion (dimensions: 12 cm (L) × 10 cm (H) × 11.5 cm (W)) represents the unstable rock mass, while the cuboidal bedrock section (30 cm edge length) provides structural constraint. Hydraulic jacks applying 2.4 MPa vertical confinement pressure ensured bedrock stability throughout testing. The failure mechanism was systematically induced through the progressive reduction in the bonded interface area (15% increment steps) via the precision machining of the controlling structural plane.

Table 1. Physical and mechanical properties of the rock blocks.

Density/(kg/m^3)	Modulus of Elasticity/GPa	Poisson's Ratio	Tensile Strength/MPa	Cohesion/MPa
2230.00	2.40	0.28	0.098	0.926

(a) (b)

Figure 5. Simulation of the collapse process of unstable rock. (**a**) Rock mass formed by pouring. (**b**) Cut of the structural plane.

3.1.2. Data Sensing Instrumentation

As shown in Figure 6a, the triaxial MEMS acceleration sensor with built-in amplifier (model INV9832-50) developed by Dongfang Institute of Technology (Hangzhou, China) was selected for this experiment. The MEMS sensing device can monitor very small impact signals with a sensitivity of 100 mV/g, and the weight of the sensing device is about 12 g, which imposes a negligible gravitational load on the rock mass, and is suitable for acquiring the dynamics index of the rock mass. As shown in Figure 6b, the INV3062C high-precision distributed data acquisition instrument of Dongfang Institute was employed in this experiment to convert the signals collected by the MEMS sensing device, and the final data were transmitted to the computer through Ethernet cables for subsequent data analysis and processing. At the same time, as shown in Figure 6c, the accuracy of the data collected by MEMS sensing was verified by the laser vibration-measuring instrument model RSV-150 from Polytec GmbH (Waldorf, Germany). The main technical parameters of the data acquisition equipment are summarized in Table 2.

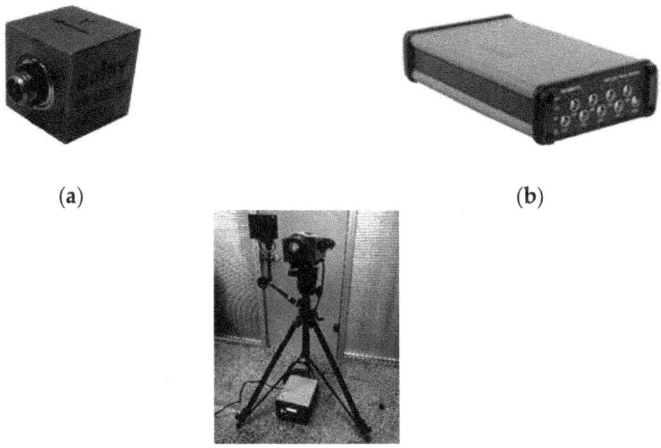

(**a**) (**b**) (**c**)

Figure 6. Data acquisition sensing equipment. (**a**) MEMS accelerometer. (**b**) Signal collector. (**c**) LDV equipment.

Table 2. Main performance parameters of the device.

Model Number	Frequency Range/kHz	Range	Resolution	Size/mm
INV9832-50	1~10	50 g	0.01 mg	19 × 19
RSV-150	0~25	1 m/s	0.5 μm/s	402 × 165 × 145 mm

3.1.3. Dynamic Simulation of Collapse Mechanisms

The collapse process of unstable rock masses was simulated through the artificially induced cutting of the dominant structural plane. As defined in Equation (9), the relative bonding length ΔL serves as a quantitative metric for structural plane degradation. Reductions in this length correlate directly with heightened deterioration severity and diminished rock mass stability.

$$\Delta L = 1 - \frac{L}{L_0} \tag{9}$$

where L is the cumulative length of artificially cut structural plane; and L_0 is the height of unstable rock mass, taking 10 cm.

The collapse simulation protocol involved progressive 0.5 cm depth cuts to structural planes until failure initiation. Table 3 quantifies the cutting length-to-relative bonding length correlation across 14 distinct working conditions for unstable rock masses.

Table 3. Simulated working conditions of unstable rock.

Condition	1	2	3	4	5	6	7	8	9	10	11	12	13	14
L/cm	0	0.5	1	1.5	2	2.5	3	3.5	4	4.5	5	5.5	6	6.5
ΔL	1	0.95	0.90	0.85	0.8	0.75	0.7	0.65	0.6	0.55	0.50	0.45	0.40	0.35

3.2. Experimental Results

3.2.1. MEMS-Laser Vibrometry Cross-Validation

To validate the MEMS sensing's efficacy in monitoring unstable rock mass vibrations under constant micro-motion conditions, a comparative analysis was conducted between high-precision LDV measurements and MEMS-acquired data. Both systems were aligned along identical measurement axes, with synchronized sampling at 1000 Hz over a 60 s acquisition period. Prior to analysis, sensitivity calibration and 5th-order Butterworth low-pass filtering (cutoff: 500 Hz) were applied to mitigate device-specific discrepancies.

As illustrated in Figure 7a, the MEMS sensor demonstrates amplitude fidelity to LDV measurements, exhibiting merely 3.4% standard deviation error with reduced amplitude variance. Figure 7b reveals that, while environmental vibrations introduce sub-100 Hz discrepancies in LDV data, both modalities show a strong spectral coherence in the range of 100–500 Hz. These results confirm MEMS technology's capability to reliably capture micro-motion vibration signatures critical for rock mass stability assessment.

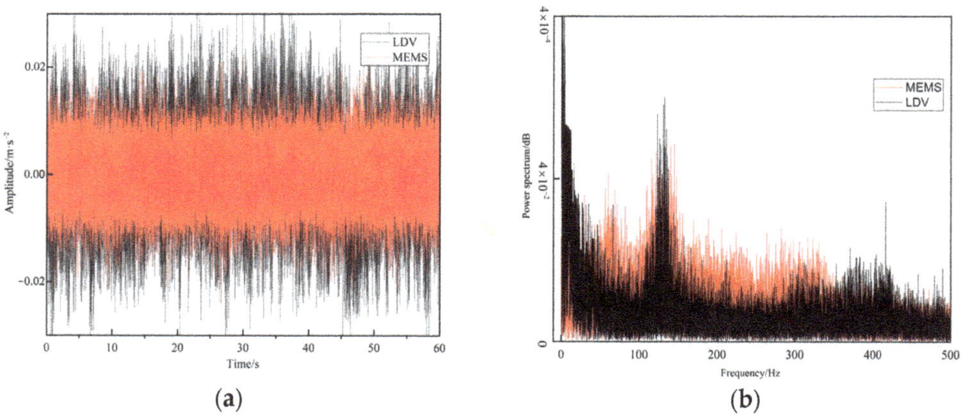

Figure 7. Comparison between LDV and MEMS sensor. (**a**) Temporal region comparison. (**b**) Frequency-domain signal comparison.

3.2.2. Correlation Analysis of Stability and NF

Following the described methodology, MEMS sensors acquired vibration curves from both bedrock and unstable rock masses, enabling the generation of frequency response function (FRF) curves across varying working conditions. The NF was derived by identifying the dominant peak within these FRF curves. As illustrated in Figure 8, the six FRF curves corresponding to distinct working conditions demonstrate a systematic leftward shift of resonance peaks (indicating frequency reduction) as the relative bonding length decreases. Quantitatively, in Condition 1 (relative bonding length = 1.0), the NF measures 641 Hz. A reduction to 0.35 in relative bonding length (Condition 14) results in a NF decline to 161 Hz, representing a 74% frequency reduction against a 65% bonding length decrease. This strong correlation confirms that diminished relative bonding lengths induce progressive declines in NF, providing a viable metric for stability assessment.

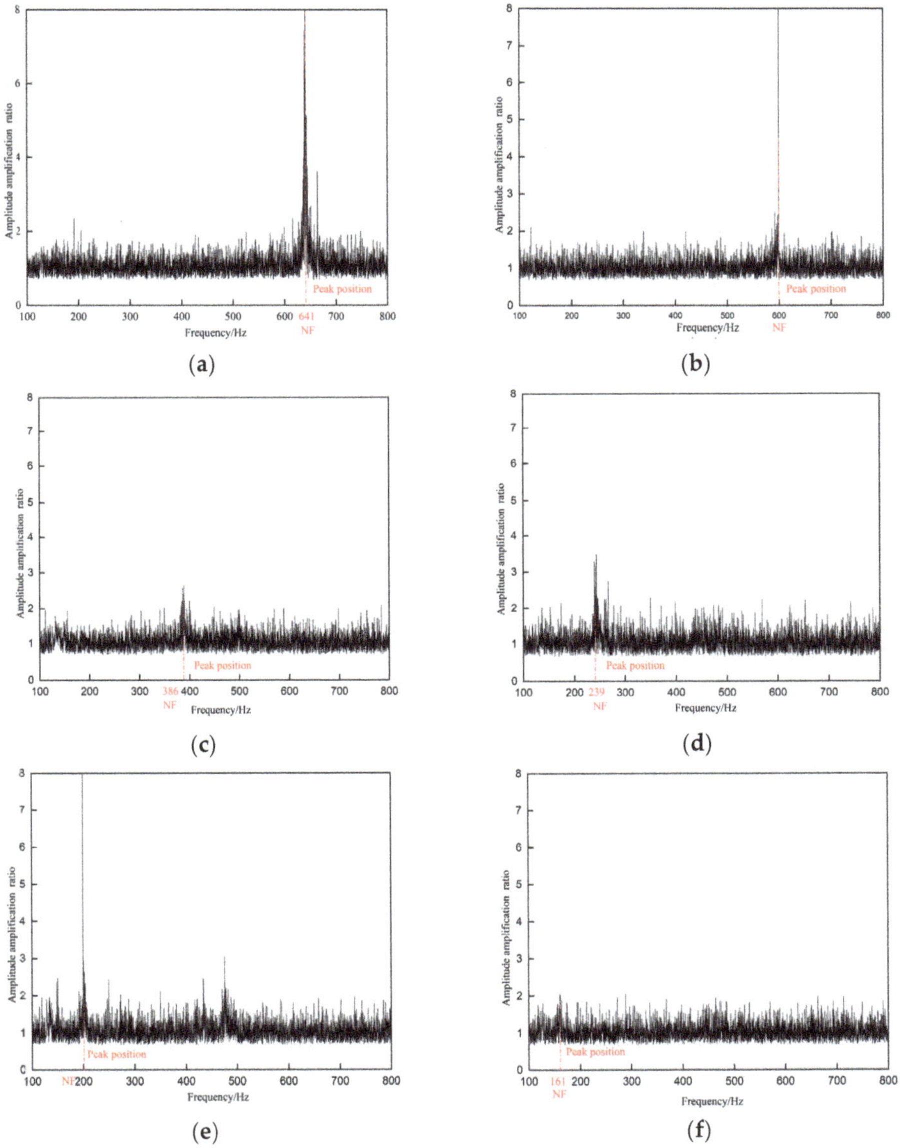

Figure 8. Frequency response curve under typical working conditions. (**a**) Frequency response curve of unstable rock mass in working condition 1. (**b**) Frequency response curve of unstable rock mass in working condition 2. (**c**) Frequency response curve of unstable rock mass in working condition 6. (**d**) Frequency response curve of working condition 9. (**e**) Frequency response curve of unstable rock mass under working condition 12. (**f**) Frequency response curve of unstable rock mass under working condition 14.

The ultimate equilibrium safety factor of unstable rock masses under each working condition was calculated using Equation (2). As shown in Figure 9, the safety factor decreases progressively with reductions in the relative bonding length of structural planes. In the initial state, the theoretical safety factor is 10.98. When the relative bonding length is reduced to 0.4, the safety factor drops to 1.75, and further decreases to 1.34 at a relative bonding length of 0.35. Both the NF and safety factor exhibit broadly consistent declining trends as the relative bonding length diminishes. However, at approximately 0.8 relative bonding length, a divergence emerges between the NF and safety factor trends. This discrepancy may stem from cutting depth inaccuracies during artificial preparation and the inherent heterogeneous anisotropic properties of the rock mass material.

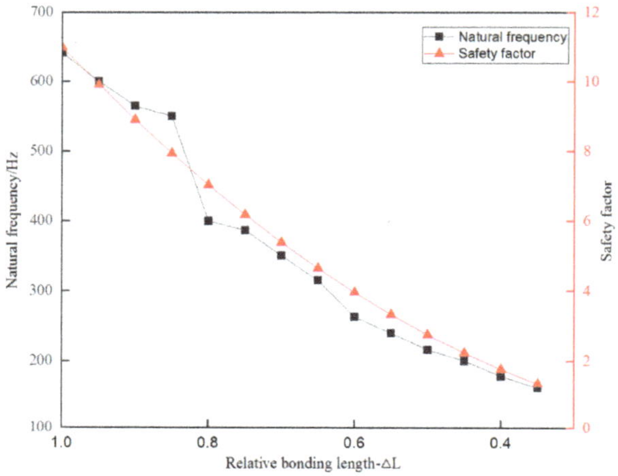

Figure 9. Relation between NF and safety factor of unstable rock.

3.2.3. Correlation Analysis of Stability and RMS-VAR

The vibration time–history curves of the bedrock and unstable rock mass, acquired by MEMS sensors, can be processed using Equation (8) to derive the dimensionless RMS-VAR. To correlate this ratio with rock mass stability, the three-axis acceleration data collected from the unstable rock mass were analyzed through their spatial geometric relationships, enabling the calculation of angular deviations toward the potential tipping direction. The computational framework is defined as follows:

$$\gamma_x = \arctan(a_x/\sqrt{(a_y^2 + a_z^2)}) \times 180/\pi \tag{10}$$

where γ_x is the MEMS sensing device x direction (tipping direction) relative to the ground horizontal tilt angle.

As shown in Figure 10, with the reduction in the relative bonding length in the structural plane of the unstable rock mass, the RMS-VAR exhibits a progressive increase, with its magnitude of change significantly exceeding that of the tilt angle. When the relative bonding length decreases from 1.0 to 0.8, the RMS-VAR from 1.02 to 1.46, while the tilt angle remains stable within this range. Upon further reduction to a relative bonding length of 0.5, the tilt angle undergoes an abrupt surge, and the RMS-VAR surpasses three times its initial amplitude. These results demonstrate that the RMS-VAR effectively captures variations in the bonding length of the dominant structural plane and provides earlier indications of stability degradation compared to detectable deformations.

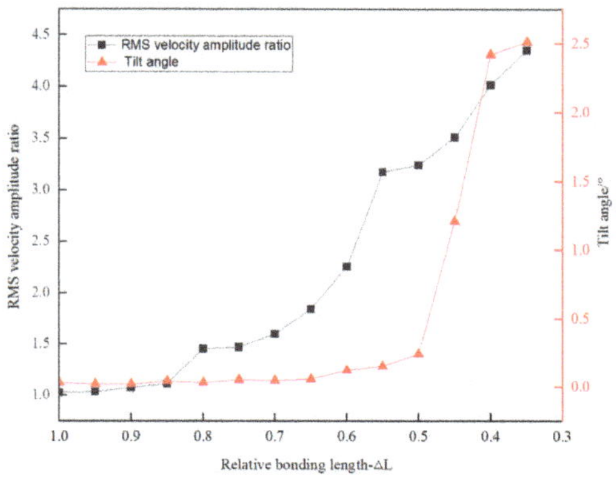

Figure 10. Relation between RMS-VAR and tilt angle.

4. Discussion

4.1. Dual-Indicator Stability Evaluation Model of Unstable Rock Mass

The collapse of unstable rock masses is a sudden geotechnical phenomenon where jointed slope rock masses gradually separate from the parent rock and undergo abrupt collapse. The stability of unstable rock masses can be assessed by monitoring the bonding state of internal structural planes. By using MEMS sensors to measure NF and RMS-VAR in real time, researchers can track the separation process between unstable rock masses and parent rock, enabling the real-time prediction of collapse timing.

As the relative bonding length of structural planes decreases, the NF values captured by MEMS sensors exhibit a declining trend, mirroring the reduction in the limit equilibrium safety factor. Simultaneously, the RMS-VAR between unstable rock masses and bedrock, measured via MEMS sensors, reflects changes in the structural plane's relative bonding length and provides earlier indications of stability degradation compared to tilt angle variations.

Based on the correlated trends of decreasing stability, declining NF, and increasing RMS-VAR, a dynamic stability assessment model (Figure 11) was established [24]. In this model, NF and RMS-VAR values in the upper-left quadrant indicate stable conditions, while migration toward the lower-right quadrant signifies destabilization. These dual indicators—NF and RMS-VAR—overcome the limitations of single-parameter assessments by enabling the real-time monitoring of stability transitions.

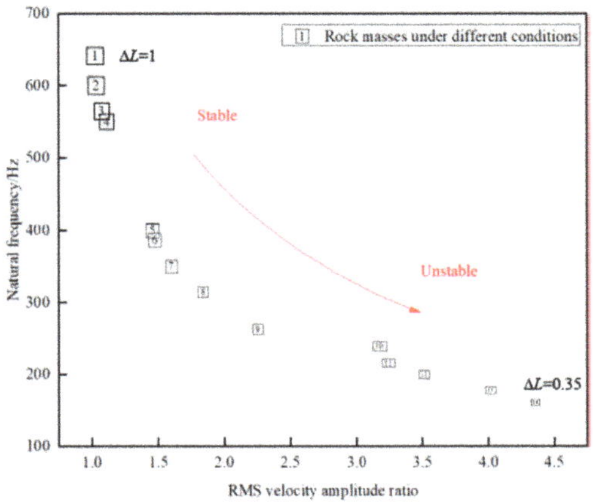

Figure 11. Dual-indicator evaluation model of unstable rock.

4.2. MEMS-Based Monitoring and Early Warning System

The proposed dual-indicator MEMS monitoring system addresses the critical challenge of real-time instability assessment in unstable rock masses by leveraging NF (NF) and RMS-VAR as core parameters. Traditional seismic sensing methods, which rely on post-trigger event analysis (e.g., collapse-induced seismic waves), suffer from delayed responses and high false-negative rates in dynamic environments. Borehole-based geotechnical investigations, while effective for structural characterization, remain cost-prohibitive for localized unstable rock masses. In contrast, the MEMS-based dual-indicator approach enables the continuous vibration monitoring of both unstable rock masses and the adjacent bedrock, exploiting their differential vibration responses to derive stability metrics with high temporal resolution.

The correlation between vibration curves from bedrock and unstable rock mass sensors is highly distance-dependent. As demonstrated in prior studies, sensor proximity (<10 m) is critical for maintaining strong signal coherence ($R^2 > 0.85$), ensuring the reliable extraction of NF and RMS-VAR. Our findings align with those of Kang et al. [25], who emphasized that signal decoupling beyond threshold distances (>15 m) introduces phase distortions, rendering response characteristics non-representative of true rock behavior. To mitigate this, we advocate for MEMS sensor deployment within optimal proximity ranges, supplemented by correlation indices to validate data fidelity—a strategy validated in slope stability monitoring.

Integrating NF and RMS-VAR with IoT cloud platforms resolves the limitations of standalone equipment performance. Localized thresholds for NF (±5% baseline), RMS-VAR (1.2–1.8), and inclined deformation (>2 mm/h) enable tiered warnings (e.g., yellow/orange/red alerts), while multi-factor big data models correlate environmental drivers (rainfall, temperature cycles) with historical instability patterns. This hybrid approach mirrors the framework proposed by Jiang et al. [26] for landslide prediction, where machine learning fused real-time sensor data with geospatial trends to achieve 92% early-warning accuracy.

However, the scalability of MEMS-based systems hinges on efficient data handling. To achieve the real-time stability monitoring of field rock masses, the dual-indicator monitoring and early-warning system based on MEMS sensors requires the integration of critical technologies including solar power supply, wireless transmission, and cloud services. While ensuring data transmission stability, it is imperative to simultaneously optimize transmission efficiency. For massive data acquisition and processing demands, the system must implement load balancing algorithms and distributed storage architecture to overcome the multi-ten-second alert issuance delays inherent in traditional warning systems. Future work should explore advanced edge computing for on-site analytics, minimizing cloud dependency in remote terrains.

5. Conclusions

Through MEMS sensing technology, the vibrational characteristic differences between bedrock and unstable rock masses were captured, yielding dynamic indices that reflect slope stability and establishing a novel methodology for collapse monitoring. Guided by the generalized spring–proton model of unstable rock masses, two monitoring parameters—NF and RMS-VAR—were proposed and validated through laboratory collapse simulations. The main conclusions are as follows:

(1) The critical structural plane length governs stability thresholds, with NF and RMS-VAR serving as effective proxies for fracture evolution. The experimental data demonstrate an inverse correlation between bonding plane length and NF, while RMS-VAR exhibit positive scaling with crack propagation rates.

(2) MEMS sensing can accurately obtain the NF of unstable rock masses and the RMS-VAR, and can reflect the stable state of unstable rock masses in real time. The NF of unstable rock masses shows a 4:3 proportionality ratio to the limit equilibrium safety factor, and the change amplitude of RMS-VAR is three times that of the tilt angle.
(3) Field implementation requires the integration of MEMS edge nodes, cloud-based analytics, and adaptive signal processing algorithms. This method provides a reliable technical guideline for rock mass monitoring in practical engineering and proposes a new concept for similar geological disaster monitoring and early warning.

Author Contributions: Conceptualization, M.X. and C.C.; methodology, C.C.; software, X.Z.; validation, C.C., M.X. and Y.D.; formal analysis, Y.D.; investigation, C.C.; resources, C.C.; data curation, Y.D.; writing—original draft preparation, Y.D.; writing—review and editing, C.C.; visualization, M.X.; supervision, Y.D. and X.Z.; project administration, X.Z.; funding acquisition, Y.D. and X.Z. All authors have read and agreed to the published version of the manuscript.

Funding: This research was funded by the National Key Research and Development Project of China grant number 2023YFC3081400, the National Natural Science Foundation of China grant numbers 42477167 and 41572274, and State Key Laboratory for GeoMechanics and Deep Underground Engineering, China University of Mining & Technology, Beijing grant number SKLGDUEK2130. The APC was funded by Fundamental Research Funds for the Central Universities grant number FRF-BD-20-01A. All authors have read and agreed to the published version of the manuscript.

Institutional Review Board Statement: Not applicable.

Informed Consent Statement: Not applicable.

Data Availability Statement: The original contributions presented in this study are included in the article. Further inquiries can be directed to the corresponding authors.

Conflicts of Interest: The authors declare no conflicts of interest.

Abbreviations

The following abbreviations are used in this manuscript:

MEMS	Micro-Electro-Mechanical System
NF	Natural Frequency
RMS-VAR	Root Mean Square Velocity Amplitude Ratio
LDV	Laser Doppler Vibrometer
SDOF	Single Degrees of Freedom
PCA	Principal Component Analysis
FRF	Frequency Response Function

References

1. Luo, G.; Cheng, Q.; Shen, W.; Ling, S.; Zhang, X.; Zou, P.; Zhao, Y. Research status and development trend of high energy rockfalls. *Earth Sci.* **2022**, *47*, 913–934. [CrossRef]
2. Du, Y.; Xie, M.W.; Jiang, Y.J.; Chen, C.; Jia, B.L.; Huo, L.C. Rock mass collapse disaster formation mechanism and early warning research review. *J. Met. Mine* **2021**, *1*, 106–119. [CrossRef]
3. Iannucci, R.; Martino, S.; Paciello, A.; D'Amico, S.; Galea, P. Investigation of cliff instability at Għajn Ħadid Tower (Selmun Promontory, Malta) by integrated passive seismic techniques. *J. Seismol.* **2020**, *24*, 897–915. [CrossRef]
4. Bottelin, P.; Baillet, L.; Larose, E.; Jongmans, D.; Hantz, D.; Brenguier, O.; Cadet, H.; Helmstetter, A. Monitoring rock reinforcement works with ambient vibrations: La Bourne case study (Vercors, France). *Eng. Geol.* **2017**, *226*, 136–152. [CrossRef]
5. Mercerat, E.D.; Payeur, J.B.; Bertrand, E.; Malascrabes, M.; Pernoud, M.; Chamberland, Y. Deciphering the dynamics of a heterogeneous sea cliff using ambient vibrations: Case study of the Sutta-Rocca overhang (southern Corsica, France). *Geophys. J. Int.* **2020**, *224*, 813–832. [CrossRef]
6. Chen, H.K.; Zhou, Y.T.; Tang, H.M. Based on the analysis of sliding type to remove unstable rock mass blasting dynamic stability calculation method. *J. Vib. Shock* **2014**, *33*, 31–34+41. [CrossRef]

7. Chen, H.K.; Zhang, R.G.; Tang, H.M.; Zhao, X.T. Pressure shear type unstable rock mass damage play rushed to study dynamic parameters. *J. Vib. Shock* **2012**, *31*, 30–33. [CrossRef]
8. Taruselli, M.; Arosio, D.; Longoni, L.; Papini, M.; Zanzi, L. Seismic noise monitoring of a small rock block collapse test. *Geophys. J. Int.* **2021**, *224*, 207–215. [CrossRef]
9. Burjánek, J.; Gischig, V.; Moore, J.R.; Fäh, D. Ambient vibration characterization and monitoring of a rock slope close to collapse. *Geophys. J. Int.* **2018**, *212*, 297–310. [CrossRef]
10. Glueer, F.; Mreyen, A.-S.; Cauchie, L.; Havenith, H.-B.; Bergamo, P.; Halló, M.; Fäh, D. Integrating Seismic Methods for Characterizing and Monitoring Landslides: A Case Study of the Heinzenberg Deep-Seated Gravitational Slope Deformation (Switzerland). *Geosciences* **2024**, *14*, 28. [CrossRef]
11. Häusler, M.; Michel, C.; Burjánek, J.; Fäh, D. Fracture Network Imaging on Rock Slope Instabilities Using Resonance Mode Analysis. *Geophys. Res. Lett.* **2019**, *46*, 6497–6506. [CrossRef]
12. Häusler, M.; Michel, C.; Burjánek, J.; Fäh, D. Monitoring the Preonzo Rock Slope Instability Using Resonance Mode Analysis. *J. Geophys. Res. Earth Surf.* **2021**, *126*, e2020JF005709. [CrossRef]
13. Wu, D.; Liang, T.; Yang, Y.; Pei, Q.; Yi, Y.; Wu, J.; Li, D. Experimental study on displacement monitoring of instable highway slope based on MEMS sensors. *Front. Earth Sci.* **2025**, *13*, 1541217. [CrossRef]
14. Praveen, H.M.; Singh, S.P. An investigation into practical implementation of mems sensors for low-speed condition monitoring applications. *Exp. Tech.* **2025**, *49*, 1–18. [CrossRef]
15. Bolla, A.; Paronuzzi, P. Numerical investigation of the pre-collapse behavior and internal damage of an unstable rock slope. *Rock Mech. Rock Eng.* **2020**, *53*, 2279–2300. [CrossRef]
16. Samadi, H.; Farrokh, E. Utilization of rock mass parameters for performance prediction of rock tbms using machine learning algorithms. *SRPH J. Fundam. Sci. Technol.* **2021**, *3*, 1–9. [CrossRef]
17. Chen, H.K.; Wang, S.J.; Chen, J. Theoretical framework of physical evolution of unstable rock mass collapse disaster and disaster outbreak time. *J. Chongqing Norm. Univ.* **2020**, *37*, 121–128. [CrossRef]
18. Colombero, C.; Comina, C.; Vinciguerra, S.; Benson, P.M. Microseismicity of an unstable rock mass: From field monitoring to laboratory testing. *J. Geophys. Res. Solid Earth* **2018**, *123*, 1673–1693. [CrossRef]
19. Liu, X.R.; Guo, X.Y.; Zhou, X.H.; Luo, X.Y.; Wang, H.; Li, P.Y.; Zhou, F.C. Study on macro-meso shear transfixion mechanisms and mechanical properties of shear band-bedrock interfaces of dangerous rock on reservoir bank. *Chin. J. Rock Mech. Eng.* **2024**, *43*, 1096–1109. [CrossRef]
20. He, Z. *Study on Instability Prediction Model of Splitting Rock Mass on Slopes*; Beijing University of Science and Technology: Beijing, China, 2023. [CrossRef]
21. Li, B.; Zhao, W.; Miao, Y.; Tian, W.; Liao, W. A method for dynamic parameter identification of an industrial robot based on frequency response function. *Int. J. Mech. Syst. Dyn.* **2024**, *4*, 461–471. [CrossRef]
22. Dahiya, N.; Pandit, K.; Sarkar, S.; Pain, A. Various aspects of rockfall hazards along the mountain roads in India: A systematic review. *Indian Geotech. J.* **2024**. [CrossRef]
23. Raja, S.; Abhishek, M.; Balegar, N.; Kristharaj, L.; Yeligar, S. Prediction of rockfall from highwall slope in an open cast mine. *J. Mines Met. Fuels* **2024**, *72*, 147–163. [CrossRef]
24. Li, L.Q.; Huang, R.Q. Steep dip bedding slope instability mechanism analysis. *J. Eng. Sci. Technol.* **2018**, *50*, 54–63. [CrossRef]
25. Kang, X.; Huang, L.; Zhang, Y.; Yun, S.; Jiao, B.; Liu, X.; Zhang, J.; Li, Z.; Zhang, H. Wearable Multi-Channel Pulse Signal Acquisition System Based on Flexible MEMS Sensor Arrays with TSV Structure. *Biomimetics* **2023**, *8*, 207. [CrossRef]
26. Jiang, H.; Wang, Y.; Guo, Z.; Zhou, H.; Wu, J.; Li, X. Landslide Displacement Prediction Stacking Deep Learning Algorithms: A Case Study of Shengjibao Landslide in the Three Gorges Reservoir Area of China. *Water* **2024**, *16*, 3141. [CrossRef]

International Journal of
Geo-Information

MDPI

Article

Exploring the Application of NeRF in Enhancing Post-Disaster Response: A Case Study of the Sasebo Landslide in Japan

Jinge Zhang [1,2], Yan Du [3], Yujing Jiang [1,*], Sunhao Zhang [1,2], Hongbin Chen [1] and Dongqi Shang [1]

[1] Department of Integrated Science and Technology, Nagasaki University, Nagasaki 852-8521, Japan
[2] School of Civil and Environmental Engineering, Nanyang Technological University, Singapore 639798, Singapore
[3] Beijing Key Laboratory of Urban Underground Space Engineering, University of Science and Technology Beijing, Beijing 100083, China
* Correspondence: jiang@nagasaki-u.ac.jp

Abstract: Rapid acquisition of 3D reconstruction models of landslides is crucial for post-disaster emergency response and rescue operations. This study explores the application potential of Neural Radiance Fields (NeRF) technology for rapid post-disaster site modeling and performs a comparative analysis with traditional photogrammetry methods. Taking a landslide induced by heavy rainfall in Sasebo City, Japan, as a case study, this research utilizes drone-acquired video imagery data and employs two different 3D reconstruction techniques to create digital models of the landslide area. Visual realism and point cloud detail were compared. The results indicate that the high-capacity NeRF model (NeRF 24G) approaches or even surpasses traditional photogrammetry in visual realism under certain scenarios; however, the generated point clouds are inferior in terms of detail compared to those produced by traditional photogrammetry. Nevertheless, NeRF significantly reduces the modeling time. NeRF 6G can generate a point cloud of engineering-useful accuracy in only 45 min, providing a 3D overview of the disaster site to support emergency response efforts. In the future, integrating the advantages of both methods could enable rapid and precise post-disaster 3D reconstruction.

Keywords: photogrammetry; neural radiance fields; landslide; 3D reconstruction

1. Introduction

Landslides, as a common geological hazard, are characterized by sudden occurrence, significant destructive power, and extensive impact areas, posing severe threats to human life, property security, and socioeconomic development [1–3]. In Japan, landslides frequently occur during the rainy summer season [4,5], and rainfall often triggers secondary landslides [6], posing significant risks to emergency response operations. Therefore, rapidly and accurately obtaining real-time conditions of landslide sites, clarifying landslide formation mechanisms and failure modes, and predicting subsequent stability are crucial for conducting effective rescue operations. To capture the spatiotemporal distribution characteristics and failure morphology of landslides, 3D reconstruction of landslide bodies is required to obtain precise terrain information, which also plays an essential role in governmental decision-making and early warning of secondary hazards [7,8].

In slope engineering, traditional 3D reconstruction methods can be broadly classified into two categories [9]: geodetic surveying techniques and remote sensing technologies. Geodetic surveying techniques, represented by total stations [10] and terrestrial laser scanners (TLS) [11], can achieve highly accurate reconstruction results. However, these instruments are complicated to operate, inconvenient to transport, and challenging to deploy in the complex terrain conditions following landslides. Airborne LiDAR scanners [12], a type of remote sensing technology, offer flexible operation and can rapidly construct high-precision 3D models. LiDAR measures distances by emitting and receiving reflected laser pulses. Its key advantage lies in its ability to penetrate vegetation canopies and capture sub-canopy terrain data, enabling the generation of accurate bare-earth digital elevation models and the detection of hazardous features such as ground fissures. For this reason, LiDAR has been widely applied in landslide detection. However, they involve high data processing costs and expensive equipment, making large-scale adoption challenging. Additionally, a limited number of researchers have applied structured-light methods [13] from remote sensing technologies for landslide disaster 3D reconstruction. Nevertheless, structured-light techniques depend on additional projection equipment and still present significant challenges for large-scale 3D reconstruction applications.

In recent years, with the rapid advancement of civilian unmanned aerial vehicles (UAVs), represented by DJI, UAV-based photogrammetric 3D reconstruction techniques have been widely adopted in the field of landslide disasters. UAV photogrammetry offers advantages such as low cost, high flexibility, high image resolution, and portability, providing efficient on-site data acquisition and accurate 3D reconstruction solutions for landslide disaster investigations. Gupta et al. [14] utilized UAVs to reconstruct two landslides in Mandi district, Himachal Pradesh, Himalaya, and compared the results with total station measurements, finding errors within 5%. Zhao et al. [15] performed detailed reconstruction of a landslide-affected area using close-range photogrammetry, obtaining precise information on surface crack development. Lucieer et al. [16] employed multi-temporal UAV models to map the displacement of the slope toe, soil blocks, and vegetation patches at the top of a landslide. Gallo et al. [17] conducted hazard assessments of rockfall disasters using UAVs, generated high-resolution point cloud models, and analyzed the morphological characteristics and fragmentation of rock blocks. Zhu et al. [18] proposed a UAV photogrammetry-based method for interpreting rock mass discontinuities and early recognition of potential hazards, utilizing UAVs to acquire the discontinuity and spatial distribution characteristics of rock slopes. Zhu et al. [19] utilized UAVs to assist emergency response by performing 3D reconstruction and investigation analysis of a rainfall-induced landslide event that occurred in Guangzhou, China, in July 2022.

Although UAV photogrammetry provides high-accuracy 3D reconstruction, the data processing phase demands substantial computational resources, leading to prolonged reconstruction times. Given the urgency during the emergency response phase following a landslide event, speed and efficiency should be prioritized. In recent years, Neural Radiance Fields (NeRF) [20], a novel 3D reconstruction technique, has gradually shown a trend toward replacing traditional photogrammetry due to its efficiency and high-quality reconstruction results. NeRF employs implicit neural representation, leveraging a multilayer perceptron (MLP) network to learn the 3D geometry and appearance information of a scene, subsequently generating high-quality novel-view images via volume rendering techniques. Compared to traditional methods, NeRF can significantly accelerate reconstruction processes and provide rapid scene representation, demonstrating considerable advantages for emergency response applications. Croce et al. [21] evaluated the application of NeRF and photogrammetric methods for three-dimensional reconstruction in the cultural heritage domain and found that NeRF trains far more rapidly than photogrammetry. Remondino et al. [22] presented a critical analysis of NeRF-based 3D reconstruction techniques and compared them with conventional photogrammetric approaches; they reported that the Nerfacto method completes modeling on the test dataset in just 15 minutes, whereas photogrammetry requires four to six hours. Moreover, comparative studies of NeRF and photogrammetric methods in reconstructing the Panagia Karmoitissa Cathedral [23] and other historic structures [24] consistently demonstrate that NeRF achieves higher efficiency than photogrammetry. However, to date, there have been no reports of applying NeRF to 3D reconstruction for landslide disaster emergency response.

This study aims to comprehensively evaluate the performance of Neural Radiance Fields (NeRF) and traditional photogrammetry methods for 3D reconstruction of landslide disasters, and explore the potential of integrating these two approaches for landslide hazard investigations. This study takes a rainfall-induced landslide in Sasebo City, Japan, as a case study, employing UAV imagery data to perform reconstructions separately using both methods. A comparative analysis of the two methods is conducted regarding visual realism and point cloud detail of the generated models.

2. Case Study: Sasebo Landslide

The case selected for this study is a landslide event that occurred in Sasebo City, located in the Kyushu region of Japan. Figure 1 shows the landslide and its surrounding area: (a) is a satellite image taken in March 2023; (b) is an aerial photograph captured on 22 September 2023. The Sasebo landslide occurred during a heavy rainfall period from 13 to 17 September 2023, representing a case of slope failure triggered by intense precipitation; uninfiltrated water was observed along the front edge of the landslide mass. According to field investigations, the landslide deposit measured approximately 89 m in length and 109 m in width, and had a volume of nearly 103,800 m^3; additionally, the Tahara–Kuroishi Line in Sasebo City was obstructed by the event.

(a) (b)

Figure 1. Overview of the Sasebo landslide: (**a**) aerial imagery, (**b**) satellite imagery.

To obtain a 3D model of the landslide site, TEC-FORCE (Technical Emergency Control FORCE) personnel in the Kyushu region used a DJI Phantom4 Pro UAV to record two videos of the site at 30 frames per second, and subsequently generated a raw dataset through processing. The dataset comprises a total of 394 images, each with a resolution of 3840 × 2160. The UAV operated at a flight altitude of 35 to 52 m, and the average overlap rate was 84.2%. It should be noted that although TEC-FORCE personnel possess extensive experience in responding to natural disasters, they are not experts in aerial surveying or 3D reconstruction. Consequently, the on-site video capture was not conducted in accordance with standard photogrammetric protocols—a point that will be further elaborated in the discussion section.

The images in the dataset capture the entire landslide body along with an expansive background, including the surrounding environment and trees. After an initial photogrammetric registration, the dataset's images are presented as shown in Figure 2. The dataset's images reveal a clear depiction of the landslide's morphology and structure, including slope deformations and fractures resulting from tensile forces acting on the soil mass—one of the key focal points in 3D reconstruction for slope engineering [25].

Figure 2. Sparse point cloud after image alignment.

3. Methodology

Three-dimensional reconstruction is a key technology in the monitoring [26,27], assessment, and post-disaster management of landslides. Conducting 3D reconstruction in landslide areas enables the accurate reproduction of the terrain and geomorphological features of the disaster site, thereby providing critical data support for landslide volume estimation, stability evaluation, and secondary hazard prediction. This section briefly introduces 3D reconstruction based on photogrammetry and that based on NeRF.

3.1. Photogrammetry-Based 3D Reconstruction

Photogrammetric 3D reconstruction technology is based on two-dimensional imagery acquired from multiple viewpoints, with its core principle lying in leveraging the geometric relationships among multi-view images to infer a scene's 3D structure. Its core workflow includes Structure from Motion (SFM) and Multi-View Stereo (MVS) [28,29]. During the SFM stage, features are first extracted and matched from the acquired oblique image sequence, employing techniques such as epipolar geometry constraints and Bundle Adjustment to simultaneously compute the exterior orientation parameters (camera poses) of all images and derive a sparse 3D structure (sparse point cloud) of the scene. Subsequently, a dense point cloud is computed using MVS, and further processes such as mesh generation, point cloud refinement, and texture mapping are performed to achieve a detailed 3D reconstruction of the target scene. Photogrammetry has been widely applied in fields such as topographic mapping [30,31], disaster identification [32], cultural heritage digitization [33], and smart cities [34].

3.2. NeRF-Based 3D Reconstruction

In computer graphics, novel view synthesis is an extremely challenging task. However, with the rapid development of deep learning technology—especially the powerful capabilities of deep neural networks in image generation and representation—new breakthroughs in novel view synthesis have emerged. Mildenhall et al. [35] introduced the concept of NeRF in 2020. NeRF adopts an entirely novel approach by forgoing the explicit storage of 3D geometric models (such as meshes or point clouds). Instead, it employs a deep neural network to implicitly represent the continuous density and color information of a scene, and utilizes volume rendering techniques to generate high-quality novel view images. Figure 3 illustrates a schematic diagram of the underlying principles of NeRF.

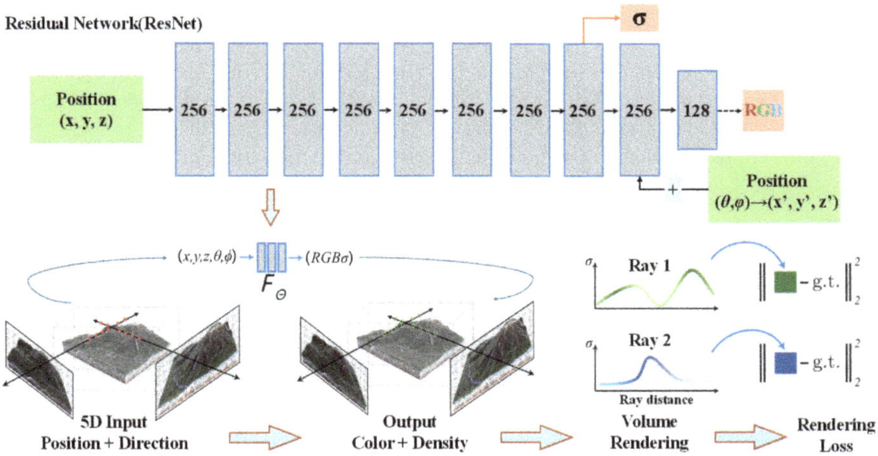

Figure 3. Diagram of NeRF, adapted from Mildenhall et al. [35].

At the core of NeRF lies a residual network (ResNet). The network takes as input the 3D coordinates (x, y, z) of a point in space along with its viewing direction (θ, φ), and outputs the corresponding volume density σ and color (RGB). This network can be represented as

$$F_\Theta : (x, y, z, \theta, \phi) \mapsto (\sigma, c) \tag{1}$$

where Θ denotes the learnable parameters of the network. σ represents the volume density at location x. c denotes the radiance (color) when observed along the viewing direction θ, ϕ. To better capture high-frequency details, NeRF typically applies a positional encoding to the input coordinates x and viewing direction θ, ϕ, mapping low-frequency information into a higher-dimensional space, thereby enabling the network to represent richer spatial variations.

Subsequently, NeRF integrates along rays using volume rendering techniques to generate the final pixel color. The pixel color $\hat{\mathbf{C}}(\mathbf{r})$ corresponding to a ray can be expressed as follows:

$$\hat{\mathbf{C}}(\mathbf{r}) = \sum_{i=1}^{N} T_i \alpha_i \mathbf{c}_i \tag{2}$$

$$T_i = \exp(-\sum_{j=1}^{i-1} \sigma_j \delta_j) \tag{3}$$

$$\alpha_i = 1 - \exp(-\sigma_i \delta_i) \tag{4}$$

where T_i represents the transmittance from the ray origin to the i sample point. α_i denotes the absorption coefficient at the i sample point. δ_i represents the distance between adjacent sample points.

Due to its ability to capture fine details and complex lighting effects, NeRF has been rapidly adopted in many fields since its introduction in 2020, such as autonomous driving [36] and cultural heritage preservation [37]. However, there are few reports on its application in rock engineering and slope engineering.

3.3. Technical Approach

The aim of this study is to explore the potential of NeRF-based 3D reconstruction in the field of geological disasters and, using traditional photogrammetry-based 3D reconstruction as a benchmark, to evaluate the advantages and limitations of NeRF. In order to achieve this, both NeRF and photogrammetry are applied to perform 3D reconstruction on the same dataset, which includes the complete landslide body along with the surrounding trees and clearly captures the landslide's features. The detailed workflow for modeling and comparing the two methods in this study is illustrated in Figure 4. All steps shown in Figure 4 were executed under the same hardware configuration, utilizing an RTX 4090 graphics card with 24 GB VRAM, a 14th-generation Intel Core i9-14900K CPU, and 128 GB of RAM.

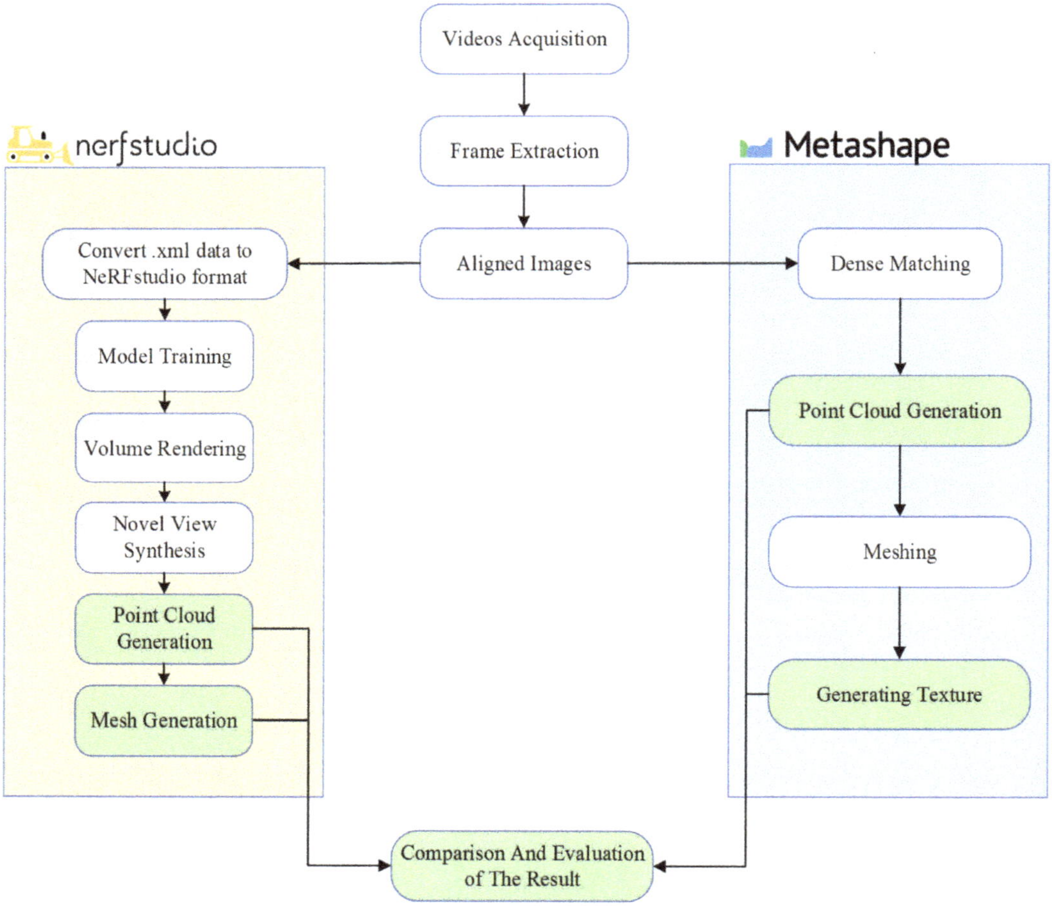

Figure 4. Flowchart of the 3D reconstruction method used in this study.

After obtaining the video and extracting frames to create the dataset, image alignment, initial point cloud generation, and camera pose estimation are required; in this study, these steps were implemented using Agisoft Metashape. For traditional photogrammetry-based 3D reconstruction, the aligned image dataset is directly used to generate a dense point cloud, create a mesh network, and finally build the model texture. All these steps were executed using the highest quality mode.

For NeRF-based 3D reconstruction, this study employed the Nerfstudio platform developed by Tancik et al. [38]. Nerfstudio also requires an aligned image dataset as the initial input. Because Nerfstudio is compatible with the image alignment performed by Metashape, the alignment results were exported in XML format and then converted for input into Nerfstudio. Subsequently, the multi-layer perceptron (MLP) is trained to generate predictions for novel viewpoints, ultimately yielding neural renderings. In this phase, Nerfstudio offers additional model options based on the required training capacity; the standard model occupies 6 GB of VRAM, while the "huge" model—employing more parameters and a denser feature network—requires 24 GB of VRAM. The huge model can capture finer scene details and high-frequency information, thereby enhancing rendering and reconstruction accuracy, albeit at the expense of slower training and rendering speeds. This study concurrently trained and generated both types of neural renderings, hereafter referred to as NeRF 6G and NeRF 24G, respectively. We employed the Nerfacto method within the Nerfstudio framework for three-dimensional reconstruction, which uses a two-stage sampling strategy consisting of an initial piecewise sampler followed by a proposal sampler. By default, the proposal network samples 64 points per ray in each of two iterations, while the main NeRF network samples 32 points per ray. Nerfacto adopts multi-resolution hash encoding with default settings including 16 encoding levels, 2 features per level, a hash map size exponent of 19, a base resolution of 16, and a maximum resolution of 2048. Since neural renderings cannot be directly compared with the meshes or point clouds produced by photogrammetry, Nerfstudio provides an API to convert neural renderings into point cloud and mesh representations. Nerfstudio leverages the depth predictions from the radiance field to reconstruct the scene's geometric structure, generating point cloud data through ray sampling and depth estimation, and producing a mesh model via Poisson reconstruction.

Finally, this study evaluated the performance of the different 3D reconstruction methods. Visually, the reconstruction results were examined from the same viewpoints. In terms of geometric accuracy, the point clouds generated by the different 3D reconstruction methods were analyzed.

3.4. Metrics

To quantitatively assess the differences between models constructed using photogrammetry and those built using NeRF relative to the actual scene—and to fully explore the potential of NeRF in post-disaster response—this study introduces various metrics, with each metric quantifying the reconstruction performance from different perspectives.

(1) Peak Signal-to-Noise Ratio (PSNR)

PSNR measures the difference between the reconstructed image and the original image using pixel-level mean squared error (MSE), thereby reflecting the accuracy of image reconstruction; a higher value indicates that the model is closer to the actual scene in terms of pixel accuracy. Its mathematical expression is as follows:

$$PSNR = 10 \cdot \lg\left(\frac{MAX_I^2}{MSE}\right) \quad (5)$$

where MSE denotes the mean squared error, and MAX is typically set to 255.

(2) Visual information fidelity (VIF)

VIF, based on natural scene statistics and human visual system theory, can measure the ability of a reconstructed image to preserve the information of the original image. VIF indicates how much visual information is retained during transmission or processing; a higher VIF value means that the distorted image maintains a higher degree of information fidelity compared to the reference image, resulting in better quality, whereas a lower VIF value suggests that significant visual information is lost, leading to poorer quality. The mathematical formulation of VIF based on a wavelet domain model is given below:

$$VIF = \frac{\sum_{j=1}^{J} \sum_{k=1}^{K} \log\left(1 + \frac{g_{j,k}^2 \sigma_{C_{j,k}}^2}{\sigma_{v_{j,k}}^2}\right)}{\sum_{j=1}^{J} \sum_{k=1}^{K} \log\left(1 + \frac{\sigma_{C_{j,k}}^2}{\sigma_{v_{j,k}}^2}\right)} \quad (6)$$

where j, k represent the scale and subband, respectively. $\sigma_{C_{j,k}}^2$ denotes the local variance (i.e., signal energy) of the reference image at scale j and subband k. $g_{j,k}$ is the gain factor. $\sigma_{v_{j,k}}^2$ is the noise variance.

(3) Feature Similarity (FSIM)

FSIM leverages the human visual system's sensitivity to critical image features—such as phase congruency and gradient magnitude—to assess the similarity between images. It captures both the structural information and details of an image, with values typically ranging from 0 to 1; the closer the value is to 1, the more similar the features are between the images, indicating higher quality. Its mathematical expression is as follows:

$$FSIM = \frac{\sum_{x \in \Omega} PC_m(x) \cdot S_L(x)}{\sum_{x \in \Omega} PC_m(x)} \quad (7)$$

where Ω represents the set of image pixels. $PC_m(x)$ denotes the minimum phase congruency index between the two images at pixel x. $S_L(x)$ represents the gradient magnitude similarity at pixel x.

(4) Learned Perceptual Image Patch Similarity (LPIPS)

LPIPS leverages pre-trained deep neural networks (e.g., VGG or AlexNet) to extract image features and compute feature differences at multiple layers between two images, thereby obtaining a perceptual similarity score. LPIPS aligns more closely with human subjective perception and is capable of capturing subtle perceptual differences that traditional metrics may fail to reflect. Unlike the previous metrics, a lower LPIPS value indicates a higher perceptual similarity between images, meaning better image quality. Its mathematical expression is as follows:

$$LPIPS(x,y) = \sum_{l} \frac{1}{H_l W_l} \sum_{h=1}^{H_l} \sum_{w=1}^{W_l} \left\| w_l \odot \left(\phi_l(x)_{h,w} - \phi_l(y)_{h,w} \right) \right\|_2^2. \quad (8)$$

where x, y represents the image under comparison; $\phi_l(x)$ denotes the feature map extracted from image I at the x layer; H_l, W_l indicate the height and width of that feature map, respectively; w_l is the weighting factor; $\odot$ stands for element-wise multiplication. Since LPIPS relies on deep networks, its formulation essentially defines a distance metric implicitly through the network architecture.

By comprehensively evaluating these four metrics, we can analyze the differences between photogrammetry-based and NeRF-based 3D reconstructions relative to the actual scene from multiple perspectives—including pixel error, information fidelity, structural consistency, and perceptual similarity.

4. Comparisons and Analyses

This chapter evaluates the performance of photogrammetry-based and NeRF-based 3D reconstructions on the Sasebo landslide dataset, and compares them with respect to the image capture angles. Figure 5 illustrates the overall view of the three constructed models, including both the landslide body and its surrounding environment.

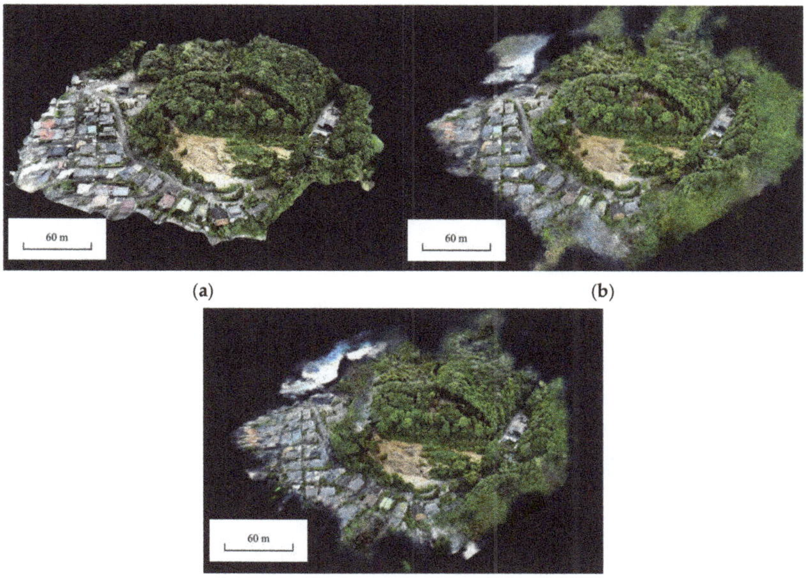

(c)

Figure 5. Overview of the three models. (**a**) Photogrammetric, (**b**) NeRF 6G, (**c**) NeRF 24G.

All three models successfully reconstructed the landslide body and the surrounding trees. The model reconstructed using photogrammetry exhibits relatively regular edges. Both NeRF-based models, however, display numerous floating artifacts along their edges. From this viewpoint, aside from the edge regions, it is difficult to discern differences in reconstruction quality. Therefore, this study selected three local regions for evaluation: the landslide body, the rear edge, and the lateral boundary of the landslide. The results are presented in Tables 1–3.

Table 1. Comparative analysis of all models for landslide crown.

Scene	Crown		
Ground Truth	**Photogrammetric**	**NeRF 6G**	**NeRF 24G**

PSNR ↑	27.9601	27.9505	27.9876
VIF ↑	0.0292	0.026	0.0321
FSIM ↑	0.6984	0.6918	0.6984
LPIPS ↓	0.6362	0.6919	0.6263

Table 2. Comparative analysis of all models for main body of landslide.

Scene	Crown		
Ground Truth	**Photogrammetric**	**NeRF 6G**	**NeRF 24G**

PSNR ↑	28.4045	28.3707	28.3841
VIF ↑	0.0695	0.0627	0.0645
FSIM ↑	0.8307	0.8225	0.8252
LPIPS ↓	0.5424	0.6183	0.5654

Table 3. Comparative analysis of all models for landslide main scarp.

Scene	Crown		
Ground Truth	**Photogrammetric**	**NeRF 6G**	**NeRF 24G**

PSNR ↑	28.5239	28.3193	28.3487
VIF ↑	0.1824	0.1914	0.2542
FSIM ↑	0.8839	0.9106	0.9339
LPIPS ↓	0.4069	0.4779	0.3685

Tables 1–3 present detailed depictions of three regions of the landslide as reconstructed by different models, along with the scores of four metrics benchmarked against the ground truth. All three models successfully reconstructed the landslide body and its surrounding environment. Although they did not achieve the fine level of detail exhibited by the ground truth, they nevertheless captured sufficient structure, details, and color. Upon close examination at high magnification, the three models differ only slightly in terms of clarity.

Through comparison, it is evident that for the landslide rear edge, the NeRF 24G model achieved the best performance—appearing most similar to the ground truth—and its scores across all four metrics were consistently higher than those of the traditional photogrammetry model. The photogrammetry model ranked second, outperforming the NeRF 6G model. Although the NeRF 6G model reconstructed the entire scene, its fewer internal parameters and relatively sparse feature network rendered it visually akin to a downsampled, lower-level version of the NeRF 24G model during the image pyramid construction, resulting in a blurred effect. For the landslide main body region, the photogrammetry model slightly outperformed the NeRF 24G model across all four metrics, while the NeRF 6G model recorded the lowest scores. This region mainly consists of post-slide soil, which features similar colors and contains numerous small soil blocks and particles, making it challenging to accurately model surface textures and subtle features. The final comparison region is the landslide lateral boundary, where the NeRF 24G model achieved the highest overall score; except for the PSNR metric, its scores on the remaining metrics surpassed those of the photogrammetry model. The PSNR metric mechanically compares the brightness and color information of the target and reference images on a pixel-by-pixel basis, whereas the other three metrics are more sensitive to information fidelity, structural similarity, and subjective human perception. Table 3 shows that the NeRF 24G model is closer to the real scene and aligns better with the notion of "similarity" from a subjective perspective.

In slope engineering, high-precision point cloud data can provide detailed three-dimensional spatial information, accurately reflecting the topography and elevation differences of slopes. By utilizing point cloud data, engineers can perform numerical simulations and finite element analyses to more accurately assess slope stability and identify potential landslide risks. Therefore, this study also compared the differences between point clouds generated using photogrammetry and those produced by the NeRF method. Figures 6 and 7 illustrate these differences.

(a) (b) (c)

Figure 6. Comparison of point clouds generated by photogrammetry and NeRF Methods. (**a**) Photogrammetry, (**b**) NeRF 6G, (**c**) NeRF 24G.

(a) (b) (c)

Figure 7. Comparison of point clouds generated by photogrammetry and NeRF Methods. (**a**) Photogrammetry, (**b**) NeRF 6G, (**c**) NeRF 24G.

Figure 6 compares the point clouds generated by the three models, with an excavator used for road clearing as the main subject in the image. It is evident that the point cloud produced by the photogrammetry method is the most detailed, faithfully representing trees, rooftops, the ground, and the excavator in its entirety. In contrast, the point cloud generated by the NeRF 6G model is noticeably sparser, losing many details, with numerous voids on the ground and unclear boundaries between the various parts of the excavator. The point cloud produced by the NeRF 24G model is slightly better than that of the NeRF 6G model, with a somewhat clearer rooftop and relatively complete excavator components; however, it still falls significantly short of the point cloud generated by the photogrammetry method.

Figure 7 presents a comparison of the point clouds of the main landslide body. Owing to the larger number of images used in the modeling process for this area, all three models produced relatively complete point clouds. The point cloud produced by the photogrammetry method remains the most detailed, with landform features such as cracks clearly rendered. The point cloud produced by the NeRF method loses many high-frequency details; however, it still effectively captures the overall undulations of the terrain and the major cracks, thereby providing an important reference for identifying macro features such as landslide deformation and soil distribution. It is worth noting that as the capacity of the NeRF model increases, improvements are observed in both detail preservation and noise control; however, the precision still falls short of that achieved by the photogrammetry method.

Figure 8 presents the three mesh models of the main landslide area. A comparison reveals that the mesh model produced by photogrammetry exhibits higher contrast and richer details, while the NeRF-generated mesh appears somewhat blurred and shows minor anomalous texture mapping errors (notably in the lower left corner of Figure 8b,c). Figure 9 illustrates the differences among the three mesh models after close-up and angle adjustments. Regarding the tensile cracks in the soil, the photogrammetry method preserves the most details—fully capturing information such as crack length and depth—whereas the NeRF 6G model loses a significant amount of detail, rendering only an approximate shape; the NeRF 24G model retains more detail than the NeRF 6G model but still falls short of the photogrammetry method. Furthermore, the photogrammetry-derived mesh preserves small surrounding stones, while the meshes generated by NeRF 6G and NeRF 24G fail to reconstruct these stones accurately, instead smoothing them out. This indicates that capturing minute objects and rough surface textures remains a challenge for NeRF.

(**a**) (**b**) (**c**)

Figure 8. Comparison of mesh generated by photogrammetry and NeRF methods. (**a**) Photogrammetry, (**b**) NeRF 6G, (**c**) NeRF 24G.

(**a**) (**b**) (**c**)

Figure 9. Comparison of detailed mesh. (**a**) Photogrammetry, (**b**) NeRF 6G, (**c**) NeRF 24G.

In practical slope engineering, surface morphology is of primary concern, as features such as undulations, slope angles, and micro-topography are key factors in determining slope stability. To evaluate this aspect, we extracted ground surface profiles from the point clouds generated by the three methods along two reference lines indicated in Figure 1. The profile derived from the photogrammetric method was used as the benchmark, and the Hausdorff Distance was adopted as the quantitative metric, where smaller values indicate closer alignment between curves. The detailed results are presented in Table 4. Key topographic features—such as large surface fissures and depressions—are highlighted with arrows of different colors in Figure 10. These highlighted sections correspond to the regions marked by arrows in Table 4.

Table 4. Comparison of cross-section.

Models	Comparative Analysis	
	Cross-Section A	**Cross-Section B**
Photogrammetric	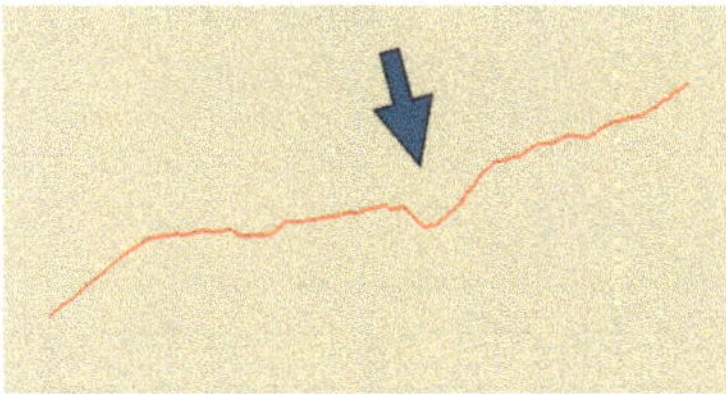	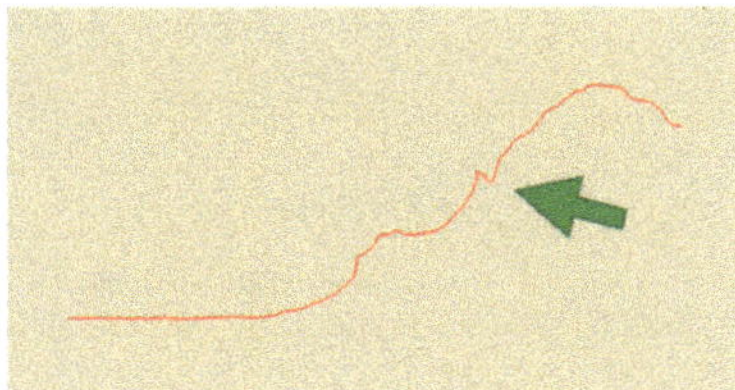
NeRF 6G	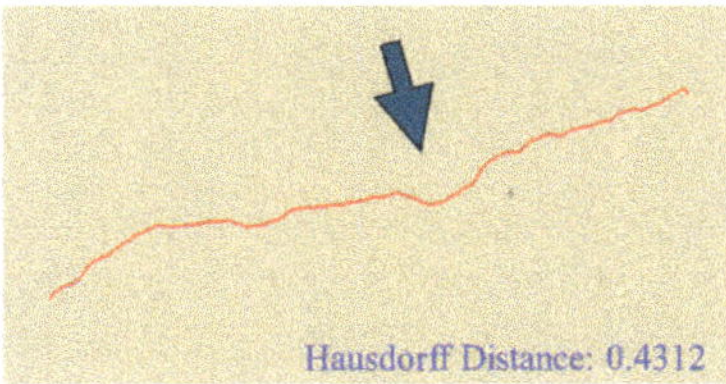	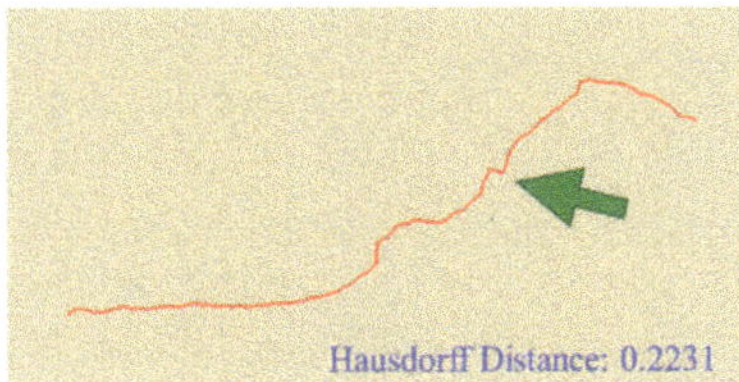
NeRF 24G	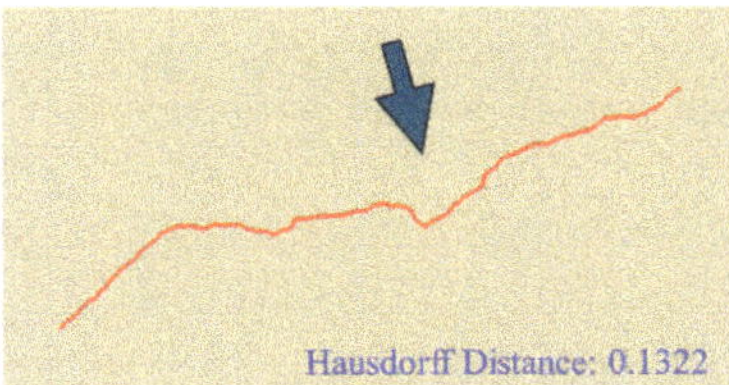	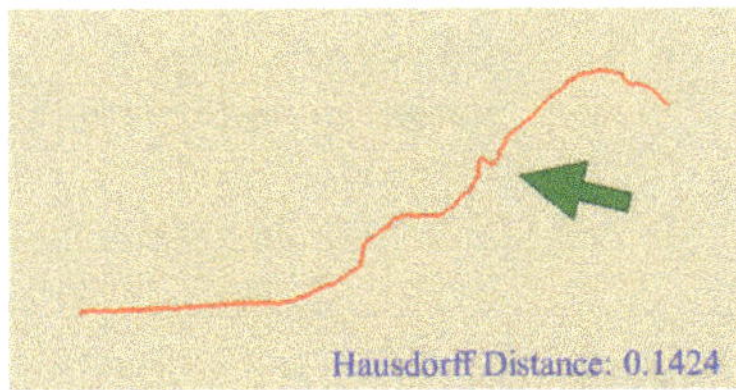

As shown by the comparison results in Table 4, both NeRF 6G and NeRF 24G achieved outcomes comparable to the photogrammetric method for Profiles A and B. Notably, NeRF 24G demonstrated superior modeling quality, offering higher geometric fidelity and closer alignment with the reference profiles. In large-scale applications such as slope engineering, the accurate reproduction of the overall terrain morphology and spatial extent is more critical than the precision of fine geometric details. From this perspective, both NeRF 6G and NeRF 24G effectively captured the primary topographic contours and variation trends. Their reconstruction accuracy is sufficient to meet the practical requirements of post-disaster emergency modeling and slope morphology interpretation. Therefore, even the lightweight NeRF 6G model shows promising feasibility and practicality for deployment in rapid response scenarios.

Figure 10. Cross-section lines.

5. Discussion

In emergency response to landslide disasters, the speed and quality of acquiring on-site 3D models play a critical role in supporting governmental emergency decision making. This paper compares the application performance of traditional photogrammetry-based modeling and NeRF-based modeling in landslide disaster response and discusses them from the following perspectives.

(1) Model Processing Time:

In terms of data processing time, there are substantial differences between NeRF and traditional photogrammetry, as shown in Table 5. From an engineering and practical application standpoint, our evaluation of time efficiency focuses on the duration required for each critical step in the 3D reconstruction process, as well as the total time needed to complete the task. This is because intermediate results at each stage may provide incremental support for emergency response operations. As illustrated in Table 5, constructing a lightweight NeRF 6G model typically takes only around 46 min, while the more detailed NeRF 24G model also requires significantly less time than the photogrammetric approach. Overall, NeRF enables faster generations of 3D models, making it feasible to complete reconstruction tasks within a single day in emergency scenarios, thereby offering timely spatial information to support on-site decision making.

(2) Rendering Effect:

In terms of rendering, traditional photogrammetry-generated texture models typically rely on image stitching, and their final appearance is easily affected by factors such as lighting conditions, viewpoint occlusions, and image stitching errors, which can result in a lack of realism. In contrast, NeRF-based modeling, by establishing an implicit volumetric representation of the scene, is capable of achieving more realistic volume rendering. Although the NeRF 6G model exhibits inferior rendering quality compared to the photogrammetry-generated texture model, the more resource-intensive NeRF 24G model demonstrates visual performance that is very close to—and even, in certain scenarios, surpasses—that of the traditional photogrammetry-generated texture model. The NeRF 24G model is more faithful to reality, providing rescue personnel with a more intuitive and realistic perception of the on-site environment, which facilitates rapid disaster assessment.

Table 5. Comparison of model generation time.

Individual Stages	Time Consumption (min)		
	Photogrammetric	**NeRF 6G**	**NeRF 24G**
Structure from Motion	4	4	4
training (only NeRF)	—	29	245
Point cloud generates and export	408	17	24
Generate texture (only photogrammetric)	14	—	—
Mesh generates and export	19	16	45
Total	455	46	318

(3) Disaster Management Data:

This study demonstrates that NeRF-based rapid reconstruction offers significant advantages for post-disaster landslide management by balancing modeling speed with the level of geometric detail required for emergency applications. While traditional photogrammetry-based modeling excels in generating highly detailed point cloud and mesh data, its time-consuming nature makes it less suitable for immediate response scenarios. In contrast, NeRF-based modeling—though inferior in geometric precision—can produce a comprehensive three-dimensional overview of the landslide site within a single day, fulfilling the core requirements of post-disaster assessment without the need for engineering-grade accuracy. Experimental results show that the point clouds generated by NeRF 6G and NeRF 24G can be effectively converted into DEMs for volume estimation, yielding values of 1102.6 m^3 and 1078.5 m^3, respectively, compared to 1082.2 m^3 obtained from photogrammetry. The small variation among these estimates confirms the reliability of NeRF in estimating landslide volume. Moreover, the NeRF-derived point clouds accurately reproduce critical terrain features, including landslide deposits, roads, and buildings along potential runout paths—elements essential for subsequent dynamic landslide simulations. Additionally, both NeRF models are capable of identifying surface features such as tensile cracks and main scarp boundaries, which are crucial indicators for early warning and monitoring deployment. Although NeRF may lack certain fine structural details, it adequately supports rapid on-site assessment and decision making, making it a valuable complementary tool to conventional photogrammetry. In summary, NeRF provides timely access to essential information for post-disaster response, including volume quantification, terrain condition analysis, and surface deformation indicators, thereby offering a practical solution for emergency 3D modeling under time constraints.

(4) Discussion on NeRF Variants:

Since the original NeRF method was introduced by Mildenhall et al. [35], a number of variants have been proposed, each making different trade-offs between speed, accuracy, and applicability. Instant-NGP employs multi-resolution hash encoding and a sparse occupancy grid to dramatically accelerate both training and rendering, but it can lose fine details in highly complex scenes [39]. Mip-NeRF adopts cone tracing and Integrated Positional Encoding (IPE) to reduce aliasing and achieve modest speed gains, yet it remains constrained to bounded scenes and often fails to converge in intricate outdoor environments [40]. Mip-NeRF 360 overcomes this limitation by using nonlinear scene parameterization to map unbounded content into a finite representation, substantially enhancing robustness for distant terrain. Zip-NeRF further combines the anti-aliasing strengths of Mip-NeRF 360 with Instant-NGP's efficient architecture, yielding approximately 22-fold faster training and lowering reconstruction error by 8% to 77% [41]. Nerfacto—the default pipeline in Nerfstudio—integrates per-image appearance embeddings, multi-resolution hash encoding, proposal-guided sampling, and scene contraction, striking an effective balance between efficiency and accuracy and making it particularly well suited for rapid-response deployments. It is important to note that these approaches have predominantly targeted indoor or small-scale scenes. Recently, large-scale NeRF variants tailored for outdoor drone imagery have begun to emerge—such as Drone-NeRF [42], UAV-NeRF [43], and FlyNeRF [44]—which employ tiling, parallel modeling, and camera-pose optimization to boost reconstruction efficiency and precision over expansive terrains. In future work, we plan to incorporate these large-scene optimization strategies to further enhance rapid 3D modeling of natural environments, including landslide scenarios.

(5) Limitations of the Study:

As mentioned in Section 2 of this study, the dataset used for modeling the Sasebo landslide was collected by TEC-FORCE personnel in the Kyushu region. Since they are not professional remote sensing practitioners, the dataset does not follow standard photogrammetric protocols. Nevertheless, traditional photogrammetry still yields clearer reconstructions of roads, buildings, and other structures—producing concentrated, uniform point-cloud density—even under sparse data (Figure 11). By contrast, NeRF point clouds appear more dispersed and "foggy" along edges, making fine structural outlines harder to discern. Future research will acquire imagery in strict accordance with photogrammetric data acquisition standards to ensure adequate coverage and consistency. Another limitation of our study is that we directly applied the Nerfacto model, even though Nerfacto is primarily optimized for smaller-scale scenes. Its default hyperparameter settings—such as hash encoding depth and sampling strategies—might not be ideally suited for large-area slope reconstruction. Future studies will therefore incorporate NeRF variants that have been specifically adapted for UAV or large-area imagery and benchmark them against Nerfacto. In addition, we used a photogrammetric model as the reference baseline, yet photogrammetry is susceptible to image quality, lens distortion and processing parameters, which can introduce geometric errors. Because laser scanning delivers more precise and stable ground truth, we plan to include high-resolution LiDAR data in future work to provide a more reliable benchmark for evaluating NeRF accuracy.

(a) Photogrammetry

(b) NeRF 24G

Figure 11. Comparison of point clouds generated by two methods under sparse image conditions.

6. Conclusions

This study addresses the needs of post-landslide emergency response by exploring the potential of Neural Radiance Field (NeRF) technology for rapid 3D reconstruction and comparing it with traditional photogrammetric methods. Using the Sasebo landslide in Japan as a case study, we acquired drone-based video imagery and applied both NeRF and photogrammetry to reconstruct 3D models of the landslide area. Two NeRF configurations—NeRF 6G and NeRF 24G—were trained to evaluate different training capacities. A comprehensive comparison of the three models demonstrates that NeRF excels in reconstruction speed and visual realism: the NeRF 24G model generates photorealistic scenes in only 318 min, significantly faster than the 455 min required by the

14. photogrammetric workflow, while NeRF 6G completes training in just 46 min and achieves visual quality close to that of photogrammetry. Although traditional photogrammetry still outperforms NeRF in geometric detail by producing the most refined point clouds and meshes, the geometric accuracy achieved by both NeRF models is sufficient for engineering-scale applications and offers considerable practical value for post-disaster assessment and rapid response.

Author Contributions: Conceptualization, Jinge Zhang and Sunhao Zhang; methodology, Jinge Zhang, Sunhao Zhang, Yan Du, and Dongqi Shang; investigation, Sunhao Zhang; writing—original draft preparation, Jinge Zhang; writing—review and editing, Jinge Zhang, Sunhao Zhang, Yujing Jiang, and Hongbin Chen; supervision, Yujing Jiang; funding acquisition, Yujing Jiang. All authors have read and agreed to the published version of the manuscript.

Funding: This work was supported by the Nagasaki University Global Human Resource Development Scholarship and the JST SPRING, Japan Grant Number: JPMJSP2172.

Data Availability Statement: Data available on request due to restrictions (e.g., privacy, legal, or ethical reasons).

Acknowledgments: We would like to thank the editor and anonymous reviewers for their constructive suggestions and comments for this article.

Conflicts of Interest: The authors declare that they have no known competing financial interests or personal relationships that could have appeared to influence the work reported in this paper.

References

1. Tanyaş, H.; Hill, K.; Mahoney, L.; Fadel, I.; Lombardo, L. The World's Second-Largest, Recorded Landslide Event: Lessons Learnt from the Landslides Triggered during and after the 2018 Mw 7.5 Papua New Guinea Earthquake. *Eng. Geol.* **2022**, *297*, 106504. [CrossRef]
2. Jain, N.; Martha, T.R.; Khanna, K.; Roy, P.; Kumar, K.V. Major Landslides in Kerala, India, during 2018–2020 Period: An Analysis Using Rainfall Data and Debris Flow Model. *Landslides* **2021**, *18*, 3629–3645. [CrossRef]
3. Li, L.; Xu, C.; Yao, X.; Shao, B.; Ouyang, J.; Zhang, Z.; Huang, Y. Large-Scale Landslides around the Reservoir Area of Baihetan Hydropower Station in Southwest China: Analysis of the Spatial Distribution. *Nat. Hazards Res.* **2022**, *2*, 218–229. [CrossRef]
4. Yano, A.; Shinohara, Y.; Tsunetaka, H.; Mizuno, H.; Kubota, T. Distribution of Landslides Caused by Heavy Rainfall Events and an Earthquake in Northern Aso Volcano, Japan from 1955 to 2016. *Geomorphology* **2019**, *327*, 533–541. [CrossRef]
5. Tsunetaka, H.; Murakami, W.; Nakao, K.; Mtibaa, S. Evaluation of the Impact of Climate Change on Rainfall for Potential Landslide Triggering in Japan. *Earth Surf. Process. Landf.* **2025**, *50*, e6050. [CrossRef]
6. Ding, Y.; Xia, Z.; Rao, Y.; Chen, H.; Gan, W. Preliminary Analysis of Wuyigou Landslide on 17 July 2024 in Zigui County, Hubei Province, China. *Landslides* **2025**, *22*, 261–266. [CrossRef]
7. Stumpf, A.; Malet, J.-P.; Allemand, P.; Pierrot-Deseilligny, M.; Skupinski, G. Ground-Based Multi-View Photogrammetry for the Monitoring of Landslide Deformation and Erosion. *Geomorphology* **2015**, *231*, 130–145. [CrossRef]
8. Fu, L.; Zhu, J.; Li, W.; You, J.; Hua, Z. Fast Estimation Method of Volumes of Landslide Deposit by the 3D Reconstruction of Smartphone Images. *Landslides* **2021**, *18*, 3269–3278. [CrossRef]
9. Yu, M.; Huang, Y.; Zhou, J.; Mao, L. Modeling of Landslide Topography Based on Micro-Unmanned Aerial Vehicle Photography and Structure-from-Motion. *Environ. Earth Sci.* **2017**, *76*, 520. [CrossRef]
10. Lachat, E.; Landes, T.; Grussenmeyer, P. Investigation of a Combined Surveying and Scanning Device: The Trimble SX10 Scanning Total Station. *Sensors* **2017**, *17*, 730. [CrossRef]
11. Telling, J.; Lyda, A.; Hartzell, P.; Glennie, C. Review of Earth Science Research Using Terrestrial Laser Scanning. *Earth-Sci. Rev.* **2017**, *169*, 35–68. [CrossRef]
12. Liu, C.; Liu, X.; Peng, X.; Wang, E.; Wang, S. Application of 3D-DDA Integrated with Unmanned Aerial Vehicle–Laser Scanner (UAV-LS) Photogrammetry for Stability Analysis of a Blocky Rock Mass Slope. *Landslides* **2019**, *16*, 1645–1661. [CrossRef]
13. Chen, Z.; Zhang, C.; Tang, Z.; Fang, K.; Xu, W. Three-Dimensional Reconstruction and Deformation Identification of Slope Models Based on Structured Light Method. *Sensors* **2024**, *24*, 794. [CrossRef]
14. Gupta, S.K.; Shukla, D.P. Application of Drone for Landslide Mapping, Dimension Estimation and Its 3D Reconstruction. *J. Indian Soc. Remote Sens.* **2018**, *46*, 903–914. [CrossRef]

15. Zhao, D.; Han, J.; Liu, X.; Zuo, Z.; Zheng, K. Exploration of the Application of Fine Reconstruction of Unmanned Aerial Vehicles in Landslide Disaster Investigation and Management. In *Hydropower and Renewable Energies*; Zheng, S., Taylor, R.M., Wu, W., Nilsen, B., Zhao, G., Eds.; Springer Nature: Singapore, 2025; pp. 489–502.
16. Lucieer, A.; Jong, S.M.d.; Turner, D. Mapping Landslide Displacements Using Structure from Motion (SfM) and Image Correlation of Multi-Temporal UAV Photography. *Prog. Phys. Geogr. Earth Environ.* **2014**, *38*, 97–116. [CrossRef]
17. Gallo, I.G.; Martínez-Corbella, M.; Sarro, R.; Iovine, G.; López-Vinielles, J.; Hérnandez, M.; Robustelli, G.; Mateos, R.M.; García-Davalillo, J.C. An Integration of UAV-Based Photogrammetry and 3D Modelling for Rockfall Hazard Assessment: The Cárcavos Case in 2018 (Spain). *Remote Sens.* **2021**, *13*, 3450. [CrossRef]
18. Zhu, Z.; Zeng, B.; Zhao, H.; Yuan, J.; Ai, D. Structure Plane Interpretation of Rockfall and Early Identification of Potential Hazards Based on UAV Photogrammetry. *Nat. Hazards* **2025**, *121*, 1779–1802. [CrossRef]
19. Zhu, Z.; Li, Q.; Zhao, L.; Huang, D.; Wu, Q.; Zuo, S. Whole-Process Survey of a Landslide Based on UAV Tilt Photography and 3D Reconstruction and Spatial Analysis Techniques. *Landslides* **2025**, *22*, 485–497. [CrossRef]
20. Šlapak, E.; Pardo, E.; Dopiriak, M.; Maksymyuk, T.; Gazda, J. Neural Radiance Fields in the Industrial and Robotics Domain: Applications, Research Opportunities and Use Cases. *Robotics and Computer-Integrated Manufacturing* **2024**, *90*, 102810. [CrossRef]
21. Croce, V.; Billi, D.; Caroti, G.; Piemonte, A.; De Luca, L.; Véron, P. Comparative Assessment of Neural Radiance Fields and Photogrammetry in Digital Heritage: Impact of Varying Image Conditions on 3D Reconstruction. *Remote Sens.* **2024**, *16*, 301. [CrossRef]
22. Remondino, F.; Karami, A.; Yan, Z.; Mazzacca, G.; Rigon, S.; Qin, R. A Critical Analysis of NeRF-Based 3D Reconstruction. *Remote Sens.* **2023**, *15*, 3585. [CrossRef]
23. Themistocleous, K.; Abate, D. The Use of Photogrammetry and NeRF Techniques to Document Built Heritage. In Proceedings of the Earth Resources and Environmental Remote Sensing/GIS Applications XV, Edinburgh, UK, 16–19 September 2024; Volume 13197, pp. 272–277.
24. Ge, Y.; Guo, B.; Zha, P.; Jiang, S.; Jiang, Z.; Li, D. 3D Reconstruction of Ancient Buildings Using UAV Images and Neural Radiation Field with Depth Supervision. *Remote Sens.* **2024**, *16*, 473. [CrossRef]
25. Yin, Y.; Deng, Q.; Li, W.; He, K.; Wang, Z.; Li, H.; An, P.; Fang, K. Insight into the Crack Characteristics and Mechanisms of Retrogressive Slope Failures: A Large-Scale Model Test. *Eng. Geol.* **2023**, *327*, 107360. [CrossRef]
26. Rapstine, T.D.; Rengers, F.K.; Allstadt, K.E.; Iverson, R.M.; Smith, J.B.; Obryk, M.K.; Logan, M.; Olsen, M.J. Reconstructing the Velocity and Deformation of a Rapid Landslide Using Multiview Video. *J. Geophys. Res. Earth Surf.* **2020**, *125*, e2019JF005348. [CrossRef]
27. Jaboyedoff, M.; Chigira, M.; Arai, N.; Derron, M.-H.; Rudaz, B.; Tsou, C.-Y. Testing a Failure Surface Prediction and Deposit Reconstruction Method for a Landslide Cluster That Occurred during Typhoon Talas (Japan). *Earth Surf. Dyn.* **2019**, *7*, 439–458. [CrossRef]
28. Tsunetaka, H.; Hotta, N.; Hayakawa, Y.S.; Imaizumi, F. Spatial Accuracy Assessment of Unmanned Aerial Vehicle-Based Structures from Motion Multi-View Stereo Photogrammetry for Geomorphic Observations in Initiation Zones of Debris Flows, Ohya Landslide, Japan. *Prog. Earth Planet. Sci.* **2020**, *7*, 24. [CrossRef]
29. James, M.R.; Robson, S. Straightforward Reconstruction of 3D Surfaces and Topography with a Camera: Accuracy and Geoscience Application. *J. Geophys. Res. Earth Surf.* **2012**, *117*, F03017. [CrossRef]
30. Lato, M.J.; Vöge, M. Automated Mapping of Rock Discontinuities in 3D Lidar and Photogrammetry Models. *Int. J. Rock Mech. Min. Sci.* **2012**, *54*, 150–158. [CrossRef]
31. Bemis, S.P.; Micklethwaite, S.; Turner, D.; James, M.R.; Akciz, S.; Thiele, S.T.; Bangash, H.A. Ground-Based and UAV-Based Photogrammetry: A Multi-Scale, High-Resolution Mapping Tool for Structural Geology and Paleoseismology. *J. Struct. Geol.* **2014**, *69*, 163–178. [CrossRef]
32. Fugazza, D.; Scaioni, M.; Corti, M.; D'Agata, C.; Azzoni, R.S.; Cernuschi, M.; Smiraglia, C.; Diolaiuti, G.A. Combination of UAV and Terrestrial Photogrammetry to Assess Rapid Glacier Evolution and Map Glacier Hazards. *Nat. Hazards Earth Syst. Sci.* **2018**, *18*, 1055–1071. [CrossRef]
33. McCarthy, J. Multi-Image Photogrammetry as a Practical Tool for Cultural Heritage Survey and Community Engagement. *J. Archaeol. Sci.* **2014**, *43*, 175–185. [CrossRef]
34. Adreani, L.; Colombo, C.; Fanfani, M.; Nesi, P.; Pantaleo, G.; Pisanu, R. A Photorealistic 3D City Modeling Framework for Smart City Digital Twin. In Proceedings of the 2022 IEEE International Conference on Smart Computing (SMARTCOMP), Helsinki, Finland, 20–24 June 2022; pp. 299–304.
35. Mildenhall, B.; Srinivasan, P.P.; Tancik, M.; Barron, J.T.; Ramamoorthi, R.; Ng, R. NeRF: Representing Scenes as Neural Radiance Fields for View Synthesis. *Commun. ACM* **2021**, *65*, 99–106. [CrossRef]
36. Shun, S.; Yuichi, O.; Yuta, S.; Ryota, K.; Kohsuke, T.; Jungpil, S.; Satoshi, N. Generating Realistic Images with NeRF for Training of Autonomous Driving Network. In Proceedings of the 2023 IEEE 16th International Symposium on Embedded Multicore/Many-Core Systems-on-Chip (MCSoC), Singapore, 18–21 December 2023; pp. 254–261.

37. Murtiyoso, A.; Grussenmeyer, P. Initial assessment on the use of state-of-the-art nerf neural network 3D reconstruction for heritage documentation. *Int. Arch. Photogramm. Remote Sens. Spat. Inf. Sci.* **2023**, *48*, 1113–1118. [CrossRef]
38. Tancik, M.; Weber, E.; Ng, E.; Li, R.; Yi, B.; Wang, T.; Kristoffersen, A.; Austin, J.; Salahi, K.; Ahuja, A.; et al. Nerfstudio: A Modular Framework for Neural Radiance Field Development. In Proceedings of the ACM SIGGRAPH 2023 Conference Proceedings, Los Angeles, CA, USA, 6–10 August 2023; Association for Computing Machinery: New York, NY, USA, 2023; pp. 1–12.
39. Arshad, M.A.; Jubery, T.; Afful, J.; Jignasu, A.; Balu, A.; Ganapathysubramanian, B.; Sarkar, S.; Krishnamurthy, A. Evaluating Neural Radiance Fields for 3D Plant Geometry Reconstruction in Field Conditions. *Plant Phenomics* **2024**, *6*, 0235. [CrossRef]
40. Lin, Z.; Xu, H. A Three-Dimensional Reconstruction Method Based on Improved Mip-NeRF. In Proceedings of the Proceedings of the 2024 International Conference on Power Electronics and Artificial Intelligence, Xiamen, China, 19–21 January 2024; Association for Computing Machinery: New York, NY, USA, 2024; pp. 483–487.
41. Barron, J.T.; Mildenhall, B.; Verbin, D.; Srinivasan, P.P.; Hedman, P. Mip-NeRF 360: Unbounded Anti-Aliased Neural Radiance Fields. In Proceedings of the 2022 IEEE/CVF Conference on Computer Vision and Pattern Recognition (CVPR), New Orleans, LA, USA, 18–24 June 2022; pp. 5470–5479.
42. Jia, Z.; Wang, B.; Chen, C. Drone-NeRF: Efficient NeRF Based 3D Scene Reconstruction for Large-Scale Drone Survey. *Image Vis. Comput.* **2024**, *143*, 104920. [CrossRef]
43. Li, L.; Zhang, Y.; Jiang, Z.; Wang, Z.; Zhang, L.; Gao, H. Unmanned Aerial Vehicle-Neural Radiance Field (UAV-NeRF): Learning Multiview Drone Three-Dimensional Reconstruction with Neural Radiance Field. *Remote Sens.* **2024**, *16*, 4168. [CrossRef]
44. Dronova, M.; Cheremnykh, V.; Kotcov, A.; Fedoseev, A.; Tsetserukou, D. FlyNeRF: NeRF-Based Aerial Mapping for High-Quality 3D Scene Reconstruction. In Proceedings of the 2024 International Conference on Unmanned Aircraft Systems (ICUAS), Crete, Greece, 4–7 June 2024; pp. 1050–1055.

International Journal of
Geo-Information

MDPI

Article

A Spatial-Network Approach to Assessing Transportation Resilience in Disaster-Prone Urban Areas

Francesco Rouhana [1,*] and Dima Jawad [2]

[1] School of Civil and Environmental Engineering, University of Connecticut, Storrs, CT 06269, USA
[2] Department of Mechanical and Manufacturing Engineering, Ontario Tech University, Oshawa, ON L1G 0C5, Canada; dima.jawad@ontariotechu.ca
* Correspondence: francesco.rouhana@uconn.edu

Abstract

Critical transportation networks in developing countries often lack structural robustness and functional redundancy due to insufficient planning and preparedness. These deficiencies increase vulnerability to disruptions and impede effective post-disaster response and recovery. Understanding how such networks perform under stress is essential to improving resilience in hazard-prone urban environments. This paper presents an integrated predictive methodology for assessing the operational resilience of urban transportation networks under extreme events, specifically tailored to data-scarce and high-risk contexts. By combining Geographic Information Systems (GISs) with complex network theory, the framework captures both spatial and topological dependencies. The methodology is applied to Beirut, the capital of Lebanon, a densely populated and disaster-prone Mediterranean city, through scenario-based simulations that account for interdependent stressors such as traffic dynamics, structural fragility, and geophysical hazards. Results reveal that the network exhibits low redundancy and high sensitivity to even minor disruptions, leading to rapid performance degradation. These findings indicate that the network should be classified as highly vulnerable. The study offers a robust framework for assessing infrastructure resilience and supporting evidence-based decision-making in critical urban network management.

Keywords: resilience; vulnerability; disaster risk reduction; graph theory; urban network analysis; network failures

1. Introduction

Civil infrastructure systems, such as transportation, communication, energy, and water networks, are complex and indispensable for providing essential goods and services in modern cities [1–4]. However, increasing urbanization makes these systems vulnerable to natural and human-induced threats [5,6]. These threats can impair their performance and cause significant socioeconomic losses [7–9]. International commitments have been made to comprehensively understand the risk and vulnerability of these systems, with the aim of reducing disaster risk and losses [10]. The overarching goal is to enhance their resilience and adaptive capacity to hazards, ensuring the safety and resilience of human settlements [10,11].

Transportation infrastructure plays a critical role in providing local communities with smooth access to their daily activities [12,13]. Disruptions to transportation networks can result in significant losses that accumulate over time and affect communities. For instance, critical failures within the network may significantly hinder the ability of emergency responders to conduct timely evacuations and provide aid to at-risk populations [14]. In seismic-prone regions, damage within these networks is closely linked to the functionality of their components, particularly where structures are susceptible to failure [15]. Hence, in the context of extreme events, it is imperative to assess the fundamental topological properties of transportation networks [16]. Equally important is the evaluation of network resilience, which enhances sensitivity to extreme events, improves preparedness for potential damage scenarios, and ensures optimal resource allocation for swift recovery [17].

This paper employs a comprehensive approach to evaluate the serviceability of transportation networks before and after extreme events. The framework identifies critical network performance measures that are essential for practical functionality. Using a combination of GIS-derived spatial data and network concepts, quantitative and statistical modeling approaches are applied to evaluate the topological network properties. Various properties of graph theory are assessed to illustrate the behavior, distribution, and efficiency of urban network components under random and targeted failures. As such, connectivity analysis and topological measures are used to identify critical infrastructure locations that can cause system failures. This study is conducted in Lebanon, a developing country highly susceptible to system failures due to extreme hazards [18]. It focuses on assessing the road network resiliency of Beirut, the highly populated capital, in the face of extreme events. The results can help policymakers prioritize critical network components, allocate resources efficiently, and minimize system downtime and associated losses.

The structure of this study is as follows: Section 2 reviews the relevant literature on resilience quantification and the application of graph theory to transportation network analysis. Section 3 outlines the data sources and introduces the proposed framework for evaluating vulnerability and redundancy in critical transportation components. Section 4 presents the case study context, while Section 5 discusses the results and their interpretation. Finally, Section 6 concludes the study and summarizes key findings.

2. Literature Review

2.1. Resilience Quantification Concept

Assessing system resilience and identifying vulnerabilities are essential steps for effective risk mitigation [19,20]. In general terms, resilience is the capability of a system to cope, absorb, and withstand the effect of a random attack and then survive and recover immediately afterward [19,21]. Figure 1 depicts the different resilience phases and their impact on network performance. The resilience process involves four stages: anticipation, absorption, adaptation, and recovery. Anticipation involves forecasting and planning for possible system failures. Absorption consists of reducing the impact of an event on the system through robustness and redundancy. The initial system percentage drop in functionality at time t_0 is based on these two interlinked aspects without differentiation. Adaptation involves resource allocation to restore functionality immediately after the event [22]. Finally, recovery consists of returning the system to an acceptable performance level at a time t_f according to predetermined standards. The final state of the system after recovery depends on the extent of repairs and improvements [23]. The area between the curves indicates the system's ability to adapt and recover from disruptions. The importance of prioritization is highlighted by the difference in performance levels between a prioritized and a non-prioritized recovery process at time t_1, which is linked to the speed of performance criteria.

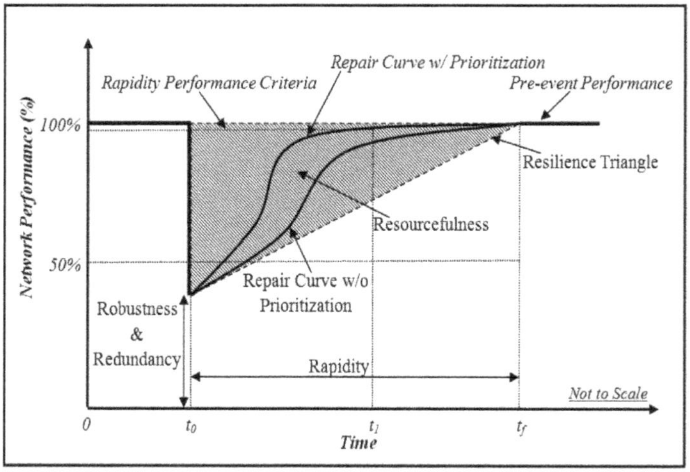

Figure 1. Role of resilience properties on network performance.

Bruneau and Reinhorn in [24] proposed measuring the degree of resilience of civil infrastructure systems by quantifying the four phases of resilience, which include inherent and adaptive capacities. Resilience can be represented as a function of vulnerability, exposure, and adaptive capacity or expressed as shown in Equation (1). The functional capacity of community infrastructure over time is characterized by its resilience, a modified integral of the $Q(t)$ function. The $Q(t)$ function refers to the quality of community infrastructure changing with time. A smaller area under the "resilience triangle" indicates greater resilience. As such, resilience is defined as the normalized area below $Q(t)$ [19,25]. For network-based infrastructure studies, evaluating network resilience involves measuring the proportion of failed links and nodes in system performance after an external shock [26,27].

$$R = \int_{t_0}^{t_f} [100 - Q(t)]dt \quad (1)$$

2.2. Graph Theory and Transportation Networks

Graph theory applies various connectivity metrics that are associated with the physical and structural characteristics of a graph. This field of network science holds significant promise for analyzing transport networks, mainly due to its suitability for managing large-scale systems effectively [28–30]. In the academic literature, the connectivity metric is one of the most used indicators to evaluate network performance [31–33]. This metric is derived from graph theory and offers a simple yet effective way to assess the resilience of transportation networks. Connectivity is intrinsically linked to the intra-dependencies of the network, which is defined by adjacency and smooth flow between vertices via edges [34]. The primary aim of analyzing network connectivity is to identify reachable pathways to prevent the possible isolation of the system, particularly for emergency services in the aftermath of an event. By analyzing the complex network structure, connectivity metrics can provide valuable insights into the network's ability to withstand disruptions and maintain functionality [28,35].

Graph theory provides measures of connectedness and redundancy, such as alpha, beta, and gamma indices, which are based on the network's number of vertices, edges, and cycles [36]. The adaptability of these indices allows for evaluating the impact of adding or removing edges or vertices on network performance. Other network topological properties that can be used to quantify the efficiency of network functionality are shown in Table 1.

Table 1. Network topological properties.

Equation	Indicator	Description
(2) $d(v) = \mid E(v) \mid$	Nodal Degree	Represents the number of edges $E(v)$ directly connected to node v. It quantifies the local connectivity of a node within the network.
(3) $d(G) = \frac{1}{\mid V \mid} \sum_{v \in V} d(v)$	Average Network Degree	Represents the mean of all node degrees $d(v)$ in the graph, where V is the set of all nodes. It indicates the overall level of connectivity in the network.
(4) $\gamma_v = \frac{E(\Gamma(v))\mid}{0.5\, d(v)(d(v)-1)}$	Clustering Coefficient	Measures the extent to which the neighbors $\Gamma(v)$ of node v are connected to each other. $E(\Gamma(v))$ is the number of edges among those neighbors. This index reflects the degree of local cohesion or transitivity in the network.
(5) $l_G = \frac{1}{n(n-1)} \sum_{i \neq j} d(i,j)$	Shortest Path	Represents the mean of the shortest path distances $d(i,j)$ between all pairs of nodes i and j in a network of n nodes. It represents the network's overall efficiency in terms of communication or flow.
(6) $b_v = \sum_{i \neq v \neq j} \frac{g(i,v,j)}{n(i,j)}$	Betweenness Centrality	Measures how often node v appears on the shortest paths between other node pairs i and j. $g(i,v,j)$ is the number of shortest paths from i to j that pass through v, and $n(i,j)$ is the total number of shortest paths between i and j. This metric identifies nodes that serve as critical bridges or bottlenecks in the network.

The network nodal degree determines the sum of vertices to the sum of edges per vertex $\mid E(v) \mid$ (Equation (2)). The average network nodal degree computes system connectivity by considering the average number of edges connected to every vertex (Equation (3)) [36,37]. Graph topology features usually count on the law followed by the degree distribution. These features depend on p_k, the ratio of vertices that possess a degree k, and the probability that a random vertex has the degree k. Furthermore, clusters are subgraphs of attached vertices whose existence reflects how perfectly the network behaves as a system. The clustering coefficients compute both local connectivity and the graph's transitivity property. The network transitivity denotes that if a vertex v_1 is joined with a vertex v_2 adjacent to a third vertex v_3, there is a high probability of adjacency between v_1 and v_3. This likelihood is designated as the clustering coefficient and rates the concentration of transitive triangles in the network. A neighborhood of vertex v is referred to as $\Gamma(v)$ and is principally a subgraph containing vertices adjacent to v but excluding it. The clustering coefficient rate between null and unity reveals the linked share of v neighbors. In Equation (4), $\mid E(\Gamma v) \mid$ represents the number of edges in the neighborhood of v, whereas the number of possible edges in $\Gamma(v)$ is denoted in the denominator [37].

Another significant concept in transportation networks is the shortest path [38]. In Equation (5), n denotes the number of network vertices. The algorithm begins at vertex i, iteratively calculating distances to connected neighbors. Once the shortest paths are determined, all vertices are evaluated to find the shortest route from i to any other vertex j. Relatedly, the most significant critical vertices can be identified by measuring their centrality with the betweenness centrality shown in Equation (6). The numerator in this equation defines the number of shortest paths from vertex i to j passing by v, whereas the denominator defines the total number of shortest paths between i and j [39]. The assessment of betweenness centrality values across all vertices in the system helps identify optimal resource allocation and guides effective interventions accordingly [40]. This measure can classify edges and vertices according to topology functions. Therefore, the likelihood of overlooking imperative vertices that may be critical bridges between regions is prevented. In this context, ranking network components form the former pace in expecting a response to the exclusion or breakdown of components.

2.3. Graph Theory and Resilience of Complex Networks

Graph-based representations have become a favored method for managing resilience, primarily due to their relative simplicity compared to alternative approaches [41]. According to Serdar et al. [42], the literature addressing the resilience of transportation networks is comparatively limited. In this regard, graph theory can effectively simplify and represent complex systems, making them more accessible for analysis and optimization. Transportation networks consist of nodes and links, which are interconnected in a complex manner. The interdependent nature of such systems gives rise to the concept of "systems-of-systems," which can enhance their functional efficiency but may also reduce their robustness, making them more susceptible to disruptions [1,15]. Therefore, graph theory can evaluate such interdependency of network properties, defining fundamental characteristics and the contribution of each component [38,41].

Graph theory principles prove highly effective in assessing network topological features and quantifying resilience and reliability [41,43]. Candelieri et al. [44] applied graph theory to analyze public transportation networks, specifically assessing vulnerability to various connectivity failure events. Zhenwu et al. [30] explored resilience in terms of connectivity with the largest cluster, employing percolation theory to identify vulnerable links in the road transportation system. Relatedly, ranking critical network components based on their importance level is crucial, as specific transportation networks lack flexibility and redundancy. Ahmed et al. [45] assessed the network credentials of infrastructure components to enhance the resilience of infrastructure systems against disruptive events. They used geographic modeling with road network data to conduct experiments, ranking bridges based on their connectivity and attributes within the road network. Chatterjee et al. [46] explored statistical properties of L-space bus networks, analyzing degree distribution, path length, and assortativity. They used a modularity-based community detection method to identify operational zones. Network resilience was assessed by removing nodes in random and targeted modes, considering degree distributions and centrality measures. Li and Han [47] studied the topological characteristics and dynamics of various network models. Applying complex network theory, they analyzed the topological features of road, bus, and track networks in urban transportation systems. Their study concluded that these features could assess the network's robustness to random attacks [47].

Several studies have applied graph theory to assess the resilience and properties of diverse transportation networks. De Bona et al. [48] analyzed bus, metro, and airport networks, finding small-world properties in degree, path length, and connectivity distributions. They noted scale-free characteristics in the initial network if the reduced network displayed a hierarchical structure. In this context, small-world networks feature high clustering and short path lengths, whereas scale-free networks are characterized by a few highly connected hubs following a power-law distribution [48]. Zhang et al. [49] applied graph theory to assess the resilience of complex metro networks. They discovered high robustness against random disruptions but relative vulnerability to intentional disruptions, aligning with characteristics of small-world and scale-free networks. Wang et al. [29] investigated railway networks, representing rail stations as vertices and transit services as edges. They analyzed complex network properties, including cumulative degree distribution, degree correlation, and betweenness centrality, revealing a core-periphery structure with small-world characteristics in the railway network. Wu and Kim [50] examined bike-sharing networks, investigating network properties and spatial autocorrelation using a network representation. Their evaluation, including nodal degree, clustering coefficient, average path length, and centrality measures, revealed a positive relationship between degree centrality and closeness centrality. The researchers suggested that high station accessibility and the small-world properties of bike-sharing networks facilitate efficient bike movement between highly clustered stations [50].

Building on the literature, this study adopts an integrated framework that combines resilience quantification and graph theory analysis to assess the vulnerability of urban transportation networks. The conceptual foundation, particularly the understanding of resilience as a dynamic, multi-phase process and the application of graph-based metrics to evaluate network topology and function, directly informs the methodological approach. Indicators such as connectivity, centrality, clustering coefficients, and shortest path measures are employed to capture structural attributes associated with robustness, redundancy, and functional importance. Applied in conjunction with resilience modeling, these metrics enable the systematic identification of critical components, and the estimation of disruption impacts on network performance. The selected methods in the next section reflect and build upon the theoretical insights reviewed, supporting a comprehensive and empirically grounded evaluation of transportation network resilience under varied hazard scenarios.

3. Materials and Methods

This section presents the proposed approach to evaluating the vulnerability of urban transportation networks under various simulated disruption scenarios. As illustrated in Figure 2, the methodology unfolds in three primary phases. The first phase involved a comprehensive survey of the study area to collect, update, and georeference relevant data. This included the identification of scenario-specific parameters such as traffic (peak traffic volumes), road class (functional classifications), geological/seismic conditions (soil and rock type-based damage increments), buildings (inventory including height, density, and locality), and natural hazards (exposure to sea-level rise, tsunamis, and river floods). While traffic and road class are line features representing network connectivity and flow, geological conditions, buildings, and natural hazards are polygon features that define spatial zones of impact and exposure. These polygons are essential for translating their effects onto the geometric structure of the transportation network and for assessing vulnerability across its overlaid nodes and links. The main road network and scenario-specific attributes were evaluated for geometric consistency, attribute completeness, and spatial accuracy using OpenStreetMap data, with visualizations conducted through Esri ArcGIS 10.8 and Google Earth platforms. Moreover, GIS was used to generate network datasets and define the entire network as $G = (V, E)$, comprising a set of vertices $V = (v1, v2 \ldots vn)$ and a set of edges $E = (e1, e2 \ldots em)$. This study considered an undirected network, assuming drivers tend to ignore traffic directions in extreme events, particularly in developing contexts. Scenario-specific weights were assigned to network components according to their functional importance and spatial exposure; for instance, major roads providing access to hospitals and links located within hazard zones were given higher weights to reflect their criticality and vulnerability.

The Urban Network Analysis (UNA) toolbox was used to evaluate network graph-based metrics, which included essential features connected to spatial network analysis such as geometry, topology, network elements (e.g., nodes and links), and spatial units (e.g., buildings) [51]. The toolbox generated adjacency matrices for nodal weight and centrality calculations. Accordingly, the degree distribution of the original graph was assessed to unveil network characteristics and vulnerability. Fundamental structural connectivity indices—including alpha (ratio of actual to maximum cycles), beta (edges-to-nodes ratio), gamma (observed-to-possible edges ratio), and the grid-tree pattern (spatial configuration)—were quantified in conjunction with density metrics (nodal, link, and network densities) to characterize the overall form and redundancy of the network. These metrics complement the more granular topological indicators presented in Table 1, which describe local connectivity and flow criticality within the network. Subsequently, local and global connectivity redundancies were defined through the average nodal degree and the clustering coefficient. The straightness property was also calculated, reflecting the degree of closeness of the shortest path between a particular node and all other nodes within a certain radius and its similarity to Euclidean distances [52]. The average betweenness centrality was computed to identify the node's significance level in the network. This centrality measure involves the number of shortest paths, which is crucial regarding redundancy and resource allocation in damaged networks.

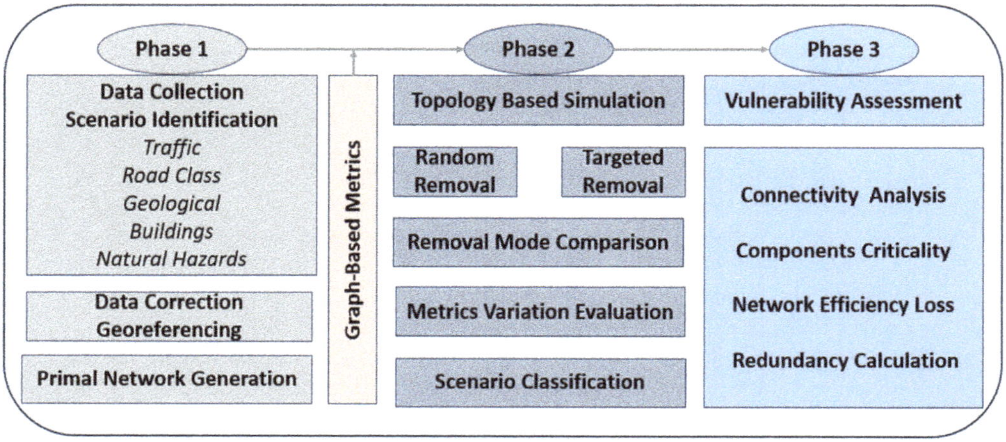

Figure 2. Methodological workflow for critical infrastructure vulnerability assessment.

In the second phase, the paper examined the behavior of the graph-based model under failure scenarios involving the random (i.e., uninformed) removal of nodes and their adjacent links at 10% intervals. The effects of these removals on various network components were simulated through random events to investigate network behavior trends. As the removal rate increased, changes in the graph's degree distribution and topological properties were analyzed. Subsequently, the removal process was repeated using a rank-ordering technique based on attributed scenarios to study the impact of targeted failure (i.e., metric-based) modes on the network. Given the significant influence of nodal failures on graph performance, a nodal betweenness rank-ordering removal was applied. This method considers both critical nodes and shortest links to identify potential cascading failure mechanisms within the system. After each removal, the highest-ranked elements in the updated centrality distribution were selected for subsequent removal fractions. Then, removal modes were compared using network metric variations, and scenario criticality was classified based on failure impacts.

In the third phase, connectivity analysis was employed to assess component criticality and to quantify potential efficiency loss under disruption scenarios. Critical components were identified using topological measures that capture their relative importance to overall system performance. Vulnerability was defined as the reduction in both connectivity and efficiency, providing a basis for identifying internal dependencies and structural weaknesses within the network. In this context, the UNA toolbox was employed to compute the redundancy index, defined as the ratio of the total length of redundant links to the length of the shortest paths for each origin–destination (OD) pair. The average redundancy index was then calculated across multiple OD pairs, specifically those connecting critical emergency centers and hospitals. The index quantifies the additional weighted distance of redundant routes compared to the shortest path, reflecting the extent of alternative connectivity available within the network. A 5% increase in the redundant paths to the shortest path was applied in the analysis, where the redundancy coefficient is set to 1.05. The analysis focused on removing top-weighted structures exhibiting the highest average betweenness centrality and nodal degree across all scenarios. Following each removal, a new network dataset was created to analyze the paths of the most weighted structures. The critical nodes were determined by their average redundancy indices to available hospitals. The network's global connectivity, percentage reach, and redundancy loss were computed after each removal fraction. Accordingly, resilience could be effectively evaluated by simulating the failure of critical network components and analyzing the resulting degradation in system connectivity.

4. Case Study

Beirut, the capital of Lebanon—a developing country located in the eastern Mediterranean region—occupies an area of approximately 20 km^2 on a peninsula along the nation's central coastline. Despite its limited size, the city is densely populated, driven in part by a substantial influx of daily commuters. Beirut also hosts Lebanon's largest and principal seaport, making it a critical urban and economic center.

Lebanon has a history of devastating earthquakes, which have resulted in large-scale infrastructure destruction and significant death tolls. Given its location between the Arabian, African, and Eurasian Plates and the presence of the 1200 km-long Levant Fault System that crosses the country, the seismic history of Lebanon is not surprising. The three main fault branches intersecting Lebanon have historically been responsible for severe earthquakes (Mw 7+) and tsunamis. Given their reoccurrence dates, a strong earthquake could occur anytime, as these faults are sufficiently mature. Earthquakes are equally probable throughout the country due to their small size relative to the critical dimension of the active faults. Huijer et al. [18] classified Lebanon as a medium-to-high seismic hazard country, with an expected peak ground acceleration (PGA) ranging from 0.2 g to 0.3 g, with a 10% probability of exceedance in 50 years. The authors suggested increasing the PGA to 0.3 g in the coastal zones.

The geological basement of Beirut dates to the Cretaceous period and is mainly composed of marly limestone coupled with conglomeratic beds. However, the sandy cover is the most significant feature of Beirut's geology, and its spatial variability can result in substantial seismic amplification [53]. While the hard rocky geology of Lebanon can attenuate seismic accelerations, the mountainous morphology along weak layer types can increase the probability of secondary hazards such as slope instabilities, rockfalls, and landslides [54,55]. In addition, the elevation of the water table and the presence of small fines in sandy areas, especially in coastal construction zones, can also lead to soil liquefaction or floods [56]. These hazards can severely damage critical infrastructure, putting the country at immediate risk.

5. Results and Discussion

The Beirut transportation network comprises 3207 nodes interconnected by 4739 unique links. Evaluating the network's degree distribution is critical for understanding its vulnerability to failure, as robustness is closely tied to the distribution of vertex degrees.

The analysis of the degree distribution in the Beirut transportation network suggests that nodes are moderately interconnected, with an average degree of three. This topology more closely resembles that of a random graph, following a Poisson distribution rather than a power law. This suggests that most nodes exhibit similar degrees, with few or no highly connected hubs—an important deviation from the heavy-tailed behavior characteristic of scale-free networks [57,58]. This structural configuration implies that the network responds to disruptions in a manner consistent with random graph behavior, wherein the impacts of random and targeted node removals are relatively comparable. Such a pattern suggests a degree of inherent robustness to targeted failures, as the absence of critical hubs reduces the likelihood of severe fragmentation following intentional attacks.

Topology-based simulation using a random removal technique was applied to assess graph theory topological and performance metrics (Figure 3). The average nodal degree, which describes the global connectivity of the system, decreased dramatically, highlighting the system's high dependence on the state of individual nodes. To assess the accessibility of alternative routes, local clustering coefficients were employed to depict the local redundancy of the system. After randomly removing 10% of the nodal network, a sharp decline in local redundancy was evident. This reflects the importance of nodal connectivity, where highly connected nodes can induce significant impacts. Following this decline, the formation of isolated groups led to a localized increase in the clustering coefficient, as smaller interconnected clusters emerged within the fragmented network. Moreover, declines in betweenness centrality and straightness degree can degrade network performance, leading to substantial delays, detours, or cancelation of trips after relatively small nodal disruptions. In parallel, trends in connectivity indices and density measures indicate reduced network efficiency, with nearly half of the network's functionality lost after just 30% of nodes were disrupted (Figure 4; see also [58]).

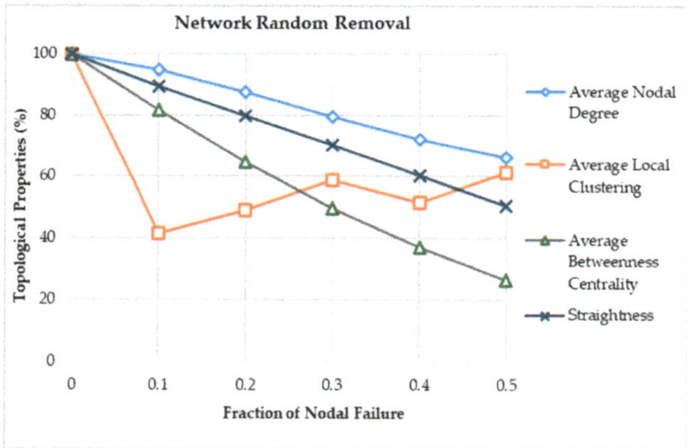

Figure 3. Random nodal removal effects on topological properties.

The integrated weighted network was evaluated to determine vulnerability measures and network effectiveness under both random and targeted removal techniques. A contextual integration approach was applied, wherein each scenario contributed unique topological properties to the network, shaping the relative importance of its components. The evaluated scenarios included: traffic data from a typical weekday noon peak (i.e., Monday) (Figure 5a); road classifications and the location of main hospitals and emergency centers in Beirut and the surrounding governorate (Figure 5b); building inventories (i.e., height, density, and locality) (Figure 5c); potential natural hazards (i.e., sea-level rise, tsunamis, and river floods) (Figure 5d); and geological characteristics (i.e., damage increment of soil and rock) (Figure 6).

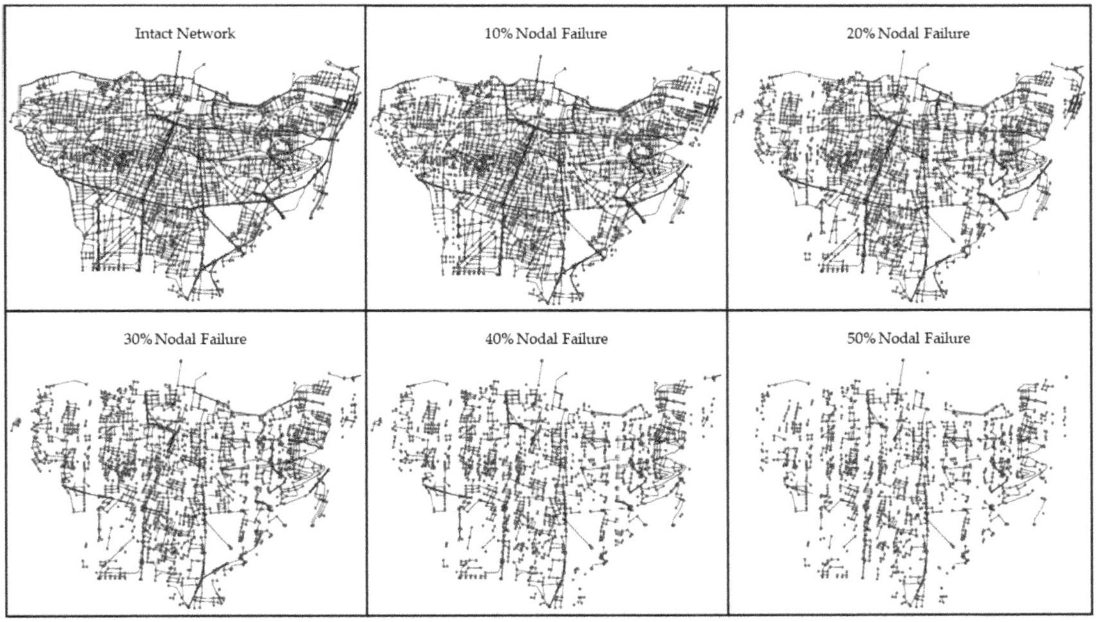

Figure 4. Nodal random removal of Beirut original graph with a 10% interval.

Scenario weights were determined through a data-driven, context-specific approach based on the relative impact of each factor on network vulnerability. Transportation data were weighted by assigning higher values to major roads (e.g., motorways and primary roads) with heavier traffic flow and direct access to critical infrastructure, particularly hospitals and emergency centers. Building-related weights reflected structural height, age, and typology, with greater emphasis placed on vulnerable low-rise and older buildings. Natural hazard risks were incorporated using spatial buffer analyses, with higher weights assigned to links located within predefined risk zones for floods, tsunamis, and sea-level rise, indicating higher exposure. Geological risk weights were derived from a seismic damage increment map, where nodes and links in areas with soft soils, elevated PGA, and strong soil-structure resonance effects received higher values. This multi-scenario weighting framework ensured that spatial risk patterns were appropriately reflected in the integrated network analysis.

A damage increment map was used to evaluate the seismic vulnerability of soil and structure data [53]. The map was based on a neural network approach that considers three typology classes: frequency ratio between soil and structure, PGA for various seismic scenarios, and horizontal-to-vertical spectral ratio interpolation, which is commonly used in seismic vulnerability and risk assessment. The parameter Dsoil–Drock represents the difference in expected structural damage between soft soil and rock under identical seismic loading. Higher values depicted in red indicate stronger amplification and thus greater damage potential (Figure 6). Links and nodes located within high-amplification zones were assigned higher weights to reflect increased vulnerability in the network analysis. Correspondingly, maps created for PGA equal to 0.25 indicated a significant damage increment for low-rise buildings (1–3 floors with an average height of 3 m each), as reported in [53,59]. The damage increment map was combined with the Beirut transportation network, along with 18,400 buildings represented by the height of their structure. Buildings were divided into three categories, where 27% of total buildings in Beirut were low-rise (i.e., 1–3 floors), 31% were medium-rise (i.e., 4–7 floors), and 42% were high-rise (i.e., >7 floors). It was considered that the low-rise buildings were the most vulnerable to seismic events as most of these structures were built before 1950 (i.e., masonry class) or between 1975 and 1990 during the civil war period, where a lack of quality control, construction, and reconstruction of structures was prevalent.

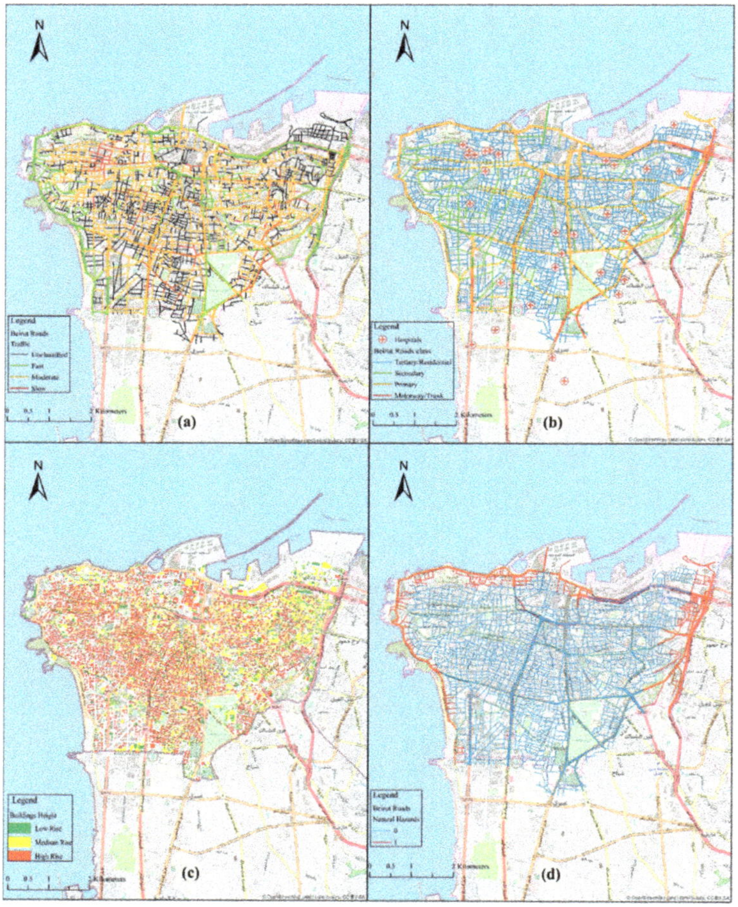

Figure 5. Beirut transportation network weighted scenarios: (**a**) typical Monday noon peak traffic; (**b**) roads class and main hospitals; (**c**) building height distribution; and (**d**) vulnerable links to natural hazards.

Based on traffic data, 16% of the network links were classified as fast-speed routes, while 74% and 10% were moderate- and slow-speed routes, respectively. Accordingly, 68% of the network nodes or intersections were deemed at a higher risk of being critical in the event of failure. As for road classification, 32% of the intersections on Beirut roads were considered essential for maintaining circulation to critical zones within the city. Moreover, 12% of network links and 20% of nodes were identified as susceptible to disturbances from natural hazards (e.g., sea-level rise and flooding), based on their proximity to designated buffer zones. Depending on the zonal geology and their susceptibility to liquefaction and landslides, 26% of network nodes were identified as vulnerable. Alternatively, the damage increment was minor in the two hills of Beirut created by rock formations, thereby having no aggravating effects on the structures. In fact, the damage on rock foundations is usually less than on soft soil due to the absence of soil amplification. Furthermore, around 25% of network links and 45% of their nodes were deemed high-risk due to their proximity to vulnerable buildings and unfavorable soil conditions (e.g., falling debris, ruptured utility lines, hazardous spills, and fires).

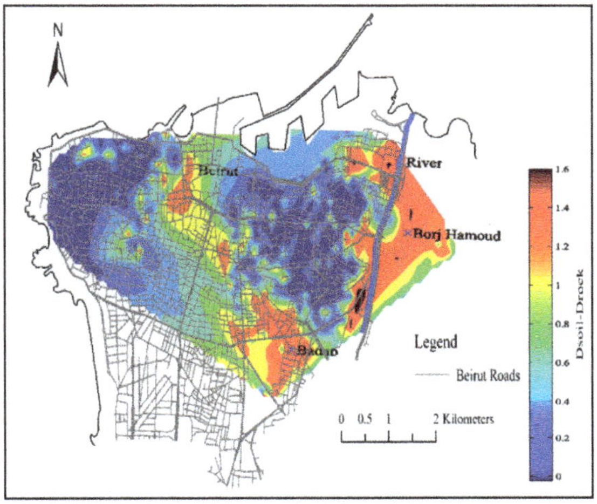

Figure 6. Interpolated map of the damage increment computed based on the neural network approach for PGA = 0.25 g, modified from [53].

The topological parameters of the weighted network were interpreted through random nodal removal after each removal fraction. The average nodal degree, straightness, and connectivity indices of all weighted scenarios showed a significant downward trend. The most noteworthy trend was the average betweenness centrality, which decreased significantly for all cases after the second random removal, as shown in Figure 7. However, nodes in the degraded network were rearranged to form newly separated clusters, as reflected by the slight increase observed in the bar chart at the third removal iteration. Overall, the substantial decrease in the topological parameter indicates a network structure failure that can lead to severe resource scarcity and hinder emergency teams. Moreover, the network exhibited a low average path length, suggesting structural characteristics similar to both random and small-world networks. In such topologies, the impact of node removal—whether random or targeted—tends to be more evenly distributed across the system [60]. This stands in contrast to scale-free networks, which, while highly resilient to random failures, are disproportionately vulnerable to targeted attacks on high-degree hubs [61]. In random networks, where most node degrees are concentrated around the average degree, the structure is practical but remains susceptible to random disruptions. Similarly, in the case of the Beirut network, system efficiency tends to decline under repeated random node removals, particularly when structurally significant nodes are impacted.

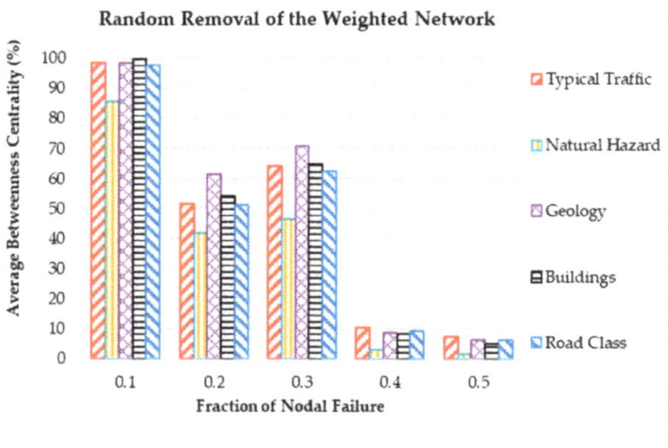

Figure 7. Random nodal removal effects on the weighted network scenarios.

The study evaluates critical network nodes and their redundancies to analyze the system's performance under different removal modes. Figure 8 illustrates the extent of disrupted links resulting from both random and targeted removal modes. The simulation results of both undisturbed and disturbed components highlighted that network performance was primarily influenced by traffic, road classification, and buildings scenarios. The comparison was based on disrupted links in the random failure removal mode over those in the targeted failure removal mode. The analysis of network traffic scenarios revealed that targeted events could result in over a 30% increase in failed links compared to random events. In road class and building scenarios, the decline persisted until the fourth removal. After that, the trend deviated, indicating the deterioration of the most highly connected intersections; however, the ratios remained lower than those in the random failure mode. Conversely, the remaining two scenarios indicated an increase in disrupted links in the random failure mode compared to the targeted one. This suggests that the weighted geology and natural hazard scenarios were not closely associated with the most connected nodes and links in the network.

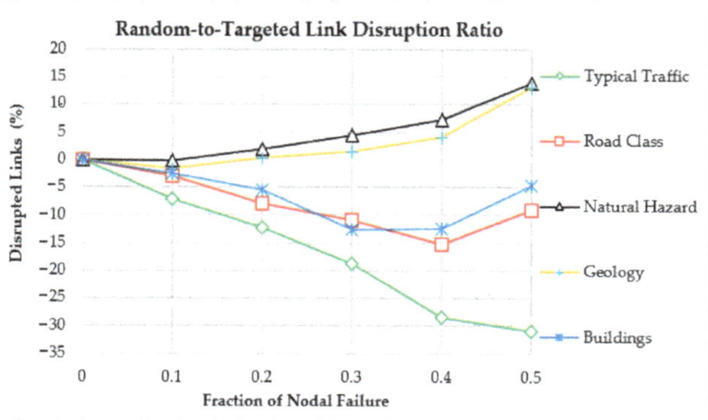

Figure 8. Comparative disruption of links under random and targeted failures.

Spatial network analysis usually assumes that transportation happens via the shortest accessible paths. However, people can experience longer routes in real networks for different subjective choices. For this reason, determining the redundant routes in a disruptive event is necessary. Alternative paths should be found between the OD matrices for a certain length quota beyond the shortest routes. In this manner, network robustness interrelated to the redundancy index can reveal the extent of additional weighted path length through OD pairs using a specified redundancy coefficient. For instance, if a 5% coefficient is applied and the redundancy output is equal to "RI = 4", four times more route lengths exist on redundant paths than on the shortest path alone. The paths obtained will increase as the redundancy coefficient rises, making the analysis more computationally intensive. In emergency and evacuation situations, time is deemed critical. In this study, the percentages of redundant nodes reachable to vital centers were computed under the targeted removal mode of top-weighted network properties across different scenarios. The results revealed that the network was shut down after disrupting only 40% of its components. Notably, the second removal fraction depicted the weakness in reaching less than half of the city's available hospitals and critical emergency centers, with only a 44% chance of reaching 13 destinations.

The average redundancy loss was assessed following each removed fraction in the weighted network. The average redundancy decreased to half its overall performance after only 20% of nodal removal. Accordingly, after evaluating the network performance and efficiency measures through different classified weighted scenarios, the critical redundant street nodes and intersections were generated and mapped as in Figure 9a. The criterion of critical nodes was mainly based on the average redundancy indices from these nodes to the available hospitals. The vulnerable key nodes were all over Beirut. However, most of the clustered nodes were located along the primary roads in the center and extending toward the capital's west side. Such points possessed low redundancy and constituted around 25% of all the network components. Correspondingly, the remaining nodes were divided into 70% with moderate and 5% with high redundancy levels, respectively. This reflects the need to focus on these exposed areas, which are more susceptible in the case of disasters (Figure 9b).

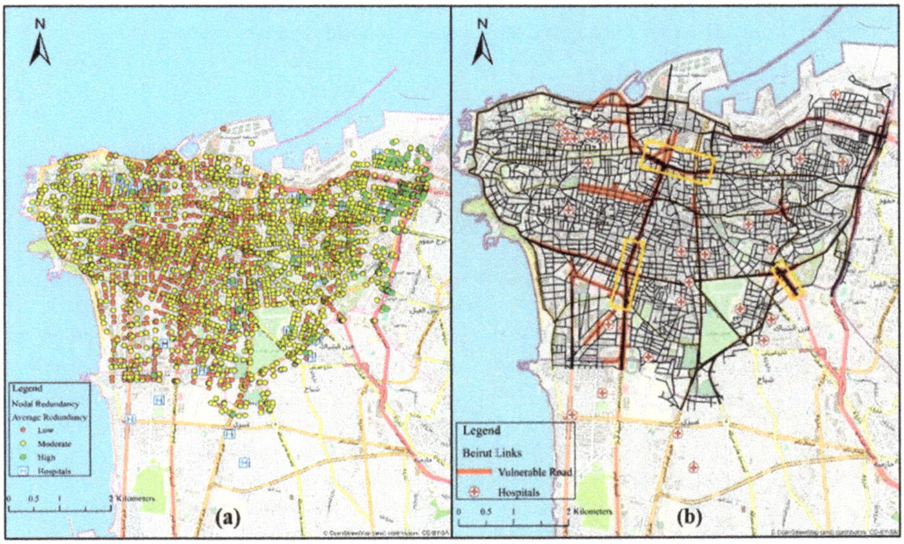

Figure 9. Network resilience and critical components: (**a**) nodal average redundancy of Beirut intersections to the accessible hospitals and (**b**) major vulnerable links and intersections.

6. Conclusions

This paper evaluates the vulnerability and criticality of the Beirut transportation network and proposes a predictive methodology to assess its operational resilience under extreme events. By integrating GIS-based spatial data with network theory, the proposed method quantifies structural vulnerabilities and functional capacities, providing targeted guidance to inform strategies for enhancing network resilience.

Engineering resilience is evaluated using centrality and redundancy metrics through quantitative and statistical modeling. The results indicate low redundancy and high sensitivity to even minor disruptions, leading to a rapid decline in overall system performance. Despite some central segments exhibiting moderate robustness, the findings suggest a lack of structurally or functionally resilient components capable of absorbing external stressors. Notably, the Beirut transportation network responds similarly to both random and targeted failure scenarios, a pattern indicative of increased vulnerability. Such findings classify the network as highly vulnerable, underscoring the critical need to understand and address the vulnerability of lifeline infrastructure for disaster preparedness, response planning, and risk mitigation.

Future research should focus on developing adaptive, dynamic vulnerability assessment methods that incorporate real-time data and account for evolving geotechnical, environmental, climatic, and socio-economic conditions. This integration with spatial-network modeling can improve the predictive capacity and operational responsiveness of transportation resilience frameworks. Identifying critical network components and classifying their risk levels will further support effective emergency routing and evacuation planning. Advancing these goals requires close collaboration among infrastructure agencies, urban planners, and emergency management authorities to enable proactive risk reduction, critical asset protection, and the long-term sustainability of urban lifeline systems.

Author Contributions: Conceptualization, Francesco Rouhana and Dima Jawad; methodology, Francesco Rouhana; software, Francesco Rouhana; validation, Francesco Rouhana and Dima Jawad; formal analysis, Francesco Rouhana; investigation, Francesco Rouhana; resources, Dima Jawad; data curation, Francesco Rouhana; writing—original draft, Francesco Rouhana; writing—review and editing, Francesco Rouhana and Dima Jawad; visualization, Francesco Rouhana; supervision, Dima Jawad. All authors have read and agreed to the published version of the manuscript.

Funding: This research received no external funding.

Data Availability Statement: Some or all data, models, or code that support the findings of this study are available from the corresponding author upon reasonable request.

Conflicts of Interest: The authors declare no conflicts of interest.

References

1. Mottahedi, A.; Sereshki, F.; Ataei, M.; Qarahasanlou, A.N.; Barabadi, A. The Resilience of Critical Infrastructure Systems: A Systematic Literature Review. *Energies* **2021**, *14*, 1517. [CrossRef]
2. Chowdhury, S.; Ren, Z.; Rouhana, F.; Wang, J.; Zhu, J. Exploring Sustainability in Resource Management: Implementation of a Water–Energy–Human Nexus Framework at a University Campus. *J. Constr. Eng. Manag.* **2024**, *150*, 10. [CrossRef]
3. Rouhana, F.; Zhu, J.; Bagtzoglou, A.C.; Burton, C.G. Analyzing Structural Inequalities in Natural Hazard-Induced Power Outages: A Spatial-Statistical Approach. *Int. J. Disaster Risk Reduct.* **2025**, *105*, 105184. [CrossRef]
4. Valencia Londoño, P.A.; Valencia Londoño, D. Persistent Vulnerability after Disaster Risk Reduction (DRR) Response: The Case of Salgar, Colombia. *Sustainability* **2024**, *16*, 4394. [CrossRef]
5. Cutter, S.L. Urban Risks and Resilience. In *Urban Informatics*; Shi, W., Goodchild, M.F., Batty, M., Kwan, M.P., Zhang, A., Eds.; The Urban Book Series; Springer: Singapore, 2021. [CrossRef]
6. Rouhana, F.; Zhu, J. Examining Rural-Urban Disparity in Disaster Impact and Recovery: Case of Tropical Storm Isaias. In *Computing in Civil Engineering 2023: Resilience, Safety, and Sustainability—Selected Papers from the ASCE International Conference on Computing in Civil Engineering 2023*; American Society of Civil Engineers (ASCE): Reston, VA, USA, 2024; pp. 219–227. [CrossRef]
7. United Nations Office for Disaster Risk Reduction (UNDRR). *Forensic Insights for Future Resilience: Learning from Past Disasters*; UNDRR: Geneva, Switzerland, 2024. Available online: https://www.undrr.org/gar/gar2024-special-report (accessed on 28 April 2025).

8. Rouhana, F.; Zhu, J.; Bagtzoglou, A.C.; Burton, C.G. Examining Rural–Urban Vulnerability Inequality in Extreme Weather-Related Power Outages: Case of Tropical Storm Isaias. *Nat. Hazards Rev.* **2025**, *26*, 03024003. [CrossRef]
9. Park, H.J.; Song, S.M.; Kim, D.H.; Lee, S.O. Socioeconomic Impact on Urban Resilience against Flood Damage. *Appl. Sci.* **2024**, *14*, 7882. [CrossRef]
10. UNDRR. Sendai Framework for Disaster Risk Reduction 2015–2030. 2015. Available online: https://www.unisdr.org/files/43291_sendaiframeworkfordrren.pdf (accessed on 25 April 2025).
11. United Nations Department of Economic and Social Affairs (UN DESA). *The Sustainable Development Goals Report 2024*; UN DESA: New York, NY, USA, 2024. Available online: https://unstats.un.org/sdgs/report/2024/ (accessed on 16 April 2025).
12. Farooq, A.; Xie, M.; Stoilova, S.; Ahmad, F.; Guo, M.; Williams, E.J.; Issa, A.M. Transportation Planning through GIS and Multicriteria Analysis: Case Study of Beijing and Xiongan. *J. Adv. Transp.* **2018**, *2018*, 1–11. [CrossRef]
13. Ntafloukas, K.; Pasquale, L.; Martinez-Pastor, B.; McCrum, D.P. A Vulnerability Assessment Approach for Transportation Networks Subjected to Cyber–Physical Attacks. *Future Internet* **2023**, *15*, 100. [CrossRef]
14. Wassmer, J.; Merz, B.; Marwan, N. Resilience of Transportation Infrastructure Networks to Road Failures. *Chaos* **2024**, *34*, 1. [CrossRef]
15. Aydin, N.Y.; Duzgun, H.S.; Wenzel, F.; Heinimann, H.R. Integration of Stress Testing with Graph Theory to Assess the Resilience of Urban Road Networks under Seismic Hazards. *Nat. Hazards* **2018**, *91*, 37–68. [CrossRef]
16. Ding, J.; Wang, Y.; Li, C. A Dual-Layer Complex Network-Based Quantitative Flood Vulnerability Assessment Method of Transportation Systems. *Land* **2024**, *13*, 753. [CrossRef]
17. Rezvani, S.M.H.S.; Silva, M.J.F.; de Almeida, N.M. Urban Resilience Index for Critical Infrastructure: A Scenario-Based Approach to Disaster Risk Reduction in Road Networks. *Sustainability* **2024**, *16*, 4143. [CrossRef]
18. Huijer, C.; Harajli, M.; Sadek, S. Re-Evaluation and Updating of the Seismic Hazard of Lebanon. *J. Seismol.* **2016**, *20*, 233–250. [CrossRef]
19. Bruneau, M.; Chang, S.E.; Eguchi, R.T.; Lee, G.C.; O'Rourke, T.D.; Reinhorn, A.M.; Shinozuka, M.; Tierney, K.; Wallace, W.A.; von Winterfeldt, D. A Framework to Quantitatively Assess and Enhance the Seismic Resilience of Communities. *Earthq. Spectra* **2003**, *19*, 733–752. [CrossRef]
20. Poulin, C.; Kane, M.B. Infrastructure Resilience Curves: Performance Measures and Summary Metrics. *Reliab. Eng. Syst. Saf.* **2021**, *219*, 107926. [CrossRef]
21. Mentges, A.; Halekotte, L.; Schneider, M.; Demmer, T.; Lichte, D. A Resilience Glossary Shaped by Context: Reviewing Resilience-Related Terms for Critical Infrastructures. *Int. J. Disaster Risk Reduct.* **2023**, *94*, 103893. [CrossRef]
22. Najarian, M.; Lim, G.J. Design and Assessment Methodology for System Resilience Metrics. *Risk Anal.* **2019**, *39*, 1885–1898. [CrossRef]
23. Ayyub, B.M. Systems Resilience for Multihazard Environments: Definition, Metrics, and Valuation for Decision Making. *Risk Anal.* **2014**, *34*, 340–355. [CrossRef]
24. Bruneau, M.; Reinhorn, A. Exploring the Concept of Seismic Resilience for Acute Care Facilities. *Earthq. Spectra* **2007**, *23*, 41–62. [CrossRef]
25. Cimellaro, G.P.; Reinhorn, A.M.; Bruneau, M. Framework for Analytical Quantification of Disaster Resilience. *Eng. Struct.* **2010**, *32*, 3639–3649. [CrossRef]
26. Liu, Y.; McNeil, S.; Hackl, J.; Adey, B.T. Prioritizing Transportation Network Recovery Using a Resilience Measure. *Sustain. Resil. Infrastruct.* **2022**, *7*, 70–81. [CrossRef]
27. Nicolosi, V.; Augeri, M.; D'Apuzzo, M.; Evangelisti, A.; Santilli, D. A Probabilistic Approach to the Evaluation of Seismic Resilience in Road Asset Management. *Int. J. Disaster Risk Sci.* **2022**, *13*, 114–124. [CrossRef]
28. Akbarzadeh, M.; Memarmontazerin, S.; Derrible, S.; Salehi Reihani, S.F. The Role of Travel Demand and Network Centrality on the Connectivity and Resilience of an Urban Street System. *Transportation* **2019**, *46*, 1127–1141. [CrossRef]
29. Wang, D.; Xue, Y.; Zhang, Q.; Niu, M. Scale-Free and Small-World Properties of a Special Hierarchical Network. *Fractals* **2019**, *27*, 1950010. [CrossRef]
30. Zhenwu, S.; Xianyu, T.; Huiyu, L.; Hui, L.; Jie, L.; Xunguo, L. Quantitative Study on the Waterlogging Resilience of Road Transportation System Based on the Validity View of System Functions. *J. Eng. Sci. Technol. Rev.* **2019**, *12*, 117–125. [CrossRef]
31. Ganin, A.A.; Kitsak, M.; Marchese, D.; Keisler, J.M.; Seager, T.; Linkov, I. Resilience and Efficiency in Transportation Networks. *Sci. Adv.* **2017**, *3*, e1701079. [CrossRef]
32. Janić, M. Modelling the Resilience of Rail Passenger Transport Networks Affected by Large-Scale Disruptive Events: The Case of HSR (High-Speed Rail). *Transportation* **2018**, *45*, 1101–1137. [CrossRef]
33. Cerqueti, R.; Ferraro, G.; Iovanella, A. Measuring Network Resilience through Connection Patterns. *Reliab. Eng. Syst. Saf.* **2019**, *188*, 320–329. [CrossRef]
34. Garrison, W.L.; Marble, D.F. Factor-Analytic Study of the Connectivity of a Transportation Network. *Pap. Reg. Sci.* **1964**, *12*, 231–238. [CrossRef]

35. Testa, A.C.; Furtado, M.N.; Alipour, A. Resilience of Coastal Transportation Networks Faced with Extreme Climatic Events. *Transp. Res. Rec.* **2015**, *2532*, 29–36. [CrossRef]
36. Albert, R.; Barabási, A.L. Statistical Mechanics of Complex Networks. *Rev. Mod. Phys.* **2002**, *74*, 47–97. [CrossRef]
37. Dueñas-Osorio, L.; Craig, J.I.; Goodno, B.J.; Bostrom, A. Interdependent Response of Networked Systems. *J. Infrastruct. Syst.* **2007**, *13*, 185–194. [CrossRef]
38. Liu, W.; Li, X.; Liu, T.; Liu, B. Approximating Betweenness Centrality to Identify Key Nodes in a Weighted Urban Complex Transportation Network. *J. Adv. Transp.* **2019**, *2019*, 9024745. [CrossRef]
39. Freeman, L.C. A Set of Measures of Centrality Based on Betweenness. *Sociometry* **1977**, *40*, 35–41. [CrossRef]
40. Reggiani, A.; Nijkamp, P.; Lanzi, D. Transport Resilience and Vulnerability: The Role of Connectivity. *Transp. Res. Part A Policy Pract.* **2015**, *81*, 4–15. [CrossRef]
41. Allen, E.; Costello, S.B.; Henning, T.F.P.; Chamorro, A.; Echaveguren, T. Integration of Resilience and Risk to Natural Hazards into Transportation Asset Management of Road Networks: A Systematic Review. *Struct. Infrastruct. Eng.* **2023**, *19*, 2238281. [CrossRef]
42. Serdar, M.Z.; Koç, M.; Al-Ghamdi, S.G. Urban Transportation Networks Resilience: Indicators, Disturbances, and Assessment Methods. *Sustain. Cities Soc.* **2022**, *76*, 103452. [CrossRef]
43. Ahmadzai, F.; Rao, K.M.L.; Ulfat, S. Assessment and Modelling of Urban Road Networks Using Integrated Graph of Natural Road Network (a GIS-Based Approach). *J. Urban Manag.* **2019**, *8*, 109–125. [CrossRef]
44. Candelieri, A.; Galuzzi, B.G.; Giordani, I.; Archetti, F. Vulnerability of Public Transportation Networks Against Directed Attacks and Cascading Failures. *Public Transp.* **2019**, *11*, 27–49. [CrossRef]
45. Ahmed, M.A.; Sadri, A.M.; Mehrabi, A.; Azizinamini, A. Identifying Topological Credentials of Physical Infrastructure Components to Enhance Transportation Network Resilience: Case of Florida Bridges. *J. Transp. Eng. Part A Syst.* **2022**, *148*, 04022071. [CrossRef]
46. Chatterjee, A.; Manohar, M.; Ramadurai, G. Statistical Analysis of Bus Networks in India. *PLoS ONE* **2016**, *11*, e0168478. [CrossRef] [PubMed]
47. Li, M.; Han, J. Complex Network Theory in Urban Traffic Network. *Atlantis Press* **2017**, *175*, 860–864. [CrossRef]
48. De Bona, A.A.; de Oliveira Rosa, M.; Ono Fonseca, K.V.; Lüders, R. A Reduced Model for Complex Network Analysis of Public Transportation Systems. *Phys. A Stat. Mech. Appl.* **2021**, *567*, 125715. [CrossRef]
49. Zhang, D.-M.; Du, F.; Huang, H.; Zhang, F.; Ayyub, B.M.; Beer, M. Resiliency Assessment of Urban Rail Transit Networks: Shanghai Metro as an Example. *Saf. Sci.* **2018**, *106*, 230–243. [CrossRef]
50. Wu, C.; Kim, I. Analyzing the Structural Properties of Bike-Sharing Networks: Evidence from the United States, Canada, and China. *Transp. Res. Part A Policy Pract.* **2020**, *140*, 52–71. [CrossRef]
51. Sevtsuk, A.; Mekonnen, M. Urban Network Analysis: A New Toolbox for ArcGIS. *Rev. Int. Géomatique* **2012**, *22*, 287–305. [CrossRef]
52. Porta, S.; Crucitti, P.; Latora, V. The Network Analysis of Urban Streets: A Dual Approach. *Phys. A Stat. Mech. Appl.* **2006**, *369*, 853–866. [CrossRef]
53. Salameh, C.; Guillier, B.; Harb, J.; Cornou, C.; Bard, P.Y.; Voisin, C.; Mariscal, A. Seismic Response of Beirut (Lebanon) Buildings: Instrumental Results from Ambient Vibrations. *Bull. Earthq. Eng.* **2016**, *14*, 2705–2730. [CrossRef]
54. Du, Y.; Xie, M.; Jia, J. Stepped Settlement: A Possible Mechanism for Translational Landslides. *Catena* **2020**, *187*, 104365. [CrossRef]
55. Chen, D.; Yin, S.; Long, W.; Zhang, Y.; Yang, Y. Heterogeneous Information Phase Space Reconstruction and Stability Prediction of Filling Body-Surrounding Rock Combination. *Int. J. Miner. Metall. Mater.* **2024**, *31*, 1500–1511. [CrossRef]
56. Lai, X.P.; Shan, P.F.; Cai, M.F.; Liu, J.; Zhang, Y. Comprehensive Evaluation of High-Steep Slope Stability and Optimal High-Steep Slope Design by 3D Physical Modeling. *Int. J. Miner. Metall. Mater.* **2015**, *22*, 1–11. [CrossRef]
57. Barthélemy, M. Spatial Networks. *Phys. Rep.* **2011**, *499*, 1–101. [CrossRef]
58. Rouhana, F.; Jawad, D. Transportation Network Resilience against Failures: GIS-Based Assessment of Network Topology Role. *Int. J. Disaster Resil. Built Environ.* **2021**, *12*, 357–370. [CrossRef]
59. Salameh, C.; Bard, P.Y.; Guillier, B.; Harb, J.; Cornou, C.; Gérard, J.; Almakari, M. Using Ambient Vibration Measurements for Risk Assessment at an Urban Scale: From Numerical Proof of Concept to Beirut Case Study (Lebanon). *Earth Planets Space* **2017**, *69*, 1. [CrossRef]
60. Watts, D.J.; Strogatz, S.H. Collective Dynamics of 'Small-World' Networks. *Nature* **1998**, *393*, 440–442. [CrossRef]
61. Holme, P.; Kim, B.J.; Yoon, C.N.; Han, S.K. Attack vulnerability of complex networks. *Phys. Rev. E* **2002**, *65*, 056109. [CrossRef]

MDPI AG
Grosspeteranlage 5
4052 Basel
Switzerland
Tel.: +41 61 683 77 34

MDPI Books Editorial Office
E-mail: books@mdpi.com
www.mdpi.com/books

www.ingramcontent.com/pod-product-compliance
Lightning Source LLC
Chambersburg PA
CBHW041934260926
51850CB00018B/133
* 9 7 8 3 7 2 5 8 6 1 2 3 1 *